Amorphous Metallic Alloys

Butterworths Monographs in Materials

The intention is to publish a series of definitive monographs written by internationally recognized authorities in subjects at the interface of the research interests of the academic materials scientists and the industrial materials engineer.

Already published

Die casting metallurgy
Control and analysis in iron and steel making
Introduction to the physical metallurgy of welding
Metals resources and energy

Forthcoming titles

Microorganisms and metal recovery
Eutectic solidification and processing
Control and analysis in steelmaking
Energy dispersive X-ray analysis of materials
Mechanical properties of ceramics
Metallurgy of high speed steels
Residual stresses in metals
Continuous casting of aluminium

Butterworths Monographs in Materials

Amorphous Metallic Alloys

Edited by
F. E. LUBORSKY, PhD
Corporate Research and Development Center, General Electric Co., Schenectady, USA

Butterworths
London Boston Durban Singapore Sydney Toronto Wellington

First published 1983

Butterworth & Co (Publishers) Ltd., 1983

British Library Cataloguing in Publication Data

Amorphous metallic alloys.—(Butterworths monographs in materials)
1. Alloys
I. Luborsky, F. E.
546′.37 TN690

ISBN 0-408-11030-9

Typeset by MS Filmsetting Ltd., Frome, Somerset
Printed and Bound in Great Britain by Butler & Tanner Ltd., London & Frome

Preface

This book on amorphous metallic alloys is intended for the student and researcher. It covers the preparation and properties of alloys produced by rapid quenching from the molten state; it does not cover in any detail alloys prepared by vacuum deposition, electro- or electroless-deposition or by sputtering. Its aim is to present an up-to-date review of the status of our knowledge in this field of materials science and technology. As such it is the first comprehensive single book on this new area of materials science. Since no field in material science today can be adequately covered in a current review by one individual, because of the diversity of talents brought to bear on the subject, this book is composed of a large set of carefully selected contributions from individual authors. Each chapter is written by an author who was selected from the world-wide pool of researchers as one of the outstanding contributors to the particular subject area in amorphous alloys. The attempt was made, and I believe successfully, to cover the entire field for these amorphous alloys. There are chapters on the history, on the fundamentals of formation, on sample preparation, on structure, on crystallization and on the unique physical, magnetic, electronic, chemical and thermal properties as well as the applications and potential applications. Each chapter has a lengthy list of references for those readers interested in more details. It is hoped that the lengthy combined subject index will be a useful appendix.

I have to thank many people, and I take pleasure in doing so. First, I must thank each of the authors of the chapters for adhering to the schedule set forth, which renders the information up to date. Thanks are also due to the publisher, Butterworths, for their great co-operation in all aspects of producing this book and, in particular, also for their adherence to a tight schedule so that this book was published promptly. I also acknowledge the inputs of my many colleagues working on amorphous metals at the General Electric Corporate Research and Development for their advice in the planning of this book. I am also extremely grateful to the General Electric Corporate Research and Development for providing the secretarial help and in particular to Cheryl Santomartino for collating the subject index from the individual chapter indices provided by each author. Finally, I am indebted to my wife for her many suggestions and for her interest in this book.

F. E. Luborsky
Schenectady

List of contributors

R. W. Cahn
Laboratoire de Métallurgie Physique, Université de Paris-Sud, Centre D'Orsay, Bâtiment 413, 91405 Orsay Cedex, France

B. Cantor
Dept. of Metallurgy and Science of Materials, University of Oxford, Parks Road, Oxford, OX1 3PH, England

H. S. Chen
Materials Research Physics Department, Bell Telephone Laboratories, Murray Hill, NJ 07974, USA

H. A. Davies
Dept. of Metallurgy, The University of Sheffield, Sheffield S1 3JD, England

T. Egami
Dept. of Metallurgy and Materials Science, K1, University of Pennsylvania, Philadelphia, PA 19174, USA

J. L. Finney
Dept. of Crystallography, Birkbeck College, Malet Street, London, WC1E 7HX, England

H. Fujimori
The Research Institute for Iron, Steel and Other Metals, Tohoku University, 1-1, Katahira-2 Chome, Sendai, Japan

K. Fukamichi
The Research Institute for Iron, Steel and Other Metals, Tohoku University, 1-1, Katahira-2 Chome, Sendai, Japan

C. F. Hague
Institut für Physik, Universitat Basil, CH-4056 Basil, Switzerland

H.-J. Güntherodt
Institut für Physik, Universitat Basil, CH-4056 Basil, Switzerland

K. Hashimoto
The Research Institute for Iron, Steel and Other Metals, Tohoku University, 1-1, Katahira-2 Chome, Sendai, Japan

H. Kronmüller
Max Planck Institute für Metallforschung, Institut für Physik, 7000 Stuttgart 80, Heisenbergstr. 1, FDR

H. Kumura
The Research Institute for Iron, Steel and Other Metals, Tohoku University, 1-1, Katahira-2 Chome, Sendai, Japan

H. H. Leibermann
Allied Corporate Technology, MetGlas Building, 6 Eastmans Rd., Parsippany, NJ 07054, USA

F. E. Luborsky
General Electric Company, Corporate Research and Development, Building K-1, Room 1C36, Schenectady, NY 12301, USA

T. Masumoto
The Research Institute for Iron, Steel and Other Metals, Tohoku University, 1-1, Katahira-2 Chome, Sendai, Japan

R. P. Messmer
General Electric Company, Corporate Research and Development, Building K-1, Room 2A30, Schenectady, NY 12301, USA

S. A. Miller
General Electric Company, Corporate Research and Development, Building K-1, Room 267MB, Schenectady, NY 12301, USA

N. Moser
Max Planck Institute für Metallforschung, Institut für Physik, 7000 Stuttgart 80, Heisenbergstr. 1, FDR

P. Oelhafen
Institut für Physik, Universitat Basil, CH-4056 Basil, Switzerland

R. C. O'Handley
Dept. Materials Science and Engineering, Building 13, Room 5065, MIT, 77 Massachusetts Ave., Cambridge, MA 02139, USA

D. G. Onn
223 Sharp Laboratory, Dept. of Physics, College of Arts and Science, University of Delaware, Newark, DE 19711, USA

S. J. Poon
Dept. of Physics, McCormack Rd., University of Virginia, Charlottesville, VA 22901, USA

K. V. Rao
Central Research Laboratory, MMM Company, P.O. Box 33221, St. Paul, MN 55133, USA

D. Raskin
Allied Corporation, Materials Research Center, Box 1021R, Morristown, NJ 07960, USA

M. Scott
Standard Telecommunication Laboratories, Ltd., Optical Waveguides Division, London Road, Harlow, Essex, CM17 9NA, England

C. H. Smith
Allied Corporation, Materials Research Center, Box 1021R, Morristown, NJ 07960, USA

F. Spaepen
Division of Engineering and Applied Physics, Harvard University, Cambridge, MA 02138, USA

K. Suzuki
Research Institute for Iron, Steel and Other Metals, Tohoku University, 1-1, Katahira-2 Chome, Sendai, Japan

A. I. Taub
General Electric Company, Corporate Research and Development, Building K-1, Room 267MB Schenectady, NY 12301, USA

C. N. J. Wagner
Dept. of Materials Science and Engineering, School of Engineering and Applied Science, University of California, Los Angeles, CA 90024, USA

E. P. Wohlfarth
Dept. of Mathematics, Imperial College of Science and Technology, Queen's Gate, London SW7 2BZ, England

Contents

Chapter 1

Amorphous metallic alloys

F. E. Luborsky

General Electric Company, Schenectady, New York, USA

1.1 Introduction

Amorphous metallic alloys are metals and metal alloys with no long range atomic order. They have also been called glassy alloys or non-crystalline alloys. They are made by a variety of techniques all of which involve the rapid solidification of the alloying constituents from the gas or liquid phases. The solidification occurs so rapidly that the atoms are frozen in their liquid configuration. There are clear structural indications and indications from their various properties that nearest neighbour, or local, order does exist in most amorphous metallic alloys, but no long range atomic order.

There are unique magnetic, mechanical, electrical and corrosion behaviours which results from this amorphous structure. For example, they behave as very soft magnetic materials, in fact magnetic losses in high magnetization alloys have been measured which are lower than those measured in any other known crystalline alloys; they are exceptionally hard and have extremely high tensile strengths and in some alloys the coefficient of the thermal expansion can be made to be zero; they have electrical resistivities which are three to four times higher than those of conventional iron or iron–nickel alloys; and finally some of the amorphous alloys are exceptionally corrosion resistant.

There are two or possibly three technologically important classes of magnetic amorphous alloy; the transition metal–metalloid (TM–M) alloys, the rare earth–transition metal (RE–TM) alloys and possibly the transition metal–zirconium or hafnium alloys. The TM–M alloys typically contain about 80 atom per cent iron, cobalt or nickel with the remainder being boron, carbon, silicon, phosphorus or aluminium and are typically prepared by rapid quenching from the melt, although other techniques such as sputtering, electrodeposition and chemical deposition have been used. The presence of the metalloids is necessary to lower the melting point making it possible to quench the alloy through its glass temperature rapidly enough to form the amorphous phase. Once made the same metalloids stabilize the amorphous phase but their presence drastically alters the magnetic, mechanical and electrical properties of the alloy by donating electrons to the *d*-band. The presumed isotropic character of the TM–M amorphous alloys had been predicted to result in very low coercivities and hysteresis loss and high permeabilities; all the properties of technological significance for application as soft magnetic materials. These good properties have been achieved in some melt-quenched alloys and we can account for their behaviour by the same models as used for conventional crystalline soft magnetic materials. The same statements can be made for the recently reported TM–Zr–Hf alloys. These normally contain about 10 atom per cent zirconium or hafnium but the addition of even a few percent of boron greatly enlarges the amorphous forming region. Because they have properties which are very similar to the TM–M alloys it is expected that they will be used in similar devices.

The RE–TM alloys, however, are normally prepared by sputter deposition and have properties especially suited to bubble memory devices, for example, low saturation magnetization and high anisotropy perpendicular to the plane. These types of amorphous alloy will not be discussed in any detail in this monograph.

Amorphous alloys have been shown to have vastly superior magnetic properties for application in large transformers and to have a combination of mechanical and magnetic properties that makes them extremely likely candidates for application in recording heads, in some electronic size transformers and in various types of sensor. Thus, the application of amorphous alloys in various magnetic devices appears to be assured. Nickel-based amorphous alloys for brazing foil have also been in use for several years. This technique provides all-metallic brazing foils with no binders, resulting in greater strength and greater assembly precision in reduced time. Many other applications have been reported in the technical literature but none has appeared on the market yet.

Until recently the major efforts in solid state physics have been confined to understanding the properties of crystalline solids. Microscopic information has been obtained from studies of the properties of single crystals. Amorphous solids now represent a new state of matter. Some of their properties are entirely as predicted. Other properties have unexpected features and ambiguities. For example, although amorphous solids consist largely of random aggregates of atoms their densities are only slightly different from the density of crystals of the same composition.

The broad theoretical question is: how does the amorphous atomic structure affect all of the characteristics, e.g. magnetic, mechanical, chemical or corrosion, and electrical. Each of these will be discussed in some detail in the forthcoming sections of this review. Much of our understanding has come from comparing the properties of the amorphous alloy with the same or a similar crystalline alloy. However, this has only limited applicability because most of the interesting amorphous alloys have no simple or single crystalline counterpart. One of the singular advantages of studies on amorphous alloys is that we can vary the composition continuously, to prepare homogenous alloys which can be studied as a function of composition and temperature without complicating interference from structural phase transitions. Although these complications do not exist, more subtle changes do occur at temperatures well below crystallization. For example, phase separation, diffusion of various species and structural relaxations all occur even though the alloy remains amorphous. That is, the amorphous phase is not a stable ground state of the solid. All of these changes can have effects on the observed properties.

The purpose of this book is to document these unique characteristics and the applications they lead to and to discuss the present status of our basic understanding of the origins of their unique behaviour. Most of the emphasis will be on melt-quenched transition metal–metalloid and transition metal–zirconium type alloys.

1.2 Historical development of amorphous metallic alloys

In the past 8000 years that humans have used metals their structure has consisted of crystalline aggregates. Historically, the first report in which a range of amorphous, i.e. non-crystalline, metallic alloys were claimed to have been made was by Kramer[1,2]. This was based on vapour deposition. Somewhat later Brenner *et al.*[3] claimed to have made amorphous metallic alloys by electrodepositing nickel–phosphorus alloys. They observed only one broad diffuse peak in the X-ray scattering pattern in the non-magnetic high-phosphorus alloys. Such alloys have been in use for many years as hard, wear and corrosion resistant, coatings. It was not until 1960 that Duwez and his coworkers discovered a method of preparing amorphous alloys by direct quenching from the melt. The story of this discovery is a fascinating one and has been told by Duwez[4]. An edited version follows.

Metallurgists are very familiar with the term 'quenching'. This is generally defined as the process of rapid cooling. The main purpose of quenching is to cool an alloy at a high enough rate so that phases stable at high temperatures are either partially retained, transformed into non-equilibrium phases, or both. Subsequent heat treatment is then used to control the relative amounts as well as the microstructure of the desired phases to achieve the optimum physical properties of the final product. In this definition of quenching nothing is said about the initial state of the material to be quenched, although it is taken for granted that it is in the solid state. This need not be the case and high cooling rates can indeed be applied to alloys in the liquid state as well.

In quenching solid alloys, the purpose is to cross the phase boundaries rapidly enough to prevent totally, or at least partially, the equilibrium reactions from taking place. In quenching from the liquid state the critical phase boundaries are the liquidus and the solidus in the phase diagram. Since the atomic mobility in a liquid is far greater than in a solid, the rates of cooling required to influence the crystallization of an alloy are obviously much greater than those necessary to prevent a phase change in the solid state. As a result, the conventional techniques of quenching a solid do not lead to any significant results when applied to a liquid alloy. Very simple techniques based on cooling by conduction of the liquid on to a solid substrate have been developed and crystallization of liquid alloys can definitely be modified and in some cases completely suppressed.

The motivation for achieving extreme rates of cooling in liquid alloys was to try to prevent the separation into two phases in binary alloy systems in which, according to the generally accepted Hume-Rothery rules, the two metals should form a complete series of solid solutions, and yet a eutectic system is found under equilibrium conditions. The test case chosen for evaluating the efficiency of the quenching techniques was the copper–silver system, and in September 1959 a complete series of solid solutions in this system was obtained at Caltech. At that time, the rate of quenching was not known, and no time was devoted to try to measure it, because more exciting results immediately followed. These unexpected results were the synthesis of a new crystalline non-equilibrium phase in the silver–germanium system which under equilibrium conditions is of the simple eutectic type. Shortly after that, the ultimate goal in quenching from the liquid state was reached when a non-crystalline (amorphous) structure was obtained in gold–silicon alloys. Total suppression of the crystallization process during solidification had been achieved.

Almost simultaneously Miroshnichenko and Salli[5] in the USSR reported on a very similar device for preparing amorphous alloys. In this technique a liquid metal alloy drop is propelled on to a cold surface where it spreads into a thin film and is thus rapidly solidified. Duwez actually propelled the liquid drop, whereas Miroshnichenko and Salli propelled two opposing pistons together with the drop in between. These techniques soon acquired, over the opposition of its inventor, the anomatopaeic designation of 'splat-cooling'. These splat-cooling techniques can generate cooling rates of greater than a million degrees per second, thus creating a completely new metallurgy of highly supersaturated solid solutions, new metastable crystalline structures and glassy alloys.

The final development in the story of the preparation of amorphous alloys was the publication by Pond and Maddin[6] of a technique for the preparation of continuous long lengths of ribbons. This opened up the possibility of large scale production and set the stage for the explosive growth of work on amorphous alloys since it was now clear that these alloys could be prepared in large quantities at low cost. Although large quantities, relative to the usual laboratory quantities of a few grams per run, are now available commercially, the price is still dropping rapidly (*Figure 1.1*). The ultimate cost

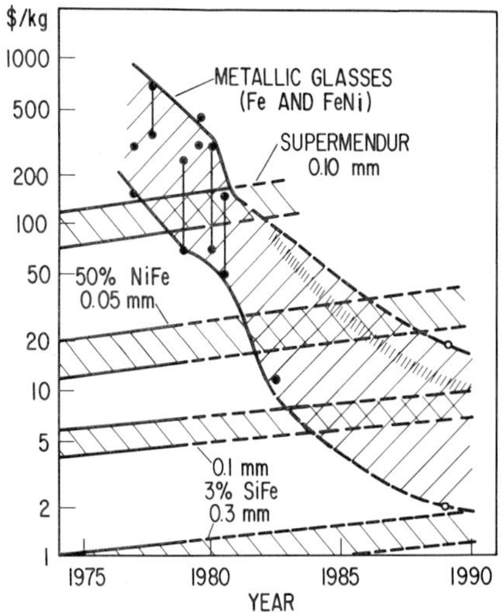

Figure 1.1 Past prices and future possible prices of metallic glasses and some competitive materials. Solid dots are prices and price ranges from Allied Chemical for METGLAS for different alloys, widths and quantities. The open circles are estimates from USA experts. The fine cross-hatched line is the prognosis from Vacuumschmelze (after Raskin and Davis[7])

(or price) has been estimated for very large quantity production to be potentially as small as 2.00 dollars/kg for the iron–boron based alloys. Thus, these alloys may become cost competitive with the oriented Fe–3.2 % Si which sells for about 1.30–1.50 dollars/kg. Concurrent with the drop in price the production quantity has been projected by Raskin and Davis[7] to rise from the present level of roughly 15 000 kg to 400 000 kg by 1985 and to 40 000 000 kg by 1987 (*Figure 1.2*). This inverse relation between price and quantity used appears to be valid for many materials and is illustrated in *Figure 1.3* for a variety of soft magnetic materials.

The other part of this fascinating story has to do with the ferromagnetic properties of amorphous metallic alloys. Because of the lack of atomic ordering it was believed for many years that ferromagnetism could not exist in amorphous solids. However, in 1960 Gubanov[8] predicted, on the basis of theoretical analysis, that amorphous solids would be ferromagnetic. This was based on evidence that the electronic band structure of crystalline solids did not change in any fundamental way on transition to the liquid state. This implies that the band structure is more dependent on short-range, rather than long-range, order so that ferromagnetism, which depends on short-range order, should not be destroyed in the corresponding amorphous solid. The theoretically expected retention of ferromagnetic behaviour in amorphous solids was first demonstrated by Mader and Nowick[9] in 1965 in work on vacuum-deposited Co–Au alloys and soon thereafter by Tsuei and Duwez[10] in work on splat-cooled Pd–20 atom % Si containing some ferromagnetic element partially substituted for the palladium. The first alloy with a substantial magnetization, further confirming

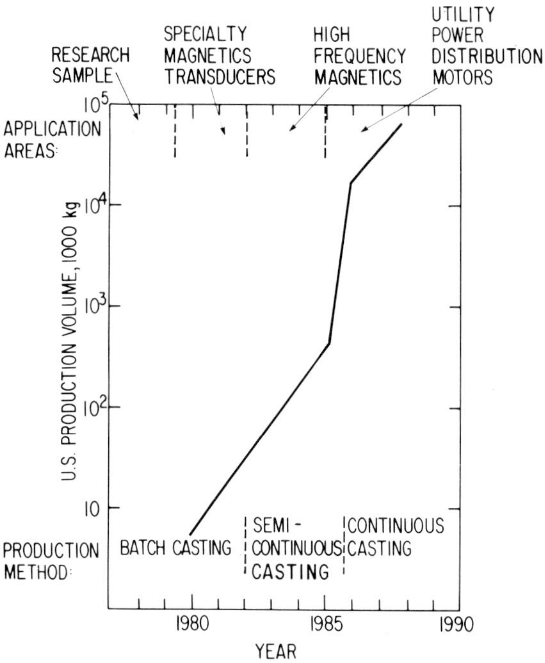

Figure 1.2 Forecast of possible production volume (after Raskin and Davis[7])

Gubanov's predictions, was $Fe_{75}P_{15}C_{10}$ reported by Duwez and Lin[11]. This appeared to be a typical soft ferromagnetic alloy with the large saturation magnetization of 7 kilogauss [0.7 T] and the relatively low coercive force of 3 oersted [240 A/m]. Simpson and Brambley[12] appear to have been the first to point out that the amorphous alloys, expected to have no magnetocrystalline anisotropy, should have very low coercivities. However, the early amorphous alloys of Co–P, prepared by deposition methods, had coercivities as high as 10—20 Oe [800—1600 A/m]. These high coercivities are now understood to arise from compositional inhomogeneities demonstrated by Chi and

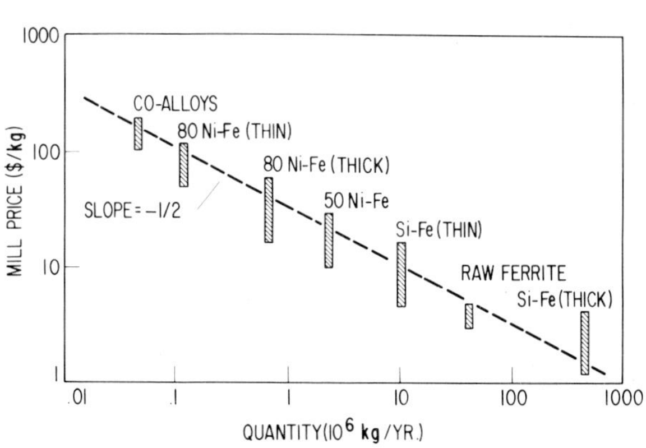

Figure 1.3 Price of various soft magnetic materials displayed as a function of the quantity sold per year. For the year 1977

Cargill[13] from small angle X-ray scattering analysis and from strain-magnetostriction anisotropy. The melt-quenched alloys of Fe–P–C appeared to be compositionally much more homogeneous but still developed coercivities of a few oersteds [160 A/m]; many orders of magnitude higher than in the commercially available Fe–Ni alloys. It is now clear that this is the result of the large strain introduced by the rapid quenching in the splat technique. This gives rise to a large strain-magnetostriction anisotropy. Amorphous alloys of Fe–Ni–P–B prepared as ribbons by the melt-quenching technique by solidification on the surface of a rapidly rotating drum exhibited even lower coercivities, in the order of 0.1 Oe [8 A/m]. Luborsky *et al.*[14] first demonstrated the reduction of coercivity in these alloys, down to less than 10 mOe [0.8 A/m] by suitable anneals and showed that the changes in properties correlated with the relief of internal strains. A new class of amorphous alloy compositions has recently been reported by Nose and Masumoto[15]. These are alloys based on the transition metal with zirconium or hafnium, e.g. $(Co,Ni,Fe)_{90}Zr_{10}$. They have somewhat lower saturation magnetizations than the $Fe_{80}B_{20}$ alloys but otherwise have very similar losses, permeabilities, magnetostrictions, etc.

1.3 Previous reviews

There have been a number of reviews covering the general phenomena of rapidly quenched metals. Duwez[4,16] reviewed in some detail most of the early work, mainly done in his laboratory. Giessen[17] alphabetically classified, by alloy, work published up to the end of 1968. Giessen and Willens[18] emphasized the underlying principles of the work reviewed and Anantharaman and Suryanarayana[19] classified the information obtained from the alloy systems studied. Giessen and Wagner[20] reviewed the structure and properties of amorphous metallic phases produced mainly by quenching from the melt. Jones[21] provided an extensive compilation of developments in splat-cooling and metastable phases. It included an analysis of the methods available for quenching from the melt and an understanding of how they work, the structural features obtained in both metastable crystalline and amorphous phases, the response of the as-quenched structures to annealing, and finally the properties and applications of splat-cooled alloys. Jones and Suryanarayana[22] published a comprehensive annotated bibliography covering the period up to near the end of 1972. This has recently been extended[23] to 1980. The status of our knowledge of the structure of metallic alloy glasses, derived both experimentally and theoretically, is covered by Cargill[24]. Our understanding of the formation and stability of amorphous structures was reviewed by Takayama[25] and the structure and crystallization by Waseda *et al.*[26]. A brief review of metallic glasses from fundamentals to practical prospects was presented by Cahn[27] as well as by Guntherodt[28] and by Gilman[29]. A complete review of the status of our understanding of the structure, transport and transformation kinetics, mechanical properities, magnetic properties, electrical properties and corrosion resistance has been published by Chen[30]. Also published in 1980 are some other rather less extensive general reviews by Chaudhari *et al.*[31], by Cahn[32], and by Luborsky[33] on magnetic properties. Furthermore, the first books are expected to appear soon[33,34].

Extensive collections of papers on glassy metals can be found in the Proceedings of four Conferences on Rapidly Quenched Metals. The first was published in *Fizika* [volume 1, supplement 2 (1970]; the second was held at Massachusetts Institute of Technology, and the Proceedings, edited by N. J. Grant and B. C. Giersen, were published in two parts, by the MIT press (1976) and in *Materials Science and Engineering* (1976). The Proceedings of the third conference, held in Brighton, were edited by B. Cantor and published by The

Metals Society, London (1978), and of the fourth, held in Japan, and edited by T. Masumoto and K. Suzuki, were published by the Japan Institute of Metals, Sendai (1982). Other extensive reports will be found in the *Scientific Reports of the Research Institutes of Tôhoku University* [Series A, No. 1, (1976); Nos. 4 and 5 (1977); No. 2 (1979) and Nos. 1 and 2 (1980)]; the book '*Metallic Glasses*', edited by J. J. Gilman and H. J. Leamy and published by the Americal Society for Metals (1978); the *Annals of the New York Academy of Sciences* (1976); the Proceedings of the Fourth International Conference on Liquid and Amorphous Metals [Supplement C8 to J. Phys (Fr.) (1980)]; Supplements to the *Scientific Reports of Tôhoku University* (1978 and 1980); the Proceedings of the Conference on Amorphous Metallic Materials, held in Smolenice, edited by P. Duhaj and P. Mrafko and published by VEDA, for the Slovak Academy of Science (1978); and the Proceedings of the Conference on Metallic Glasses: Science and Technology, held in Budapest and published by the Central Research Institute for Physics, Budapest (1980).

References

1. Kramer, J., *Annln Phys.*, **19.** 37 (1934)
2. Kramer, J., *Z. Phys.*, **106,** 675 (1937)
3. Brenner, A., Couch, D. E. and Williams, E. K., *J. Res. natn. Bur. Stand.*, **44,** 109 (1950)
4. Duwez, P., *Trans. Am. Soc. Metals*, **60,** 607 (1967)
5. Miroshnichenko, I. S. and Salli, I. V., *Ind. Lab.*, **25,** 1463 (1959) in English; *Zav. Lab.*, **25,** 1398 (1959)
6. Pond, R. (jr.) and Maddin, R., *TMS-AIME*, **245,** 2475 (1969)
7. Raskin, D. and Davis, L. A., *IEEE Spectrum*, **18,** 28 (1982)
8. Gubanov, A. I., *Fizika*, **2,** 502 (1960)
9. Mader, S. and Nowick, A. S., *J. Vac. Soc.& Technol.*, **2,** 35 (1965)
10. Tsuei, C. L. and Duwez, P., *J. appl. Phys.*, **37,** 435 (1966)
11. Duwez, P. and Lin, S. C. H., *J. appl. Phys.*, **38,** 4096 (1967)
12. Simpson, A. W. and Brambley, D. R., *Phys. Status Solidi*, **43,** 291 (1971)
13. Chi, G. C. and Cargill, G. S. (III), *Mater. Sci. Engng*, **23,** 155 (1976)
14. Luborsky, F. E., Becker, J. J. and McCary, R. O., *IEEE Trans. Magn.*, MAG–11, 1644 (1975)
15. Nose, M. and Masumoto, T., *Sci. Rep. Res. Insts Tôhoku Univ.*, **28,** 232 (1980)
16. Duwez, P., *'Progress in Solid State Chemistry'*, Ed. Riess, H., vol. 3, p. 377 Pergamon Press, Oxford (1966)
17. Giessen, B. C., *'Developments in Structural Chemistry of Alloy Phases'*, Ed. Giessen, B. C., p. 277, Plenum Press, New York (1969)
18. Giessen, B. C. and Willens, R. H., *'Use of Phase Diagrams in Ceramic, Glass and Metal Technology*, Ed. Alper, A. M., vol. 3, p. 103, Academic Press, New York (1970)
19. Giessen, B. C. and Wagner, C. N. J., *'Liquid Metals'*, Ed. Beer, S. Z., p. 633, Marcel Dekker, New York (1972)
20. Anantharaman, T. R. and Suryanarayana, C., *J. mater. Sci.*, **6,** 1111 (1971)
21. Jones, H., *Rep. Prog. Phys.*, **36,** 1425 (1973)
22. Jones, H. and Suryanarayana, C., *J. mater Sci.*, **8,** 705 (1973)
23. Suryanarayana, C., Plenum Publ. Corp., New York (1980)
24. Cargill, G. S. (III), *Solid State Phys. 30*, 227 (1975)
25. Takayama, S., *J. mater. Sci.*, **11,** 164 (1976)
26. Waseda, Y., Okazaki, H. and Masumoto, T., *J. mater. Sci.*, **12,** 1927 (1977)
27. Cahn, R. W., *Metall. Mater. Technol.*, p. 309 (June, 1978)
28. Guntherodt, H. J., *Adv. Solid State Phys.*, **17,** 25 (1977)
29. Gilman, J. J., *Science, N.Y.*, **208,** 856 (1980)
30. Chen, H. S., *Rep. Prog. Phys.*, **43,** 353 (1980)
31. Chaudhari, P., Giessen, B. C. and Turnbull, D., *Scient. Am.*, 98 (April 1980)
32. Cahn, R. W., *Contemp. Phys.*, **21,** 43 (1980)
33. Luborsky, F. E., *'Ferromagnetic Materials'*, Ed. Wohlfarth, E. P., vol. 1 chap. 6, North-Holland Publ. Co., Amsterdam (1980)
34. Hasegawa, R. (Editor), *Publ. CRC*, in the press
35. Luborsky, F. E., Chap. 19, this volume

Chapter 2

Metallic glass formation

H. A. Davies

Dept. of Metallurgy, University of Sheffield, Sheffield, UK

2.1 Introduction

A non-crystalline solid phase represents the ultimate state of solid metastability and may be formed by a number of different routes, including solidification from the liquid or vapour states, deposition from a chemical solution or an electrolyte and by high-energy ion- or neutron-bombardment of crystalline materials. A non-crystalline solid formed by continuous cooling from the liquid state is known as a glass in the original sense of the word. The vitrification of a melt requires that it be cooled at a rate sufficiently high to avoid a significant degree of crystallization, so that the 'disordered' atomic configuration of the liquid state is frozen-in, or at least largely so. Glass formation occurs easily in some familiar classes of non-metallic material such as silicates and organic polymers. In these, the nature of the bonding places severe limits on the rate at which the atomic or molecular rearrangements, necessary for maintaining thermodynamic equilibrium during cooling, can occur; thus the melt solidifies to a glass, even at low rates of cooling often less than 10^{-2} K/s. Metallic melts, in contrast, have non-directional bonding, so that atomic rearrangements occur very rapidly, even at high degrees of undercooling below their equilibrium freezing temperatures. Hence, very high cooling rates ($>10^5$ K/s) must generally be imposed to form metallic glasses, and it was not until the development of a splat cooling technique that the first metallic glass phase was formed by continuous cooling from the melt by Klement, Willens and Duwez[1] in 1960.

Although this Au–25 atom %Si glass was produced only in very small quantity and proved to be unstable to room temperature, the achievement added an important new dimension to both metal science and glass science. Subsequent studies have shown that metallic glass phases can be formed at cooling rates of 10^6 K/s or less in a wide variety of alloy systems, albeit over limited composition ranges. Some of these have technologically attractive soft magnetic and mechanical properties and excellent corrosion resistance. These properties have stimulated widespread scientific study and, together with the development of processes for continuous casting of alloy glasses as uniform thin strip, are leading to significant commercial exploitation of this new class of material. Section thicknesses for most of the potentially useful metallic glasses are however presently limited to less than about 80 μm because of their high critical cooling rates $\dot{T}_c$ for glass formation. In fact, $\dot{T}_c$, which is a convenient measure of glass forming ability (GFA), spans a range of at least seven orders of magnitude for metallic alloys—from about 100 K/s to more than

10^9 K/s and corresponding to a range of critical thicknesses x_c from about 1 mm down to below 0.1 μm^2. An overview is presented of the present understanding of the glass forming behaviour of metallic alloys and of its scientific and technological significance. The formation of amorphous alloys by techniques other than quenching from the melt will not be considered here, except where appropriate for comparison purposes.

2.2 Undercooling of the melt and glass formation

When a metal or alloy melts, the three dimensional lattice arrangement of atoms, that is, the long-range order, is destroyed and, in the liquid, the atoms vibrate about positions that are constantly and rapidly interdiffusing. During melting, the crystal and liquid phases are in equilibrium and, for pure metals, the volume, enthalpy and entropy undergo discontinuous change; the enthalpy and entropy increase, and the volume usually does also, except where the atomic packing in the crystal is relatively open, as for semimetals. A liquid, at temperatures above the melting point, is in a state of internal equilibrium and its structure and properties are independent of its thermal history. It is characterized by inability to resist a shear stress. Metallic liquids have high fluidity because of the non-directional nature of the metallic bond while, in contrast, molten silicates, borates and similar substances, in which the atomic bonding is strongly covalent, are characterized by very low fluidity or, conversely, high viscosity.

On cooling, a liquid must undercool below the equilibrium crystallization temperature before crystallization can begin because of the existence of an energy barrier to the formation of nuclei. The degree of undercooling that occurs depends on several factors, including the initial viscosity of the liquid, the rate at which the viscosity increases with decreasing temperature, the temperature dependence of the free energy difference between the undercooled liquid and the crystal phase, the interfacial energy between melt and crystal, the volume density and the efficacy of heterogeneous nucleating particles and the imposed cooling rate. Liquid transition metals such as iron, nickel and cobalt, in which the heterogeneous nucleants have been largely removed with a flux, can be undercooled in bulk by more than 200 K under slow cooling conditions[3]. Similarly, when metallic melts are subdivided into very fine discrete droplets, suspended in a liquid medium, at least some droplets will be devoid of any heterogeneous nucleating particles and nucleation occurs homogeneously at relative undercoolings typically 0.2—$0.3T_m$, where T_m is the equilibrium freezing temperature[4]. However, the growth rates for crystals in metallic melts are very rapid once nucleated and, where the rate of heat removal to the surroundings is small, rapid recalescence occurs.

If, however, the liquid is cooled rapidly by heat removal to a sink, the influence of heterogeneous nucleants is increasingly delayed for kinetic reasons and, as the cooling rate is increased, the undercooling is enhanced and the recalescence decreased. Thus, the temperature range over which crystallization proceeds becomes increasingly depressed, leading to structural modifications; this leads initially to the refinement of microstructural features then, for alloys, depending on the composition, to extension of solute solubility and the formation of metastable crystalline phases. Eventually, when the cooling rate is sufficiently high, crystallization is suppressed because of insufficient time for significant growth or, in the extreme, for nucleation. In this case, the shear viscosity of the melt η increases continuously, as shown in *Figure 2.1*.

Although the driving force for nucleation is continually increasing, this is opposed by the rapidly decreasing atomic mobility which, at very high undercoolings, dominates. Eventually, the atomic configuration of the liquid departs from equilibrium and then shortly thereafter becomes homogeneously frozen, at the so-called glass transition

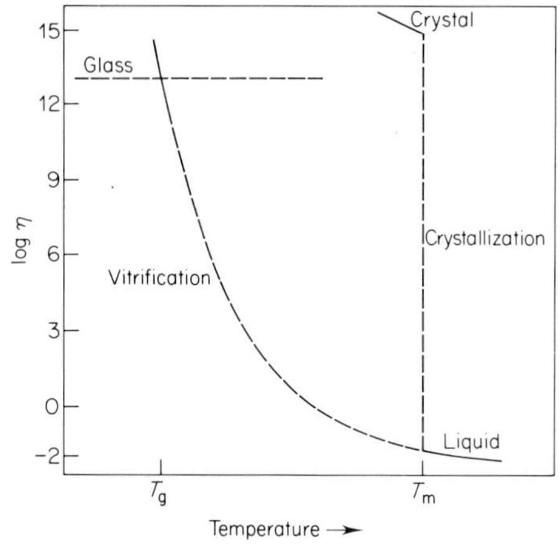

Figure 2.1 Viscosity (η)–temperature (T) behaviour corresponding to crystallization and vitrification of a metallic melt

temperature T_g. This structural freezing to the glass state is, by convention, considered to occur when η is about 10^{13} poise. However, in practice, there is no unique glass structure for a given material since the temperature and the structure, at which departure from configurational equilibrium occurs, depends on the cooling rate. Thus T_g and the glass structure are both cooling-rate dependent, the variation in the latter manifesting itself in different structural relaxation behaviour during subsequent annealing of the glass. At T_g, the temperature dependence of the volume and of the enthalpy decrease discontinuously, the marked decrease in specific heat C_p reflecting the fewer degrees of freedom resulting from the freezing of the structure (*Figure 2.2*). Whereas the difference ΔC_p in the specific heat of the undercooled liquid and the crystalline state increases with falling temperature, it becomes very small once vitrification has taken place. On the free volume model of atomic transport, the glass transition corresponds to the elimination of most of the free volume[5].

Analysis of diffraction data[6] suggests that the atomic configuration of a metallic alloy glass bears a rather close resemblance to the static structure of the equilibrium liquid near its freezing temperature, but with the atoms in the glass vibrating about positions that are more condensed and now frozen. The glass structure has been proposed as approximating to an assemblage of dense random-packed hard spheres[7] of two or more different sphere diameters. For alloys where the solute atoms are sufficiently concentrated and are sufficiently different in size from those of the solvent, these configurations are stable against crystallization. As would be expected for the dense

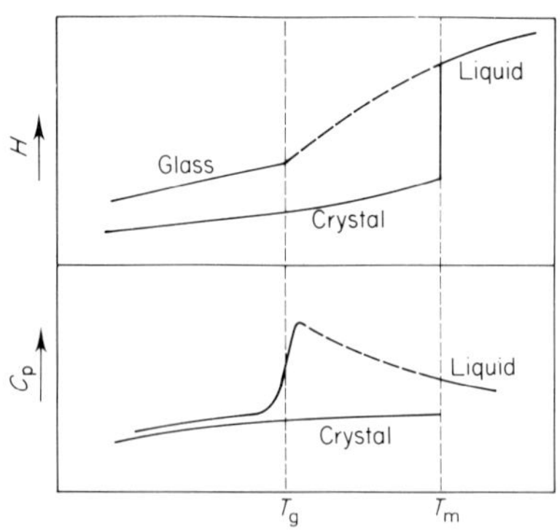

Figure 2.2 Temperature dependence of the enthalpy H and specific heat C_p corresponding to crystallization and vitrification of a metallic melt

packing of atoms approximating to hard spheres, there is a significant degree of topological short-range order. There is also clear evidence of *chemical* short-range order, i.e. non-random packing of atoms of different species, that is carried through from the liquid, and, indeed, enhanced[8]. It has been further suggested[9] that the local structure in a metallic glass resembles that in the derivative 'inter-metallic' crystalline phase with an essentially disordered arrangement of these structural units, similar to network glasses.

The process of vitrifying a liquid is therefore, strictly, one of avoiding the nucleation of crystals during cooling, although a more practical definition would be the avoidance of a just detectable fraction of crystal. The glass forming ability of a material is then determined by the kinetics of the nucleation process and, for practical purposes, of the early stages of crystal growth.

2.3 Glass forming ability in metallic materials

2.3.1 Nominally pure metals

In accordance with a kinetic view of glass formation, it is likely that any metallic alloy containing more than, at most, a few atomic per cent of solute, and, possibly as little as a fraction of a percent, could be quenched to a glass, given a sufficiently high cooling rate $\dot{T}$ and, importantly, given that it is cooled to a temperature sufficiently low to avoid spontaneous crystallization. The situation of pure metals and very dilute alloys is less clear. Several studies have been made of thin films produced by condensation from the vapour phase at very low temperatures. This is a process that gives a much higher effective quench rate (of the order of 10^{13} K/s) than is normally attainable from the melt; since deposition occurs atom by atom and the heat of condensation is rapidly removed through the thin film. Amorphous thin films of several transition metals, of generally unspecified purity, have been produced by vapour deposition[10,11] and, for instance, crystallization temperatures T_x of up to 530 K[10] have been reported for amorphous nickel films. In contrast, subsequent careful study of the influence of gaseous impurity on the condensed film structure indicated that amorphous films of nickel could not be produced at 4 K when the impurity content was 0.07 %[12]. (On the other hand, films of another close-packed metal, cobalt, and of chromium, iron and manganese, of similar nominal impurity content, condensed[13] in the amorphous state at 4 K but it is possible that they were more susceptible to impurity stabilization than nickel).

In the case of quenching from the melt, a glassy phase was observed in very thin sections (about 100 nm thick) at the edges of splat foils of nickel, quenched in air[14]. This phase was found to crystallize at about 420 K and it is likely to have been stabilized by dissolved gaseous impurity, possibly up to 2 atom %. The cooling rate $\dot{T}$ corresponding to the 100 nm splat quenched thickness is estimated to be of the order of 10^9—10^{10} K/s[15]. Pure germanium was found to vitrify at a similar $\dot{T}$, both by splat quenching in argon[16] and by dispersion of melt as 30 nm particles in argon[17]. However, although germanium is metallically bonded in the liquid state, it assumes tetrahedral, covalent bonding on vitrification and the structural rearrangement required for crystallization stabilizes the glass to high temperatures, even in the absence of impurity atoms. Similarly, pure silicon has also recently been vitrified in very thin sections by use of a pulsed laser surface melting technique.

Notwithstanding the various reports of amorphous films of nominally pure metals, the balance of evidence suggests that a minimum concentration of impurity atoms is required to promote the vitrification of a *metal* but which may depend on the chemical nature of the impurity[2,18,19]. For pure metals having close- or densely-packed crystal structures, crystallization of a glassy phase would require little structural adjustment and no thermal activation, which, in effect, means that the presence of a free surface, where atomic packing is relatively loose, would lead to spontaneous crystallization.

2.3.2 Influence of alloying

The effect of adding solute atoms to a pure metal, and especially if they are of different size and chemical character from the host atoms, is to require that significant diffusive rearrangement occur during crystallization of the glass phase, i.e. to stabilize the glass. As the solute concentration increases, T_g generally increases and T_x, when measured dynamically under conditions of relatively rapid heating (more than about 1 K/s), approximates more closely to T_g. On the other hand, on alloying, the temperature at which freezing begins under equilibrium conditions, i.e. the liquidus temperature T_l, often decreases. In such cases, the interval between T_l and T_g generally decreases with increasing solute concentration, so that the probability of being able to cool through the interval without inducing crystallization is enhanced, i.e. the glass forming ability GFA is increased. (The GFA is, as indicated earlier, measured by the magnitude of $\dot{T}_c$ and thus of x_c. In practice, however, the observed x_c is always smaller than the theoretical limit because of resistance to heat flow at the interface between the melt and the heat sink.)

Accordingly, the alloy systems for which glass formation occurs most readily, i.e. having the highest GFA, are those that manifest either one or more *deep* eutectics, or a steep and substantial decrease in T_l with increasing percentage of solute, to a 'plateau' over which T_l is low in comparison with the melting point T_m of the host metal. Thus, there is a *tendency* for the GFA to vary inversely with the ratio T_g/T_l, which is referred to as the reduced glass temperature T_{rg} [19,20]. The lower is T_l and the higher is T_g, then the higher tends to be the GFA. The highest known values of T_{rg} for metallic alloys are in the range 0.66—0.69, e.g. 0.69 for $Ca_{65}Al_{35}$ [21].

The meaning of 'readily glass forming' is somewhat arbitrary; by convention, it has been taken to indicate completely glass forming either when splat-quenched as foils about 10—20 μm thick or, more recently, with the advent of melt spinning processes, when continuously cast as ribbon or strip 10—20 μm thick. Thus, the glass forming range (GFR) for an alloy system is conventionally defined as that within which x_c is more than about 10—20μm. In reality of course, the GFR will depend on the value of x_c chosen, the smaller is x_c the wider being the GFR. Values of x_c of 10—20 μm correspond to $\dot{T}_c$ of the order of 10^6—10^7 K/s, for most of the melt quenching processes commonly employed, and to T_g/T_l generally about 0.45 [14]. In practice, also, there is the additional requirement that T_g be higher than room temperature, unless the alloy is quenched below ambient.

For condensation processes such as vapour deposition and cathodic sputtering, the ranges of alloy systems and compositions over which amorphous phases are formed[22,23] are wider than for liquid quenching at 10^6 K/s, because of the higher effective quench rate, though care may be required to control the rate of deposition and the substrate temperature, to avoid crystallization. Amorphous film formation by electrolytic or electrodeless deposition has been reported for both metal–metalloid alloys (e.g. Ni–P, Co–P[24]) and metal–metal alloys (e.g. Cr–W, Cr–Mo[25]). Interestingly, for the former, it is at compositions close to eutectics that amorphous films form most easily.

2.3.3 Classification of readily glass forming (RGF) alloy systems

There are a number of distinct categories of alloy system within which glass formation from the melt occurs readily. Lists of RGF compositions and composition ranges have been presented by Takayama[26] and by the present author[15,27] and for binary metal–metal systems by Suryanarayana[28]; further compilations are contained within two bibliographies [29,30]. The number and diversity are continually increasing, however, and no attempt is made here to present an exhaustive list because of the complexity in the case of ternary, quaternary and higher order alloys.

The first class to be discovered and investigated were the late transition metal–metalloid (TL–M) type; the TL include Group VIIB, Group VIII and, here also, Group IB noble metals. They remain the most important, technologically, especially in the case of glasses based on iron, cobalt and nickel for soft ferromagnetic applications. It was at one time believed that the GFR was, in each case, rather narrow and centred around a deep eutectic at approximately 20 atom % metalloid, i.e. in the range around 13—25 atom %. Indeed, for many binary systems, e.g. Pd–Si[31] and Fe–B[32], this is the case, but recent work, especially, has shown that there are many exceptions, e.g. Ni–B_{31-41}, in addition to Ni–$B_{17-18.5}$ [33]; Co–B_{17-41} [34] and Pt–$Sb_{34-36.5}$ [35].

When further solute species are substituted, either one or more metalloids or transition metals, the GFT may be considerably enhanced, and the GFR may broaden or change considerably. For instance, $Pd_{78}Cu_6Si_{16}$ [36] and $Pd_{40}Ni_{40}P_{20}$ [37] are much easier glass-formers than $Pd_{84}Si_{16}$ and $Pd_{80}P_{20}$, respectively, and, for a second metalloid addition, $Ni_{92-x}Si_8B_x$ alloys are readily glass forming where x is between about 10 and 39 [33], the technologically useful Fe–Si–B alloys between 13 and 35 atom % (Si + B)[38] and Co–Al–B alloys between 35 and 82 atom % Co[39] (Al is apparently acting as a metalloid in the latter case). Substituting the TL by only 6—22 atom % of an early transition metal may shift the GFR very markedly, for instance down to 2—8 atom % of boron in (Fe, Co, Ni)–TE–B alloys, where TE is a group IVB, VB or VIB metal[40]. Alternatively, substitution by TE solutes may render a very difficult glass former RGF, e.g. Fe–C_{15-24} when Fe is partly substituted (up to 50 %) by various combinations of molybdenum, chromium and/or tungsten[41].

Recent work has shown that a number of TE–M alloys are also RGF, close to primary eutectic compositions which lie mostly in the range 15—30 atom % metalloid. The eutectic temperatures for binary eutectic alloys of this type are generally much higher than those for the TL–M type, at least partly reflecting the higher T_m for the TE elements. It appears that the GFR are also narrower than for the TL–M alloys but partial substitution of the metal solvent by a second TE is possible. Examples of this type include Ti–Si_{15-20} [42,43]; Nb_{80}–Si_{20}, (W or Mo)$_{70}Si_{20}B_{10}$ and $Ti_{50}Nb_{35}Si_{15}$ [43]; Re_{65}–Si_{35} and $W_{60}Ir_{20}B_{20}$ [44].

The second major group consists of alloys of the type TE–TL. The TE have high T_m and addition of a TL (or a Group IB metal) generally leads to very rapid decrease in T_l, down to a eutectic; T_l then remains relatively low across one or more intermetallic phases of relatively low stability, and other eutectic reactions, so that the GFR is usually quite wide, even for binary systems. Examples of this type are Cu–Ti_{35-70} [45]; Cu–$Zr_{27.5-75}$ [46], Ni–$Zr_{33-42,60-80}$ [47]; Nb–Ni_{40-66} and Ta–Ni $_{40-70}$ [48]: for the Group IV metal–TL alloys, such as (Co, Ni or Fe)Zr, there is also a narrow RGF region at exceptionally low solute contents, e.g. Co–Zr_{9-16} and Fe–Zr_{9-11} [49]; these have potentially useful soft ferromagnetic properties. Also included in this group are alloys where the TL is a platinum group metal; here, the eutectic temperatures are generally very high (above 1500 °C) and only a few specific compositions appear to have been investigated, e.g. $Nb_{55}Ir_{45}$, $Ta_{55}Rh_{45}$ [44]. They have very high crystallization temperatures (above 1000 K).

Most of the binary alloy systems of lanthanide metals with late transition and Group IB metals (RE–TL) have deep eutectics and, accordingly, several have been shown to be RGF, centred approximately about the eutectic composition in each case. (Yttrium is also included with the RE metals.) Their GFR lie on the RE-rich side of the system in

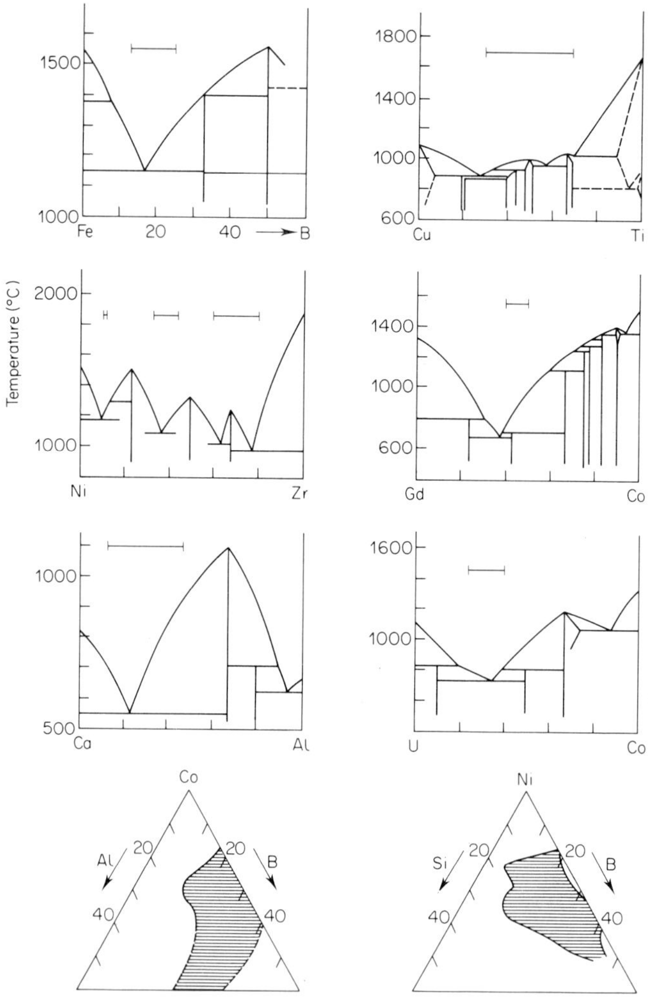

Figure 2.3 Composition dependence of the liquidus temperatures and glass forming ranges corresponding to a cooling rate of about 10^6 K/s for six types of binary alloy system and the glass forming range for two ternary systems

each case. Examples of this type include La–Au_{18-26} and $La_{78}Ni_{22}$ [50]; Gd–Fe_{32-50} and $Er_{68}Fe_{32}$ [51]; Gd–Co_{40-50} [52]. Some lanthanum-based alloys containing a B sub-group metal or a metalloid are also glass-forming, e.g. La–Al_{18-34} and La–Ge_{17-22} [53]; (Ce or $Pr)_{89}Sn_{11}$ [54]. The GFR for these systems do not overlap with the composition ranges of interest for possible magnetic bubble applications and samples of the latter have been prepared by sputtering[55].

Another major class of RGF alloys is represented by binary and multicomponent alloys of the Group IIA alkaline earth (AE) metals with certain B sub-group metals, with Group IV TE or with TL and Group IB noble metals. The AE metals, in common with RE metals, have extremely high affinities for oxygen; consequently, they must be vitrified in a controlled atmosphere but the resulting glasses are also generally reactive in air. In some cases, the GFR are exceptionally wide. Examples of this class include

Ca–Al$_{12.5–47.5}$ [56,57]; Ca–Cu$_{12.5–62.5}$ [56]; Ca–Pd[58]; Mg–Zn$_{25–32}$ [59]; Be–Zr$_{50–70}$ [60]; $Sr_{70}Ga_{30}$ and $Sr_{70}Mg_{30}$ [57] (in the latter case the alloy involves two Group IIA elements).

Finally, several alloys based on the actinide metals, thorium, uranium, neptunium and plutonium for which the eutectic temperatures are substantially depressed, have been established as RGF[61–63]. Examples include U–Co$_{24–40}$ [61], Np–Ga$_{30–40}$ [63], and Pu–Ni$_{\sim 12–30}$ [63].

The composition ranges for RGF for selected binary alloy examples of each type of system are shown in *Figure 2.3* together with the respective equilibrium phase diagram. Also given are two examples of ternary systems indicating the range of RGF. The binary alloy data serve to emphasize the generally close correlation between the GFR and the regions of lowest T_l, in each system.

2.4 Theory of metallic glass formation

2.4.1 Semiempirical criteria

As was indicated in an earlier section, the interval between T_l and T_g, in relation to T_l, and thus the magnitude of the ratio T_g, would be expected to have a substantial influence on the kinetics of glass formation in alloys. But since T_g has been found to vary only relatively slowly with composition in most systems investigated (*see*, for instance, refs. 31 and 32) the composition-dependence of the GFA is often dominated by the variation of T_l. There is a more direct correlation than for network glasses, for instance; this arises from the relative constancy of the shear viscosity η of metallic melts at their respective freezing temperatures (of the order of a few cP), since they are structurally more homogeneous than network glass-forming melts. The viscosities of the latter at T_l may vary with composition by several orders of magnitude, depending on the strength and directionality of the covalent bonding, and thus the GFA does not scale closely with T_l[64].

Accordingly, various semiempirical methods have been proposed for relating the GFA for metallic systems to the depression of T_l below some mean value. In one approach, Marcus and Turnbull[65] used the normalized parameter $\Delta T/T_l^{\circ}$, where ΔT is expressed as the deviation of T_l from the ideal solution liquidus temperature T_l°, which is given by:

$$T_l^{\circ} = \frac{\Delta H_f^A T_m^A}{\Delta H_f^A - R \ln (1 - x) T_m^A} \tag{2.1}$$

ΔH_f^A and T_m^A are the heat of fusion and melting point, respectively, of the solvent metal and x is the mole fraction of solute. Large positive values of the parameter, it was proposed, should indicate a high GFA. Although some of the known RGF TL–M binary alloys were found to have high values (about +0.4), the approach did not successfully predict RGF for many TL–TE alloys, for which the parameter had a negative value.

It was observed [66] that, for many of the easier glass forming eutectic alloys, both metal–metal and metal–metalloid, $T_e/T_m^A \sim 0.6$, where T_e is the eutectic temperature and T_m^A the melting point of the major component. Donald and Davies[67] proposed that the GFA of alloys could be related to the simple parameter:

$$\Delta T^* = (T_l^{mix} - T_l)/T_l^{mix} \tag{2.2}$$

which represents the fractional departure of T_l from the simple rule of mixtures melting temperature T_l^{mix}, where

$$T_l^{mix} = \sum^{n} x_i T_m^i$$

x_i and T_m are the mole fraction and melting point, respectively of the ith component of an n-component alloy. It was found[67] that, in an assessment of 68 melt-quenched binary and ternary alloys, those that were RGF, i.e. could be fully vitrified at $\dot{T} \sim 10^6$ K/s, had values of $\Delta T^* > 0.2$, with only eight exceptions to the rule. This provides a reasonably reliable indication of whether an alloy is RGF, providing its liquidus temperature is known. Several binary alloys whose GFA had, at that time, not been verified experimentally, were identified as being potential RGF[67] and subsequently further systems have been suggested[68]. The criterion has been employed successfully to predict RGF in a number of beryllium-based alloys[69]. In systems where exceptionally high melting point phases occur, ΔT^* is more appropriately calculated with respect to T_m of the pure element and T_l of the intermetallic phase which bound the composition in question.

As was noted earlier, substitution of a second, or more, solutes will often render a difficult binary glass former RGF (e.g. $Fe_{80}P_{13}C_7$ or $Fe_{62}Mo_{20}C_{18}$, cf. $Fe_{83}C_{17}$) or a RGF alloy more easily glass forming (e.g. $Pd_{78}Cu_6Si_{16}$. cf. $Pd_{82}Si_{18}$). It is possible to rationalize the decreases in $\dot{T}_c$ largely in terms of further depression of T_l, i.e. an increased ΔT^*, arising from the establishment of ternary eutectics. However, other factors such as an enhanced T_g or less quantifiable effects, related to changes in the liquid–nucleus interfacial energy and in the probability of heterogeneous nucleation, may also play significant roles.

Other criteria proposed include a difference in atomic diameters of at least 15% and a difference in group number Δn of at least 5 for the two constituents of a binary alloy[70]. In fact, it is possible in selected cases to obtain RGF in systems for which Δn has values down to zero, e.g. Be–(Ti, Zr); where $\Delta n = 2$; Ca–Mg where $\Delta n = 0$.

Some recent studies have concentrated on the construction of two-dimensional maps to establish criteria for RGF in alloys, based on two parameters. The use of the unlike atom bond strength, expressed by the magnitude of the excess negative enthalpy of mixing ΔH_m, combined with the melting point depression ($T_l^\circ - T_l$), was proposed by Zielinski *et al.*[71]; this gave good separation of RGF and non-RGF metal–metal alloys into discrete zones based on theoretically derived ΔH_m. It is not clear that ΔH_m and ($T_l^\circ - T_l$) are independent parameters; recent studies by Güntherodt *et al.*[72] support the correlation of increasing GFA with increasing negative ΔH_m. This is to be expected on kinetic grounds since deep depressions of T_l are invariably associated with strong interaction on mixing of alloy components. Giessen[73] proposed, for binary alloys, the combined use of ΔH_m and the component atomic radius ratio r_A/r_B, where A represents the smaller atom, and constructed a so-called 'Hume-Rothery' map. These give satisfactory delineation of RGF and non-RGF systems, with the former, as expected, concentrated in the regions having coordinates of high ΔH_m and low r_A/r_B. However, the survey also shows that RGF can occur in alloys where r_A/r_B is nominally as high as 0.94, e.g. $Nb_{55}Ir_{45}$.

2.4.2 Kinetics of glass formation

Though the one- and two-parameter representations described in the previous section are useful predictive and illustrative aids, they are to varying degrees oversimplifications. A quantitative rationalization of GFA is most usefully approached in terms of the kinetics of crystallization, since we are interested in the avoidance of detectable crystallization.

The first such treatment, by Turnbull[19], considered the process of crystal nucleation. It was assumed, for simplicity, that the crystals forming have the same composition as the liquid, that nucleation transients are unimportant and that the bulk free enthalpy change associated with transformation of the undercooled liquid to the crystal phase, ΔG_v, is given by the linear approximation $\Delta G_v = \Delta H_f \Delta T_r$, where ΔH_f is the molar enthalpy of fusion and ΔT_r the reduced undercooling, $(T_l - T)/T_l$. The homogeneous nucleation frequency I_v was shown to be given by:

$$I_v = \frac{k_n}{\eta(T)} \exp\left[-16\pi\alpha^3\beta/3T_r\Delta T_r^2\right] \tag{2.3}$$

where k_n is a kinetic constant, $\eta(T)$ is the shear viscosity at temperature T (which is inversely related to the atomic diffusivity D), T_r is the reduced temperature T/T_l and α and β are dimensionless parameters related, respectively, to the liquid–crystal interfacial energy σ and to the molar entropy of fusion, ΔS_f, thus:

$$\alpha = (N\bar{V}^2)^{1/3}\sigma/\Delta H_f$$
$$\beta = \Delta S_f/R$$

N is Avogadro's number and $\bar{V}$ the molar volume of the crystal. It is evident from equation 2.3 that, as the magnitude of $\alpha\beta^{1/3}$ increases, I_v decreases very steeply, for a given temperature and melt viscosity.

For metallic melts, in common with other simple materials, experimental derivations of σ indicate $\alpha\beta^{1/3} \approx 0.5$; this is consistent with their high resistance to homogeneous nucleation up to large undercoolings, ΔT_r, of about 0.2. An approximately constant value of $\alpha\beta^{1/3}$ being assumed, the principal variable governing I_v and its temperature dependence then becomes the function $\eta(T)$. Since $\eta(T_l)$ is approximately constant for alloy melts at, typically, a few centipoise, the reduced glass temperature T_{rg} becomes the controlling factor. As ΔT_r increases, I_v initially increases steeply, reflecting the increasing driving force, until, eventually, it peaks and undergoes a sharp decrease, as the increasing η becomes dominating at large ΔT_r (*see Figure 2.1*). As T_{rg} is increased, η increases more rapidly with increasing ΔT_r and the I_v–T_r peak is rapidly lowered and shifted to higher T_r. Thus it becomes easier to avoid nucleation, i.e. $\dot{T}_c$ decreases and the volume in which a single nucleus can be avoided, for a given imposed $\dot{T}$, increases. It was predicted, for instance, that an alloy having $T_{rg} = 0.5$ would vitrify in droplets of diameter below about 60 μm at $\dot{T}$ around 10^6 K/s, which was in good accord with experimental observation.

A related kinetic approach was taken by Uhlmann[74] for a number of non-metallic glass formers, but account was also taken of a limited degree of crystal growth; this provides a convenient and practically useful method of quantitatively estimating $\dot{T}_c$. This was adopted by Davies *et al.*[75] for some selected metallic glass formers and was subsequently extended and generalized for metallic systems[15,20,76,77]. Following the Johnson–Mehl–Avrami treatment of transformation kinetics, the fraction of transformed phase x in time t, for small x, is given by:

$$x \sim \pi I_v u_c^3 t^4/3 \tag{2.4}$$

where u_c, the crystal growth velocity, can be expressed as:

$$u_c = \frac{kT_f}{3\pi a_0^2 \eta}\left[1 - \exp(-\Delta T_r \Delta H_f/RT)\right] \tag{2.5}$$

f is the fraction of sites at the crystal surfaces where atomic attachment can occur (=1 for close packed crystals and $0.2\,\Delta T_r$ for faceted crystals) and a_0 the mean atomic diameter.

Initially, the form of the expression used [2] for I_v was that based on the Hoffman model for the free enthalpy of crystallization, for which $\Delta G = T_r \Delta T_r \Delta H_f$. (Again, it is assumed for this purpose that transformation is to a single crystalline phase.) This gives

$$I_v = \frac{kT\bar{N}_v}{3\pi a_0^3 \eta} \exp\left[\frac{-1.07}{\Delta T_r^2 T_r^3}\right] \tag{2.6}$$

The pre-exponential constants in equations 2.5 and 2.6 are derived on the assumption that the atomic diffusivity is identical across the liquid–nucleus and liquid–crystal interfaces and related to η through the Stokes–Einstein equation. $\bar{N}_v$ is the average volume concentration of atoms.

Thus, on substitution for I_v and u_c in equation 2.4, the time, t, to achieve a small fraction of crystal, X, is given by:

$$t \sim \frac{9.3\eta a_0^2 X}{kTf^3\bar{N}_v}[\{\exp(1.07/\Delta T_r^2 T_r^3)\}/\{1 - \exp(-\Delta H_f \Delta T_r/RT)\}^3]^{1/4} \tag{2.7}$$

Any effects of nucleation transients and of heterogeneous nucleation are neglected, though there are grounds for assuming that these assumptions are reasonable for RGF alloys[2,15].

A time–temperature–transformation (T–T–T) curve can be computed by use of equation 2.7, the time t being expressed as a function of T_r to transform to a barely detectable fraction of crystal, which we take arbitrarily to be 10^{-6}. Computed T–T–T curves for $X = 10^{-6}$ for nickel and for three alloys for which η at temperatures above T_l has been determined experimentally, $Pd_{84}Si_{16}$, $Pd_{78}Cu_6Si_{16}$ and $Au_{78}Ge_{14}Si_8$, are given in *Figure 2.4*. For the nickel and palladium based alloys, interpolated η were based on the Vogel–Fulcher expression

$$\eta = A \exp\left[\frac{B}{(T - T_0)}\right]$$

where A, B and T_0 are constants[15,27], and for the AuGeSi alloy the η–T relation predicted by a free volume analysis[78] was used. A rough crystal–melt interface was assumed for pure nickel ($f = 1$) and a smooth interface ($f = 0.2\Delta T_r$) for the alloys which crystallize to 'intermetallic' phases.

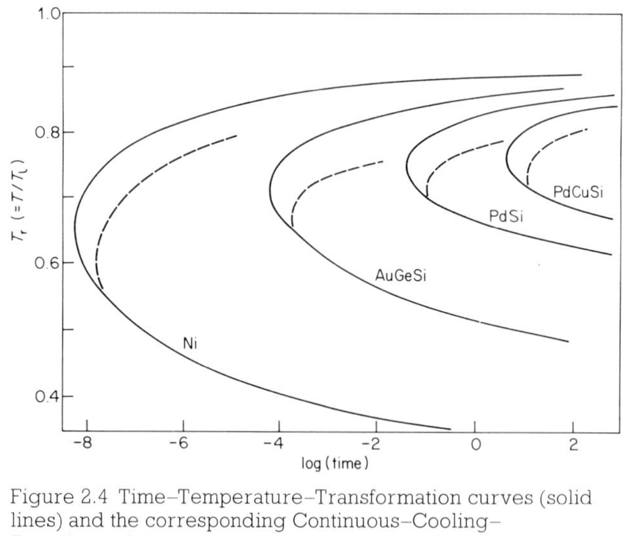

Figure 2.4 Time–Temperature–Transformation curves (solid lines) and the corresponding Continuous–Cooling–Transformation curves (dashed lines) for a fraction crystal of 10^{-6} for Ni, $Au_{78}Ge_{14}Si_8$, $Pd_{82}Si_{18}$ and $Pd_{78}Cu_6Si_{16}$

The form of the T–T–T curves and the magnitude of the time at the nose, t_n, is largely a reflection of the competition between the increasing driving force for nucleation and the decreasing atomic mobility with decreasing T_r. The large depressions of T_l, and also, to some extent, the smaller increases in T_g on alloying, increase T_{rg} and result in the substantial retardations of the crystallization process for the alloys, in comparison with the pure nickel. Correction of these isothermal transformation curves for continuous cooling conditions[79] yields C–C–T curves, also given in *Figure 2.4*. Thus the predicted $\dot{T}_c$ is given, on this basis, by the cooling curve that just avoids interception of the nose of the C–C–T curve, i.e. $\dot{T}_c \approx (T_l - T_n)/t_n$ where T_n is the temperature at the nose. The values for nickel and the AuGeSi, PdSi and PdSiCu alloys are $\sim 10^{10}$, $\sim 10^6$, ~ 3000 and ~ 35 K/s, respectively. These are broadly in agreement with experimental observation (considering the numerous uncertainties and assumptions involved in the computations); for instance, impurity-stabilized amorphous nickel was formed in thicknesses of about 100 nm, corresponding to a $\dot{T}$ of about 10^{10} K/s[14], and the PdCuSi alloy can be fully vitrified, as may some PdNiP and PtNiP alloys, as water quenched rods of diameter about 2 mm, corresponding[80] to $\dot{T}_c$ around 100 K/s, and the work of Boettinger[81] indicates that $\dot{T}_c$ is 125 K/s for $Pd_{77}Cu_6Si_{17}$. Naka *et al.*[82] have estimated $\dot{T}_c$ for $Pd_{78}Cu_6Si_{16}$ experimentally as about 500 K/s, which is significantly greater than expected from the observations of Chen[80]. The large decrease in $\dot{T}_c$ between $Pd_{82}Si_{18}$ and $Pd_{78}Cu_6Si_{16}$ can, on a purely kinetic basis, be ascribed[75] to the 56 K decrease in T_l and corresponding increase in T_{rg}.

The nickel, with an estimated T_{rg} of about 0.25, and the $Pd_{78}Cu_6Si_{16}$ alloy, with $T_{rg} = 0.63$, represent close to opposite ends of the established range of T_{rg} for metallic materials, with $Au_{78}Ge_{14}Si_8$ having $T_{rg} = 0.47$ and close to the lower limit of RGF. The kinetically predicted $\dot{T}_c$ is plotted in *Figure 2.5* against T_{rg} for several metallic materials[15,82,83], in addition to those discussed above, and for pure tellurium and germanium. The technologically useful iron-, nickel- and cobalt-based transition metal–metalloid (T–M) alloys have $\dot{T}_c$ generally within the range $\sim 3 \times 10^4$—10^6 K/s and T_{rg} 0.5—0.58. (Where a glass transition is not thermally manifest, the crystallization temperature T_x, measured dynamically at a high $\dot{T}$, is taken as an acceptably close estimate of T_g.) For most of the examples given here, the predicted $\dot{T}_c$ accord well with experimentally derived estimates [15,82,83], the agreement being particularly good for a number of palladium- and nickel-based alloys[82,83]. However, Nb–Ni alloys, and also

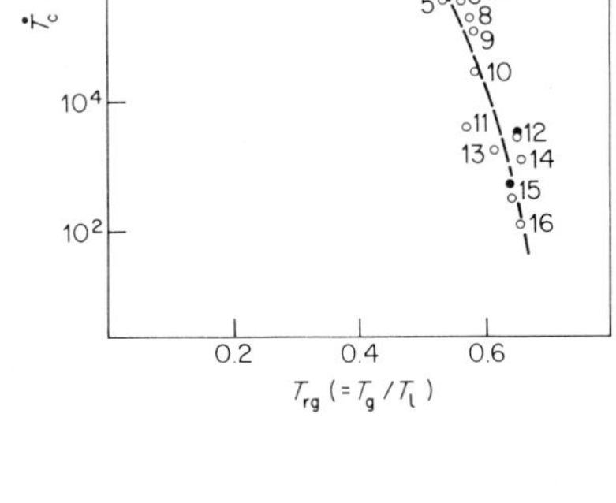

Figure 2.5 Predicted (○) and experimentally measured (●) critical cooling rate for glass formation $\dot{T}_c$ (based on C–C–T curves for the formation of a fraction crystal of 10^{-6}) plotted against reduced glass temperature T_{rg} (= T_g/T_l) for a representative range of elements and alloys. 1, Ni; 2, Te; 3, $Au_{78}Ge_{14}Si_8$; 4, $Fe_{83}B_{17}$; 5, $Fe_{41.5}Ni_{41.5}B_{17}$; 6, $Co_{75}Si_{15}B_{10}$; 7, Ge; 8, $Fe_{79}Si_{10}B_{11}$; 9, $Ni_{75}Si_8B_{17}$; 10, $Fe_{80}P_{13}C_7$; 11, $Pt_{60}Ni_{15}P_{25}$; 12, $Ni_{53}Pd_{27}P_{20}$ (ref. 83); 13, $Pd_{82}Si_{18}$; 14, $Ni_{62.4}Nb_{37.6}$; 15, $Pd_{78}Cu_6Si_{16}$ (ref. 82); 16, $Pd_{40}Ni_{40}P_{20}$

calcium- and strontium-based alloys[56,57] appear to vitrify less readily than would be predicted from their very high values of T_{rg}. This may result from heterogeneous nucleation, possibility on oxide particles, or alternatively because the liquid–crystal interfacial energy σ, and thus $\alpha\beta^{1/3}$, are unusually low. There is further evidence for anomalous glass-forming behaviour of niobium- and tantalum-containing alloys from a study of the casting of nickel-based ternary alloy glasses[84]. Good correlation between $\dot{T}_c$, x_c and T_{rg} has been observed for several T–Zr[85] and FeSiB and CoSiB[86] systems.

Comparison between measured x_c and predicted x_c, derived from a modified equation 2.7, which is based on a refined model of the free enthalpy of crystallization[87] (where $\Delta G_v = 2T_r . \Delta T_r . \Delta H_f/(1 + T_r)$), has been made for beryllium-based[88] and NiSiB and other NiB-based glasses[78]. The predicted $\dot{T}_c$ based on this model are about an order of magnitude greater than those based on the Hoffman model[77]. However, substantial discrepancies were found between predicted and experimentally estimated x_c for some of the nickel-based ternary glasses[77], associated with scatter in the relation between measured x_c and T_{rg}, which was also observed previously for NiSiB glasses[89], possibly for the reasons proposed for NiNb above. But, in contrast to the NiSiB compositions well within the RGF range, most alloys along the RGF boundary (for melt spun ribbon approximately 17 μm thick, equivalent to a $\dot{T}$ of about 10^6 K/s) corresponded closely with the $T_{rg} = 0.52$ isometric (which is equivalent to a $\dot{T}_c$ of approximately 10^6 K/s)[33]. A similar study[90] for the technologically important FeSiB system gave satisfactory quantitative agreement between the predicted and measured isometrics of x_c. Here, the measured x_c were determined by a magnetic coercivity technique[91]. This was a more sensitive indicator of the first appearance of crystals than the optical microscopy technique of Hagiwara *et al.*[92] which reported maximum x_c greater by a factor of 4.

A significant limitation to any theoretical prediction based on equation 2.3, equations 2.6 and 2.7, or the subsequent modification [77] to equation 2.7, is the assumption that the crystallization process is partitionless, i.e. to a single phase of the same composition as the melt. In cases where the liquidus depression on alloying is very steep, as for many RGF systems, the so-called T_0 curves for the eutectic phases, it has been proposed, drop catastrophically well before the eutectic composition is reached. Since T_0 represents the temperature above which the free enthalpy of the crystalline solid solution phase is greater than that of the melt, partitionless crystallization would be precluded over a composition range spanning the eutectic. This point has been discussed recently[81,93,94] with particular reference to PdCuSi. In support of this argument, it was shown experimentally[81] that solidification for $Pd_{77}Cu_6Si_{17}$ changed directly from the eutectic to single phase glass at a critical growth velocity, corresponding to a $\dot{T}$ of 125 K/s. Although the sharp discontinuity in solute diffusion distance represented by the transition from eutectic to glass may indicate significant local recalescence effects during eutectic solidification, the theoretical treatment of glass forming kinetics needs to be modified in detail to take account of the additional diffusion involved in partitioning, particularly for the more readily glass forming systems. Nevertheless, the broad agreement that exists with experimental observation lends support to the general validity of the kinetic approach.

2.4.3 Alloy chemical considerations

The evidence that ready glass formation in metallic alloy systems is associated with highly depressed liquidus temperatures is clear and the correlation between GFA and the parameter T_{rg} is strong. As discussed in the previous section, this can be rationalized on purely kinetic grounds of rate of increase of viscosity with increasing ΔT_r, although there are other quantifiable but lesser factors, such as the effect of temperature in the denominator of the pre-exponential term of equation 2.7, which would predict increasing GFA with decreasing T_l, irrespective of the magnitude of T_{rg}. There have been numerous discussions of metallic glass formation, directly in terms of alloy chemical factors[70–73,95–97], such as enthalpy of formation, relative sizes of the component atoms, electron concentration and electronegativity differences between the components. It is appropriate, alternatively, to consider briefly the influence of these various factors on the terms of the equations that govern the kinetics of the complex process of crystallization (cf. equation 2.7).

It is well-established that as the size and valence differences between component atoms increase, and thus as the electronegativity difference increases, the atomic interaction, expressed by the negative excess enthalpy and free enthalpy of mixing, increases. Strong unlike atom interaction (high negative ΔH_m) leads to the formation of stable intermetallic compound phases between which eutectic reactions occur at low temperatures; for RGF alloy systems, these stoichiometric intermetallic compounds are frequently Frank–Kasper phases (μ, σ and Laves phases) for metal–metal systems or cementite-type phases for metal–metalloid systems. When the magnitudes of the excess thermodynamic quantities are known for the liquid and crystalline phases, it is possible to predict[98] precisely the composition dependence of T_l. This has been performed [99] for the RGF system Au–Si. Hafner[100,101] has shown that it is possible to explain alloy constitution diagrams for RGF systems based on simple metals (e.g. Ca–Al, Mg–Zn, Ca–Zn) by a first-principles approach. It was shown that bonding in stable phases arises from matching of the nearest neighbour distances or, in the case of the liquid phase, the maxima of the pair distribution functions, with the attractive minima of the interatomic pair potentials. Hafner argues that the geometrical basis of these stable crystalline and liquid structures is tetrahedral close packing with the liquid phase being stabilized for a majority concentration of the larger atoms and Frank–Kasper phases stabilized for a majority concentration of the smaller atoms.

The magnitude of the glass transition temperature T_g for an alloy should be dependent largely on the slopes of the repulsive parts of the interatomic potentials, which will determine the atomic diffusivity in the glass phase. This, in turn, will be reflected in the magnitude of the cohesive energy ΔH_c for the alloy. A relation between T_g and ΔH_c was proposed earlier[102] and a fairly good correlation established for a wide range of metal glasses between T_g and ΔH_s, the 'linear weighted' sublimation enthalpy, which approximates[2,67] to ΔH_c. A detailed investigation for a range of T–M glasses based on iron, nickel and cobalt indicated that changes in T_g on alloying correlated well with atomic size differences and changes in electron density[103].

The possible influence of alloy chemistry on other terms in the kinetic equations have yet to be considered in detail. For instance, it is assumed, for simplicity and tractability, that the free enthalpy of crystallization and other nucleation parameters can be treated in a uniform way and the validity of these assumptions can be questioned. The melt viscosities of some of the very easy glass forming palladium-based alloys at their T_l are extremely high (above 200 poise)[82], in comparison with those of other metallic melts, including a few RGF alloys, that have been measured experimentally. Indeed they are substantially higher than would be expected on the basis of Arrhenius extrapolation from higher temperatures and this factor alone contributes significantly to their high GFA. The reason for these high viscosities is not clear but there is some evidence that the interfacial tension σ is also very enhanced in these systems.

2.5 Influence of process variables

The techniques available for producing metallic glasses in various forms, ranging from powder, through fibre and wire to strip, and the major process variables, have been reviewed[15]. The principal developments since then have been the evolution of the planar flow casting process for wide amorphous alloy strip production[104] and of an improved method for casting amorphous alloy wire[105]. The aim in all the techniques is to bring the melt into clean contact with (a) cool and highly conducting heat sink(s) at a high relative velocity, in order to promote efficient and uniform melt spreading or subdivision and intimate thermal contact. The higher is $\dot{T}_c$ for an alloy, then the smaller must be the section thickness x, i.e. the length of the heat path.

The cooling rate achieved is also clearly a function of the heat transfer coefficient h (i.e. the efficiency with which heat is conducted) across the interface(s), with the heat sink. Where no interfacial resistance to heat flow exists ('ideal cooling conditions'), the cooling rate is then determined by the conductivity and thickness of the liquid medium (for a highly conducting sink). In practice, ideal cooling conditions are not achieved, except in the case of 'laser glazing' where a self-substrate exists. For chill block melt spinning, h has been estimated experimentally to be approximately $10^5\ \mathrm{W\,m^{-2}\,K^{-1}}$, with a variation from system to system of about half an order of magnitude[106–107], and it is presumed to be similar for planar flow casting. The computed variation of $\dot{T}$ with x for a RGF alloy $Fe_{40}Ni_{40}P_{14}B_6$, for ideal cooling, for the range of h appropriate to melt spinning and to 'gun' splat quenching, where h is higher at about 3×10^6 K/s [108], is given in *Figure 2.6*.

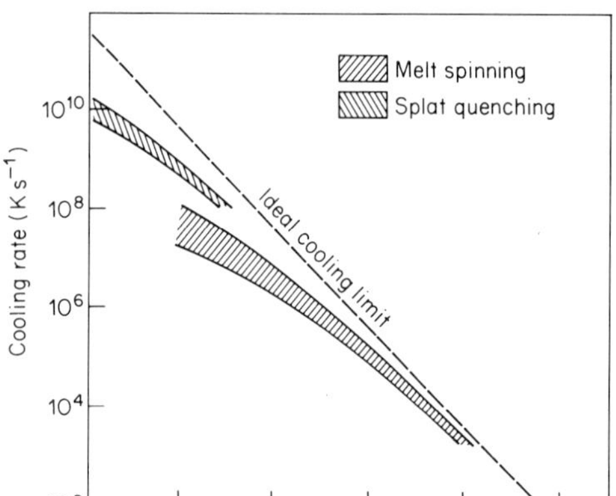

Figure 2.6 Cooling rate $\dot{T}$ as a function of section thickness x computed for the glass forming alloy $Fe_{40}Ni_{40}P_{14}B_6$, based on experimentally derived values of heat transfer coefficient (cross hatched areas) and ideal cooling, i.e. no interface resistance (dashed line), for splat quenching and for chill block melt spinning. Unidirectional cooling of a rectangular plate section to one heat sink is assumed in all cases

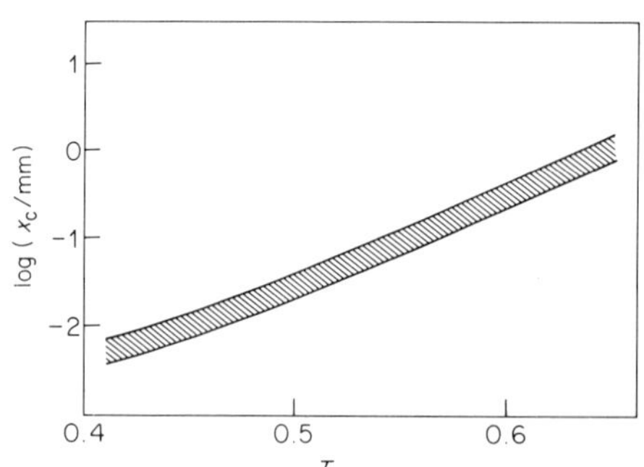

Figure 2.7 Estimated range of maximum section thickness of glass phase as a function of T_{rg}, derived from data in *Figures 2.5* and *2.6* for chill block melt spinning

The estimated maximum glass thickness x_c derived from *Figures 2.5* and *2.6* is then plotted against T_{rg} in *Figure 2.7*.

Also of importance in a casting technique is the duration of contact t_c with the heat sink. Clearly, t_c should be sufficiently large that the temperature of the material being quenched has decreased to close to, or below, T_g before contact is lost. On loss of contact $\dot{T}$ can decrease catastrophically, depending on the relative velocity of the ambient gas and the section thickness. If t_c is too short either the material will not vitrify or it will devitrify during the low $\dot{T}$ phase, after removal from the heat sink. The magnitude of t_c can also determine the state of quench of the glass, since a glass that leaves a substrate at higher temperature will tend to undergo more relaxation during the low $\dot{T}$ phase, down to ambient temperature.

Acknowledgements

The author is grateful to the Science and Engineering Research Council, The Royal Society and the Procurement Executive of the Ministry of Defence for financial support.

References

1. Klement, K., Willens, R. H. and Duwez, P., *Nature*, **187**, 869 (1970)
2. Davies, H. A., *Phys. Chem. Glasses*, **17**, 159 (1976)
3. Colligan, G. A. and Bayles, S., *Acta metall.*, **10**, 895 (1962)
4. Turnbull, D. and Cech, R. E., *J. appl. Phys.*, **21**, 805 (1950)
5. Turnbull, D. and Cohen, M. H., *J. chem. Phys.*, **34**, 120 (1961)
6. Waseda, Y. and Chen, H. S. *Sci. Rep. Res. Insts Tôhoku Univ.*, **28**, 143 (1980)
7. *See* a review by Cargill, G. S. (III), *Solid St. Phys.*, **30**, 227 (1975)
8. Sakata, M., Cowlam, N. and Davies, H. A., (a) *J. Phys. Colloque*, **41**, C8, suppl. 8, 190 (1980); (*b*) *J. Phys. F*, **11**, L157 (1981)
9. Gaskell, P. H., *J. Non-cryst. Solids*, **32**, 207 (1979)
10. Tamura, K. and Endo, H., *Phys. Lett.*, **29A**, 52 (1969)
11. Fujime, S., *Jap. J. appl. Phys.* **5**, 1029 (1966); **6**, 305 (1967)
12. Bennett, M. R. and Wright, J. G., *Phys. Status Solidi*, (*a*) **13**, 135 (1972)
13. Leung, P. K. and Wright, J. G., *Phil. Mag.* **30**, 185, 995 (1974)
14. (*a*) Davies, H. A., Aucote, J. and Hull, J. B., *Nature*, **246**, 13 (1973); (*b*) Davies, H. A. and Hull, J. B., *J. mater. Sci.*, **11**, 215 (1976)
15. Davies, H. A., *'Proc. 3rd Int. Conf. Rapidly Quenched Metals'*, Ed. Cantor, B., vol. I, p. 1, The Metals Society, London (1978)
16. Davies, H. A. and Hull, J. B., *Scripta Met.*, **7**, 637 (1973)
17. Vučić, Z., Etlinger, B. and Kunstelj, D., *J. Non-cryst. Solids*, **20**, 451 (1976)
18. Turnbull, D., *J. Phys. Colloque*, **35**, 4, 1 (1974)
19. Turnbull, D., *Contemp. Phys.*, **10**, 473 (1969)
20. Davies, H. A. and Lewis, B. G., *Scripta Met.*, **9**, 1107 (1975)
21. Giessen, B. C., Hong, J., Kadakoff, L., Polk, D. E., Ray, R. and St. Amand, R., *'Proc. 3rd Int. Conf. Rapidly Quenched Metals'*, Ed. Cantor, B., vol. I, p. 249, The Metals Society, London (1978)
22. *E.g.*, Mangin, P., Marchal, G., Rodmaq, B. and Janot, C., *Phil. Mag.*, **36**, 643 (1977)
23. *E.g.*, McGuire, R. Mizoguchi, T., Gambino, R. J. and Kirkpatrick, S., *J. appl. Phys.*, **49**, 1689 (1978)
24. *E.g.*, Brenner, A. E., Couch, D. E., and Williams, R. E., *J. Res. natn. Bur. Stand.*, **44**, 109 (1950)
25. Furuya, H., Hasegawa, N., Watanabe, T. and Tanabe, Y., *'Proc. 4th Int. Conf. Rapidly Quenched Metals'*, Ed. Masumoto, T. and Suzuki, K. vol. I, p. 93, Japan Institute of Metals, Sendai (1982)
26. Takayama, S., *J. mater. Sci.*, **11**, 164 (1976)
27. Davies, H. A., *'Proc. Conf. Amorphous Metallic Materials, Smolenice, 1978'*, Ed. Duhaj, P. and Mrafko, P., p. 107, Veda, Bratislava (1980)
28. Suryanarayana, C., *Sci. Rep. Res. Insts Tôhoku Univ.*, **28A**, Suppl., 143 (1980)
29. Jones, H. and Suryanarayana, C., *J. mater. Sci.*, **8**, 705 (1973)
30. Suryanarayana, C., *'Rapidly Quenched Metals—a Bibliography 1973–1979'*, Plenum Press, New York (1980)
31. Lewis, B. G. and Davies, H. A., *Mater. Sci. Engng*, **23**, 179 (1976)

32. Ray, R., Hasegawa, R., Chou, C.-P. and Davis, L. A., *Scripta Met.*, **11**, 973 (1977)
33. Donald, I. W., and Davies, H. A., *J. mater. Sci.*, **15**, 2754 (1980)
34. Inoue, A., Kitamura, A. and Masumoto, *Trans. Japan Inst. Metals*, **20**, 404 (1979)
35. Srivastava, P. K., Giessen, B. C. and Grant, N. J., *Metall. Trans.*, **3**, 977 (1972)
36. Chen, H. S. and Turnbull, D., *Acta. metall.*, **17**, 1021 (1969)
37. Maitrepierre, P. L., *J. appl. Phys.*, **41**, 498 (1970)
38. Hagiwara, M., Inoue, A. and Masumoto, T., *Sci. Rep. Res. Insts Tôhoku Univ.*, **29**, 351 (1981)
39. Inoue, A., Kitamura, A. and Masumoto, *J. mater. Sci.* **16**, 1895 (1981)
40. Inoue, A., Kobayashi, K., Nose, M. and Masumoto, *J. Phys. Colloque*, **41**, C8, suppl. 8, 831 (1980)
41. Inoue, A., Masumoto, T., Arakawa, S. and Iwadachi, T., *'Proc. 3rd Int. Conf. Rapidly Quenched Metals'*, Ed. Cantor, B., vol. I, p. 265, The Metals Society, London (1978)
42. Polk, D. E., Calka, A. and Giessen, B. C., *Acta metall.*, **26**, 1097 (1978)
43. Inoue, A. and Masumoto, T., *Sci. Rep. Res. Insts. Tôhoku Univ.*, **28A**, Suppl., 165 (1980)
44. Fischer, M., Polk, D. E. and Giessen, B. C., *'Proc. 1st Int. Conf. Rapid Solidification Processing, Reston, 1977'*, Eds. Mehrabian, R., Kear, B. H. and Cohen, M., p. 140, Claitor's, Baton Rouge (1978)
45. Sakata, M., Cowlam, N. and Davies, H. A., *'Proc. 4th Int. Conf. Rapidly Quenched Metals'*, Eds. Masumoto, T. and Suzuki, K., vol. I, p. 327, Japan Institute of Metals, Sendai (1982)
46. (*a*) Ray, R., Giessen, B. C. and Grant, N. J., *Scripta Met.* **2**, 359 (1968); (*b*) Raman, R. V., Ph.D. Thesis, Northeastern University, Boston (1977)
47. Gregan, G. P. J., Dong, Y. D. and Scott, M. G., *'Proc. Int. Conf. Metallic Glasses, Budapest*, 1980, vol. II, p. 197, Kultura, Budapest (1981)
48. Giessen, B. C., Madhava, M., Polk, D. E. and Vander Sande, J. B., *Mater. Sci. Engng*, **23**, 145 (1976)
49. Nose, M. and Masumoto, T., *Sci. Rep. Res. Insts Tôhoku Univ.*, **28A**, Suppl., 222 (1980)
50. Johnson, W. L., Poon, S. J. and Duwez, P., *Phys. Rev.*, **B11**, 150 (1975)
51. Buschow, K. H. J., *J. less-Common Metals*, **66**, 89 (1979)
52. Fukamichi, K., Kikuchi, M., Masumoto and Matsura, M., *Phys. Lett.*, **A73**, 436 (1979)
53. Agyeman, K., Muller, R. and Tsuei, C. C., *Phys. Rev.*, **B19**, 193 (1979)
54. Kawano, K., Ino, H., Nishikawa, S. and Nihei, *'Proc. 4th Int. Conf. Rapidly Quenched Metals*, Ed. Masumoto, T. and Suzuki, K., vol. I, p. 371, Japan Institute of Metals, Sendai (1982)
55. Chaudhari, P., Cuomo, J. J. and Gambino, R. J., *IBM Jl Res. Dev.*, **11**, 66 (1973)
56. St. Amand, R., and Giessen, B. C., *Scripta Met.* **12**, 1021 (1978)
57. Sommer, F., Duddek, G. and Predel, B., *Z. Metallk.*, **69**, 587 (1978)
58. Quoted in ref. 21
59. Catka, A., Madhava, M., Polk, D. E., Giessen, B. C., Matyja, H. and Vander Sande, J. B., *Scripta Met.*, **11**, 65 (1977)
60. Tanner, L. E. and Ray, R., *Acta metall.*, **27**, 1727 (1979)
61. Ray, R. and Musso, E., *U.S. Pat.* 3 981 722 (1976)
62. Giessen, B. C. and Elliott, R. O., ref. 15, vol. I, p. 406
63. Elliott, R. O. and Giessen, B. C., *Acta metall.*, **30**, 785 (1982)
64. Rawson, H., *'Inorganic Glass-Forming Systems'*, Academic Press, London–New York (1967)
65. Marcus, M. and Turnbull, D., *Mater. Sci. Engng* **23**, 211 (1976)
66. Chen, H. S. and Jackson, K. A., *'Metallic Glasses'*, Eds. Gilman, J. J., and Leamy, H. J., p. 77, American Society of Metals, Metals Park, Ohio (1978)
67. Donald, I. W. and Davies, H. A., *J. Non-cryst. Solids*, **30**, 77 (1978)
68. Nielsen, H. J. V., *J. Non-cryst. Solids*, **33**, 285 (1979)
69. Holt, J. B., Ankeny, D. C. and Cline, C. F., *Scripta Met.* **14**, 959 (1980)
70. Polk, D. E. and Giessen, B. C., ref. 66, p. 1
71. Zielinski, P. G., Ostalek, J., Kijek, M. and Matyja, H. ref. 15, vol. I, p. 337
72. Güntherodt, H. J. *et al.*, *J. Phys Colloque*, **41**, C8, 381 (1980)
73. Giessen, B. C., ref. 25, vol. I, p. 213
74. Uhlmann, D. R., *J. Non-cryst. Solids*, **7**, 337 (1972)
75. Davies, H. A., Aucote, J. and Hull, J. B., *Scripta Met.* **8**, 1179 (1974)
76. Davies, H. A., *J. Non-cryst. Solids*, **17**, 266 (1975)
77. Reeve, J., Davies, H. A. and Donald, I. W., ref. 25, vol. I, p. 221
78. Ramachandrarao, P., Cantor, B. and Cahn, R. W., *J. Non-cryst. Solids*, **24**, 109 (1977); *J. mater. Sci.*, **12**, 2488 (1977)
79. Grange, R. A. and Keifer, J. M., *Trans. Am. Soc. Metals*, **29**, 85 (1941)
80. Chen, H. S., *Acta metall.*, **22**, 1505 (1974)
81. Boettinger, W. J. ref. 25, vol. I, p. 99
82. Naka, M., Nishi, Y. and Masumoto, T., ref. 15, vol. I, p. 231

83. Nishi, Y., Suzuki, K. and Masumoto, T., ref. 25, vol. I, p. 217
84. Reeve, J. and Davies, H. A., to be published
85. Nishi, Y., Morohoshi, T., Kawakami, M., Suzuki, K. and Masumoto, T., ref. 25, vol. I, p. 111
86. Hagiwara, M., Inoue, A., Tomioka, H. and Masumoto, T. ref. 25, vol. I, p. 115
87. Thompson, C. V. and Spaepen, F., *Acta metall.*, **27,** 1855 (1979)
88. Tanner, L. E., *Acta metall.*, **28,** 1805 (1980)
89. Hagiwara, M., Inoue, A. and Masumoto, *Metall. Trans., A,* **12,** 1031 (1981)
90. Luborsky, F. E., Reeve, J., Davies, H. A. and Liebermann, H. H., *IEEE Trans. Magn.*, **MAG-18,** in the press
91. Luborsky, F. E., Lie· ermann, H. H. and Walter, J. L., ref. 47, vol. I, p. 203
92. Hagiwara, M., Inoue, A. and Masumoto, T., *Sci. Rep. Res. Insts. Tôhoku Univ.*, **A29,** 351 (1981)
93. Boettinger, W. J., Biancaniello, F. S., Kalonji, G. M. and Cahn, J. W., *'Rapid Solidication Processing II'*, p. 50, Claitor's Publishing Divn., Baton Rouge (1980)
94. Massalski, T. B., ref. 25, vol. I, p. 203
95. Giessen, B. C. and Whang, S. H., ref. 8(*a*) p. C-8, 95
96. Sommer, F., *Z. Metallk.*, **72,** 219 (1981)
97. Sommer, F., Fripan, M. and Predel, B., ref. 25, p. 209
98. Ansara, I., *Int. Metals Reviews,* **24,** 20 (1979)
99. Castanet, R., Chastel, R. and Bergmann, C., *Mater. Sci. Engng* **32,** 93 (1978)
100. Hafner, J. and von Heimendahl, L., *Phys. Rev. Lett.*, **42,** 386 (1979)
101. Hafner, J., *Phys. Rev.*, **B21,** 406 (1980)
102. Cohen, M. H. and Turnbull, D., *J. chem. Phys.*, **34,** 120 (1960)
103. Donald, I. W. and Davies, H. A., *Phil. Mag.*, **A42,** 277 (1980)
104. Narasimhan, M. C., *U.S. Pat.* 4 142 571 (1979)
105. Ohnaka, I., Fukusako, T., Ohmichi, T., Masumoto, T., Inoue, A. and Hagiwara, M., ref. 25, vol. I, p. 31
106. Davies, H. A., Shohoji, N. and Warrington, D. H., ref. 93, p. 153
107. Davies, H. A., Lewis, B. G. and Donald, I. W., ref. 44, p. 78
108. Davies, H. A. and Hull, J. B., *J. mater. Sci.*, **9,** 707 (1974)

Chapter 3

Sample preparation: methods and process characterization

H. H. Liebermann

Allied Corporate Technology, Parsippany, New Jersey, USA

3.1 Introduction

Alloys produced by rapid quenching from the melt have been shown to exhibit metallurgical characteristics not obtainable by conventional casting methods. Such features include refined grain size, extended solid solutions[1,2], metastable phases[3,4], and metallic glasses[5–7]. The quench rate achieved in the rapid solidification is primarily controlled by specimen thickness and the nature of interfacial contact between substrate and sample[8–9]. Calculations by Ruhl[10] have shown that average quench rate varies inversely with the square of specimen thickness for ideal cooling or as the inverse of thickness for Newtonian (interface-controlled) cooling. Aside from high quench rate, another important factor in the preparation of rapidly quenched samples is the ability to undercool the melt to such an extent that subsequent recalescence does not cause morphological destabilization of the product[11,12].

A high average sample quench rate can be achieved by promoting rapid heat removal from the bulk. Perhaps the simplest means by which this may be achieved is by maximizing the contact area between the melt and cooling medium by rapidly increasing the liquid alloy surface area. This may be effected either by altering the shape of the melt during processing (splat quenching, melt-spinning, etc.) or by physically segmenting the melt by various means (atomization, etc.). Of course, atom-by-atom preparation methods such as sputtering and vapour deposition have maximum sample surface-to-volume ratios on quenching. An alternative means by which high quench rates can be achieved is by briefly subjecting the sample surface to a high power input, causing localized melting and rapid solidification of the melt using the bulk of the specimen as a substrate (self substrate quenching). The morphology and characteristics of some rapidly quenched amorphous and crystalline alloys has been shown to depend significantly on the sample preparation method used[13–16]. Comprehensive reviews describing both the fabrication and physical properties of amorphous alloys have been written[6,17–24]. The following sections will describe various sample preparation techniques as well as characterization and limitations of these methods.

3.2 Sample preparation techniques

3.2.1 Sputtering and evaporation methods

These specimen fabrication techniques are based on an atom-by-atom constitution of the product and require that the kinetic energy of the condensing atom exceed the binding energy of the atoms in the substrate surface[25]. An arrangement of atoms having low mobility at the substrate surface will promote metallic glass formation by subsequently depositing atoms. Both the sputtering and evaporation methods of sample preparation are very sensitive to deposition conditions. For example, the presence of an impurity atom species such as oxygen has been found to facilitate the formation of an amorphous atomic structure[26–28]. The substantially high quench rate inherent in sputtering and evaporation methods (over 10^8 K/s) allows the formation of some metallic glasses which cannot be produced by melt quenching techniques. In fact, high melting temperature and chemically reactive elements [such as the rare earths (lanthanides)] may be combined with other elements at low temperatures to form amorphous alloys[29]. While the use of sputtering and evaporation methods results in samples made at high quench rates, the rate of sample mass accumulation is very slow compared with that of the techniques described in the following sections.

3.2.2 Chemical deposition and electrodeposition

While various transition metal–metalloid amorphous alloy systems have been prepared by chemical deposition and electrodeposition, the precise composition of the product depends strongly on the deposition conditions and bath composition during sample formation. In contrast, sample fabrication by other methods such as melt quenching, sputtering, etc. results in a product having a composition very nearly that of the melt from which it is formed.

Typical bath compositions and deposition parameters have been described[30]. In particular, some effects of bath composition and electrodeless plating parameters on the formation of Ni–B amorphous alloy have been discussed by Watanabe and Tanabe[31]. Amorphous alloys may be formed by such deposition methods if the bath used is able to supply 10—30 atom % of phosphorus or boron. The tendency for the incorporation of hydrogen into the sample atomic structure may have an adverse effect on the properties of amorphous alloys made by chemical and electrodeposition methods.

3.2.3 Ion implantation

The broad variety of atomic species which can be introduced into the surface of many materials by ion implantation has generated recent interest in the use of this method for the fabrication of unique crystalline[32] and amorphous[33] surface layers. Structural measurements of samples made suggest that the implanted alloys are identical to alloys of the same nominal composition made by conventional techniques. While the implantation method allows substantially greater freedom from the compositional restrictions imposed by constitutional phase diagrams, similar compositional rules for the formation of metallic glasses seem to be operative for both implanted and conventionally quenched alloys. Consideration of an individual atom implantation event as a thermal spike phenomenon results in an effective quench rate of about 10^{14} K/s. Ion implantation must necessarily be considered as a surface modification technique only because of the energy loss sustained by the penetrating ion on successive collisions with atoms in the surface of the target material. Therefore, the likelihood of bulk sample modification in a single stage operation by ion implantation seems unlikely except for thin specimens.

3.2.4 Splat quenching and spray deposition

Perhaps the earliest melt quenching method used in the rapid solidifcation of metastable crystalline and amorphous materials was the gun technique[34] of Duwez in which molten alloy is atomized by expulsion from a shock tube and propelled against a substrate surface having a high thermal conductivity. Numerous variations of the principles operative in this splat quenching method have evolved over the years. For example, the piston and anvil[35] and twin piston[36] methods cause the rapid flattening and quenching of a molten alloy droplet between impacting substrate surfaces. Modifications of the original twin piston method include increased substrate impact by using magnetically-driven substrate platens[4,37] and also limited twin substrate proximity on impact to reduce mechanical deformation of the splat during quenching[38]. The use of diamond as a low-temperature substrate material has been shown to have advantages over that of the conventionally used copper[39]. Merits of low temperature quenching techniques have been discussed by Leontic *et al.*[40]. In contrast, it has been reported that the use of a hot substrate[41] promotes enhanced melt wetting on the substrate surface and thereby enhances splat quench rate. The arc-hammer method[42] for producing splats of rapidly quenched metals consists of arc-melting a button of material on a water-cooled hearth and then striking this button with a hammer. Although foils made by splat quenching techniques typically have an irregular shape and variable thickness, the thinnest parts of such samples are typically suitable for characterization by transmission electron microscopy without the need for thinning.

Several variations of melt droplet spraying and deposition have been used to fabricate rapidly solidified samples. One such variation has been to inject molten alloy into a hot plasma flame and cause the formation of rapidly quenched splats on impact with a high conductivity substrate surface[43–45]. Modification of this basic process to cause plasma jet impingement on to a rotating substrate wheel[46,47] has resulted in greater sample yields. The use of a plasma or other melt spray operation advantageously combines melting, quenching, and compaction into a single operation[48–50].

3.2.5 Fabrication of powders

Power is a convenient material form for compaction into near-final shapes. Considerable effort has been expended to prepare amorphous powders and rapidly solidified crystalline powders having refined microstructure and grain size as well as homogeneous, stable phases. Details of amorphous alloy powders will be discussed in Chapter 24.

Gas atomization

This method is the earliest used for the rapid solidification of powder and involves the atomization of molten alloy stream by gas jets. The quench rate prevalent during processing depends largely on melt droplet size and on the type of atomization gas used. For example, the use of helium results in microcrystalline superalloy powder having a finer microstructure than an equivalent size sample made using argon[51]. It is assumed that helium provided a more rapid cooling than did argon in the gas atomization. Various characteristics and operating parameters of gas atomization have been discussed by Grant[52].

Gas–water atomization

In an effort to increase further the cooling rate of fine powders, the molten alloy droplets formed by gas atomization are quenched in streams of high velocity water[53]. The alloy $Cu_{60}Zr_{40}$ has been made into amorphous powder particles up to 100 μm in diameter by this atomization technique. Secondary dendrite arm spacing measurements of crystalline aluminium–copper alloy powder made on gas–water atomized powder particles yielded an average quench rate of 10^5—10^6 K/s. Although gas–water atomized powder typically exhibits a quench rate greater than that achieved with gas atomization, the maximum individual powder particle size tolerable is still limited by the critical cooling rate of a given glass-forming alloy composition[16,54].

Centrifugal atomization

This powder fabrication method is based on a stream of molten alloy which is directed on to the face of a spinning disc. The droplets formed on impact with the spinning disc are quenched by forced convective cooling in a gaseous atmosphere such as argon or helium[55]. Flow rates of about 500 g/s coupled with disc speeds of 100 m/s can result in the production of powder particles less than 200 μm in diameter. Experimentally derived quench rates obtained from dendrite arm spacing measurements of crystalline materials are about 5×10^5 K/s for this atomization process.

Spark erosion

Spark erosion was first described by Svedberg as a method for producing fine particles[56]. The process involves the use of a repetitive spark which is maintained between two electrodes of the material to be quenched while immersed in a dielectric fluid. The spark discharge produces localized meltings or vaporization of the electrode material, causing the rapid solidification of these in the dielectric fluid. Details of the spark erosion method for the fabrication of alloy powders depend on the process parameters and choice of dielectric[57], which dielectric is usually an organic fluid. Physical property measurements of amorphous alloy powders made by the spark erosion method under various conditions have been reported[16] and suggest that the average quench rate of this process can exceed that of the other atomization processes.

3.2.6 Filaments, ribbons and tapes

While atomized powder particles have been made amorphous for certain glass-forming alloy compositions, each run yields a range of particle sizes which is a consequence of the mechanics of melt droplet formation[52]. This particle size spread results in a quench rate difference between powder particles, causing some non-uniformity of the product. Alternatively, the fabrication of rapidly quenched materials by continuously making the product from the melt results in improved sample geometry and quench rate uniformity. Some of these continuous melt solidification processes are described in the following subsections.

Filaments

While powder particles generally have point symmetry, a filament is characterized by an axis of symmetry along its length. The melt quenching of filaments having a round cross-section is restricted to a limited number of processes. The early method of Taylor[58,59] involves a rapid increase in melt-container surface area by quickly drawing out the glass tube in which the molten alloy is contained. A more widely used principle for the fabrication of filaments have a circular cross-section is the expulsion of molten alloy through a circular orifice. Melt jet stabilization, solidification, and cooling are provided by various quenching media[60] such as gases [61,62] or liquids[63–66]. The use of a solid substrate as a quenching medium for the molten alloy jet results in filaments having a quasi-circular cross-section. Such processes include melt drag[67], crucible [68,69] and pendant drop[68,70] melt extraction.

Ribbons

The fabrication of thin metallic ribbons by bringing a stream of molten alloy in contact with a rapidly moving substrate surface is not a recently devised process[71]. The method of pressurized orifice chill block melt-spinning (CBMS)[72–76] involves the formation of a melt jet by the expulsion of molten alloy through an orifice and the impingement of this jet against a rapidly moving substrate surface. The puddle, which results from continued impingement of the melt jet, serves as a local reservoir from which ribbon is continuously formed and chilled. Both the initial quench rate (about 10^6 K/s) in processing and the final ribbon geometry depend upon the nature of the melt puddle. For example, a puddle which is smooth and unchanging during the course of CBMS typically results in ribbons with smooth edges and surfaces[77,78]. The most common substrate surfaces described in the literature are the inside of drums[79] or wheels[80–83], the outside of wheels[74–76], between twin rollers[84–90], and on belts[91]. Various schemes for increasing ribbon–substrate sticking distance during CBMS have been employed[92,93] in order to enhance the ribbon average quench rate. The CBMS process has been extensively used for the fabrication of rapidly quenched materials because of its inherent simplicity and facility for making uniform, continuous ribbons.

Tapes

The need for wide amorphous alloy ribbons or tapes in prototype devices, such as transformer cores, has spurred process development for the fabrication of samples wider than the typical few millimetres obtained by the CBMS method described in the previous section. A method for making tape several centimetres wide would require an elongated molten alloy reservoir which is transversely maintained in contact with the moving substrate surface. The use of a slit rather than a circular orifice in conventional CBMS has been shown to be suitable for making wide ribbon or tape in a reduced atmosphere[94]. The use of a planar melt jet, however, is made difficult because of stream geometry destabilization caused by the high surface tension and low viscosity of molten alloys[95]. Therefore, the planar flow[91,96] or strip casting method for the fabrication of rapidly quenched tapes has been developed in which the melt nozzle is held very close to the moving substrate surface. This close proximity dampens perturbation of the melt on the substrate and thereby allows the formation of rapidly quenched tape under a laboratory atmosphere.

3.2.7 Advanced melt-spinning techniques

Some new and useful variations of the basic CBMS processes described have been developed to make amorphous alloy ribbons and tapes for specific applications. For example, it is well known that glass-forming alloys quenched at less than the critical cooling rate will result in a brittle and probably partially crystalline sample[97]. This knowledge has been applied to the fabrication of amorphous alloy tapes having highly embrittled lines, shown in *Figure 3.1*, by using a substrate surface having a predetermined low conductivity pattern[98]. A transverse micrograph showing locally extensive

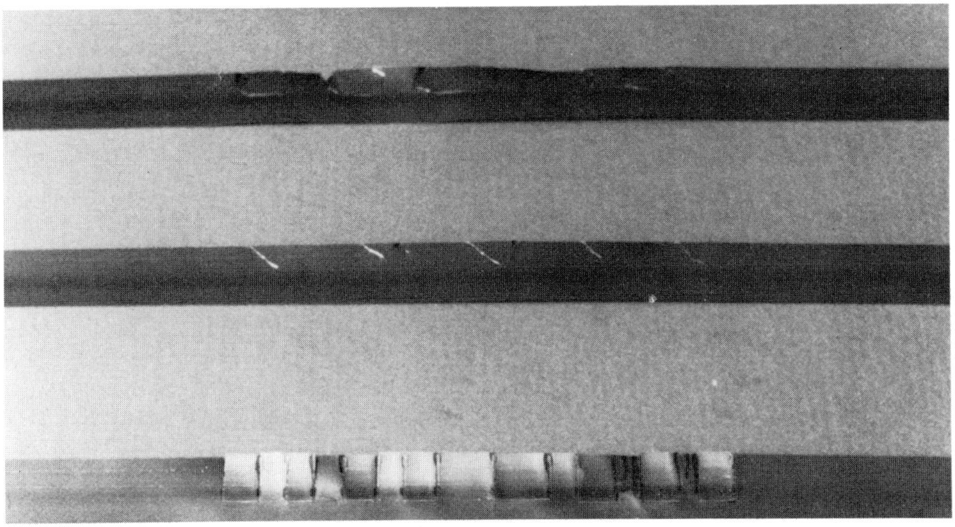

Figure 3.1 $Fe_{81.5}B_{14.5}Si_4$ amorphous alloy tapes (13 mm wide) made using a substrate wheel bearing scribings of the shown patterns

crystallization in the embrittled line of the tape appears in *Figure 3.2*. This approach may be used to produce amorphous alloy ribbons having predetermined cut-outs by simply causing breakage along the embrittled lines. Therefore, graded quench rates and sample microstructures can be generated in as-made ribbons and tapes by the use of graded sample–substrate interfacial characteristics.

Another innovation in the chill block melt-spinning of amorphous alloys is the fabrication of helical ribbons and tapes; that is a product having a predetermined, continous in-plane curvature[99], as shown in *Figures 3.3* and *3.4*. The key to the fabrication of these kinds of ribbon and tape is the use of a prolonged sample-substrate sticking distance. Such a ribbon or tape would clearly be self-nesting during assembly into a bulk form and would undergo less property degradation due to stresses when compared with straight ribbon wound into the same bulk configuration. Combination of the two new chill block melt-spinning technologies discussed so far can result in helical amorphous alloy tape having a predetermined pattern of cut-outs for possible use in electric motor stators and rotors[100]. Other applications of rapidly quenched helical ribbons are expected.

A development which allows the chill block melt-spinning of composite ribbons and

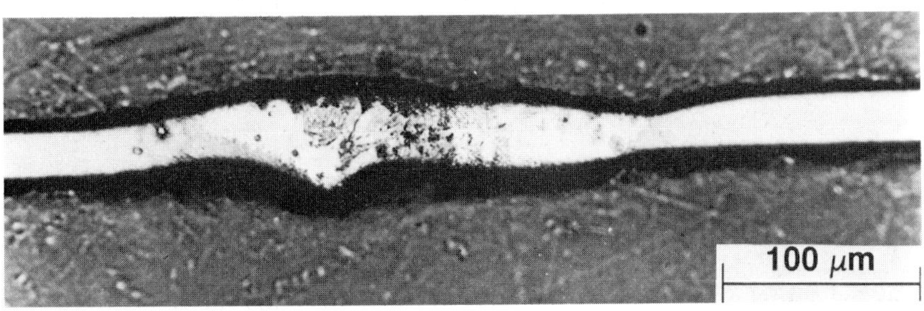

Figure 3.2 Transverse section micrograph showing partial crystallization of an embrittled line in the tape shown in *Figure 3.1*

Figure 3.3 Chill block melt-spinning of helical $Fe_{40}Ni_{40}B_{20}$ amorphous alloy ribbon on the flat face of a 12.5 cm diameter copper wheel

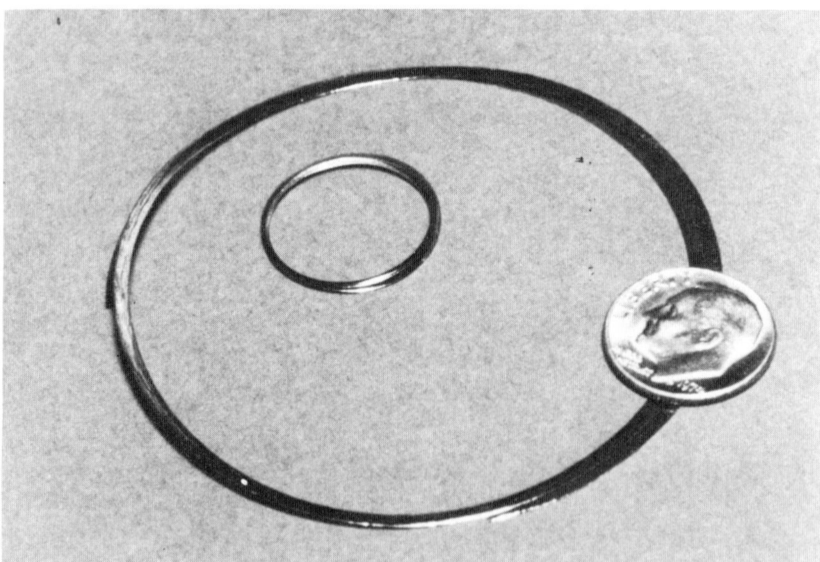

Figure 3.4 Typical $Fe_{40}Ni_{40}B_{20}$ amorphous alloy helical ribbons with average diameters of 2 cm and 6 cm

Figure 3.5 Photograph of a strip casting run in which a two-layer $Fe_{81.5}B_{14.5}Si_4$ amorphous alloy ribbon is formed

tape is gaining interest[101–105]. In this process, successive layers of rapidly quenched material are created by deposition of melt atop layers already formed. As an example, *Figure 3.5* shows the fabrication of two-layer thick $Fe_{81.5}B_{14.5}Si_4$ amorphous alloy tape. The key to the prevention of ribbon crystallization on deposition of the melt to form the second layer is to balance the kinetics of heat transfer against those of structural relaxation and crystallization of the alloy[102]. *Figure 3.6* shows a transverse section micrograph of a two layer $Fe_{40}Ni_{40}B_{20}$ amorphous alloy ribbon made in a vacuum. Note that while the presence of two layers is seen at the ribbon edge, there is complete absence of a demarcation line between layers. Thus, such procedures allow the fabrication of microstructurally uniform composite ribbon and tape. Similar procedures have been used to make composite ribbons of two or more different compositions in amorphous–amorphous[101–104], amorphous–crystalline[102], and crystalline–crystalline[102,105] combinations.

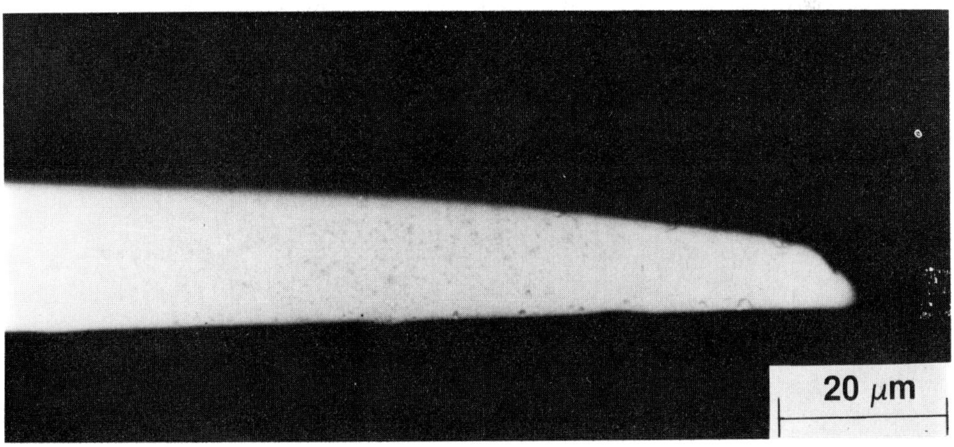

Figure 3.6 Transverse section micrograph showing no visible boundaries between the two layers in a $Fe_{40}Ni_{40}B_{20}$ composite amorphous alloy ribbon cast in a vacuum (after Liebermann[102])

3.2.8 Electron beam and laser surface melting

The use of directed energy and ion beam processing methods for surface modification of metallic materials has been reviewed by Kear *et al.*[106]. Rapid quench rates can be obtained by melting a thin layer on the surface of a substrate material. Electron and laser beams are particularly suitable for melting small surface areas rapidly and continuously because of their high energy densities (about $100\,kW/cm^2$). As with other rapid solidification methods, cooling rates have been measured by cell diameter or secondary dendrite arm spacing. Typical quench rates thus derived range[107] from 4×10^4 K/s to 5×10^6 K/s. This range of cooling rate is sufficiently great to have formed a 400 μm thick amorphous layer on the surface of a $Pd_{91.7}Cu_{4.2}Si_{5.1}$ substrate[108]. In addition to forming amorphous surface layers, surface melting can also be used to form new surface compositions by melting thin films[109] or by continuously feeding alloy wire or powder to the substrate surface[110]. A variation of these methods is the injection of hard particles into the melt track generated by the laser or electron beam to improve mechanical wear resistance of the surface[111].

3.3 Process characterization

Mathematical models of rapid solidification processes having various characteristic quench rates have been based both on heat transfer and fluid dynamic considerations. For example, a finite difference calculation has been used to describe the solidification of spherical droplets in atomization[55] and of fibres in melt extraction[112]. Equations for melt extraction have been generated in which both fluid and thermal aspects of the process have been considered[113]. Heat flow calculations have also been done to model self substrate quenching[55,114]. The following sections will briefly discuss some of the characterization and modelling which has been done for some selected melt-quenching processes.

3.3.1 Splat quenching

The first detailed mathematical modelling of splat quenching by Ruhl[10] distinguished between ideal and interface-controlled (Newtonian) cooling. All results were based on solutions to finite difference equations which contained variables corresponding to adjustable experimental parameters. Basic conclusions were that the most significant variables in splat cooling are the splat thickness, splat–substrate interfacial heat transfer coefficient, and instantaneous splat temperature. Similar

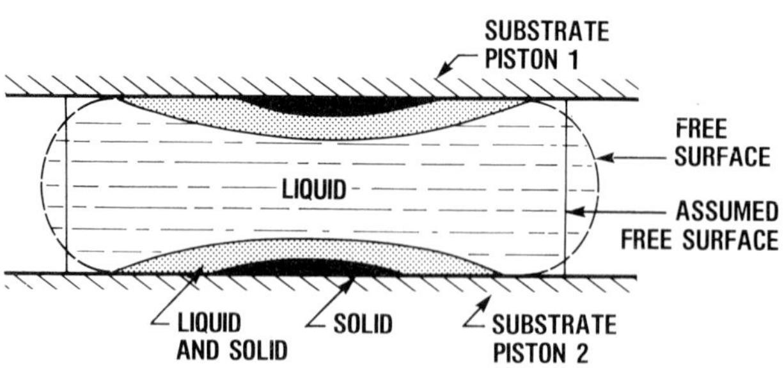

Figure 3.7 Schematic representation of the deforming splat during quenching (after Miyazawa and Szekely[116])

calculations on the splat quenching process were later done by Jones[9]. Direct measurement of splat cooling rates has also been used to characterize the process[115]. Recently, much more detailed modelling of the splat cooling process has been done and includes factors such as melt droplet spreading, convective heat flow, and actual movement of the substrate piston after impact [116,117]. An example given in this work, of the complexity in modelling twin substrate piston splat quenching, is schematically shown in *Figure 3.7*. Note that mechanical deformation effects on the splat after solidification make a detailed description of the process physics difficult. A virtual adjunct method calculation describing heat flow in unidirectional solidification has been applied to splat cooling processes[118]. Application of this model to normal freezing for conventional casting has already shown agreement with experimental data.

3.3.2 Chill block melt-spinning

An accurate description of this process is made difficult by the simultaneous occurrence of heat and mass transfer. Since ribbons used in applications may typically be stacked or wound into some bulk form, the as-made ribbon must be controlled, during its fabrication, to produce a smooth, uniform and reproducible product in order to maximize the spatial filling factor. To this end, empirical relations have been established between process parameters and ribbon width and thickness[74–76,119–121]. *Figures 3.8* and *3.9* show two examples of experimental results on the width of ribbons made by chill block melt-spinning. The small degree of data scatter about the least-squares fitted lines indicates good agreement of these experimental data with the respective models proposed. Details pertaining to various aspects of chill block melt-spinning have been modelled. For example, the fluid dynamics and mass transfer of an axisymmetric jet impinging on a barrier have been calculated[122] as have the heat transfer characteristics of such a jet on impingement[123]. Models describing the freezing of a liquid on to a moving sheet (dip-forming)[124,125] describe a physical situation very similar to that which prevails in the melt puddle during melt-spinning. Recently, however, some mathematical models have been established which describe the chill block melt-spinning process exclusively. All of these models are based on the propagation of a boundary layer which determines ribbon thickness. While thermal and momentum boundary layers occur simultaneously during ribbon formation, Kavesh[75] has calculated that the thermal boundary layer penetration exceeds that of the momentum boundary layer, thereby allowing modelling

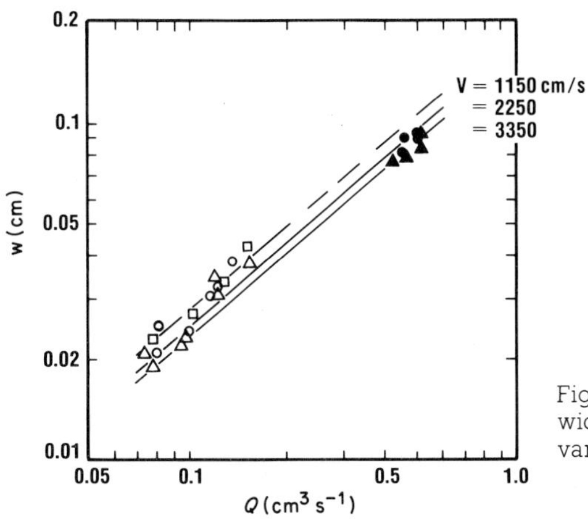

Figure 3.8 Experimental CBMS ribbon width as a function of material flow at various substrate speeds (after Kavesh[75])

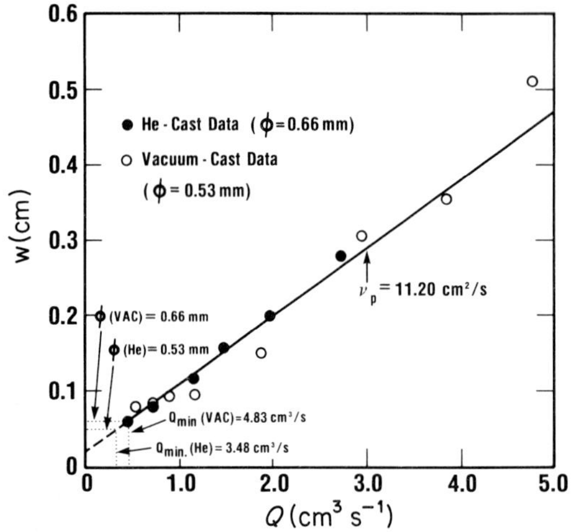

Figure 3.9 Experimental CBMS ribbon width as a function of material flow for melt-spinning in helium and in vacuum (after Liebermann[76])

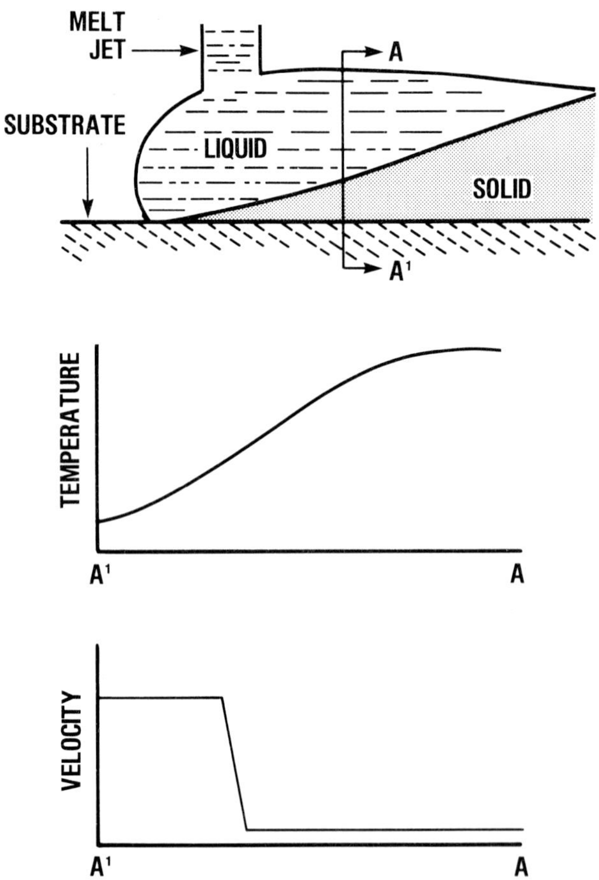

Figure 3.10 Schematic diagram of the molten alloy puddle on the rapidly moving substrate surface with temperature and velocity profiles along section A–A' of the melt puddle (after Kavesh[75])

of the process in terms of thermal aspects alone. Schematic temperature and velocity profiles, suggested by Kavesh, through a section *A–A′* of a melt puddle during chill block melt-spinning are schematically shown in *Figure 3.10.* Note that while temperature varies continuously along section *A–A′*, there is an abrupt velocity change in the solidification front, taken as the glass transition isotherm. In contrast to this ribbon formation modelling, based on thermal boundary layer propagation, more recent works contend that the momentum transport mode is predominant[126–128]. Furthermore, some recent modelling work has considered a combined thermal–momentum boundary layer effect[129,130] as well as temperature-dependent melt viscosity[131] in order more realistically to simulate the melt puddle during processing. Thus, a variety of approaches to describing the process physics of chill block melt-spinning have evolved. All are successful in describing experimental observations to some extent. However, the complex nature of ribbon formation and the factors which control it remain yet to be described by a physically consistent model.

3.3.3 Twin roller quenching

The inception of this ribbon processing method was spurred by the quest for greater average cooling rates. It turns out that while twin roller quenching has a slightly higher initial quench rate, its short ribbon–substrate contact time causes a sharp quench rate reduction after departure from the substrate[132]. Mathematical modelling for twin roller quenching is more difficult than for single roller methods because inadvertent mechanical working of the ribbon product between the roller may occur during processing. In spite of this, a model has been established which describes the heat transfer, liquid and solid phase flow, as well as the hydrostatic pressure distribution in the solid for twin roller quenching[133]. In addition, another modelling work has used experimental data to estimate the temperature, flow fields and average cooling rates[89]. While the somewhat higher initial quench rate and improved ribbon surface microgeometry make twin roller quenching an attractive process, the demanding apparatus geometry constraints in the melt pool region make this a difficult process to control.

3.3.4 Planar flow casting

The process differs from chill block melt-spinning in that the nozzle which feeds molten alloy is situated very close to the moving substrate surface[91]. Therefore, perturbation of the melt on the substrate by the gas boundary layer, substrate surface roughness, etc. is substantially reduced owing to mechanical constraint of the melt by the crucible nozzle walls, as shown[91] schematically in *Figure 3.11.* The effects of the gas

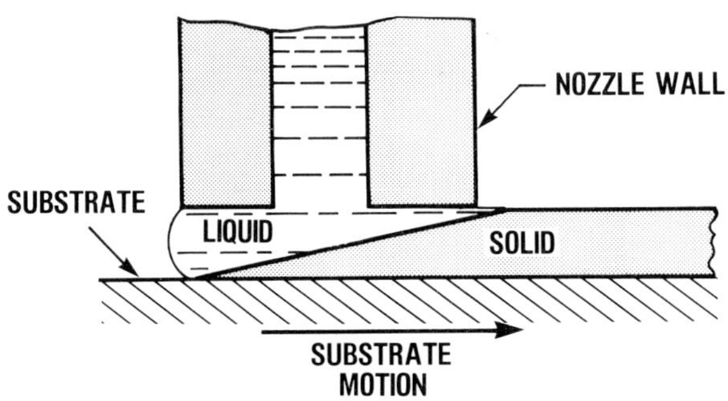

Figure 3.11 Shematic side view of nozzle–substrate geometry in planar flow casting (after Narasimhan[91])

boundary layer on melt puddle stability and melt–substrate interfacial contact have been studied by Huang and Fiedler[134]. A clear advantage of this rapidly quenched alloy tape process is that the nozzle–substrate geometry shown in *Figure 3.11* can be extended indefinitely normal to the page, thereby allowing the production of wide tapes. In reality, care must be exercised to maintain a fixed nozzle–substrate gap because it significantly affects ribbon dimensions, smoothness, and quench rate. Narasimhan describes the fluid dynamics of the melt in the planar flow casting process as being laminar[91]. Empirical characterization of the planar flow processes for the fabrication of rapidly quenched metallic ribbons has been performed[96,135,136]. Calculations by Stokes have shown the melt flow to be highly laminar and approximated by lubrication theory[137]. Caution must be exercised with this approach because the temperature dependence of melt viscosity has not been considered in the calculations. In a more conventional mathematical model, it has been found that planar flow casting is similar to chill block melt-spinning in that the propagation of a momentum rather than a thermal boundary layer controls ribbon dimensions[138].

3.4 Conclusion

The sample preparation methods described all have limitations as to quench rate, specimen geometry, etc. In fact, the non-equilibrium nature of amorphous alloys makes the atomic short range order and free volume dependent on the preparation method used and on experimental variables within a given process. For example, a comparison of non-equilibrium alloys made by vapour and by liquid quenching shows differences which are attributable to the nature of each preparation method used[139]. Differences in the as-prepared state of amorphous alloys are noted even for samples made by melt quenching alone[14,15,93,132,140]. This suggests that secondary effects associated with the details of sample preparation can have a significant effect on the structure and properties of rapidly quenched samples.

While all of the sample preparation methods discussed have been used to produce amorphous metallic ribbons in the laboratory, the commercialization of a particular process depends on the nature of the final product form desired. For example, methods such as sputtering, evaporation, implantation, and spray deposition are visualized as commercially viable processes for surface modification and coating. Although very high quench rates (above 10^8 K/s) can be achieved with processes such as deposition etc., the rate of amorphous metal mass generation is relatively low. This would probably limit such processes to the fabrication of small quantities of special compositions which could otherwise not be formed by use of other processing techniques. Various powder and ribbon fabrication methods have been commercialized. While quench rates for these processes (10^5—10^6 K/s) are not as high as those of some other methods, the existing melt-quenching technology allows the fabrication of amorphous metallic alloys at high mass production rates (about 20 cm^3/s). Methods for consolidation of the as-made powder and ribbon may be necessary in order to obtain bulk forms made from as-cast materials.

Acknowledgments

I am grateful to M. M. Dean and R. P. Laforce, both of General Electric Corporate Research and Development, for typing the manuscript and providing the photographs of *Figures 3.1*, and *3.3*, *3.4* and *3.5*, respectively.

References

1. Giessen, B. C., in *'Developments in the Structural Chemistry of Alloy Phases'*, Ed. Giessen, B. C., p. 227, Plenum Press, New York (1979)
2. Jones, H. and Suryanarayana, C., *J. mater. Sci.*, **8,** 705 (1975)
3. Ramachandrarao, P., Scott, M. G. and Chadwick, G. A., *Phil. Mag.*, **25,** 961 (1972)
4. Cahn, R. W., Krishananand, K. D., Laridjani, M., Greenholz, M. and Hill, R., *Mater. Sci. Engng*, **23,** 83 (1976)
5. Polk, D. E. and Giessen, B. C. in *'Metallic Glasses'*, p. 1, American Society of Metals, Metals Park, Ohio (1978)
6. Cahn, R. W., *Contemp. Phys.*, **21,** 43, (1980)
7. Suryanarayana, C., *Sci. Rep. Res. Insts Tôhoku Univ.*, **A28,** 143 (1980)
8. Jones, H., *Rep. Prog. Phys.*, **36,** 1425 (1973)
9. Jones, H. in *'Rapid Solidification Processing Principles and Technologies'*, Eds. Mehrabian, R., Kear, B. H. and Cohen, M., p. 28 Claitor's Publishing Division, Baton Rouge, LA (1978)
10. Ruhl, R. C., *Mater. Sci. Engng*, **1,** 313 (1967)
11. Perepezko, J. H., in *'Rapid Solidification Processing Principles and Technologies II'*, Eds. Mehrabian, R., Kear, B. H. and Cohen, M., p. 56, Claitor's Publishing Division, Baton Rouge, LA (1980)
12. Cohen, M., Kear, B. H. and Mehrabian, R., in ref. 11, p. 1
13. Wood, J. V. and Honeycombe, R. W. K., *J. mater. Sci.*, **9,** 1189 (1974)
14. Chi, G. C., Chen, H. S. and Miller, C. E., *J. appl. Phys.*, **49,** 1715 (1978)
15. Calvayrac, Y., Harmelin, M., Quivy, A., Chevalier, J. P. and Bigot, J., *J. Phys.* (Fr.), **C8,** 114 (1980)
16. Liebermann, H. H. and Walter, J. L. in *Rapidly solidified Amorphous and Crystalline Alloys*, Eds, Kear, B. H., Giesen, B. C. and Cohen, M., p. 111, North Holland (1982)
17. Takayama, S., *J. mater. Sci.*, **11,** 164 (1976)
18. Davies, H. A., *Rev. Chim. Min.*, **16,** 349 (1979)
19. Chen, H. S., Leamy, H. J. and Miller, C. E., *A. Rev. mater. Sci.*, **10,** 363 (1980)
20. Chen, H. S., *Rep. Prog. Phys.*, **43,** 353 (1980)
21. Leontic, B., in *'Proc. Conf. Amorphous Metallic Materials, Smolenice, 1978'*, Eds. Duhaj, P. and Mrafko, P., p. 225, Veda, Bratislava, Yugoslavia (1980)
22. Davis, L. A., DeCristofaro, N. and Smith, C. H. in '*Metallic Glasses: Science and Technology*', vol. 1, p. 1, Budapest, Hungary (1980)
23. Hilzinger, H. R. and Hock, S., ref. 22, p. 71
24. Luborsky, F. E., in *'Ferromagnetic Materials'*, Ed. Wohlfarth, E. P., vol. 1, p. 451, North-Holland Publishing Company, New York (1980)
25. Wehner, G. K. and Anderson, G. S., in *'Handbook of Thin Film Technology'*, chap. 3, McGraw-Hill, New York (1970)
26. Bennett, M. R. and Wright, J. G., *Phys. Status Solidi (a)*, **13,** 135 (1972)
27. Leung, P. K. and Wright, J. G., *Phil. Mag.*, **30,** 995 (1974)
28. Polk, D. E., Dube, C. E. and Giessen, B. C. in *'Proc. 3rd Int. Conf. Rapidly Quenched Metals'*, Ed. Cantor, B., vol. I, p. 220, The Metals Society, London (1978)
29. Leamy, H. J. and Dirks, A. G., *J. Phys. D*, **10,** L95 (1977)
30. Simpson, A. W. and Brambley, D. R., *Phys. Status Solidi (b)*, **43,** 291 (1971)
31. Watanabe, T. and Tanabe, T., *Mater Sci. Engng*, **23,** 97 (1976)
32. Poate, J. M. and Gullis, A. G., in '*Ion Implantation*', Ed. Hirvonen, J. K., vol. 18, chapt 4, Materials Science and Technology (1980)
33. Grant, W. A., Ali, A., Chadderton, L. T., Grundy, P. J. and Johnson, E., in ref. 28, p. 63
34. Duwez, P. and Willens, R. H., *Trans.* TMS-AIME, **227,** 362 (1963)
35. Pietrokowsky, P., *Rev. scient. Instrum.*, **34,** 455 (1963)
36. Harbur, D. R., Anderson, J. W. and Maraman, W. J., *Trans. TMS-AIME*, **245,** 1055 (1969)
37. Guntherodt, H. J., Institut fur Physik, Universitat Basel, Basel, Switzerland, personal communication (1981)
38. Nayar, P. K. K., *Trans. Indian Inst. Metals*, **33,** 233 (1980)
39. Ramachandrarao, P., Laridjani, M. and Cahn, R. W., *Z. Metallk.*, **63,** 43 (1972)
40. Leontic, B., Lukatela, J., Babic, E. and Ocko, M. in ref. 28, p. 41
41. Togano, K., Kumakura, H. and Tachikawa, K., '*Proc. 4th Int. Conf. Rapidly Quenched Metals*', vol. II, Eds, Masumoto T. and Suzuki, K., p. 125, Japan Institute of Metals, Sendai (1982)
42. Ohring, M. and Haldipur, A., *Rev. scient. Instrum.*, **42,** 530 (1971)
43. Moss, M., Smith, D. L. and Lefever, R. A., *Appl. Phys. Lett.*, **5,** 120 (1964)
44. Yamauchi, K. and Nakagawa, Y., *Jap. J. appl. Phys.*, **10,** 1730 (1971)
45. Rairden, J. R. and Liebermann, H. H., General Electric Corporate Research and Development, unpublished research (1978)

Chapter 4

Modelling the atomic structure

J. L. Finney

Dept. of Crystallography, Birkbeck College, London, UK

4.1 Introduction

Amorphous alloys were prepared first by the rapid cooling of the melt[1], bypassing crystal nucleation. An obvious parallel exists with the preparation of silicate glasses, in which the essential structure of the liquid is retained, although some structural relaxation will probably occur during cooling. When polycrystalline, distorted lattice, dislocation and other models of amorphous alloys failed to account for the experimental structural data[2] (the radial distribution function or the pair correlation function; *see* Chapter 5), it seemed sensible to lcok towards the parent liquid structure to assist our understanding of these new materials, although non-liquid related preparation techniques (Chapter 3) leave open the possibility of the existence of structural variants ('amorphous polymorphism' [3]).

Unfortunately although we know in great quantitative detail the structures of *crystalline* materials, our ability to handle the structures of *non-crystalline* condensed liquid phases is limited. Assuming we know how the N constituent particles interact (the potential function), we require to construct the configurational integral:

$$Q = \frac{1}{V^N N!}\int \cdots \int e^{-\varphi/kT} dV_1 \cdots dV_N \tag{4.1}$$

where ϕ is the potential energy of the assembly with atom 1 in volume element dV_1 etc. Once Q is known, we can calculate the thermodynamics of the assembly by purely formal manipulations.

Although the high coordination and density of the crystal means we must take into account a relatively large number of interactions in calculating ϕ, the crystal lattice allows us to make a drastic simplification: for example, for a primitive ideal crystal (one asymmetric unit per unit cell), Q can be written as the product of N identical partition functions, which can be calculated if the atomic positions are known. For a *real* crystal, we can use the ideal crystal as a reference state for a perturbation-type analysis. At the other end of the spectrum, a perfect gas, there are no interactions between atoms except on collision, hence $\phi \equiv 0$, making the evaluation of Q trivial. For a dilute real gas, as contributions to ϕ arise from the small number of close interatomic approaches, Q can be evaluated in terms of pairs of atoms. As the density increases, we must consider larger clusters of atoms (three, four, ...). This 'cluster expansion' procedure[4], however, becomes mathematically intractable for reasonably dense gases, and fails to converge before we reach liquid densities.

Calculation of Q for a dense assembly is made difficult by the need to consider a relatively large number of intermolecular interactions. For a crystal, the approximate identity of local environments allows us drastically to simplify the problem by making use of lattice symmetry; hence the popularity until relatively recently of lattice models of liquids (and glasses). *Where there is no underlying lattice structure, this approximation cannot be made*, and we are unable to calculate the configurational integral unless the number of interactions is very small, as in a dilute gas. Thus, other methods have to be developed, and this leads us into liquid state theory[5,6]. Even here, unphysical approximations must be made to make calculations tractable; there is still no adequate theory of the liquid state[7].

In the absence of an adequate theory, we are driven to an alternative structural approach which constructs approximate *models* either in the laboratory or on the computer which are arguably representative of the structural arrangement in the liquid or glass. In the liquid case, our model (a set of positional coordinates of atom centres) represents an *instantaneous snapshot* of atomic positions in the real liquid. For an amorphous solid, translational motions will have been frozen out, and our stationary model will relate directly to possible atom positions in the real assembly. Once such a model has been constructed, its structural properties, in particular its radial distribution function (RDF) and density, can be calculated and compared with experiment.

Such *liquid* models are now constructed routinely in the computer, using *computer simulation* techniques which, within the limitations of the assumed potential function, result in sample molecular assemblies which are consistent with the required equilibrium Gibbs ensemble. For non-crystalline solids (which are *non-equilibrium* assemblies) the use of such techniques is problematical (*see* below). Initially, we prefer to consider simpler approaches which illustrate more clearly the structural principles of non-crystalline dense structures. Earlier detailed reviews of such model-building techniques can be found in the literature[3,8–12].

4.2 The ideal hard-sphere glass: the Bernal model

Bernal argued[13] that the structure of a simple liquid (one of spherical molecules with an angle-independent potential function) is determined by volume exclusion. As the liquid density is about the same as (though generally a few per cent lower than) the crystal density, the constituent atoms must have high coordination numbers, with surrounding atoms generally in contact, an argument consistent[5,6,14] with experimental coordination numbers of around 8—12. Hence, he proposed a zeroth order model in which the atoms were considered as hard spheres, and their local structure determined by the restrictions on space-filling consequent upon the inability of two atoms to approach more closely than one diameter. The structure of *crystals* of spherical molecules such as the inert gases and simple metals had long been rationalized in this way, the familiar f.c.c. (face-centred cubic) and h.c.p. (hexagonal close packed) structures being the consequence of high density and *the presence of a crystal lattice*. What Bernal did was to extend this approach to liquids, for which theoretical[15] and experimental evidence argued against lattice organization. The ideal liquid might therefore have a structure similar to that realized when equal hard spheres are packed together at high density *without allowing any 'crystallization' to occur*.

Such models were physically constructed in the laboratory, and their structural properties examined, independently by Bernal and coworkers[13,16] and Scott[17].

Bernal's models were built [16,18] by pouring several thousand steel balls into a suitable flexible container (originally a football bladder) placed within a previously-prepared irregular surface to prevent surface crystallization. The assembly was tightly bound by several meters of rubber strip to exert a compressive force, and then 'kneaded' to facilitate sphere rearrangements to maximize the density. The mass was 'fixed' by pouring through it a molten mineral wax (earlier experiments[16] used paint), and after setting, the exterior skins were peeled off. The coordinates of each sphere were measured by use of a converted toolmaker's bench[19]. Scott's procedure[17] involved filling a beaker or spherical flask with spheres, densification being achieved by mechanical vibration.

Both procedures appear somewhat *ad hoc*, yet the results from the two different laboratories were in excellent agreement. In particular, both concluded that the maximum density of a hard-sphere non-crystalline packing was 0.6366, to within 0.1 %. Moreover, recent attempts to reproduce this structure on the computer[20] yield an essentially indistinguishable radial distribution function.

Figure 4.1 A laboratory-constructed random close-packing of hard-spheres, showing the absence of lattice ordering

This model 'random close-packed structure' is shown in *Figure 4.1*, and its lack of crystallinity is immediately apparent. Calculations on the sphere coordinates demonstrated that the model possessed many of the structural features of simple liquids, and gave appealing qualitative explanations of supercooling, nucleation, melting, fluidity and diffusion[13,21]. Although very simple, this structure contains within it the essential physics of simple liquids[7], and provides an illuminating structural framework within which to consider the structures of other simple non-crystalline systems.

Because the model is static, and of maximum density (although lower density — 'random loose' — packings can be constructed[17]), it is particularly appealing as a model of a zero temperature ideal *glass* of spherical molecules. Its specific application to amorphous metallic alloys was proposed in the early 1960s by Cohen and Turnbull[22], although it was nearly ten years later before both model and experimental data were adequate to justify Cargill's making a quantitative comparison using data from Bernal's laboratory. He concluded that the experimental RDF of an amorphous $Ni_{76}P_{24}$ bore a strong resemblance to that of a dense random packing[2]. The two RDFs are shown in *Figure 4.2(a)*; the random packing was the only model that could reproduce the experimental splitting of the second peak.

There are, however, two major problems with this identification between

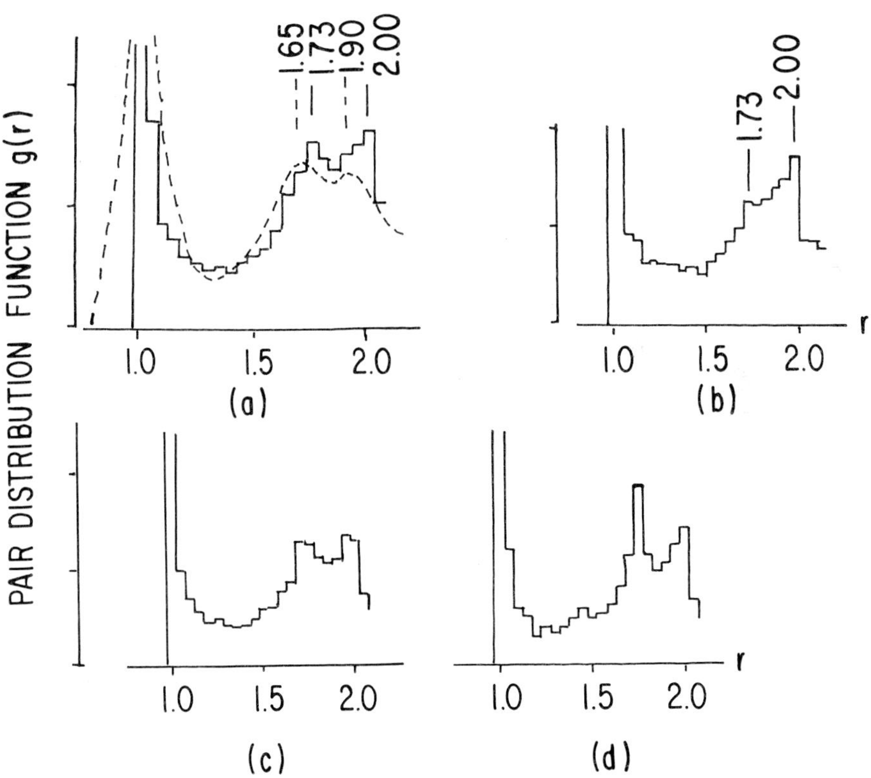

Figure 4.2 Pair distribution functions out to about two diameters of several single component hard-sphere models. (*a*) Hard-sphere dense packing[21] ($\rho = 0.6366$) compared with experimental $Ni_{76}P_{24}$ data[2] (dashed line); (*b*) computer-constructed model using Bennett's 'global' criterion[24]; (*c*) and (*d*) 'gas-compression' models of densities (*c*) 0.639 and (*d*) 0.664. Distance *r* is measured from a given centre in units of first peak position

4.3.2 Collective rearrangements

Sequential addition procedures can produce different 'random' structures depending on the pocket criterion. All, however, are 'loose-packed' structures which fail to show the structural details found in real amorphous alloys: to reproduce those, collective rearrangements are required to increase the density.

These movements are easily made on laboratory models by *ad hoc* shaking and kneading. It is difficult (and *very* expensive) to program the computer to make them. Two approaches have been followed for hard-sphere structures. The first[34–36] simulates the compression of a hard-sphere gas. N positions are chosen at random in a box (usually spherical) and a small sphere assigned to each point. Checks are made that no two spheres overlap; when an overlap occurs, the two spheres are moved apart along their line of centres until they just touch (overlaps created thereby being ignored). When all overlaps have been cleared, the sphere radius is increased and the overlap-clearance procedure repeated. This method had been attempted by Bernal[13] in the late 1950s, but his computer was not up to the task, and the method was first used successfully by Mason[34]. Several such 500-sphere models of density between 0.57 and 0.67 have been built, and their RDFs and local structures examined[35]. Strong similarities were found between the laboratory-built dense packings and computer-built assemblies of similar density [*Figures 4.2(a)* and *4.2(c)*]. This suggests (but not conclusively) that the 'statistical structure' of a hard-sphere random packing above the loose packing density (~ 0.60) may be unique for that density. Below ρ about 0.60, this is not so. The main point of interest, however, concerns our apparent ability to build a model with a density some 4 per cent higher than the apparent laboratory limit of 0.637. Moreover, in the RDF of such a high density packing the relative intensities of the split peak components have reversed [*Figure 4.2(d)*].

Both these effects can be put down to a further failure to take full account of packing constraints, this time at the boundary. Ideally, our finite model should be a valid sample from an effectively infinite statistically-equivalent assembly. The laboratory model attempted, apparently successfully, to do this by using irregular surfaces as model boundaries, while both the sequential computer building and the hard-sphere gas compression methods do away with the boundary altogether. This is particularly serious for the hard-sphere gas compression method, in that the absence of a boundary, by removing a strong packing constraint, allows the spheres to rearrange unrestricted by the requirement that the model should be capable of being embedded in an equivalent assembly. Thus, although this method can give apparently good agreement with both the experimental RDF and density, it can do so only at the expense of a relaxation of the required packing constraints. The local structures of these high density clusters also show significant differences from the laboratory dense packing[35].

We could largely solve this 'embedding problem' by performing the collective rearrangements on an assembly using periodic boundary conditions, in which the (in this case cubic) box containing our hard-sphere gas is surrounded by identical 'image' boxes. This has been done successfully in two[37], but apparently not in three, dimensions for hard spheres*. Woodcock[38] has, however, performed a lengthy molecular dynamics calculation in which the density of an assembly of 500 spheres with periodic boundary conditions was slowly increased from a low density fluid state to almost close-packing.

Note added in proof.* A recent paper (Jodrey, W. S. and Tory, E. M., *Powder Tech.*, **30, 111 (1981)) uses periodic boundary conditions in an implementation of the gas compression algorithm. Again, the maximum density achieved was 0.6366, which provides further evidence of this upper limit to density.

The projected maximum density limit was at 0.637 ± 0.002, indistinguishable from that of the laboratory model (0.6366 ± 0.0004). Moreover, the RDF at the highest density achieved (at 0.628 still significantly below the apparent maximum) was identical to that of the hard-sphere model within computational and experimental uncertainties. Despite the apparent *ad hoc* nature of the laboratory-built model, its density and structure appear to be indistinguishable from that of this 'configurationally-arrested' hard-sphere fluid, supporting our tentative assignment of the hard-sphere dense packing as a zero temperature ideal glass. The collective rearrangements produced by either molecular dynamics, or laboratory kneading and shaking, appear very similar.

4.3.3 Summary

Computer construction methods raise a basic problem of constructing a model consistent with the packing constraints imposed by the high density of the real systems we are trying to model. Sequential addition procedures necessarily result in lower density models which do not show the characteristic RDF features found in real systems: collective rearrangements are apparently necessary to increase the density adequately to show a clear second peak splitting. The addition of a sphere on to a partly-built model is constrained only by the packing constraints exerted by the already-existing model on one side only of the new sphere. Similarly, collective rearrangements with free boundaries take inadequate cognisance of the packing constraints that would be effective were the cluster consistent with being embedded in a statistically-equivalent ensemble. The additional freedom allowed by a free boundary facilitates very high density packings with apparently improved (with respect to experimental) RDFs. Just as low density models with good RDFs must be considered inadequate, so must high density models if they are incapable of being extended in size indefinitely without the consequent exertion of packing constraints changing the density and RDF. The icosahedral model of Sadoc is an example of the embedding problem on a finer scale: the density and RDFs of the subunit clusters are reasonable, but they cannot be packed together without drastically reducing the density.

Although these remarks refer to hard-sphere packings they apply also to more 'realistic' 'soft' sphere models. Experimental comparison must be made with both density and RDF *at least*: we can fit either with many models, but to fit both by one fully embedded model is difficult, but necessary. Otherwise, we would have failed to consider sufficiently the packing constraints which, in dense systems, are strong determinants of the structure.

4.4 'Real' amorphous alloys

No real alloy consists either of hard-spheres, or of only one type of atom. We need therefore to examine the effects of softer (ideally realistic) potentials on the ideal hard-sphere structure, and to see how this structure changes when two (or more) components and their chemistry are considered.

4.4.1 Potential functions

Two main routes are followed to probe the effects of soft potentials. First, a previously-constructed hard-sphere assembly is rescaled to a soft potential, and energy minimization calculations performed. Most work on single component systems has used simple potentials such as those of the Lennard-Jones or Morse variety, and conjugate gradient minimization methods have allowed relatively large structures (of a thousand or more atoms) to be handled[39]. Secondly, the use of computer simulation — molecular dynamics (MD) and Monte Carlo (MC) — techniques has been explored.

In the molecular dynamics procedure, an assembly of atoms is set up in a (usually cubic) box, with periodic boundary conditions to minimize boundary problems. The atoms are each assigned a velocity at random, ensuring the average kinetic energy corresponds to the required temperature. The forces acting on each atom are calculated, and the atoms allowed to move according to Newton's laws. The temporal development of the system is followed until the assembly has apparently equilibrated, whereupon its structural and dynamic properties can be extracted. The Monte Carlo procedure works rather differently but produces sample configurations which are in the limit consistent with the molecular dynamics assemblies. Again starting from an initial configuration of the required density, a randomly chosen atom is moved in a random direction by a (modulated) random amount and the potential energy change ΔU calculated. If ΔU is negative, the move is accepted and the process repeated; otherwise, the move is accepted with a Boltzmann weighting. It can be shown that both MC and MD methods converge towards configurations that are consistent with a Gibbs ensemble of the model molecules chosen.

These computer simulation techniques have been applied very successfully to liquids, and in fact were used early on to demonstrate the similarity between inert gas liquids and Bernal's model[21]. Provided the calculation is pursued for long enough (though how long is arguable, especially for non-simple molecules) the validity of the resultant structure (MC and MD) and dynamics (MD only) *depends on the adequacy of the potential functions used.* The method is often used to test proposed potential functions.

When applied to amorphous solids, the reservation with respect to potential functions remains, and we have an additional problem: we are trying to simulate a *non-equilibrium* assembly. Using molecular dynamics, we start with a liquid assembly and then reduce the temperature progressively. However, computational restrictions force us to reduce the temperature at a rate which is unrealistically high (perhaps about 10^{11} K s^{-1}) compared with laboratory quench rates of less than about 10^{6} K s^{-1}. The resultant structure is really a 'configurationally-arrested' liquid, rather than an amorphous solid which has been allowed to adjust its structure during the quench. Monte Carlo suffers similar problems in that temperature must be reduced (or pressure increased) unrealistically rapidly.

Relaxation calculations[3,23,39,40], molecular dynamics[20,38,41,42] and Monte Carlo[43–46] simulations have been reported on single component systems using largely Lennard-Jones or Morse potentials. Many of these calculations were performed under different conditions; as a result, it is possible to draw only tentative conclusions concerning the variability of the resulting amorphous structure with respect to the different conditions used. We give here a brief summary relevant to the binary alloy work discussed below, fuller details having been published elsewhere[3,40].

Potential function and model boundaries

Both RDF and internal structure can vary with potential function[3]. For a very soft potential such as a Morse 3, structural differences are also induced by the different truncation procedures used in energy calculations[3]. These latter differences are particularly evident when free boundaries are used, and might be attributed to a long-range compression effect which would be much reduced for an embedded model.

The same starting configuration relaxed under free and confined boundary conditions showed structural differences. However, the boundary problem may be less critical for soft potential assemblies than for hard-spheres. In the hard-sphere case, the 'free volume' available for internal rearrangement depends strongly on the presence of an unconstrained boundary; for soft spheres, internal adjustments are facilitated in addition by the softness of the repulsive core. Nevertheless, as such calculations require only a minor increase in computer time, it would seem sensible to use periodic boundary conditions to reduce the free boundary problem.

Starting configuration for energy minimization

If we could show that the relatively low density sequentially constructed assemblies led to relaxed structures indistinguishable from those obtained from dense, collectively-rearranged hard-sphere starting configurations, then the objections raised earlier to sequential construction methods could be ignored. The limited evidence on this point suggests[3] that, although resultant RDFs of relaxed models using starting configurations of densities 0.637 (close-packed) and 0.60 (loose-packed) are statistically indistinguishable, there are measurable differences in internal structure. Which starting point is more realistic is unclear, as is the effect on resultant structure of using even lower density starting configurations. Although one might argue that the detected local structural differences are physically unimportant, the remaining uncertainty advises prudent use of several different starting configurations.

Computer simulation.

Just as the hard-sphere MD calculation of Woodcock[38] tended towards a structure whose RDF is apparently identical to that of the hard-sphere dense random packing[38,47] so MD calculations on rapidly quenched Lennard-Jones atoms yield RDFs very similar to those obtained from free boundary relaxations[41,42]. Thus, despite the reservations raised earlier about the identity of structures produced at realistic laboratory quench rates and at accessible computer quench rates, indistinguishable RDFs can be obtained. We should stress several points, however. First an amorphous Lennard-Jones system has never been experimentally prepared, and it may well be that the unrealistically high quench rates used in the simulations may be theoretically necessary to produce one. In this case, the apparent structural similarity between MD and relaxation calculations would be expected although we could not justifiably generalize this similarity to alloys which are produced at quench rates low enough to allow internal structural rearrangements on the time scale of the quench. Secondly, the RDF is a relatively crude measure of structure, and similar RDFs can hide subtle structural differences. Thirdly, some MD work suggests there may be a significant structural dependence on quench rates[42], and structural data exist to allow a detailed investigation of this point[48].

4.4.2 Binary amorphous alloys

Each of the model-building techniques discussed above has its disadvantages. When we consider trying to model real amorphous *alloys*, the two procedures that look most attractive are (i) *computer simulations* and (ii) *relaxation calculations*, preferably using periodic boundaries. Monte Carlo or molecular dynamics quenching of a previously set-up liquid configuration will produce a 'configurationally-arrested liquid' which will necessarily ignore those rearrangements that will occur on the time-scale of the laboratory quench. The relative importance of such processes to the final structure is unclear, and detailed examination of particular cases is required. The possibility of identifying these processes (or perhaps relaxations from sequentially constructed computer models) with structural relaxation of as-quenched glasses could be pursued (*see* Chapter 10).

A major problem with relaxation calculations is the choice of starting configuration. For single component models, we have evidence for a dependence of structure on starting configuration. For a multicomponent system, this dependency will be much greater. In addition to the packing constraints that dominate the single-component structure, we must consider the 'dispersion' of the two components in the alloy. This is generally termed 'chemical ordering', and refers to the relative arrangement of A atoms around B, B around A, etc. Relaxing a sequentially-constructed dense starting configuration will not significantly perturb the topology of the local chemical ordering: if it is incorrect at the start, it will remain incorrect; the structure will be locked into an incorrect local energy minimum. 'Correct' topology of the starting configuration is therefore essential; subsequent relaxation is merely a refinement, though one which (on the evidence of single component hard→soft sphere relaxations) will significantly change the model RDF.

Thus, either method must be used with care, and the possible consequences of variations in procedure explored fully. For sequential model-building followed by relaxation, alternative starting configurations should be used, while for simulation calculations, quench-rate dependences must be examined. A hybrid of the two approaches seems attractive, though has yet to be attempted. On the reasonable assumption that (provided the potential functions are 'adequate') a computer-simulated liquid alloy has the correct liquid 'chemical ordering', then that retained on rapid quenching would be a reasonable starting configuration for a relaxation calculation[10]. Within the constraints of available computer time, quench rate dependences of the starting structure could be examined, and the detailed local structures of the final 'relaxed' models compared.

At this point, it is perhaps useful to restate the three major problems facing the would-be serious modeller of binary or more complex amorphous alloys. First, *the potential functions*. We know single component amorphous structures depend upon the potential, and this dependence is probably stronger for the *binary metal* alloy. For any resemblance of reality, therefore, we must abandon the oversimplified Lennard-Jones and Morse forms. This raises additional problems. In metals, we cannot ignore many-body effects from the delocalized electrons, and hence density-dependent potentials must be used. When we considered the second component B, we need a description of the AB interaction which is consistent with the resulting 'chemical ordering' found in the liquid. For 'simple' metals, some progress is possible by use of established techniques in metal physics, as in the fundamental work of Hafner[49]. Where metalloids are involved, we have even bigger problems: directionality of the metal–metalloid interaction, which would have consequences for 'chemical ordering', must somehow be considered.

In the light of these rather basic problems of potentials, it could be argued that more *ad hoc* model potentials could be used, provided the model potential is designed to reproduce successfully (or at least be consistent with) the observed local chemical ordering. This raises our second problem, that of *knowing what the chemical ordering is*. For single component models, one RDF is the maximum structural information that can immediately be obtained from a diffraction experiment. For two components, we require *three partial* RDFs, one for each of the three possible pairs AA, AB, BB (*see* (Chapter 5). Merely obtaining these data for a given alloy is a tall order: three independent scattering experiments are required in which the scattering power of the atoms is different. In principle this can be done for a limited number of suitable alloys by use of isotope substitution in neutron work, and anomalous dispersion for X-rays. Only very limited partial RDF data are currently available, though work in progress should lead to much better local ordering characterization in a few alloys. Other techniques such as EXAFS, Mössbauer and NMR have also been used to try to obtain similar local information, although the interpretations of especially the last two are strongly model-dependent[10].

Thirdly, even when the 'chemical ordering' is known, it is not a trivial task to construct an amorphous alloy model with the two components suitably dispersed to reproduce it. For the single-component models, *packing constraints* are strong structure determinants, yet they are poorly understood. For the binary, within this poorly-understood framework, we must build in the additional constraint of a prescribed 'chemical ordering'.

In this final section, we can do little more than indicate the approaches that have been tried to tackle these problems. This discussion could be framed in terms of how simultaneous satisfaction of these two (competing?) *packing* and *'chemical ordering'* constraints is attempted. All methods have used subsequent relaxation of the initial configuration, using potentials (generally Lennard-Jones variants) which are sometimes designed to bias the assembly towards the desired local structure.

'Dense random packing' models

Under this heading we consider model-building approaches which *satisfy packing constraints preferentially to 'chemical ordering'*. Although earlier laboratory models were built by Scott and Kovacs[50] and Cherry and Finney[51], no attempt was made to bias the model towards a particular local ordering. Boudreaux and Gregor[52–54] applied Bennett's sequential addition procedure to (initially) hard-spheres of size and composition ratio designed to mimic $Fe_{80}P_{20}$ or $Pd_{80}Si_{20}$. Technical modifications were made to the building procedure to ensure compositional homogeneity[54], and an attempt made to input some chemical order constraints by forbidding near neighbour P–P contacts. The initial hard-sphere models (of several compositions to allow investigation of structural variations with composition) brought out a general defect of the sequential addition procedure: the structure was highly anisotropic. The addition procedure (probably partly because a site closest to the centre would tend to be planar) showed a characteristic RDF [*see Figure 4.2(c)*] only in the radial direction. Collective rearrangements would be expected to take full account of the packing constraints in all directions and thus remove this anisotropy. This was successfully done by relaxing the assemblies under Lennard-Jones potentials. As no attempt was made to design potentials which might bias the relaxation towards either retaining or enhancing ordering, the 'chemical ordering' was fed in solely through the exclusion of near-neighbour small (metalloid) atoms. Relaxation increased Fe–Fe and P–Fe coordination numbers by 30 per cent and 41 per cent respectively, to values in reasonable agreement with experiment[54]. Related structures have been built to model $Fe_{85}B_{15}$[55] and $Fe_{\sim 80}P_{\sim 20}$[56] and the results compared only partially successfully with limited partial RDF data.

The main problem with sequential addition procedures for modelling binaries with 'chemical order' is feeding in local ordering constraints which are consistent with the packing constraints *at each stage during construction*. The free boundary forming half the neighbourhood of each atom means *only half the constraints are defined at each addition stage*; therefore decisions must necessarily be made on limited information. Forbidding close contacts of particular types of atom is straightforward, but the more complex constraints that will probably be required fully to reproduce required chemical ordering seem far more difficult to feed in to the sequential addition process.

Built-in chemical ordering

An alternative is *first* to build in the desired chemical ordering (perhaps to excess) and to allow the relaxation process to bring in the dense packing constraints, perhaps using model potential functions designed to favour the original locally-ordered structures. This approach is strongly related to established model-building techniques for network glasses such as vitreous silica and amorphous silicon and germanium, although there the absence of significant packing constraints means that the prescribed chemical ordering remains. Gaskell first explored the possibilities of the method in his work on $Pd_{80}Si_{20}$ and related transition metal–metalloid alloys[57] for which experimental coordination numbers measured by EXAFS and neutron scattering gave results which were similar to those found in corresponding crystalline modifications which showed a trigonal prismatic coordination [*Figure 4.3*(*a*)]. The widespread occurrence of this kind of coordination polyhedron over a wide composition range suggests that it is reasonable to expect it to occur frequently in the amorphous alloy.

A model was therefore built (initially in the laboratory, later transferred to the computer), by joining trigonal prismatic units in the manner of *Figure* 4.3(*b*). This emphasizes the strong trigonal prismatic chemical ordering at the expense of packing constraints (the model in fact contains large cavities). The packing constraints are then brought more into play by computer-relaxing the assembly. Again Lennard-Jones potentials were used, though for one model the second neighbour palladium atom is treated differently to try to emphasize the trigonal prismatic ordering in the relaxed structure. The resultant structure shows encouraging agreement with experimental total RDFs: no partial RDF data are available for comparison.

The interplay between the initial trigonal prismatic coordination and the increasingly strong effect of the packing constraints as the relaxation proceeds is a particularly interesting problem which requires attention. At experimental densities,

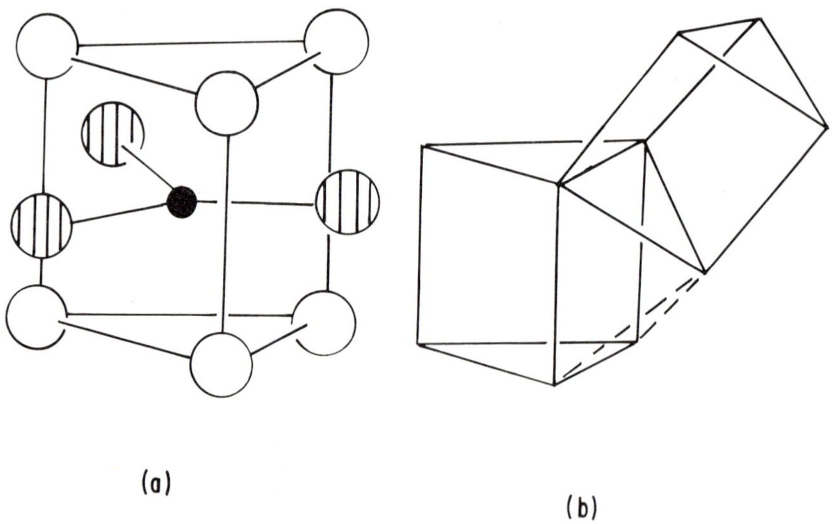

Figure 4.3 (*a*) Trigonal prismatic units used in building Gaskell's starting configuration by adding together units as in (*b*). The small black circle represents the silicon atom, the open circles first neighbour palladium atoms and the shaded circles second neighbour palladium atoms

widespread perfect trigonal prismatic order appears not to be possible in a non-crystalline assembly, but how small the distortions need be to accommodate such local ordering is unclear. Using recently developed techniques[58] we can begin to probe the degree of 'trigonal prismaticity' remaining in these structures. Initial results show that many such units do indeed remain, although generally distorted, and there is evidence that significant movements are forced on some atoms, in particular several silicon atoms become close to being first neighbours. We could speculate on the possibility of constructing a more perfectly trigonal prismatic (yet densely packed) non-crystalline assembly from a totally different starting point, with such an assembly inaccessible by relaxation from a configuration conforming to the starting hypothesis of *Figure 4.3(b)*.

There is also the interesting question of how far the Boudreaux and Gaskell approaches might tend towards similar relaxed structures. Fujiwara and Ishii[56] constructed Boudreaux-type models of $Fe_{100-x}P_x$ alloys, and attempted to characterize the local ordering. Although it is not clear how capable their analysis techniques were of describing unambiguously the phosphorus coordination geometry, they argued for a strong tendency to trigonal prismatic organization. Further work using more sensitive analysis techniques is under way to examine this point in more detail[59].

Other methods

In addition to computer simulated quenching suggested above, relying on realistic potential functions to reproduce the desired chemical ordering in the liquid structure starting configuration, one further approach seems worth pursuing. Boudreaux's approach puts primary emphasis on the packing, (rather weak) constraints being added to try to mimic some of the features of expected chemical ordering (although the sequential nature of the building procedure means that the packing constraints cannot be fully considered every time a sphere is added). Gaskell's method which emphasizes local ordering primarily, leaving relaxation to sort out packing constraints secondarily, seems a more flexible approach. An interesting third possibility currently being tried[60] starts from an already-constructed densely-packed single component model, and feeds into it chemical ordering information. The philosophy behind this approach argues that a pre-existing single component model contains the essential packing constraints in a dense non-crystalline structure. The (variable) topology is precisely defined, and therefore complete data are available for investigating the possibilities of labelling the atoms A or B consistent with the desired local ordering. Allowing a degree of statistical variation in the topological and metrical perfection of the local orderings we are trying to insert, many different possible combinations of A and B labellings can be explored and screened; relaxation calculations can then be performed in the usual way, using periodic boundary conditions. The simplest case of such an approach was used by von Heimendahl and Hafner[61] for $Mg_{70}Zn_{30}$, using what were hoped to be realistic potentials developed from pseudo-potential theory[49]. As zero chemical ordering was assumed, A or B (Mg or Zn) labels were assigned randomly (subject to composition). The related procedure of Kobayashi *et al.*[62] includes much weaker packing constraints, their starting configuration being set up by a random *parking* algorithm.

Extending this 'seeding' approach to the non-random chemical ordering case is far from trivial. It attempts to partly decouple the topological disorder (packing constraints) from chemical ordering so as to allow their (partially) separate consideration, by using a completely defined non-crystalline structure within which to probe the combinational possibilities. As such a single-component 'random lattice' (!) is a better approximation to the packing constraints operating in a binary where the two components are nearly equal in size, it seems sensible to start by modelling a weakly-ordered alloy of approximately equal size components.

4.5 Conclusions

The major problem with modelling real amorphous alloys is to construct a non-crystalline assembly that is consistent with both local chemical ordering and the overall packing constraints of high density structures. Network glasses such as vitreous silica present much less of a problem as the chemistry (linked SiO_4 tetrahedra) dominates the structure, which exhibits very strong chemical ordering with only minor considerations of packing. Single component 'ideal' glasses, be they of hard or soft spheres, provide the opposite extreme, where packing constraints dominate and the chemistry of molecules with spherical symmetry is not a significant consideration. The real problems lie in between where the interplay between 'packing' and 'chemistry' is crucial in determining the structure.

We are handicapped in tackling this problem in two major ways. First, we lack good, *model-independent* data on chemical ordering, though current work will remove this problem for selected alloys. Secondly, our understanding of the operation of packing constraints in dense non-crystalline assemblies is poor. Even for single-component 'ideal' model systems, we have no adequate theory, and we are forced to model-building studies to gain some (albeit restricted) insight. Even here, there are complications in that different construction procedures can lead to structurally different model assemblies ('amorphous polymorphism')[3], although there is evidence to suggest that at the higher densities of interest for real systems, a single ideal hard-sphere (and perhaps soft-sphere) structure can be (statistically) defined.

Although we have much experience of single-component models, which have been useful in helping at least to see (if not to understand) how variable model construction parameters can affect final structure, only a limited amount of work has been done to model binary amorphous alloys. Two main approaches put different stresses on chemical ordering and on packing constraints in the starting structure, and have raised interesting questions concerning the interplay between these two constraints which are crucial in determining the model structure. Despite the severe problems in modelling effectively the structures of real binary, and eventually ternary, amorphous alloys, progress is expected also via new 'topological seeding' techniques. Although we might make use of computer simulation to set up liquid configurations with chemical ordering determined by the assumed potential functions, neither MC or MD methods will solve all problems, being restricted to unrealistically high quench rates. Simpler, cruder methods are still appropriate, and may do more in the long term to improve our understanding of these fascinating structures.

References

1. Duwez, P., Williams, R. J. and Klement, W., *J. appl. Phys.*, **31,** 1136 (1966)
2. Cargill, G. S. (III), *Acta Crystallogr.*, **A25,** (S3), 595 (1969); *J. appl. Phys.*, **41,** 12 (1970)
3. Finney, J. L., *'Diffraction Studies of Non-Crystalline Substances'*, Eds. Hargittai, I., and Orville-Thomas, W. Y., p. 440, Elsevier, Amsterdam (1981)
4. Mayer, J. E. and Mayer, M. G., *'Statistical Mechanics'*, Wiley, New York (1940)
5. Croxton, C. A., *'Liquid State Physics'*, Cambridge University Press (1974)
6. Hanson, J. P. and McDonald, I. R., *'The Theory of Simple Liquids'*, Academic Press, New York (1976)
7. Ziman, J. M., *'Models of Disorder'*, Cambridge University Press (1979)
8. Finney, J. L., *Nature*, **266,** 309 (1977)
9. Cargill, G. S. (III), *Solid St. Phys.*, **30,** 227 (1975)
10. Finney, J. L., Gellatly, B. J. and Wallace, J., *'Metallic Glasses: Science and Technology'*, Eds. Hargittai, C., Bakonyi, I. and Kemeny, T., p. 55, Akademiai Kiado, Budapest (1981)

11. Boudreaux, D. S., in *'The Magnetic, Chemical and Structural Properties of Glassy Metallic Alloys'*, Ed. Hasegawa, R., chap. 2, Chemical Rubber Co., Boca Raton, Florida, USA (1981)
12. Cargill, G. S. (III), *'Atomic Energy Review'*, Suppl. 1, p. 63, International Atomic Energy Agency, Vienna (1981)
13. Bernal, J. D., *Proc. R. Soc.*, **A284**, 299 (1964)
14. Watts, R. D. and McGee, I. J., *'Liquid State Chemical Physics'*, Wiley, New York (1976)
15. Pippard, A. B., *'Elements of Classical Thermodynamics'*, p. 122, Cambridge University Press (1966)
16. Bernal, J. D., *Nature*, **188**, 910 (1960)
17. Scott, G. D., *Nature*, **188**, 908 (1960)
18. Finney, J. L., PhD Thesis, University of London (1968)
19. Bernal, J. D., Cherry, I. A., Finney, J. L. and Knight, K. R., *J. Phys. E*, **3**, 388 (1970)
20. Angell, C. A., Cheeseman, P. A., Clarke, J. H. R. and Woodcock, L. V., in *'The Structure of Non-Crystalline Materials'*, Ed. Gaskell, P. H., p. 191, Taylor and Francis, London (1977)
21. Finney, J. L., *Proc. R. Soc.*, **A319**, 495 (1970)
22. Cohen, M. H. and Turnbull, D., *Nature*, **203**, 964 (1964)
23. Barker, J. A., Finney, J. L. and Hoare, M. R., *Nature*, **257**, 120 (1975)
24. Bennett, C. H., *J. appl. Phys.*, **43**, 2727 (1972)
25. Adams, D. J. and Matheson, A. J., *J. chem. Phys.*, **56**, 1989 (1972)
26. Norman, L. D., Maust, E. E. and Skolnick, L. P., *Bull, U.S. Bur. Mines*, 658 (1971)
27. Ichikawa, T., *Phys. Status Solidi (a)*, **29**, 293 (1975)
28. Mrafko, P. and Durhaj, P., *J. Non-cryst. Solids*, **17**, 143 (1975)
29. Sadok, J. F., Dixmier, J. and Guinier, A., *J. Non-cryst. Solids*, **12**, 46 (1973)
30. Visscher, W. M. and Bolsterli, M., *Nature*, **239**, 504 (1972)
31. Matheson, A. J., *J. Phys C*, **7**, 2569 (1974)
32. Bonissent, A. and Mustaftschiev, B., *Phil. Mag.*, **35**, 65 (1977)
33. Finney, J. L., *J. Phys (Fr.)*, **36**, C2-1-11 (1975)
34. Mason, G., *Faraday Soc. Discuss.*, **43**, 75 (1967)
35. Finney, J. L., *Mater. Sci. Engng*, **23**, 199 (1976)
36. Koskenmaki, D. C., *Mater. Sci. Engng*, **23**, 207 (1976)
37. Kobe, S., *'Proc. Int. Symp. on Amorphous Magnetism'*, Troy, New York (1976)
38. Woodcock, L. V., *J. chem. Soc. Faraday II*, **72**, 1667 (1976)
39. von Heimendahl, L., *J. Phys. F*, **5**, L141 (1975)
40. Lançon, F., Billard, L., Laugier, J. and Chamberod, A., *J. Phys. F*, **12**, 259 (1982)
41. Rahman, A., Mandell, J. J. and McTague, J. P., *J. chem. Phys.*, **64**, 1564 (1976)
42. Damgaard Kristensen, W., *J. Non-cryst. Solids*, **21**, 203 (1976)
43. Street, W. B., Raveché H. J. and Mountain, R. D., *J. chem. Phys.*, **61**, 1960 (1974)
44. Wendt, H. R. and Abraham, F. F. *Phys. Rev. Lett.*, **41**, 1244 (1978)
45. Abraham, F. F., *J. chem. Phys.*, **72**, 359 (1980)
46. Bennett, C. H. and Finney, J. L., *'Rapidly Quenched Metals'*, Eds. Grant, N. and Giessen, B. C., p. 231, Massachusetts Institute of Technology Press, Cambridge, Mass. (1976)
47. Finney, J. L., *Proc. R. Soc.*, **A319**, 479 (1970)
48. Finney, J. L., unpublished results
49. Hafner, J., *Phys. Rev.*, **B21**, 406 (1980)
50. Scott, G. D. and Kovacs, G. J., *J. Phys. D*, **6**, 1007 (1973)
51. Cherry, I. A. and Finney, J. L., unpublished results
52. Boudreaux, D. S. and Gregor, J. M., *J. appl. Phys.*, **48**, 152 (1977)
53. Boudreaux, D. S. and Gregor, J. M., in ref. 46, p. 323.
54. Boudreaux, D. S. and Gregor, J. M., *J. appl. Phys.*, **48**, 5057 (1977)
55. Fujiwara, T., Chen, H. S. and Waseda, Y., *J. Phys. F*, in the press
56. Fujiwara, T. and Ishii, Y., *J. Phys. F*, **10**, 1901 (1980)
57. Gaskell, P. H., *J. Non-cryst. Solids*, **32**, 207 (1979)
58. Gellatly, B. J. and Finney, J. L., *J. Non-cryst. Solids*, **50**, 313 (1982)
59. Finney, J. L., Gellatley, B. J. and Bouquiere, J. P., *Proc. TSM Symp. Modelling the Structure and Properties of Amorphous Materials*, St. Louis, Mo, Oct. 1982
60. Bushnell-Wye, G. and Finney, J. L., unpublished results
61. von Heimendahl, L. and Hafner, J., *Phys. Rev. Letters*, **42**, 386 (1979)
62. Kobayashi, S., Maeda, K. and Takeuchi, S., *J. phys. Soc. Japan*, **48**, 1147 (1980)

Chapter 5

Experimental determination of atomic scale structure of amorphous alloys by scattering experiments

C. N. J. Wagner

Dept. of Materials Science and Engineering, University of California, Los Angeles, USA

5.1 Introduction

The atomic-scale structure of amorphous metallic alloys can be evaluated from diffraction data gathered in X-ray, neutron, or electron scattering experiments. However, the analyses of their scattering patterns yield only one-dimensional descriptions of the three-dimensional distributions of the atoms in amorphous alloys[1–3]. The structural data deduced from the scattering experiments have recently been complemented by detailed analyses of the extended X-ray absorption fine-structures (EXAFS)[4] of amorphous metals.

The conventional scattering experiments of multicomponent, amorphous samples permit us to evaluate the total structure factor (or interference function)* $I(K)$ which represents a weighted sum of the partial atomic pair (AP) structure factors $I_{ij}(K)$, i.e.,

$$I(K) = \sum_i \sum_j W_{ij}(K) I_{ij}(K) \tag{5.1}$$

where

$$W_{ij}(K) = \frac{c_i c_j f_i(K) f_j^*(K)}{\langle f(K) \rangle^2} \tag{5.2}$$

c_i and f_i being the atomic concentration and scattering amplitude, respectively, of

* In crystalline solids, the terms structure factor and interference function are referring to the scattering of the atoms in the unit cell and the interference due to the three-dimensional arrangements of the unit cells, respectively. The interference function yields the three Laue equations. In amorphous metals, one can only describe the arrangement of the atoms in terms of an atomic distribution function, and a separation of the scattered intensity into structure factor and interference function is not possible. Historically, the term interference function was used first for $I(K)$ (*see* Guinier[5]), but Faber and Ziman[6] introduced the term structure factor for $I(K)$ because it is related to the structure of the liquid or amorphous sample. Since the scattered intensity is the square of the scattering amplitude, thus a factor, von Laue[7] suggested to refer to intensities as factors in reciprocal space, and to atomic distributions as functions in real space. Thus, we will adopt the term structure factor for $I(K)$ of liquid or amorphous samples.

element i, $K = (4\pi/\lambda)\sin\theta$ is the length of the diffraction vector, and $\langle f \rangle = \sum_i c_i f_i(K)$.

The partial AP structure factor is the Fourier transform of the partial atomic pair (AP) distribution function $\rho_{ij}(r)$ which describes the number of j-type atoms per unit volume at the distance r from an i-type atom. In an n-component system, there are $n(n+1)/2$ partial AP distribution functions, because it can be shown[1] that $c_i\rho_{ij}(r) = c_j\rho_{ji}(r)$, i.e., the number of ij pairs must be equal to the number of ji pairs. Thus, in a binary alloy, three AP partial structure factors $I_{11}(K)$, $I_{22}(K)$ and $I_{12}(K)$ must be determined which can be accomplished by carrying out three separate scattering experiments in each of which the weight factor $W_{ij}(K)$ (equation 5.2) has been changed by one of the following methods:

1. *Isomorphous substitution*: One or both elements are partially or totally replaced by physically and chemically similar elements in the amorphous sample.
2. *Isotopic substitution*: In neutron scattering experiments, the scattering amplitude of element i can be changed by choosing different isotopes, when available.
3. *Anomalous dispersion of the scattering amplitude*: Close to the absorption edge of element i, the scattering amplitude is sufficiently changed owing to resonance effects.
4. *Three radiation experiment*: The weight factors $W_{ij}(K)$ are different for X-rays, and for neutron and electron diffraction.
5. *Polarized neutrons*: In ferromagnetic alloys, polarized neutrons will interact with the magnetic moments of the atoms.

Unfortunately, the changes in $W_{ij}(K)$ by any of the above methods are usually relatively small, and careful experiments must be carried out to obtain reliable partial AP structure factors $I_{ij}(K)$. Some favourable cases exist when one of the alloying elements (or isotopes) has a negative scattering amplitude (e.g. ^{7}Li, nat Ti, nat Mn, and ^{62}Ni) in neutron scattering.

In EXAFS experiments, the local structure about an element i, whose absorption fine structure is measured, can be probed. This local structure corresponds to the atomic distribution function $\rho_i(r)$ defined as

$$\rho_i(r) = \sum_{j=1}^{n} \rho_{ij}(r) \tag{5.3}$$

or the coordination number N_i, given by:

$$N_i = \sum_{j=1}^{n} N_{ij} \tag{5.4}$$

where N_{ij} is the number of j-type neighbours about the i-type atom, i.e.,

$$N_{ij} = \int_{r_l}^{r_u} 4\pi r^2 \rho_{ij}(r)\,\mathrm{d}r \tag{5.5}$$

where r_l and r_u are the lower and upper radius, respectively, of the first coordination shell.

Local fluctuations in atomic density and/or concentration might be observable by small-angle scattering which is sensitive to electron density variation on the scale of more than 20 Å (2.0 nm) in diameter. Fluctuations in structure larger than the diameter of the probing electron beam might be detectable in the transmission electron microscope.

5.2 Scattering theory of amorphous metals

Before we discuss the experiments which might yield information about the atomic-scale structure, we must define some of the terms which are needed to convert the experimental intensity to the atomic distribution function $\rho_{ij}(r)$, which is the Fourier transform of the partial structure factor $I_{ij}(K)$ (equation 5.1), i.e.

$$G_{ij}(r) = 4\pi r\left[\frac{\rho_{ij}(r)}{c_j} - \rho_0\right] = \frac{2}{\pi}\int_0^\infty K[I_{ij}(K) - 1]\sin Kr\,\mathrm{d}K \tag{5.6}$$

The total intensity, scattered by N atoms in the sample of volume V, is given for a binary alloy by[2]

$$I_N(K) = N\{\langle f^2\rangle + c_1^2 f_1^2[I_{11}(K) - 1] + 2c_1c_2 f_1 f_2[I_{12}(K) - 1] + c_2^2 f_2^2[I_{22}(K) - 1]\} \tag{5.7}$$

which can also be expressed as[8]

$$I_N(K) = N\{\langle f^2\rangle + \langle f\rangle^2[S_{NN}(K) - 1] + 2\Delta f\langle f\rangle S_{NC}(K) + c_1c_2(\Delta f)^2\left[\frac{S_{cc}(K)}{c_1c_2} - 1\right] \tag{5.8}$$

where

$$\langle f^2\rangle = c_1 f_1^2 + c_2 f_2^2, \langle f\rangle = c_1 f_1 + c_2 f_2, \quad \text{and} \quad \Delta f = f_1 - f_2$$

The partial number-concentration (NC) structure factors $S_{N-C}(K)$ are related to the partial AP structure factors $I_{ij}(K)$. Introducing the relation

$$I_i(K) = \sum_j c_j I_{ij}(K) \tag{5.9}$$

which is equivalent to equation 5.3, we can write

$$S_{NN}(K) = c_1 I_1(K) + c_2 I_2(K) \tag{5.10}$$

$$S_{NC}(K) = c_1c_2[I_1(K) - I_2(K)] \tag{5.11}$$

$$S_{CC}(K) = c_1c_2[1 + c_2 I_1(K) + c_1 I_2(K) - I_{12}(K)] \tag{5.12}$$

Since $I_{ij}(K)$ will modulate about unity for large values of K, it follows that $S_{NN}(K)$ will also tend towards unity, whereas $S_{NC}(K)$ will modulate about zero and $S_{CC}(K)$ about c_1c_2. Consequently, the total scattering $I_N(K)$ will be $N\langle f^2\rangle$ for large values of K.

It has been customary to introduce the total structure factor, which is an intensity function, normalized in such a way that it also modulates about unity. This can be accomplished in two ways[2]:

$$I(K) - 1 = \frac{[I_N(K)/N] - \langle f^2\rangle}{\langle f\rangle^2} \tag{5.13}$$

or

$$S(K) = \frac{I_N(K)}{N\langle f^2\rangle} \tag{5.14}$$

Combination of equations 5.7 and 5.13 yields immediately equation 5.1. If we use the definition of $S(K)$ (equation 5.14), we obtain from equation 5.7:

$$S(K) - 1 = \sum_i \sum_j W'_{ij}(K)[I_{ij}(K) - 1] \tag{5.15}$$

where

$$W'_{ij}(K) = \frac{c_i c_j f_i(K) f_j^*(K)}{\langle [f(K)]^2 \rangle} \tag{5.16}$$

The Fourier transforms of the total structure factor, i.e. the total reduced atomic distributions functions $G(r)$, are related to the partial atomic distribution functions $G_{ij}(r)$ (equation 5.6):

$$G_I(r) = \frac{2}{\pi} \int_0^\infty K[I(K) - 1] \sin K_r \, \mathrm{d}K = \sum_i \sum_j W_{ij}(0) G_{ij}(r) \tag{5.17}$$

$$G_S(r) = \frac{2}{\pi} \int_0^\infty K[S(K) - 1] \sin K_r \, \mathrm{d}K = \sum_i \sum_j W'_{ij}(0) G_{ij}(r) \tag{5.18}$$

It is readily seen that the differential correlation function $D(r)$ is given by:

$$\begin{aligned} D(r) &= \langle f(0) \rangle^2 G_I(r) = \langle [f(0)]^2 \rangle G_S(r) \\ &= \left\{ \sum_i \sum_j 4\pi r \, c_i c_j f_i(0) f_j(0) \rho_{ij}(r)/c_j \right\} - 4\pi r \, \langle f(0) \rangle^2 \rho_0 \end{aligned} \tag{5.19}$$

The Fourier transforms $\rho_{N-C}(r)$ of the number-concentration structure factors $S_{N-C}(K)$, defined as:

$$4\pi r[\rho_{NN}(r) - \rho_0] = \frac{2}{\pi} \int_0^\infty K[S_{NN}(K) - 1] \sin K_r \, \mathrm{d}K \tag{5.20}$$

$$4\pi r \, \rho_{NC}(r) = \frac{2}{\pi} \int_0^\infty K[S_{NC}(K)] \sin K_r \, \mathrm{d}K \tag{5.21}$$

$$4\pi r \, \rho_{CC}(r) = \frac{2}{\pi} \int_0^\infty K \left[\frac{S_{CC}(K)}{c_1 c_2} - 1 \right] \sin K_r \, \mathrm{d}K \tag{5.22}$$

are related to $\rho_i(r)$ and $\rho_{ij}(r)$, i.e.[1,2]

$$\rho_{NN}(r) = c_1 \rho_1(r) + c_2 \rho_2(r) \tag{5.23}$$

$$\rho_{NC}(r) = c_1 c_2 [\rho_1(r) - \rho_2(r)] \tag{5.24}$$

$$\rho_{CC}(r) = c_2 \rho_1(r) + c_1 \rho_2(r) - \frac{\rho_{12}(r)}{c_2} = [c_2 \rho_1(r) + c_1 \rho_2(r)] \, \alpha(r) \tag{5.25}$$

where $\alpha(r)$ is the generalized Warren–Cowley chemical short-range order (CSRO) parameter[1,2]:

$$\alpha(r) = 1 - \frac{\rho_{12}(r)}{c_2[c_2 \rho_1(r) + c_1 \rho_2(r)]} \tag{5.26}$$

The atomic distribution function $\rho_{NN}(r)$ describes the global or topological short-range order of the amorphous alloys. The term $\rho_{NC}(r)$ expresses the correlation between density and concentration fluctuation and is zero when $\rho_1(r) = \rho_2(r)$ which is the case in the substitutional alloy consisting of atoms of identical size. The concentration correlation function $\rho_{CC}(r)$ is of practical interest because it describes the chemical short-range order in the substitutional alloy. Since only the first peak in $\rho_{NN}(r)$ or $4\pi r^2 \rho_{NN}(r)$ is usually well defined, we will rewrite the CSRO-parameter in terms of the number of atoms in the first coordination shell. It follows from equations 5.4, 5.5 and 5.26 that

$$\alpha_1 = 1 - \frac{N_{12}}{c_2(c_2 N_1 + c_1 N_2)} \tag{5.27}$$

which reduces to the Warren–Cowley parameter only if $N_1 = N_2 = N_{NN}$, i.e. $\rho_{NC}(r) = 0$.

However, important information can be deduced directly from the radial concentration correlation function $4\pi r^2 \rho_{CC}(r)$. It modulates about zero; negative peaks (or minima) correspond to distances with preferred unlike atom pairs, whereas positive peaks (maxima) indicate preference for like atom pairs. However, the maxima and minima in $4\pi r^2 \rho_{CC}(r)$ do not always correspond to the maxima in the topological distribution function $4\pi r^2[\rho_{NN}(r) - \rho_0]$ which yields the total coordination number $N_{NN} = c_1 N_1 + c_2 N_2$.

5.3 Experimental methods

The structure of amorphous alloys can be deduced from the X-ray and neutron diffraction patterns. The basic experimental requirement consists of measuring the scattered intensity as a function of the length of the diffraction vector K, i.e.

$$K = \left(\frac{4\pi}{\lambda}\right) \sin\theta = \left(\frac{4\pi}{hc} \sin\theta\right) E = \left(\frac{4\pi m}{h} \sin\theta\right) \frac{L_0 + L}{t} \tag{5.28}$$

where λ is the wavelength of X-rays or neutrons, E is the energy of X-rays, and t is the time-of-flight of neutrons. L_0 and L are the flight paths of the incident and scattered neutrons, respectively, m is the mass of the neutrons, h is Planck's constant, and c is the velocity of the photons. It is readily seen that two experimental conditions can be employed.

1. Variable 2θ-method: K is varied by changing the angle 2θ between the incident and scattered beam using monochromatic X-rays or neutrons.
2. Variable λ-method: K is varied by using polychromatic X-rays or neutrons and measuring the energy of the photons[9] or the time-of-flight of the neutrons[10,11] at fixed scattering angle 2θ.

A second condition must be satisfied for a successful evaluation of the structure of amorphous alloys from their scattering patterns, i.e. the scattered intensity must be measured to values of K beyond which there exist no modulations in $I(K)$ about unity or in $I_N(K)$ about $\langle f^2 \rangle$. In the case of amorphous alloys, structure in $I(K)$ has been observed[12] to values of K beyond 25 reciprocal Ångströms. Thus, short-wavelength X-rays and epithermal neutrons must be used to reach such high values of K. Otherwise, one has to terminate the data at a finite K-value which introduces the so-called termination-of-series error which causes a broadening of the peaks in the radial distribution function with a corresponding loss of details in the short-range atomic-scale structure of the amorphous alloys[13].

Because of the relatively high absorption cross-sections of most metals for X-rays, the slab or thin-plate geometry for a specimen is preferred in X-ray diffraction experiments. In this case, the sample will intersect the entire primary beam at any scattering angle 2θ. The scattered intensity can be measured in reflection or in transmission. The transmission geometry is advantageous for intensity measurements at small scattering angles. In this case, errors in precise positioning of the sample into the diffractometer axis are less serious than in the reflection geometry.

The rod geometry is usually used in neutron diffraction experiments where the incident beam is wider than the diameter of the cylindrical sample. This is particularly true for metallic glasses where the amount of material available for a diffraction experiment might be limited.

Variable 2θ-method

In the conventional diffraction experiment, the scattered intensity from the amorphous sample is measured by a continuous or step-scanning of the detector as a function of the scattering angle 2θ. This represents a time-consuming and inefficient process since the intensity is measured only at one given value of 2θ at any time and the monochromatic X-ray or neutron beam from conventional sources (X-ray generator or nuclear reactor) is relatively weak. Even with the synchrotron radiation, one runs into intensity problems when using a high-resolution monochromator to select the incident monochromatic radiation and a medium-resolution monochromator in the diffracted beam to reject the fluorescence from the sample, present in an anomalous dispersion experiment.

The use of a curved, one-dimensional position-sensitive detector for X-rays and neutrons should greatly reduce the counting time. A curved proportional counter has been built covering an angular range of 45 degrees in 2θ, with a position resolution of 0.08 degree in 2θ and a quantum efficiency of 50 per cent for 60 keV photons[14]. This position-sensitive detector will be used to measure the scattered intensity of amorphous samples, and should permit us to cover a K-range from 0.5 to 30 Å^{-1}.

Variable λ-method

With the development of solid state detectors with excellent energy-resolution for X-ray photons (less than 200 eV for 6 keV photons), and pulsed sources for neutrons, it became possible to measure the scattered intensities from an amorphous sample as a function of photon energy in X-ray experiments[9,15] or the neutron time-of-flight in pulsed neutron experiments[10,11]. Thus, a single 2θ position of the detector permits the coverage of a large K range. For example, setting the detector at $2\theta = 40$ degrees and using neutrons with wavelengths λ ranging from 0.25 to 5 Å, we can measure the scattered intensity in the K range from 0.9 to 17 Å^{-1}. In the case of X-rays, the useful wavelength range is much more limited, from about 0.15 to 0.7 Å, and several 2θ positions must be chosen to cover the same range in K from 0.9 to 17 Å^{-1}.

A typical experimental set-up for X-ray diffraction studies of amorphous samples is shown in *Figure 5.1*. It consists of a 100 kV X-ray tube, a sample vacuum enclosure, and a hyperpure germanium detector. With the appropriate electronics, it can be used for application of the variable-λ and -2θ methods.

In principle, neutron diffraction should be more advantageous for structural studies of amorphous metals because the coherent scattering amplitude of neutrons does not fall off with increasing K which is so troublesome with X-rays where it reduces the intensity scattered at higher K to rather small values. On the other hand, the primary intensity emanating from a conventional X-ray tube is much higher than that from a conventional neutron source, and this fact should offset the rapid fall-off of the scattered X-ray intensities at large K to some extent. The new radiation sources, synchrotron for X-rays and spallation source for neutrons, will certainly increase the primary beam flux,

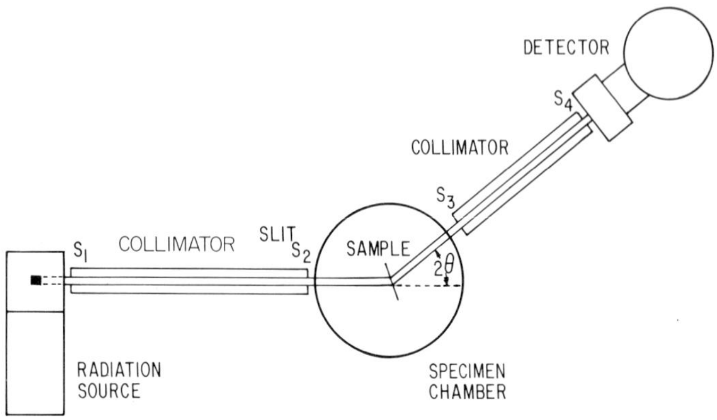

Figure 5.1 Experimental set-up for X-ray scattering investigations of amorphous metals

thus reducing the counting time for a given statistical accuracy or improve the counting statistics for a given counting time.

The data reduction is much simpler for the variable-2θ method than for the variable-λ method, because for the latter technique one must know the variation of the primary beam intensity and of the absorption coefficient with wavelength λ. This follows readily from the expression for the X-ray intensity scattered by an amorphous sample:

$$I_s(2\theta, E) = \{I_p(E)A(2\theta, E, E)P(2\theta)[I_a(2\theta, E) + I^{ms}_{coh}(2\theta, E)] + I_p(E')A(2\theta, E', E)P(2\theta)I_{inc}(2\theta, E')\} * M(E) \quad (5.29)$$

where

$I_p(E)$ = intensity of the primary beam as a function of the photon energy $E = hc/\lambda$;
$A(2\theta, E, E')$ = absorption factor of the primary X-rays with energy E' and of the scattered X-rays with energy E;
E' = initial energy of the primary photons which is reduced to $E = E' - \Delta E$ only after inelastic Compton scattering, i.e. $E' = E(1 - 0.00392E\sin^2\theta)$;
$P(2\theta)$ = polarization factor;
$I_a(2\theta, E)$ = coherent scattering per atom from the amorphous sample;
$I^{ms}_{coh}(2\theta, E)$ = double scattering of the coherent radiation;
$I_{inc}(2\theta, E')$ = incoherent Compton scattering;
$M(E)$ = energy-resolution function of the detecting system.

The symbol (*) represents the convolution product, i.e. $I(E)*M(E) = \int I(U)M(E - U)\,dU$.

Usually the resolution function $M(E)$ is much sharper than $I_a(2\theta, E)$ so that $I_a(2\theta, E)*M(E) \simeq I_a(2\theta, E)$. But the convolution product $I_{inc}(2\theta, E')*M(E)$ cannot be readily simplified even though $I_{inc}(2\theta, E')$ is a slowly varying function because the response function $M(E)$ is centred about E, and $I_{inc}(2\theta, E')$ is spread out at E for each particular value of E'. This problem has been discussed by Ruland[16] who approximated the convolution $I_{inc}(2\theta, E')*M(E)$ by the function $Q(K)I_{inc}(K)$, where $Q(K)$ is a slowly varying function, equal to unity at $K = 0$ and zero at large K when a monochromator is used in the diffracted beam or a narrow energy window is applied to the energy response of a solid state detector.

Thus, for the conventional 2θ-X-ray method we find

$$I_s(2\theta, E) = I_p(E)A(2\theta, E, E)P(2\theta)$$

$$\left\{ I_a(K) + I^{ms}_{coh}(K) + \frac{A(2\theta, E', E)}{A(2\theta, E, E)} Q(K)I_{inc}(K) \right\} \quad (5.30)$$

The expressions for the absorption factor can be found in several publications[1,17] for the different methods and geometries. The polarization factor depends on the kind of X-rays which are used in the experiment, i.e. characteristic or continuous radiation from an X-ray tube or synchrotron radiation.

Since we know that $I_a(Kr) = I_N(K)/N$ will tend towards $\langle f^2 \rangle$, we can use this fact to normalize the measured intensity $I_s(2\theta, E)$ from the sample and evaluate the coherent intensity $I_a(K)$ per atom, expressed in electron units. It is also possible to use the sum rule to normalize the data[18]. It follows from equations 5.17 and 5.19 that

$$\int_0^\infty K^2[I(K) - 1]\,dK = -2\pi^2\rho_0 \quad (5.31)$$

The data reduction for the variable-λ X-ray method is much more involved and has been discussed in detail by Egami[9] and Wagner *et al.*[15]. The neutron diffraction technique is dealt with by Suzuki[13].

5.4 Evaluation of the partial structure factors in binary amorphous alloys

The structure of binary substitutional alloys is characterized by three atomic distribution functions $\rho_{11}(r)$, $\rho_{12}(r)$ and $\rho_{22}(r)$ which are the Fourier transforms of the partial structure factors $I_{ij}(K)$ (equation 5.6). In order to determine these three partial structure factors, at least three independent scattering experiments must be carried out. Since the total structure factor $I(K)$ (equation 5.1) or $S(K)$ (equation 5.15) is the weighted sum of the partial factors $I_{ij}(K)$, we must change the weighting factors $W_{ij}(K)$ (equation 5.2) or $W'_{ij}(K)$ (equation 5.16) in each experiment. This can be accomplished by any of the methods (1) to (5) outlined in the Introduction, which permit us to vary the scattering amplitude f_i of element i without affecting the structure. Thus, we have to solve a set of linear equations in $I_{ij}(K)$, which follows from equation 5.1 or 5.15:

$$i^1(K) = W^1_{11}(K)i_{11}(K) + 2W^1_{12}(K)i_{12}(K) + W^1_{22}(K)i_{22}(K)$$
$$i^2(K) = W^2_{11}(K)i_{11}(K) + 2W^2_{12}(K)i_{12}(K) + W^2_{22}(K)i_{22}(K)$$
$$i^3(K) = W^3_{11}(K)i_{11}(K) + 2W^3_{12}(K)i_{12}(K) + W^3_{22}(K)i_{22}(K)$$

where $i(K) = I(K) - 1$ and $i_{ij}(K) = I_{ij}(K) - 1$ are the minor interference functions. This set of equations can be written in matrix notation[19], with $J(K) = i^{(n)}(K)$ and $P(K) = i_{ij}(K)$

$$[J(K)] = [V(K)][P(K)] \quad (5.32)$$

The solution of equation 5.32, i.e. the functions defined in $[P(K)]$, is simply given by:

$$[P(K)] = [[V(K)]^{-1}][J(K)] \quad (5.33)$$

A unique solution is found for $P(K)$ if the determinant of $[V(K)]$ is different from zero, i.e. if the three equations are linearly independent. However, the normalized determinant of the matrix $[V(K)]$, i.e.

$$|V(K)|_n = |V_{ij}(K)/\{\sum_j [V_{ij}(K)]^2\}^{1/2}| \tag{5.34}$$

is normally very small, i.e. the set of equations (equation 5.32) is ill-conditioned. (Well-conditioned equations yield $|V|_n$ of the order ± 1.)

A more quantitative measure of conditioning has been proposed by Livesey and Gaskell[20], who introduced the figure of merit T (called Turin's number), defined as:

$$T = \| V(K) \| \, \| V^{-1}(K) \| \simeq \| V(K) \|_E \| V^{-1}(K) \|_E \tag{5.35}$$

where $\| V(K) \|$ is the norm of the matrix $[V(K)]$ and $\| V(K) \|_E$ is the Euclidian norm of the matrix $[V(K)]$, i.e.

$$\| V(K) \|_E = \left(\sum_i \sum_j |V_{ij}^2| \right)^{1/2} \tag{5.36}$$

If $\Delta i(K)$, $\Delta W_{ij}(K)$ and $\Delta i_{ij}(K)$ are the errors with the minor interference function $i(K)$, the weighting factors $W_{ij}(K)$ and the partial interference function $i_{ij}(K)$, respectively, one finds that

$$\frac{\| \Delta P \|}{\| P \|} \leqslant T \frac{\| \Delta J \|}{\| J \|} \tag{5.37}$$

and

$$\frac{\| \Delta P \|}{\| P \|} \leqslant \frac{T \dfrac{\| \Delta V \|}{\| V \|}}{1 - T \dfrac{\| \Delta V \|}{\| V \|}} \tag{5.38}$$

The figure of merit T has been found to lie in the range between 30 and 200 for metallic glass systems whose partial structure factors have been evaluated so far (*Table 5.1*). Even if one can achieve an accuracy of $\Delta i/i = \Delta J/J$ of ± 1 per cent one encounters a relative error of ± 100 per cent for the partial structure factors when $T = 100$. Thus, one must resort to additional conditions resulting from the fact that equation 5.7 is a quadratic function[21] in f_1 and f_2, i.e.

$$c_1 + c_1^2 i_{11}(K) > 0 \tag{5.39}$$

$$c_2 + c_2^2 i_{22}(K) > 0 \tag{5.40}$$

$$[c_1 + c_1^2 i_{11}(K)] - [c_2 + c_2^2 i_{22}(K)] + [c_1 c_2 i_{12}(K)]^2 > 0 \tag{5.41}$$

Similarly, one finds from equation 5.8 that[22]

$$S_{NN}(K) > 0 \tag{5.42}$$

$$S_{CC}(K) > 0 \tag{5.43}$$

$$S_{NN}(K) S_{CC}(K) - [S_{NC}(K)]^2 > 0 \tag{5.44}$$

In addition, the sum rules must be satisfied, i.e.

$$\int_0^\infty K^2 i_{ij}(K)\, dK = \int_0^\infty K^2 [S_{NN}(K) - 1]\, dK = -2\pi^2 \rho_0 \tag{5.45}$$

TABLE 5.1 Values of the normalized determinant $|V|_n$ of the weight factor matrix $[V(K)]$ (equation 5.34) and Turia's number T (equation 5.35)

Alloy	*Method*	$|V|_n$	T	*Ref.*
$Ni_{75}Si_{25}$	X-Ray anomalous scattering using Mo-, Cu-, and Co-K_α radiation	−0.00014	1600	*a*
$Co_{81}P_{19}$	X-Ray nuclear and magnetic neutron scattering	−0.0345	30	*b*
$Fe_{80}B_{20}$	X-Ray, and neutron scattering with nat Fe and ^{57}Fe	0.0028	230	ref. 24
$Ni_{81}B_{19}$	Neutron isotopic substitution using ^{62}Ni, ϕNi and nat Ni	−0.505	6	ref. 23
$Cu_{57}Zi_{43}$	Neutron isotopic substitution using nat Cu, Cu and Cu	0.00275	225	*c*
$Ni_{35}Zr_{65}$	X-Ray isomorphous substitution with 0 and 30% Hf and neutron	0.015	75	*d*
$La_{80}Ga_{20}$	X-Ray isomophous substitution with Al assuming I_{GaGa} to be negligible	0.15	13	ref. 28
$Ni_{35}Ti_{65}$	X-Ray and neutron scattering assuming $S_{NC}(K)$ = hard sphere	0.998	2	ref. 27
$Be_{43}Zr_{57}$	X-Ray isomorphous substitution with Hf assuming I_{BeBe} to be negligible	−0.07	30	*e*
$Cu_{35}Ti_{65}$	X-Ray and neutron scattering assuming $S_{NC}(K) \sim 0$	0.999	9	*f*

[a]Y. Waseda and S. Tamaki, *Z. Phys.* **B83,** 315 (1976). [b]J. F. Sador and J. Dixmier, *J. mater. Sci. Engng* **23,** 187 (1976). [c]T. Mizoguchi, T. Kudo, T. Irisawa, N. Watanabe, N. Niimura, M. Misawa and K. Suzuki, '*Rapidly Quenched Metals III*', The Metal Society, London, (1978), p. 415. [d]Dokyol Lee, Alfred Lee, C. N. J. Wagner, L. E. Tanner and S. K. Soper, *Proc. 5th Int. Conf. Physics of Non-Crystalline Solids*, Montpelier, France (to be published). [e]M. Maret and C. N. J. Wagner (to be published). [f]M. Sakata, N. Cowlam and H. A. Davies, *J. Phys. (Fr.)* **41,** C8—190 (1980).

and

$$\int_0^\infty K^2 S_{NC}(K)\,dK = \int_0^\infty K^2[S_{CC}(K) - c_1c_2]\,dK = 0 \tag{5.46}$$

Several metal–metalloid and metal–metal glasses have been subjected to an analysis of the partial structure factors. The pertinent data are summarized in *Table 5. 1.* From the small values of the normalized determinant of the weighting factors $|W_{ij}(K)|_n = |V(K)|_n$ and the large values of Turin's number it can be judged that all investigators encountered difficulties in the evaluation of the partial functions, i.e. a direct calculation of the partial functions using Cramer's rule did not yield physically meaningful functions in all cases without invoking the constraints expressed in equations 5.39 and 5.40 or equations 5.42 to 5.44.

Recently, the method of isotopic substitution was applied by Lamparter *et al.*[23] to evaluate the partial AP structure factors $I_{ij}(K)$ for $Ni_{81}B_{19}$. The availability of the ^{62}Ni isotope with negative scattering length ($f = -0.87 \times 10^{-12}$ cm) and the ^{60}Ni isotope with small but positive f ($= 0{\cdot}28 \times 10^{-12}$ cm) permitted the preparation of nickel isotope mixture with zero scattering length (null matrix: ϕNi). Thus, the neutron diffraction experiment of the alloy $\phi Ni_{81}B_{19}$ yielded directly the partial function $I_{BB}(K)$. In addition, alloys were prepared with natural Ni and with ^{62}Ni, respectively. From the diffraction patterns, the partial AP structure factors could be readily determined, which are shown in *Figure 5.2.* The corresponding Fourier transforms, i.e. the reduced AP distribution functions $G_{ij}(r)$, are shown in *Figure 5.3.* The boron–boron correlation function exhibits a double first peak and strong oscillations up to 15 Å. This indicates that the arrangement of the boron atoms shows strong chemical short-range order.

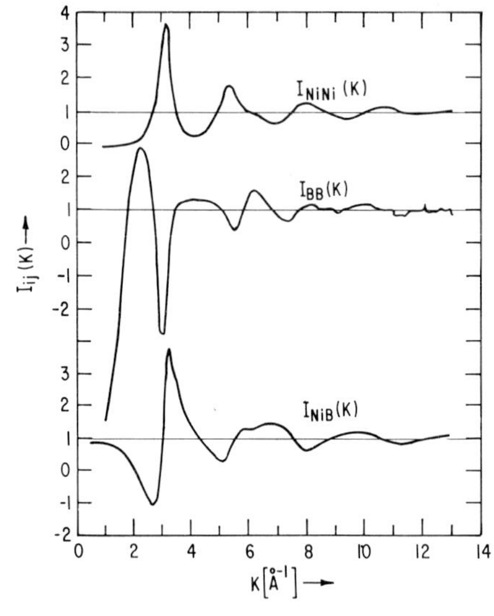

Figure 5.2 Partial atomic pair structure factors $I_{ij}(K)$ of amorphous $Ni_{81}B_{19}$ (after Lamparter *et al.*[23])

Similar results were found[23,24] for $Fe_{80}B_{20}$ where an X-ray experiment and two neutron experiments with natural Fe and the isotope ^{57}Fe, respectively, were carried out. The partial AP structure factors $I_{ij}(K)$ are shown in *Figure 5.4*, and the corresponding reduced distribution functions in *Figure 5.5*. The metal–metal and metal–metalloid partials in $Fe_{80}B_{20}$ are quite similar to their respective counterparts in the $Ni_{81}B_{19}$ glass whereas $I_{BB}(K)$ and $G_{BB}(r)$ show less structure in the amorphous $Fe_{80}B_{20}$. Whether this is a real effect is not certain at the present time since great difficulties were encountered in the evaluation of $I_{BB}(K)$ for $Fe_{80}B_{20}$ whereas this function could be directly determined with the null Ni alloy.

That chemical short-range order is present in the metal–metalloid alloys can be

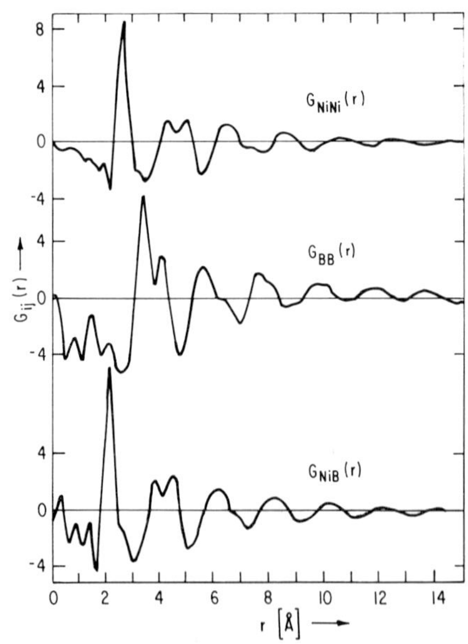

Figure 5.3 Partial reduced pair distribution functions $G_{ij}(r)$ of amorphous $Ni_{81}B_{19}$ (after Lamparter *et al.*[23])

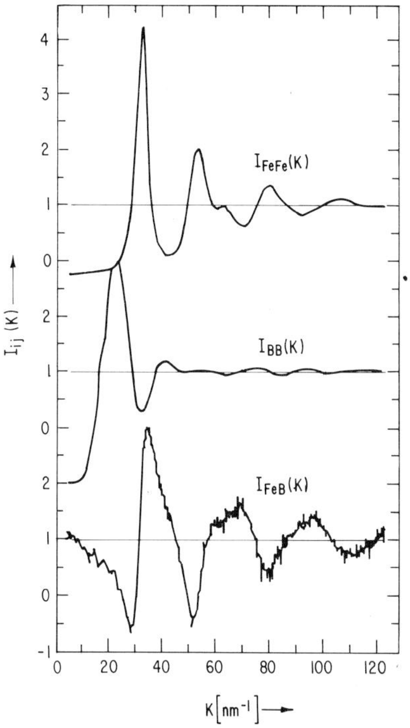

Figure 5.4 Partial atomic pair structure factors $I_{ij}(K)$ of amorphous $Fe_{80}B_{20}$ (after Nold *et al.*[24])

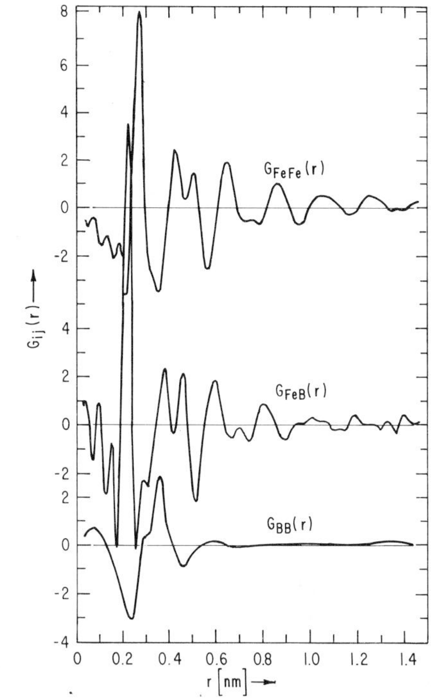

Figure 5.5 Partial reduced pair distribution functions $G_{ij}(r)$ of amorphous $Fe_{80}B_{20}$ (after Nold *et al.*[24])

readily seen from the NC structure factors $S_{N-C}(K)$ of $Fe_{80}B_{20}$ shown in *Figure 5.6*. The modulation of $S_{CC}(K)/(c_1c_2)$ about unity is much stronger than would be expected from a hard-sphere model consisting of atoms of two different sizes, whereas the size factor $S_{NC}(K)$ is very close to the prediction of hard-sphere model[25,26].

Because of the fact that natural titanium has a negative scattering length ($f = -0.34 \times 10^{-12}$ cm) for neutrons, it became advantageous to evaluate the number-concentration partial function $S_{N-C}(K)$ and their corresponding Fourier transform, i.e. $G_{N-C}(r)$ for $Ni_{35}Ti_{65}$ glass[27] from an X-ray and a neutron diffraction experiment assuming that the third partial function $S_{NC}(K)$ can be approximated by the Percus–Yevick hard-sphere model. It is readily seen that the weight factors of $S(K)$ (equation 5.14), when used in the expression for $I_N(K)$ given in equation 5.8, have the following values for $Ni_{35}Ti_{65}$: $a = \langle f \rangle^2/\langle f^2 \rangle = 0.986$ and $1 - a = c_1c_2(\Delta f)^2/\langle f^2 \rangle = 0{\cdot}014$ for X-rays and $a = 0.046$ and $1 - a = 0.954$ for neutrons. The magnitudes of the terms $2\Delta f\langle f \rangle/\langle f^2 \rangle = 2[a(1-a)/(c_1c_2)]^{1/2} = 0.49$ for X-rays and 0.88 for neutrons are appreciable, but $S_{NC}(K)$ is usually quite small. Therefore, the X-ray pattern $S^X(K)$ of $Ni_{35}Ti_{65}$ is dominated by $S_{NN}(K)$ and the neutron pattern $S^N(K)$ by $S_{CC}(K)$.

Combination of the two experimental structure factors $S^N(K)$ and $S^X(K)$ with the theoretical $S^{PY}_{NC}(K)$ makes it possible to evaluate $S_{NN}(K)$ and $S_{CC}(K)$ which are shown in *Figure 5.7* together with the theoretical $S_{NC}(K)$. The Fourier transforms of the partial *NC* structure factors, i.e., $G_{NN}(r)$, $G_{CC}(r)$ and $G_{NC}(r)$ are also presented in *Figure 5.6*. It is readily visible that $G_{CC}(r) = 4\pi r\rho_{CC}(r)$ exhibits a negative first peak which indicates preference for unlike nearest neighbours at $r_1 = 2.58$ Å. The distance is smaller than the average of the atomic sizes of nickel ($r_{Ni} = 2.49$ Å) and titanium ($r_{Ti} = 2.89$ Å), but

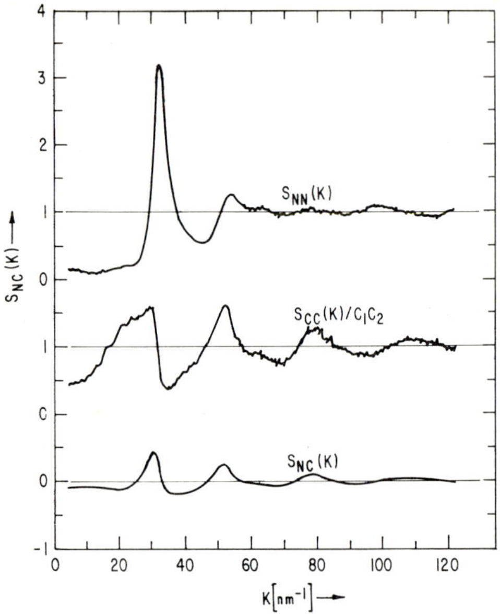

Figure 5.6 Partial number-concentration structure factors $S_{NC}(K)$ of amorphous $Fe_{80}B_{20}$ (after Nold *et al.*[24])

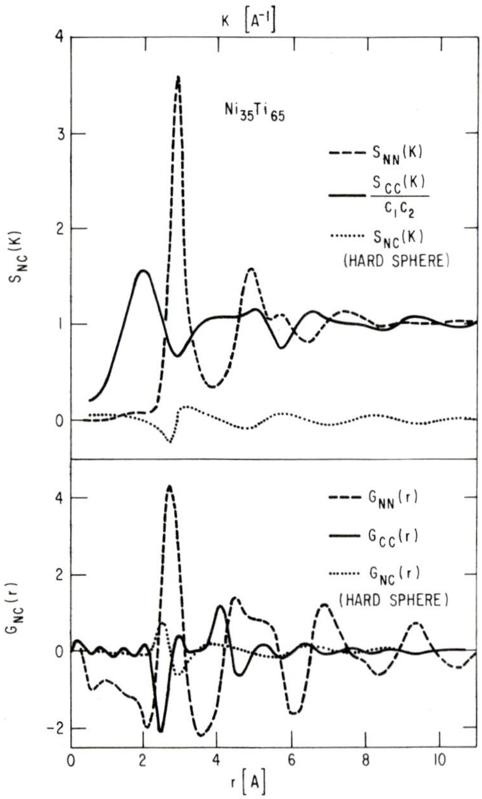

Figure 5.7 Partial number-concentration structure factors $S_{NC}(K)$ and reduced pair correlation functions $G_{NC}(r)$ of amorphous $Ni_{35}Ti_{65}$ (after Wagner *et al.*[27])

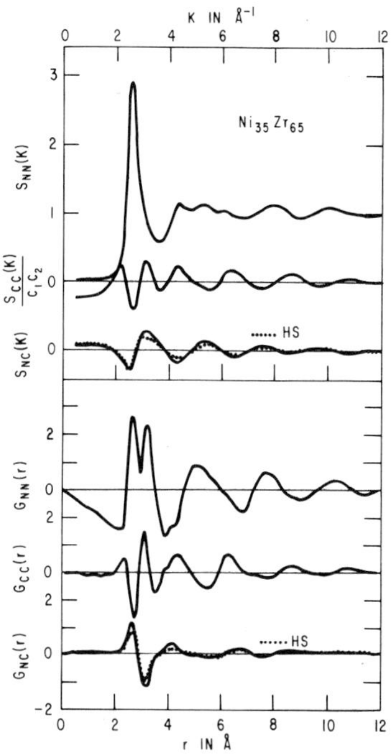

Figure 5.8 Partial number-concentration structure factors $S_{NC}(K)$ and reduced pair correlation functions $G_{NC}(r)$ of amorphous $Ni_{35}Zr_{65}$ (after Wagner[2])

is similar to those found in NiTi (r_{NiTi} = 2.58 Å with 8 neighbours) and $NiTi_2$ (with an average r_{NiTi} = 2.5 Å with six Ti about one Ni atom).

The isomorphous substitution method has been used to evaluate the partial structure factors in amorphous $Ni_{35}(Zr\text{-}Hf)_{65}$ alloys by Wagner and Lee[2] and in amorphous La–Ga–Al alloys by Williams[28]. In this method, it has been tacitly assumed that the substitution of Zr by Hf and Ga by Al does not change the structure of the glass. This might not be a bad assumption because crystalline $NiZr_2$ and $NiHf_2$, and La_3Ga and La_3Al are crystallographically isomorphous. The partial *NC* structure factors $S_{N-C}(K)$ and the reduced partial correlation functions $G_{N-C}(r)$ are shown in *Figure 5.8*. It should be noted that $S_{NC}(K)$ and $G_{NC}(r)$ agree very well with the corresponding theoretical Percus–Yevick partial functions[25,26].

5.5 EXAFS studies of the local atomic arrangements in amorphous metals

Extended X-ray absorption fine structure (EXAFS) spectroscopy has proven to be a powerful tool to provide information about atomic scale structure of amorphous alloys which complements data from X-ray and neutron scattering experiments. Since EXAFS is determined from the absorption edge of one particular element of the alloys, one actually determines the local atomic arrangement about that element.

The quantity of interest in EXAFS spectroscopy is the function $X(E)$ which is obtained from the absorption coefficient $\mu(E)$. It is defined as[4]

$$X(E) = \frac{\mu(E) - \mu_0(E)}{\mu_0(E)} \qquad (5.47)$$

where $\mu_0(E)$ is the absorption edge of the isolated atom, i.e. it corresponds to the smoothly varying part of $\mu(E)$. The energy E above the absorption threshold E_0 can be converted to k, the momentum of the photoexcited core level electrons, i.e.

$$k = \sqrt{\frac{2m}{\hbar^2}(E - E_0)} = \frac{K}{2} \tag{5.48}$$

where K is the length of the diffraction vector (equation 5.28).

In the single scattering approximation[29,30,31] the fine structure above the K or L_1 edge can be described as:

$$X(k) = \sum_j \frac{N_j}{kr_j^2} \exp(-2\sigma_j^2 k^2) F_j(k) D_j \sin[2kr_j + \phi_j(k)] \tag{5.49}$$

where $F_j(k)$ is the backscattering amplitude from each of the N_j neighbours of type j, $\phi\,(k)$ is the total phase shift experienced by the photoelectron, σ_j^2 is the mean square width of a Gaussian distribution of j-type atoms in a single shell of radius r_j about the absorbing atom, and $D_j \lesssim 1$ is a factor which takes into account the photoelectron losses due to inelastic scattering processes. Values for $F_j(k)$ and $\phi_j(k)$ have been calculated by Teo and Lee[32] for many elements.

The Fourier transform of the EXAFS function $X(k)$ is a radial structure function, i.e.

$$\Phi_n(r) = \frac{1}{\sqrt{(2\pi)}} \int_{k_{min}}^{k_{max}} k^n X(k) \exp(2iKr) W(k)\, dk \tag{5.50}$$

$\Phi_n(r)$ has a real part $\Phi_{R,n}(r)$ and an imaginary part $\Phi_{I,n}(r)$. Various choices of exponents $n = 1, 2$, and 3 have been suggested[32] for backscatterers with $Z > 57$, $36 < Z < 57$, and $Z < 36$, respectively. k_{min} and k_{max} are the minimum and maximum k-values, respectively, of the usable experimental data. $W(k)$ is a window function which is used to reduce the $X(k)$ function smoothly to zero at k_{min} and k_{max}.

It has been noted[5,33] that the peaks in $|\Phi_n(r)|$ are displaced to smaller r-values from the actual interatomic distances r_j in the material being examined because of the k-dependent phase shift $\phi_j(k)$ in equation 5.49. The magnitude of the displacement $\Delta r_j = r_j - r$ depends upon the chemical nature of the absorber and backscattering atoms, and on the details of the Fourier transformation, e.g. on the parameters n, $W(k)$ and the integration limits of the Fourier integral. Values of Δr_j are required to calculate the actual separation between neighbouring atoms. Methods to evaluate Δr_j have been discussed in the literature[33].

By analysing EXAFS from crystalline samples of iron–phosphorus and cobalt–phosphorus alloys, Cargill[33] concluded that the position displacements Δr for Fe–Fe and Co–Co pairs, caused by phase shift terms, are in the range of 0.3 to 0.4 Å. This is much smaller than the value 0.61 Å used by Wong[29] for the Fe–Fe. Because of the large value of Δr, the strong first peak at $r \simeq 2$ Å in $|\Phi_n(r)|$ of metal–metalloid samples has been attributed to metal–metal pairs ($r \simeq 2.5$ Å), whereas Cargill[33] suggests that it might correspond to metal–metalloid pairs at $r \sim 2.3$ Å. Further studies are required to fully test Cargill's hypothesis.

Acknowledgement

Part of this research was supported by grant DMR80-07939 from the National Science Foundation.

References

1. Wagner, C. N. J., *J. Non-cryst. Solids,* **31,** 1 (1978)
2. Wagner, C. N. J., *J. Non-cryst. Solids,* **42,** 3 (1980)
3. Cargill, G. S., *Atomic Energy Review, Suppl.* **1,** 63 (1981)
4. Wong, J., *'Glassy Metals I',* Eds. Guntherodt, H. J. and Beck, H., p. 45, Springer Verlag, Berlin (1981)
5. Guinier, A., *Theory et Technique de la Radio-Crystallographie',* Dunod, Paris (1956)
6. Faber, T. E. and Ziman, J. E., *Phil Mag.,* **11,** 153 (1965)
7. Hosemann, R. and Bagchi, S. M., *'Direct Analysis of Diffraction by Matter',* North-Holland Publ. Co. Amsterdam (1962)
8. Bhatia, A. B. and Thornton, D. E., *Phys. Rev.,* **B2,** 3004 (1970)
9. Egami, T., *'Glassy Metals I',* Eds. Guntherodt, H. J. and Beck, H., p. 25, Springer Verlag, Berlin (1981)
10. Sinclair, R. N. and Wright, A. C., *J. Nucl. Instrum. Meth.* **114,** 451 (1974)
11. Suzuki, K., Misawa, M., Kai, K. and Watanabe, N., *J. Nucl. Instrum. Meth.,* **147,** 519 (1977)
12. Suzuki, K., Fukunaga, T., Misawa, M. and Masumoto, T., *Mater. Sci. Engng.,* **23,** 215 (1976)
13. Suzuki, K., Chapter 6, this volume
14. Sleaford, B., Perez-Mendez, V. and Wagner, C. N. J., *Adv. X-Ray Analysis,* in the press
15. Wagner, C. N. J., Lee, D., Tai, S. and Keller, L., *Adv. X-Ray Analysis,* **24,** 245 (1981)
16. Ruland, W., *Br. J. appl. Phys.,* **15,** 1301 (1964)
17. Paalman, H. H. and Pings, C. A., *J. appl. Phys.,* **33,** 2635 (1962)
18. Wagner, C. N. J., *'Liquid Metals, Chemistry and Physics',* Ed. Beer, S. Z., p. 257, Marcel Dekker, New York (1972)
19. Edwards, F. G., Enderby, J. E., Howe, R. A. and Page, D. I., *J. Phys. C,* **8,** 3483 (1975)
20. Livesey, A. K. and Gaskell, P. H., *'Proc. 4th Int. Conf., Rapidly Quenched Metals',* Eds. Masumoto, T. and Suzuki, K., vol. I, p. 335, Japan Institute of Metals, Sendai (1982)
21. Enderby, J. A., North, D. M. and Egelstaff, P. A., *Phil. Mag.,* **14,** 961 (1966)
22. Bletry, J., *Z. Naturf.,* **33a** 327 (1978)
23. Lamparter, P., Sperl, W., Nold, E., Reiner-Harbach, G. and Steeb, S., ref. 20, p. 343
24. Nold, E., Lamparter, P., Rainer-Harbach, G. and Steeb, S., *Z. Naturf.* **36a,** 1032 (1981)
25. Ashcroft, N. W. and Lengreth, D. C., *Phys. Rev.,* **156,** 685 (1967)
26. Enderby, J. E. and North, D. M., *Phys. Chem. Liq. Metals,* **1,** 1 (1968)
27. Wagner, C. N. J., Lee, D., Keller, L., Tanner, L. E. and Ruppersberg, H., ref. 20, p. 331
28. Williams, A., *J. Non-cryst. Solids,* **45,** 183 (1981)
29. Sayers, D. E., Lytle, F. W. and Stern, E. A., *Adv. X-Ray Analysis,* **13,** 248 (1970)
30. Stern, E. A., *Phys. Rev.,* **B10,** 3027 (1974)
31. Haensel, R., Rabe, P., Tolkiehn, G. and Werner, A., *Liquid and Amorphous Metals,* Eds. Luscher, E. and Coufal, H., p. 459, Sijthoff and Noordhoff, Amsterdam (1980)
32. Teo, B. K. and Lee, P. A., *J. Am. chem. Soc.,* **101,** 2815 (1978)
33. Cargill, G. S. (III), ref. 20, p. 389

Experimental determination of short-range structure of amorphous alloys by pulsed neutron scattering

Kenji Suzuki

The Research Institute for Iron, Steel and Other Metals, Tohoku University, Sendai, Japan

6.1 Introduction

Among the various kinds of experimental technique, the radiation scattering experiment is one of the most direct methods for studying the spatial arrangement and motion of constituent atoms in amorphous metals. In this experiment, we measure the momentum and energy transferred between radiation and target sample during the scattering process. The radiation scattering experiments so far performed for condensed material targets using conventional radiations are summarized in *Table 6.1*.

The atomic spacing is several Ångström units and the interaction energy between atoms is of the order of 100 meV in amorphous metals. The ideal choice of the

TABLE 6.1 Comparison of experiments with various types of radiation

Radiation	*Range*	*Wavelength*	*Energy*	*Experiment**
Photon	Microwave	1 ~ 100 cm	$10^{-4} \sim 10^{-6}$ eV	NMR, ESR
	Infrared	> 7700 Å	< 1.6 eV	IR, Raman
	Visible	3800 ~ 7700 Å	1.6 ~ 3.3 eV	V, Raman
	Ultraviolet	< 3970 Å	> 3.1 eV	UV, Raman
	X-Ray	0.01 ~ 100 Å	1240 keV ~ 124 eV	Diffraction XPS, Emission, EXAFS
	γ-Ray	< 1 Å	> 12.4 keV	Mössbauer Compton, γ–γ
Electron		1 ~ 0.037 Å	150 eV ~ 100 keV	Diffraction
Neutron	Cold	> 4 Å	< 5 meV	SAS, INS Diffraction
	Thermal	0.5 ~ 4 Å	330 ~ 5 meV	Diffraction INS
	Epithermal	< 0.5 Å	< 330 meV	Diffraction INS

*NMR: Nuclear magnetic resonance
ESR: Electron spin resonance
IR: Infrared spectroscopy
V: Visible spectroscopy
UV: Ultraviolet spectroscopy
XPS: X-Ray photoelectron spectroscopy
EXAFS: Extended X-ray absorption fine structure
γ–γ: Time differential gamma–gamma perturbed angular correlation
SAS: Small angle scattering
INS: Inelastic neutron scattering

wavelength and energy of the radiation used must be of this order. When electrons and X-rays have a wavelength comparable with the average atomic spacing in amorphous metals, their energy is too high to examine the dynamics of amorphous metals. The wavelength of light is too large compared with the atomic distance in amorphous metals, but its energy resolution is in the correct range for finding the atomic vibration in amorphous metals. Thermal neutrons are a unique and powerful probe for investigating simultaneously the structure and dynamics of amorphous metals, because their wavelength and energy satisfy well the condition of ideal choice mentioned above.

Thermal neutrons are scattered by the nucleus and unpaired electrons in the target sample. The nuclear scattering is usually used to investigate the atomic structure and dynamics in condensed materials. Since the neutron scattering length of an individual nucleus varies with different isotopes and their spin states, coherent and incoherent scattering takes place according to the existence and absence of interference occurring between neutron waves scattered by different nuclei. The coherent neutron scattering provides the information of collective structure and motion of atoms such as atomic pair distribution function, phonon dispersion and so on. On the other hand, the localized vibration and atomic diffusion in terms of individual motion of atoms are observed by the incoherent scattering experiment.

This chapter describes the high resolution observation of the short-range structure of amorphous alloys by the coherent total scattering and incoherent inelastic scattering using short wavelength pulsed neutrons generated from accelerators.

6.2 Neutron total scattering experiment

The double differential cross-section[1] for the coherent scattering $\mathrm{d}^2\sigma_{\mathrm{coh}}/\mathrm{d}\Omega\,\mathrm{d}E'$, which represents the fraction of neutrons scattered per nucleus per unit area into a solid angle $\Delta\Omega$ and with neutron energy in the range $\Delta E'$, is proportional to the Fourier transform of the Van Hove[2] space-time correlation function $\mathfrak{G}(\vec{r}, t)$:

$$\frac{\mathrm{d}^2\sigma_{\mathrm{coh}}}{\mathrm{d}\Omega\,\mathrm{d}E'} = N\frac{k'}{k}\frac{\langle b^2\rangle}{2\pi\hbar}\iint \mathrm{d}\mathbf{r}\,\mathrm{d}t\,\mathrm{e}^{\mathrm{i}(\mathbf{Q}\cdot\mathbf{r}-\omega t)}\mathfrak{G}(\mathbf{r}, t) \qquad (6.1)$$

where N is the number of nuclei in the target sample, $\mathbf{k}$ and $\mathbf{k}'$ are the wave vectors of incident neutrons with the energy E and scattered neutrons with the energy E', $\hbar\mathbf{Q} = \mathbf{k} - \mathbf{k}'$ is the momentum transfer, $\hbar\omega = E - E'$ is the energy transfer, b is the neutron scattering length of the nucleus fixed at the position in the target and $\langle b\rangle$ is the average scattering length per nucleus.

The Van Hove space-time correlation function $\mathfrak{G}(\vec{r}, t)$ is written as the sum of two terms in the classical limit:

$$\mathfrak{G}(\mathbf{r}, t) = \mathfrak{G}_{\mathrm{s}}(\mathbf{r}, t) + \mathfrak{G}_{\mathrm{d}}(\mathbf{r}, t) \qquad (6.2)$$

where $\mathfrak{G}_{\mathrm{s}}(\mathbf{r}, t)$ is the 'self, correlation function representing the probability that if an atom is located at the origin at time zero, the same atom is moved to position $\mathbf{r}$ at time t, and $\mathfrak{G}_{\mathrm{d}}(\mathbf{r}, t)$ is the 'distinct' correlation function which means the probability that there is a different atom at position $\mathbf{r}$ at time t, when an atom is found at the origin at time zero.

The total structure factor $S(Q)$ is defined as[3]

$$S(Q) = \frac{1}{2\pi}\iiint \mathrm{d}\mathbf{r}\,\mathrm{d}t\,\mathrm{e}^{\mathrm{i}(\mathbf{Q}\cdot\mathbf{r}-\omega t)}\{\mathfrak{G}(\mathbf{r},t)-\rho\}|_{\mathrm{const.}\,Q}\,\mathrm{d}\omega \tag{6.3}$$

$$= \iint \mathrm{d}\mathbf{r}\,\mathrm{d}t\,\mathrm{e}^{\mathrm{i}\mathbf{Q}\cdot\mathbf{r}}\,\delta(t)\{\mathfrak{G}(\mathbf{r},t)-\rho\}$$

$$= 1+\rho\int \mathrm{e}^{\mathrm{i}\mathbf{Q}\cdot\mathbf{r}}\{g(r)-1\}\,\mathrm{d}\mathbf{r} \tag{6.4}$$

where ρ is the average number density of the sample and the relation of equation 6.2 with $t = 0$ is used:

$$\mathfrak{G}(r,0) = \mathfrak{G}_s(r,0) + \mathfrak{G}_d(r,0)$$
$$= \delta(r) + \rho g(r)$$

Thus, integrating $\mathrm{d}^2\sigma_{\mathrm{coh}}/\mathrm{d}\Omega\,\mathrm{d}E'$ over all energy transfers at a constant value of the momentum transfer gives the pair distribution function $g(r)$ which represents the probability of finding a pair of atoms with a spacing of r at the same instant of time.

In the experimental measurement of the angular differential cross-section $(\mathrm{d}\sigma_{\mathrm{coh}}/\mathrm{d}\Omega)_{\mathrm{total}}$ by coherent neutron total scattering, the energy integration corresponding to equation 6.3 is automatically performed by a detector with a detecting efficiency $f(E')$ at a constant scattering angle 2θ using a finite energy E of incident neutrons:

$$\left(\frac{\mathrm{d}\sigma_{\mathrm{coh}}}{\mathrm{d}\Omega}\right)_{\mathrm{total}} = \int_{-E}^{\infty} \frac{\mathrm{d}^2\sigma_{\mathrm{coh}}}{\mathrm{d}\Omega\,\mathrm{d}E'}\,f(E')|_{\mathrm{const}\cdot 2\theta}\,\mathrm{d}E' \tag{6.5}$$

The total structure factor $S(Q)$ defined by equation 6.3 is accessible from equation 6.5, if we use incident neutrons with energy high enough to realize the static approximation ($k \approx k'$) and the constant detector efficiency ($f(E') \sim$ constant). The high flux of epithermal neutrons in an electron LINAC (γ, n) source as shown in *Figure 6.1* is quite effective to obtain unambiguously the instantaneous configuration of atoms in liquids and amorphous solids[4]. *Figure 6.2* shows the schematic view of a neutron total

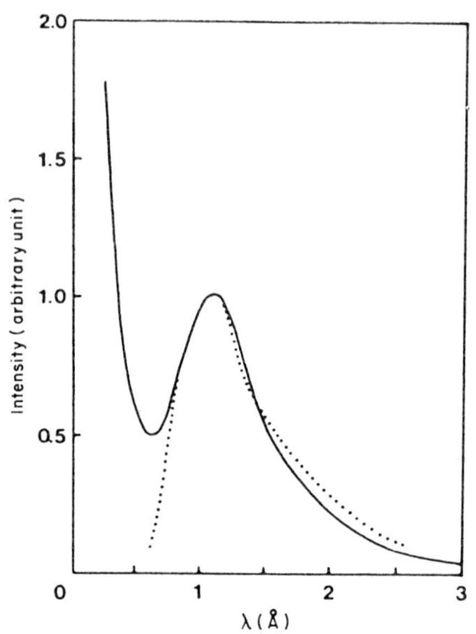

Figure 6.1 Wavelength spectrum of (γ, n) pulsed neutron source installed at Tohoku University 300 MeV electron LINAC[4]. Dotted line is a typical spectrum of conventional steady state nuclear reactors

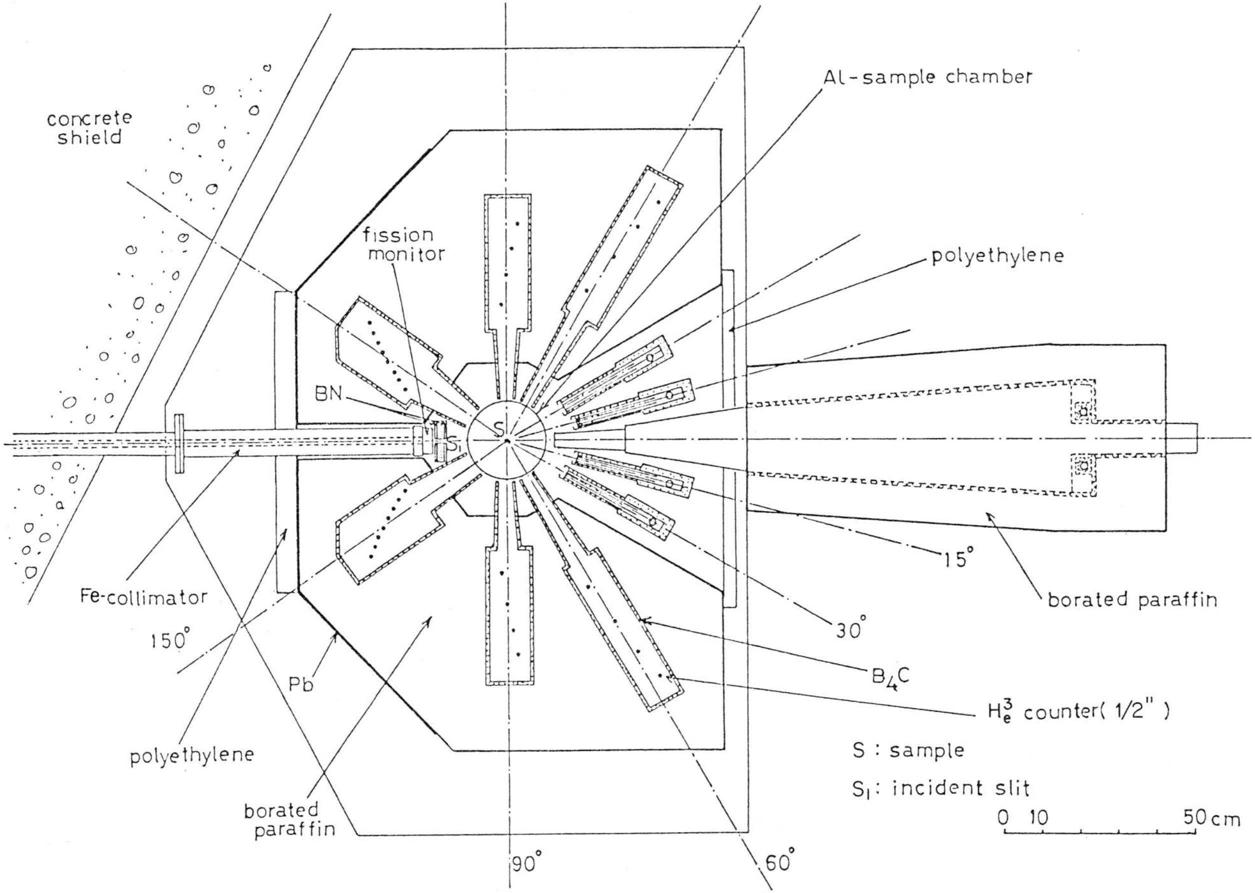

Figure 6.2 Schematic view of neutron total scattering spectrometer Mark II installed in Tohoku University Electron LINAC Laboratory (after Ueno *et al.*[5])

scattering spectrometer Mark II installed at the Tohoku University 300 MeV electron LINAC. The essential features are found in shieldings located inside a beam tube collimator and around a specimen, a small angle scattering system and time-focusing arrangement of detectors. Details of the spectrometer specifications and performances have been described in a previous report[5].

6.3 High-resolution observation of short-range structure

In contrast to liquid metals, the atomic arrangement in amorphous metals still conserves the polyhedral unit structures with less fluctuations in the bond length and angle[6]. Therefore, it may be expected that to distinguish the effect of the local atomic

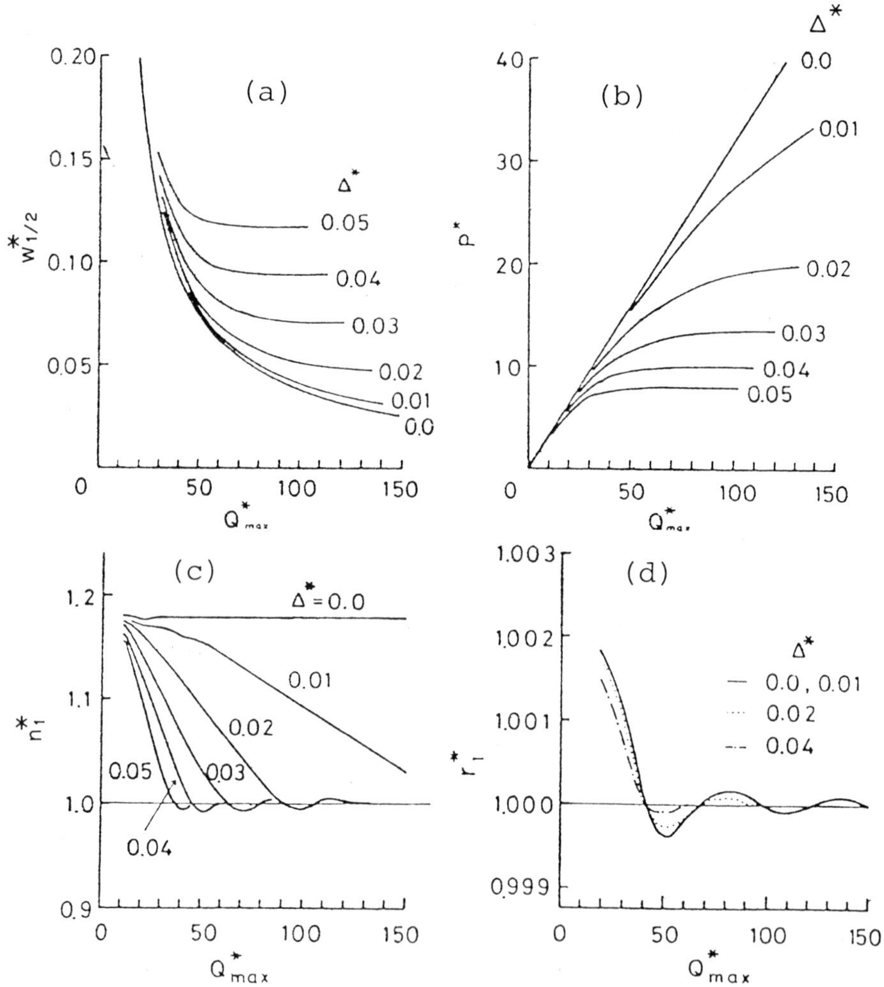

Figure 6.3 Variation in the profile of single peak RSDF as a function of Δ^* and Q^*_{max}. (*a*) FWHM $w^*_{1/2}$; (*b*) peak height p^*; (*c*) area n^*_1 and (*d*) position r^*_1. Asterisk superscript means the non-dimensional reduced quantity, $Q^* = Qr_1$, $\Delta^* = \Delta/r_1$; Q is the scattering vector, r_1 is the atomic distance and $\Delta^2 = \langle(r_1 - \langle r_1\rangle)^2\rangle$ (after Misawa[7])

coordinations in amorphous alloys on their total pair distribution functions, without the difficulties often involved in the partial structure separation, the experimental resolution in real space is much improved by extending the scattering vector up to an extraordinarily high value in the structure factor observation.

Here, we demonstrate how the short-range fine structure in amorphous solids is resolved by measuring the high momentum transfer structure factor with short wavelength neutrons, using a model function:

$$S_M(Q) = 1 + \sum_{i=1}^{N} n_i \frac{\sin(Qr_i)}{Qr_i} \exp\left(-\frac{\Delta_i^2}{2} Q^2\right) \tag{6.6}$$

where n_i is the number of ith atoms at a distance r_i from the origin, Q is the scattering vector and $\Delta_i^2 = \langle (r_i - \langle r_i \rangle)^2 \rangle$ is the mean square fluctuation of atomic spacing.

The Fourier transform of equation 6.6 truncated at Q_{max} gives the real space distribution function (RSDF):

$$\text{RSDF} = \frac{2}{\pi} \int_0^{Q_{max}} Qr\{S_M(Q) - 1\} \sin(Qr)\, dQ \tag{6.7}$$

which approaches the Gaussian function of equation 6.8 by extension of Q_{max} to infinity:

$$\lim_{Q_{max} \to \infty} \text{RSDF} = \sum_{i=1}^{N} \frac{n_i}{\sqrt{2\pi\Delta_i^2}} \exp\left\{-\frac{(r - r_i)^2}{2\Delta_i^2}\right\} \tag{6.8}$$

Misawa *et al.*[7] have discussed the effect of truncating the scattering vector on the position, height, width and area of a single peak RSDF for the case of $N = 1$. The results are summarized in *Figure 6.3* where the fluctuation in the atomic spacing is used as a parameter. The behaviour of RSDF with small Δ^2 is sensitively influenced by the truncation of the scattering vector in the $S_M(Q) \rightarrow$RSDF Fourier transformation. On the other hand, Fukunaga *et al.*[8] have examined how much the mutual separation between two peaks in RSDF is modified by truncating the Fourier transformation of $S_M(Q) \rightarrow$RSDF, when two Gaussian functions come near each other for the case of $N = 2$. *Figure 6.4* shows a comparison between two different models of $S_M(Q)$:

Model 1	$N = 1$	$r = 2.9$ Å	$2\sqrt{\text{-}2 \ln 2)}\Delta_1 = 0.08$ Å	$n_1 = 2$
Model 2	$N = 2$	$r_1 = 2.8$ Å	$2\sqrt{(2 \ln 2)}\Delta_2 = 0.08$ Å	$n_1 = 1$
		$r_2 = 3.0$ Å	$2\sqrt{(2 \ln 2)}\Delta_2 = 0.8$Å	$n_2 = 1$

There are found only small differences between amplitudes of the oscillatory $S_M(Q)$'s over the low Q region, i.e less than 10 Å^{-1}. The phases in both the oscillations, however, are clearly reversed in the high Q region above 18 Å^{-1}.

The RSDF's obtained by truncating the Fourier transformation of $S_M(Q) \rightarrow$RSDF at several different values of Q_{max} are compared in *Figure 6.5*. The RSDF's approach gradually to the original Gaussian functions with increasing Q_{max}. Particularly, the reappearance of the complete separation between both Gaussian functions needs an extraordinarily high value of Q_{max} of more than 40 Å^{-1}. If the value of Q_{max} is truncated at less than 20 Å^{-1}, we cannot determine whether the RSDF has a single peak or double peak. The measurement of the high momentum transfer structure factor is necessary to find the fine short-range structure as well as the precise position and coordination number among nearest neighbour atoms in amorphous metals. Therefore, short-range wavelength neutrons in the epithermal energy region generated

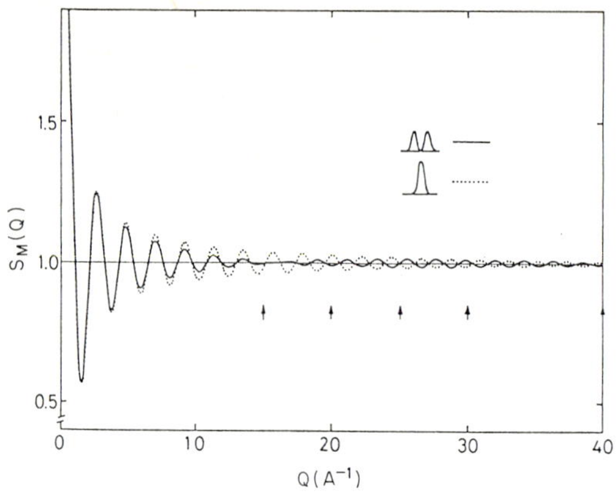

$$S(Q)=1+\sum_i n_i \frac{\sin(Q\cdot r_i)}{Q\cdot r_i}\exp\left(-\frac{\Delta_i^2}{2}Q^2\right)$$

Figure 6.4 Model structure factors $S_M(Q)$ corresponding to single and double peak RSDF's as shown in the bottom of *Figure 6.5*

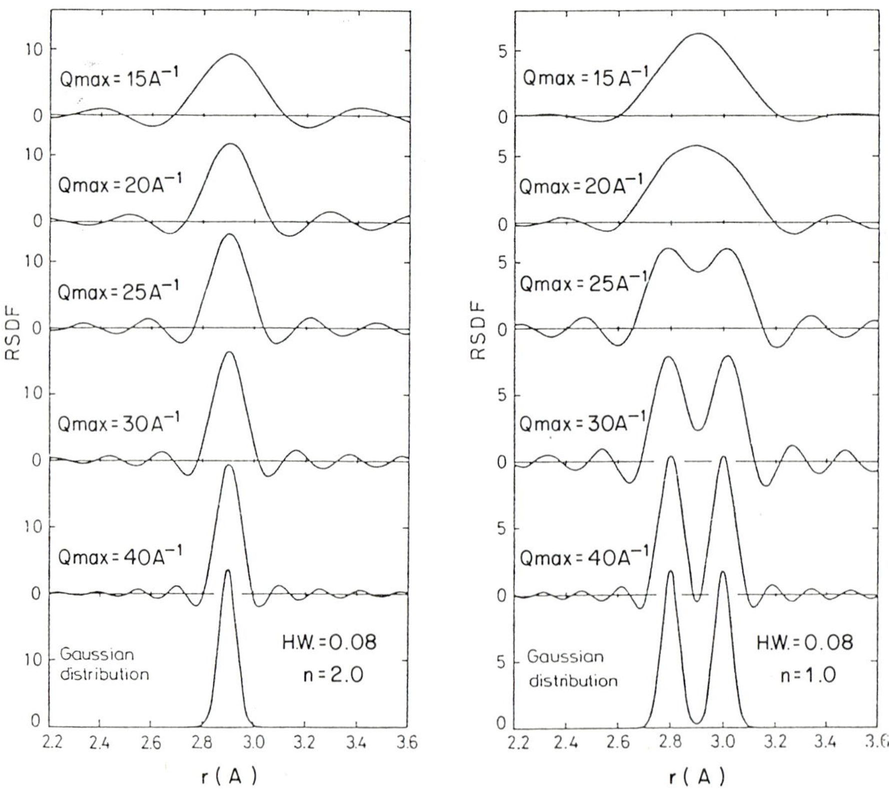

Figure 6.5 Real space distribution functions RSDF with single and double peak, and effect of truncating the $S_M(Q) \rightarrow$ RSDF Fourier transformation on their peak profiles. Values of $Q_{max}/\text{Å}^{-1}$ are shown on each curve (after Fukunaga[8])

from accelerator sources are a quite powerful probe for obtaining the highly resolved short-range structure[9].

6.4 Metal–metalloid amorphous alloys

Palladium–silicon amorphous alloy system is a good candidate to examine the short-range structure of metal–metalloid amorphous alloys by neutron total scattering, because the nuclei of palladium and silicon atoms are excellent coherent neutron scatterers and additionally have no contribution from magnetic scattering.

Figure 6.6 shows the experimental $S(Q)$'s of $Pd_{80}Si_{20}$ alloy in the glass and liquid state measured by pulsed neutron total scattering using the Tohoku University electron LINAC (γ, n) source[10]. The oscillation in $S(Q)$ of $Pd_{80}Si_{20}$ glass definitely persists up to the value of $Q \geqslant 20\,Å^{-1}$. The shoulder on the high Q side of the second peak, which is the famous feature common in $S(Q)$ of amorphous metals, disappears in the liquid state, but the second peak itself still conserves an asymmetrically distorted profile.

The effect of truncating the $S(Q)\rightarrow g(r)$ Fourier transformation on the resolution of the first and second peak profiles of the radial distribution function (RDF) $4\pi r^2 \rho g(r)$ of $Pd_{80}Si_{20}$ glass[11,12] is illustrated in *Figure 6.7*. The RDF of $Pd_{80}Si_{20}$ glass displays clearly the first peak split into two peaks and the second peak having three small humps. The subpeak on the low r side ($r \sim 2.4$ Å) of the first peak corresponds to the Pd–Si correlation and the main peak at $r \sim 2.8$ Å is almost entirely contributed by the Pd–Pd correlation. This assignment is confirmed with the comparison of the RDF between $Pd_{80}Si_{20}$ glass and Pd_3Si crystalline compound, as shown in *Figure 6.8*. Such a first peak splitting is not found in $g(r)$ of $Pd_{80}Si_{20}$ liquid[10], although *Figure 6.9* shows that a slight hump appears at the position corresponding to the Pd–Si correlation of the $Pd_{80}Si_{20}$ glass. We think that the short-range structure of $Pd_{80}Si_{20}$ alloy is essentially equivalent in the glass and liquid state.

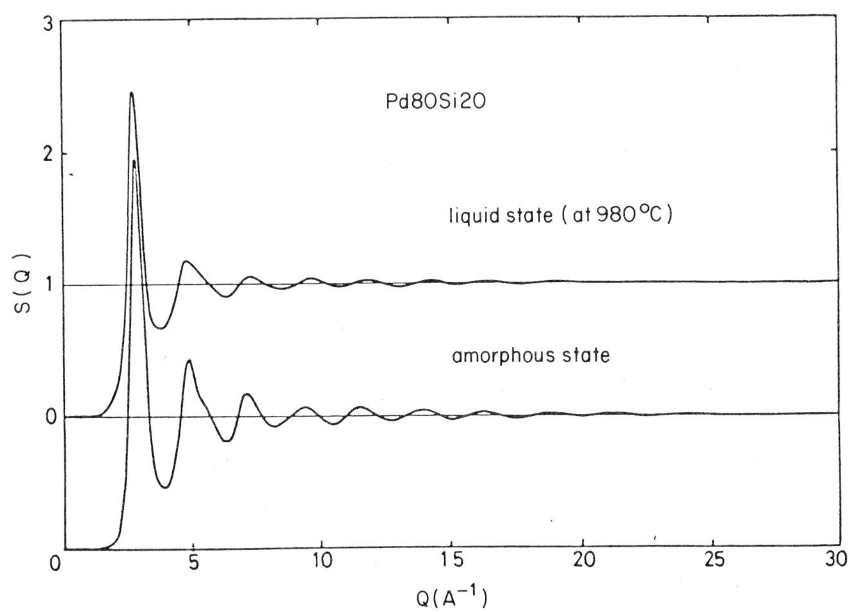

Figure 6.6 Neutron total structure factors $S(Q)$ of $Pd_{80}Si_{20}$ alloy in the glass (room temp.) and liquid 980 °C) state (after Fukunaga and Suzuki[10])

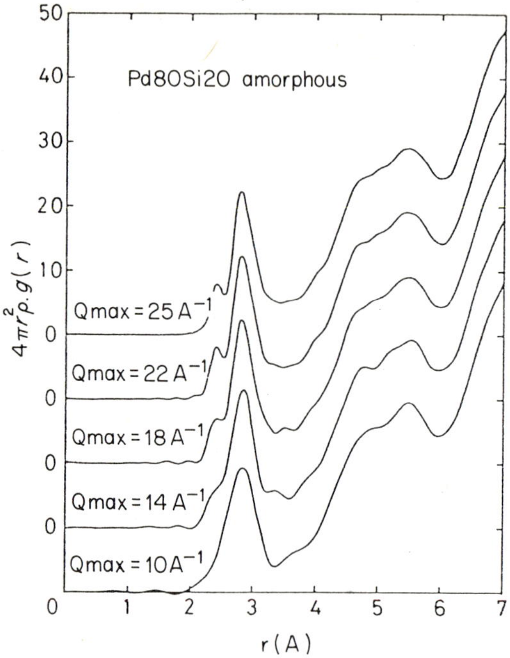

Figure 6.7 Effect of truncating the $S(Q) \rightarrow g(r)$ Fourier transformation on the RDF of $Pd_{80}Si_{20}$ glass[11,12]. Values of $Q_{max}/\text{Å}^{-1}$ are on the curves

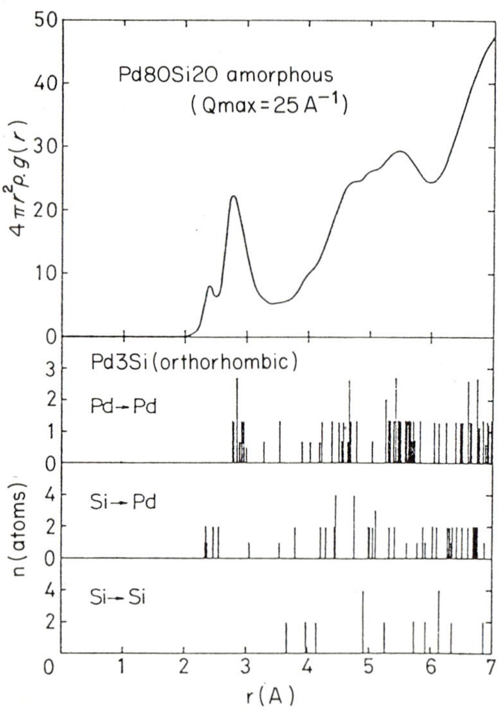

Figure 6.8 Comparison between neutron RDF of $Pd_{80}Si_{20}$ glass ($Q_{max} = 25\text{Å}^{-1}$) and the crystal structure of the orthorhombic Pd_3Si compound (after Suzuki *et al.*[11,12])

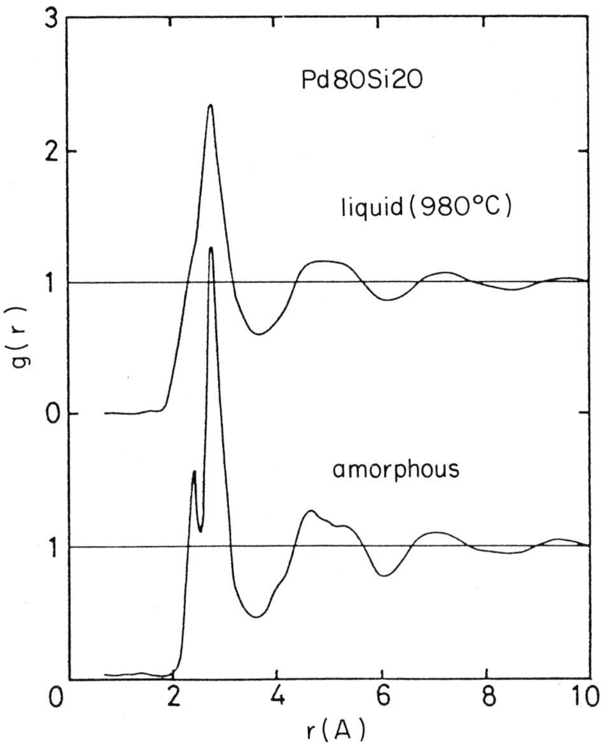

Figure 6.9 Neutron total pair distribution functions $g(r)$ of $Pd_{80}Si_{20}$ alloy in the glass (room temp.) and liquid (980 °C). state (after Fukunaga and Suzuki[10])

To examine further the relationship of the short-range order in Pd–Si alloys between the glass and crystal state, Fukunaga *et al.*[10,13] have carried out the neutron total scattering measurement of Pd–Si glasses with three different silicon contents. *Figure 6.10* shows that the height of the first peak in $S(Q)$'s decreases, the width of the first peak increases and the shoulder of the second peak becomes less conspicuous with increasing silicon content.

Figure 6.11 shows the pair distribution functions $g(r)$'s which are the Fourier transforms of the $S(Q)$'s truncated at $Q_{max} = 25\,Å^{-1}$. The peak widths and coordination numbers corresponding to Pd–Si and Pd–Pd correlations in the first peak of the RDF's are shown in *Figures 6.12* and *6.13* as a function of silicon content. The peak width of the Pd–Si correlation is much narrower than that of the Pd–Pd correlation and both widths are independent of silicon content. The coordination number of palladium atoms around a palladium atom (n_{Pd-Pd}) is about 10.6 Pd atoms at the silicon content below 20 atom % and then approaches the value found in Pd_3Si crystalline compound with increasing silicon content. The coordination number of palladium atoms surrounding a silicon atom (n_{Si-Pd}) linearly decreases and approaches again the value in Pd_3Si crystalline compound with increasing silicon content. The numerical parameters characterizing the first peak profile in the RDF's described above are summarized in *Table 6.2* together with the density and the first peak parameters of the $S(Q)$'s.

The second peak profiles in the $g(r)$'s of Pd–Si glasses, as found in *Figure 6.11*, show a characteristic variation that the first subpeak located at a lower r position

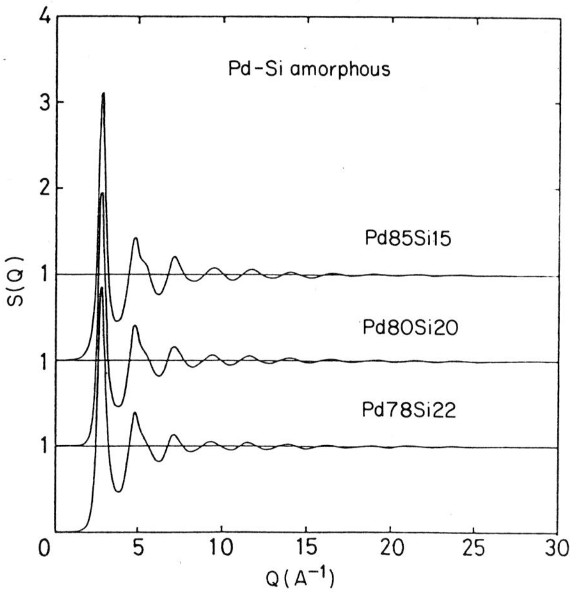

Figure 6.10 Neutron total structure factors $S(Q)$ of Pd–Si glasses (after Fukunaga *et al.*[10,13])

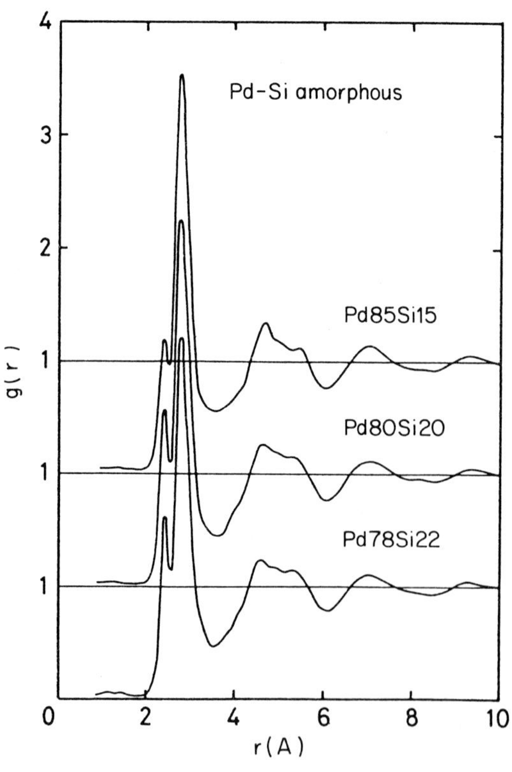

Figure 6.11 Neutron total pair distribution functions $g(r)$ of Pd–Si glasses, where $S(Q) \rightarrow g(r)$ Fourier transformation is truncated[10,13] at $Q_{max} = 25$ Å^{-1} (after Fukunaga *et al.*[10,13])

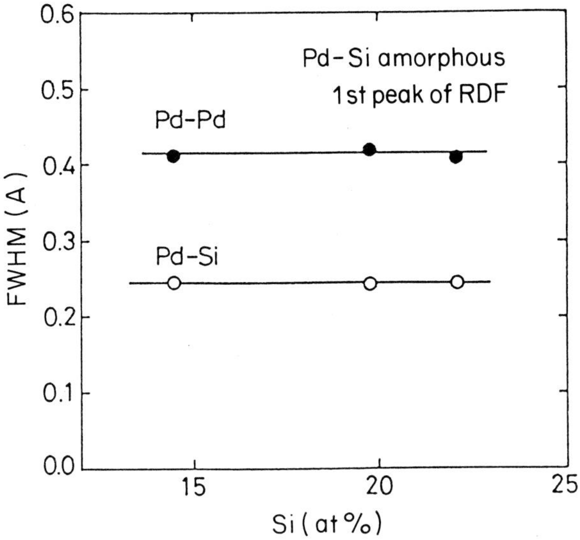

Figure 6.12 First peak widths of Pd–Si and Pd–Pd correlations in neutron RDF's of Pd–Si glasses (after Fukunaga and Suzuki[10])

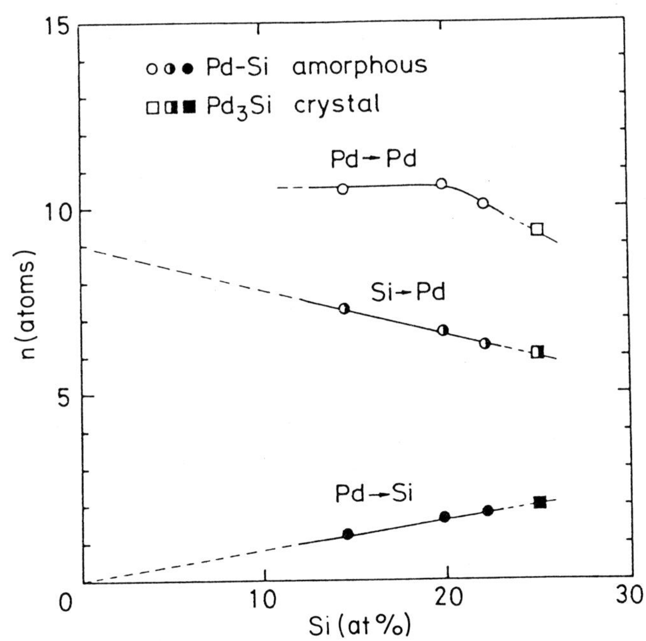

Figure 6.13 Coordination numbers obtained from neutron RDF's of Pd–Si glasses, where n_{A-B} means the number of B atoms around an A atom (after Fukunaga *et al.*[13])

TABLE 6.2 Structure parameters of Pd–Si glasses and liquid

Glass sample	$Pd_{85}Si_{15}$	$Pd_{80}Si_{20}$	l-$Pd_{80}Si_{20}$*	$Pd_{78}Si_{22}$
Si content (atom % Si)	14.5	19.8	19.8	22.1
Density/g cm^{-3}	10.83	10,54	(10.0)	10.31
First peak in $S(Q)$				
Position/Å	2.85	2.83	2.76	2.84
FWHM/Å^{-1}	0.515	0.567	0.66	0.577
First peak in RDF				
Pd–Si Position/Å	2.42	2.42	(2.42)	2.43
FWHM/Å^{-1}	0.246	0.242	—	0.243
n_{Si-Pd}	7.25	6.64	6.96	6.31
n_{Pd-Si}	1.23	1.64	(1.72)	1.79
Pd–Pd Position/Å	2.78	2.80	(2.82)	2.81
FWHM/Å^{-1}	0.409	0.420	—	0.406
n_{Pd-Pd}	10.5	10.6	(10.5)	10.1

*l-$Pd_{80}Si_{20}$ means the liquid state at 980 °C

becomes smaller and shifts toward a lower r region with increasing silicon content but both of the second subpeaks at the middle r position and the third subpeak at the higher r position have no significant changes in their profiles.

Unfortunately, it is very difficult to identify the Si–Si correlation in the total RDF of Pd–Si glasses measured by neutron scattering. Therefore, we require the experimental derivation of three partial structures describing the Pd–Pd, Pd–Si and Si–Si correlations in Pd–Si glasses.

The experimental total structure factor $S(Q)$ of an A–B amorphous alloy is generally represented as the weighted sum of the three partial structure factors $S_{AA}(Q)$, $S_{AB}(Q)$ and $S_{BB}(Q)$ associated with A–A, A–B and B–B correlations in the alloy:

$$S(Q) = w_{AA} S_{AA}(Q) + 2w_{AB} S_{AB}(Q) + w_{BB} S_{BB}(Q) \tag{6.9}$$

where the $S(Q)$ is defined as

$$\frac{d\sigma_{coh}}{d\Omega} = N\langle b\rangle^2 S(Q) \tag{6.10}$$

the weighting factors w_{ij} for the i–j correlation is written as

$$w_{ij} = \frac{c_i c_j b_i b_j}{\langle b\rangle^2} (i, j = \text{A or B}) \tag{6.11}$$

c_i and b_i are the concentration fraction and atomic scattering length of ith atom in the alloy and $\langle b\rangle = c_A b_A + c_B b_B$.

The three partial structure factors $S_{ij}(Q)$ can be resolved from equation 6.9, if three independent measurements of $S(Q)$ are carried out using three different values for each w_{ij}. Equation 6.11 shows that the variation in w_{ij} values is achieved by changing the value of b_A and/or b_B through the utilization of anomalous scattering in X-ray diffraction, isotope substitution in neutron diffraction and/or a combination of different kinds of radiation.

Figure 6.14 shows the partial structure factors and partial pair distribution functions of Pd–Pd, Pd–Si and Si–Si correlations in $Pd_{80}Si_{20}$ glass, which were separated using the combination of X-ray, electron and neutron diffraction by Fukunaga and Suzuki[14].

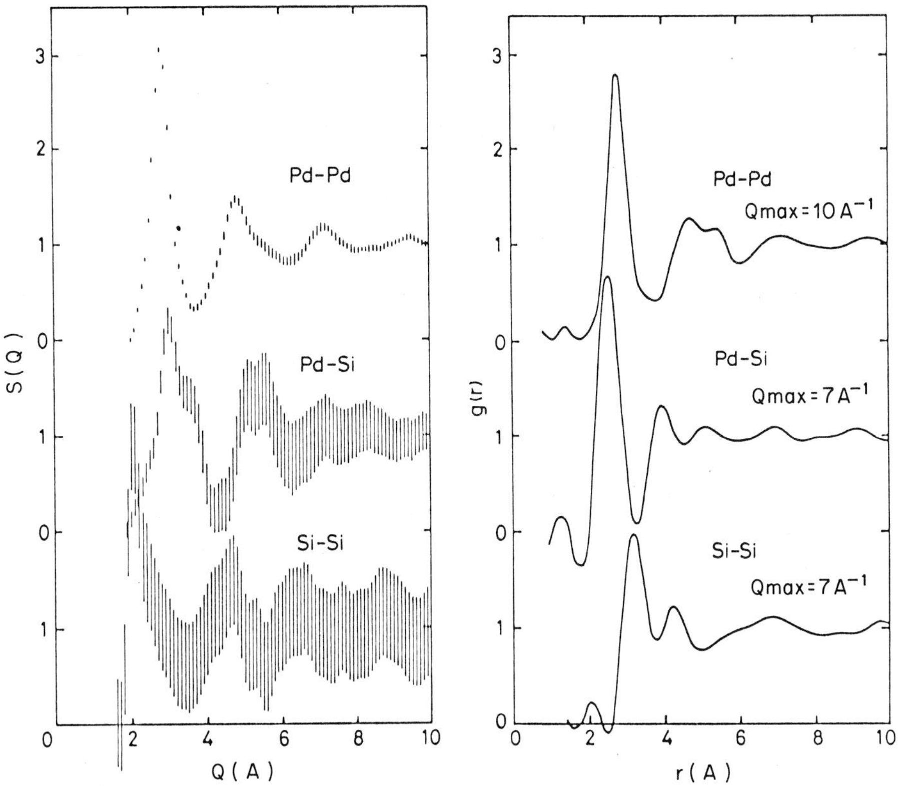

Figure 6.14 Partial structure factors and partial pair distribution functions for Pd–Pd, Pd–Si and Si–Si correlations in $Pd_{80}Si_{20}$ glass (after Fukunaga and Suzuki[14]). For Pd–Pd $Q_{max} = 10Å^{-1}$ and for Pd–Si and Si–Si $Q_{max} = 7Å^{-1}$

The overall behaviour of $S_{PdPd}(Q)$ is quite analogous to that of the experimental $S(Q)$. This means that the total structure factor of $Pd_{80}Si_{20}$ glass is predominantly dominated by the Pd–Pd correlation even in neutron scattering experiment. The $S_{PdSi}(Q)$ has a shoulder on the high Q side of the first peak. Such a shoulder or asymmetrical profile of the first peak extended to the high Q region is commonly found in the metal–metalloid partial structure factor of amorphous alloys including transition metal (Fe, Co, Ni)–metalloid(B, P) glasses[15–18]. This characteristic behaviour may suggest a well-defined local coordination of metal atoms surrounding a metalloid atom in metal–metalloid amorphous alloys. The first peak of $S_{SiSi}(Q)$ appears in the low Q region compared with those of $S_{PdPd}(Q)$ and $S_{PdSi}(Q)$. This observation is in good agreement with the behaviour commonly found in the metalloid–metalloid partial structure factor of various kinds of metal–metalloid amorphous alloys[14–18].

The first neighbour position for Pd–Pd correlation in $4\pi r^2 \rho g_{PdPd}(r)$ is 2.85 Å, which is a little larger than the value (2.80 Å) obtained from the high resolution neutron total RDF. The first peak of $g_{PdSi}(r)$ is located at 2.60 Å. This distance is larger than the value (2.41 Å) in the high resolution neutron total $g(r)$. These discrepancies are caused by the truncation effect in the $S(Q)$–$g(r)$ Fourier transformation, mentioned above in *Figure 6.3*. The first peak in $g_{SiSi}(r)$ is located at about 3.2 Å and the small second peak appears at the position close to the first peak. Such a behaviour is also found in the

$g_{BB}(r)$ of $Ni_{81}B_{19}$ glass[18], while the first peak in $g_{BB}(r)$ of $Fe_{80}B_{20}$ glass[18] shows a sharp single peak without the small second peak closely adjacent to the first peak.

The observation on the total RDF, as shown in *Figure 6.8*, that the short-range order existing in Pd_3Si crystalline compound is essentially preserved in that of Pd–Si glasses, is verified by the experimental $g_{SiSi}(r)$ indicating that silicon atoms are never at the nearest neighbour position even in $Pd_{80}Si_{20}$ glass as well as in Pd_3Si crystalline compound.

The high resolution observation of the short-range structure supports Gaskell's proposal[19] that the structure of metal–metalloid amorphous alloys is constructed from polyhedral unit structures with a well-defined local coordination around a central metalloid atom, which are almost similar to the polyhedral unit structures existing in the corresponding crystalline compound.

The high resolution observation of the total $S(Q)$ and RDF of Pd–Ge glasses performed by Hayashi *et al.*[20] using the electron LINAC (γ, n) neutron source indicates more obviously that the local environment surrounding a germanium atom consists of about six palladium atoms over all the glass-forming composition range and consequently the Ge–Ge correlation describes the centre–centre correlation of the polyhedral unit structures existing in Pd–Ge glasses.

The most important key specifying the structure of an individual metal–metalloid amorphous alloy must be found in the method of connecting the polyhedral unit structures in the alloy. A large difference has been already discovered between $g_{BB}(r)$ of $Fe_{80}B_{20}$ glass and that of $Ni_{81}B_{19}$ glass in a pioneering work by Lamparter *et al.*[18]. The precise determination of the metalloid–metalloid correlation in various kinds of amorphous alloy is awaited.

6.5 Metal–metal amorphous alloys

It is interesting to examine whether some kinds of chemical short-range order still exist in metal–metal amorphous alloys as well as in metal–metalloid amorphous alloys. Bhatia and Thornton[21] have proposed that the total structure factor $S_{BT}(Q)$ can be divided into the three partial structure factors representing the number density fluctuation $S_{NN}(Q)$, the concentration fluctuation $S_{CC}(Q)$ and the cross-correlation between density and concentration $S_{NC}(Q)$:

$$S_{BT}(Q) = \frac{1}{\langle b^2\rangle}\{\langle b\rangle^2 S_{NN}(Q) + 2\Delta b\langle b\rangle S_{NC}(Q) + (\Delta b)^2 S_{CC}(Q)\} \tag{6.12}$$

where the experimental angular differential cross-section is related to $S_{BT}(Q)$ as follows:

$$\frac{d\sigma_{coh}}{d\Omega} = N\langle b^2\rangle S_{BT}(Q) \tag{6.13}$$

$$\langle b^2\rangle = c_A b_A^2 + c_B b_B^2, \quad \langle b\rangle = c_A b_A + c_B b_B \quad \text{and} \quad \Delta b = |b_A - b_B|$$

The physical meaning of $S_{NN}(Q)$ corresponds exactly to the topological short-range order in terms of the spatial correlation between the number density fluctuations, while $S_{CC}(Q)$ describes only the chemical short-range order defined as the atomic scale spatial correlation between the concentration fluctuations. In contrast to X-ray diffraction, neutron scattering makes it possible to measure selectively $S_{NN}(Q)$ and $S_{CC}(Q)$, because there are some nuclei with negative neutron scattering lengths.

If we have an amorphous alloy with an average neutron scattering length zero ($\langle b\rangle = 0$), called a neutron zero alloy, equation 6.12 denotes that only $S_{CC}(Q)$ of the alloy can be exclusively measured. In fact, Fukunaga *et al.*[22] prepared the $Ni_{26}Ti_{74}$ glass, which has an average neutron scattering length of zero because of the negative value of $b_{Ti} = -0.34 \times 10^{-12}$ cm and positive value of $b_{Ni} = 1.03 \times 10^{-12}$ cm [23].

The neutron $S_{BT}(Q)$'s of $Ni_{26}Ti_{74}$ and $Ni_{33}Ti_{67}$ glasses[22] are shown in *Figure 6.15*, together with X-ray $S(Q)$ defined by equation 6.10. The oscillation in $S_{BT}(Q)$'s of both the glasses persists up to a very high Q region beyond 25 Å^{-1}. *Figure 6.16* shows the reduced atomic distribution function $G(r)$ obtained as the Fourier transform of $S_{BT}(Q)$:

$$G(r) = 4\pi r\rho(r) = \frac{2}{\pi}\int_0^\infty Q\{S_{BT}(Q) - 1\}\sin(rQ)\,dQ \tag{6.14}$$

The neutron $G(r)$ of the $Ni_{26}Ti_{74}$ glass is exactly equal to the reduced concentration correlation function $G_{CC}(r) = 4\pi r\rho_{CC}(r)$, which has a sharp and negative minimum at $r = 2.56$ Å and two positive peak maxima around $r \sim 3$ and 4 Å. The negative peak maximum denotes that the Ni–Ti unlike atom pair is preferentially formed at the nearest neighbour position in the glass. From the comparison between the crystal structure of the $NiTi_2$ compound (Fe_3W_3C-type)[24] and $G_{CC}(r)$ of the $Ni_{26}Ti_{74}$ glass, we find that the Ni–Ti unlike-atom correlations appear at the same position in both the crystal and glass state of Ni–Ti alloys. The two positive peaks around $r \sim 3$ and 4 Å in $G_{CC}(r)$ are predominantly contributed from the Ti–Ti and Ni–Ni like-atom correlations in the $NiTi_2$ crystal structure. Therefore, the chemical short-range order in Ni–Ti glasses is quite analogous to that in the $NiTi_2$ crystalline compound.

The neutron $S_{BT}(Q)$'s of Cu–Ti glasses with three different titanium contents[22]

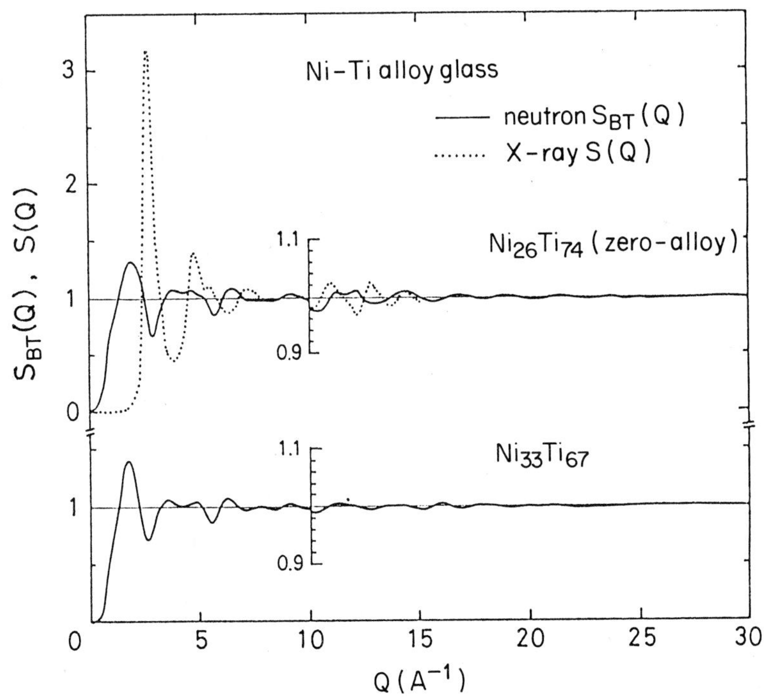

Figure 6.15 Bhatia–Thornton neutron structure factors $S_{BT}(Q)$ (full line) of $Ni_{26}Ti_{74}$ and $Ni_{33}Ti_{67}$ glasses, and X-ray total structure factor $S(Q)$ (broken line) of $Ni_{26}Ti_{74}$ glass. $S_{BT}(Q)$ of $Ni_{26}Ti_{74}$ glass is exactly equal to the $S_{CC}(Q)/c_{Ni}c_{Ti}$ due to a neutron zero scattering alloy (after Fukunaga *et al.*[22])

are shown in *Figure 6.17*. The high Q oscillation in $S_{BT}(Q)$ is definitely preserved up to the value of $Q \geqslant 25\,Å^{-1}$. With increasing titanium content, the main peak at $Q \sim 3.2\,Å^{-1}$ is drastically diminished and the pre-peak around $Q \sim 1.7\,Å^{-1}$ is modified from a single peak at 30 atom % of titanium to a double-headed one at 60 atom % of titanium. *Figure 6.18* shows $G(r)$'s for Cu–Ti glasses which are the Fourier transforms of $S_{BT}(Q)$'s truncated at $Q_{max} = 26\,Å^{-1}$. These $G(r)$'s have a narrow and positive peak maximum at $r = 2.45\,Å$, and a broad and negative peak minimum at $r = 2.75\,Å$. The positive peak at $r = 2.45\,Å$ rapidly increases with increasing copper content.

The X-ray total radial distribution functions RDF^X for $Cu_{40}Ti_{60}$ and $Cu_{50}Ti_{50}$ glasses are represented by three partial radial distribution functions, RDF_{CuCu}, RDF_{CuTi} and RDF_{TiTi} as follows:

$$RDF^X(Cu_{40}Ti_{60}) = 0.219RDF_{CuCu} + 0.498RDF_{CuTi} + 0.283RDF_{TiTi} \tag{6.15}$$

and

$$RDF^X(Cu_{50}Ti_{50}) = 0.323RDF_{CuCu} + 0.491RDF_{CuTi} + 0.186RDF_{TiTi} \tag{6.16}$$

Since the weighting factors of RDF_{CuTi} for Cu–Ti unlike atom pairs have nearly the same value in both glasses as shown in equations 6.15 and 6.16, the characteristic feature of Cu–Cu and Ti–Ti like atom pair correlations in the glasses can be found in the differential radial distribution function defined as

$$\begin{aligned}\Delta RDF &= RDF^X(Cu_{40}Ti_{60}) - RDF^X(Cu_{50}Ti_{50}) \\ &\approx 0.1(RDF_{TiTi} - RDF_{CuCu})\end{aligned} \tag{6.17}$$

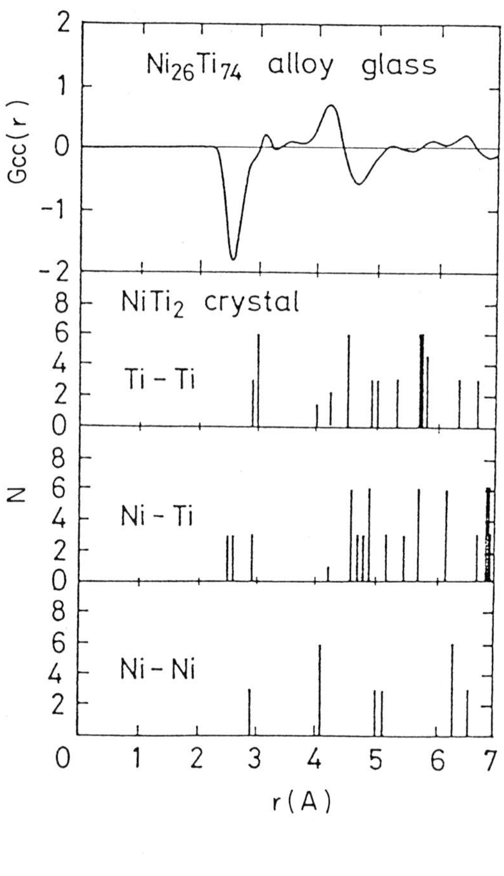

Figure 6.16 Reduced concentration correlation function $G_{CC}(r)$ of $Ni_{26}Ti_{74}$ glass, and Ti–Ti, Ni–Ti and Ni–Ni correlations in $NiTi_2$ crystalline compound (after Fukunaga *et al.*[22])

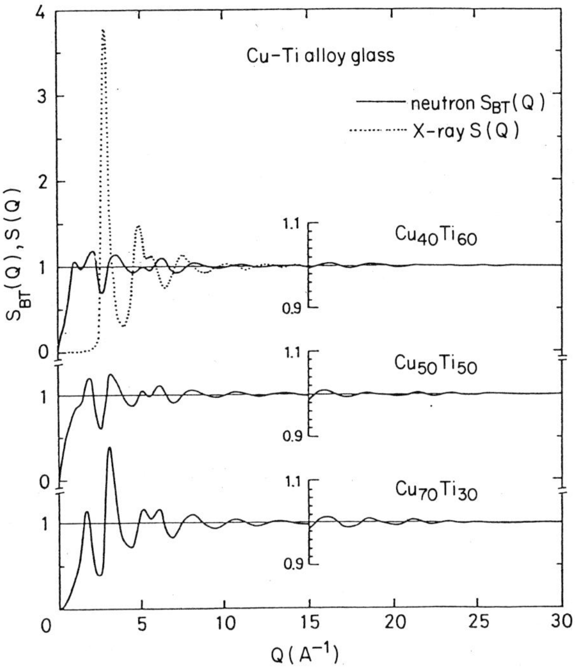

Figure 6.17 Bhatia–Thornton neutron structure factors $S_{BT}(Q)$ (full line) of Cu–Ti glasses and X-ray total structure factor $S(Q)$ (broken line) of $Cu_{40}Ti_{60}$ glass (after Fukunaga *et al.*[22])

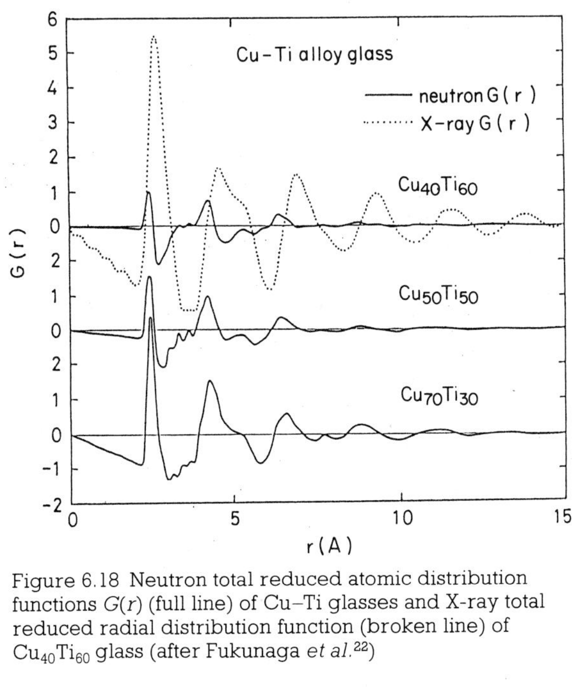

Figure 6.18 Neutron total reduced atomic distribution functions $G(r)$ (full line) of Cu–Ti glasses and X-ray total reduced radial distribution function (broken line) of $Cu_{40}Ti_{60}$ glass (after Fukunaga *et al.*[22])

if the local atomic arrangement is not drastically modified between the two glasses. As shown in *Figure 6.19*, the ΔRDF definitely assigns the Cu–Cu nearest neighbour correlation at $r = 2.50$ Å because of the negative peak, and the Ti–Ti correlation predominantly contributes to the positive peak around $r = 3.0$ Å in Cu–Ti glasses, while the Cu–Ti correlation for ΔRDF $=0$ exists at the distance of $r = 2.70$ Å. Therefore, the sharp and positive peak found at $r = 2.45$ Å in neutron $G(r)$ is concluded

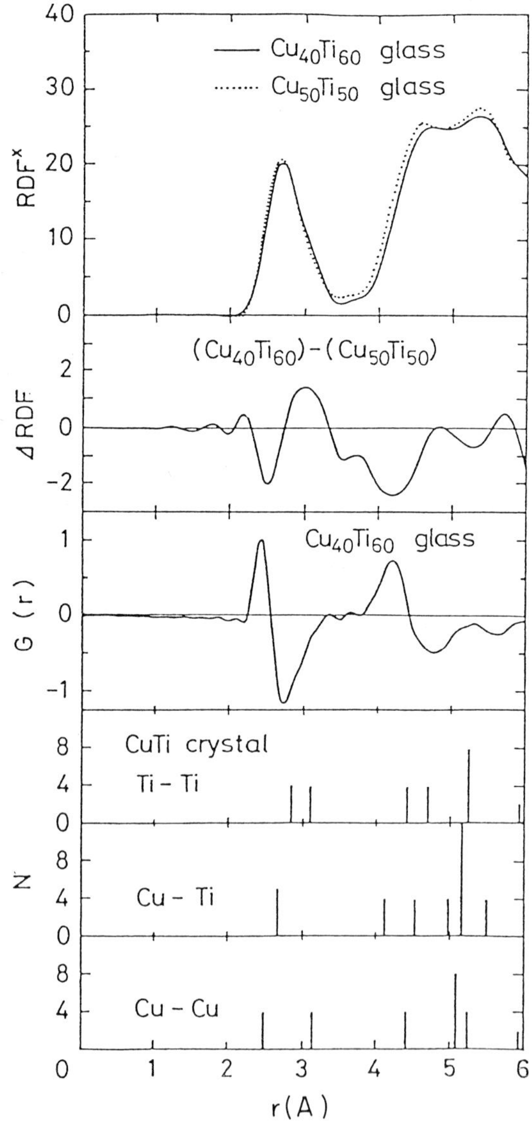

Figure 6.19 X-Ray total radial distribution functions RDF^X, differential radial distribution function ΔRDF and neutron total reduced atomic distribution function $G(r)$ for Cu–Ti glasses, and Ti–Ti, Cu–Ti and Cu–Cu correlations in CuTi crystalline compound (after Fukunaga *et al.*[22])

to be ascribed to the Cu–Cu correlation, while the positive contribution from Ti–Ti correlation is completely hidden behind the broad negative peak minimum around $r = 2.75$ Å corresponding to the Cu–Ti correlation. Such a short-range structure between nearest neighbour atoms in Cu–Ti glasses is found in the crystal structure of the CuTi compound (γ-CuTi type)[25] as shown in *Figure 6.19*.

There are definitely some differences in the chemical short-range order between Ni–Ti glasses and Cu–Ti ones, which correspond exactly to the difference of the crystal structure between $NiTi_2$ and CuTi. The Cu–Cu correlation in Cu–Ti glasses arises from the nearest neighbour position available for direct contact as found in the CuTi crystal structure, while the Ni–Ni correlation in Ni–Ti glasses does not exist at the direct contact position as also in the $NiTi_2$ crystalline compound. Therefore, we conclude that metal–metal amorphous alloys still preserve the chemical short-range structure analogous to that found in the corresponding crystalline compounds as well as metal–metalloid amorphous alloys.

It is noteworthy that Cu–Ti glasses are formed over a wide composition range from 30 to 65 atom% of titanium by the melt-quenching method, because Cu–Ti glasses prefer to have the short-range structure similar to the dense random packing of binary hard spheres. Since there is a severe restriction on the short-range structure for the prohibition of Ni–Ni contact in Ni–Ti glasses, Ni–Ti alloys can be vitrified only within a narrow composition range from 60 to 74 atom % of titanium.

6.6 Local environment around hydrogen atoms in amorphous alloys

Hydrogen and deuterium atoms are easily absorbed into metal–metal amorphous alloys consisting of a combination of early and late transition metals in the periodic table. Therefore, hydrogen and deuterium atoms are expected to be a promising probe of the atomic scale structure and dynamics in amorphous metals as well as in crystalline metals. Neutron scattering experiments have the unique advantage of detecting sensitively hydrogen and deuterium atoms in amorphous alloys, since the nucleus of the hydrogen atom has a quite large value for the incoherent neutron scattering cross-section (79.7 barns) and the deuterium atom contains a nucleus with a reasonable magnitude for the coherent neutron scattering cross-section (5.6 barns)[23].

The atomic spacing between deuterium atoms and metallic atoms is fairly short and the vibrational energy of hydrogen atoms is considerably higher in amorphous alloys, because the size and mass of hydrogen and deuterium atoms are extremely small. Therefore, short wavelength neutrons with epithermal energy are necessary as incident neutrons to obtain well-resolved experimental observations about the local environment around a deuterium atom and the localized vibration of hydrogen atoms in amorphous alloys. Concerning this point, the use of pulsed neutrons with a high flux above the thermal energy region generated from accelerators[26] is a more prudent choice than the use of steady state reactor neutrons.

Figures 6.20 and *6.21* show the total structure factor $S(Q)$'s and the RDF's of $Pd_{0.35}Zr_{0.65}D_x$ ($x = 0 \sim 1.25$) glasses obtained by Kai *et al.*[27] using a spallation neutron source installed at the 500 MeV proton booster synchrotron in the Japanese National Laboratory for High Energy Physics. With increasing deuterium content, the profiles of the first and second peak in the $S(Q)$'s are rapidly modified and the high Q oscillation is rather enhanced. Since the Fourier transformation of $S(Q) \rightarrow g(r)$ is truncated at $Q_{max} = 30$ Å^{-1}, the first peak around $r = 2$ Å corresponding to the D–M (M = Pd, Zr) correlation is clearly distinguished from the higher order peaks of

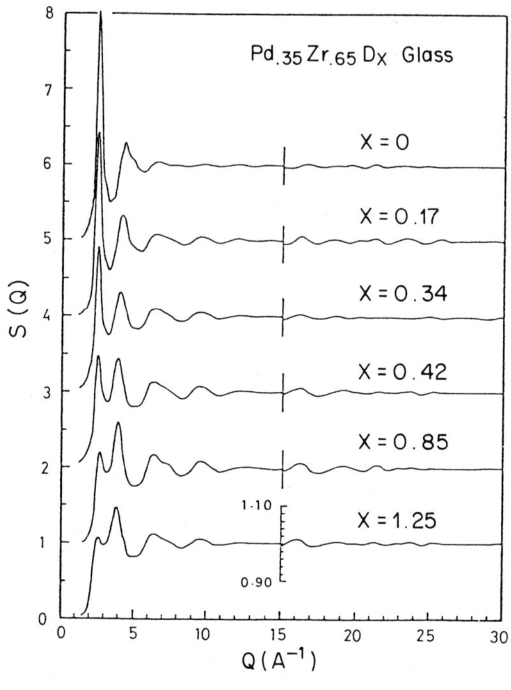

Figure 6.20 Neutron total structure factors $S(Q)$ of $Pd_{0.35}Zr_{0.65}D_x$ ($x = 0 \sim 1.25$) glasses for values of x shown on the curves (after Kai *et al.*[27])

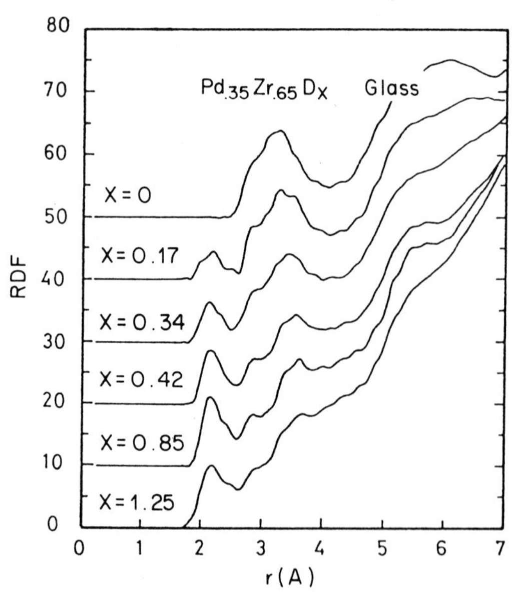

Figure 6.21 Neutron total radial distribution functions RDF of $Pd_{0.35}Zr_{0.65}D_x$ ($x = 0 \sim 1.25$) glasses for values of x shown on the curves. The Fourier transformation of $S(Q) \rightarrow g(r)$ is truncated at $Q_{max} = 30 \text{Å}^{-1}$ (after Kai *et al.*[27])

the M–M correlations in the RDF's. The first peak in the RDF's is separated into the two Gaussian peaks indicating the D–M correlations at $r^1_{DM} \sim 2.1$ Å and $r^2_{DM} \sim 2.4$ Å by a least squares fitting.

The high r Gaussian peak is assigned to the D–Zr correlation and the low r Gaussian peak is contributed to from both the D–Pd and D–Zr correlations. *Figure 6.22* shows the coordination numbers of metallic atoms around a deuterium atom in $Pd_{0.35}Zr_{0.65}D_x$ glasses as a function of deuterium content, calculated from the area under the Gaussian peaks.

We can find that deuterium atoms initially prefer to occupy a tetrahedral-like site consisting of four metal atoms in $Pd_{0.35}Zr_{0.65}D_x$ glasses with low deuterium content. With increasing deuterium content, deuterium atoms are located at an octahedral-like site having six metal atoms around a deuterium atom on average near the deuterium content of $x = 0.42$. Finally, the average local environment around a deuterium atom approaches a hexahedral-like site consisting of five metal atoms at the deuterium content of $x = 1.25$. These types of polyhedral site are known to exist often in crystalline transition metal hydrides such as PdH_x[28], ZrH_x[29] and $NiZrH_x$[30]. It is noteworthy that the octahedral and hexahedral-like sites in $Pd_{0.35}Zr_{0.65}D_x$ glasses may indicate Jahn–Teller splitting because of the existence of two different D–Zr spacings.

In the RDF of the $NiZrD_{1.8}$ glass, where the $S(Q) \rightarrow g(r)$ Fourier transformation is truncated at $Q_{max} = 25$ Å^{-1}, the clear separation between D–Ni and D–Zr correlations is found, as shown in *Figure 6.23*, because of a large difference of atomic size between nickel and zirconium atoms. The small peak located at $r = 1.7$ Å denotes the D–Ni correlation, while the large peak appearing around $r = 2.1$ Å corresponds to the D–Zr correlation in the glass. These atomic spacings of D–Ni and D–Zr correlations in

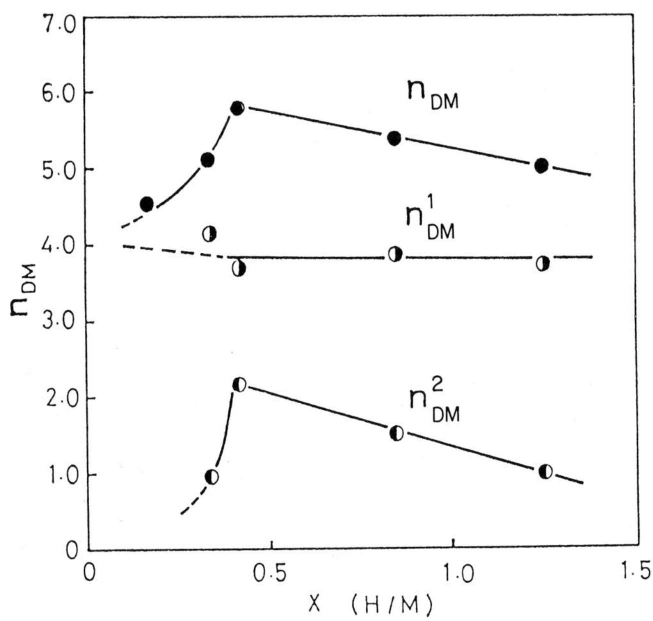

Figure 6.22 Coordination numbers of metallic (M) atoms surrounding a deuterium atom in $Pd_{0.35}Zr_{0.65}D_x$ ($x = 0 \sim 1.25$) glasses. n^1_{DM} is the number of M atoms located at small D–M spacing, n^2_{DM} means the number of M atoms with large D–M spacing and $n_{DM} = n^1_{DM} + n^2_{DM}$ (after Kai *et al.*[27])

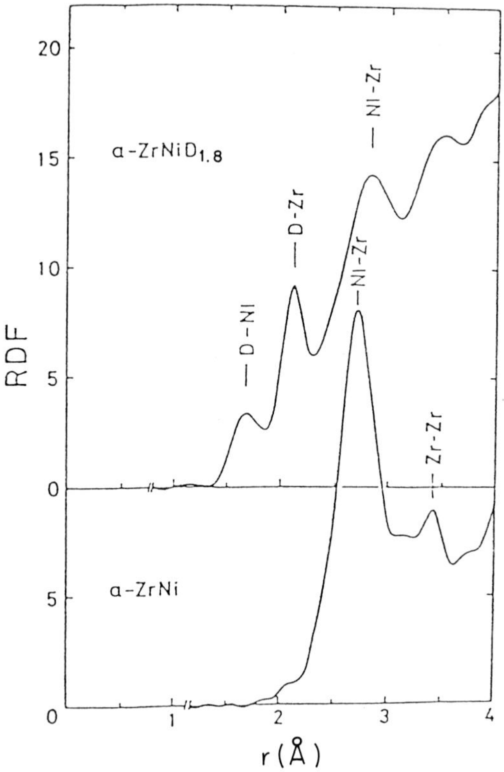

Figure 6.23 Neutron total radial distribution functions RDF of NiZr and $NiZrD_{1.8}$ glasses (after Kaneko *et al.*[31])

$NiZrD_{1.8}$ glass are nearly equal to those of H–Ni ($r = 1.66$ Å) and H–Zr ($r = 2.1$ Å) correlations in the tetrahedral site in the $NiZrH_3$ crystalline compound[30]. The coordination numbers of metallic atoms surrounding a deuterium atom are calculated to be one nickel and three zirconium atoms from the area under both peaks. This result supports again the fact that a deuterium atom prefers to occupy the tetrahedral site consisting of one nickel and three zirconium atoms on the average.

The energy spectrum of the localized hydrogen vibration provides knowledge of the topological distortion and chemical fluctuation of the local environment around a hydrogen atom in amorphous alloys. As the hydrogen atom has a large incoherent neutron scattering cross-section ($\langle b^2 \rangle - \langle b \rangle^2$) as well as a high vibrational frequency ($h\omega = E - E'$) owing to its small mass, the frequency distribution $g(\omega)$ of hydrogen atoms in amorphous alloys can be obtained directly from the measurement of the double differential cross-section for the incoherent single-phonon scattering process[1] by use of epithermal energy pulsed neutrons:

$$\frac{d\sigma_{\text{incoh}}}{d\Omega dE'} = \frac{k'}{k}(\langle b^2 \rangle - \langle b \rangle^2)\frac{Q^2\langle u^2 \rangle}{2m}\left\{\frac{1}{[\exp(h\omega/k_B T) - 1]} + \left(\frac{1}{2} \pm \frac{1}{2}\right)\right\} N \exp(-2W)\frac{g(\omega)}{\omega} \quad (6.18)$$

where $\exp(-2W)$ is the Debye–Waller factor and $\langle u^2 \rangle$ is the mean square of hydrogen vibration amplitude.

Figure 6.24 shows the final $g(\omega)$'s of a $NiZrH_{2.8}$ crystal and a $NiZrH_{1.8}$ glass

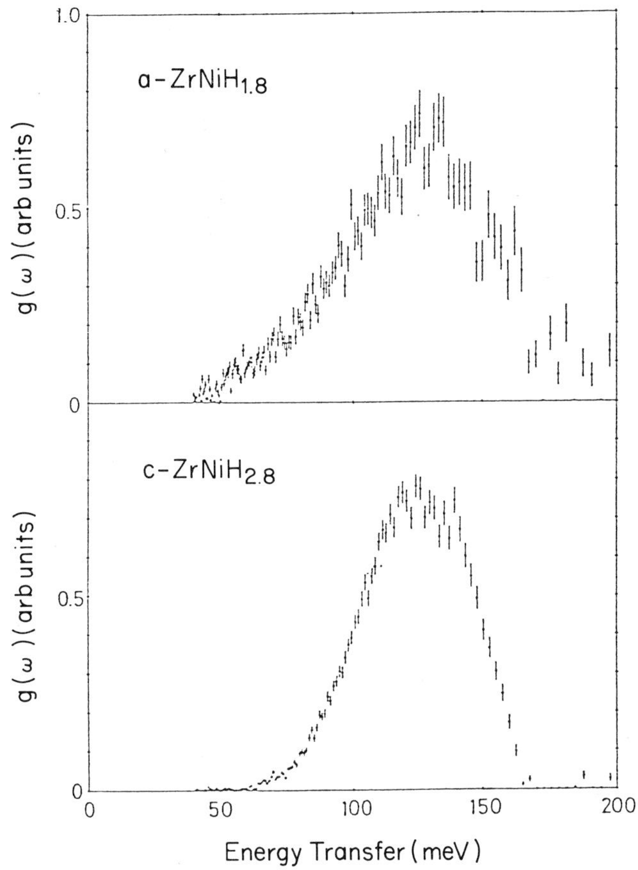

Figure 6.24 Localized vibration frequency spectra $g(\omega)$ of hydrogen atoms absorbed in $NiZrH_{1.8}$ crystal and $NiZrH_{2.8}$ glass (after Kaneko *et al.*[31])

measured by Kaneko *et al.*[31] using a time-focusing crystal analyser spectrometer[26] installed at a (γ, n) neutron source in the Tohoku University 300 meV Electron LINAC Laboratory. The peak centres of the $g(\omega)$'s are located at 130 ± 3 meV in both the crystal and glass. The fact that there is no variation in the peak centre position implies that the sites occupied predominantly by hydrogen atoms are equivalent for the crystal and glass state of the NiZr alloy. However, the width of the $g(\omega)$ for the $NiZrH_{1.8}$ glass (FWHM $= 70 \pm 5$ meV) increases by about 40 per cent compared with that of the $NiZrH_{2.8}$ crystal (FWHM $= 50 \pm 5$ meV), although the glass includes less hydrogen content than the crystal. Particularly, $g(\omega)$ of the $NiZrH_{1.8}$ glass has wide tails on both extreme sides of the energy range below 100 meV and above 150 meV. The root mean square of the hydrogen vibration amplitude obtained from the Debye–Waller factor in equation 6.18 shows no significant difference between the $NiZrH_{2.8}$ crystal [$\sqrt{(\langle u_x^2 \rangle)} = 0.192$ Å] and the $NiZrH_{1.8}$ glass [$\sqrt{(\langle u_g^2 \rangle)} = 0.205$ Å].

Based on the observation described above, we conclude that hydrogen and deuterium atoms absorbed into NiZr glass are preferentially located at the tetrahedral-like sites in the glass, where topological distortion is rather restricted but the chemical order in the tetrahedral unit structures fluctuates considerably around the average composition of one nickel and three zirconium atoms. The tetrahedral site consisting of two nickel and two zirconium atoms may enhance the contribution to $g(\omega)$ from the lower energy vibration, while hydrogen atoms sitting in the tetrahedron constructed from four zirconium atoms shift the centre position of $g(\omega)$ toward the higher energy region.

6.7 Conclusions

The well-resolved short-range structure of amorphous alloys is determined by the measurements of the high momentum transfer structure factor and hydrogen vibration frequency spectrum using epithermal energy pulsed neutrons generated from accelerators.

The structure of amorphous alloys is constructed from the polyhedral unit structures of which topological distortion and chemical disorder are rather modest. Particularly, the chemical short-range order of metal–metalloid amorphous alloys is quite close to that of the corresponding crystal structure. Even in metal–metal amorphous alloys the chemical short-range order is still preserved without significant deviations from the crystal structure.

An essential key for characterizing the individual structure of amorphous alloys subject to different preparation and treatment methods is to find the way of connecting the polyhedral unit structures under a certain condition. Based on this point of view, the precise determination of the metalloid–metalloid correlation is important in metal–metalloid amorphous alloys and also hydrogen atoms absorbed into metal–metal amorphous alloys play a unique role in providing an insight into their atomic scale structure and dynamics.

EXAFS[32] and some nuclear techniques[33] such as the Mössbauer effect and nuclear magnetic resonance make it possible to observe selectively the local atomic configuration around a specific atom in amorphous alloys. High resolution electron microscope experiments provide the direct observation of the arrangement of atoms in amorphous alloys. The experimental determination of the structure of amorphous metals is certainly making progress with the development of excellent experimental techniques.

References

1. Windsor, C. G., *'Chemical Applications of Thermal Neutron Scattering'*, Ed. Willis, B. T. M., p. 1, Oxford University Press, London (1973)
2. Van Hove, L., *Phys. Rev.*, **95**, 249 (1954)
3. Egelstaff, P. A., *'An Introduction to the Liquid State'*, Academic Press, London (1967)
4. Suzuki, K., Misawa, M., Kai, K. and Watanabe, N., *J. Nucl. Instrum. Meth.*, **147**, 519 (1977)
5. Ueno, M., Misawa, M. and Suzuki, K., *Res. Rep. Lab. nucl. Sci. Tôhoku Univ.*, **13**, 254 (1980)
6. Suzuki, K., *Ber. Bunsenges. phys. Chem.*, **80**, 689 (1976)
7. Misawa, M., Doctoral Thesis, Tohoku University (1976)
8. Fukunaga, T., Doctoral Thesis, Tohoku University (1978)
9. Suzuki, K., *'Proc. 4th Int. Conf. Rapidly Quenched Metals'*, Eds. Masumoto, T. and Suzuki, K., vol. 1, p. 309, Japan Institute of Metals, Sendai (1982)
10. Fukunaga, T. and Suzuki, K., *Sci. Rep. Res. Insts Tôhoku Univ.*, **A29**, 153 (1981)
11. Suzuki, K., Fukunaga, T., Misawa, M. and Masumoto, T., *Mater. Sci. Engng*, **23**, 215 (1976)
12. Suzuki, K., Fukunaga, T., Misawa, M. and Masumoto, T., *Sci. Rep. Res. Insts., Tôhoku Univ.*, **A26**, 1 (1976)
13. Fukunaga, T., Misawa, M., Fukamichi, K., Masumoto, T. and Suzuki, K., *'Proc. 3rd Int. Conf. Rapidly Quenched Metals'*, Ed. Cantor, B., vol. II, p. 325, The Metals Society, London (1978)

14. Fukunaga, T. and Suzuki, K., *Sci. Rep. Res. Insts Tôhoku Univ.*, **A28,** 208 (1980)
15. Waseda, Y., Okazaki, H. and Masumoto, T., *Sci. Rep. Res. Insts Tôhoku Univ.*, **A26,** 202 (1977)
16. Waseda, Y. and Tamaki, S., *Z. Phys.*, **B23,** 315 (1976)
17. Sadoc, J. F. and Dixmier, J., *Mater. Sci. Engng*, **23,** 187 (1976)
18. Lamparter, P., Sperl, W., Nold, E., Rainer-Harbach, G. and Steeb, S., '*Proc. 4th Int. Conf. Rapidly Quenched Metals*', Eds. Masumoto, T. and Suzuki K., vol. I, p. 343, Japan Institute of Metals, Sendai (1982)
19. Gaskell, P. H., *J. Non-cryst Solids,* **32,** 207 (1979)
20. Hayashi, N., Fukunaga, T. Ueno, M. and Suzuki, K., ref. 18, p. 355
21. Bhatia, A. B. and Thornton, D. E., *Phys. Rev.,* **B2,** 3004 (1970)
22. Fukunaga, T., Kai, K., Naka, M., Watanabe, N. and Suzuki, K., ref. 18, p. 347
23. Bacon, G. E., 'Neutron Diffraction', 2nd edn., p. 31, Oxford University Press, London (1967)
24. Rostoker, W., *Trans. Am. Inst. Min. metall. Engrs,* **194,** 209 (1952)
25. Karlsson, N., *J. Inst. Metals* **79,** 391 (1951)
26. Suzuki, K., Niimura, N., Kajitani, T., Kai, K., Endoh, Y., Tomiyoshi, S., Yamada, Y. and Yamaguchi, S., *'Proc. 4th Meeting of ICANS',* Ed. Ishikawa, Y., p. 485, National Laboratory for High Energy Physics, Tsukuba (1980)
27. Kai, K., Fukunaga, T., Nomoto, T., Watanabe, N. and Suzuki, K., *'Proc. 4th Int. Conf. Rapidly Quenched Metals',* Eds. Masumoto, T. and Suzuki, K., vol. II, p. 1609, Japan Institute of Metals, Sendai (1982)
28. Worsham, J. E., Wilkinson, M. K. and Shull, C. G., *J. Phys. Chem. Solids,* **3,** 303 (1957)
29. Slaggie, E. L., *J. Phys. Chem. Solids,* **29,** 923 (1968)
30. Westlake, D. G., *J. less-Common Metals,* **75,** 177 (1980)
31. Kaneko, H., Kajitani, T., Hirabayashi, M., Ueno, M. and Suzuki, K., ref. 27, p. 1605
32. Wong, J., '*Glassy Metals I*', Ed. Güntherodt, H.-J. and Beck, H., p. 45, Springer-Verlag, Berlin (1981)
33. *Proc. Int. Conf. Amorphous Systems Investigated by Nuclear Methods,* Hungary, 1981, Eds. Kajcsos, Zs., Dezsi, I., Horvath, D., Kemeny, T., Marcziz, L. and Nagy, D. L., Budapest (1982)

Chapter 7

Atomic short-range ordering in amorphous metal alloys

T. Egami

Dept. of Metallurgy and Materials Science, University of Pennsylvania, Philadelphia, USA

7.1 Atomic short-range order in amorphous and liquid alloys

Anyone who studies or produces the metallic glasses would be first surprised, and may even be somewhat puzzled, by the degree of variability in many properties of the as-quenched metallic glasses. Most of the metallic glasses containing iron or cobalt are excellent soft magnetic materials, but their magnetic permeability varies from one run of production to the other, and from one portion of the ribbon to another. Furthermore, these properties can be altered significantly by post-quench treatments, such as annealing and mechanical deformation. In early times, researchers, including the present author, attributed this variance to difference in the 'amorphousness', implying that some of, or at least portions of, metallic glasses are not truly amorphous, but merely microcrystalline. This view, however, turned out to be incorrect. The reversibility of many of the properties[1,2] and the structural evidence[3] indicate that these effects are due to some atomic ordering within the amorphous phase, rather than due to phase transformation into crystallinity. In other words, the amorphous state is not a unique state, but contains a considerable and continuous variation of its microscopic state. Such a microscopic state is described by the atomic short range order (SRO), but since it is difficult to determine the SRO's directly by experiment, there is a fair amount of confusion about how they behave, or even how they are defined[4,5]. In this chapter, we summarize our latest view on this subject, in an attempt to seek simplicity and coherency in the seemingly complex subject. We will first define the *compositional or chemical short range order* (CSRO) in terms of the *atomic pair distribution functions* (PDF), and consider physical properties related to CSRO. We then define the geometrical or physical short range order (GSRO) in terms of the atomic level local parameters, with an emphasis on the difference between the density fluctuations and the shear stress fluctuations.

Finally, we discuss the thermal properties of the SRO's, both in equilibrium and non-equilibrium states. Although the research in this field, particularly those related to the last subject, is admittedly still incomplete, and sizable portions of the descriptions are conjecture, it is hoped that the concepts developed here would assist in many ways to the understanding of various properties of metallic glasses at a microscopic level.

7.2 Chemical short-range order (CSRO)

7.2.1 Definition of chemical short-range order

Since all the metallic glasses which are stable at room temperature are alloys, the state of mixture of the alloy components is an important parameter of the system. In most cases, a completely random mixture is not attained, so that the chemical composition around the atoms of each alloying component is different from the average. The system then has chemical short-range order (CSRO). Since the effect of CSRO on observed properties is limited in most cases to within the first nearest neighbour shell, the simplest way to define CSRO is to use the deviation in the nearest neighbour composition from the average as a parameter, which for a binary (A–B) system is

$$\alpha_{\mathrm{p}} = 1 - \frac{Z_{\mathrm{AB}}}{\langle Z \rangle C_{\mathrm{B}}} = 1 - \frac{Z_{\mathrm{BA}}}{\langle Z \rangle C_{\mathrm{A}}} \tag{7.1}$$

where $Z_{\alpha\beta}(\alpha\beta = \mathrm{A, B})$ is the number of β coordination of an α atom, C_α is the concentration of α, and $\langle Z \rangle$ is the total coordination number. This parameter is known as the *Warren–Cowley short range order parameter*[6,7] (*see also* ref. 8), and provides an excellent measure of chemical affinity. However, for the determination of this parameter, and for more involved studies, we need to know CSRO as a function of distance from atoms. This is best described by the composition fluctuation atomic pair correlation function used by Bhatia and Thornton[9]:

$$\rho_{\mathrm{CC}}(r) = \frac{1}{4\pi r^2 N} \sum_{i,j} \delta(r - r_{ij})(\eta_i^{\mathrm{A}} - C_{\mathrm{A}})(\eta_j^{\mathrm{B}} - C_{\mathrm{B}}) \tag{7.2}$$

where r_{ij} is the distance between the ith and jth atoms, and

$$\begin{aligned} \eta_i^{\mathrm{A}} &= 1 \quad \text{if the } i\text{th site is occupied by A atom} \\ &= 0 \quad \text{otherwise} \end{aligned} \tag{7.3}$$

and likewise for η_j^{B}. These parameters, however, are non-directional, while for some properties such as the magnetic anisotropy, we need to describe *directional CSRO*. There is no established method for this, but a parameter,

$$I_n^{l,m}(A) = \frac{\sqrt{4\pi}}{C_{\mathrm{A}} Z_{\mathrm{A}} N} \sum_{i,j} \left(\frac{r_{ij} - \bar{a}_{\mathrm{A}}}{\bar{a}_{\mathrm{A}}} \right)^n \eta_i^{\mathrm{A}} \left(\frac{C_{\mathrm{B}} Z_{\mathrm{B}}}{\langle Z \rangle} - \eta_j^{\mathrm{B}} \right) Y_l^m\left(\frac{\vec{r}_{ij}}{r_{ij}} \right) \tag{7.4}$$

where Z_{A} is the total coordination number of an A atom, $\bar{a}_{\mathrm{A}}$ is the average nearest neighbour distance from an A atom, and Y_l^m are the spherical harmonics, should be able to serve the purpose. Equation 7.4 is an extension of 7.1, and α_{p} is approximately given by

$$\alpha_{\mathrm{p}} \sim I_0^{0,0}(A) \frac{\langle Z \rangle}{C_{\mathrm{B}} Z_{\mathrm{B}}} = I_0^{0,0}(B) \frac{\langle Z \rangle}{C_{\mathrm{A}} Z_{\mathrm{A}}} \tag{7.5}$$

$I_n^{l,m}(A)$ can be determined, in principle, when equation 7.2 is known for all directions of r.

The CSRO of the system can thus be defined unequivocally. Unfortunately, however, the experimental determination of these parameters is a very difficult task. While a considerable amount of work has been done for liquid alloys[10], relatively few attempts have been made for metallic glasses, by neutron diffraction using the isotopic substitution technique[11], by X-ray diffraction making use of the anomalous dispersion[12], or using isomorphous substitution[13,14] by combining neutron and X-ray diffraction results [15–17], and by EXAFS measurement[18–20]. Furthermore, the accuracy of these measurements is not yet high enough to provide an explanation for physical and chemical properties in terms of the measured CSRO except for a few cases[12]. It is possible, however, to conjecture on the basis of circumstantial considerations such as the composition dependence, that some important changes in the properties due to heat treatment or mechanical deformation are due to changes in CSRO. We will discuss two such cases below, namely the changes in the Curie temperature and induced magnetic anistropy.

7.2.2 Curie temperature and CSRO

Soon after the extensive research on the magnetic properties of metallic glasses started in 1974—75, it was recognized that the Curie temperature, T_C, of ferromagnetic metallic glasses is not so stable, but is susceptible to heat treatments[21–23]. Typical results of the change in T_C with annealing temperature are shown in *Figure 7.1.*

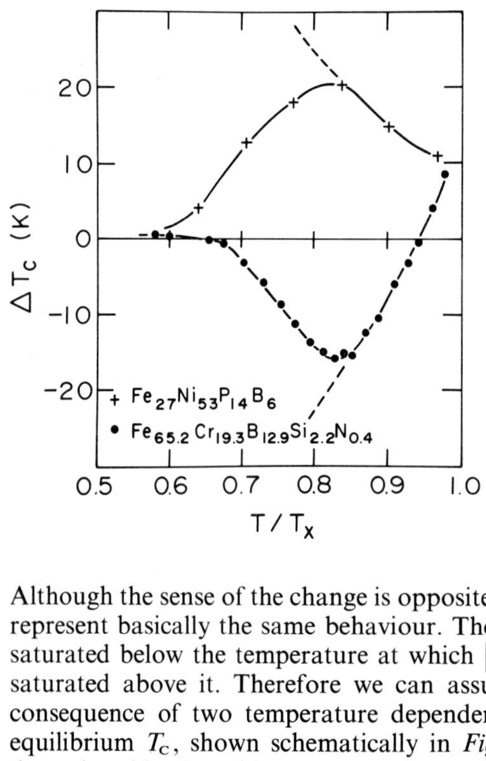

Figure 7.1 Changes in the Curie temperature T_C with annealing temperature T_a for (+) $Fe_{27}Ni_{53}P_{14}B_6$ (ref. 23) and (●) $Fe_{65.2}Cr_{19.3}B_{12.9}Si_{2.6}$ (ref. 24). The values of ΔT_C in (+) are obtained by 180 min. annealing of as-quenched samples at T_a, while those in (●) are obtained by successive annealing for 5 min. at each temperature T_a. However, they represent essentially the same treatment. T_a is scaled by the crystalline temperature T_x [360 °C for (+) and 465 °C for (●)] which is approximately equal to the glass transition temperature T_g. Dashed lines indicate the extrapolation of the equilibrium T_C at each annealing temperature (after Liebermann and co-workers[23,24])

Although the sense of the change is opposite for Fe–Ni[23] and Fe–Cr[24] base alloys, they represent basically the same behaviour. The isothermal studies show that ΔT_C has not saturated below the temperature at which $|\Delta T_C|$ is maximum in *Figure 7.1*, while it is saturated above it. Therefore we can assume that this behaviour is produced as a consequence of two temperature dependent phenomena[2]; a temperature dependent equilibrium T_C, shown schematically in *Figure 7.1* by dashed lines, and temperature dependent kinetics which become faster at higher temperature, with the time constant obeying the Arrhenius activation law with a wide distribution of the activation energy[25]. That T_C is saturated to an equilibrium value at high temperatures can readily be shown by the nearly complete reversibility of T_C during cyclic annealing[2,24].

This temperature dependence in the equilibrium value of T_C is best explained in terms of the changes in CSRO. In general the equilibrium CSRO of the system is dependent on temperature, and it decreases with increasing temperature because of the effect of entropy[26]. In the magnetic alloys, the change in the CSRO around the magnetic ions will produce the change in T_C. In Fe–Ni alloys, a close contact of iron and nickel atoms will raise T_C[27], while in FeCr that of iron and chromium will lower T_C[24]. Such changes in the CSRO do not have to involve diffusional exchange of atoms, but will take place also via more collective atomic rearrangements discussed later.

7.2.3 Field induced magnetic anisotropy

Since the glassy structure by itself is macroscopically isotropic, magnetic amorphous alloys should have no intrinsic magnetic anisotropy. If the alloy is annealed in a magnetic field, however, it develops a macroscopic magnetic anisotropy[1,28–30] (*see* Chapter 16). This phenomenon is also characterized by a temperature dependent saturation value $K_u(\infty)$ and temperature dependent kinetics of approach to saturation, $K_u(t)/K_u(\infty)$. Typical examples of these temperature dependences[29] are shown in *Figures 7.2* and *7.3*. The kinetics of this phenomenon are almost identical to that of

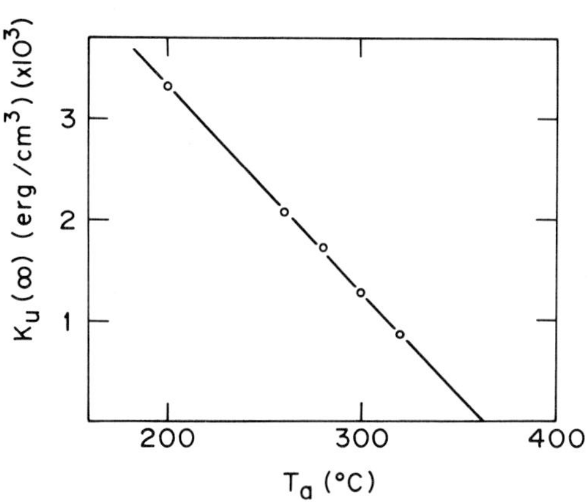

Figure 7.2 Temperature dependence of $K_u(\infty)$ for $Fe_{40}Ni_{38}Mo_4B_{18}$ alloy (after Chambron and Chamberod[29])

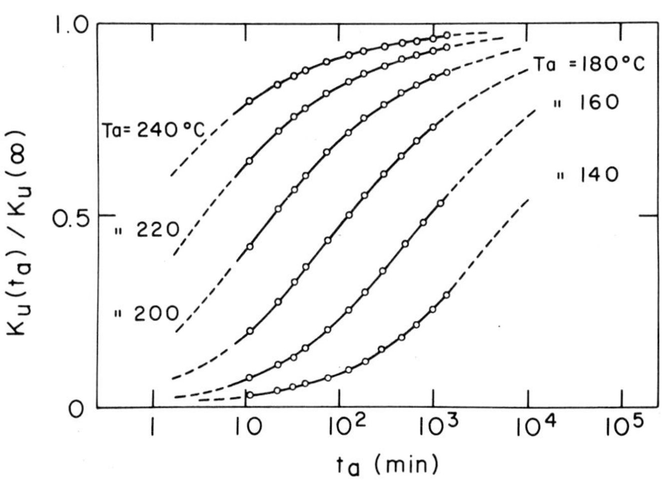

Figure 7.3 Kinetics of saturation $K_u(t_a)/K_u(\infty)$ for $Fe_{40}Ni_{40}P_{14}B_6$ alloy. Values of T_a are shown on the curves in degrees Celsius (after Chambron and Chamberod [229])

ΔT_C. The field induced anisotropy is a well known phenomenon for crystalline alloys, and is explained in terms of the *directional CSRO* of magnetic ions[31–33]. It appears that the same mechanism explains the bulk of the induced anisotropy in metallic glasses, judging from the composition dependence[34,35]. However, direct measurement of the anisotropic SCRO such as given by equation 7.4 by diffraction experiments is all but impossible, since the structural anisotropy would be very small[31,32]. This induced anisotropy is quite important for practical applications, since it determines the permeability[36], after-effect[37,38], and in some cases magnetic losses[39,40] of the magnetic amorphous alloys.

7.3 Geometrical short-range order (GSRO)

7.3.1 Geometrical short-range order defined in terms of the atomic level local structural parameters

While many properties depend upon CSRO, which describes the state of mixture of alloying elements as discussed above, they also depend upon the details of the geometry of the structure itself, or the degree of randomness of the structure. This is a more subtle kind of short range order, and is best studied in a single elementary glass, in order to be free from the effect of CSRO, even though such a glass is not stable at room temperature. We have been calling this short range order *topological short range order* (TSRO)[2]; however, a recent study showed that geometrical distortions which are independent of topology are also important, as will be discussed below. Therefore, the SRO which characterizes local structure of an elementary glass may be called geometrical short range order (GSRO), which includes both TSRO and *distortional short range order* (DSRO).

In network glasses, such as amorphous silicon or germanium, oxides and polymers, the topology of the structure is well defined in terms of connectivity of bonded atoms, and many physical and chemical properties are determined largely by TSRO[5]. In metallic glasses, the metallic nature of the atomic bonding makes the definition of atomic connectivity ill-defined. However, since the first nearest neighbour is reasonably well defined owing to the presence of a deep valley in the PDF between the first and the second peaks, it is still possible to define topology of the system in terms of the topology of *Voronoi polyhedra* which is equivalent to Wigner–Sietz cells in crystals (*see* Chapter 4). Nevertheless, since the distribution in the nearest neighbour distances, or the width of the first peak of the PDF, is significant ($\Delta r/r \sim 0.15$), a more detailed account of the state of packing around atoms becomes necessary in order to relate the local structure of metallic glasses to the observed properties. A method of describing the local structure, or GSRO, in terms of a set of local parameters was recently proposed by Egami, Maeda and Vitek[41].

One of the parameters is the *atomic level stresses*, defined by

$$\sigma_i^{\alpha\beta} = \frac{1}{2\Omega_i} \sum_j f_{ij}^{\alpha} r_{ij}^{\beta} \tag{7.6}$$

where $\alpha\beta$ denote Cartesian components, Ω_i is the local volume of the ith atoms, and f_{ij} is the two-body force between the ith and jth atoms. For a system interacting via a two-body central force potential $\phi(r_{ij})$, equation 7.6 becomes

$$\sigma_i^{\alpha\beta} = \frac{1}{2\Omega_i} \sum_j \frac{1}{r_{ij}} \frac{d\phi}{dr_{ij}} r_{ij}^{\alpha} r_{ij}^{\beta} \tag{7.7}$$

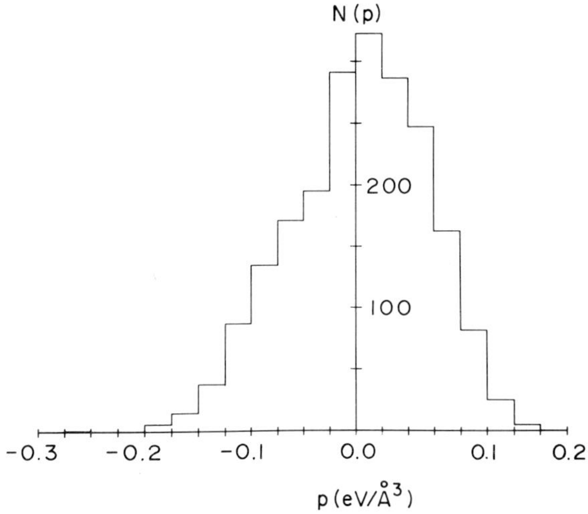

Figure 7.4 Distribution of atomic level hydrostatic stress p, calculated for model amorphous iron (after Srolovitz *et al.*[43])

Since the macroscopic average of the local environment is spherically symmetric, we may use the spherical harmonics equivalents,

$$\sigma_i^{l,m} = \frac{1}{2\Omega_i} \sum_j r_{ij} \frac{\mathrm{d}\phi}{\mathrm{d}r_{ij}} Y_l^m \left(\frac{\vec{r}_{ij}}{r_{ij}} \right) \qquad (l = 0, 2) \tag{7.8}$$

and in particular the rotational invariant,

$$p = \frac{\sqrt{4\pi}}{3} \sigma_i^{0,0} = \tfrac{1}{3}(\sigma_i^{XX} + \sigma_i^{YY} + \sigma_i^{ZZ}) \tag{7.9}$$

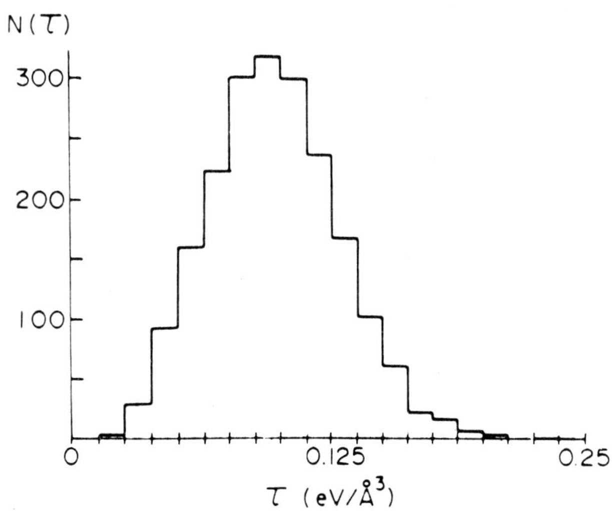

Figure 7.5 Distribution of atomic level shear stress τ calculated for model amorphous iron (after Srolovitz *et al.*[43])

which is the hydrostatic stress, and

$$\tau = \sqrt{\frac{4\pi}{15}}\left[\sum_m (\sigma^{2,m})^2\right]^{1/2} = \left[\tfrac{1}{6}\left\{(\sigma^{XX} - \sigma^{YY})^2 + (\sigma^{YY} - \sigma^{ZZ})^2 + (\sigma^{ZZ} - \sigma^{XX})^2\right\}\right]^{1/2} \tag{7.10}$$

(X, Y, Z in principal directions)
which is the von Mises shear stress. An example of the distribution of p and τ, calculated for a model of amorphous iron, is shown in *Figure 7.4* and *7.5*. The hydrostatic stress p is closely related[42] to the local volume Ω_i, so that it describes the fluctuation in the local density. The shear stress τ, on the other hand, describes the local distortion of the atomic environment from spherical symmetry. In addition, the site symmetry coefficients α_0 and β, which are closely related to p and τ respectively[43], were proposed as the local structural parameters[41].

7.3.2 The local structural parameters and topology

It is natural to expect the local structural parameters introduced above to be related, at least in part, to the local topology of the structure. A detailed study[44,45] showed that indeed the hydrostatic stress p is linearly related to the local coordination number Z_i, as shown in *Figure 7.6*. Therefore, the *local pressure fluctuations* are related to topological irregularities, which are essentially *disclinations* considered for instance by Spaepen[46] and Halperin and Nelson[47] in the case of two-dimensional systems, and for the three-dimensional case by Kléman and Sadoc[48,49]. Note that compressed atoms tend to have Z_i smaller than the average. This is because when Z_i is small, the atoms in the nearest neighbour shell have a surface tension which results in compression at the central atom. Similarly a large Z_i produces a surface compression and dilatational stress at the central atom. A local compressive or dilatational stress itself is elastically unstable, but it is sustained by the surface tension or compression among the nearest neighbour atoms[45]. Therefore it is appropriate to think of the local hydrostatic stress as

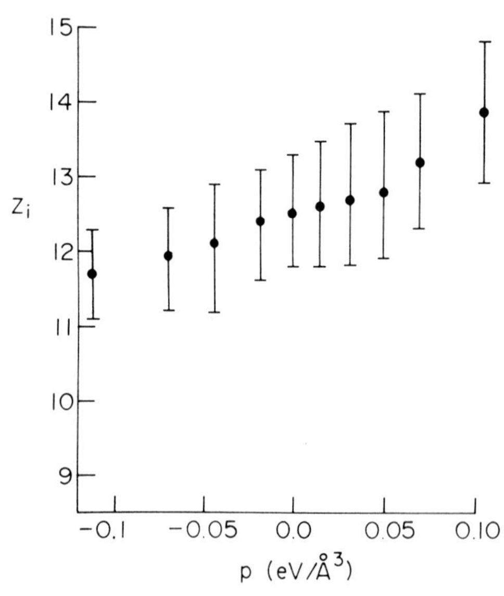

Figure 7.6 Relation between the local hydrostatic stress p and the local coordination number Z_i, calculated for model amorphous iron (after Egami and Srolovitz[45])

being produced when an atom is fitted to a nearest neighbour shell whose size is smaller or larger than the atom, very much like an inclusion in an elastic medium[45,50].

While the hydrostatic stress p is closely related to Z_i, it is not significantly affected by other details of the topology of the Voronoi polyhedra. The shear stress τ shows even weaker correlation with topology. It is almost independent of Z_i, and the average values of τ for each type of *Voronoi polyhedon* are virtually equal[44]. Therefore, while the hydrostatic pressure p describes TSRO, the shear stress τ describes a different kind of SRO, which may be called *distortional* SRO (DSRO). Thus, GSRO includes both TSRO and DSRO. It should be noted that the atomic level stresses can be defined equally well in multicomponent alloy systems, and the distribution of the stresses is also very similar to the single component system, when appropriately scaled with the elastic constants[51].

7.3.3 GSRO and pair distribution function

Although at present it is not possible to measure directly the local structural parameters by experiment, their distribution should be reflected in the pair distribution function of the system. More specifically, if the distribution of p or τ is wide, the first peak of the PDF should also be wide. The relation, in fact, is relatively simple. It was found[52] that the local PDF depends upon p by

$$G_p(r) = G_0(r + \gamma p) \tag{7.11}$$

where

$$G_p(r)\,\delta p = \frac{1}{r} \cdot \frac{\sum_{ij} \delta(r - r_{ij})}{\sum_i} \tag{7.12}$$

i satisfies the condition

$$p - \frac{\delta p}{2} < p_i < p + \frac{\delta p}{2} \tag{7.13}$$

and γ is the phase shift factor which is constant beyond the second peak of the PDF. Therefore the total PDF is

$$\begin{aligned} G(r) &= \langle G_p(r) \rangle \\ &= G_0(r) + \frac{\gamma^2}{2!} \frac{d^2 G_0}{dr^2} \langle p^2 \rangle + \cdots \end{aligned} \tag{7.14}$$

where $\langle \ldots \rangle$ denotes the volume average, since $\langle p \rangle = 0$. As expected, the peak widths of the PDF are increased as $\langle p^2 \rangle$ is increased. However, both the coordination number and the peak positions of $G(r)$ are largely independent of $\langle p^2 \rangle$. Therefore, the effect of changing $\langle p^2 \rangle$ is similar to what one might expect as the effect of changing temperature. While equation 7.14 makes it possible to calculate the changes in $G(r)$ due to the change in $\langle p^2 \rangle$, it is not possible to estimate $\langle p^2 \rangle$ from a measured $G(r)$, since we do not know $G_0(r)$ of the system with sufficient accuracy.

On the other hand, the relation between $G(r)$ and τ is more complex, and cannot be expressed analytically. In particular, the change in $\langle \tau \rangle$ produces shifts in the peak positions of the PDF[52].

7.3.4 GSRO and physical properties

One of the most important properties of the atomic level stresses is that their volume average $\langle \sigma^{\alpha\beta} \rangle$ are always zero, since they are equal to the external stress. Therefore the stress state is characterized by their second moments $\langle p^2 \rangle$ and $\langle \tau^2 \rangle$. Note that τ is an r.m.s. of the shear stress, therefore it represents essentially the second moment. Therefore the bulk properties to which all the atoms contribute more or less equally, such as density, resistivity, bulk magnetic properties and the PDF, are relatively weakly affected by the changes in the atomic stress distribution, since it reflects only second order changes. The effect is much more pronounced for properties which are generally considered to be 'structure-sensitive' by metallurgists, such as diffusivity and mechanical properties, since these properties are largely controlled by relatively few atoms in extreme environments, which are the lattice defects in the case of crystalline solids. The distinction between p and τ is also important here. The changes in p would result in atomic transport, since it changes the topology and the local density; however, the changes in τ would not produce atomic transport. The reorientation of τ, however, would produce the anelasticity, which is an important characteristic of the glassy system. The changes in τ under the applied stress play a large role in mechanical deformation and fracture. The changes in p and τ in the alloy system provide the mechanism by which the local CSRO is altered. In the following, we examine some of these relations between the observable properties and the local stress fluctuations, particularly after heat treatment.

7.4 Thermal effects on SRO

7.4.1 SRO in thermal equilibrium

As has been mentioned earlier, the equilibrium CSRO depends upon temperature owing to the entropy effect, in very much the same way as does the CSRO in crystalline solids. Therefore the Bragg–Williams approximation should give a good first order description of the phenomenon[31–33]. The order–disorder temperature, however, will be entirely smeared out because of the fluctuation in the interaction parameters. The equilibrium GSRO is more difficult to calculate, but a recent theory[45] predicts that

$$\langle p^2 \rangle = (B/K^{\alpha}\Omega)kT \qquad (T > T_g) \tag{7.15}$$

$$\langle \tau^2 \rangle = (2G/K^{\gamma}\Omega)kT \qquad (T < T_{g'}) \tag{7.16}$$

where B is the bulk modulus, G is the shear modulus, and K^{α} and K^{γ} are constants of the order of unity. They are estimated in a continuum approximation to be

$$K^{\alpha} = 9/4$$
$$K^{\gamma} = 45/23 \tag{7.17}$$

The two equations 7.15 and 7.16 extrapolate to zero at 0 K, while we know that $\langle p^2 \rangle$ and $\langle \tau^2 \rangle$ are finite at 0 K from the computer simulation results shown in *Figures 7.4* and *7.5*. Therefore $\langle p^2 \rangle$ and $\langle \tau^2 \rangle$ have to freeze at certain temperatures at which expressions 7.15 and 7.16 coincide with the 0 K values as shown in *Figure 7.7*. These temperatures define the *glass temperatures*. For the model amorphous iron, they are

$$T_g = 760\,\mathrm{K} \qquad T_{g'} = 2120\,\mathrm{K} \tag{7.18}$$

Although the accuracy of this estimate is not high owing to the various approximations involved, the calculated value of T_g, at which the pressure fluctuation freezes, is in good agreement with the experimental values for iron-based metallic glasses. The *upper glass temperature* $T_{g'}$, at which the shear fluctuation freezes, is very close to the melting

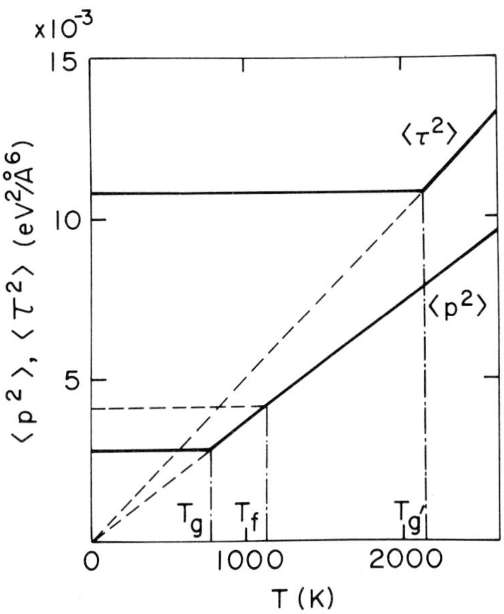

Figure 7.7 Temperature dependence of $\langle p^2 \rangle$ and $\langle \tau^2 \rangle$ calculated for model amorphous iron. T_g, glass transition temperature; T_f, fictive temperature; $T_{g'}$, upper glass transition temperature (after Egami and Srolovitz[45])

temperature T_m. According to this theory, a liquid at temperatures above T_g has two regimes, atomic fluid and molecular liquid. The normal atomic fluid, in which each atom moves in the Brownian fashion, exists only above $T_{g'}$, while below $T_{g'}$ the atomic motions become more collective, so that the system appears to be made of many molecules with finite life-times, and the viscosity increases rapidly with decreasing temperature. Although there is no experimental confirmation of the existence of a well defined upper glass transition at $T_{g'}$, the overall behaviour that the theory predicts is in accord with experimental observations.

7.4.2 Irreversible structural relaxation

Since metallic glasses are obtained by rapid quenching from the melt, the structure of the as-quenched glass is not in thermal equilibrium, and a subsequent annealing produces *structural relaxation*. The effects of annealing are wide range and complex (*see* Chapters 10 and 11), but some of them are irreversible below T_g, while others are reversible. The irreversible structural relaxation is best represented by the viscosity which increases linearly with time[53] (*see* Chapter 13) and the changes in the PDF observed by X-ray diffraction[3]. It was found recently[52] that the changes in the PDF can be quite satisfactorily explained in terms of the change in $\langle p^2 \rangle$, by equation 7.14, as shown in *Figure 7.8*. On the other hand, the changes in the PDF produced by the reduction in $\langle \tau^2 \rangle$ did not agree with the observation; therefore it is concluded that *the irreversible structural relaxation results largely from the reduction in* $\langle p^2 \rangle$ *alone*[52]. This conclusion is not surprising if we consider the temperature dependence of $\langle p^2 \rangle$ and $\langle \tau^2 \rangle$ shown in *Figure 6.7*. Since $\langle \tau^2 \rangle$ freezes at a high temperature where the atomic relaxation time is of the order of 10^{-11} to 10^{-12} sec, $\langle \tau^2 \rangle$ can relax to its minimum value during the cooling at a normal rapid quenching rate of around 10^5 to 10^6 K/s. By the time $\langle p^2 \rangle$ freezes, however, the viscosity and the relaxation time are much more increased, so that $\langle p^2 \rangle$ can be frozen at a value larger than the equilibrium value. This value can be characterized by the temperature at which the frozen value of $\langle p^2 \rangle$ is in equilibrium, the so called *fictive temperature*, T_f. With subsequent annealing, $\langle p^2 \rangle$

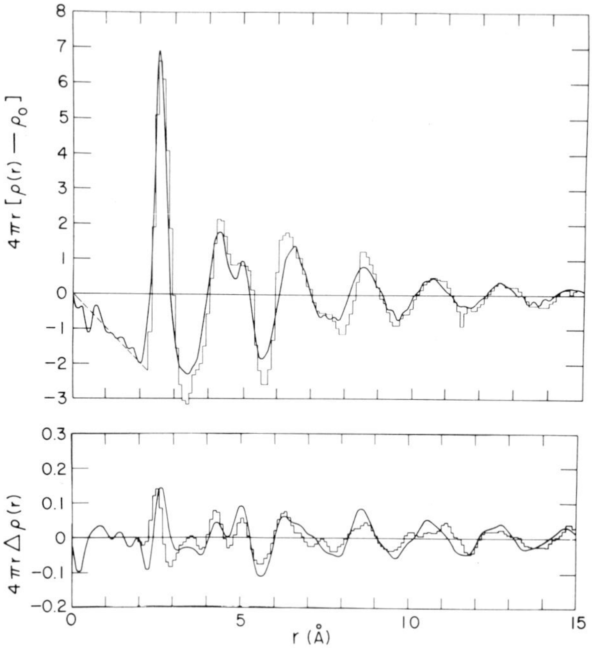

Figure 7.8 Top: pair distribution function (PDF) of $Fe_{40}Ni_{40}P_{14}B_6$ (solid line) and PDF of model amorphous iron (histogram). Bottom: change in the PDF of $Fe_{40}Ni_{40}P_{14}B_6$ produced by annealing at 350 °C for 30 min. (solid line, Egami[3]) and the change in PDF predicted by the decrease in $\langle p^2\rangle$ (histogram, Srolowitz *et al.*[52])

relaxes toward its low temperature value, while $\langle \tau^2\rangle$ does not change any further. The change in $\langle p^2\rangle$ should affect viscosity and diffusivity, but no detailed theory to relate $\langle p^2\rangle$ to diffusivity has yet been developed. If we use the free volume theory, the diffusivity D should scale with

$$D \sim \exp(-B_p/(\langle p^2\rangle - \langle p^2\rangle_{eq.})) \tag{7.19}$$

where $\langle p^2\rangle_{eq.}$ is the equilibrium value of $\langle p^2\rangle$, since the change in $\langle p^2\rangle$ is proportional to the change in volume[42]. Thus D is decreased as $\langle p^2\rangle$ is decreased by relaxation.

7.4.3 Anelasticity

It has been known for some time that a part of the creep deformation recovers, indicating the presence of *anelasticy* in metallic glasses[54–59]. The total creep therefore is given by

$$\varepsilon_t = \varepsilon_{visc} + \varepsilon_{anel} \tag{7.20}$$

where ε_{visc} is the creep due to viscous flow. Unlike in crystalline solids, ε_{visc} is not a steady state flow, but since the viscosity increases linearly with time it is given by

$$\varepsilon_{visc} = \sigma b \ln(t/t_0 + 1) \tag{7.21}$$

where b and t_0 are constants. The anelastic part is given by

$$\varepsilon_{anel} = \sigma b' \int (1 - e^{-t/\tau_0}) N(\tau_0)\, d\tau_0 \tag{7.22}$$

where $N(\tau_0)$ is the density of the relaxation time τ_0, and is given[58] by a wide distribution of the activation energies around 2 eV. The most conspicuous point about the anelasticity in metallic glasses is that it does not depend strongly upon thermal history, and persists even after long annealing which reduces the diffusivity to very small values. Therefore the microscopic mechanism of anelasticity must be distinct from that of structural relaxation. There are two types of possible mechanism which would produce the observed anelasticity. One is the local shear transformation which changes only the local shear stress τ, but breaks no atomic bonds, therefore leaving the topology unchanged. In other words it is a pure DSRO transformation. The other is the local DSRO transformation which involves the local change in TSRO, but leaves the overall stress distribution unchanged. The activation energy of the former process would be smaller than that of the latter, so that it may be important at low temperatures. The anelasticity observed at high temperatures probably involves the latter process. The details, however, can only be clarified by a suitable computer simulation study. The persistence of anelasticity is one of the most important properties of the amorphous alloys and is most probably the direct consequence of the inherent mismatch of the atomic structure.

7.4.4 Kinetics of change in CSRO

All the atomic rearrangements discussed above, namely irreversible structural relaxation, diffusion and anelasticity, can change the local CSRO of the alloys. It is readily recognized that the CSRO defined by the *Warren–Cowley parameter* α_p can be changed only when the local pressure p is changed. On the other hand, the CSRO defined by the more generalized parameters of expression 7.4 can be altered by changes in either pressure p or shear stress τ. When the local topology is changed, the value of α_p is changed only at the atoms involved directly in the topological change, namely forming or breaking of the bond. However, $I_n^{l,m}(A)$ is changed at most of the neighbouring atoms. Since the mechanism by which the CSRO is changed is not unique, the kinetics of the change also should be mixed. Indeed a part of the change in T_C appears to be irreversible, and it may be a consequence of irreversible structural relaxation. A large part of the change in T_C, however, is reversible. Furthermore, the *field induced magnetic anisotropy* has no irreversible part, and can be completely wiped out by annealing at above T_C. These reversible changes are most likely to be directly related to *anelasticity*. Although a careful experimental comparison by simultaneous measurements of anelasticity and T_C or anisotropy has not been made, the kinetics of these three properties appear to be very similar, suggesting that they result from the same type of atomic rearrangements. The anelastic shear transformation can be coupled to T_C via $I_1^{0,0}(A)$, if the exchange constant is strongly dependent upon the atomic separation within the first peak of PDF. This mechanism explains well the changes in T_C in Fe–Ni and Fe–Cr alloys shown in *Figure 7.1*. The field induced magnetic anisotropy is related to $I_n^{2,m}(A)$ $(n = 0, 1)$ which can also be changed by the anelastic shear transformation. It is noteworthy that $I_0^{2,m}(A)$ is changed only when the local TSRO is altered, while the local pure DSRO change which leaves TSRO unchanged affects only $I_1^{2,m}(A)$. The magnetic after-effect observed at low temperature [60,61] suggests that there are anisotropy reorientation mechanisms with small activation energies. It is likely that they are the pure DSRO reorientation processes.

There is evidence that the kinetics of the T_C change and the field induced anisotropy are dependent upon thermal history[25,29]. The dependence is, however, rather weak, and is definitely distinct from irreversible structural relaxation. This behaviour may be explained by invoking a two-level system model similar to the one used to explain the low temperature anomalies[62–64], in order to describe the anelasticity.

7.5 Concluding remarks

Unlike the crystalline structure, the atomic structure of amorphous alloys is highly degenerate and non-unique. As a consequence it has many internal degrees of freedom which can be easily changed by heat treatment or mechanical deformation. This variability of the structure and properties is an important problem both for those who want to use the amorphous alloys for application and for those who study them scientifically. These degrees of freedom are described in terms of the short range order, although its definition is sometimes vague. In this chapter, we presented an unambiguous definition of SRO's, both for CSRO and GSRO, and considered their effects on a few of the physical properties. Other physical properties clearly influenced by SRO's include electrical resistivity[65,66], specific heat[67,68], elastic constants[69,70] and superconductive transition temperatures[71,72]. All of them are now known to change both irreversibly and reversibly, reflecting various mechanisms of the changes in SRO's touched upon in this chapter. Although some parts of the description are still conjecture, the concepts introduced here explain the observed phenomena rather well. Further research would make the picture more detailed and quantitative.

Acknowledgement

A significant part of this chapter is based upon the results of collaborations involving the author and Drs V. Vitek, D. Srolovitz and C. D. Graham, Jr. Their contributions are gratefully acknowledged. The author is also thankful to Drs A. C. Anderson and S. J. Poon for making their manuscripts available before publication. Part of this work was supported by NSF–MRL Grant DMR79–23647.

Note added in proof. Recent measurements[73,74] have shown that the kinetics of change in the Curie temperature, the reorientation of the field-induced magnetic anisotropy, and the change in the anelasticity measured as internal friction, are all linearly correlated, suggesting strongly that they indeed share the identical microscopic mechanism, which is presumably the local shear transformation discussed in this chapter.

References

1. Berry, B. S. and Pritchet, W. C., *Phys. Rev. Lett.*, **34**, 1022 (1975)
2. Egami, T., *Mater. Res. Bull.*, **13**, 557 (1978)
3. Egami, T., *J. mater. Sci.*, **13**, 258 (1978)
4. Chaudhari, P. and Turnbull, D., *Science*, N.Y., **199**, 11 (1978)
5. Messmer, R. P. and Wong, J., *J. Non-cryst. Solids*, **45**, 1 (1981)
6. Warren, B. E., Averbach, B. L. and Roberts, B. W., *J. appl. Phys.*, **22**, 1493 (1951)
7. Cowley, J. M., *J. appl. Phys.*, **21**, 24 (1950)
8. Cargill, C. S. and Spaepen, F., *J. Non-cryst. Solids*, **43**, 91 (1981)
9. Bhatia, A. B. and Thornton, D. E., *Phys. Rev.*, **B2**, 3004 (1970)
10. Chieux, P. and Ruppersberg, H., *J. Phys. (Paris)*, **41**, C8, 145 (1980)
11. Mizoguchi, T., Kudo, T., Irisawa, T., Watanabe, N., Niimura, N., Misawa, M. and Suzuki, K., *'Proc. 3rd Int. Conf. on Rapidly Quenched Metals'*, Ed. Cantor, B., vol. 2, p. 384. The Metals Society, London (1978)
12. Waseda, Y. *'The Structure of Non-Crystalline Materials'*, McGraw Hill, New York (1980)
13. Cargill, G. S. and Tsuei, C. C., in *'Proc. 3rd Int. Conf. on Rapidly Quenched Metals'*, Ed. Cantor, B., vol. 2, p. 337, The Metals Society, London, 1978
14. Williams, A., *J. Non-Cryst. Solids*, **45**, 183 (1981)
15. Sadoc, J. F. and Dixmier, J., *Mater. Sci. Engng*, **23**, 187 (1976)
16. Sakata, M., Cawlam, N. and Davies, H. A., *J. Phys. F*, **9**, L235 (1979)
17. Ruppersberg, H., Lee, D. and Wagner, C. N. J., *J. Phys. F*, **10**, 1645 (1980)

18. Wong, J., in '*Glassy Metals I*' Ed. Güntherodt, H. J., p. 45, Springer-Verlag, Berlin (1981)
19. Chen, H. S., Teo, B. K. and Wang, R., *J. Phys. (Paris)*, **41,** C8, 254 (1980)
20. Raoux, D., Sadoc, J., Lagarde, P., Sadoc, A. and Fontaine, A., *J. Phys. (Paris)*, **41,** C8 207 (1980)
21. Chen, H. S. Sherwood, R. C., Leamy, H. J. and Gyorgy, E. M., *IEEE Trans. Magn.*, **MAG-12,** 433 (1976)
22. Chen, H. S., Sherwood, R. C. and Gyorgy, E. M., *IEEE Trans. Magn.*, **MAG-13,** 1538 (1977)
23. Liebermann, H. H., Graham, C. D. (jun.), and Flanders, P. J., *IEEE Trans. Magn.* **MAG-13,** 1541 (1977)
24. Kudo, T., Egami, T. and Liebermann, H. H., '*Proc. 4th Int. Conf. on Rapidly Quenched Metals*', Eds. Masumoto, T. and Suzuki, K., vol. 2, p. 1187, Japan Institute of Metals, Sendai (1982)
25. Chen, Y.-N. and Egami, T., unpublished work
26. Sakata, M., Cawlam, N. and Davies, H. A., *J. Phys. F*, **11,** L157 (1981)
27. Egami, T., *IEEE Trans. Magn.*, **MAG-17,** 1600 (1981)
28. Luborsky, F. E. and Walter, J. L., *Mater. Sci. Engng*, **28,** 77 (1977)
29. Chambron, W. and Chamberod, A., *Solid St. Communs*, **33,** 157 (1980); **35,** 61 (1980)
30. Ho, K.-Y., Flanders, P. J. and Graham, C. D. (jun.), *J. appl. Phys.*, **53,** 2279 (1982)
31. Chikazumi, S. and Graham, C. D. (jun.), in '*Magnetism and Metallurgy*', Eds. Berkowitz, A. and Kneller, E., p. 577, Academic Press, New York (1969)
32. Neel, M. L., *J. Phys. Radium*, **15,** 225 (1954)
33. Taniguchi, S., *Sci. Rep. Res. Insts Tôhoku Univ.*, **A7,** 269 (1955)
34. Luborsky, F. E. and Walter, J. L., *IEEE Trans. Magn.* **MAG-13,** 953 (1977)
35. Fujimori, H., Morita, H., Obi, Y. and Ohta, S., in '*Amorphous Magnetism II*', Eds. Levy, R. A. and Hasegawa, R., p. 393, Plenum Press, New York (1977)
36. Masumoto, T. and Egami, T., *Mater. Sci. Engng*, **48,** 147 (1981)
37. Ohta, K., Matsuyama, T., Kajiura, M. and Teranishi, T., *J. Magnetism magn. Mater.*, **19,** 765 (1980)
38. Jagielinski, T., *J. appl. Phys.*, **53,** (1982)
39. Luborsky, F. E., McCary, R. O. and Becker, J. J., in '*Rapidly Quenched Metals*', Eds. Grant, N. J. and Giessen, B. C., p. 476, Massachusetts Institute of Technology Press, Cambridge, Mass. (1976)
40. Fujimori, H., Yoshimoto, H., Masumoto, T. and Mitera, T., *J. appl. Phys.*, **52,** 1893 (1981)
41. Egami, T., Maeda, K. and Vitek, V., *Phil. Mag.*, **A41,** 883 (1980)
42. Egami, T., Maeda, K., Srolovitz, D. and Vitek, V., *J. Phys. (Paris)*, **41,** C8, 274 (1980)
43. Srolovitz, D., Maeda, K., Vitek, V. and Egami, T., *Phil. Mag.*, **A44,** 847 (1981)
44. Srolovitz, D., Maeda, K., Takeuchi, S., Egami, T. and Vitek, V., *J. Phys. F*, **11,** 2209 (1981)
45. Egami, T. and Srolovitz, D., *J. Phys. F.*, **12,** 2141 (1982)
46. Spaepen, F., *J. Non-cryst. Solids*, **31,** 207 (1978)
47. Halperin, B. I. and Nelson, D. R., *Phys. Rev. Lett.*, **41,** 121 (1978)
48. Kleman, M. and Sadoc, J. F., *J. Phys. (Paris) Lett.*, **40,** L569 (1979)
49. Sadoc, J. F., *J. Non-cryst. Solids*, **44,** 1 (1981)
50. Eshelby, J. D., *Proc. R. Soc.*, **A241,** 376 (1957)
51. Srolovitz, D., Thesis, University of Pennsylvania (1981)
52. Srolovitz, D., Egami, T. and Vitek, V., *Phys. Rev.*, **B24,** 6936 (1981)
53. Taub, A. I. and Spaepen, F., *Acta metall.*, **28,** 1781 (1980)
54. Berry, B. S., in '*Metallic Glasses*', Eds. Gilman, J. J. and Leamy, H. J., p. 161, American Society of Metals, Metals Park, Ohio (1978)
55. Maddin, R. and Masumoto, T., *Mater. Sci. Engng*, **4,** 153 (1972)
56. Logan, J. and Ashby, M. F., *Acta metall.*, **22,** 1047 (1974)
57. Murata, T., Kimura, H. and Masumoto, T., *Scripta Met.*, **10,** 705 (1976)
58. Argon, A. S. and Kuo, H. Y., *J. Non-cryst. Solids*, **37,** 241 (1980)
59. Taub, A. I. and Spaepen, F., *J. mater. Sci.*, **16,** 3087 (1981)
60. Moser, N. and Kronmuller, H., *J. Magnetism magn. Mater.*, **19,** 275 (1980)
61. Allia, P. and Vinai, F., *J. Phys. (Fr.)*, **41,** C8 654 (1980)
62. Anderson, P. W., Halperin, B. I. and Varma, C. M., *Phil. Mag.*, **25,** 1 (1972)
63. Phillips, W. A., *J. Low Temp. Phys.*, **7,** 351 (1972)
64. Black, J. L., in ref 18, p. 167
65. Lin, C.-H., Bevk, J. and Turnbull, D., *Solid St. Communs*, **29,** 641 (1979)
66. Balanzat, M., Mairy, C. and Hillairet, J., *J. Phys. (Paris)*, **41,** C8 854 (1980)
67. Onn, D. G., *J. appl. Phys.*, **52,** 1788 (1981)
68. Chen, H. S., *J. appl. Phys.*, **52,** 1868 (1981)
69. Chen, H. S., *J. appl. Phys.*, **49,** 3289 (1978)
70. Kursumovic, A., Scott, M. G., Girt, E. and Cahn, R. W., *Scripta Meta.*, **14,** 1303 (1980)
71. Anderson, A. C., Koch, C. C. and Scarbrough, J. O., *Phys. Rev.*, **B26,** 1156 (1982)
72. Poon, S. J., Chap. 22 (this volume)
73. Flanders, P. J., Morito, N. and Egami, T., to be presented at the Intermag Conf. (1983)
74. Morito, N. and Egami, T., to be presented at the Intermag Conf. (1983)

Chapter 8

Local electronic structure theory of amorphous metals

R. P. Messmer

General Electric Company, Schenectady, New York, USA

8.1 Introduction

The physical and chemical properties of a material are determined, in the final analysis, by the electronic structure of the material. As a consequence, much useful insight into the properties of a material can be gained by determining the electronic structure theoretically through quantum mechanical calculations and by performing experiments which probe the electronic structure.

On the theoretical side, considerable information has been obtained for crystalline metals and metallic alloys via quantum mechanical studies which employ energy band calculations[1,2]. However, in order to perform such band strcture calculations one must have a regular periodic structure and a unit cell which is not too large (e.g., 12 atoms or less). The latter is not a theoretical restriction, only a practical one, as regards computational tractability on commonly available computers. When no long range order exists, as in amorphous metals, the band theory approach cannot be used and a serious problem presents itself to those who wish to obtain theoretical information regarding electronic structure.

At present, there is no general theoretical approach to obtaining the electronic structure of a disordered or non-periodic solid. However, if one is willing to replace the quest for a general theory by some pragmatism, an avenue remains open to gaining some information. In particular if one focuses on a local atomic environment in the solid, the use of clusters can provide a practical method for obtaining certain aspects of the electronic structure. This approach has been used with considerable success to gain knowledge of electronic effects associated with chemisorption[3–5], defects in semiconductors[6–8], impurities in metals[9] and other phenomena[10,11].

An essential prerequisite to employing this approach, is the definition of a relevant local region of the solid so that one may choose a finite cluster of atoms to model the local environment. The second step is to choose a method of calculation to solve for the electronic structure of the cluster. Usually, one uses molecular orbital theory, which is the finite analogue of band theory. Two examples of the choice of clusters are:

1. The site of an impurity or defect and one or more shells of neighbours to

investigate the electronic effects of defects and impurities in metals[9] or semiconductors[6–8],

2. An atom or molecule together with a group of metal atoms having the same geometry as the surface chemisorption site[3–5] to investigate the bonding of an adsorbate to a substrate.

When one thinks of this approach as applied to amorphous metal–metalloid alloys, chemical intuition would suggest that the focus be on the metalloid and its immediate vicinity. In fact, there is now substantial evidence for short-range chemical order in such glasses due to metal–metalloid interactions[12–14]. Thus it seems quite reasonable to use a cluster approach to study this aspect of electronic structure in metal–metalloid glasses. However, one must question how well a cluster of atoms can simulate the electronic structure to be found in a bulk metal. This question has been studied extensively and in the next section we compare bulk and cluster models of crystalline metals so as to have a proper calibration on what one can realistically hope to achieve in studying glassy alloys by this approach.

8.2 Comparison of bulk and cluster models of electronic structure

Here we will consider the results of cluster calculations for metals and compare them with the results of band theory. In particular we will compare the calculated densities of states and the magnetic behaviour. In doing the cluster calculations, the form of molecular orbital theory which is chosen is the so-called self-consistent-field Xα-scattered wave (SCF-Xα-SW) method[15–17]. It is the complete analogue for clusters of the Korringa–Kohn–Rostoker (KKR) method of band theory[18–20] for bulk metals.

One of the most readily accessible theoretical characteristics of a metal cluster is its density of states (DOS). The DOS is a function which tells one about the number of energy levels per unit energy as a function of energy. It is usually calculated for bulk metals from band theory computations and has been important in comparing theory with photoemission experiments on metals (*see* Chapter 9). The ability to obtain such information from cluster calculations allows a comparison to be made between the DOS of clusters and that of the bulk metal.

The first calculations which made such comparisons were for clusters of nickel, palladium, platinum and copper[21]. However, we will discuss more recent results on iron clusters[22] as an illustrative example.

The calculated energy level diagram for a Fe_{15} cluster which contains a central atom together with the eight nearest neighbours and six next nearest neighbours fixed at the bulk interatomic separation, is shown in *Figure 8.1*. The Fermi level, ε_F, separates the unoccupied spin orbitals from the occupied ones and is indicated by a dashed line in *Figure 8.1*. The calculations were carried out in a spin-polarized fashion, allowing different molecular orbitals for different spins. The spin-polarized energy levels in *Figure 8.1* are displayed in two different ways. In the central panel all of the energy levels are displayed, in the left and right panels the up-spin and down-spin levels are shown respectively, broken down into their symmetry types. The symmetry types are given at the top of the figure and are the labels of the irreducible representations under which the orbitals transform according to the O_h point group of the cluster. When each of the discrete energy levels of *Figure 8.1* is broadened by a Gaussian of width parameter 0.2 eV, the DOS shown in *Figure 8.2* is generated. Note that the energy scale

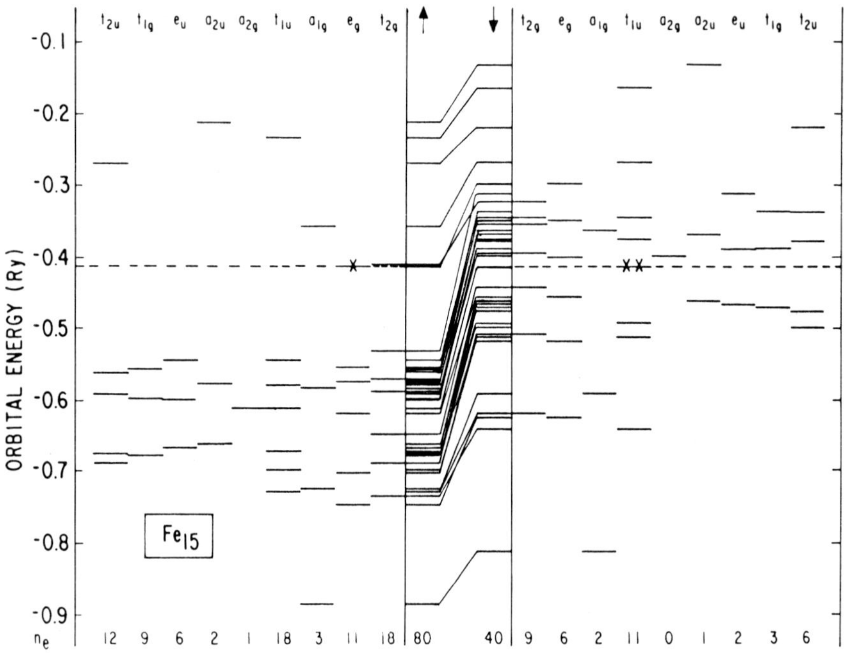

Figure 8.1 Orbital eigenvalues of a spin-polarized calculation for Fe_{15}. The dashed line denotes the position of the Fermi level. At the bottom of the figure, the number of electrons of each spin and symmetry type are shown as well as the total number of majority- and minority-spin electrons

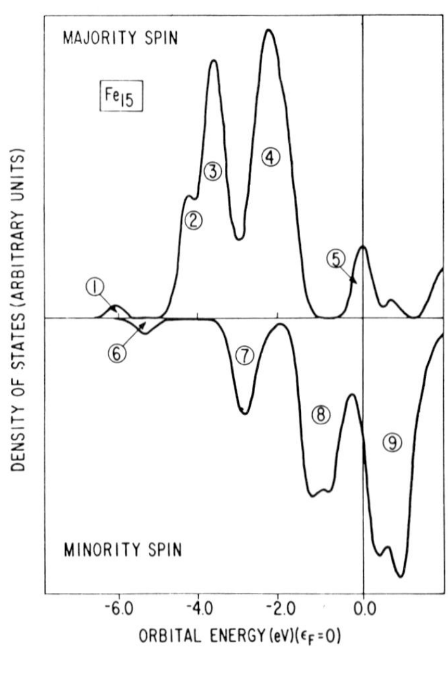

Figure 8.2 The majority- and minority-spin DOS for the Fe_{15} cluster. The curves were generated by replacing each discrete energy level of *Figure 8.1* with a Gaussian of width parameter 0.2 eV

of *Figure 8.1* is in rydbergs (1 Ryd = 13.60583 eV), whereas the scale in *Figure 8.2* is in electron volts.

In *Figure 8.3*, the DOS from a recent band structure calculation for ferromagnetic iron is shown[1]. For the majority-spin (up-spin) DOS of Fe_{15}, five peaks, labelled 1—5 are obtained below ε_F, which may be associated with the features labelled 1—5 in the bulk DOS results. The cluster and the band DOS show a reasonable degree of correspondence and, for both, the most prominent features are the three peaks which increase in magnitude as one goes from peaks 1 to 3. Relative to bulk iron, peaks 2 and 3 for Fe_{15} are at about the right energy but somewhat too close to each other and peak 4 is about one eV too low. The large separation between peaks 4 and 5 for the cluster is the most noticeable difference between the cluster and the band DOS. For the minority-spin (down-spin), the peaks labelled 6—9 are the most prominent features of both results. Again the relative magnitude and disposition of the peaks is quite similar in the two calculations. Both results indicate that the majority- and minority-spin DOS are not simply shifted mirror images of one another, i.e. there are significant departures from the rigid-band model. A detailed discussion of other similarities and differences between the cluster and bulk electronic structures has been presented elsewhere[22]. The most significant point, however, as regards the present discussion is the overall similarity between the two DOS results which is achieved with only a 15-atom model of iron.

We now turn to a discussion of the magnetic properties obtained from cluster calculations. In the first Xα-SW work on transition metal clusters[21], it was found that Ni_8 and Ni_{13} clusters were magnetic while Pd_{13} and Pt_{13} were not. For the Ni_8 cluster the calculated average spin magneton number per atom was 0.25, whereas for the Ni_{13} cluster the value was 0.46. These may be compared with the bulk value of 0.54. As

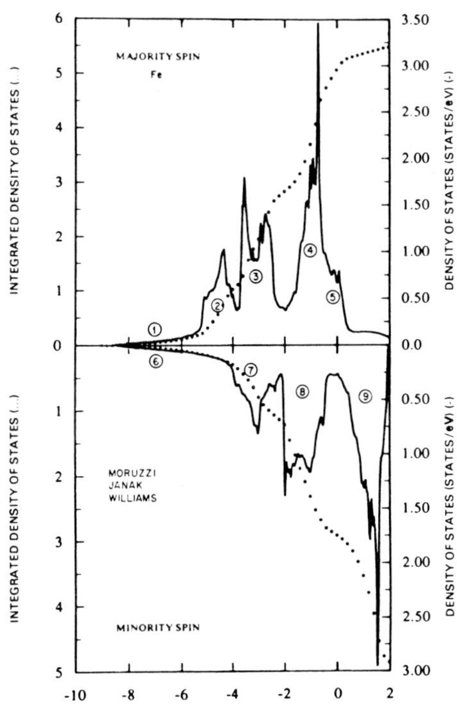

Figure 8.3 The majority- and minority-spin DOS for bcc iron from the band structure calculations of Moruzzi, Janak and Williams[1]

would be expected, the spin density on the central atom is quite different from that found on the 12 surface atoms.

Although it is rather remarkable that the 13-atom clusters of nickel, palladium and platinum which all have the same number of valence electrons turn out to be magnetic for nickel and non-magnetic for palladium and platinum, and thus mimic the behaviour found in the bulk metals, one cannot expect that this will be a general phenomenon. In fact, recent calculations[23] on a 15-atom chromium cluster have shown that such clusters carry a net magnetic moment, while it is well known that the bulk metal is antiferromagnetic. Nonetheless, the Cr_{15} cluster results try to mimic the antiferromagnetic behaviour to a large degree, but owing to surface effects there is not an exact cancellation of moments of opposite spin orientation. It is interesting to consider the Cr_{15} results in a broader context, however; namely, to study the magnetic properties of clusters of three different elements, which have the same crystal structure in the bulk (body centred cubic, bcc) and exhibit three different types of magnetic ordering. The metals are vanadium (non-magnetic), chromium (antiferromagnetic) and iron (ferromagnetic).

Figure 8.4 shows the orbital energies from spin-polarized Xα-SW calculations for V_{15}, Cr_{15} and Fe_{15}. The cluster geometry is shown at the bottom of the figure and consists of 15 atoms representing the bcc structure with interatomic distances

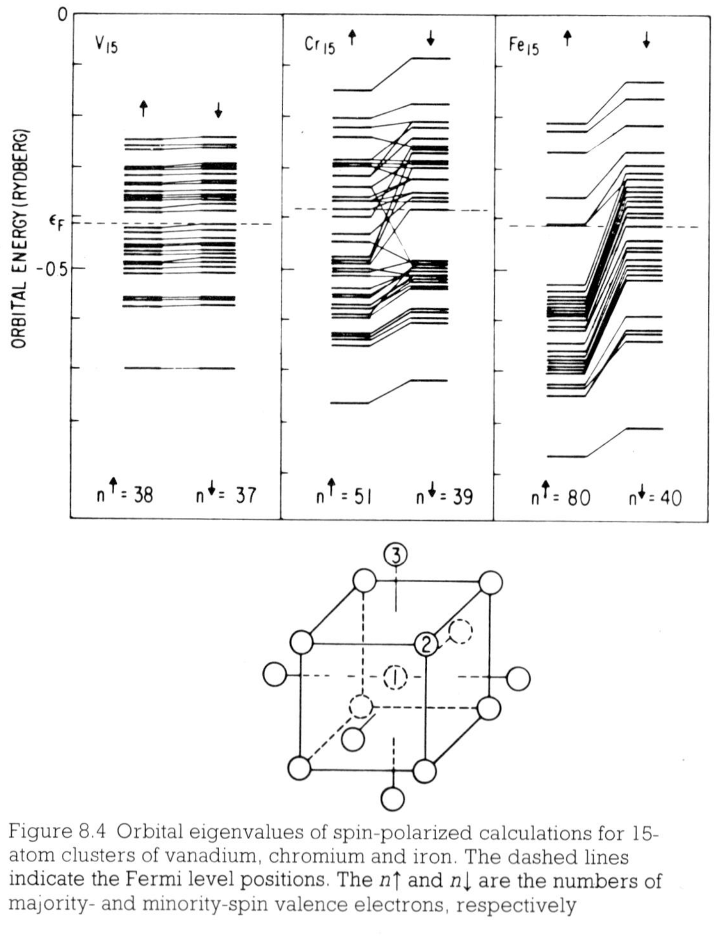

Figure 8.4 Orbital eigenvalues of spin-polarized calculations for 15-atom clusters of vanadium, chromium and iron. The dashed lines indicate the Fermi level positions. The $n\uparrow$ and $n\downarrow$ are the numbers of majority- and minority-spin valence electrons, respectively

appropriate to the observed bulk values. The lines connecting up-spin and down-spin energy levels of *Figure 8.4* merely relate the same symmetry types in sequential order.

The calculations were carried out fully self-consistently, allowing a flow of electrons between the two spin manifolds such that at convergence Fermi statistics are satisfied. In the case of V_{15}, the added degree of freedom provided by spin-polarization has a negligible effect. The converged result has a single extra majority spin electron and the exchange splitting is very small. Thus V_{15} mimics, as far as possible for a system with an odd number of electrons, the non-magnetic situation of bulk vanadium.

For Cr_{15}, an approximate calculation of the magnetic moment[23] for the various atoms, yields $-0.7\,\mu_B$ for the central atom, $+4.1\,\mu_B$ for the first neighbours and $-3.4\,\mu_B$ for the second neighbours, which may be compared with $+0.7\,\mu_B$ for bulk antiferromagnetic chromium as determined from neutron diffraction data[24]. The near perfect agreement of the central atom value with experiment must be considered largely fortuitous. Thus while the alternation in sign characteristic of bulk chromium is well represented by the results, they also indicate that the 15 atom cluster in the bulk geometry should have a permanent magnetic moment and also that the moments for the surface atoms may be quite different from those of atoms in the interior. Adding further shells of atoms will undoubtedly lead to equivalent spin-up and spin-down potentials in the interior, but the overall magnetic ordering will remain, as it is largely determined by local short-range interactions.

Turning finally to the case of iron, one sees from *Figure 8.4* that Fe_{15} shows a very large spin polarization and has an excess of 40 majority spin electrons. If the contribution to the magnetization from orbital angular momentum (which is small for bulk iron[17]) is neglected, this yields an average magnetic moment per iron atom of $2.7\,\mu_B$ which compares with the experimental value of $2.2\,\mu_B$ for bulk iron. However, the three groups of symmetry-equivalent iron atoms in the Fe_{15} cluster are in different environments and one would not expect that the charge and spin densities for them would be the same. An approximate calculation[22] of magnetic moments yields $1.1\,\mu_B$ for the central atom, $2.7\,\mu_B$ for the first neighbours and $2.8\,\mu_B$ for the second neighbours; thus the cluster is ferromagnetic. One sees again that while the qualitative features of magnetism of bulk iron are reasonably well interpreted with the results of an Fe_{15} cluster, there are definite effects which arise from the finite size of the cluster (surface effects).

Summarizing cluster magnetism for 15-atom clusters of three bcc transition metals, one finds that the clusters appear to be sufficiently large to exhibit magnetic ordering which is qualitatively similar to that shown by the bulk metals and hence insight into certain aspects of bulk properties can be obtained. However, the clusters also exhibit characteristic differences as compared with the bulk and these intrinsic differences are important and of interest in their own right.

8.3 Cluster models of transition metal–metalloid systems

From the results discussed in the previous section, one can see that a cluster of atoms can give a surprisingly good representation of the qualitative features of the DOS and some of the magnetic characteristics of bulk metals. In this section we will consider the chemical bonding between metal atoms and metalloid atoms in clusters and discuss the implications of such results for gaining insight into the nature of metal–metalloid glasses. Owing to space limitations and simplicity of presentation, only four-atom metal clusters (of tetrahedral geometry) with a metalloid atom at the centre, will be discussed. Calculations on six- and nine-atom clusters[25] only reinforce the general qualitative conclusions to be drawn from the smaller clusters, although the larger clusters provide more detailed information and are better models.

In *Figure 8.5*, the results for a Fe_4B cluster are shown[25]. To the left of the figure the calculated orbital energies are displayed. The highest occupied up-spin level is the $2t_1$ with three electrons, the highest occupied down-spin orbital is the $1t_1$ with one electron. A comparison of results for Fe_4 and Fe_4B (the former results are not shown here) demonstrates that the addition of the boron atom to the cluster results in a decrease of the magnetic moment of the Fe_4 cluster, which is consistent with known experimental information. The replacement of the boron atom with a phosphorus atom decreases the magnetic moment even further, again consistent with experiment. At the upper right of *Figure 8.5* is a sketch of the geometry of the cluster, showing the plane in which the molecular orbital wave functions are plotted. The molecular orbital wave functions responsible for Fe–B bonding are shown in the lower right of *Figure 8.5* and are the $1t_2$ and $1a_1$ orbitals. The remainder of the occupied orbitals are essentially pure metal in character. The $1a_1$ orbital shows the interaction of B $2s$ with Fe $3d$ character in a bonding fashion. The $1t_2$ orbital (only one component is shown) displays the bonding interaction between B $2p$ and Fe $3d$ atomic orbitals.

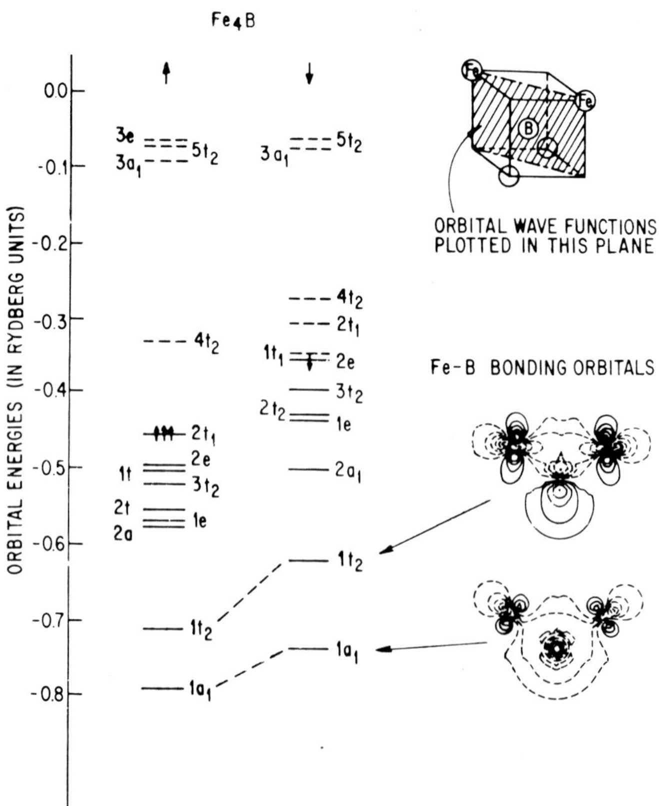

Figure 8.5 Orbital eigenvalues of a spin-polarized calculation for Fe_4B. At the upper right, the geometry of the cluster is shown. At the lower right, the orbitals responsible for Fe–B bonding are shown

Significant bonding interactions between metalloids (boron, phosphorus, carbon) and metals (iron, nickel, chromium, manganese) have been found [25,26] for *all* systems investigated. These include a variety of metals and metalloids, with different cluster sizes and different cluster geometries. These results clearly support arguments put forth on the basis of experimental information[12–14] that there are significant metal–metalloid bonding interactions in these systems.

Figure 8.6 shows the orbital energy levels which arise from a calculation of the Ni_2Fe_2B cluster[27]. The levels which are predominantly Ni or Fe in the content of their wave functions are so labelled. The effect of the boron atom is to spread out the levels as compared with those of Ni_2Fe_2. It should also be noted that the nickel levels are more spread out in energy than the iron levels as a consequence of the interaction with the boron atom. This suggests that the boron atoms may be interacting more strongly with the nickel atoms than with the iron atoms, an effect which will be discussed later. Two valence levels are not shown in *Figure 8.6*; they are the up- and down-spin $1a_1$ levels which occur at -11.0 and -10.6 eV, respectively. These levels correspond to the 2*s* level of boron with a small admixture of nickel and iron 4*s* and 3*d* character. In *Figure 8.7* contour plots of the up-spin $1a_1$ and $5a_1$ orbitals are shown, from which it can be seen that the $1a_1\uparrow$ level is predominantly B 2*s* although there is a small bonding interaction with the iron atoms [*see Figure 8.7*(*b*)], while the $5a_1\uparrow$ orbital shows a bonding interaction between the B $2p_z$ and the Ni *d* orbitals and a non-bonding or slightly antibonding interaction with the iron atoms. The other orbitals with significant B 2*p* character (noted in *Figure 8.6*) also have bonding interactions with the metal atoms.

The highest occupied levels for Ni_2Fe_2B are the $6b_2\uparrow$ and $7a_1\downarrow$ which have the same energy and an occupancy of 0.6 and 0.4, respectively. For this case n_B (the effective magneton number) is equal to 1.71. No SCF solution was found which had

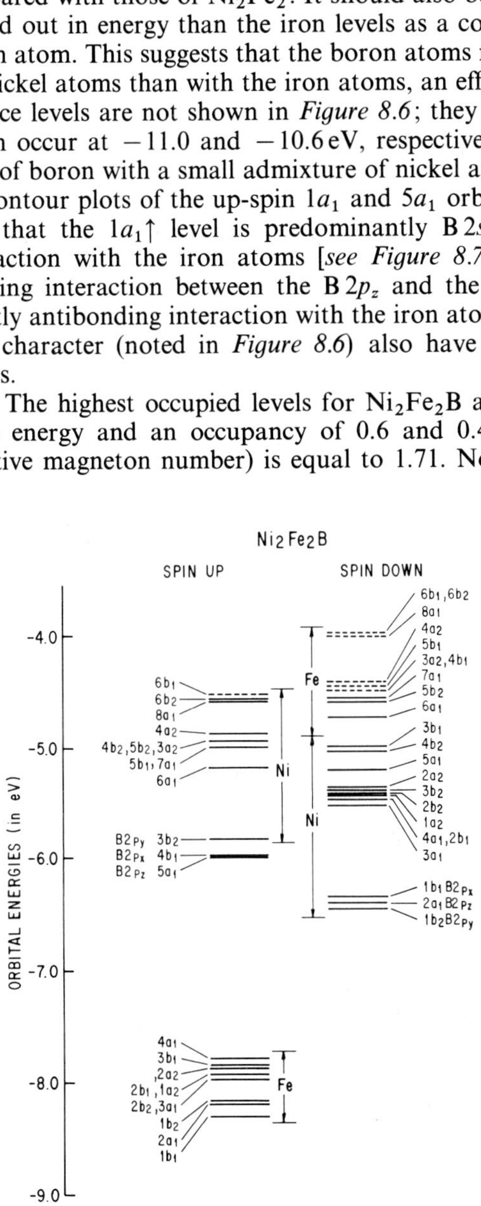

Figure 8.6 Orbital eigenvalues of a spin-polarized calculation for Ni_2Fe_2B. The levels which have orbitals containing significant boron character are labelled

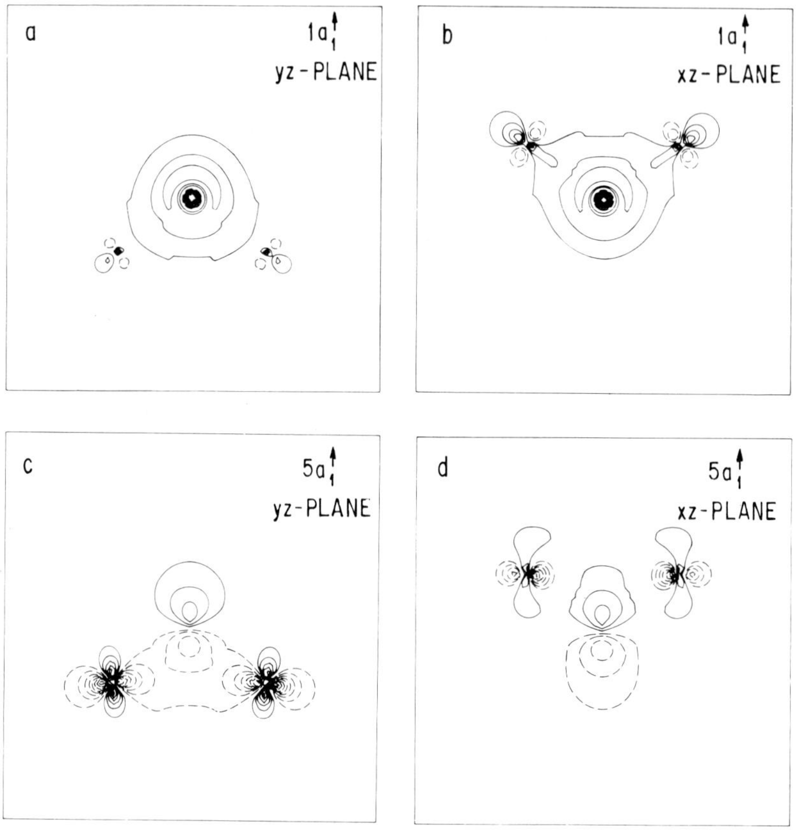

Figure 8.7 Orbital contour plots of the up-spin $1a_1$ and $5a_1$ orbitals of Ni_2Fe_2B. The *yz*-plane shows the interaction of boron with the nickel atoms. The *xz*-plane shows the interaction of boron with the iron atoms

integral occupation numbers and simultaneously satisfied Fermi statistics. The integral occupation number solution, which is very close to the level diagram of *Figure 8.6*, has the $6b_2\uparrow$ level empty and the $7a_1\downarrow$ containing one electron with the $6b_2\uparrow$ level at a slightly lower energy (0.15 eV) than the $7a_1\downarrow$ level (i.e., Fermi statistics are not satisfied). To determine reliably such small differences in energy levels is beyond the inherent accuracy of the calculations and thus one cannot really decide which is the more appropriate solution. Fortunately, as far as the chemical bonding aspects are concerned, the two solutions are virtually identical. The only significant difference is clearly in the calculation of the magnetic moment which depends very sensitively on the change in occupancy from 0.6 electron in an up-spin orbital and 0.4 electron in a down-spin orbital to one electron in a down-spin orbital. Thus the value of $n_B = 1.38$ is obtained for the integral occupation case. These values of n_B may be compared with the experimental value of 1.31 for the $Ni_{40}Fe_{40}B_{20}$ amorphous alloy[28]. The calculated value of n_B for a shortened metal–metalloid distance of 1.53 Å is 0.83.

An interesting question with regard to analysing these results is whether there is a preferential interaction between a given metalloid and the iron or nickel atoms in the cluster. If such preferential interactions do occur the view that the metalloid atoms simply fill in the Bernal holes of a DRPHS structure of metal atoms becomes rather

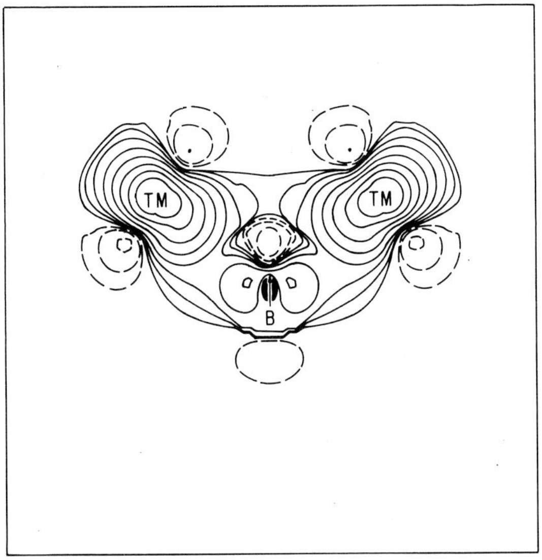

Figure 8.8 Plot of difference in bonding charge density for Ni_2Fe_2B. Positive contours about boron B and the transition-metal (TM) sites show a preferential Ni–B bonding interaction

tenuous. Rather, one would have to admit to the possibility of a *local atomic scale chemistry* arising from such electronic interactions.

In order to ascertain whether there are specific interactions between metal atoms and metalloid atoms, an analysis of bonding charge densities between the metalloid atom and the metal atoms should be particularly useful[27]. This bonding charge density can be plotted in the plane containing the metalloid and the nickel atoms on the one hand, and in the plane containing the metalloid and the iron atoms on the other hand. In order to see if there is a preferential interaction, a difference density can be obtained by aligning the metal atoms from the two planes such that the nuclei coincide. Thus the difference charge density is defined as the difference between the two bonding charge densities once the metal atoms have been properly aligned. If there is a preferential interaction between the boron atom and the nickel atoms, positive contour lines of the difference charge density (DCD) should surround the boron atom, whereas negative contour lines of the DCD should surround the boron atom if the preferential interaction is between boron and the iron atoms[27]. A contour plot of the DCD for the Ni_2Fe_2B cluster is shown in *Figure 8.8*. The contours around the boron atom are positive, indicating a preferential B–Ni interaction.

Recently, experimental evidence for chemical ordering between nickel and boron in $Ni_xFe_yB_z$ alloys has been accumulating[29–32]. The experimental data are from such diverse techniques as differential scanning calorimetry[29], extended X-ray absorption fine structure[30], and Mössbauer spectroscopy[31,32]. The theoretical results discussed above suggest that the nature of this chemical ordering is a preferential bonding interaction between boron and nickel atoms. Other results[27] suggest that a preferential iron-phosphorus interaction also exists in Fe–Ni–B–P alloys.

Although the theoretical results discussed here are for rather small clusters, they do demonstrate the utility of cluster investigations and the type of information they can provide. Furthermore, the implications of such calculations transcend the problem of amorphous alloys. They also are of potential value in discussing impurities (metalloids) which segregate to grain boundaries[10,11].

In summary, we have found evidence for the following general features regarding metal–metalloid glasses:

1. There is definite bonding between metal and metalloid atoms for all systems investigated.
2. There are preferential bonding interactions between metalloid atoms and certain metal atoms of the alloy.
3. It appears that the strengths of the preferential interactions are related to the differences in electronegativities between the metalloid and metal atoms, although further work is needed to make this quantiative.

Acknowledgments

This work was partially supported by the U.S. Department of Energy under Contract No. DE-AC02-79ER-10382. The author is grateful to Drs J. Wong, J. L. Walter and F. E. Luborsky for discussions.

References

1. Moruzzi, V. L., Janak, J. F. and Williams, A. R., '*Calculated Electronic Properties of Metals*', Pergamon, New York (1978)
2. Williams, A. R., Gelatt, C. D. and Janak, J. F., in '*Theory of Alloy Phase Formation*', Ed. Bennett, L. H., p. 40, The American Institution of Mining, Metallurgical and Petroleum Engineers, Warrendale, PA, (1980)
3. Bennett, A. J., McCarroll, B. and Messmer, R. P., *Phys. Rev.*, **B3,** 1397 (1971)
4. Messmer, R. P. and Salahub, D. R., *Phys. Rev.,* **B16,** 3415 (1977)
5. Messmer, R. P., Lamson, S. H. and Salahub, D. R., *Solid St. Communs.*, **36,** 265 (1980)
6. Messmer, R. P. and Watkins, G. D., *Phys. Rev. Lett.,* **5,** 503 (1970)
7. Messmer, R. P. and Watkins, G. D., *Phys. Rev.,* **B7,** 2568 (1973)
8. Watkins, R. P. and Messmer, R. P., *Phys. Rev. Lett.,* **32,** 1244 (1974)
9. Johnson, K. H., Vvedensky, D. D. and Messmer, R. P., *Phys. Rev.*, **B19,** 1519 (1979)
10. Briant, C. L. and Messmer, R. P., *Phil. Mag.*, **B42,** 569 (1980)
11. Messmer, R. P. and Briant, C. L., *Acta metall.*, **30,** 457 (1982)
12. Chen, H. S., *Acta metall.,* **22,** 897 (1974)
13. Hayes, T. M., Allen, J. W., Tauc, J., Giessen, B. C. and Hauser, J. J., *Phys. Rev. Lett.,* **40,** 1282 (1978)
14. Johnson, W. L. and Williams, A. R., *Phys. Rev.,* **B20,** 1640 (1979)
15. Slater, J. C. and Johnson, K. H., *Phys. Rev.,* **B5,** 844 (1972)
16. Johnson, K. H. and Smith, F. C. (jun.), *Phys. Rev.* **B5,** 831 (1972)
17. Slater, J. C., '*Quantum Theory of Molecules and Solids*', vol. 4, McGraw-Hill, New York (1974)
18. Korringa, J., *Physica,* **13,** 392 (1947)
19. Kohn, W. and Rostoker, J., *Phys. Rev.*, **94,** 1111 (1954)
20. Segall, B. and Ham, F. S., '*Methods in Computational Physics*', vol. 8, Academic Press, New York, (1968)
21. Messmer, R. P., Knudson, S. K., Johnson, K. H., Diamond, J. B., and Yang, C. Y., *Phys. Rev.,* **B13,** 1396 (1976)
22. Yang, C. Y., Johnson, K. H., Salahub, D. R., Kaspar, J. and Messmer, R. P., *Phys. Rev.*, **B24,** 5673 (1981)
23. Salahub, D. R. and Messmer, R. P., *Surf. Sci.,* **106,** 415 (1980)
24. Shull, C. G. and Mook, H. A., *Phys. Rev. Lett.,* **16,** 184 (1966)
25. Messmer, R. P., unpublished results
26. Messmer, R. P. and Briant, C. L., unpublished results
27. Messmer, R. P., *Phys. Rev.,* **B23,** 1616 (1981)
28. Becker, J. J., Luborsky, F. E. and Walter, J. L., *IEEE Trans. Magn.,* **MAG-13,** 988 (1977)
29. Walter, J. L., *Mater. Sci. Engng,* **39,** 95 (1979)

30. Wong, J., Lytle, F. W., Greegor, R. B., Liebermann, H. H., Walter, J. L. and Luborsky, F. E., in *'Proc. 3rd Int. Conf. on Rapidly Quenched Metals'*, Ed. Cantor, B., vol. 2, p. 345, The Metals Society, London (1978)
31. Vincze, I., Van der Woude, F., Kemeny, T. and Schaafsma, A. S., *J. Magnetism magn. Mater.*, **15–18,** 1336 (1980)
32. Vincze, I., Boudreaux, D. S. and Tegze, M., *Phys. Rev.* **B19,** 4896 (1979)

Chapter 9

Electronic structure determination

C. F. Hague*, P. Oelhafen and H.-J. Güntherodt

Institut für Physik, Universität Basel, Basel, Switzerland

9.1 Introduction

An important aim of modern solid state physics is to obtain information on the electronic structure of crystalline materials by means of experiment and theoretical calculations. Powerful tools for this purpose have been developed over the past decade. The band structure of many crystalline materials can now be calculated from realistic models and can be confronted with experiment. The calculation of the electronic structure of amorphous materials, however, is not as yet a tractable problem because of the loss of periodicity and this highlights the urgent need for measurements on structurally disordered systems.

Of course information could be obtained from the liquid state but the technological difficulties involved in very high temperature measurements have so far limited the attempts. Only recently have the metallic glasses opened up a new field for the systematic study of the amorphous phase. Much effort has gone into trying to elucidate whether the measured densities of states can be related to the glass-forming ability of alloys, but it has to be said that there are still insufficient data to give a clear answer. In addition to this aspect, however, metallic glasses also offer the opportunity of analysing new alloy concentrations which were not available in the crystalline state and with often quite surprising results.

Since we will be dealing with multicomponent materials it is appropriate straightaway to make the distinction between techniques which probe the total density of states (TDOS) and those which probe the local densities of states (LDOS). By TDOS we mean the contributions from all atomic sites put together and by LDOS we mean the contribution from each type of atomic site, taken separately.

Unquestionably the most direct tools for probing the occupied part of the TDOS are ultraviolet photoelectron spectroscopy (UPS) and X-ray photoelectron spectroscopy (XPS). Another useful but less direct method is optical spectroscopy. The unoccupied TDOS, lying above the Fermi energy E_F, can be analysed by use of Bremsstrahlung isochromat spectroscopy (BIS), also called inverse photoemission.

The main probes of the LDOS are soft X-ray spectroscopy (SXS), Auger electron

* Permanent address: Laboratoire de Chimie Physique, Université Pierre et Marie Curie, F-75231 Paris Cedex 05.

spectroscopy (AES), and XPS core-level analysis. The LDOS above E_F may be probed by soft X-ray appearance potential spectroscopy (SXAPS) which is one of the many possible modulation techniques here applied to an isochromat spectroscopy or, if the samples are available as thin films (a few hundred nm thick), by SXS in the absorption mode.

The preparation techniques used to obtain metallic glasses are dealt with in Chapter 3 but a word of warning is needed: all the techniques we describe for probing the electronic structure are surface sensitive so only contamination-free specimens prepared specially for such experiments should be used.

9.2 Experiments on the electronic structure

This section is meant for readers who are not familiar with the spectroscopy techniques used in electronic structure determinations. No attempt will be made to give a theoretical description of the physical processes involved. There are many excellent textbooks and reviews which can be consulted for a thorough treatment of the theory and the experiments[1–12]. Our purpose here is to make the non-specialist aware of the tools available and help him to see why different techniques measuring apparently identical properties produce different data.

9.2.1 Photoelectron spectroscopy

The principle is that a photon of well defined energy is used to impart enough energy to an electron to eject it from the atom, to which it is bound, into the vacuum [*Figure 9.1(a)*]. The vast majority of experiments on alloys use the He I or He II ultraviolet resonance lines for UPS and the Mg-K_α or Al-K_α X-ray emission lines in the case of XPS (*see Table 9.1* which gives the energy and the width of the main photon emissions used in electron spectroscopy). The difference between the energy of the incident photon and the kinetic energy of the emitted electron measured by the spectrometer is a measure of how tightly the electron was bound to the atom. The spectra are normally represented with a binding energy scale set relative to the Fermi energy (E_F or ε_F) and not relative to the vacuum level.

The number of electrons emitted per unit energy at a given kinetic energy will be related to the number of occupied electron states per unit energy at a given binding energy in the specimen, i.e. to the density of occupied states. We should note that *a priori* it is not possible to distinguish electrons according to the type of site they come from in the alloy.

TABLE 9.1 The four most common line sources used in photoemission experiments (*see* for instance ref. 6)

Source	*Energy*/eV	*Relative intensity*	*Linewidth*/eV
He I	21.22	100	0.003
Satellites	~23.5	~5	
He II	40.82	100	0.017
Satellites	~52	~3	
Mg-$K_{\alpha 1,2}$	1253.6	100	0.680
Satellites $K_{\alpha 3,4}$	~1263	~14	
Al-$K_{\alpha 1,2}$	1486.6	100	0.830
Satellites $K_{\alpha 3,4}$	~1497	~10	

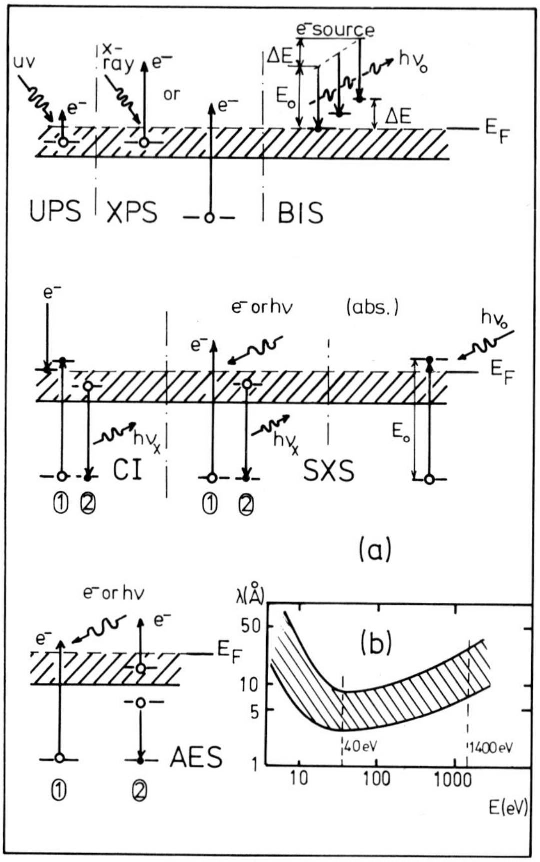

Figure 9.1 (*a*) Schematic representations of photon and photoelectron processes which can be used to probe the electronic structure of materials. The final state only is depicted for UPS, XPS, BIS, and SXS (absorption). Both the initial ionization process (1) and final state (2) are shown for CI, SXS (emission), and AES. For simplicity the work function has been neglected. (*b*) The mean free-path (λ) as a function of electron energy (E) above E_F lies within the hatched region for most materials in this universal curve

Photoemission experiments have in common with the other techniques described below that they in fact describe the convolution of two states. There is a transition from an initial state (it is best thought of as a hole in the continuum above the vacuum level) to a final state in which an electron has filled the hole in the continuum but has left behind a hole in the valence band. The transition probability is subject to a dipole matrix element. If we assume that the matrix element is constant over the energy range of interest, we can also consider that we have measured the convolution of initial and final states. In fact the spectrum will differ from the occupied TDOS, only if there is structure in the continuum states (or if the matrix element varies with energy). Structure exists fairly close to the vacuum level but even this will be attenuated in amorphous systems because of the absence of long range order. It is generally accepted that with the He II excitation the electrons are ejected into a structureless part of the continuum and more obviously so if the spectra are induced by Al- or Mg-K_α.

There is another important factor which must be taken into account in the interpretation of photoemission valence band spectra, and that is the mean free path of an excited electron in the specimen under study. The energy dependence of the mean free path is given in *Figure 9.1(b)*. For a given material the mean free path is much longer for an electron having a kinetic energy of 1400 eV than 40 eV which means that the bulk will contribute more to the intensity of a spectrum if XPS is used rather than UPS. It implies of course that in addition XPS is less surface sensitive than UPS. We note that if the mean free path is long, the electron wavevectors will remain well defined and we will expect selection rules to appear in the matrix elements of transitions from the occupied states to the empty states. In XPS this limitation is compensated by the increase in the number of possible final states as the energy of these states above the vacuum level increases.

In summary, XPS will provide a closer picture of the bulk TDOS, while UPS remains invaluable because of its intrinsically excellent resolving power. Monochromatization of Al-K_α for XPS will lead to significant improvement in resolution because only part of the width of the X-ray emission is used and also because satellite lines are eliminated but at the cost of signal strength.

XPS applied to the study of core-level binding energy shifts and line shapes can also provide indirect information on electronic structure. The core-line shifts encountered on going from the pure elements to alloys have been under investigation for some time, but as yet no consensus exists on the subject. The interpretation given to a core shift will depend on whether one views it in terms of charge transfer from one atomic site to another, or whether one considers that an adjustment in atomic volume has led to a reorganization between on-site *s-p-d*-electrons. Progress will doubtless be made as a result of repeated confrontations between experiment and theory.

Apart from conceptual differences we must also face quite complex relaxation processes in core-level excitations[7,8]. Here we will limit ourselves to two factors of particular interest to measurements on alloys. To begin with all the electrons with radial distributions extending beyond the core-level from which the electron has been ejected will see the core of a $(Z + 1)$ atom: this atomic site has become a $(Z + 1)$ impurity in the host.

This consideration has been used to advantage since a correlation has been established between core-line shifts in dilute alloys and the energy of solution of an element in a host metal. Agreement between experiment and semi-empirical energy of solution values is good if the energy of solution of the $(Z + 1)$ atom in each constituent is taken into account[13,14]. A relationship between core-line shifts in concentrated alloys and the heat of formation has also been established independently for substitutional alloys[15] and glassy metals[16].

The other aspect of interest in the analysis of core-lines results from the formation of low energy electron–hole pairs as the conduction electrons attempt to screen-out the charge created by the ejection of a core electron. There is now ample theoretical and experimental evidence that this process modifies the line shapes in a way that is very much dependent on the local density at the Fermi energy[7,17,18].

9.2.2 Isochromat spectroscopies

Although these techniques have been little used in the analysis of amorphous alloys they will be described briefly as they are a natural complement to photoemission experiments and as such are likely to be developed[9]. BIS requires a soft X-ray spectrometer set at a fixed Bragg reflection so that it will pass photons of one energy $h\nu_0$

9.3.1 Early transition metal–late transition metal glasses

The first results from photoemission experiments on TE–TL metallic glasses came as a surprise[27–29]. For instance, *Figure 9.2* shows the UPS ($h\nu = 40.8\,\text{eV}$) spectra for pure palladium and zirconium and three Pd_xZr_{100-x} glasses ($x = 25, 30$ and 35). We notice two well separated peaks in the density of states of the alloys: one lying close to E_F, the other some 4 eV below E_F. The way in which the relative intensities of the peaks vary with composition shows that the peak at the higher binding energy is dominated by Pd 4*d* states, while the Zr 4*d* states lie close to E_F.

Other measurements[30,31] on TE–TL metallic glasses confirm this splitting into two components with a large shift in the TL component away from E_F (*Figure 9.3*). We see that the separation in the peak positions depends on the relative position of the TE and TL group numbers in the periodic classification.

An analysis of the core-line shapes confirms that the TE component contributes a high density of states at E_F and that the TL component contributes little: the TL core levels are distinctly more symmetric in the alloys than in the pure elements and the TE core levels become more asymmetric[25].

It was already known that substitutional alloys fell into the split-band category[32] but in such cases no large shifts in the peak position of one of the constituents relative to the pure element had been observed or predicted[20,33,34]. Steiner, Schmidt, and Hüfner have proposed[35] that the TE–TL alloys, also clearly split band alloys, are no different basically from the standard *d*-metal alloys like Cu_xNi_{100-x} and Ag_xPd_{100-x}. They argue that the TL shift can be explained phenomenologically from Meidema theory[36] and that the XPS spectra recorded over the complete concentration range for coevaporated Pd_xZr_{100-x} and Cu_xZr_{100-x} crystalline films bear quite a close resemblance to similar measurements on, say, Ag_xPd_{100-x}.

Certainly the large TL shifts are not a result of the amorphous phase. For instance the XPS spectra of crystalline Cu_3Zr_2 and $PdZr_2$ are very similar to glassy $Cu_{60}Zr_{40}$ and $Pd_{35}Zr_{65}$ respectively[24]; and other TE–TL alloys with the CsCl structure have been observed with equally large TL shifts[37]. The questions which need to be answered all the same are: what is the physical reason for the TL peak position relative to E_F, and

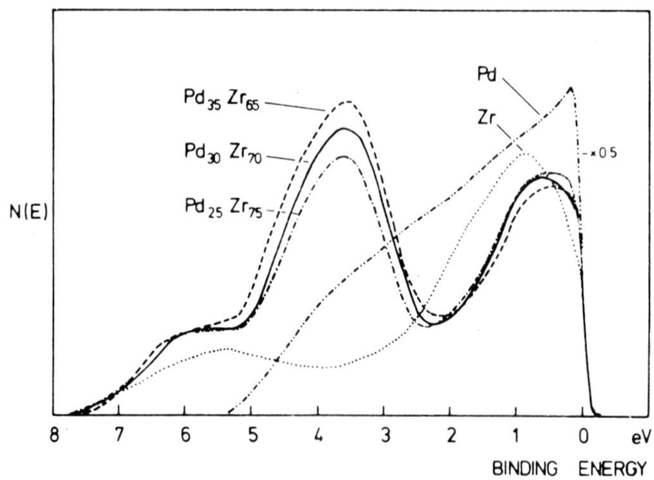

Figure 9.2 UPS valence band spectra of glassy Pd_xZr_{100-x} as a function of composition ($x = 25, 30, 35$) and the crystalline pure elements palladium and zirconium

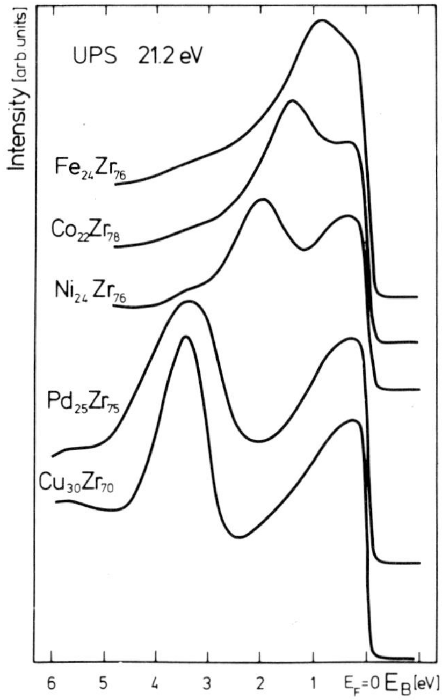

Figure 9.3 UPS valence band spectra of glassy $Fe_{24}Zr_{76}$, $Co_{22}Zr_{78}$, $Ni_{24}Zr_{76}$, and $Cu_{30}Zr_{70}$ at 21.2 eV showing the variation in peak separation as a function of group number difference. Also shown is $Pd_{25}Zr_{75}$ for comparison

what, if any, is the relation between the electronic structure and the glass-forming ability?

Further information is available from an investigation of the LDOS by SXS[38]. *Figure 9.4* shows the $L\beta_{2,15}(4d \rightarrow 2p_{3/2})$ X-ray emission bands for palladium and zirconium in the pure elements and $Pd_{30}Zr_{70}$. Bearing in mind that the resolution is an order of magnitude inferior to that in UPS spectra, we note that the zirconium peak has narrowed in the region of its maximum intensity and that a shoulder has appeared on the low energy side of the emission. The Pd 4*d* distribution has also narrowed and become distinctly asymmetric with a shift of 0.5 eV in the peak position. The two

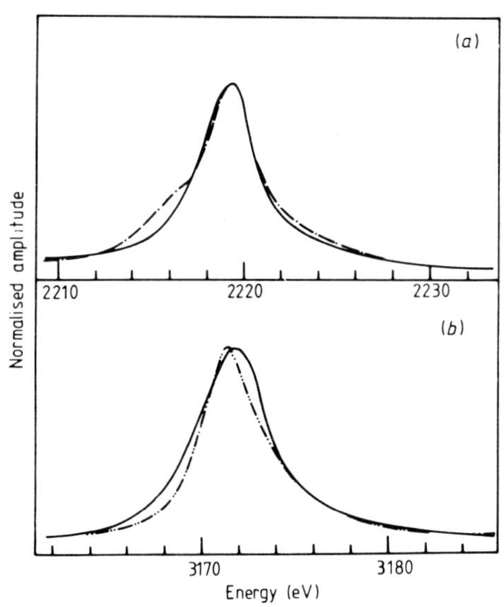

Figure 9.4 SXS valence band spectra for glassy $Pd_{30}Zr_{70}$. The $L\beta_{2,15}(4d \rightarrow 2p_{3/2})$ emission bands for palladium (full line) and zirconium (broken line) in (*a*) the alloy and (*b*) the pure element. The spectra are set against the X-ray transition energy

emissions from the alloy are situated relative to E_F in *Figure 9.5* by means of the appropriate core-level binding energies measured on the same specimen.

The data presented in *Figures 9.2* and *9.5* can be interpreted on an intuitive basis along the lines set out by Moruzzi, Williams and Janak[39]. They showed that band narrowing in a *d*-metal intermetallic compound occurs when the energy overlap is small between the constituent densities of states. This is because electron tunnelling between neighbouring atomic sites is then reduced. They also show that if the *d*-band splitting is of the order of the *d*-band widths, the energy levels are repelled by interaction.

Narrowing is discernible in the peaks of all these alloys,[40] but at first sight the narrowing and the shoulder in the zirconium LDOS are in contradiction since the latter coincides with the energy of the maximum in the palladium LDOS. The explanation is that the zirconium states consist of a narrow band near E_F and split-off hybridized states in coincidence with the palladium LDOS peak. In contrast the overlap between silver and palladium is small in Ag_xPd_{100-x} so that little hybridization is present in this substitutional alloy[41].

There are two ways of approaching the problem from a theoretical point of view. One is to consider that a start can be made at understanding amorphous alloys if the electronic structure of the crystalline alloys of similar composition are calculated and understood. For this purpose, the augmented-spherical-wave (ASW) method[42] has been used to calculate a number of alloys in the TE–TL glass-forming category[26,40]. This type of computation is self-consistent and provides calculated heats of formation in excellent agreement with experiment. Moreover the only inputs required are the atomic numbers and the crystal structure. An ASW calculation of the *d* density of states is presented in *Figure 9.6* for $PdZr_3$.

The other approach is to look at relatively crude but quantitative information on the density of states of the *disordered* phase (as pointed out in the Introduction the calculations will be relatively crude because quite important approximations have to be made). Fairlie, Temmerman and Gyorffy have constructed 19 atom clusters representative of the $Pd_{30}Zr_{70}$ glassy alloy and also other metallic glasses[38,43] for which they have solved the Schrödinger equation exactly. By varying both the substitutional disorder and the structural disorder of these clusters, they were able to show that the way the atoms are distributed relative to each other has far more impact on the density

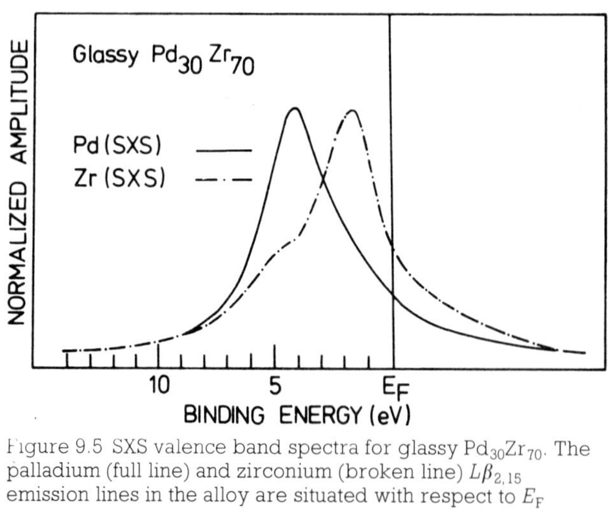

Figure 9.5 SXS valence band spectra for glassy $Pd_{30}Zr_{70}$. The palladium (full line) and zirconium (broken line) $L\beta_{2,15}$ emission lines in the alloy are situated with respect to E_F

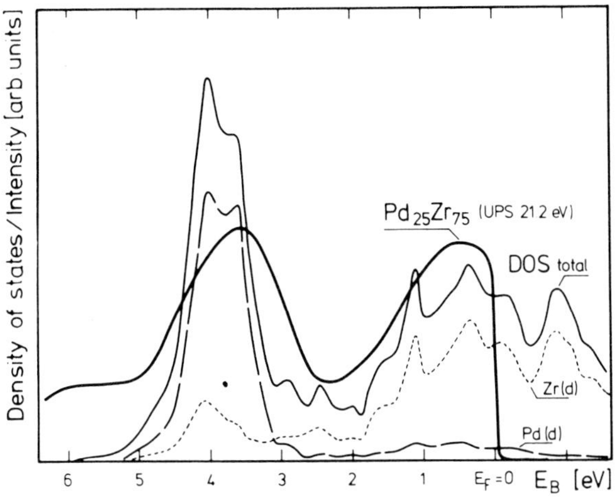

Figure 9.6 One theoretical description of Pd–Zr alloys: the electronic structure of $PdZr_3$ with the Cu_3Au symmetry showing the local 4*d* densities of states

of states than the details of the structural disorder. This supports the idea that information may be gained by studying the crystalline phase in which the main features of the density of states have already been established. The calculated $L\beta_{2,15}$ emission bands of palladium and zirconium in $Pd_{30}Zr_{70}$ are given in *Figure 9.7* for one of the clusters. Unfortunately this type of calculation is still insufficiently detailed to reveal small changes in the densities of states when the structural order is varied.

From both types of calculation the splitting in the zirconium density of states is apparent. It seems that the relatively localized Pd 4*d* band has repelled the Zr 4*d* states (and hence pushed up E_F) but has hybridized with the rather delocalized states at the bottom of the broad zirconium band to form a bonding state[43].

Optical spectra experiments on $Cu_{50}Zr_{50}$ have confirmed a narrowing of the Cu *d*

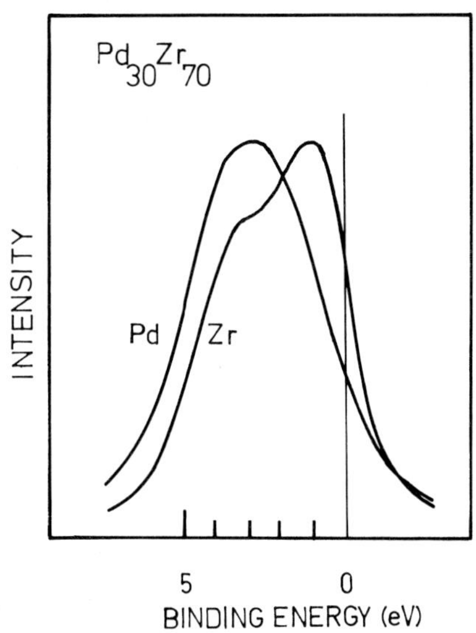

Figure 9.7 A theoretical description of glassy Pd–Zr alloys: the palladium and zirconium $L\beta_{2,15}$ emission bands in amorphous $Pd_{30}Zr_{70}$ as obtained by means of a 19-atom cluster calculation

component relative to the pure element and are also in general agreement with photoemission experiments as concerns the shifts encountered in the constituent *d*-band peaks[44].

9.3.2 Transition metal–polyvalent metal glasses

This category of metallic glasses is the one which has been most studied to date both experimentally and theoretically. The arguments put forward by Nagel and Tauc[45] establishing a correlation between glass stability and a *local* minimum in the density of states at E_F at the glass-forming composition, were tested on these glasses with some success. By *local* minimum it is meant that any rigid-band shift in the position of the Fermi level due to a change in the composition will tend to increase the density of states at E_F.

The first combined UPS and XPS experiments were carried out on $Pd_{77.5}Cu_6Si_{16.5}$ (a particularly easy glass former)[46]. They tended to support the presence of a local minimum at E_F, but the authors' conclusions were that more results were needed before the local minimum hypothesis could be considered confirmed. A number of other UPS and XPS measurements were then undertaken independently on $Pd_{100-x}Si_x$ [47–50] (*see*, for instance, *Figure 9.8*), but they led to different conclusions.

Three of the experiments presented evidence in favour of an appreciable contribution to the density of states at E_F from the Pd 4*d* states. One of them established a relation[49] between the Pd 4*d* peak position relative to E_F and the composition in the region $15 \leqslant x \leqslant 25$.

The fourth experiment[47] added AES as a local probe of the Pd 4*d* and Si 3*s*–3*p* states to the UPS and XPS measurements. The conclusions are in contradiction with the other experiments as concerns the presence of 4*d* states at E_F. These authors interpreted their results as showing that there were no Pd 4*d* states at E_F, but excluded the possibility of charge transfer from silicon to palladium to explain the filling of the *d* band. The latter was attributed to on-site charge redistribution.

The use of Auger spectroscopy to study the silicon LDOS is quite favourable because it is well established that for Si the $L_{2,3}VV$ spectrum is a self convolution of the valence band. Instead of a single maximum as in pure silicon, the Auger spectrum was observed to consist of three peaks. The spectrum was interpreted with the help of a theoretical calculation for the density of states of a silicon substitutional impurity in a

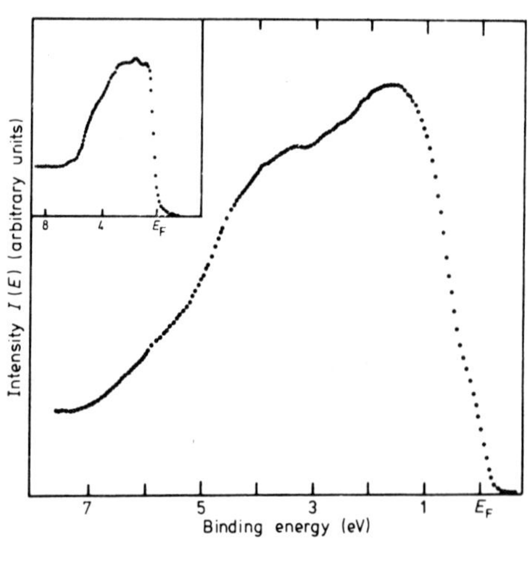

Figure 9.8 Experimental valence band determinations from Pd–Si alloys: high resolution XPS measurements on glassy $Pd_{80}Si_{20}$. The inset shows the XPS valence band of pure palladium

palladium matrix[51]. It was concluded that in fact there was a local maximum in the density of states at E_F due to Si 3*p* states, the main Si 3*p* peak lying 5 eV below E_F and the Si 3*s* states being split off and lying 10 eV below E_F. It is interesting to note that there is quantitive agreement with a very similar experiment[52] performed on Pd_2Si (*see Figure 9.9*).

According to SXS measurements[53] on liquid $Pd_{100-x}Si_x$ ($5 \leqslant x \leqslant 56$) the Si 3*p* contribution to the valence band has a double peaked structure which one can attribute to bonding and antibonding states. In *Figure 9.10* we show the Si-$K_\beta(3p \rightarrow 1s)$ X-ray emission band for two extreme concentrations in the liquid phase and also for $Pd_{80}Si_{20}$ in the glassy phase. For low silicon content ($\lesssim 10$ atom %) only the bonding states are occupied, but as the silicon content increases the antibonding states begin to fill up and a peak at E_F becomes apparent. Terakura[54] explained the shape of the 3*p* states in terms of an anti-resonance between the Pd 4*d* states and the Si 3*p* states resulting in a minimum in the 3*p* density of states at the peak position of the 4*d* density of states. *Figure 9.11* shows the SXS LDOS of copper in $Pd_{77.5}Cu_6Si_{16.5}$ and liquid $Cu_{30}Ge_{70}$ (unpublished results). It clearly demonstrates that the *d* states for a *d*-metal dissolved in a polyvalent metal become highly localized owing to reduced electron tunnelling to neighbouring sites, except when another *d*-metal is present.

Formally, the evidence is against the Nagel and Tauc theory as we will again see in the next section. However, it is interesting to note that any reduction in the *d* density of states at E_F will result in an increase in the 3*p* density of states at E_F and vice versa. If

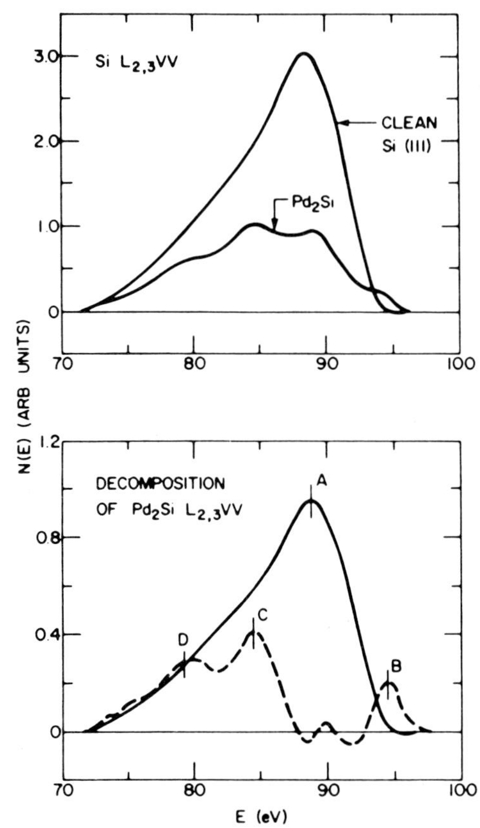

Figure 9.9 Experimental valence band determinations from Pd–Si alloys: Top: AES measurement on crystalline Pd_2Si, compared with pure silicon. Bottom: change in the AES spectrum (dashed curve) as a result of alloying

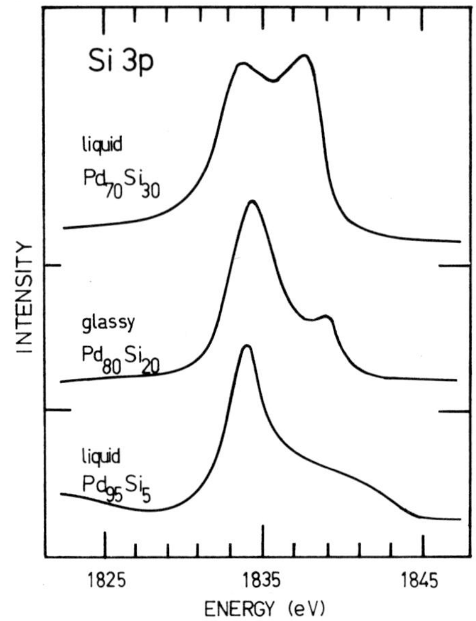

Figure 9.10 SXS local densities of states for various liquid or glassy alloys with a polyvalent metal: Si 3*p* electron distributions

the glass forming ability of an alloy does correlate with electronic structure, it can only depend on quite subtle changes[55] which are likely to escape detection by photoemission experiments. We have picked just one example amongst many, but experiments on other glassy alloys[56] such as Ni–P show that the electronic structure is grossly the same as that encountered in Pd–Si. It is interesting to note that differences in unoccupied densities of states have been observed in Co–P and Ni–P thin films by soft X-ray absorption spectroscopy[57] and in Vitrovac 6010 ($Co_{58}Ni_{10}Fe_5B_{16}Si_{11}$) by SXAPS[58] according to whether the specimens were amorphous or crystalline. The changes are quite small and seem to concern the *s*-states on the transition metal sites.

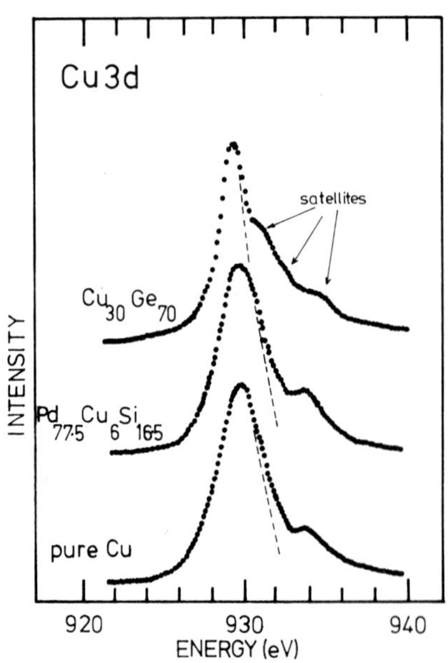

Figure 9.11 SXS local densities of states for various liquid or glassy alloys with a polyvalent metal: Cu 3*d* electron distributions

9.3.3 Simple metal–simple metal glasses

We have seen that the metalloids have a rather complex behaviour as a result of alloying with transition metals but we might expect that an ideal case to study would be one where the glass is made up of only free electron-like metals. Glasses have in fact been realized from Ca–Al, Ca–Mg, Ca–Zn, Mg–Zn etc. There has been some delay in undertaking electronic structure investigations, in fact, because of the formidable task of preparing oxygen-free specimens. The problem was finally solved for Ca–Al by preparing polycrystalline ingots of appropriate amounts of vacuum distilled calcium and high purity aluminium. The glasses were then made by vacuum splat-cooling in the case of $Ca_{70}Al_{30}$ and melt-spinning[59] under an argon atmosphere for $Ca_{50}Al_{50}$.

In view of the fact that this is the only set of experiments to have been performed on this type of material so far, *Figures 9.12, 9.13* and *9.14* are taken from ref. 59 as are most of the other data presented in this section.

Three interesting features should be noted in the UPS ($h\nu = 21.2\,eV$) results presented in *Figure 9.12*. First, the density of states is relatively low at E_F, secondly the main peak is narrow, and thirdly a shoulder is apparent towards the high binding energies. All these points are observed in both alloys.

Pure calcium and aluminum, being free electron-like, have an approximately parabolic increase in the density of states starting at the bottom of the valence band. There is little structure in the occupied part of the band. The band widths are 3 eV and 11 eV for calcium and aluminium respectively (*see* for example ref. 60). It is a surprise therefore to observe a narrow peak in these spectra having a width, apparently, of only 3 or 4 eV. We notice that the width of this peak seems to be reduced as aluminium is added which is in contradiction with the rigid band approach, since adding more electrons should shift E_F away from the bottom of the band. In addition, the sharp drop on the high binding energy side of the peak and the intense shoulder suggest that there is probably a splitting of the band. On p. 130 we pointed out that valence band states at high binding energies were likely to be smeared by Auger broadening which would explain the lack of detail.

An SXS measurement on $Ca_{70}Al_{30}$ is also reported in ref. 59 and is given in *Figure 9.13*. It concerns the Al-$K_\beta(3p \rightarrow 1s)$ emission band (this X-ray transition was also

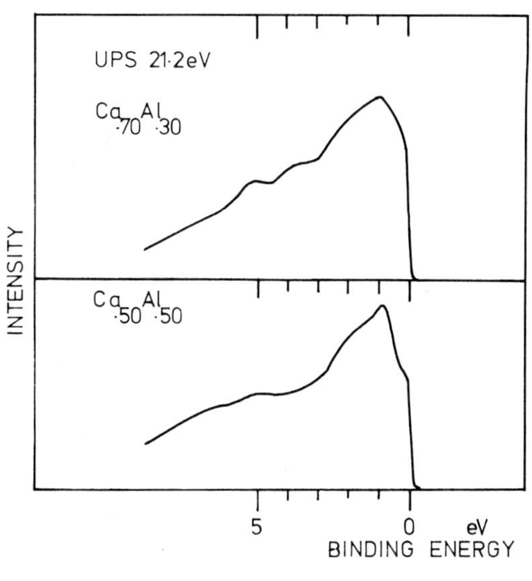

Figure 9.12 Experimental density of states in Ca–Al alloys: UPS ($h\nu = 21.2\,eV$) valence band measurements

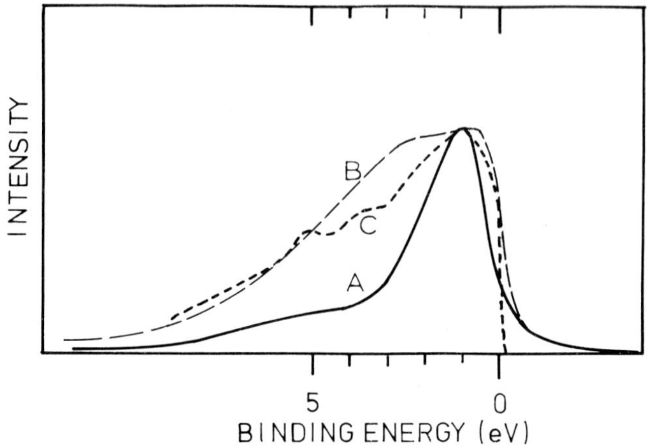

Figure 9.13 Experimental density of states in Ca–Al alloys: *A*, SXS $K_\beta(3p \rightarrow 1s)$ emission band of Al in $Ca_{70}Al_{30}$; *B*, K_β emission for pure aluminium; *C*, UPS curve. The X-ray emission is situated relative to E_F by means of core-level measurements

the one used to probe the Si 3*p* states in $Pd_{100-x}Si_x$ and presented in *Figures 9.10* and *9.11*). The UPS measurements for the alloy and the X-ray emission for pure aluminium are also shown. The relative energies are determined from core-level measurements. A striking reduction in the Al 3*p* states from the 11 eV of pure aluminium to approximately 3.5 eV in the alloy is observed. Here too there is structure on the high binding energy side of the emission but this is attributed to oxygen contamination during the experiment (further details on typical experimental conditions will be found in ref. 61).

Here again unexpectedly strong hybridization is observed and it is instructive to try to explain it qualitatively on an intuitive basis. In free electron-like metals the *s*-states, to an appreciable extent, lie at the bottom of the band, but the density of *p* states is low at the bottom of the band and increases rapidly towards E_F (the latter can be observed in a slightly accentuated way owing to matrix element effects in the K_β emission of pure aluminium given in *Figure 9.13*). As aluminium is dissolved in calcium there is reduced energy overlap for the aluminium *s*-states because the next nearest neighbours are mainly calcium atoms tending to form a narrower valence band. This explains why the 3*s* states must become more localized and split-off from the 3*p* states at aluminium sites. The situation seems to be the same even for 50 atom % of aluminium, but in the absence of data concerning the calcium LDOS it is not possible to push this rough analysis further.

Fortunately it has been possible to perform an ASW calculation for Ca_3Al which confirms very nicely the split between *s* and *p* states on aluminium sites and also reveals that the calcium valence states have hybridized with the aluminium states so that a mixed *sp* peak is localized at the Al 3*s* energy (about 5 eV below E_F) on calcium sites and there is also a main *sp* peak near E_F in coincidence with the 3*p* peak on aluminium sites. There is also very satisfactory quantitative agreement between theory and experiment. The LDOS are shown in *Figure 9.14* for each site; they are the sum of the partial *s* and *p* densities of states, and the TDOS is also given.

The density of states at E_F seems to be higher for the Ca_3Al calculation than for the photoemission measurements on glassy $Ca_{70}Al_{30}$. In addition, although SXS is not a very sensitive method for estimating densities of states, it seems that in comparison the

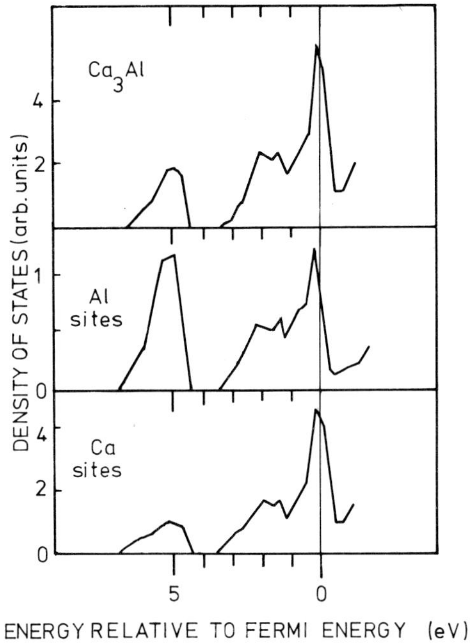

Figure 9.14 Calculated electronic structure of Ca_3Al in the Cu_3Au structure: total density of states (top), LDOS Al sites (middle), LDOS on Ca sites (bottom). The peak at 5 eV binding energy on Al sites is almost entirely due to Al 3*s* states; the peak just below E_F is the 3*p* contribution

calculation overestimates the Al 3*p* contribution to the density of states at E_F. The explanation might be that the slightly different compositions are responsible. The addition of aluminium to bring the composition up from 25 to 30 atom % could entail a further drop in both the Al 3*p* density of states and the total density of states at E_F by a rigid band-like shift of E_F away from the bottom of the band. Such an extrapolation is reasonable over such a small composition change since the $Ca_{50}Al_{50}$ UPS spectrum is quite similar. We also note that the density of states at E_F has lowered in $Ca_{50}Al_{50}$ but this glass was difficult to make.

Once again there seems to be no formal concordance with the Nagel and Tauc criterion. In the particular context of such simple metal–simple metal alloys it has been pointed out by Hafner[62] that many alloys which do fit this criterion do not in fact form glasses; for instance, Ca–Li, Li–Mg, Al–Mg etc. So it can be instructive to look at the electronic structure of these alloys also. SXS data are already available for several of these alloys and show that they do not fall into the rigid band category.[63,64] However, there is nothing to suggest that band splittings of the kind encountered in Ca–Al occur. The Korringa–Kohn–Rostoker coherent-potential-approximation (KKR–CPA) method has been employed with success on substitutional alloys[20,41], and has also been applied[64] to Li_xMg_{100-x}. The Li and Mg *p* states are almost unchanged relative to the pure elements and only the *s*-states are slightly modified. Coming back to the Ca–Al system, $CaAl_2$ is a Laves phase lying outside the glass-forming region. It has been studied[65] by SXS and a band structure calculation is also available[66]. The Al 3*p* states have already narrowed to some extent but the hybridization effects are less marked than in the glass-forming composition region.

9.4 Conclusions

Electronic structure determinations have made little headway in picking out particular features which could lead to increased stability against crystallization. The experiments lack sensitivity to the small changes in the density of states brought about by structural disorder. At present the same may be said of the theoretical models. Calculations on amorphous alloys are still not sufficiently realistic to feed back information to the experimentalist indicating where modifications between ordered and amorphous phases may be expected.

One particularly striking characteristic of a large number of glass-forming alloys is that they are subject to large hybridization effects. This means that such investigations should contribute significantly to the understanding of alloys. Until recently we had to make do with oversimplified notions such as rigid bands and charge-transfer between constituents. The fact that theoretical justification can be found for redistributions in the density of states of the size discussed here, implies that rapid progress will now be made to understand both crystalline and amorphous alloying.

Acknowledgements

We thank B. L. Gyorffy, E. Hauser, J. Kübler, J.-M. Mariot, V. L. Moruzzi, S. R. Nagel, W. M. Temmerman, and A. R. Williams for discussions and collaboration. We are indebted to all the members of Experimentelle Physik der Kondensierten Materie group in Basle and in particular U. Gubler, G. Indlekofer, J. Krieg and R. Lapka.

One of us (CFH) thanks the University of Basle, the Swiss National Science Foundation, and the Centre National de la Recherche Scientifique for making his stay in Basle possible.

Financial support from the Swiss National Science Foundation is gratefully acknowledged.

References

1. Sevier, K. D., *'Low Energy Electron Spectroscopy'*, Wiley, New York (1972)
2. Fadley, C. S., in *'Electron Spectroscopy'*, Eds. Brundle, C. R. and Baker, A. D., vol. 2, p. 1, Academic Press, London (1979)
3. Urch, D. S., ref. 2, vol. 3, p. 1
4. McGuire, G. E. and Holloway, P. H., ref. 2, vol. 4, p. 2
5. Fuggle, J. C., ref. 2, vol. 4, p. 86
6. Cardona, M. and Ley, L., in *'Photoemission of Solids I*, Eds. Cardona, M. and Ley, L., p. 1. Topics in Applied Physics, Vol. 26: Springer-Verlag, Berlin (1978)
7. Wertheim, G. K. and Citrin, P. H., ref. 6, p. 197
8. Wendin, G., *'Breakdown of the One Electron Pictures in Photoelectron Spectra'* (Structure and Bonding 45), Springer-Verlag, Berlin (1981)
9. Baer, Y., in *'Emission and Scattering Techniques'*, Ed. Day, P., NATO Advanced Study Institute, Reidel, Dordrecht (1981)
10. Rooke, G. A., in *'X-Ray Spectroscopy'*, Ed. Azaroff, L. V., p. 173, McGraw-Hill, New York (1974)
11. Ulmer, K., in *'Band Structure Spectroscopy of Metals and Alloys'*, Eds. Fabian, D. J. and Watson, L. M., p. 521, Academic Press, London (1973)
12. Nagel, D. J., ref. 11, p. 457
13. Johansson, B. and Mårtensson, N., *Phys. Rev.*, **B21**, 4427 (1980)
14. Steiner, P., Hüfner, S., Mårtensson, N. and Johansson, B., *Solid St. Communs*, **37**, 73 (1981)
15. Mårtensson, N., Nyholm, R., Calen, H., Hedman, J. and Johansson, B., *Phys. Rev.*, **B24**, 1725 (1981)
16. Oelhafen, P., *J. Phys. F: Metal Phys.*, **11**, L41 (1981)
17. Shevchik, N. J., *Phys. Rev. Lett.*, **33**, 1336 (1974)
18. Folmer, J. C. W. and de Boer, D. K. G., *Solid St. Communs*, **38**, 1135 (1981)
19. O'Brian, H. M. and Skinner, H. W. B., *Phys. Rev.*, **45**, 379 (1934)
20. Durham, P. J., Ghaleb, D., Gyorffy, B. L., Hague, C. F., Mariot, J.-M., Stocks, G. M. and Temmerman, W. M., *J. Phys. F: Metal Phys.*, **9**, 1719 (1979)
21. Von Barth, V. and Grossman, G., *Solid St. Communs*, **32**, 645 (1979)
22. Güntherodt, H.-J., in '*Festkörperprobleme*' (Advances in Solid State Physics, vol. 17) Ed. Treusch, J., Vieweg, Braunschweig (1977)

23. Güntherodt, H.-J., Oelhafen, P., Lapka, R., Künzi, H.-U., Indlekofer, G., Krieg, J., Laubscher, T., Rudin, H., Gubler, U., Rösel, F., Ackermann, K. P., Delley, B., Fischer, M., Greuter, F., Hauser, E., Liard, M., Müller, M., Kübler, J., Bennemann, K. H. and Hague, C. F., *J. Phys. (Fr.)*, **41,** C8 381 (1980)
24. Güntherodt, H.-J., Oelhafen, P., Hauser, E., Greuter, F., Lapka, R., Rösel, F., Jacobs, R., d'Albuquerque E Castro, J., Kübler, J., Hague, C. F., Bennemann, K. H., Fairlie, R. H., Temmerman, W. M. and Gyorffy, B. L., p. 619, Institute of Physics Conference Series No. 55 (1981)
25. Oelhafen, P., Hauser, E., and Güntherodt, H.-J., in '*Inner-Shell and X-Ray Physics of Atoms and Solids*', Eds. Fabian, D. J., Kleinpoppen, H. and Watson, L. M., p. 575, Plenum Press, New York (1981)
26. Oelhafen, P., in '*Glassy Metals II*' Eds. Beck, H. and Güntherodt, H.-J., Springer Verlag, Berlin (1983)
27. Nagel, S. R., Tauc, J. and Giessen, B. C., *Solid St. Communs,* **22,** 471 (1977)
28. Amamou, A. and Krill, G., *Solid St. Communs,* **28,** 957 (1978)
29. Oelhafen, P., Hauser, E., Güntherodt, H.-J. and Bennemann, K. H., *Phys. Rev. Lett.,* **43,** 1134 (1979)
30. Amamou, A., *Solid St. Communs,* **33,** 1029 (1980)
31. Oelhafen, P., Hauser, E. and Güntherodt, H.-J., *Solid St. Communs,* **35,** 1017 (1980)
32. Seib, D. H. and Spicer, W. E., *Phys. Rev.,* **B2,** 1676 (1970)
33. Stocks, G. M., Williams, R. W. and Faulkner, J. S., *J. Phys. F: Metal Phys.,* **3,** 1668 (1978)
34. Hüfner, S. Wertheim, G. K. and Wernick, J. H., *Phys. Rev.,* **B8,** 4511 (1973)
35. Steiner, P., Schmidt, M. and Hüfner, S., *Solid St. Communs,* **35,** 493 (1980)
36. Miedema, A. R., de Châtel, P. F. and de Boer, F. R., *Physica,* **100B,** 1 (1980)
37. Battye, F. L., Schultz, H., Goldmann, A., Hüfner, S., Seipler, D. and Eschner, B., *J. Phys. F: Metal Phys.,* **8,** 709 (1978)
38. Hague, C. F., Fairlie, R. H., Temmerman, W. M., Gyorffy, B. L., Oelhafen, P. and Güntherodt, H.-J., *J. Phys. F: Metal Phys.,* **11,** L95 (1981)
39. Moruzzi, V. L., Williams, A. R. and Janak, J. F., *Phys. Rev.,* **B10,** 4856 (1974)
40. Kübler, J., Bennemann, K. H., Lapka, R., Rösel, F., Oelhafen, P. and Güntherodt, H.-J., *Phys. Rev.,* **B23,** 5176 (1981)
41. Durham, P. J., Gyorffy, B. L., Hague, C. F., Mariot, J.-M., Pindor, A. J. and Temmerman, W. M., p. 145, Institute of Physics Conference Series No. 55 (1981)
42. Williams, A. R., Kübler, J. and Gelatt, C. D. (jun.), *Phys. Rev.,* **B19,** 6094 (1979)
Fairlie, R. H., Temmerman, W. M. and Gyorffy, B. L., *J. Phys. F: Metal Phys.,* **12,** 1641 (1982)
44. Rivory, J., Frigerio, J. M. and Rebouillat, J. P., *J. Phys. (Paris),* L481 (1981)
45. Nagel, S. R. and Tauc, J., *Phys. Rev. Lett.,* **35,** 380 (1975)
46. Nagel, S. R., Fisher, G. B., Tauc, J. and Bagley, B. G., *Phys. Rev.,* **B13** 3284 (1976)
47. Riley, J. D., Ley, L., Azoulay, J. and Terakura, K., *Phys. Rev.,* **B20,** 776 (1979)
48. Oelhafen, P., Liard, M., Güntherodt, H.-J., Berresheim, K. and Polaschegg, H. D., *Solid St. Communs,* **30,** 641 (1979)
49. Waclawski, B. J. and Boudreaux, D. S., *Solid St. Communs,* **33,** 589 (1980)
50. Cartier, E., Baer, Y., Liard, M. and Güntherodt, H.-J., *J. Phys. (Fr.) F: Metal Phys.,* **10,** L21 (1980)
51. Asada, T., and Terakura, K., *J. phys. Soc. Japan,* **47,** 1495 (1979)
52. Ho, P. S., Rubloff, G. W., Lewis, J. E., Moruzzi, V. L. and Williams, A. R., *Phys. Rev.,* **B22,** 4784 (1980)
53. Hague, C. F., *J. Phys.* (*Paris*), **41,** C8 433 (1980)
54. Terakura, K., *J. Phys. F: Metal Phys.*, **6,** 1385 (1976)
55. Gaspard, J. P., in '*Electronic Structure of Crystal Defects and of Disordered Systems,* Les Editions de Physique, Les Ulis (1981)
56. Domashevskaya, E. P., Therekhov, V. A., Ugai, Ya.A., Nefedov, V. I., Sergushin, N. P. and Firsov, M. N., *J. Electron Spectrosc. and relat. Phenom.,* **16,** 441 (1979)
57. Belin, E., Bonnelle, C., Flechon, J. and Machizaud, F., *J. Non-cryst. Solids*, **41,** 219 (1980)
58. Dose, V. and Haertl, A., *Phys. Rev. Lett.*, **47,** 132 (1981)
59. Nagel, S. R., Gubler, U. M., Hague, C. F., Krieg, J., Lapka, R., Oelhafen, P., Güntherodt, H.-J., Evers, J., Moruzzi, V. L. and Williams, A. R., *Phys. Rev. Lett.*, **49,** 575 (1982)
60. Moruzzi, V. L., Janak, J. F. and Williams, A. R., '*Calculated Electronic Properties of Metals*', Pergamon Press, New York (1978)
61. Hague, C. F., *Phys. Rev.*, **B25,** 3529 (1982)
62. Hafner, J., *Phys. Rev.*, **B21,** 406 (1980)
63. Neddermayer, H., in ref. 11
64. Stocks, G. M., Tagle, J. A., Calcott, T. A. and Arakawa, E. T., in ref. 25, p. 619
65. Wiech, G., and Zöpf, E., in ref. 11, p. 173
66. Radwan, A. M., and Taut, M., *Phys. Status Solidi* (*b*) **76,** 605 (1976)

Chapter 10

Crystallization

M. G. Scott

Standard Telecommunications Laboratories Ltd., Harlow, Essex, UK
[Formerly, University of Sussex, Brighton]

10.1 Introduction

A glass is defined as any solid formed by the continuous cooling of a liquid such that detectable crystallisation is avoided[1]. That this is possible is illustrated simply by *Figure 10.1*, which shows schematically as a function of temperature the time taken for a small amount of crystallinity to form in an undercooled melt. An essential feature of this (TTT) diagram is that as the liquid is undercooled below T_m, the equilibrium melting point, the crystallization rate does not continue to rise indefinitely but reaches a

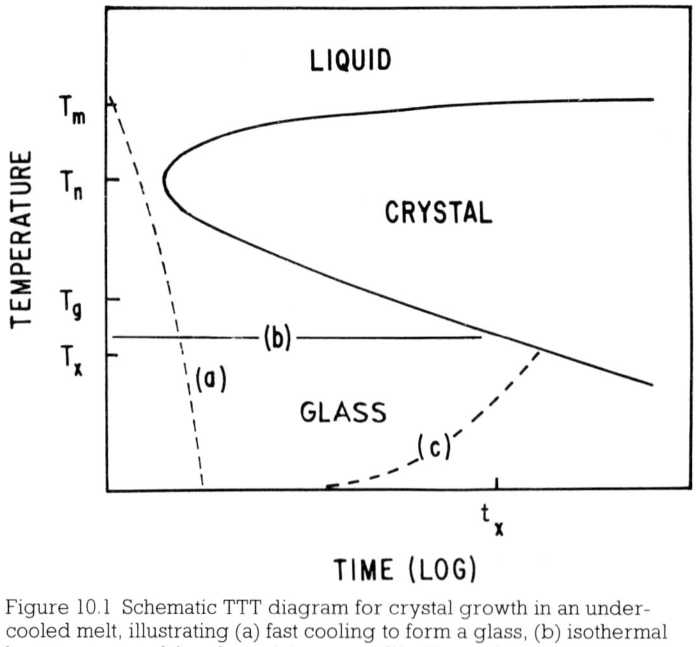

Figure 10.1 Schematic TTT diagram for crystal growth in an undercooled melt, illustrating (a) fast cooling to form a glass, (b) isothermal heat treatment of the glass giving crystallization at time t_x, and (c) slow heating of the glass producing crystallization at T_x

maximum at a temperature T_n. Consequently if the liquid is quenched from above T_m to well below T_n in a time less than t_{min}, the minimum time for crystallization, the undercooled liquid is retained and, at T_g, the glass transition temperature, the configuration is frozen to form a glass. For metallic alloys this requires[2] a cooling rate of between 10^2 and 10^{10} K/s. It also follows from *Figure 10.1* that if the glass is subsequently heated to a temperature $T < T_n$ and held isothermally for a time t_x it will begin to crystallize. Alternatively if the glass is heated slowly from room temperature, crystallization begins at a temperature, T_x, which increases with increasing heating rate. (Of course if it were possible to heat sufficiently quickly to pass the nose of the diagram crystallization could be totally prevented.) The ease with which this devitrification occurs varies widely between different glassy systems. The familiar oxide glasses transform so sluggishly that extreme measures (such as the deliberate introduction of nuclei) are often needed to induce crystallization even at T_n; on the other hand, as will be shown in this chapter, metallic glasses transform so rapidly that it is rarely possible to heat them much above $0.5T_m$ before spontaneous crystallization occurs.

In practice the above description is simplistic in two respects. In the first place a single isothermal transformation diagram has been used to describe both isothermal and dynamic heat treatments whereas accurate description of the latter requires either continuous cooling (CCT) or continuous heating (CHT) transformation diagrams. However, the only difference to the isothermal diagram is a shift of the transformation to longer times at a given temperature[3]. Secondly, only one potential crystallization product has been considered. As will be shown later, there are a number of possible crystallization reactions, each of which should be represented on the transformation diagram. *Figure 10.2* illustrates one possibility. Just below T_m the undercooled liquid crystallizes to the equilibrium phase, α, whilst at lower temperatures a second, metastable phase β can form. As will be shown in this chapter most metallic glasses

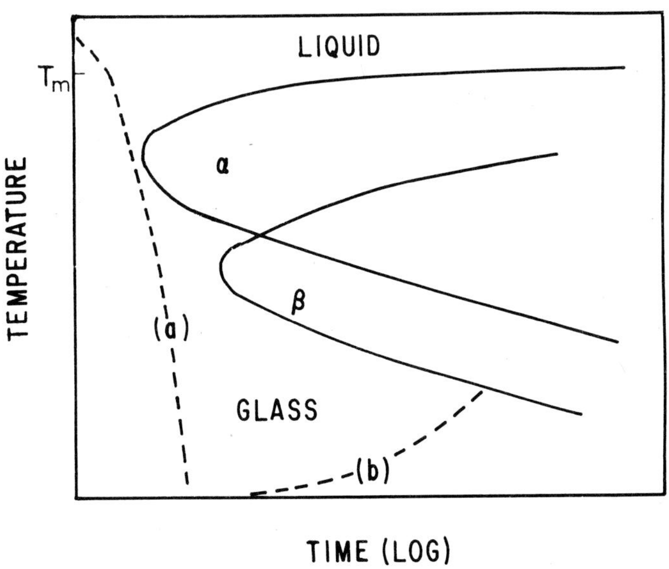

Figure 10.2 Modification of the TTT diagram of *Figure 10.1* to show equilibrium (α) and non-equilibrium (β) phases. (a) During rapid quenching to form a glass crystallization of α must be avoided; (b) heat treatment of the glass generally produces β

crystallize initially to one or more metastable phases. An important consequence of this is a non-dependence of the kinetics of crystallization (in this case of β) on those of glass formation (i.e. avoidance of α).

The crystallization of metallic glasses is accompanied by a densification [4] of about 1 per cent, a decrease in enthalpy[5] equivalent to about half the latent heat of melting (i.e. a few kJ/mol) and significant changes in most physical properties. It occurs by a nucleation and growth process identical to that in liquids just below the melting point, and as such offers a unique opportunity to study under controlled conditions the growth of crystals into an isotropic medium and to test the applicability of classical nucleation and growth theories at large undercoolings. Much of this chapter is devoted to a discussion of these aspects. However, aside from justifiable scientific curiosity, the devitrification of metallic glasses has a number of practical implications. In the first place, the loss of many of the desirable properties on crystallization imposes a strict limit on the operating times at elevated temperatures. Secondly, it is becoming increasingly clear that the partial or full crystallization of metallic glasses can be used to produce novel and useful microstructures unobtainable by other means. Some of these are now exploited commercially[6] and are discussed at the end of the chapter.

10.2 Experimental techniques

Any of the property changes described above may be used to monitor the crystallization of a metallic glass. Amongst the easiest to measure is electrical resistivity; *Figure 10.3*, curve (a), shows its variation when the glass $Fe_{40}Ni_{40}B_{20}$ is heated at a uniform rate from room temperature. The onset of crystallization is marked by a sharp decrease in resistivity. Similar sharp changes have been observed in elastic modulus[7], magnetic coercivity[8], saturation magnetization[9], and all have been used to monitor the crystallization process. However, such methods are indirect and considerable caution should be exercised in using them to determine the kinetics of crystallization. An essential prerequisite is that the property should change linearly with the fraction of crystallinity, a fact which has often been assumed but not established.

An almost essential tool in any laboratory studying metallic glasses is differential

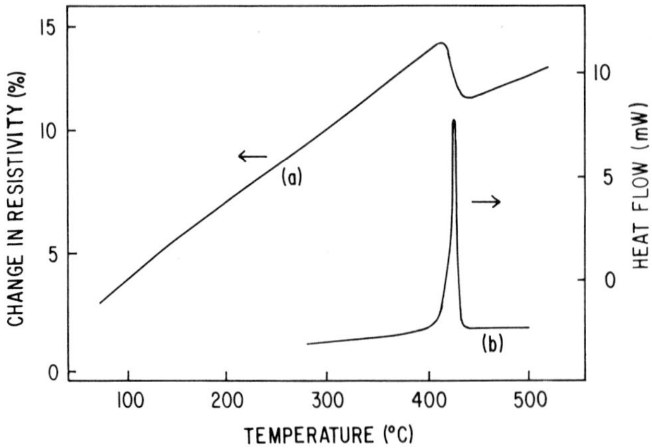

Figure 10.3 (a), Resistivity curve obtained by heating the glass $Fe_{40}Ni_{40}B_{20}$ at 5 K/min from room temperature (courtesy of J. Rout). (b), DSC trace obtained under the same conditions

scanning calorimetry (DSC). Here the sample (typically a few mg) and an inert reference of similar mass are subjected to identical thermal programmes and the heat flow between the two is monitored. This technique, which was originally developed for polymer research, is ideally suited to metallic glasses on account of their large enthalpy of crystallization. *Figure 10.3*, curve (b), shows the thermogram obtained by heating the glass $Fe_{40}Ni_{40}B_{20}$ at 5 K/min. Crystallization is marked by a large exotherm from which it is possible to define a dynamic 'crystallization temperature', either as the onset temperature, T_x, or peak temperature, T_p. The dependence of both T_x and T_p on heating rate (β) has been used frequently to determine an activation energy for crystallization, E_x, though there is still disagreement about the relative merits of the methods of Ozawa[10], in which log T_x/β is plotted against $1/T_x$ and Kissinger[11], in which log T_x^2/β is plotted against $1/T_x$. Both give an approximately straight line from the slope of which E_x can be found. Since the instantaneous heat flow is directly proportional to the rate of transformation, isothermal DSC is frequently used to deduce the crystallization kinetics. Some examples will be quoted later.

Both X-ray diffraction and Mössbauer spectroscopy[12] have been used in conjunction with the techniques described above to identify the crystallizing phases. In many cases this task is complicated because the phases produced are not usually the equilibrium ones. For a complete understanding of the crystallization process, however, direct microstructural observations are essential, and all forms of microscopy have a role to play. *Figure 10.4* shows typical micrographs of the partially crystallized glass

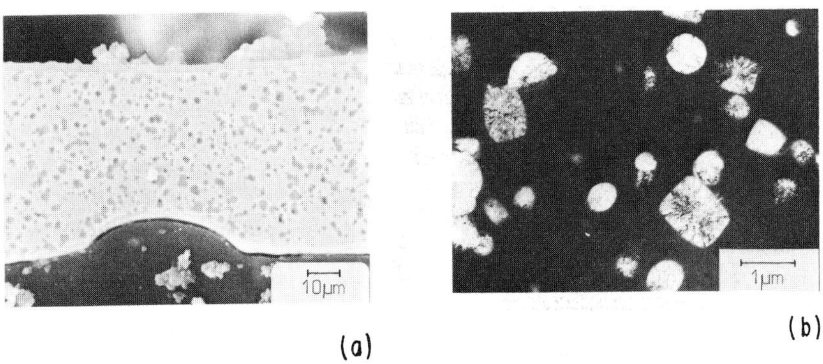

Figure 10.4 (a) SEM micrograph of glass $Fe_{40}Ni_{40}P_{14}B_6$ annealed for 3 h at 365 °C; (b) TEM micrograph of the same glass annealed for 11 min at 385 °C

$Fe_{40}Ni_{40}P_{14}B_6$ obtained by (a) scanning electron (SEM) and (b) transmission electron microscopy (TEM)[13]. Because of the fine scale of the microstructures produced TEM has been used extensively; together with electron diffraction it can provide full information about the morphology, crystallography and orientation of the phases and, after suitable correction for sectioning effects, direct measurements of nucleation and growth rates. The addition of energy dispersive X-ray (EDX) analysis allows *in-situ* measurement of the chemical compositions of the crystals and untransformed glass. In all cases electron microscopy should be performed on specimens thinned after annealing. *In-situ* heat treatment of thinned specimens can give misleading results owing to surface nucleation and diffusion.

10.3 Crystallization temperatures and their compositional dependence

It should be clear from *Figure 10.1* that, unlike melting and other changes of state, crystallization of a glass does not occur at a unique temperature; rather given sufficient time devitrification can occur at any temperature, though the time–temperature relationship is such that most metallic glasses can be considered as stable indefinitely at room temperature. Nevertheless the literature on metallic glasses abounds with quotations of crystallization temperatures, T_x, usually measured at a practical heating rate (5—50 K/min) in a differential scanning calorimeter as described above. Such values depend not only on the heating rate but many other factors such as the thermal history of the glass, its method of preparation, and amount of trapped gas. Moreover, T_x measured in this way should never be taken as an indicator of the safe operating temperature of a metallic glass; for example, whereas many iron-based glasses have T_x in the region of 400 °C, their maximum long term operating temperatures are only of the order of 150 °C[14]. However, armed with these qualifications, the dynamic crystallization temperatures do provide a useful and quick means to compare the relative thermal stabilities of different glasses.

For most metallic glasses T_x is between 0.4 and 0.6 T_m; in absolute terms, therefore it can vary from as little as 100 °C for some simple metal glasses[15] to in excess of 1000 °C for glasses based on refractory metals[16]. In many binary metal–metalloid glasses T_x is a maximum near the equilibrium eutectic composition (*see Figure 10.5* for Fe–B alloys)[17]. The same does not appear to be the case[18] in all-metal glasses, such as Ni–Zr (*Figure 10.6*), which show a monotonic decrease of T_x with increasing zirconium content despite the existence in the equilibrium diagram of two eutectics and an intermetallic compound. Numerous attempts have been made to correlate the relative thermal stabilities of different glasses to thermodynamic, crystallographic and electronic factors. Amongst the earliest were those of Chen and colleagues[19–21] who showed that in many ternary metal–metalloid systems the most stable glasses were those in which the transition metal constituents (*a*) had different atomic sizes, (*b*) had

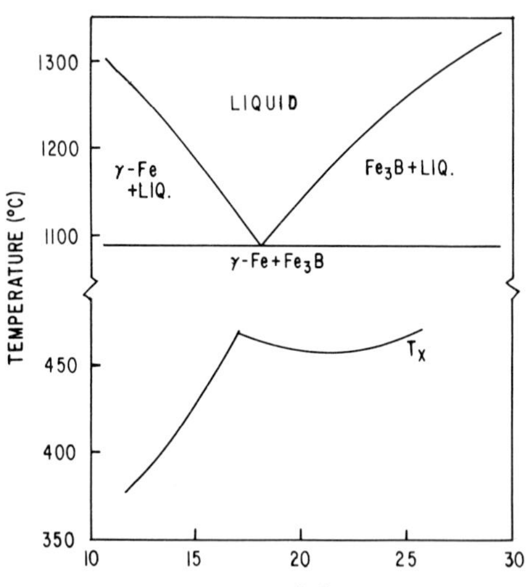

Figure 10.5 Liquidus temperature and T_x as a function of composition for Fe–B

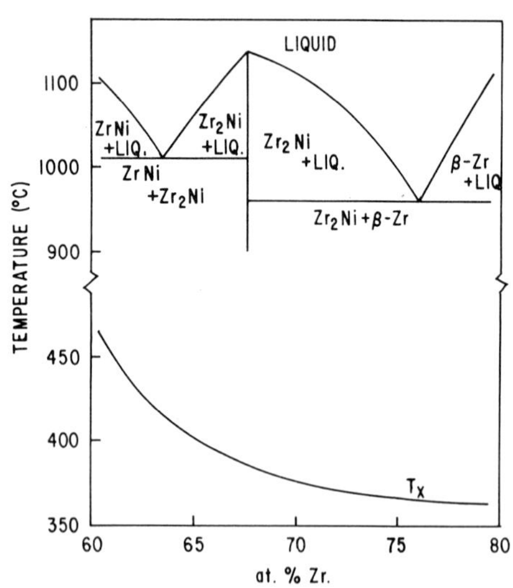

Figure 10.6 Liquidus temperature and T_x as a function of composition for Zr–Ni

strong interactions with the metalloid (as measured by the heat of mixing) and (*c*) formed intermetallic compounds of different structures with the relevant metalloid. The rôle of atomic size has been taken up subsequently in a number of cases. Walter[22], Donald *et al.*[23], and Lovas *et al.*[24] have all shown independently that in glasses of the type (Fe, Ni)–M–B–Si where M is selected from a large number of metallic elements T_x is increased when the atomic size of M is larger than that of iron or nickel and decreased when it is smaller. Conversely Giessen and Elliott[25] showed that in binary actinide–transition metal glasses T_x increases with increasing fraction of the smaller atom. It is by no means clear why, in the absence of other factors, the addition of an outsize atom should change the thermal stability of a binary glass. Walter[22] has proposed that in Fe–B alloys substitution of some of the iron by a larger atom disturbs the close packing, thereby reducing the free volume and the diffusion coefficient. Such an approach, based on a 'confusion' principle, predicts that ternary glasses will be more resistant to crystallization than the constituent binaries. In Fe–Ni–B, at least, this is true[9]. A similar but more quantative approach has been proposed by Bushow *et al.*[26,27] to account for the composition dependence of T_x in a number of binary intertransition metal and rare earth–transition metal glasses. Using tabulated data for the enthalpies of formation of monovacancies in pure metals and an appropriate correction factor for alloying, they calculated the energy of formation of a hole equal in size to the radius of the smaller atom. T_x scales well with this energy which is assumed to be proportional to the activation energy for viscous flow[28] and therefore diffusion. The two assumptions implicit in this argument, namely that the activation energy for crystallization E_x, is the same as that for viscous flow[28] and that T_x scales[24,29] as E_x, have been verified independently, albeit for different glasses. Nevertheless the above arguments assume implicitly that the crystallization is controlled by long range diffusion; as shown later this is not always the case.

All of the correlations described so far ignore the dependence of conduction electron density on alloying. Both Naka *et al.*[30] and Donald and Davies[31] have shown that when iron is partially substituted by various transition metals in iron–metalloid glasses there is a scaling of T_x and other properties with electron/atom ratio. In a similar vein Inoue *et al.*[32] have shown that in (Fe, Co, Ni)–M–B glasses, where M is any transition metal, T_x increases with the difference in electronegativity between (Fe, Co, Ni) and M. One of the most significant attempts at an electronic description of thermal stability was that of Nagel and Tauc[33]. By treating the glass as a nearly free electron solid they argued that the glassy phase should be most stable at compositions where the Fermi level is at a minimum in the density of states. Mathematically this condition is met when $2k_f = k_p$ where k_f is the wave vector at the Fermi surface (assumed spherical) and k_p is the reciprocal space vector of the first peak in the structure factor of the glass. This model has met with some success in predicting both the composition of maximum stability in some metal–metalloid glasses and the large glass forming range in some intertransition metal glasses; however, it does not appear to be universally applicable. An important consequence of Nagel and Tauc's hypothesis is that at compositions near $2k_f = k_p$ the temperature coefficient of resistance of the glass should be negative[34]. Significantly good correlations between dR/dT and T_x have been found in Ni–Zr[35], $Fe_{80-x}Ni_xB_{20}$, $(Fe_{0.5}Ni_{0.5})_{1-y}B_y$[36] and $Fe_{80}M_3B_{17}$[24]. In all cases T_x increases as dR/dT becomes more negative. Moreover Nagumo[37] has indicated that when the Si/B ratio is varied in $Ni_{75}(Si\text{–}B)_{25}$ the maximum in T_x coincides with the minimum in thermoelectric power. It is clear therefore that the conduction electrons play a significant role in the stabilization of metallic glasses and further correlations of this type are to be expected.

10.4 Thermodynamics of crystallization: crystallization reactions

Metallic glasses crystallize by a nucleation and growth process. The driving force is the free energy difference between the glass and the appropriate crystalline phase(s). The possible devitrification reactions may be illustrated, therefore, by the hypothetical free energy diagram[38] in *Figure 10.7*, which represents the variation of free energy with composition of the glass and various crystalline phases at the chosen annealing temperature. Assuming that there is no tendency to clustering, the curve for the glass, G, will be concave downwards as shown and, since the glass is metastable, will always be higher than that of at least one crystalline phase. In this case two stable phases are included: a terminal solid solution α and an intermetallic phase θ. Depending on the composition these can exist either independently or in equilibrium with compositions c_α and c_θ respectively, as given by the well known common tangent construction. Also shown is a phase M which, although it has a lower energy than the glass, is metastable with respect to the equilibrium mixture of α and θ. It can co-exist in metastable equilibrium with α of composition c'_α. The following crystallization processes can now be identified. In each case the numbers in brackets refer to those labelling the arrows in *Figure 10.7*.

(*a*) Polymorphous transformation of the glass to a crystalline phase of the same composition. The product may be either θ (1) or M (2) or a supersaturated solid solution (3). In the latter two cases subsequent decomposition can occur to the equilibrium mixture of α and θ (2′ and 3′).

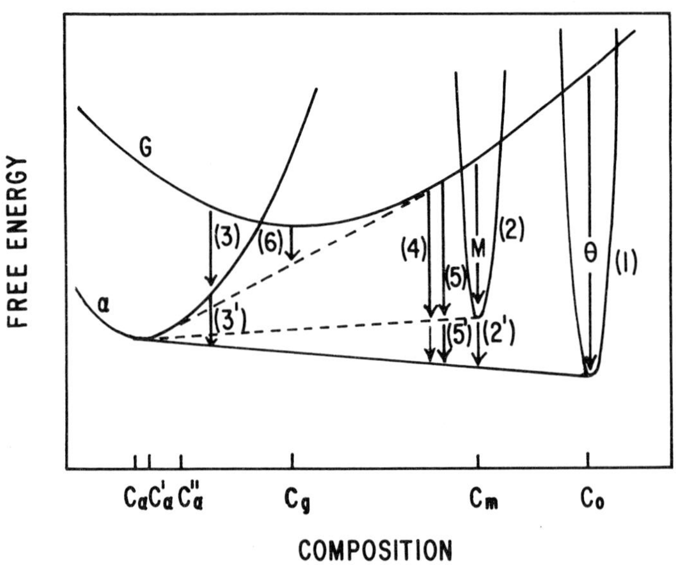

Figure 10.7 Hypothetical free energy diagram to illustrate the crystallization of a metallic glass. G, α, θ, M are respectively the free energy curves of the glass, a terminal solid solution, a stable intermetallic phase, and a metastable phase. Stable equilibrium is indicated by the solid line; metastable equilibrium by the broken lines. The numbered arrows refer to the devitrification processes described in the text

(*b*) The glass can reduce its free energy to a point on the common tangent between either α and θ (4) or α and M (5). In these cases the two crystalline phases grow in a coupled fashion analogously to the *eutectic crystallization* of liquids. In the case of the metastable eutectic between α and M subsequent further decomposition to α and θ can occur. (There is a case for describing such a reaction as 'eutectoid' since it occurs entirely within the solid state; however, the term 'eutectic' seems to have established itself, presumably because the parent material is liquid-like.)

(*c*) *Primary crystallization* of supersaturated solid solution (6). Since the α has a composition c''_α which is less than that of the glass c_g solute is rejected from the growing crystals into the glass. Ultimately the untransformed, enriched glass transforms by one of the other mechanisms discussed above.

All metallic glasses crystallize by one of the above mechanisms. Which of the reactions occurs in a given case will depend not only on the thermodynamic driving force but on the activation barrier to, and therefore kinetics of, each. In most cases the equilibrium phases do not form directly from the glass but crystallization proceeds by way of one, and sometimes several, metastable phases. For example in the Fe–B system shown in *Figure 10.5* the stable phases are α–Fe and Fe_2B, yet the glass always crystallizes to the metastable Fe_3B with the equilibrium phases forming only at high temperatures[39]. In the following sections the morphologies and kinetics of the various crystallization mechanisms are described in detail.

10.5 Growth rates and morphologies

10.5.1 Polymorphous crystallization

The polymorphous transformation of a glass to a single crystalline phase of the same composition is the simplest of the crystallization reactions described above. Yet, except possibly at low annealing temperatures where it is claimed that some glasses transform to a supersaturated solid solution, it is the least common. In metal–metalloid alloys, where glass formation is restricted to narrow composition ranges near equilibrium eutectics, polymorphous crystallization to an equilibrium phase has never been reported; instead in a few cases, metastable phases are formed, usually within very limited composition ranges. A good example of this is the formation of the tetragonal phase Fe_3B in Fe–B glasses containing[39] between 24 and 26 atom % of boron. On the other hand in metal–metal glasses, where the glass forming range is frequently much larger, the same constraints do not apply and equilibrium phases such as Zr_2Ni [40] and Mg_7Zn_3 [15] grow from the glass without composition change and often well away from stoichiometry. The crystals usually nucleate randomly within the glass and grow until they impinge and consume all of the matrix. Their morphologies vary from unfaulted equiaxed single crystals to heavily twinned polycrystals. *Figure 10.8* shows two examples: (a) Ni_3P [41] and (b) Zr_2Ni [40]. The origins of the fine structure in the Zr_2Ni crystals are unclear, but it is believed that the crystals are highly twinned. Such highly faulted morphologies are frequently observed in the crystallization products of metallic glasses. Since the crystals are growing with no long range diffusion into an isotropic medium their external shapes should be a function of only the crystallographic dependence of growth rate. It is, therefore, somewhat surprising that crystals grown in the glass often have different morphologies from those of the same phase grown directly from the melt. *Figure 10.9* shows an example[18] in the glass $Ni_{56}Zr_{42}$. In both cases the crystals are the equilibrium phase $Ni_{10}Zr_7$; the large spherical crystals [*Figure 10.8*(*a*)] were produced during quenching of the ribbon and presumably at undercoolings less

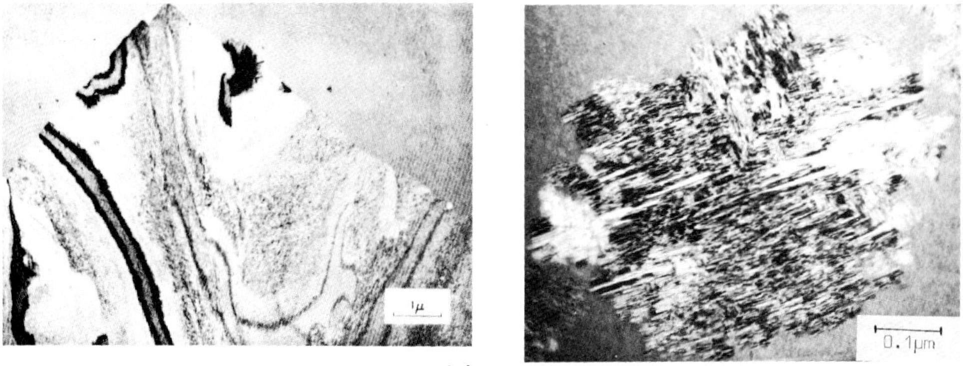

(a) (b)

Figure 10.8 Morphologies of polymorphous crystallization: (a) equiaxed crystal in Ni_3P (courtesy of D. Turnbull); (b) highly faulted crystal in Zr_2Ni

than 150 °C; the smaller ellipsoidal crystals were obtained at large undercoolings during subsequent heat treatment of the glass. The origins of this difference in morphology are not known.

Figure 10.10 shows the growth kinetics of the Zr_2Ni crystals shown in *Figure 10.8(b)*. As expected for polymorphous transformation the growth is linear with time. Turnbull[42] has shown that for such transformations the growth rate, u, is given by:

$$u = a_0 \nu_0 [\exp(-\Delta F_a/kT)][1 - \exp(-\Delta F_v/kT)] \tag{10.1}$$

where a_0 is the atomic diameter, ν_0 is the atomic jump frequency, ΔF_v is the molar free energy difference between the two phases and ΔF_a is the activation energy for an atom to leave the matrix and attach itself to the growing phase. A plot of u against T therefore has a maximum as described in the introduction to this chapter. Since the crystallization of metallic glasses is always studied at large undercoolings (the growth rate data in *Figure 10.10* were obtained at $\Delta T \sim 800$ °C), $\Delta F_v \gg kT$ and equation 9.1 reduces to:

$$u = a_0 \nu_0 \exp(-\Delta F_a/kT) \tag{10.2}$$

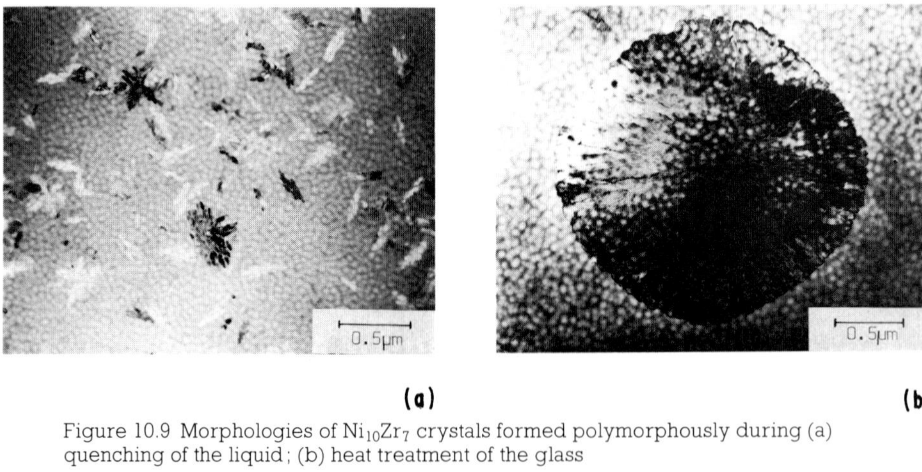

(a) (b)

Figure 10.9 Morphologies of $Ni_{10}Zr_7$ crystals formed polymorphously during (a) quenching of the liquid; (b) heat treatment of the glass

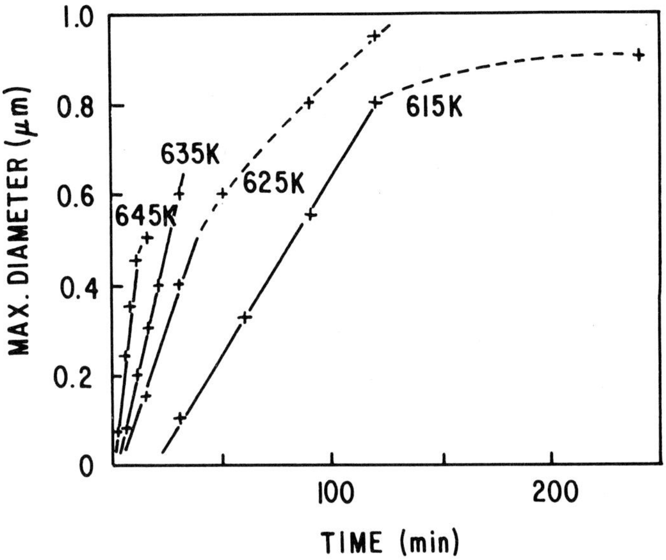

Figure 10.10 Growth kinetics of Zr_2Ni crystals in glass of same composition. The broken lines indicate crystal impingement

This Arrhenius dependence of growth rate on temperature has been confirmed experimentally for both the polymorphous and eutectic crystallization of metallic glasses. Both the activation energies ΔF_a, typically a few hundred kJ/mole, and the pre-exponentials, typically $10^9 \, ms^{-1}$, are high and inconsistent with a growth mechanism involving activated jumps of single atoms across the glass–crystal interface. However, no attempts have been made to investigate the detailed mechanisms of crystal growth, and there is clearly a need for high resolution studies of the nature of the glass–crystal interfaces and their mobilities.

10.5.2 Eutectic crystallization

At large undercoolings the composition range over which eutectic growth can occur (coupled-zone) is often large, e.g., from 17 to 24 atom % of boron in Fe–B[39], and it is, therefore, a common decomposition mechanism in metallic glasses. Nevertheless it has been studied in detail in only a handful of alloys and, as shown below, the details of the growth processes are still not fully understood. The microstructural aspects of eutectic crystallization are well illustrated by the much studied[13,43,44] glass $Fe_{40}Ni_{40}P_{14}B_6$ (METGLAS 2826). Devitrification occurs by the apparently homogeneous nucleation and growth within the amorphous matrix of two phase barrel-shaped crystals (*Figures 10.4, 10.11* and *10.12*) which grow unchanged until they impinge. This morphology has been observed over the temperature range 630 to 830 K though there is some evidence that it may change during isothermal crystallization at lower temperatures[13]. Each crystal contains two finely-spaced crystalline phases: a continuous phase, which has been identified as a metastable tetragonal compound $(FeNi)_3(PB)$ and is isomorphous with Fe_3P; and rods of Fe–Ni austenite which grow radially[44] outwards from the centre of the crystal as indicated in *Figure 10.12*. The austenite phase additionally contains some fine structure which has been attributed to either micro-twinning[13] or micro-inversion domains[45]. The two phases have the simple orientation relationship $\langle 100\rangle(FeNi)_3PB \| \langle 110\rangle$austenite: $\langle 001\rangle$tetragonal$\| \langle 112\rangle$austenite such

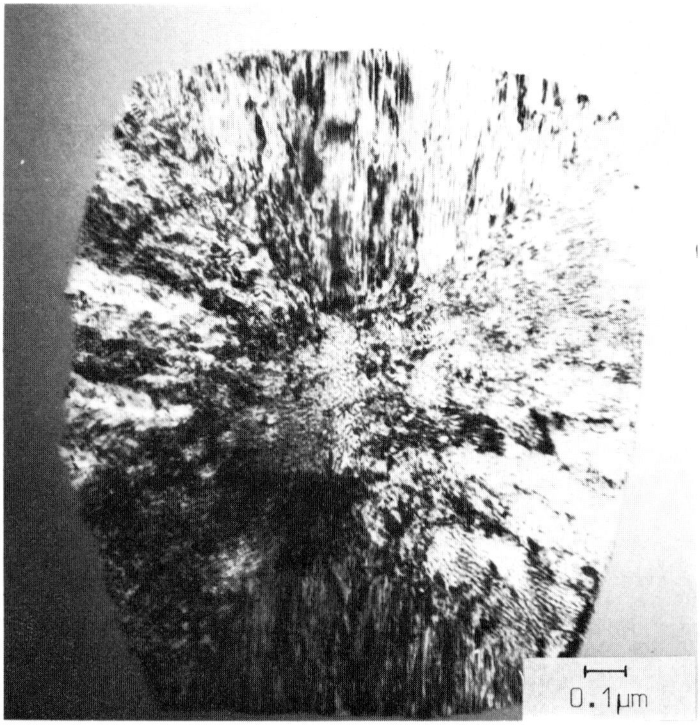

Figure 10.11 Barrel shaped eutectic crystal in $Fe_{40}Ni_{40}P_{14}B_6$ annealed for 13 min. at 385 °C

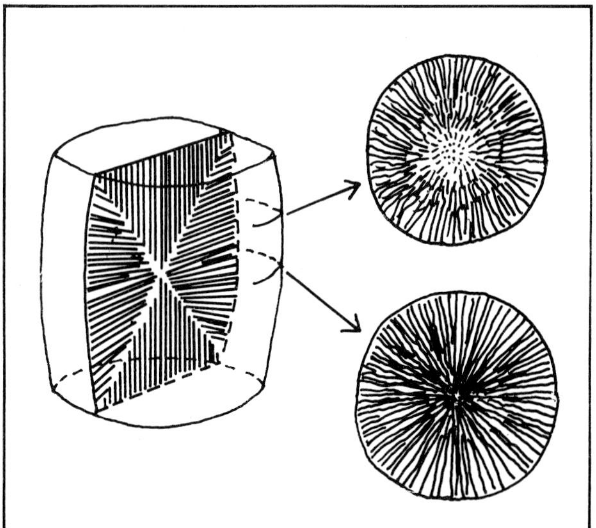

Figure 10.12 Schematic to show internal structure of crystal. The rods are austenite; the matrix is $(FeNi)_3PB$ (courtesy of M. V. Heimendahl)

that the *c* axis of the tetragonal phase is parallel to the barrel axis[13]. The overall crystal shape is determined by the growth morphology of the tetragonal phase and is highly sensitive to metalloid content; if the phosphorus is replaced by boron to produce the glass $Fe_{40}Ni_{40}B_{20}$ the crystals become ellipsoidal in shape[46]. Similar eutectic morphologies with well defined orientation relationships between the phases are produced by the devitrification of a number of other metallic glasses[47–49].

Like polymorphous crystallization the eutectic transformation is a discontinuous reaction; the overall composition of the crystal and glass are the same and the matrix remains unchanged until the glass–crystal interface sweeps past it. For such reactions the growth rate of a crystal should be independent of time until hard impingement with another crystal occurs[50]. Measurements of eutectic growth rates in several metallic glasses have confirmed the linear kinetics[44,48,49–51,52]. In the limited temperature range (typically 50 K) over which such measurements are generally made the growth rate increases linearly with the exponential of inverse temperature. However, Morris[43] has confirmed that in $Fe_{40}Ni_{40}P_{14}B_6$ the growth rate–temperature relationship deviates from Arrhenius behaviour at high temperature, indicating that the nose of the TTT diagram is being approached. Unlike polymorphous crystallization, eutectic crystallization requires diffusion to enable the partition of solute between the two growing phases. This diffusion may occur either in the glass ahead of the growing crystal or in the glass–crystal interface. For steady-state growth, therefore, a balance will be established between the crystal growth rate (u), the interphase spacing (λ) and the diffusion coefficient (D). For crystalline eutectics and eutectoids the following diffusion equations can be set up[53]:

$$D_m = \lambda u / 2\pi\alpha \text{ for diffusion in the matrix} \qquad (10.3)$$

$$D_i = \lambda^2 u / 4\pi^2 \alpha\delta \text{ for diffusion in the interface} \qquad (10.4)$$

In these equations α, which is not independent of λ, is a dimensionless term describing the compositions of the matrix and growing phase at the interface and δ is the thickness of the interface layer in which the diffusion is assumed to occur. Conventional eutectic and eutectoid reactions are usually studied at low undercoolings where u increases and λ decreases with *decreasing* temperature. By contrast the crystallization of metallic glasses is always carried out at large undercoolings where growth rate u increases with *increasing* temperature. In only four cases[39,44,48,49] have interphase spacings been measured as a function of temperature; in all four the spacing decreased with increasing temperature consistent with equations 10.3 and 10.4. Unlike conventional eutectics ($u\lambda^2$ = constant[54]) however, there appears to be no universal relationship of u with λ. In $Fe_{40}Ni_{40}P_{14}B_6$ a decrease in temperature of 50 K produced a six-fold increase in eutectic spacing; using these values and equation 9.3 Tiwari *et al.*[44] deduced a diffusion coefficient D_m which was in remarkably good agreement with that measured independently by Cahn *et al.*[55] for boron in amorphous $Fe_{40}Ni_{40}B_{20}$. They conclude therefore that the necessary solute redistribution occured by diffusion of boron in the glass ahead of the interface. Conversely application of the same equation by Boswell and Chadwick[48] to the cellular decomposition of Pd–Cu–Si in which λ was almost invariant of u gave a D_m which was several orders of magnitude larger than that expected for metal atom diffusion in the glass. Moreover in both this glass and Ni–Pd–P[49] the activation energy for growth was close to that for viscous flow and consistent with Chen and Turnbull's[56] model of interface controlled growth. Such inconsistencies between different glasses are not entirely unexpected and may be a consequence of different diffusion mechanism of large and small atoms. Analogous behaviour is found in conventional eutectoid decomposition; pure Fe–C austenites decompose primarily by bulk diffusion of carbon; in alloy steels and other substitutional systems such as Cu–In the diffusion is at the pearlite–matrix interface[53].

It is clear from the above that further careful measurements of interphase spacings and growth rates are necessary to establish beyond doubt the mechanism of eutectic crystallization in metallic glasses. In crystalline eutectic and eutectoid systems such measurements are conventionally made by directional decomposition in which the sample is pulled at a constant rate through a negative temperature gradient[57]. Similar experiments on oxide glasses with use of a positive temperature gradient have confirmed[58] the inverse relationship between u and λ. In principle the same technique could be applied to metallic glasses. However, in practice this is unlikely to be successful since the temperature gradients required are high, and can only be achieved by localized heating by use of electron beams or lasers[59].

10.5.3 Primary crystallization

Primary crystallization of a phase of different composition from that of the glass is the initial stage of crystallization of many metallic glasses. The primary phase may be either a terminal solid solution, e.g. α–Fe in many iron–metalloid glasses[51], or an intermediate phase, e.g. $Cu_{10}Zr_7$ in $Cu_{56}Zr_{44}$ [60]. Because of the inapplicability of EDX analysis to light elements such as boron and carbon direct verification of the composition difference between matrix and crystal has been obtained in only a few cases. For example, von Heimendahl and Oppolzer[61] showed that the primary fcc crystals in $Fe_{32}Ni_{36}Cr_{14}P_{12}B_6$ were depleted in phosphorus and chromium with respect to the overall composition of the glass. Other measurements of this type have been restricted to glasses of the all-metal type such as Cu–Zr[60] and Ni–Zr[40] in which compositional analysis of all elements is possible.

The growth of primary crystals, whether they be from liquid, glass or crystalline solid solution, may be limited by either (*a*) the rate of transfer of atoms across the advancing interface (interface control) or (*b*) the rate of diffusion of atoms towards or away from the growing phase (diffusion control). All of the evidence to date is that the crystallization of metallic glasses is diffusion controlled. Under such circumstances the composition profile, through a growing crystal, is expected to be as shown in *Figure 10.13* where c_0, c_α, and c_{eq} are respectively the compositions of the glass, the crystal and that of the glass in equilibrium with the crystal at the interface[62]. They are related by the appropriate free energy diagram as shown in *Figure 10.14*. (Strictly c_α and c_{eq} will

Figure 10.13 Composition profile expected through primary crystal

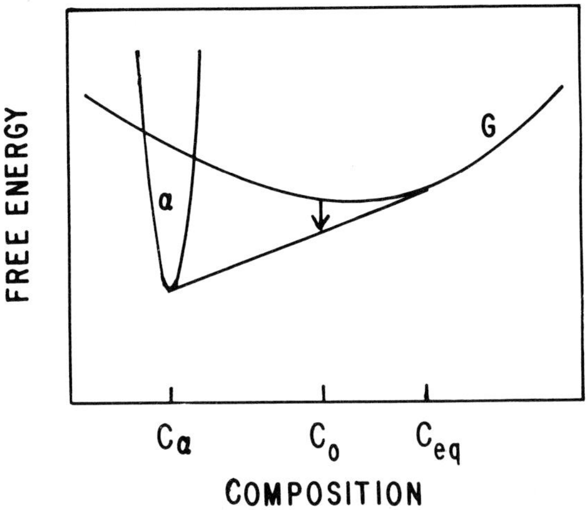

Figure 10.14 Free energy diagram to show relationship of C_{eq}, C_o and C_α.

depend on the size of the crystal through the Gibbs–Thomson correction for curvature. For small crystals, less than 1 μm, this effect may be considerable). Surprisingly, the composition gradient in the matrix ahead of the crystal has been confirmed experimentally only once, by Freed and Vandersande[60] for roughly spherical $Cu_{10}Zr_7$ crystals growing in a $Cu_{58}Zr_{42}$ glass; the measured composition profile (*Figure 10.15*) was qualitatively similar to that given in *Figure 10.13*. As a consequence of this composition gradient the region ahead of the crystal is constitutionally supercooled. For the growth of crystals from liquids such conditions lead to instabilities of the liquid–solid interface and consequently to dendritic growth[63]. For solid state transformations, however, Doherty[64] has shown that a highly anisotropic interfacial energy and the high misfit between the matrix and parent phases usually stabilizes the interface and prevents dendritic growth. The crystallization of metallic glasses, where crystals grow into an isotropic medium in the solid state, is intermediate to the above two cases and the morphologies of the primary crystals range from spherical to highly

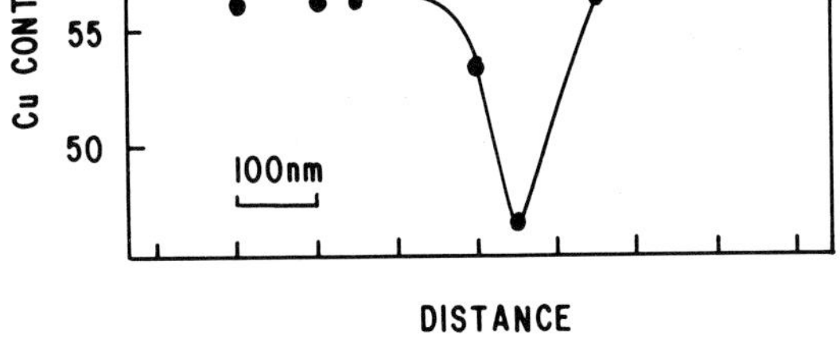

Figure 10.15 Measured composition profile through crystal in $Cu_{56}Zr_{44}$ glass (courtesy of J. B. Vander Sande)

dendritic. The morphology is strongly dependent on alloy composition; for example, Shimomura *et al.*[65] have shown that the progressive replacement of carbon by boron in the glass $Fe_{80}(CB)_{20}$ leads to a change in the primary α–Fe crystals from small spheres to large dendrites (*Figure 10.16*). They attribute this behaviour to a reduction of the interfacial energy by boron, but this has not been verified independently.

The growth rate of primary crystals depends on their morphology. In the absence of interfacial instabilities the growth is parabolic with time, i.e., r (the mean radius of the crystals) increases linearly with the square root of the time. *Figure 10.17* shows an example for the growth of Zr_2Ni crystals in a matrix of composition $Zr_{63}Ni_{37}$. The diffusion coefficient, D, for the growth process can be deduced from:

$$r = \sqrt{\alpha D t} \qquad (10.5)$$

where α is a dimensionless expression involving the compositions of the crystal and of the matrix both at the interface and well away from the crystal[66]. Koster[67], in particular, has used measurements of parabolic growth rates in a number of metal–metalloid glasses to deduce diffusion coefficients and their compositional dependence. In the few cases where independent measurements of D are available the agreement with these is reasonable. (For a full discussion of diffusion in metallic glasses see Chapter 25.) However, it should be borne in mind that the above equation assumes

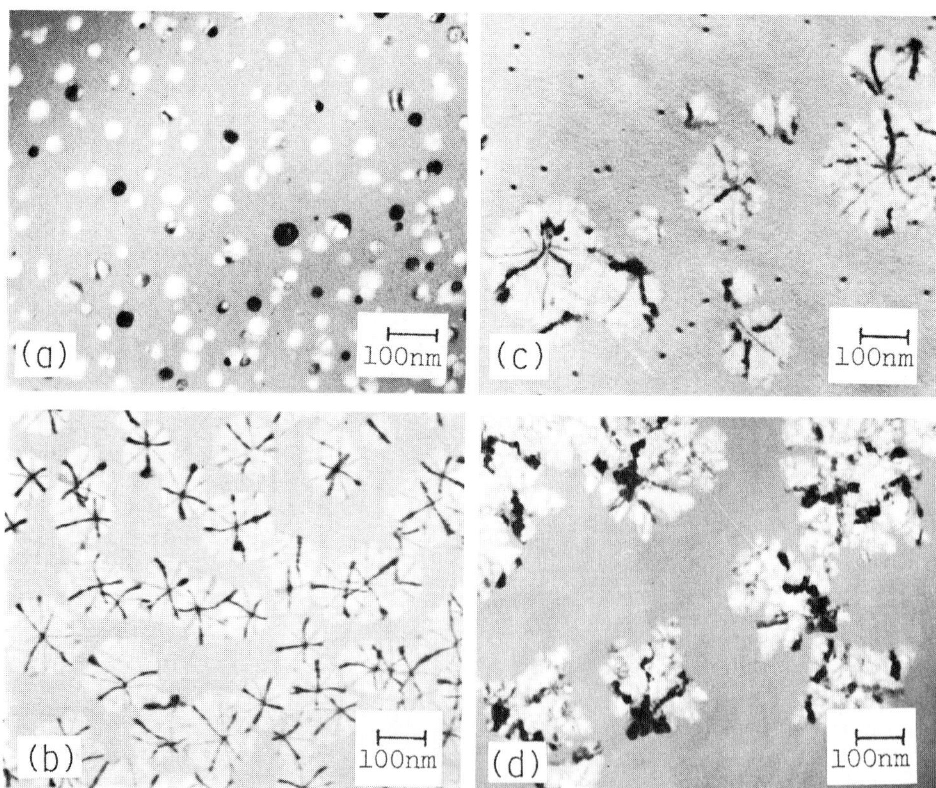

Figure 10.16 Change in morphology of primary α-Fe crystals as function of boron to carbon ratio in $Fe_{80}(CB)_{20}$ glasses. (a) $Fe_{80}C_{19}B_1$ (b) $Fe_{80}C_{15}B_5$ (c) $Fe_{80}C_{10}B_{10}$ (d) $Fe_{80}C_5B_{15}$ (courtesy of P. Shingu)

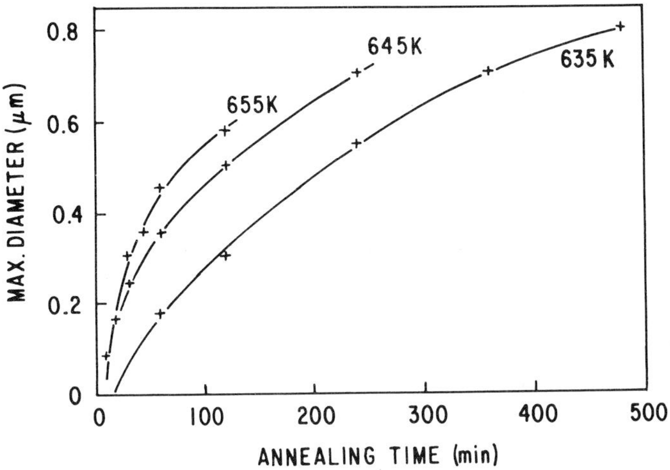

Figure 10.17 Growth kinetics of primary Zr_2Ni crystals in $Zr_{63.5}Ni_{36.5}$ glass

that the crystals remain spherical throughout their growth. Inspection by the author of most of the published micrographs of primary crystals shows there is invariably some tendency towards directional or dendritic growth. The diffusion coefficients deduced from the parabolic increase of mean radius should therefore be treated with some caution. For dendritic growth the primary arms are expected to increase linearly with time[64]. A detailed study[8,68] of the growth of α–Fe dendrites in $Fe_{81}B_{13.5}C_{3.5}Si_2$ (*Figure 10.18*) has confirmed this behaviour. Moreover, the dendritic growth direction, ⟨100⟩, is exactly that expected for growth of a bcc phase into an isotropic medium.

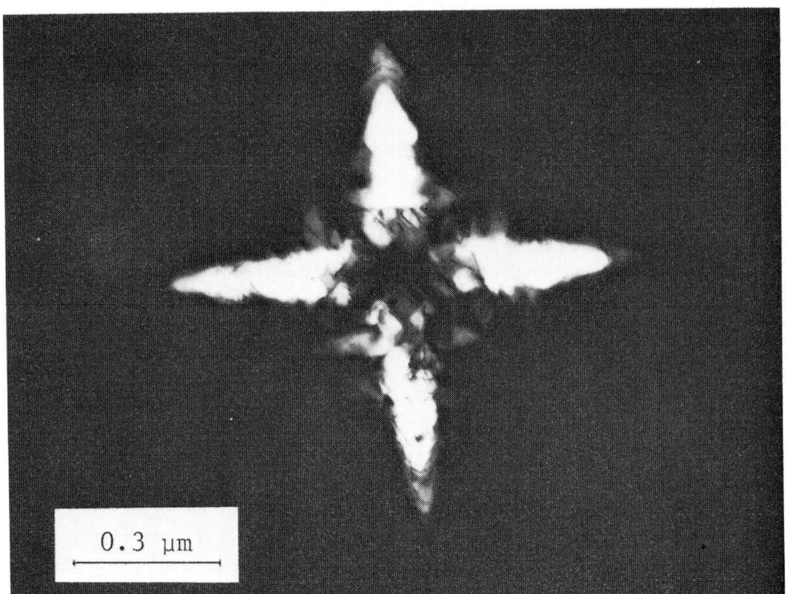

Figure 10.18 Dendritic growth of primary α-Fe crystals in $Fe_{81}B_{13.5}C_{3.5}Si_2$ (courtesy of J. C. Swartz)

10.6 Nucleation

The question of nucleation in metallic glasses has been considered seriously only recently[69] and there are still arguments about the relative roles of true nucleation within the glass and the athermal growth of pre-existing or 'quenched-in' nuclei. Added to the difficulties of interpretation are those of measurement of nucleation rates. Either the number density of crystals as a function of time can be counted from micrographs or the nucleation rate can be calculated from measurements of the growth rate and overall transformation kinetics. In the first case errors can be made unless proper allowance is made for sectioning effects and in both cases the results are unreliable during the early stage of crystallization when the numbers and sizes of crystals are small.

From classical nucleation theory[70] the critical radius, r_c for homogeneous polymorphous nucleation in an undercooled melt, is given by

$$r_c = 2\sigma T_m / H_v \Delta T \tag{10.6}$$

where σ is the solid–liquid interfacial energy, T_m is the liquidus temperature, ΔT the undercooling and H_v the crystallization enthalpy. An estimate of r_c for the crystallization of the glass Fe–B can be obtained by taking σ as equal to the solid–liquid interfacial energy[71] in iron, i.e. $0.25\,\mathrm{J\,m^{-1}}$ and H_v as the heat of crystallization, $80\,\mathrm{J\,g^{-1}}$. At a typical crystallization temperature, $T_x = 0.5T_m$, r_c is of the order of 1 nm; this value is only slightly greater than the unit cell size[52] of Fe_3B and may well explain why the phases produced by crystallization of metallic glasses usually have simple crystal structures. More seriously it throws doubts on the applicability of classical nucleation theory at such large undercoolings.

The steady state nucleation rate, I_s, is given by[70]

$$I_s = I_0 \exp(-\Delta G_c/RT) \exp(-Q_n/RT) \tag{10.7}$$

where

$$\Delta G_c = (16\pi\sigma^3/3)(\Delta F_v)^2 \tag{10.8}$$

ΔG_c is the energy required to form a critical sized nucleus, Q_n is the activation energy for transfer of an atom from the matrix to the embryo and ΔF_v is the free energy difference between the glass and the crystal. *Figure 10.19* shows the temperature dependence of I_s, calculated from equation 10.7 for a number of metallic glass formers[72]. By studying the crystallization of $Fe_{40}Ni_{40}P_{14}B_6$ over a wide temperature range Morris has confirmed that I_s does indeed have a maximum. However in most cases crystallization has been studied in a narrow temperature range, well below the maximum in the curve, where ΔF_v is large, $\Delta G_c \gg Q_n$ and I_s decreases in an Arrhenius fashion with decreasing temperature. *Figure 10.20* shows some typical measurements[40] for $Ni_{36.5}Zr_{63.5}$. After an initial incubation period the nucleus density increases linearly with time with no evidence of site saturation. The nucleus densities, typically $10^{18}\,\mathrm{m^{-3}}$, are much higher than those obtained during solidification just below T_m and, more importantly, than the density of the heterogeneous sites. The activation energies for nucleation Q_n are typically 500—600 kJ/mol, which are considerably higher than those for growth in the same glass.

For steady state nucleation an equilibrium concentration of embryos must exist at all times. In practice this cannot be the case and after the specimen is brought to the annealing temperature there must be an initial time period during which the population of embryos is established and the nucleation rate rises from zero to its steady state value. The role of this transient nucleation in devitrification of non-metallic glasses has been discussed in detail by Gutzow[73], who has shown that the nucleation rate $I(t)$ at

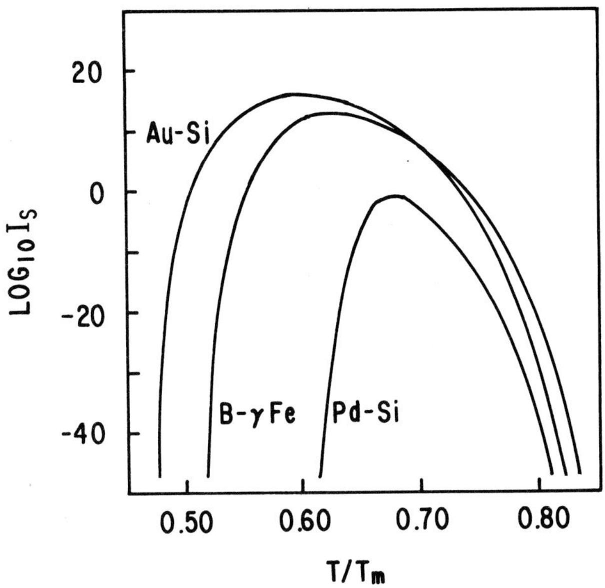

Figure 10.19 Calculated steady state homogeneous nucleation frequency, I_s, as a function of reduced temperature, T/T_m, for 3 metallic glass formers (courtesy of A. L. Greer)

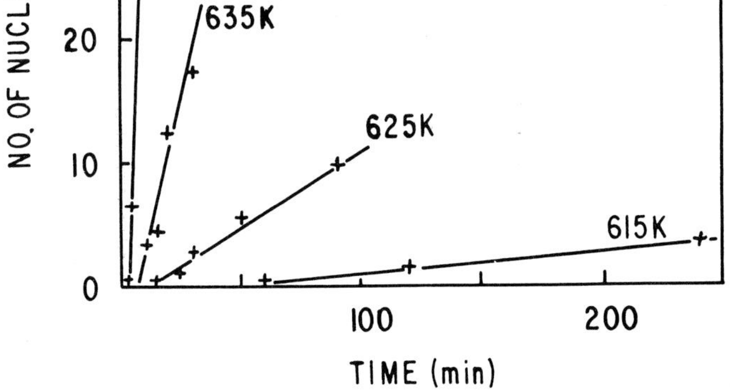

Figure 10.20 Measured nucleation rate for polymorphous crystallization of the glass Zr_2Ni

time t may be approximated by:

$$I(t) = I_s \exp(-\tau/t) \qquad (10.9)$$

where τ scales as the shear viscosity. At small undercoolings τ is negligible and steady state nucleation is established almost instantly. At a typical devitrification temperature however, τ is expected to be of the order of 10^2 to 10^3s, and give rise to measurable incubation times as shown in *Figures 10.10* and *10.20*. Below T_g, τ is expected to become so long that steady state nucleation is unlikely to be achieved. Under such circumstances, athermal growth of the quenched-in nuclei may predominate as follows. During glass formation the undercooled liquid is quenched through the temperature range in which I_s is large and τ is small (*Figure 10.19*). Although there is insufficient time for their growth, nuclei formed in this temperature interval are retained in the glass at room temperature. Since r_c decreases with decreasing temperature many of these nuclei will be supercritical at the annealing temperature and on subsequent heat treatment may grow with no activation barrier. Under such circumstances the overall transformation kinetics will be identical to those for growth and at any time the partially transformed glass will contain a fixed number of equal sized crystals all of which started to grow at time zero. Evidence for this process has been found in a number of metallic glasses[53,73,74], in particular the much studied[75] $Fe_{80}B_{20}$ for which Greer[52] has shown that the DSC curves are consistent with a population density of quenched-in nuclei which varies with the quenching rate, an observation which may go some way to explaining the wide variation of transformation kinetics found by different investigators of this glass[76]. In general, growth of 'quenched-in' nuclei is expected to be the dominant mechanism at low annealing temperatures and in difficult glass formers. Conversely, high annealing temperatures and high quenching rates will favour transient and steady state nucleation.

This far only homogeneous nucleation has been considered and the role of potential heterogeneous sites has been ignored. In view of the large difference between glass–crystal and crystal–crystal interfacial energies heterogeneous nucleation at surfaces or interfaces is unlikely to occur to the same extent as in crystalline solid state transformations. Nevertheless preferential crystallization at ribbon surfaces has been reported in a number of glasses, though this may be a consequence of compositional inhomogeneities rather than genuine heterogeneous nucleation[39], e.g. the glass $Fe_{40}Ni_{40}B_{20}$ is believed to crystallize from the surface because preferential oxidation of iron destablizes the glass. Similarly the segregation of metalloids to cracks and other defects in heavily deformed glasses[77] may lead to crystallization on these sites. Pre-existing crystals, either quenched-in or produced by primary crystallization, in some cases act as nucleation sites[8,78]; this may be a consequence of solute rejection and hence a reduction of stability in the vicinity of the crystals.

10.7 Overall crystallization kinetics

Since the crystallization of metallic glasses is a nucleation and growth process the overall rate of transformation will reflect the time and temperature dependence of both. As shown in the introduction the overall shape of the TTT diagram is a C. However, most measurements of transformation kinetics are made well below the nose of the diagram in the linear region. Solid state nucleation and growth transformations are usually described by the Johnson–Mehl–Avrami equation[79]:

$$\chi(t) = 1 - \exp[-b(t - t_0)^n] \qquad (10.10)$$

where $\chi(t)$ is the fraction transformed after time t, t_0 is the incubation time, b is a rate constant and n is an exponent which need not be an integer. Transformation data for numerous metallic glasses have been fitted successfully to this equation[80-85], even though it is strictly valid only for spherical particles[52]. The exponent n, which generally lies between 1.5 and 4, may be written as $n = n_n + n_g$, where n_n describes the time dependence of the nucleation rate ($0 \geqslant n_n \geqslant 1$) and n_g that of the growth rate ($1.5 \geqslant n_g \geqslant 4$). Unambiguous interpretation of the value of n can be obtained only by parallel microstructural investigations and independent measurements of the nucleation and growth rates as described in the last two sections. Typically n_n should be 0 for athermal growth of quenched-in nuclei and unity for steady state nucleation; values of 1.5, 3 and 3 for n_g are expected for primary, eutectic and polymorphous growth respectively, though these values will be reduced if the crystals show pronounced directionality of growth. Reported values of n greater than 4 are thought to arise from errors in analysis, such as the failure to subtract incubation times[81]. The alternative explanation of an autocatalytic acceleration of nucleation and growth rates[86] seems improbable since the temperature rise ahead of a growing crystal has been calculated to be extremely small[52]. The activation energy, E_x, for the overall crystallization process can be obtained from either the time taken for a certain fraction of crystallinity or from the temperature dependence of the rate constant:

$$b = b_0 \exp(E_x/kT) \tag{10.11}$$

or from continuous heating experiments using Kissinger's[11] or Ozawa's[10] method. Typically E_x is of the order of a few hundred kJ/mol, i.e. close to that for viscous flow. However E_x will reflect the activation energies for both nucleation (E_n) and growth (E_g) and it is more meaningful to determine these separately. Ranganathan and von Heimendahl[87] have shown that they are related by:

$$E = (n_n E_n + n_g E_g)/n \tag{10.12}$$

where n, n_n and n_g have the meanings described above. The significance of E_n and E_g have been discussed in earlier sections.

10.8 Phase separation

In *Figure 10.7* it was assumed that the heat of mixing in the undercooled liquid is negative at all temperatures, giving rise to a free energy versus composition curve which is always concave downwards. Any clustering in the liquid, however, will lead to a positive heat of mixing and a free energy curve which at a low temperature is as shown in *Figure 10.21*. At compositions between c and c' the initially single phase glass can then lower its free energy by separating into two glassy phases with compositions given by the common tangent construction. Two distinct microstructural mechanisms are possible: in the composition intervals c to c_s and c'_s to c' there is an activation barrier to phase separation which must therefore occur by a nucleation and growth mechanism; between c_s and c'_s, i.e. where d^2G/dc^2 is negative, no activation barrier exists and spinodal decomposition occurs. In the first case discrete particles of the second phase nucleate and grow with fixed composition; in the second sinusoidal fluctuations of composition appear and grow in intensity at a wavelength which depends on temperature.

Although phase separation is well known in oxide glasses[88] and is the basis of Vycor and many glass ceramics it is relatively scarce in metallic systems. A double glass transition in Ni–Pd–P[89] and Ti–Be–Zr[90] has been taken as evidence for two glassy

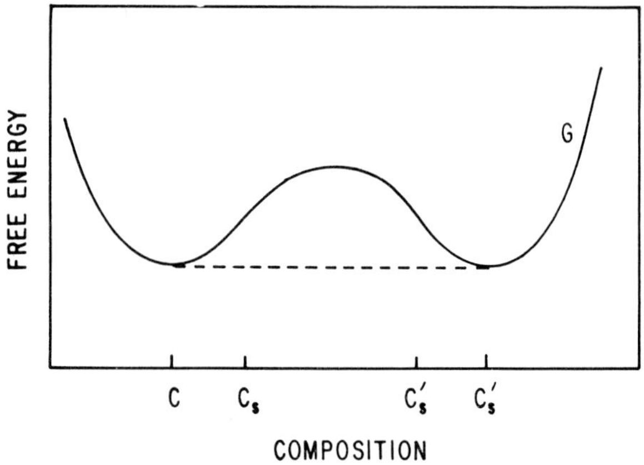

Figure 10.21 Schematic free energy versus composition diagram to illustrate phase separation of a glass. Phase separation can occur between c and c'; between c_s and c_s' the mechanism is spinodal

phases in these alloys. The microstructures of some as-quenched Fe–B[91] and Ti–Be–Zr[90] glasses show two interconnected glassy phases. The glass $Pd_{74}Au_8Si_{18}$ is single phase after quenching but phase-separates[56] on annealing near 400 °C. Chou and Turnbull[82] have used small angle X-ray scattering to show that, in the early stages at least, the decomposition is spinodal, with the wavelength of the composition fluctuations of the order of 20 nm. The scattering intensity at selected wavelengths initially increases linearly with time consistent with Cahn's linear theory of spinodal decomposition[92]. At longer times the rate of growth decreases and there is some sharpening of the interface between the regions of different compositions.

10.9 Influence of external factors

Thus far it has been assumed that crystallization has been studied at atmospheric pressure and in the absence of constraints such as applied stress and magnetic field. Since the crystallization of metallic glass involves a densification high pressures might be expected to accelerate the process. In fact a number of studies have shown unambiguously that the reduced atomic mobility at high pressures retards the crystallization[93–98]; in Pd–Ag–Si glasses for example T_x is increased by 1.4 K/kbar[96]. At the same time crystallization sequences are changed to favour those crystalline phases which have maximum density, e.g. above 100 kbar primary crystallization of bcc Fe in Fe–B glasses is replaced by the hcp ε phase[98]; in Nb–Si glasses the cubic A15 Nb_3Si is formed in preference to the normal tetragonal form[97]. Conversely to the effect of pressure, application of a tensile stress increases the free volume and therefore the atomic diffusivity[99]. The resulting acceleration of crystallization has been noted in both $Fe_{40}Ni_{40}P_{14}B_6$ [100] and $Fe_{32}Ni_{36}Cr_{14}P_{12}B_6$ [101]. Von Heimendahl *et al.*[101] have shown that, whilst the nucleation rate is always increased, the growth rate increases significantly only in the case of primary crystallization. Deformation followed by subsequent heat treatment has no influence on the crystallization behaviour because the structural changes induced by deformation are recovered at temperatures well below those necessary to induce crystallization[102].

The potential use of metallic glasses in nuclear reactor environments has led to interest in their susceptibility to radiation damage. The crystallization behaviour has been observed after bombardment by neutrons[103–105], ions[106,107] and fission fragments[108]. Two opposing effects might be expected: the destruction of short range order might decrease the number of available nuclei; the creation of free volume will accelerate crystal growth. The results to date are ambiguous and often contradictory; both increases[103,107,108] and decreases[104–106] in T_x have been reported. Some of these discrepancies may arise from the difficulty in preventing temperature rises in the specimen during radiation. Finally Kopewicz *et al.*[109] have noted that the application of a high frequency field can induce crystallization in metallic glasses even when they are held at cryogenic temperatures.

10.10 Technical implications

The properties of metallic glasses are highly sensitive to small amounts of crystallinity and as these materials find an increasing number of applications the dangers of using them without a detailed knowledge of their long-term thermal stability cannot be overstressed. The core of a power distribution transformer, for example, must operate near 100 °C for at least 20 years with no significant degradation of properties[14]. Unfortunately all of the crystallization studies described in this chapter have been carried out within 100 K of T_g and there is no public knowledge of the long term stability of metallic glasses at lower temperatures. Near T_g the crystallization process is likely to involve both nucleation and growth whereas at lower temperatures growth of quenched-in nuclei may predominate. Extrapolation of high temperature data through several hundred degrees is therefore extremely unreliable and no substitute for long term stability tests that all manufacturers and users must feel obliged to run.

The heat treatment of metastable phases is a cornerstone of physical metallurgy and it is now becoming clear that metallic glasses are highly suitable candidates. In some cases the introduction of a controlled amount of crystallinity can actually enhance the properties of a metallic glass, e.g. Fe–B–C–Si alloys where the eddy current losses at high frequencies are reduced by the precipitation of a few percent of primary α-Fe crystallites which refine the domain structure by providing nucleation sites for domain walls[110–112]. The reduction in losses depends on the volume fraction, size and spacing of the crystals and therefore critically on the heat treatment conditions. Contrast this with the low frequency behaviour of the same alloys which is impeded by small amounts of surface crystallinity introduced during the melt-quenching process[110,113,114]. The search for high T_{cr}, high field, high current superconducting devices has provoked an interest in the fine grain crystalline microstructures produced on devitrification of metallic glasses. Hf–Zr–V glasses, for example, are superconducting only below 2 K; yet when heat treated to produce a 20 nm dispersion of C-15 Laves phases T_{cr} is increased significantly and the small grain size provides the desired flux pinning[115]. Moreover, since the superconducting properties are unaffected by exposure to fast neutrons, such materials are good candidates for a fusion reactor environment. In a similar vein, ductile tapes superconducting up to 18 K may be made by crystallizing sputtered Nb–Ge films[116]. More recently Ray[6,117] has shown that a large category of complex transition metal–refractory metal–boron glasses can be heat treated to produce high strength, high hardness, ductile microcrystalline phases. These materials which are now available commercially[118] may be pulverized and subsequently hot-consolidated to produce complex shapes such as machine tools.

Acknowledgements

This chapter is dedicated to the former members of the Sussex Metastable Alloys Group with all of whom it was a pleasure to work. In particular I thank Drs Watanabe, Dong and Gregan who did much of the experimental work and Drs Doherty, Cantor and Cahn for discussions. My work was generously supported by both the UK Science and Engineering Research Council and the US Office of Naval Research. I thank Shirley Stay for word-processing the manuscript.

References

1. Jones, G. O., *'Glass'*, p. 1, Chapman and Hall, London (1971)
2. Davies, H. A., *'Proc. 3rd Int. Conf. on Rapidly Quenched Metals'*, Ed. Cantor, B., vol. I, p. 1, The Metals Society, London (1978)
3. Anderson, P. M., Steinberg, J. and Lord, A. E., *J. Non-cryst. Solids*, **34**, 267 (1979)
4. Chen, H. S., *J. appl. Phys.*, **49**, 3289 (1978)
5. Polk, D. E. and Chen, H. S., *J. Non-cryst. Solids*, **15**, 165 (1974)
6. Ray, R., *J. mater. Sci.*, **16**, 2924 (1981)
7. Kursumovic, A. and Scott, M. G., *Appl. Phys. Lett.*, **37**, 620 (1980)
8. Swartz, R. C., Kossowsky, R., Haugh, J. J. and Krause, R. F., *J. appl. Phys.*, **52**, 3324 (1981)
9. Luborsky, F. E., *Mater. Sci. Engng*, **28**, 139 (1977)
10. Ozawa, T., *J. therm. Anal.*, **2**, 301 (1970)
11. Kissinger, H. E., *Analyt. Chem.*, **29**, 1702 (1957)
12. Kemeny, T., Vincze, I., Fogarassy, B. and Arajs, S., ref. 2, vol. I, p. 291
13. Watanabe, T. and Scott, M. G., *J. mater. Sci.*, **15**, 1131 (1980)
14. Luborsky, F. E., *Proc. 2nd Int. Symp. Amorphous Magnetism*, Eds. Levy, R. and Hasegawa, R., Plenum Press, New York (1976)
15. Calka, A. Madhava, M., Polk, D. E., Giessen, B. C. and Vandersande, J., *Scripta Met.*, **11**, 65 (1977)
16. Johnson, W. L., Poon, S. J., Durand, J. and Duwez, P., *Phys. Rev.* **B18**, 201 (1978)
17. Ray, R., Hasegawa, R., Chou, C.-P. and Davis, L. A., *Scripta Met.*, **11**, 973 (1977)
18. Dong, Y. D., Gregan, G. and Scott, M. G., *J. Non-crystal. Solids*, **43**, 403 (1981)
19. Chen, H. S. and Park, B., *Acta metall.*, **21**, 393 (1973)
20. Chen, H. S., *Acta metall.*, **22**, 897 (1974)
21. Chen, H. S., *Acta metall.*, **24**, 153 (1976)
22. Walter, J. L., *Mater. Sci. Engng*, **50**, 137 (1981)
23. Donald, I. W., Ward, K. D., Davies, H. A. and Crangle, J. C., ref. 40
24. Lovas, A., Granasy, L., Zambo-Balla, K. and Kiraly, J., *'Proceedings of Conference on Metallic Glasses; Science and Technology*, Eds. Hargitai, C., Bakonyi, I. and Kemeny, T., vol. II, p. 291 Kultura, Budapest (1981)
25. Giessen, B. C. and Elliott, R. O., ref. 2, vol. I, p. 406
26. Buschow, K. H. J. and Beekmans, N. M., *Phys. Rev.*, **B19**, 3843 (1979)
27. Buschow, K. H. J. and Dirks, A. G., *J. appl. Phys. D*, **13**, 251 (1980)
28. Chen, H. S., *J. Non-cryst. Solids*, **27**, 257 (1978)
29. Chen, H. S., *Appl. Phys. Lett.*, **29**, 12 (1976)
30. Naka, M., Tomizawa, S., Masumoto, T. and Watanabe, T., *'Rapidly Quenched Metals'*, Eds. Giessen, B. C. and Grant, N. J., p. 273 Massachusetts Institute of Technology Press, Cambridge (1976)
31. Donald, I. W. and Davies, H. A., ref. 2, vol. I, p. 273
32. Inoue, A., Kobayashi, K., Kanehira, J. and Masumoto, T., *Sci. Rep. Res. Insts Tôhuku Univ.*, **A29**, 331 (1981)
33. Nagel, S. R. and Tauc, J., *Phys. Rev. Lett.*, **35**, 380 (1975)
34. Cote, P. J. and Meisel, L. V., ref. 2, vol. I, p. 62
35. Pekala, K., Trykozko, R., Ostatek, J. and Matyja, H., *J. appl. Phys.*, **22**, 369 (1980)
36. Hillmann, H. and Hilzinger, H. R., ref. 2, vol. I, p. 371
37. Nagumo, M., Takahashi, T., Arai, T. and Hasegawa, T., ref. 2, vol. I, p. 383
38. Koster, U. and Weiss, P., *J. Non-cryst. Solids*, **17**, 359 (1975)
39. Koster, U. and Herold, U., *'Glassy Metals I'*, Eds. Güntherodt, H. J. and Beck, H., p. 225, Springer-Verlag, Heidelberg (1981)

40. Scott, M. G., Gregan, G. and Dong, Y. D., *'Proc. 4th Int. Conf. on Rapidly Quenched Metals'*, Eds. Masumoto, T. and Suzuki, K., Japan Institute of Metals, Sendai (1982)
41. Bagley, B. G. and Turnbull, D., *Acta metall.*, **18,** 857 (1970)
42. Turnbull, D., *Solid St. Phys.*, **3,** (1956)
43. Morris, D. G., *Acta metall.*, **29,** 1213 (1981)
44. Tiwari, R. S., Ranganathan, S. and Von Heimendahl, M., *Z. metallk.*, **72,** 563 (1981)
45. Wu, Y. K., Laing, J. Z. and Kuo, K. H., *Phys. Status Solidi A*, **64,** 113 (1981)
46. Walter, J. L. and Bartram, S. F., ref. 2, vol. I, p. 307
47. Duhaj, P., Barancok, D. and Ondreika, A., *J. Non-cryst. Solids*, **21,** 411 (1976)
48. Boswell, P. G. and Chadwick, G. A., *Scripta Met.*, **10,** 509 (1976)
49. Boswell, P. G. and Chadwick, G. A., *Scripta Met.*, **11,** 701 (1977)
50. Fine, M. E., *'Phase Transformations in Condensed Systems'*, p. 81, Macmillan, New York (1964)
51. Herold, U. and Koster, U. ref. 2, vol. I, p. 281
52. Greer, A. L., *Acta metall.*, **30,** 171 (1982)
53. Cahn, J. W., and Hagel, W. C., *'Decomposition of Austenite by Diffusional Processes'*, Eds. Zackay, V. F. and Aaronson, H. I., p. 131 Interscience, New York (1962)
54. Chadwick, G. A., *'Metallography of Phase Transformations'*, p. 119, Butterworths, London (1972)
55. Cahn, R. W., Evetts, J. E., Patterson, J., Somekh, R. E. and Kenway-Jackson, G., *J. mater. Sci.*, **15,** 702 (1980)
56. Chen, H. S. and Turnbull, D., *Acta metall.*, **17,** 1020 (1969)
57. Livingston, J. D., *Mater. Sci. Engng*, **7,** 61 (1971)
58. Carpay, F. M. A. and Cense, W. A., *J. Cryst. Growth*, **24/25,** 551 (1974)
59. Cense, W. A. and Albers, W., *J. Non-cryst. Solids*, **28,** 391 (1978)
60. Freed, R. L. and Vandersande, J. B., *Acta metall.*, **28,** 103 (1980)
61. Von Heimendahl, M. and Oppolzer, H., *Scripta Met.*, **12,** 1087 (1978)
62. Burke, J., *'The Kinetics of Phase Transformations in Metals'*, p. 160, Pergamon, London (1965)
63. Shewmon, P. G., *'Transformations in Metals*, p. 170, McGraw-Hill, New York (1969)
64. Doherty, R. D., *'Crystal Growth'*, Ed. Pamplin, B. R., p. 485, Pergamon, London (1980)
65. Shimomura, K., Shingu, P. and Ozaki, R., *J. mater. Sci.*, **15,** 1175 (1980)
66. Aaron, H. B., Fainstein, D. and Kotter, G. R., *J. appl. Phys.*, **41,** 4404 (1970)
67. Koster, U., Herold, U. and Becker, A., ref. 40
68. Swartz, J. C., Haugh, J. J., Krause, R. F. and Kossowsky, R., *J. appl. Phys.*, **52,** 1908 (1981)
69. Chalmers, B., *'Physical Metallurgy*, p. 235, Wiley, New York (1962)
70. Fine, M. E., ref. 50, p. 7
71. Woodruff, D. P., *'The Solid-Liquid Interface'*, p. 28, Cambridge University Press (1973)
72. Thompson, C. V., Greer, A. L. and Drehman, A. J., ref. 40
73. Gutzow, I., *Contemp. Phys.*, **21,** 121 (1980)
74. Boswell, P. G., *J. mater. Sci.*, **15,** 1939 (1980)
75. Schaafsma, A. F., Snijders, H., Vanderwoude, F., Drijver, J. W. and Radelaar, S., *Phys. Rev. B*, **20,** 4423 (1979)
76. Leake, J. A. and Greer, A. L., *J. Non-cryst. Solids*, **38/39,** 735 (1980)
77. Walter, J. L., Bacon, F. and Luborsky, F. E., *Mater. Sci. Engng*, **24,** 239 (1976)
78. Banerjee, S. and Scott, M. G., unpublished work
79. Burke, J., ref. 62, p. 192
80. Scott, M. G., *J. mater. Sci.*, **13,** 291 (1978)
81. Scott, M. G. and Ramachandrarao, P., *Mater. Sci. Engng*, **29,** 137 (1977)
82. Chou, C.-P. and Turnbull, D., *J. Non-cryst. Solids*, **17,** 169 (1975)
83. Bagley, B. G. and Vogel, E. M., *J. Non-cryst. Solids*, **18,** 29 (1975)
84. Funakoshi, N., Kanamori, T. and Manabe, T., *Jap. J. appl. Phys.*, **16,** 515 (1977)
85. Burton, J. J. and Ray, R., *J. Non-cryst. Solids*, **6,** 393 (1971)
86. Funakoshi, N., Kanamori, T. and Manabe, T., *Jap. J. appl. Phys.*, **17,** 11 (1978)
87. Ranganathan, S. and Von Heimendahl, M., *J. mater. Sci.*, **16,** 2401 (1981)
88. Doremus, R. H., *'Glass Science'*, p. 44, Wiley, New York (1973)
89. Chen, H. S., *Mater. Sci. Engng*, **23,** 151 (1976)
90. Tanner, L. E., *Scripta Met.*, **14,** 657 (1980)
91. Walter, J. L. and Bartram, S. F., ref. 2, vol. I, p. 307
92. Cahn, J. W., *Trans. Metall. Soc. Am. Inst. Min. metall. Petrol. Engrs.*, **242,** 166 (1968)
93. Wang, W. K., Iwasaki, H. and Fukamichi, K., *J. mater. Sci.*, **15,** 2701 (1980)
94. Iwasaki, H. and Masumoto, T., *J. mater. Sci.*, **13,** 2171 (1978)
95. Cedergren, M. and Backstrom, G., *J. Non-cryst. Solids*, **30,** 69 (1978)
96. Emmens, W., Vrijen, J. and Radelaar, S., *J. Non-cryst. Solids*, **18,** 299 (1975)
97. Suryanarayana, C., Wang, W. K., Iwasaki, H. and Masumoto, T., *Solid St. Communs*, **34,** 861 (1980)

98. Ogawa, Y., Nunogaki, K., Endo, S., Kiritani, M. and Fujita, E., ref. 40
99. Spaepen, F., ref. 2, vol. II, p. 253
100. Patterson, J. and Jones, D. R. H., *Scripta Met.*, **13,** 947 (1979)
101. Von Heimendahl, M., Tiwari, R. S. and Claus, J.-C., ref. 40
102. Krishnanand, K. D. and Cahn, R. W., *Scripta Met.*, **9,** 1259 (1975)
103. Doi, K., Ayano, T. and Kawamura, K., *J. Non-cryst. Solids,* **34,** 405 (1979)
104. Gerling, R. and Wagner, R., ref. 40
105. Cahn, R. W., Toloui, B., Akhtar, D, and Thomas, M., ref. 40
106. Azam, N., Lenaour, L., Rivera, C., Grosjean, P., Sacovy, P. and Delaplace, J., *J. nucl. Mater.*, **83,** 298 (1979)
107. Ayano, T., Ohno, H., Utumi, K., Kawamura, K. and Furukawa, K., *J. nucl. Mater.*, **79,** 430 (1979)
108. Elliott, R. O., Koss, D. A. and Giessen, B. C., *Scripta Met.*, **14,** 1061 (1980)
109. Kopcewicz, M., Gonder, U. and Wagner, H. G., *J. appl. Phys.*, **23,** 1 (1980)
110. Davis, L. A., Cristofaro, N. and Smith, C. H., ref. 24, vol. I, p. 1
111. Hasegawa, R., Fish, G. E. and Ramanan, V. R. V., ref. 40
112. Datta, A., Decristofaro, N. and Davis, L. A., ref. 40
113. Fiedler, H. C., Livingston, J. D. and Huang, H. C., *J. Magnetism magn. Mater.*, to appear
114. Fiedler, H. C., *'Proc. Third Joint INTERMAG — MMM Conf.*, Montreal, 1982
115. Tenhover, M., *IEEE Trans. Magn.*, **17,** 1021 (1981)
116. Tsuei, C. C., *Appl. Phys. Lett.*, **33,** 262 (1978)
117. Ray, R., *J. mater. Sci.*, **16,** 2927 (1981)
118. Devitrium, Marketed by Allied Corporation, N.J., USA

Chapter 11

Structural relaxation in metallic glasses

H. S. Chen

Bell Laboratories, Murray Hill, New Jersey, USA

11.1 Introduction

A metallic glass can be regarded as a metallic solid with a frozen-in melt structure which lacks three-dimensional atomic periodicity. In general, it is not in an internal equilibrium state and relaxes structurally to the latter whenever atoms attain an appreciable mobility. Associated with the structural relaxation, many physical properties change; some drastically and others only slightly.

For the past few years, great advances have been made in the fabrication and property optimization for the application of metallic glasses as electromagnetic devices. Concurrently the understanding of the stability and structure relaxation processes have drawn great interest, and a number of papers on this subject have appeared[1–3]. These investigations often lead to new controversy and new problems which remain to be solved.

This chapter reviews past observations and present understanding of the relaxation process and its influence on physical properties. Section 11.2 describes ordering parameters to characterize different states of glassy structure. Section 11.3 presents briefly the effects of relaxation on physical properties, and Section 11.4 describes the general characteristics of structure relaxation. Emphasis is laid particularly on the different effects of relaxation exerted upon macroscopic properties, such as volume, elastic constants and enthalpy; structurally sensitive properties, such as resistivity, lifetime of positron annihilation and microhardness, and other properties. Section 11.5 lists annealing effects on properties such as internal friction, magnetization processes and embrittlement, and section 11.6 kinetics of relaxation processes in as-quenched and stabilized samples. The reversibility and cross-over phenomena are critically examined. Section 11.7 discusses the influence of deformation and irradiations, from which investigations the understanding of microstructure is richly revealed, and Section 11.8 presents recent investigations of enthalpy relaxation which reveal many significant novel features. A new model glass transition in the framework of percolation theory is proposed in section 11.9. The model predicts the localized modes of structure relaxation in sub-sub-T_g annealling and the concept of a distribution in glass transitions. Additional remarks are made in Section 11.10 on the small volume contraction as compared with large changes in other properties, and different relaxation modes influencing in various ways some properties.

11.2 Glassy state ordering parameters

In the region of the glass transition, the properties of glass forming liquid are not simply dependent on temperature and pressure but also on time. For a simple case, as shown[4] in *Figure 11.1*, for $Au_{77}Si_9Ge_{14}$ at each cooling rate, if crystallization is by-passed below the melting point T_m, a liquid structure freezes at different temperatures T_g's, determined by the relaxation time constant, τ, and time of measurement, t_{meas} [1]

$$\tau(T_g) = t_{meas} \tag{11.1}$$

and

$$t_{meas} \approx (k_B T_g^2/Q_a)\dot{T}^{-1} \tag{11.2}$$

where k_B is Boltzmann's constant. Q_a is the apparent activation energy and $\dot{T} = dT/dt$ is the quenching rate. The factor $k_B T_g^2/Q_a$ is 5 to 10 K. At fast quenching of 10^6 K/s, commonly experienced during melt spinning of metallic glass formation, the glass freezes at T_{g1}, at which $\tau \approx 10^{-5}$ s into a glass denoted by G_1. At slower cooling rates of about 1 K/s (in practice, it is done by cooling the liquid after reheating the fast quenched glass just above T_{g2}), the glass freezes at a lower temperature T_{g2}. The resultant glass G_2 as compared with glass G_1 has lower enthalpy, H (and volume, V), by one third of the heat of fusion (the volume contraction) compared with the liquid–crystal transformation, and higher viscosities by six orders of magnitudes. At extremely slow cooling rates, the enthalpy decreases following path 'd'. In practice the length of path 'd' is limited by the increasingly slow rates of cooling required. Extrapolation of H (or V) against T would equal and become below the crystalline value, at $T_0 \lesssim 0.8T_g$, which is the well known Kauzeman's paradox[5].

It is obvious that to specify the glassy state, an additional set of parameters, beside temperature and pressure, is required. Use is often made of the concept of fictive temperature, T_f, first proposed by Tool[6] and some macroscopic properties such as free volume and configurational entropy[7]. T_f is defined as the temperature at which a glass would be in equilibrium if brought very rapidly to that temperature. In the case of simple cooling, T_f may be approximated by T_g, and the concept of T_f describing a different glassy state is appropriate. When two glasses G_1 and G_2 with different fictive temperature are annealed at the intermediate temperature T_a, they relax from both sides towards the state corresponding to the fictive temperature T_a, (*Figure 11.1*). However, if glass G_1 is annealed at temperature T_a' below T_{g2}, until it relaxes to the

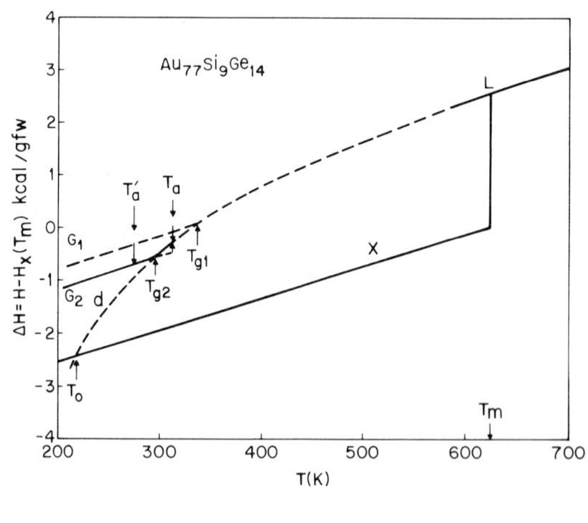

Figure 11.1 Enthalpies H of $Au_{77}Si_9Ge_{14}$ in various states; L = liquid, G = glass, X = crystal. Experimental data are shown by solid lines (after Chen and Turnbull[4]). $H_x(T_m)$ is set as a reference

state corresponding to the fictive temperature and then brought up rapidly to T_{g2}, one may expect that nothing would occur. In fact, it has been seen in many glasses that the volume[8], the index of refraction[9], and Curie temperature[10] initially increase (or decrease) rapidly then approach the equilibrium value. This is known as the memory effect or cross-over phenomenon. Apparently the single parameter T_f is not sufficient to describe the structural state of a glass. Two relaxation times and distribution in relaxation times have been employed in explaining this effect as applied to the relaxation process near the glass transition or sub-T_g region[11].

Davies and Jones[12] showed that if a single ordering parameter is sufficient to determine the glassy state, the following relationship should hold:

$$\Delta\beta\Delta C_p/(\Delta\alpha)^2 TV = 1 \qquad (11.3)$$

where $\Delta\beta$, $\Delta\alpha$ and ΔC_p respectively are changes in compressibility, thermal expansion and specific heat at T_g. Experimental data in most glassy systems, except metallic glasses which have not been tested, show that equation 11.3 becomes the inequality[13]

$$\Delta\beta \cdot \Delta C_p/(\Delta\alpha)^2 TV > 1 \qquad (11.4)$$

indicating that more than one order parameter is required. It is clear then that fictive temperature, specific volume or refractive index alone is not sufficient to specify the relaxation behaviour of a glass. We will show further in the following that the approach of using a distribution in relaxation times combined with the relaxation time depending on macroscopic properties, e.g. V_f, S_g or T_f, is also not adequate to describe the low temperature (sub-sub-T_g) relaxation in metallic glasses.

11.3 Relaxation phenomena

Metallic glasses usually obtained at high quenching greater than 10^6 K/s possess a large frozen-in structural disorder and thus have high atomic diffusivities. Many metallic glasses, e.g. Pd–Si, Fe–B, and Zr–Cu, undergo structural relaxation just above room temperature; several hundred degrees below T_g. Associated with the structural relaxation, many physical properties, such as atomic diffusivity[14], viscosity[15], mechanical ductility[16] and magnetic anisotropy[17] in particular, alter drastically. Although the change in density associated with structural relaxation is small[18], $\sim 0.5\%$, an appreciable change has been observed in many other properties, e.g. Young's modulus[18] by $\sim 7\%$, enthalpy[19] by 200 ~ 300 cal/mole, Curie temperature[20,21] by as much as 40 K, and electrical resistivity[22,23] by $\sim 2\%$. Annealing also reduces the internal friction[24,25], rates of stress relief[26], magnetization processes[17], and magnetic disaccommodation effect[27], and shifts these processes to higher temperature. Some structure sensitive properties, e.g. the lifetime and angular correlation of positron annihilation[28], mechanical hardness[29] and electrical resistivity[23] exhibit opposite isochronal annealing effects.

The effect of annealing on physical properties may be found in the related chapters, and will not be detailed here. The reader may also wish to refer to chapters 6 and 8 in Treatise on Materials Science and Technology[30]. In the following we will centre our discussion on the general characteristics of relaxation processes. Emphasis will also be laid on the description of recent novel observations and their interpretations and on the presentation of a model for the glass transition.

11.4 Low temperature (sub-sub-T_g) and high temperature (sub-T_g) relaxation

Available data suggest that the relaxation processes at low temperature ($T_a < T_g - 150\,K$) is distinguishable from that at high temperature near T_g. In the following, manifestation of these two distinguishable processes will be exemplified in three typical cases: (*a*) macroscopic properties, e.g. volume, enthalpy and Young's modulus; (*b*) structure sensitive properties, e.g. the lifetime of positron annihilation and microhardness; and (*c*) semimicroscopic properties, e.g. electrical resistivity and Curie temperature.

11.4.1 Volume, Young's modulus and enthalpy relaxation

As-quenched metallic glasses, upon heating, exhibit volume contraction, exothermic reaction and an increase in Young's modulus. The changes are slow initially and become more rapid at temperature about 100 K below T_g. The rates of change indicate two distinguishable broad distributions as exemplified in *Figure 11.2* in which the differences in apparent specific heat, ΔC_p ($\equiv C_p - C_{p,s}$) are shown, where C_p and $C_{p,s}$ are respectively the apparent specific heat of virgin samples prepared at different rates of quenching and the reference sample cooled slowly[19] (=20 K/min) through T_g. The heat evolution curves indicative of structural relaxation clearly show two broad distributions; the low temperature or sub-sub-T_g($T_g - 200\,K < T < T_g - 100\,K$) (II), and a high temperature one near T_g or sub-T_g ($T \geqslant T_g - 100\,K$) (III) with apparent peak temperatures at T_2 and T_3. The intensity of the relaxation decreases with decreasing

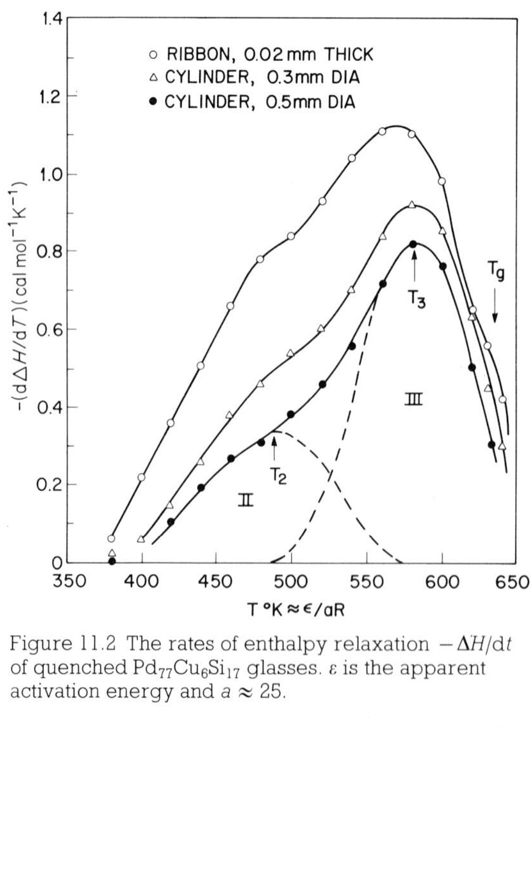

Figure 11.2 The rates of enthalpy relaxation $-\Delta H/dt$ of quenched $Pd_{77}Cu_6Si_{17}$ glasses. ε is the apparent activation energy and $a \approx 25$.

quenching rate, and the ratio of the sub-sub-T_g relaxation to the total one also decreases from 0.4 for the ribbon sample ($-\dot{T} \equiv -dT/dt \geqslant 10^5$ K/s) to about 0.2 for the water quenched cylindrical sample ($-\dot{T} \leqslant 10^3$ K/s). The activation energies for the relaxation (II) are low, ~1.0 to 2.0 eV. It was noted that with the decreasing quenching rates of the samples, not only the onset but also the peak temperatures in sub-sub-T_g relaxation process shift to higher temperatures by 30 K.

11.4.2 The lifetime of positron annihilation and microhardness

Figure 11.3 illustrates the change[28] in the lifetime τ_1 and microhardness[29] H_v for the $Pd_{77}Cu_6Si_{17}$ glass. The annealing times were 30 min and 10 min respectively for τ_1 and H_v measurements. As shown in curve (*a*), τ_1 increases by about 5 ps immediately following an anneal at $T_a \geqslant 350$ K and remains nearly constant for $T_a = 350$ K to 500 K with a plateau at $T_2' = 450$ K. With further annealing, τ_1 decreases approaching a value being about 5 ps lower than the initial value at $T_3' \approx 600$ K. τ_1 appears to increase

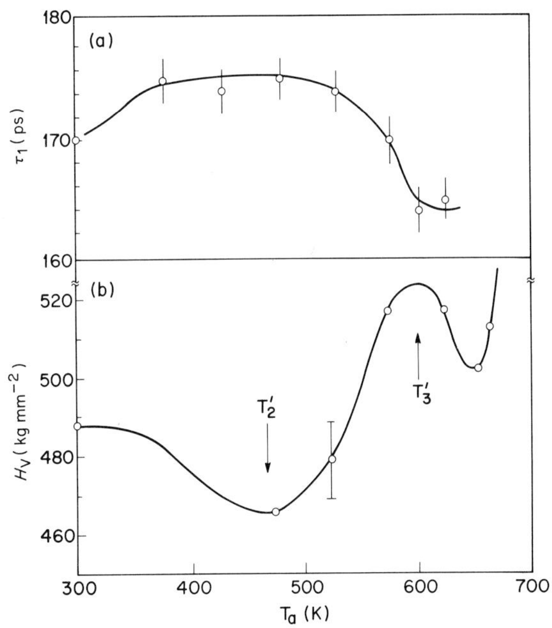

Figure 11.3 (*a*) Isochronal lifetime τ_1 of positron annihilation, and (*b*) microhardness H_v of the PdCuSi glass. Measurements are taken at room temperature

again when T_a approaches T_g. Similar behaviour was reported[31] recently for $Co_{73}Fe_5Si_{12}B_{10}$ and $Pd_{80}Si_{20}$[32]. A plausible explanation is that during the low temperature structural relaxation, some atomic order (or clustering) occurs locally in a more or less rigid matrix, and acts as additional positron trapping sites, whereas high temperature relaxation is a long-range cooperative process which reduces the structural heterogeneity. For comparison in curve (*b*), the sub-sub-T_g anneals lower microhardness H_v resulting from induced structural heterogeneity acting as stress concentration centres, while high temperature homogenization and stabilization increase H_v.

11.4.3 Curie temperature T_C and Electrical Resistivity

During sub-sub-T_g anneal, T_C of many metallic glass ferromagnets increases[20,21] by 5—10 per cent. On further annealing at higher temperature, T_C exhibits a plateau or a peak depending on composition and duration of annealing[21,33]. For short-time anneal in series of $(Fe_{1-x}M_x)_{75}P_{16}B_6Al_3$ with M = Ni, Mo and Cr[20], T_C of the glasses increases with T_a and tends to saturate at T_g. In simpler binary and ternary alloys of $Fe_{80}G_{20}$ (G = B, P, C and/or Si)[33], the variation of T_C with T_a is very sensitive to composition, e.g. the alloys containing phosphorus and carbon, such as $Fe_{80}P_{10}C_{10}$, $Fe_{80}B_{10}C_{10}$ and $Fe_{77}P_{20}Al_3$, exhibit a T_C peak just above T_{C0} (the lowest T_C observable), while glasses containing silicon and boron, such as $Fe_{100-y}B_y$ (y = 15—25), $Fe_{80}B_{10}Si_{10}$ and $Fe_{80}P_{10}Si_{10}$ show T_C increasing monotonously with T_a. In a few cases, e.g. $Fe_{65}Cr_{19}B_{14}Si_2$[34] and $Fe_{90}Zr_{10}$[35] glasses, opposite behaviour is shown, i.e. T_C decreases with T_a.

Figure 11.4 shows the T_C of $Fe_{40}Ni_{40}P_{14}B_6$ subjected to isochronal annealing with t_a = 10 and 60 min. T_C increases monotonically with annealing temperature T_a and

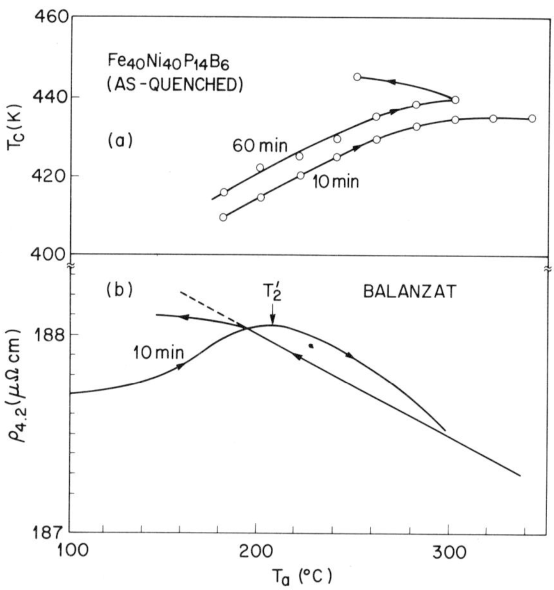

Figure 11.4 (*a*) The isochronal Curie temperature T_C and (*b*) resistivity at 4.2 K $\rho_{4.2}$ for as-quenched $Fe_{40}Ni_{40}P_{14}B_6$ glass. $T_g(\tau^* = 10\,min.) = 370\,°C$

shows saturation only in the region of $T_g(t_a = 10\,min) = 370\,°C$ which is contrary to a T_C peak observed by Greer and Spaepen[36]. In comparison the resistivity at 4.2 K, $\rho_{4.2}$, of the same alloy glass[23] increases with T_a initially and attains a maximum at $T_2' \approx 200\,°C \approx T_2$ the peak temperature in the enthalpy relaxation. $\rho_{4.2}$ decreases at higher temperature, and subsequent cooling increases it. The ρ peak temperature $T_2' = 200\,°C$ is well below $T_g(t_a = 10\,min) = 370\,°C$ and apparently is not in equilibrium state for those T_a at or above 200 °C as was claimed. In fact a continuous and gradual drift in ρ and T_C can be seen after repeated cyclic heating in the region of $T_a \approx 300\,°C$. The difference in annealing effects on T_C of different compositions and on the ρ of a corresponding glassy alloy may be attributed to the different coupling constants to sub-sub-T_g and sub-T_g relaxation processes.

11.5 Influence of annealing on other properties

Annealing reduces the room temperature internal friction Q^{-1}, and shifts the loss spectrum to high temperature. Q^{-1} increases exponentially with temperature with apparent activation energies $A = 0.5$—$1.0\,eV$[24,25]. Although the parameter A has the dimension of an energy, it has no direct physical meaning because of the broad distribution in both frequency and activation energies[37,38].

In simple alloy systems such as binary Fe–B, Fe–P[39], and Zr–Cu[40] the glasses embrittle as a result of crystallization, but those glasses based on nickel, palladium and platinum, whose *d*-shell is nearly filled, exhibit a high ductility even in a partially crystalline state[41,42]. However, in multicomponent systems, particularly iron based alloys, such as $(Fe_{1-x}M_x)_{75}P_{16}B_6Al_3$ (M = Ni, Co, Mn, Mo and Cr)[16], $Fe_{40}Ni_{40}P_{14}B_6$, $Fe_{50}Ni_{30}P_{14}B_6$, $Fe_{40}Ni_{40}B_{20}$[43], and $Fe_{80}P_{13}C_7$[44] ductility is lost at relatively low temperatures without crystallization. The cause of embrittlement is not clear yet, but based on composition dependence of the mechanical ductility, it was suggested[16,39] that the enhanced embrittlement in alloy glasses may be attributed to the same local atomic clustering affecting the lifetime of positron annihilation τ_1, resistivity, ρ, and microhardness, H_v. The enhancement in small angle X-ray scattering (SAXS) observed[45] upon annealing accords with this proposition.

The DC magnetization and AC permeability, μ_{ac}, deteriorate drastically if the metallic glass magnets are annealed between 400 K and T_C in zero magnetic field, owing to a random self-induced magnetic anisotropy. If annealing is conducted above T_C and cooling is fast enough to prevent further atomic rearrangements, substantial improvement in both DC and AC characteristics is observed. Annealing below T_C in a saturation magnetic field in the direction of ribbons leads to lower coercive force and higher remanence ratio but drastically lowers μ_{ac} owing to the formation of large magnetic domains along the ribbons[46]. Partial crystallization is shown to increase μ_{ac}, as discussed in Chapter 19.

Annealing reduces the disaccommodation effects in zero-magnetostrictive cobalt based alloy glasses[27]. The reversible part D^{I} decreases and its peak shifts from 270 K for the as-quenched state to about 370 K, while the irreversible part D^{II} decreases in magnitude but the relaxation peak remains near 470 °K. This sort of magnetic after-effect arises from atomic rearrangement near domain walls.

As expected, annealing drastically decreases some atomic diffusivities[14] and reduces creep rates[15] by as much as six orders of magnitude. However, little effect was observed[47] on diffusivity of the small metalloid atoms boron in $Ni_{60}Nb_{40}$. The 'iso-configurational' creep rate is seen to increase with annealing time[15]. For detail, the reader may refer to Chapters 12 and 24.

11.6 Kinetics of relaxation processes

Isothermal relaxation processes of as-quenched metallic glasses below and near T_g have been investigated from the change in enthalpy recovery, Curie temperature[48], X-ray intensity function[49], volume contraction[50], and Young's modulus[51]. The rates of equilibration obey approximately a $\log t$ law. Most analysis of the logarithmic t recovery is based on variable activation energies, e.g. the linear increase in activation energy either with decreasing number of defects with decreasing disorder[48,49], and on free volume models[50,51]. In cases of stress relief and magnetic ageing performed well below T_g, the kinetics of annealing are given by the first order rate reaction with a single thermally activated process, which contrasts with the $\log t$ law observed. The causes of the different behaviour are not clear yet. The stress relief exhibits two distinguishable processes, the initial short-time relaxation with low activation energies $Q_{re} \sim 0.5$ eV, and the final long time relaxation with $Q_{re} \approx 1$ eV. The rates of stress relief are reduced in samples produced by a slower rate of cooling[52].

In samples preconditioned, or 'stabilized', by annealing at higher temperatures, the reorientation of magnetically induced anisotropy, measured by the remanence to saturation ratio for $Fe_{40}Ni_{40}P_{14}B_6$ [17], is seen to obey simple first order rate reaction kinetics. Later studies[53] on $Fe_{40}Ni_{40}P_{14}B_6$ and $Fe_{75}P_{15}C_{10}$ glasses reveal similar behaviour deduced from the change in ΔE to monitor the kinetics of directional ordering. Q_{re} is approximately 1.4 eV for the FeNi and $\sim$1 eV for the FePC. These values are only about half of $Q(\approx 2.5$ eV) found from the frequency–time shift in internal friction measurements[54]. The latter yields the frequency factors ($10^{20} \sim 10^{25}$ s^{-1}), much greater than the Debye frequency $Q_D \approx 10^{13}$ s^{-1}, suggesting that the first order rate reaction with a single activation energy is not appropriate.

An important feature of the magnetic anneal is that the induced magnetic anisotropy is reversible, suggesting directional ordering processes involving thermally activated local atomic movements. Egami[55] and Kudo and Egami[34] found that T_C of a stabilized glass was reversible and could be cycled between two temperatures (*see Figure 11.8*). Similar behaviour has been seen for Young's modulus[56] and resistivity[23]. Based on the low Q_{re} (≈ 1 eV) observed, Berry and Prichet suggested that magnetic ordering involved movement of metalloid atoms, phosphorus, boron, or carbon, while Egami and others suggest that the reversible behaviour is due to chemical short-range order (CSRO). These interpretations, however, are not well substantiated and require further examination (see p. 178).

11.7 Deformation and irradiation effects

Preferential etching is seen to occur at the sites of shear deformation bands in PdCuSi, NiPB and FePSi glasses subjected to bending[57] and in PdCuSi[58] subjected to compression. This etching demonstrates clearly the change in local chemical potential of the deformed zones. On heating of cold rolled PdCuSi, stored energy is released over a broad temperature range[59]. The fact that the enthalpy relaxation, occurring with a slight shift to lower temperature, resembles that found for the as-quenched glass, suggests that the structural disorder incurred during deformation is similar to the frozen-in disorder in the as-quenched state. Only about 4 per cent of the expended energy, e.g. about 50 cal/mole at 40 per cent thickness reduction, is stored in the slip band as compared with 250 cal/mole for the as-quenched sample. Consequently, the overall changes due to deformation would be small in volume ($\leqslant 0.1$%), in X-ray scattering intensities (~ 2%) and in Young's modulus (~ 1%). These changes thus should be barely detectable, although much larger changes were reported[60,61]. Because of the heterogeneous anisotropic deformation, the magnetization behaviour is drastically modified, e.g. coercive force increases[62] from about 0.01 Oe before rolling to 20 Oe after 40 per cent cold rolling in $Fe_{40}Ni_{40}P_{14}B_6$. Annealing of the deformed sample near T_g practically eliminates the deformation effects through structural recovery in thermally stable glasses. Thus cold rolling as well as irradiations are often found to have little effect on T_g and crystallization temperature.

The microstructure of cold rolled PdCuSi has been investigated by use of positron annihilation methods[63]. Both the lifetime and angular correlation of the glass exhibit a negligible change upon cold rolling at room temperature as exemplified by the angular correlation in *Figure 11.5*. In comparison the angular distribution in crystalline

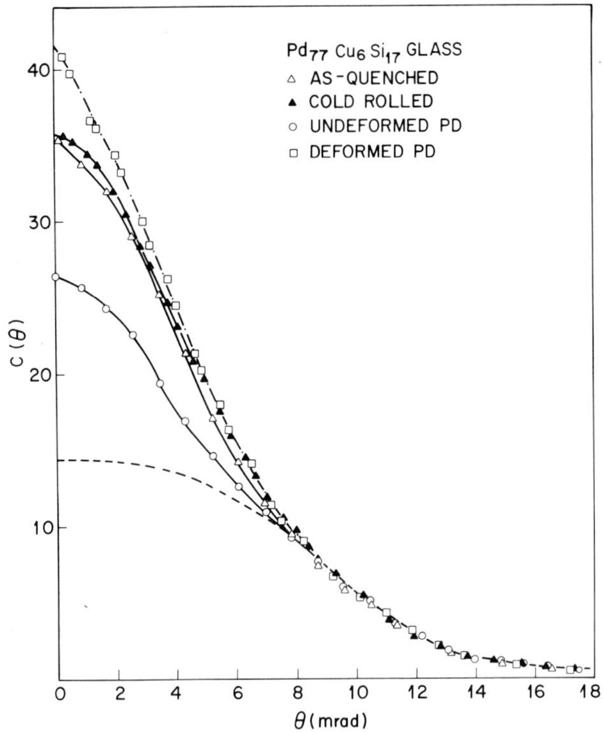

Figure 11.5 The angular correlation of positron annihilation in deformed and undeformed $Pd_{77}Cu_6Si_{17}$ glass and palladium metal: △, as quenched; ▲, cold rolled; ○, undeformed palladium; □, deformed palladium

palladium becomes sharper, and the intensity of the narrow component increases drastically. The lifetime and angular correlation of the glassy samples fall between that of undeformed and deformed palladium metal. 1 MeV electron irradiation on PdNiSi[64] and $Fe_{40}Ni_{40}P_{14}B_6$[65] reveals the same results which lead to the proposition that the plastic deformation as well as the irradiation induce no 'vacancy-like' defects. It is plausible, however, that the vacancy–interstitial pairs created during these processes recombine through local atomic regrouping because of high atomic mobilities at room temperature. Recent measurements[66] reveal, in fact, that at low enough temperatures (below 77 K), the vacancy-type defects can be induced and retained in irradiated glassy samples. Evidence for this is the increase in positron-lifetime, τ_1, which approaches the values corresponding to vacancies. Both τ_1 and ρ increase with irradiation dose at 77 K. Upon heating, both τ_1 and ρ decrease monotonously with temperature, recovering the initial values near room temperature. τ_1 of a number of metallic glasses falls between that of corresponding metals without and with vacancies. These observations are in harmony with the earlier room temperature measurements, and in metallic glasses, if they contain defects, the defects would be smaller in volume than vacancies.

As with the internal friction measurements, the magnitude of dispersion in the cold-rolled sample is three-fold greater than in the as-quenched sample[67]. Annealing reduces both the velocity dispersion and internal friction. The absorption peak and velocity dispersion are broad and occur at a temperature near 20 K for a number of metallic glasses at frequencies of about 1 kHz. It was noteworthy that the data exhibit a large asymmetry in the low temperature regime. Interestingly, a similar peak is observed near 55 K ($\tau = 10^{-7}$ s) in the temperature dependence of the spin–lattice relaxation time T_1 on ^{31}P and ^{11}B in a $Ni_{78}P_{14}B_8$ glass from NMR measurements[68]. A similar asymmetry in the absorption peak is also seen[69] in metallic glasses which have absorbed hydrogen occurring near 200 K.

11.8 A new aspect of structural relaxation

Kinetics of structural relaxation near T_g in non-metallic glasses are often analysed in the framework of a distribution in relaxation times, $H(\ln \tau)$:

$$p - p_\infty = (p_0 - p_\infty)\int_0^\infty H(\ln \tau)\exp(-t/\tau)\mathrm{d}\ln\tau \tag{11.5}$$

where p_0, p and p_∞ are values of properties at time $t = 0$, t and ∞ respectively. While the description of the relaxation process in terms of a distribution in times is appropriate, the following empirical expression

$$p - p_\infty = (p_0 - p_\infty)\exp[-(t/\tau)^n] \tag{11.6}$$

with $0 \leqslant n \leqslant 1$ is often employed in practice. To account for the non-linear behaviour, the characteristic relaxation time $\tau(T, \sigma)$ is assumed to be not only temperature dependent but also dependent on a structural order parameter σ, defined by 'macroscopic' properties[70] such as T_f, V_f or S_{conf}. No detailed studies of relaxation kinetics have been made in metallic glasses, but it is clear from the above-mentioned phenomena that the methods of analysis employed for sub-T_g relaxation is not applicable for the sub-sub-T_g relaxation. In the following, a brief description of recent structural relaxation studies[37,38,71] and interpretation of data are presented. A new model glass transition based on percolation theory will then be proposed and its implications on structural models of relaxation processes discussed.

Figure 11.6 shows the thermograms of as-quenched $Pd_{48}Ni_{32}P_{20}$ glass subjected to annealing at 380 K for 210 hours. After the annealing (curve 1), the sample exhibits an excess endothermic reaction ($\Delta C_p \equiv C_p - C_{p,s} > 0$), above $T_a(=380\,K)$ peaking at T'_m. ΔC_p then decreases and is followed by a broad exothermic peak ($\Delta C_p < 0$) which merges with the thermogram of the as-quenched sample. Both the temperature T'_m and

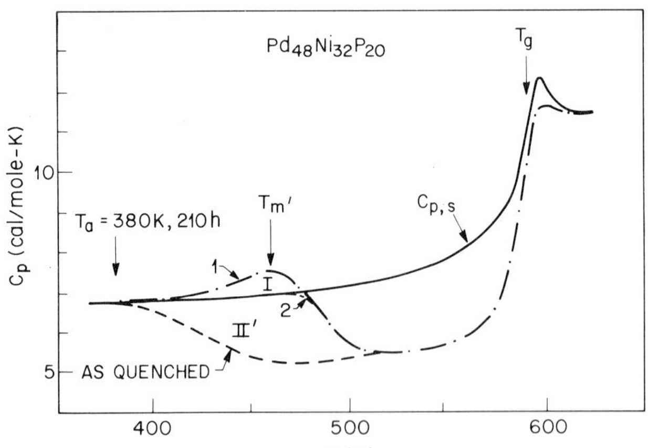

Figure 11.6 The thermograms of the $Pd_{48}Ni_{32}P_{20}$ glass after various heat treatments. The rate of scan is 20 K/min.

the magnitude $\Delta C_{p,max}$ of the endothermic peak increases with annealing time t, approximately as log t, but a slight decrease in $\Delta C_p(T)$ at corresponding temperatures is seen. Similar behaviour has been found also in $Ni_{36}Fe_{32}Cr_{14}P_{12}B_6$ by Drijiver *et al.*[72]. Upon rescanning (curve 2), after the annealed sample was scanned to T'_m then cooled to room temperature, the endothermic peak diminishes and C_p falls on $C_{p,s}$ for $T \lesssim T_m$ and then follows the same exothermic reaction at higher temperatures. Clearly, the relaxation process contributing to the endothermic peak (I) is reversible, while the stabilization leading to the exothermic reaction (II′) is irreversible. The reversible relaxation (I) is believed to arise from short-range localized structural relaxation (SLSR) in the stabilized portion of the more or less rigid matrix (II′) (*see Figure 11.9*). The irreversible structural stabilization (II′) results from the annihilation of defects leading to the reduction in free volume which is medium-range cooperative structural relaxation (MCSR) involving only a portion of the glassy material. As the duration of anneal increases and T_a approaches T_g the stabilized regions expand over the whole system (III in *Figure 11.2*), and the relaxation process becomes the long-range cooperative structural relaxation (LCSR) commonly observed for the sub-T_g anneal.

To facilitate the understanding of the relaxation processes, a systematic detailed study has been conducted on preconditioned samples to avoid the complication arising from superposition of reversible and irreversible processes[38]. *Figure 11.7* illustrates thermograms of annealed samples at 420 K after stabilization via slow cooling (20 K/min) through the glass transition ($T_g = 582$ K). At the same rates of heating and cooling a hysteresis phenomenon in the region of glass transition, i.e. $T \geqslant 530$ K, is seen as expected. The annealed sample, upon being heated, shows a $C_p(T)$ behaviour which follows closely that of $C_{p,s}(T)$ up to 420 K, then exhibits an excess endothermic specific heat ΔC_p, before merging again on $C_{p,s}$ at a temperature below T_g. This indicates that C_p in the range about 420 K is dependent on thermal history and consists of a configurational contribution $C_{p,\sigma}$ in addition to the vibrational specific heat $C_{p,v}$. The values of $C_{p,v}$ then should be extrapolated from values of C_p below 420 K as shown in *Figure 11.7*.

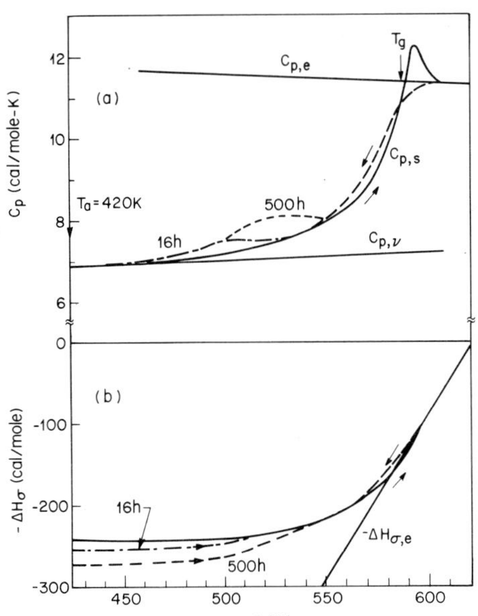

Figure 11.7 The apparent specific heats, C_p, and corresponding configurational enthalpy $-\Delta H_\sigma$ for the PdNiP glassy samples

To examine the significance of the endothermic reaction, the configurational enthalpy

$$\Delta H_\sigma(T) = -\int_{620}^{T} (C_p - C_{p,v})\mathrm{d}T' \qquad (11.7)$$

is shown in *Figure 11.7(b)*. Here $\Delta H_{\sigma,e}(T)$ is the ΔH_σ of the undercooled liquid, and is set to be zero at 620 K. As seen, $-\Delta H_\sigma(420\,\mathrm{K})$ falls progressively with annealing time as $\ln t^n$ with $1 \leqslant n \leqslant 3$, as the glassy structure stabilizes. It may be noted that upon heating, the $-\Delta H_\sigma(T)$ of the annealed samples increases towards the reference value and merges with it well below T_g. This implies that, as a result of structural relaxation, which occurs during the sub-sub-T_g anneal, it is possible to recover the initial structure of glass without reheating through the glass transition. The $-\Delta H_\sigma(T)$ curves thus tend to stay to the left of the equilibrium line $-\Delta H_{\sigma,e}(T)$. These features differ significantly from the phenomena commonly observed in the usual sub-T_g anneal for the glass materials.

The same calorimetric behaviour found in the sub-sub-T_g anneal was seen for samples annealed at higher temperatures $T_g - 100 < T_a < T_g$. The endothermic reaction, however, sets in at $T = T_a$. This phenomenon becomes less pronounced as T_a is raised closer to T_g, and, depending on the length of annealing time, a delayed C_p peak characteristic of sub-T_g anneal begins to appear at the transition. For detailed description and data analysis, the reader may refer to recent papers[37,38,71]. Only significant features of the results are summarized in the following.

1. The relaxation spectrum $H(\ln \tau)$ for the sub-sub-T_g anneal is seen to have a broad distribution in both activation energies ($Q = 1$ eV to 2.5 eV) and frequency factors ν_0, varying from $10^{14}\,\mathrm{s}^{-1}$ to $10^{22}\,\mathrm{s}^{-1}$. The enthalpy coupling constant is found to depend on the annealing temperature; $\gamma(T_a) \propto (T_m - T_a)$. This T_a dependence of $\gamma(T_a)$ and relatively large ν_0 compared with the Debye frequency $\nu_0 = 10^{13}\,\mathrm{s}^{-1}$, is reminiscent of the relaxation kinetics commonly observed in the sub-T_g anneal, and the relaxation spectrum may be considered phenomenologically as a distribution of characteristic glass transitions.
2. The sub-sub-T_g anneals stabilize a portion of the relaxation spectrum. Upon heating, destabilization occurs above the annealing temperature and the initial glass structure is recovered. The effect of initial annealing is eliminated upon heating but will reappear upon repeated annealing. The effects of low temperature annealing are thus reversible and additive. The reversibility and additivity facilitate the interpretation of the 'reversibility' in Curie temperature and cross-over phenomena[37]. *Figure 11.8* shows the change in T_C of a $Fe_{40}Ni_{40}P_{14}B_6$ glass subjected to a cyclic ageing between 473 K and 523 K ($t_a = 30$ min) and corresponding endothermic specific heat ΔC_p after each anneal. The main features of the results are: (1) Each increase in temperature (to 523 K) eliminates (or reduces) the structural relaxation incurred during the previous low temperature ($T_a = 473$ K) anneal, i.e. the disappearance of the low temperature spectrum '1', and a rapid drop in T_C by nearly same amount of about 4 K. It is then followed by a gradual increase in T_C, reminiscent of memory phenomenon. The effect of the high temperature anneal is represented by the ΔC_p peak '2'. (2) Each decrease in anneal temperature (to 473 K) recreates the low temperature spectrum superimposed on the high temperature one (curve 3), and T_C is raised gradually by about 4 K. (3) Each successive annealing at 523 K leads to a gradual development of the ΔC_p spectrum and an increase in T_C but with a much slower rate, i.e. $\Delta T_C \propto \ln \sum_i \Delta t_i$, becoming negligibly small at later stages of cyclic ageing. Thus changes in Curie temperature upon cyclic ageing appear reversible. It may be

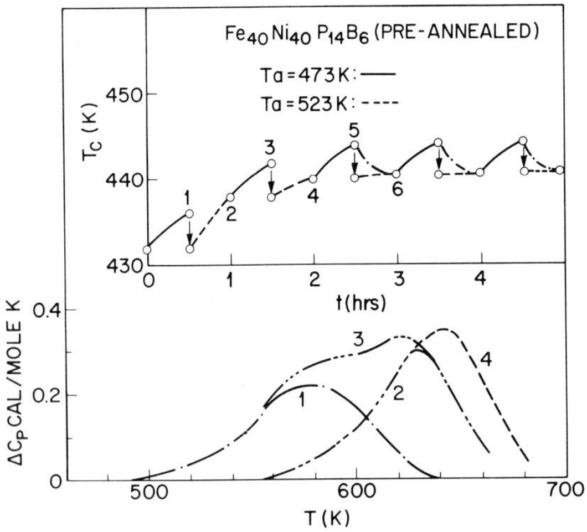

Figure 11.8 The changes in T_C (inset) and corresponding enthalpy revolution of the pre-annealed $Fe_{40}Ni_{40}P_{14}B_6$ glass

worth mentioning that in previous reversibility studies, it was arbitrarily assumed that upon step heating T_C[34,55], ρ[23] and E[56] gradually approached the high temperature 'equilibrium' values monotonically as shown schematically by dash-dot lines in the inset to *Figure 11.8*. These assumptions are not truly valid, and the interpretations of this phenomenon with the concept of chemical short-range ordering are not well substantiated.

11.9 A new model glass transition

A new model glass transition has been proposed using the framework of percolation theory[38]. It is assumed that

1. The structure of a glass and a liquid is heterogeneous. The inhomogeneity arises in monatomic materials from fluctuation in density, and in alloys from concentration fluctuations as well. It is visualized[73,74] that a glass consists of liquid-like regions of large free volume or high local free energy, and solid-like regions with small free volume or low free energy.
2. Each region undergoes infrequent transitions $k_1 \ll \nu_D (=10^{13}\,s^{-1})$, between local minima well separated by energy barriers $\Delta\varepsilon_i$, in configurational space[75]. Each local minimum represents a different structure. A distribution in the heights of $\Delta\varepsilon_i$ is responsible for the relaxation time τ_i being proportional to $\exp(-\Delta\varepsilon_i/k_B T)$.

The relaxation spectra are shown schematically in *Figure 11.9*. It consists of an asymmetric distribution around a main peak and a long tail in the short times. At temperatures $T_1 > T_g$[curve (*a*)], the whole spectrum lies to the left of the time of measurement τ^* ($=30$ sec for example), so that the whole system undergoes frequent configurational transformation and is liquid-like, or in an equilibrium state. During the cooling, the whole spectrum shifts to longer times such that small portion of regions with $\tau > \tau^*$ behave as isolated solid-like clusters embedded in a liquid-like matrix. As

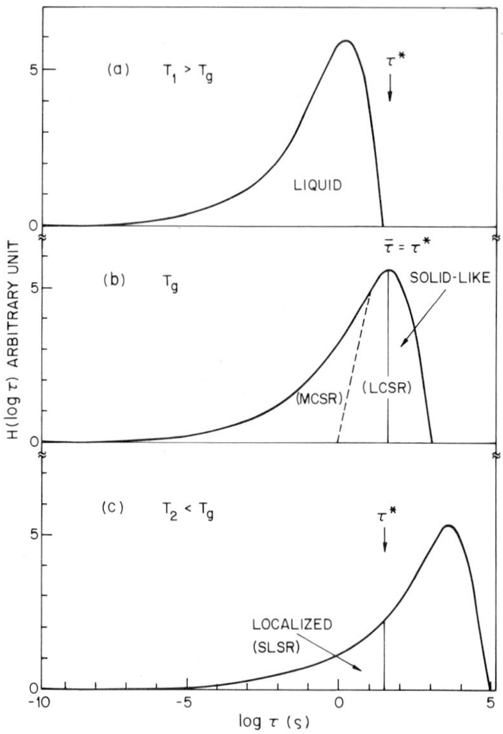

Figure 11.9 Schematic relaxation time spectrum $H(\log \tau)$ for a metal glass at various temperatures

the temperature approaches T_g [curve (*b*)] the solid-like clusters increase in number and grow in size to such an extent that an infinite cluster is formed. This results in a drastic decrease in macroscopic flow and liquid–glass transition occurs. The glass transition $T_g(\tau^*)$ is then related to a percolation process. It is noteworthy that the solid-like regions occupy only a small fraction, $t \approx 1/4$ to $1/3$, of the system[76], and are surrounded by liquid-like matrix. The relaxation processes near T_g are thus cooperative in nature (CLSR). At a temperature T_2 about 100 K below T_g [curve (*c*)], the solid-like regions grow and the fraction of liquid-like regions decrease to below the percolation limit t' ($\approx t$). The liquid-like regions now are isolated from each other and are embedded in the rigid matrix. Relaxation then occurs locally in the liquid regions independent from each other (SLSR). It may be emphasized that SLSR in each cluster (probably about 10 Å or greater in domain size) involving a number of atoms is cooperative in nature, and is capable of undergoing a localized liquid–glass transition when τ_i exceeds approximately τ^*. In the region $T_2 < T_a < T_g$, structural relaxation will occur in the liquid matrix in a cooperative manner, involving only a portion of the material (MCSR). It is distinguished from LCSR occurring at T_g [*see* curve (*b*)].

The present model predicts localized modes of relaxation in sub-sub-T_g anneal and upon heating the glass recovers the initial state without reheating through T_g. It also predicts the concept of a distribution of glass transition temperatures. The model reproduces fairly well many ageing behaviours mentioned in previous sections. An as-quenched sample shows shifts of the whole spectrum $H(\ln \tau)$ to shorter times (with slight modification in distribution) and exhibits structural relaxation at lower temperatures. It is likely that the magnetic disaccommodation and anelastic effects involving SLSR may occur at temperatures well below the temperatures extrapolated from the viscosity and diffusion data which are believed to involve MCSR and long-range cooperative structural relaxation LCSR.

11.10 Concluding remarks

The sub-sub-T_g relaxation of an as-quenched sample consists of reversible short-range localized structural relaxation (SLSR) and medium-range cooperative structural relaxation (MCSR). The ratio of SLSR to MCSR is dependent on annealing temperature and duration of annealing. It is believed that SLSR contributes to the increase in the lifetime of positron annihilation τ_1 and resistivity ρ, and causes the drastic loss of ductility in many metallic glasses. The effects of MCSR on these properties are however not clear and require further investigation. As the annealing temperature approaches T_g and as the duration of anneal increases, the relaxation proceeds with long-range cooperative structural relaxation (LCSR) and leads to a decrease in τ_1 and ρ owing to the reduction in the structural heterogeneity incurred during sub-sub-T_g anneals. It is obvious that changes in τ_1, ρ and Curie temperature T_C with annealing are greater in alloy systems because of additional concentration fluctuation.

The equilibrium glassy state cannot be attained in the experimental time of weeks at 100 K below the thermally manifested $T_g (\tau^* = 30\,\text{s})$. In many cases, the 'reversibility' in T_c, ρ and E on cycling between two temperatures, and the cross-over behaviour reported, in fact never attained equilibrium state values (*see Figure 11.4*). It is our opinion that the assignment of the 'reversible' relaxation as chemical short range order (CSRO) and the irreversible one as topological short-range order (TSRO) is not well founded. The reversibility and cross-over phenomena can be interpreted in terms of the creation and annihilation, respectively, of SLSR and MCSR, and even LCSR if T_a approaches T_g, during annealing.

It may be remarked that the volumetric relaxation of metallic glasses is small, of the order $\Delta v/v \approx (0.3$—$0.5) \times 10^{-2}$, but other macroscopic properties change by an order of magnitude. The relatively small change in volume as compared with other properties is rather straightforward, based on solid mechanics, since the change in volume arises from the anharmonic term or second order elastic constants. It was shown by Zener[77] that the mean dilation of a *self-stressed body* depends on the quadratic terms in the stress–strain law, and may be written

$$\bar{\Theta} = c_d \bar{W}_d + c_s \bar{W}_s \tag{11.8}$$

with coefficients

$$c_d = -(1 + \mathrm{d}\ln K/\mathrm{d}\ln V)/K_0 \tag{11.9}$$

$$c_s = -(1 + \mathrm{d}\ln \mu/\mathrm{d}\ln V)/K_0 \tag{11.10}$$

where $\bar{W}_d$ and $\bar{W}_s$ are the mean energy stored in unit volume by dilation strains and shear strains respectively. The values in the parentheses in equations 11.9 and 11.10 are about twice the Grüneisen constant (2 to 3). Taking the bulk modulus $K_0 \approx 2 \times 10^{12}$ dyne/cm^2 and $\bar{W}_d + \bar{W}_s = 300$ cal/mole $\approx 1.5 \times 10^9$ erg/cm^3 for the enthalpy of relaxation of as-quenched glasses it yields, from equations 11.8—11.10:

$$\Delta\bar{\Theta} \approx (4 \sim 6) \times 1.5 \times 10^9/2 \times 10^{12} \approx (3 \sim 5) \times 10^{-3}$$

In comparison, the dilation energy on an element under hydrostatic pressure p is

$$W_d = K_0\Theta^2/2 = p^2/2K_0 \tag{11.11}$$

while shear stress contributes to distortion (or spherical symmetry) but exerts little on the interatomic fluctuation; (Θ_s in fact being of the order shown in equations 11.8 and 11.10). The dilational fluctuation arises mainly from hydrostatic pressure and compression. We arrive at

$$W_d \cong \frac{K_0}{2}\Sigma\Theta^2 = K_0\langle\Theta^2\rangle/2 \qquad (11.12)$$

Taking $W_d \approx W_s$, and the heat of crystallization $\Delta H_{cry} = 1.5$ kcal/mole $= 8 \times 10^9$ erg/cm^3, we find that the average dilational energy for an element, $W_d \approx 4 \times 10^9$ erg/cm^3 and (4—0.75) $\times 10^9$ erg/cm^3 in the as-quenched and relaxed state respectively. We have the mean square interatomic fluctuation in the as-quenched specimen $\sqrt{(\langle\Theta^2\rangle_q)} \approx 14\%$ and in the annealed specimen

$$\sqrt{\langle\Theta^2\rangle_a} = \sqrt{\frac{3.25}{4}} \cdot \sqrt{\langle\Theta^2\rangle_q} \approx 0.90\sqrt{\langle\Theta^2\rangle_q}$$

a 10 per cent structural relaxation upon annealing. These values are in fair agreement with volumetric, X-ray RDF measurements and computer simulations (*see also* Chapter 7).

Lastly, it may be noted that the structural relaxation affects glass–crystal transformation processes, and is related to embrittlement behaviour. If a glass is crystallized at low temperatures, it crystallizes by precipitation of a metastable phase embedded in a glassy matrix with an activation energy of about 1 eV, probably dominated by MCSR. However, at higher temperature near T_g, the processes of crystallization are determined by atomic diffusivity being dominated by LCSR. The activation energies for both crystallization and embrittlement in a series of metallic glasses are seen to increase linearly from 1 eV at low temperatures 150 K or more below T_g to 2.5 eV at about 50 K below T_g, and then increase rapidly to 4 eV to 6 eV at T_g, i.e. to the values found from atomic diffusivities and viscosity measurements. This temperature dependence of Q indeed is reminiscent of the activation energy spectrum for the structural relaxation[78].

References

1. For a review, *see* Chen, H. S., *Rep. Prog. Phys.*, **43,** 353 (1980)
2. Collection of papers in Proc. LAM4, Grenoble 1980; *J. Phys. (Paris)*, **41,** C8 (1980)
3. Proc *4th Int. Conf. Rapidly Quenched Metals*, Eds. Masumoto, T. and Suzuki, K., Japan Institute of Metals, Sendai (1982)
4. Chen, H. S. and Turnbull, D., *J. chem. Phys.*, **48,** 2560 (1968)
5. Kauzeman, W., *Chem. Rev.*, **43,** 219 (1948)
6. Tool, A. M., *J. Am. Ceram. Soc.*, **29,** 240 (1946)
7. Turnbull, D. and Cohen, M. H., *J. chem. Phys.*, **29,** 1049 (1958); also Adam, G. and Gibbs, J. H., *J. chem. Phys.*, **43,** 139 (1965)
8. Kovacs, A. J., *Fortschr. Hochpolymforsch.*, **3,** 394, (1963)
9. Ritland, H. N., *J. Soc. Glass Technol.*, **39,** 99 (1955)
10. Greer, A. L. and Leak, J. A., *J. Non-cryst. Solids*, **33,** 291 (1979)
11. Aklonis, J. J. and Kovacs, A. J., *Contemp. Top. Polym. Sci.*, **3,** 267 (1979)
12. Davies, R. O. and Jones, G. O., *Adv. Phys.*, **2,** 370 (1953)
13. Goldstein, M., *J. chem. Phys.*, **64,** 4767 (1976)
14. Chen, H. S., Kimerling, L. C., Poate, J. M. and Brown, W. L., *Appl. Phys. Lett.*, **32,** 461 (1978)
15. Taub, A. I. and Spaepen, F., *Acta metall.*, **28,** 1781 (1980); Taub, A. I. and Luborsky, F. E., *Acta metall.*, **29,** 1939 (1981)
16. Chen, H. S., *Mater. Sci. Engng.*, **26,** 79 (1976)
17. Luborsky, F. E., *'American Institute of Physics Conf. Proc.'*, **29,** 209 (1976)
18. Chen, H. S., *J. appl. Phys.*, **49,** 3289 (1978)
19. Chen, H. S. and Coleman, E., *Appl. Phys. Lett.*, **28,** 245 (1976)

20. Chen, H. S., Sherwood, R. C., Leamy, H. J. and Gyorgy, E. M., *IEEE Trans. Magn.*, **MAG-12,** 933 (1976)
21. Liebermann, H. H., Graham, C. D. and Flanders, P. J., *IEEE Trans. Magn.*, **MAG-13,** 1541 (1977)
22. Chen, H. S., *Solid St. Communs,* **33,** 915 (1980)
23. Balanzat, M., *Scripta Met.* **14,** 173 (1980)
24. Chen, H. S., Leamy, H. J. and Barmatz, M., *J. Non-cryst. Solids,* **5,** 444 (1971)
25. Berry, B. S. and Pritchet, W. C., *J. appl. Phys.*, **44,** 3122 (1973)
26. Williams, R. S. and Egami, T., *IEEE Trans. Magn.*, **MAG-12,** 927 (1976)
27. Matsuyama, T., Ohta, K. and Kajiura, M., *Jap. J. appl. Phys.*, **19,** 55 (1980); Moser, N. and Kronmüller, H., *J. Magnetism magn. Mater.*, **19,** 275 (1980)
28. Chen, H. S. and Chuang, S. Y., *Appl. Phys. Lett.*, **31,** 255 (1977)
29. Chen, H. S. and Lo, C. C., *'Proc. 2nd Int. Conf. on Rapidly Quenched Metals'*, p. 413, Massachusetts Institute of Technology Press, Cambridge, Mass. (1976)
30. Lasocka, M. and Matyja, H., in *'Treatise on Materials Science and Technology'*, vol. 20, p. 261; Li, J. C. M., *ibid,* p. 325, Academic Press, New York (1981)
31. Shima, K., Tanigawa, S. and Waseda, Y., ref. 3, p. 543
32. Tanigawa, S., Shima, K. and Masumoto, T., ref. 3, p. 501
33. Chen, H. S., Sherwood, R. C. and Gyorgy, E. M., *IEEE Trans. Magn.*, **MAG-13,** 1538 (1977)
34. Kudo, T. and Egami, T., ref. 3, p. 1187
35. Ohnuma, S., Shirakawa, K., Nose, M., Kanehira, J. and Masumoto, T., ref. 3, p. 1203
36. Greer, A. L. and Spaepen, F., *Ann. N.Y. Acad. Sci.*, **371,** 218 (1981)
37. Chen, H. S., *J. appl. Phys.*, **52,** 1868 (1981)
38. Chen, H. S., *J. Non-cryst. Solids,* **46,** 289 (1981)
39. Chen, H. S., *Scripta Met.*, **11,** 367 (1977)
40. Pratten, N. A. and Scott, M. G., *Scripta Met.*, **12,** 137 (1978)
41. Chen, H. S., Krause, J. T. and Coleman, E., *J. Non-cryst. Solids,* **18,** 157 (1975)
42. Chou, P. C. P. and Spaepen, F., *Acta metall.*, **23,** 609 (1975)
43. Walter, J. L., Bacon, F. and Luborsky, F. E., *Mater. Sci. Engng.*, **24,** 239 (1976)
44. Masumoto, T., *Sci. Rep. Res. Insts. Tôhoku Univ.*, **A26,** 246 (1977)
45. Walter, J. L., LeGrand, D. G. and Luborsky, F. E., *Mater. Sci. Engng,* **29,** 161 (1977); Osamura, K., Shibue, K., Shingu, P. H. and Murakami, Y., *J. mater. Sci.*, **14,** 945 (1979)
46. Gyorgy, E. M., Leamy, H. J., Sherwood, R. C. and Chen, H. S., *American Institute of Physics Conf. Proc.* **29,** 198 (1975)
47. Kijek, M., Akhtar, D., Cantor, B. and Cahn, R. W., ref. 3, p. 573
48. Chen, H. S., *J. appl. Phys.*, **49,** 462 (1978)
49. Egami, T. and Ichikawa, T., *Mater. Sci. Engng.*, **32,** 293 (1978)
50. Kursumovic, A., Cahn, R. W. and Scott, M. G., *Scripta Met.*, **14,** 1245 (1980)
51. Kursumovic, A. and Scott, M. G., *Appl. Phys. Lett.*, **37,** 620 (1980)
52. Chi, G. C., Chen, H. S. and Miller, C. E., *J. appl. Phys.*, **49,** 1915 (1978)
53. Berry, B. S. and Pritchet, W. C., *J. appl. Phys.*, **47,** 3295 (1976)
54. Berry, B. S. and Pritchet, W. C., *American Institute of Physics Conf. Proc.*, **34,** 292 (1976)
55. Egami, T., *J. mater. Sci.*, **13,** 2587 (1978)
56. Kursumovic, A., Scott, M. G., Grit, E. and Cahn, R. W., *Scripta Met.*, **14,** 1303 (1980)
57. Chen, H. S., Leamy, H. J. and O'Brien, M. J., *Scripta Met.* **7,** 115 (1973)
58. Pampillo, C. A. and Chen, H. S., *Mater. Sci. Engng,* **13,** 181 (1974)
59. Chen, H. S., *Appl. Phys. Lett.*, **29,** 328 (1976)
60. Masumoto, T. and Maddin, R., *Mater. Sci. Engng.*, **19,** 1 (1975)
61. Waseda, Y., Aust, K. T. and Masumoto, T., *Scripta Met.*, **13,** 187 (1979)
62. Luborsky, F. E., Walter, J. L. and LeGrand, D. G., *IEEE Trans. Magn.*, **MAG-12,** 930 (1976)
63. Chen, H. S. and Chuang, S. Y., *Appl. Phys. Lett.*, **27,** 316 (1975)
64. Chen, H. S., *Phys. Status Solidi,* **(a) 34,** K127 (1976)
65. Howell, R. H. and Hopper, R. W., *Scripta Met.*, **13,** 367 (1979)
66. Moser, P., Hautojarvi, P., Yli-Knuppila, J. and Chamberod, A., ref. 3, p. 759
67. Barmatz, M., Wyatt, K. W. and Chen, H. S., *'Rapidly Quenched Metals'*, p. 431, Massachusetts Institute of Technology Press, Cambridge, Mass. (1976)
68. Aliaga-Guerra, D., Panissod, P. and Durand, J., *J. Phys. (Paris),* **41,** C8 674 (1980)
69. Berry, B. S. and Pritchet, W. C., *Scripta Met.*, **15,** 637 (1981)
70. Uhlmann, D. in *'Metallic Glasses'*, p. 128, American Society of Metals, Metals Park, Ohio, (1978)
71. Chen, H. S., ref. 3, p. 495
72. Drijver, J. W., Mulder, A. L., Emmens, W. C. and Radelaar, S., *'Proc. 3rd Int. Conf. on Rapidly Quenched Metals'*, p. 363, Ed. Cantor, B., vol. I, The Metals Society, London (1978)
73. Cohen, M. H. and Greer, G. S., *Phys. Rev., B,* **20,** 1077 (1979)

74. Cryst, M., *J. Phys. (Paris)*, **41,** C8, 107 (1980)
75. Goldstein, M., *J. chem. Phys.*, **51,** 3728 (1969)
76. Shante, K. K. S. and Kirkpatrick, S., *Adv. Phys.*, **20,** 325 (1971)
77. Zener, C., *Trans. Am. Inst. Min. metall. Engrs,* **147,** 361 (1942)
78. Chen, H. S., ref. 3, p. 555

Chapter 12

Strength, ductility and toughness — a study in model mechanics

H. Kimura and T. Masumoto

The Research Institute for Iron, Steel and Other Metals, Tohoku University, Sendai, Japan

12. 1 Introduction

This chapter proposes a model mechanics from which it is possible to set up the phenomenology of the inhomogeneous plastic flow and fracture for amorphous alloys, rather than reviewing experimental results and details of various theoretical micro-mechanisms*. We employ this model in an attempt to organize the current knowledge of flow and fracture events but to assess this field critically. We intend, in this chapter, to gain a better understanding of the actual strength, inhomogeneous flow and related phenomena of amorphous alloys, with particular attention to their quantitative description; i.e., we will not be concerned with the micro-mechanisms or with homogeneous flow.

The basic idea and procedure should be described in more detail. Model mechanics, as we use this term, is a theoretical and experimental approach in which, through an ideal mechanical system using a model material, one presents a theory, new concepts and parameters describing the origin of the phenomena under consideration. So far, there seems to be no fruitful route to systematize the strength, using parameters and variables with a clear physical justification–certainly none capable of giving attention to the *existence* of mechanics of these phenomena. In the present status of the research on the strength of amorphous alloys, a model mechanics study has the potential to complete the phenomenology of flow and fracture. We conveniently set up a sharp division, in both experimental and theoretical fields, of the flow between the static (topological) description of yield strength and an atomistic mechanism, and a dynamical formulation of mobility and kinetics of the shear bands and of flow units involved. This procedure is suitable to our purpose, rather than, for example, to make a division between theory and experiment. On the other hand, we divide the study of the fracture into two categories. We deal mainly with the fracture events, correlated with the origin of the strength of amorphous alloys; e.g. how ductile is an amorphous alloy, what is the nature of the embrittlement process. We further describe a modelling of mechanical processes for various fracture phenomena in terms of an ideal plastic solid.

* See references 1—6 for a detailed collection of papers on mechanical properties.

This chapter comprises three sctions discussing:

1. The statics of plastic deformation
2. The dynamics of inhomogeneous plastic flow
3. The mechanics of fracture

In each section, we first propose a model mechanics or a concept which could be essential to a physical process and phenomenon underlying the strength of an amorphous alloy, and then, with a model, derive basic rules which can generally apply to various alloy compositions. At the same time, we present testing methods and their principles to deal with thin specimens when possible. We then try to present an overview of the various complicated, established events of the flow and the fracture from a logical extension of our concepts. Finally, we examine different implications, if any, to the micro-mechanism involved in flow and fracture, and describe future needs.

12.2 Statics of plastic deformation

12.2.1 Deformation mechanics

A thin, amorphous specimen, as tested under tension in a conventional testing machine, exhibits a brittle behaviour before a well defined yielding in the stress–strain curve[7–10]. This behaviour may be referred to as brittle, in the engineering sense of the term. The energy-balance in a machine–specimen system can permit a 'bulk' specimen, or a specimen in a machine with a very high rigidity, both to avoid such instability in tension, and to reveal macroscopic yielding. For a round bar of amorphous $Pd_{78}Cu_6Si_{16}$ alloy at room temperature and at 77 K, *Figure 12.1* shows the tensile stress–extension curves; and *Figure 12.2* shows tensile plastic deformation after general yielding[11]. The specimen deforms elastically before a deviation from linearity in the stress–strain curve; there is no trace of plastic deformation at available resolutions. The plastic flow occurs locally (local yielding) in the form of slip bands and then proceeds through the entire cross-section of the round specimen (general yielding). The angle of the slip band to the tensile axis has the constant value of 55 degrees. General yielding appears to occur before the application of maximum load; the general yield stress is thus somewhat lower (1 or 2 per cent) than the tensile yield stress deduced from the

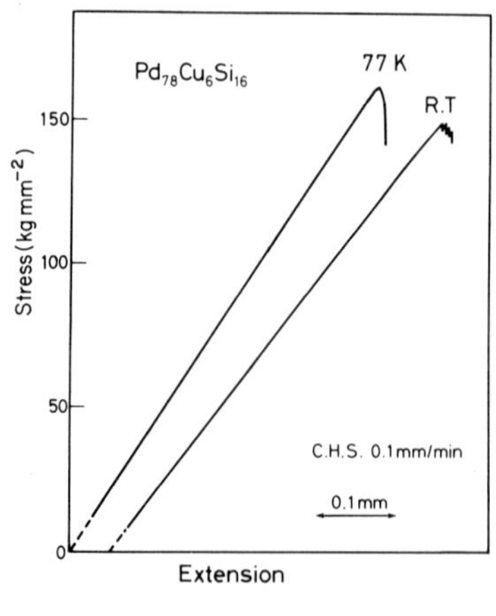

Figure 12.1 The tensile stress–extension curves at room temperature (R.T.) and 77 K for a round bar of amorphous $Pd_{78}Cu_6Si_{16}$ alloy

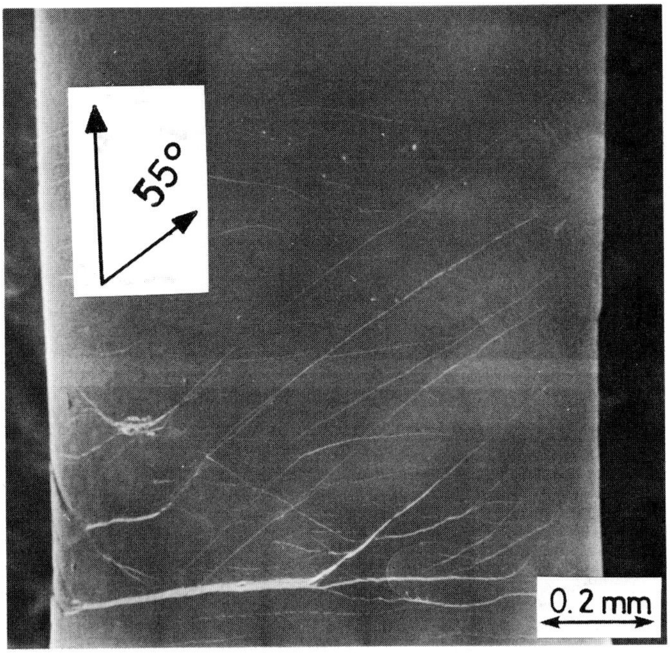

Figure 12.2 Tensile plastic deformation after general yielding, showing the slip band angle of 55 degrees

maximum load (apparent strain hardening). In this stage, the multiplication and intersection of shear bands can be seen. In the further stage wherein the load decreases, following a maximum, plastic deformation occurs by serrated flow and appears to proceed mainly on a single slip plane inclined at 55 degrees to the tensile axis, resulting in a decrease in the load-bearing area (plastic instability). The level of yield stress for a tensile specimen is conveniently derived from the maximum load, $148 \pm 2\,\mathrm{kg/mm^2}$. The tensile deformation for such a bulk specimen reveals the important characteristics of flow for an amorphous alloy. Now, we attempt to determine, more rigoriously by way of continuum theory, the yield characteristics and their related variables. These variables include general yielding, quantitative descriptions of strain hardening and softening, localization of shear slip, slip-band angle, intersection and multiplication of slip bands, serrated flow, and the onset of plastic instability. Here, we are going to present material tests which adequately describe the physical characteristics of static flow.

Evidence of an ideal plastic solid

The plastic behaviour of amorphous $Pd_{78}Cu_6Si_{16}$ is examined in terms of reasonable macroscopic variables in continuum plastic theory. *Figure 12.3* shows the torsional moment (M_T) as a function of torsional angle (θ), and *Figure 12.4* shows the torsional plastic deformation at the specimen surface[11,12]. Simple torsion theory gives the calculation of the pure shear stress (τ):

$$\tau = \frac{3M_T + \theta \dfrac{dM_T}{d\theta}}{2\pi d^3} \tag{12.1}$$

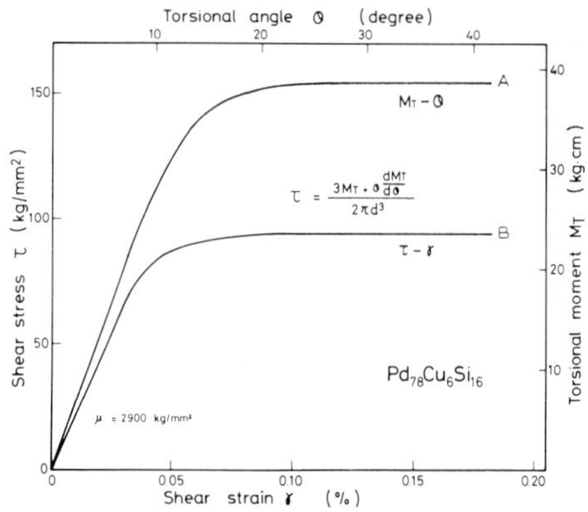

Figure 12.3 Torsional moment (M_T)-torsional angle (θ) curve for amorphous $Pd_{78}Cu_6Si_{16}$ ($\mu \approx 2900\,kg/mm^2$)

where d is the diameter of a round-bar specimen. The torsional deformation–angle curve A in *Figure 12.3* clearly shows non-strain hardening ($dM/d\theta = 0$). Assuming an ideally plastic solid, we derive from equation 12.1 a pure shear stress of $91 \pm 9\,kg/mm^2$ for amorphous $Pd_{78}Cu_6Si_{16}$ (curve B). The torsion moment–torsion angle curve in the region of the elastic–plastic flow allows us to adjust to a simple torsion theory prediction for an elastic–perfectly plastic solid. The morphology of slip bands in torsion (*Figure 12.4*) is characterized by the two sets of intersecting straight slip-bands along planes of maximum shear stress, as predicted in simple torsion theory. We also find that the density of slip bands increases (i.e. the interval between the slip bands decreases) with increasing torsion strain, indicating the stable deformation of torsional flow. In

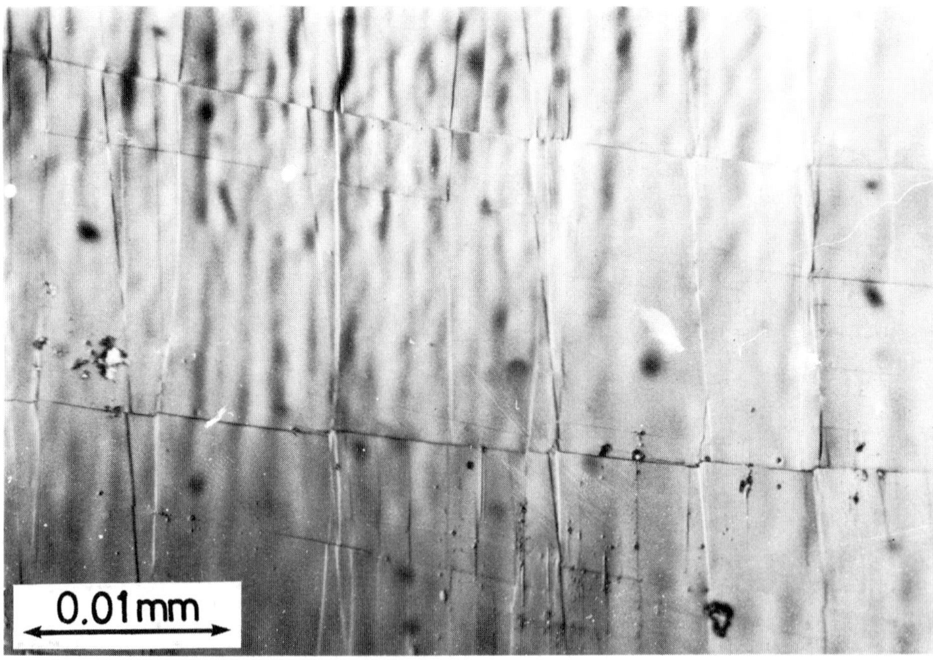

Figure 12.4 Torsional plastic deformation at surface of the specimen used for *Figure 12.3*

conclusion, the specimens of amorphous $Pd_{78}Cu_6Si_{16}$ under torsion give evidence of elastic–perfectly plastic deformation. The nucleation and propagation of slip bands at a stress concentration of the specimen is responsible for local yielding appearing in tension, but an inherent micro-yielding behaviour in uniaxial tension is not expected.

For a thin specimen, ideal plastic behaviour was confirmed by Krenitsky and Ast[13] who analysed the compression of a loop specimen (i.e., the bending) of amorphous $Fe_{40}Ni_{40}P_{14}B_6$.

Antiplane strain mode in plane stress

Figure 12.5 shows the tensile plastic deformation of an amorphous $Pd_{78}Cu_6Si_{16}$ thin plate specimen[11]. The plastic deformation proceeds by a local necking mode; this is a tearing mode, known as an antiplane strain mode by which tensile failure of unnotched amorphous thin sheets typically occurs under the conditions of plane stress.

Notched specimens of amorphous $Pd_{78}Cu_6Si_{16}$ are used to analyse the deformation mechanics of the antiplane strain mode. *Figure 12.6* illustrates tearing in an antiplane strain mode ahead of a notch tip[14]. Dugdale predicted for a rigid–perfectly plastic solid that the size (R) of the two-dimensional plastic zone is related to the

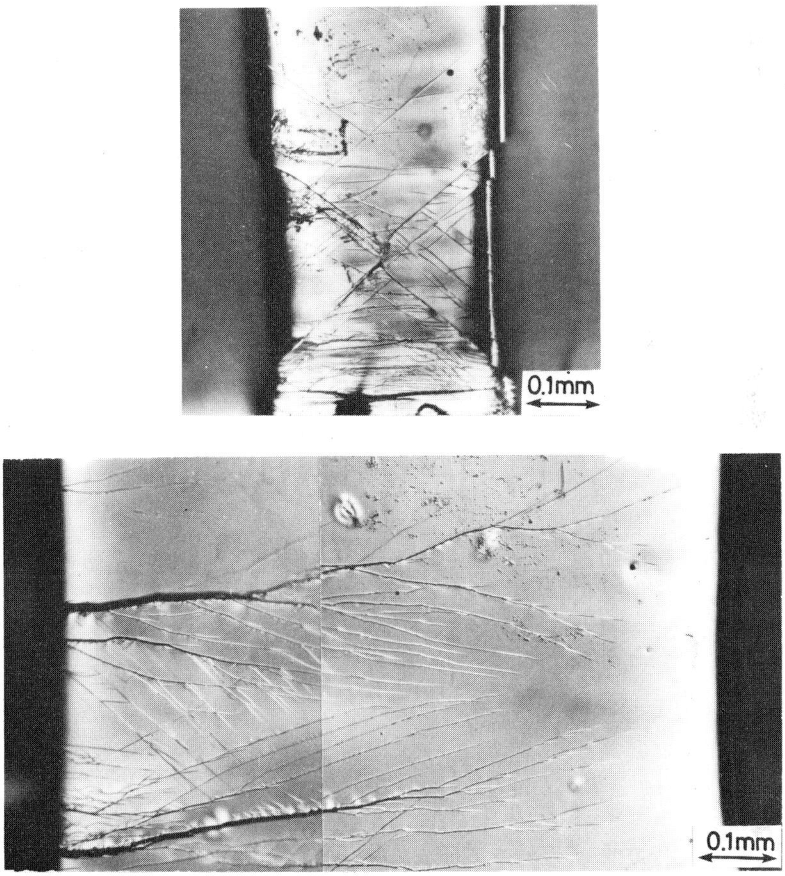

Figure 12.5 Tensile plastic deformation of amorphous $Pd_{78}Cu_6Si_{16}$ of thin plate specimen

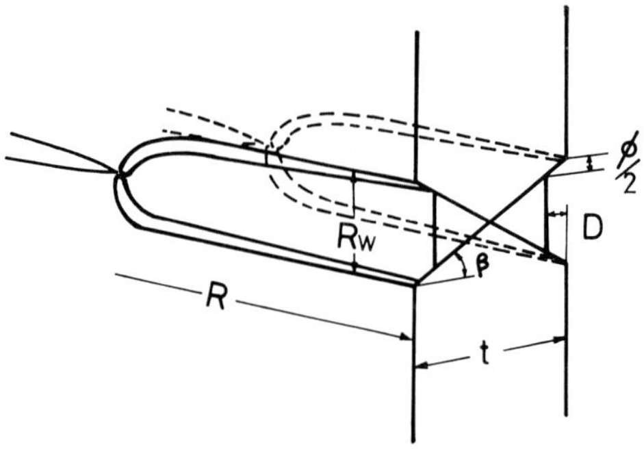

Figure 12.6 An antiplane-strain mode

applied stress (σ) and the crack length (a) by the relationship[15]

$$R = a\left\{\sec\left[\frac{\pi\sigma}{2\sigma_Y}\right] - 1\right\} \tag{12.2}$$

The notch root contraction (D), accommodated by local necking in the antiplane strain mode (as a function of bend angle (θ_b)), is formulated with the J integral for a rigid–perfectly plastic solid (after Kimura and Masumoto[14]):

$$D = C_1\theta_b \tag{12.3}$$

where $C_1 = (1/\sigma_Y)J/(2\sin\beta)$ (β is the angle of the slip plane to the thickness direction). For amorphous $Pd_{78}Cu_6Si_{16}$ plate specimens of two different thicknesses, *Figure 12.7* shows tensile plastic deformation ahead of a notch tip, and *Figure 12.8* shows the plastic zone size as a function of the ratio of net stress to yield stress[11]. The experimental results with two specimens before general yielding are in good agreement with Dugdale's relationship; the solid line in *Figure 12.8* denotes prediction of the Dugdale

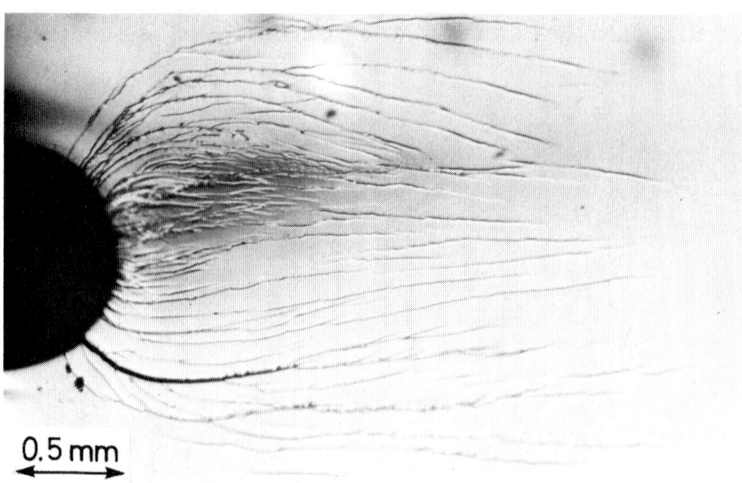

Figure 12.7 Tensile plastic deformation ahead of a notch tip for plate specimen of amorphous $Pd_{78}Cu_6Si_{16}$

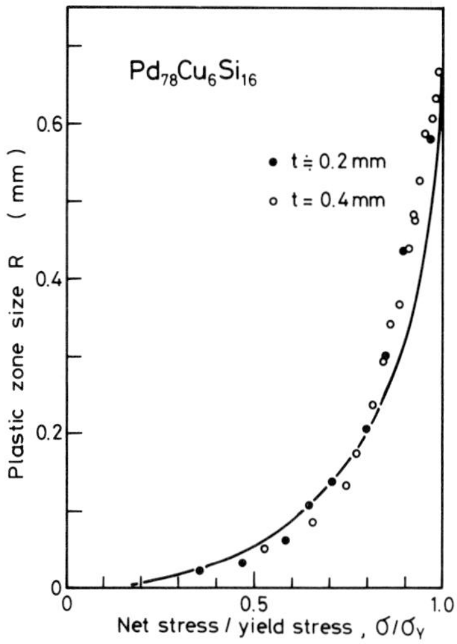

Figure 12.8 The plastic zone size as a function of the ratio of net stress to yield stress in amorphous $Pd_{78}Cu_6Si_{16}$

model, corrected for a finite width (W) of the specimen:

$$R = a\left\{\frac{W}{\pi a}\arcsin\left[\sin\left(\frac{\pi a}{2W}\right)\sec\left(\frac{\sigma}{2\sigma_Y}\right)\right] - 1\right\}$$

Figure 12.9 shows out-of-plane contraction (D) at the notch root; and *Figure 12.10* shows measured contraction of two different specimens as a function of measured bend angle[14]. The linear relation between the bend angle and the contraction, predicted by equation 12.3, is in excellent agreement with the experimental results obtained with specimens of different thickness. We may conclude that the inhomogeneous plastic flow in two and three dimensions of the antiplane-strain mode is that of a non-strain-hardening, rigid, perfectly plastic solid. *Figure 12.11* shows the variation[11] of the general yielding stress in tensile-tested specimens with plate thickness, t. At the critical plate thickness to net section (W − a) ratio, $t/(W - a) = 0.6$, a transition between plane stress and plane strain occurs in the deformation mode[11,14]. Note that the net stress at general yielding in the antiplane-strain mode increases with plate thickness. Based on

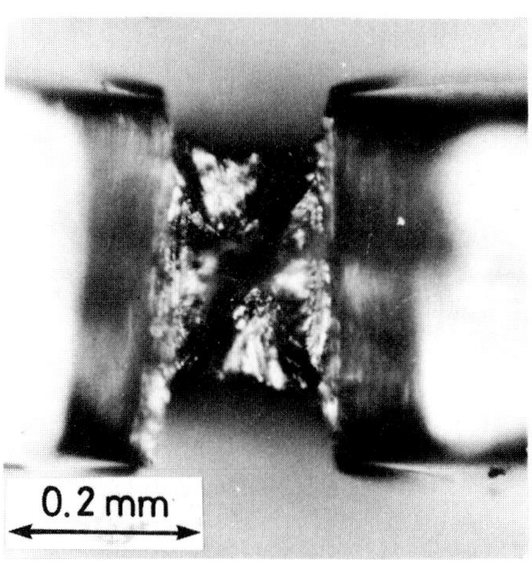

Figure 12.9 Out-of-plane contraction at notch root in antiplane-strain mode for amorphous $Pd_{78}Cu_6Si_{16}$ in V-notch bending

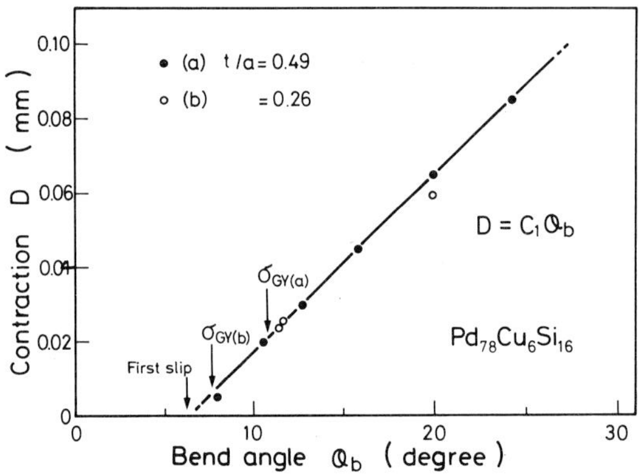

Figure 12.10 The measured contraction of two different specimens as a function of measured bend angle

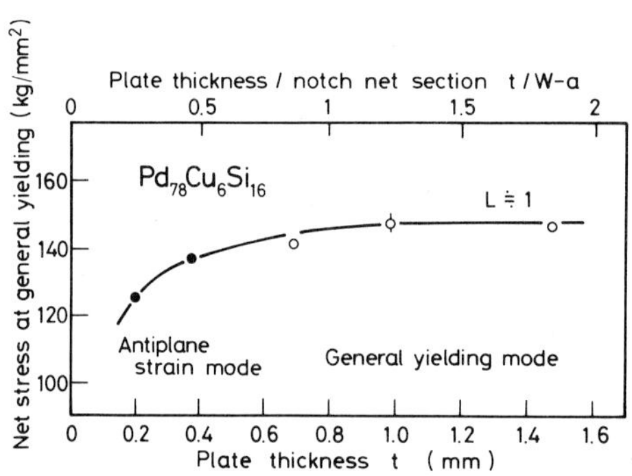

Figure 12.11 The general yield stress as a function of plate thickness in V-notch tensile specimens of amorphous $Pd_{78}Cu_6Si_{16}$

equations 12.2 and 12.3, the J integral at general yielding (J_{GY}) for the antiplane-strain mode in three dimensions is given by[14]:

$$J_{GY} = (\sigma_Y \phi_0)t \tag{12.4}$$

where ϕ_0 is the dimensionless crack opening displacement. Since the J integral and net stress at general yielding are related, equation 12.4 explains the rapid increase in the net stress at general yielding shown in *Figure 12.11*. This conclusion is important in an engineering sense where one is interested in the fracture strength of an unnotched specimen. The plastic instability is likely to occur before general yielding in the antiplane-strain mode; this fact promotes a further decrease in the observed fracture strength of the antiplane-strain mode.

Plate specimens with thickness–net section ratios greater than 0.6 show the general yielding mode as shown in *Figure 12.11*. For a ribbon specimen, a transition in the deformation mode occurs at the thickness ratio of 1/8 as reported by Davis who tested the amorphous $Fe_{40}Ni_{40}P_{14}B_6$ specimens with a gauge length, revealing a general yielding mode[16].

12.2.2 Yield criterion

One may reasonably expect that the deviatoric stress plays a primary role in plastic flow in an amorphous alloy undergoing shear slip. von Mises' criterion requires that yielding occurs when deviatoric stress (i.e., the root mean-square shear stress) (k)

equals the plastic shear resistance [$\tau(p)$] for the given pressure component of the principal stress:

$$k = \frac{1}{\sqrt{6}}\{(\sigma_1 - \sigma_2)^2 + (\sigma_2 - \sigma_3)^2 + (\sigma_3 - \sigma_1)^2\}$$
$$= \tau(p) \qquad (12.5)$$

where σ_1, σ_2 and σ_3 are components of the principal stress.

Yielding under various stress states

Figure 12.12 shows compressive deformation[11,12] for amorphous $Pd_{78}Cu_6Si_{16}$. The slip band angle is 45 degrees, in agreement with the results of Pampillo and Chen[17]. *Figure 12.13* shows the compressive yield stress as a function of the aspect ratio

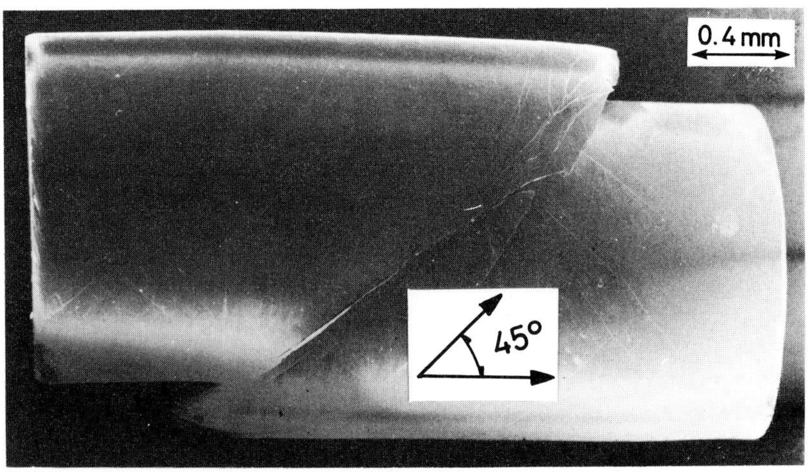

Figure 12.12 The compressive deformation for amorphous $Pd_{78}Cu_6Si$, indicating the slip band angle of 45 degrees

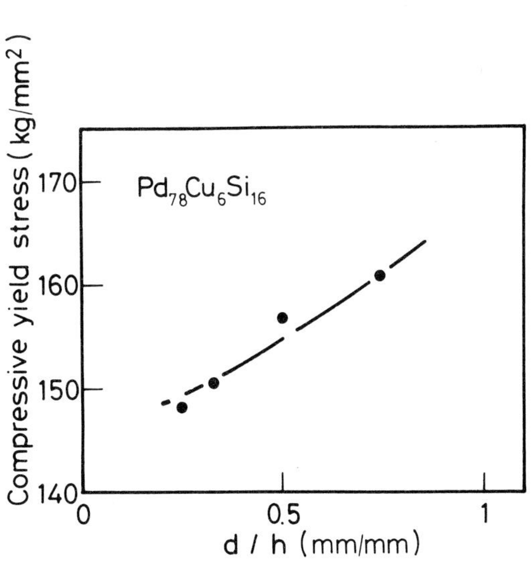

Figure 12.13 The variation of compressive yield stress with aspect ratio

(height/diameter) of the round bar[11]. The compressive stress for amorphous $Pd_{78}Cu_6Si_{16}$ decreases with decreasing aspect ratios. The compressive yield stress for the aspect ratio of 0.8 is higher than that of tensile yield stress (148 kg/mm^2), which is in agreement with the previous results.

The ratio of compressive yield stress in plane strain (σ_{YP}) on the modified von Mises' criterion, to the uniaxial yield stress (σ_Y) is [11,12]:

$$\frac{\sigma_{YP}}{\sigma_Y} = \frac{2}{\sqrt{3}}\frac{1 + k\xi_0/\sqrt{3}}{1 - k\xi_0} \tag{12.6}$$

where $\sigma_{YP} = -2k/(1 - k\xi_0)$, $\sigma_Y = (1\sqrt{3})k/(1 - k\xi_0/\sqrt{3})$ and ξ_0 is a pressure coefficient [i.e., $\tau(p) = k + \xi_0 p$]. According to the Mohr–Coulomb criterion, the yield stress in plane strain and in uniaxial are identical. Therefore, the decrease in compressive yield stress as shown in *Figure 12.13* may be ascribed to plastic deformation under a plastic constraint, depending on the aspect ratio of height (h) to diameter (d), i.e., under a plane strain condition. The smooth extrapolation of the dependence of yield stress on the aspect ratio of a uniaxial specimen gives the compressive yield stress of 148 kg/mm^2.

Figure 12.14 shows the yield loci (σ_1–σ_2 yield surface) for samples of amorphous $Pd_{78}Cu_6Si_{16}$, using the macroscopic yield stresses for compressive, tensile and torsional deformation[12]. We can see that the sample obeys von Mises' criterion

$$k = [\tfrac{1}{3}(\sigma_1^2 + \sigma_2^2 - \sigma_1\sigma_2)]^{1/2}$$

not a Mohr–Coulomb criterion nor Tresca's criterion in which the dilatational component of the stress tensor affects the shear stress. The relationship between shear stress (τ) and tensile stress (σ) shown in *Figure 12.15* defines the yield surface in a combined torsional–tensile test[12]. This relationship fits a prediction based on von Mises' criterion [$\sigma^2 + 3\tau^2 = (2k)^2$] showing clear evidence that under combined stresses

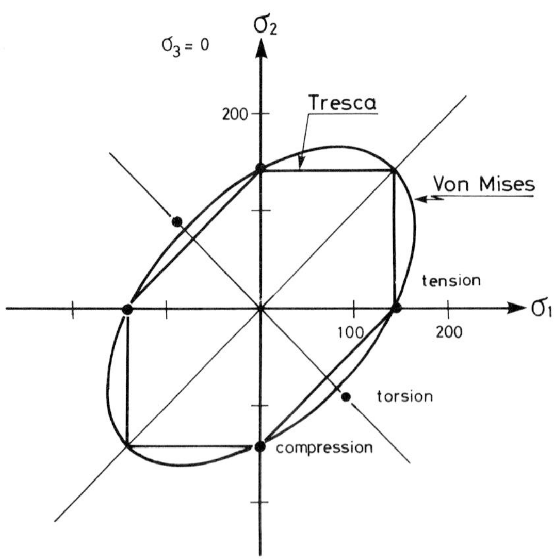

Figure 12.14 The yield loci (σ_1–σ_2 yield surface) for samples of amorphous $Pd_{78}Cu_6Si_{16}$, using the compressive, tensile and torsional shear stresses

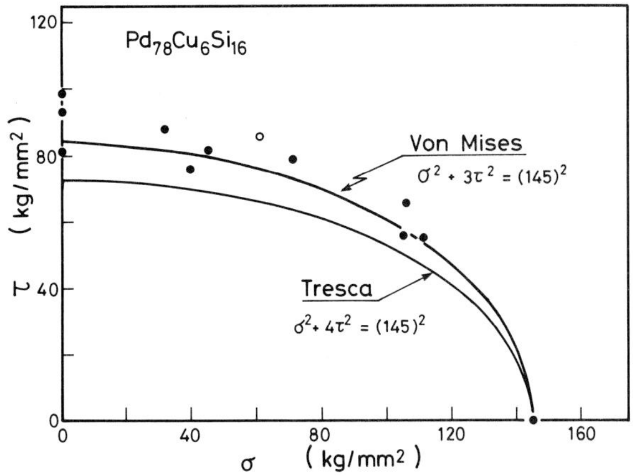

Figure 12.15 τ–σ yield surface in a combined torsional–tensile test for amorphous $Pd_{78}Cu_6Si_{16}$

amorphous $Pd_{78}Cu_6Si_{16}$ is a von Mises' material. *Figure 12.16* illustrates Green's plane-strain slip-line field for general yielding of 45 degree V-notch bent specimens*, as derived for a rigid perfectly plastic solid[18]. The plastic constraint factor (L) under full plane strain is predicted[18]:

$$L = \sigma_{GY}/\sigma_Y \tag{12.7}$$

$$= 1.26 \tag{12.7'}$$

Figure 12.17 shows the general yielding pattern[14] in three-point bending for a standard 45 degree V-notch specimen of amorphous $Pd_{78}Cu_6Si_{16}$. The general yielding process for a V-notch specimen is well defined experimentally; it can be followed easily *in situ* with an optical microscope by observing the formation of slip bands on the specimen surface. The general yield stress (σ_{GY}) for amorphous $Pd_{78}Cu_6Si_{16}$ is derived as 178 kg/mm^2, by assuming von Mises' criterion; the plastic constraint factor of 1.20 is deduced for standard specimens, using the uniaxial yield stress (σ_Y) of 148 kg/mm^2. The

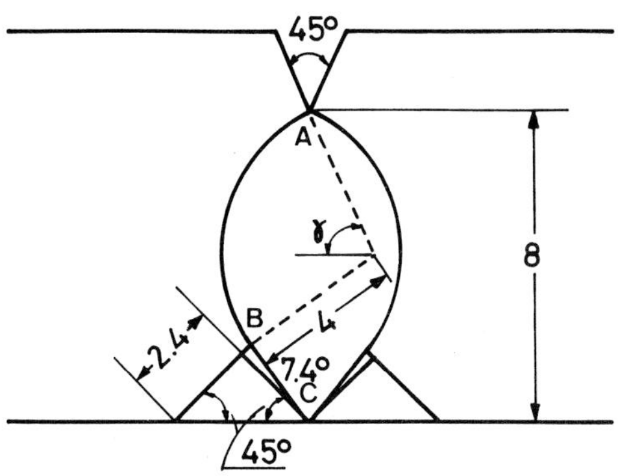

Figure 12.16 Green's slip line field for the V-notch with a circular fillet in plane strain

*The standard specimen used is a rectangular prismatic bar of plate thickness and plate width of 1 mm, and a notch depth of 0.2 mm.

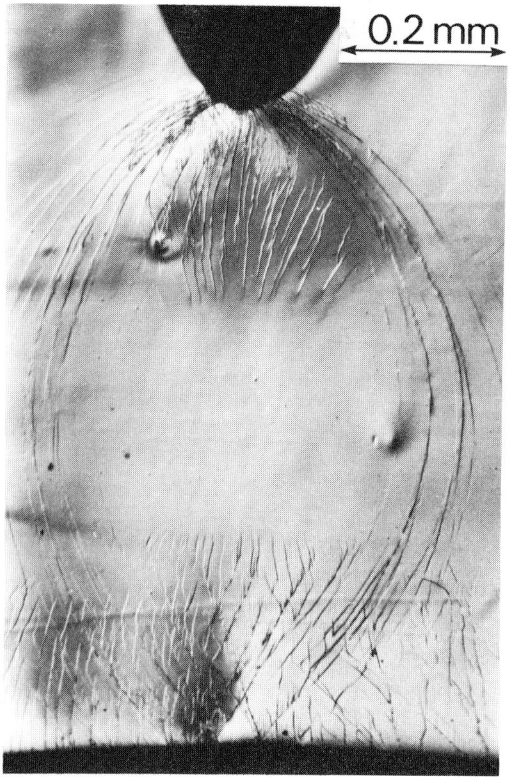

Figure 12.17 The general yielding pattern in three-point bending for a standard 45 degree V-notch specimen of amorphous $Pd_{78}Cu_6Si_{16}$

derived value accords with the value of 1.22 found for acute notches in three-point bending by Ewing[19], in agreement with equation 12.7′. The central angle (γ) of the 65 degrees for the standard specimen with a 45 degree notch agrees with that predicted by Green's slip line field. These agreements between theory and experiment confirm the applicability of von Mises' criterion to plastic flow in an amorphous alloy.

A pressure modified yield criterion

It is convenient to express the function of equation 12.5 explicitly as:

$$\tau(p) = \tau_0(1 + \xi P) \tag{12.8}$$

where τ_0 is the shear stress at zero pressure, and ξ is the pressure coefficient. A method is developed[20] with which one can obtain a pressure dependence of shear yield stress under very much higher pressure for thin specimens of various amorphous alloys. The specimen is compressed in a diamond cell. With a ruby-fluorescence method, the pressure distribution on the specimen surface is measured as a function of position. The function of equation 12.8 being assumed, the pressure as a function of position (r) is[20]

$$P_{(r)} = \frac{1}{\xi}\left[\left(\exp\frac{2\tau_0 \cdot \xi \cdot v}{h[1 - (2/3)\tau_0\xi]}\right) - 1\right] \tag{12.9}$$

where h is the specimen thickness under compression.

Figure 12.18 shows the measured pressure as a function of distance on the surface of a specimen of amorphous $Fe_{40}Ni_{40}P_{14}B_6$ [20]. The solid line is that predicted by equation 12.9, with $\xi = 0.0039\,\text{kbar}^{-1}$, $\tau_0 = 8.31\,\text{kbar}$ ($84.7\,\text{kg/mm}^2$). The pressure

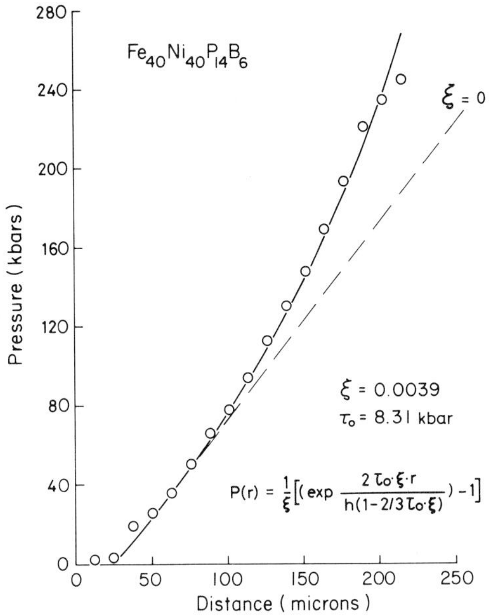

Figure 12.18 The measured pressure as a function of distance on the surface of a specimen of amorphous $Fe_{40}Ni_{40}P_{14}B_6$

dependence of yield stress for this alloy is normal up to 250 kbar and follows the explicit form of equation 12.8.

Davis and Kavesh were the first to measure the influence of hydrostatic pressure on flow in compression for amorphous $Pd_{77.5}Cu_6Si_{16.5}$ at hydrostatic pressures up to 6.2 kbar[21]. They derived the pressure coefficient of 0.005 ± 0.0007 per kilobar of applied pressure for amorphous $Pd_{77.5}Cu_6Si_{16.5}$. These values of pressure coefficients ($\xi = 0.0039$ for $Fe_{40}Ni_{40}P_{14}B_6$, and $\xi = 0.005$ for $Pd_{77.5}Cu_6Si_{16.5}$) are considerably smaller, compared with that of a glassy polymer in which a volume dilatation plays a role in yielding, but are close to the pressure coefficient measured in a crystalline material.

Having determined the pressure-dependence, we can finally get a yield criterion for an amorphous alloy by combining von Mises' criterion and the pressure-dependence function of equation 12.8.

Slip-band angle

The angle of the 'visible' slip bands is one of the characteristics of flow for an amorphous alloy. The concept of the strain-rate component can describe plastic flow of a highly local nature by noting that 'local' deformation can only occur on planes in which, on general yielding, the extension rate is zero[22]. According to this idea, the observed angle of 45 degrees for compression of a round bar (*Figure 12.12*), corresponds to that for plane-strain deformation (which does not contradict the previous discussion on the aspect-ratio dependent compressive yield stress); the observed 55 degrees for tension (*Figure 12.2*) is equal to the predicted value for uniaxial extension. The difference between slip band-angle in tension and compression is not due to a normal stress induced variation of 45 degrees shear yielding (i.e., plane strain extension with a volume dilatation). The small pressure coefficient of shear stress as described in the above section also does not account for such a large deviation from 45 degrees as pointed out by Davis and Kavesh[21].

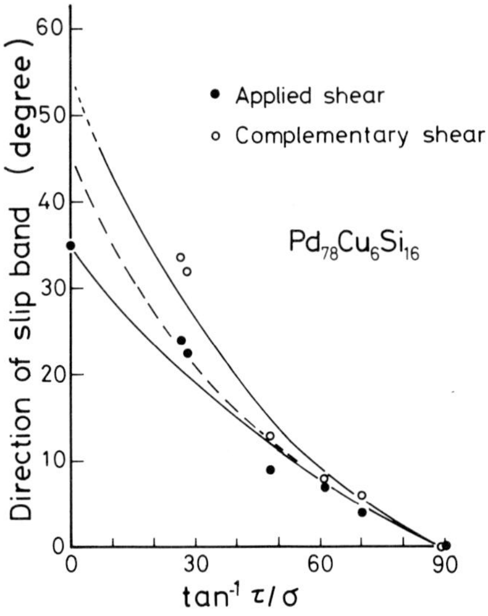

Figure 12.19 The angle of applied shear band and of complementary band observed on the surface of combined torsional–tension specimen as a function of the ratio of shear stress to tensile stress for $Pd_{78}Cu_6Si_{16}$

We try to examine explicitly the variation of slip-band angle with a systematic change of stress state. *Figure 12.19* shows the angle of the applied shear band, and of the complementary band, observed on the surface of a combined torsional–tension specimen as a function of the ratio of shear stress to tensile stress[12]. The measured values fall fairly well on the two solid curves which are based on the concept of strain rate component.

12.2.3 Micro-mechanics of plastic deformation

The slip band is the most basic flow unit for amorphous alloys at available resolutions. It is of significant importance in the formulation of a theory that accounts for phenomena resulting from the inhomogeneity of plastic flow in amorphous alloys.

Critical shear stress on a slip band

A theory is developed for a simple tear test from which one can determine the critical shear stress (τ) acting on a single shear band[23]. Assuming an ideally plastic solid, the critical shear stress in pure shear is given by

$$\Gamma/2 = \tau t \qquad (12.10)$$

where Γ is the tear energy, and t is the plate thickness.

According to Rivlin and Thomas, the tear energy (fracture energy per unit area) for shear crack propagation was originally defined as[24]:

$$\Gamma = 2F/t \qquad (12.11)$$

where F is tearing force. The critical discrepancy between the definition of equation 12.10 and equation 12.11 is their thickness dependence of tear energy: Γ in equation 12.10 is proportional to plate thickness, while equation 12.11 assumes the constancy of Γ for plate thickness. Note that the shear yield stress in tearing can be derived only by applied parameters, plate thickness and tearing force, avoiding a need of Young's modulus in the bend equation.

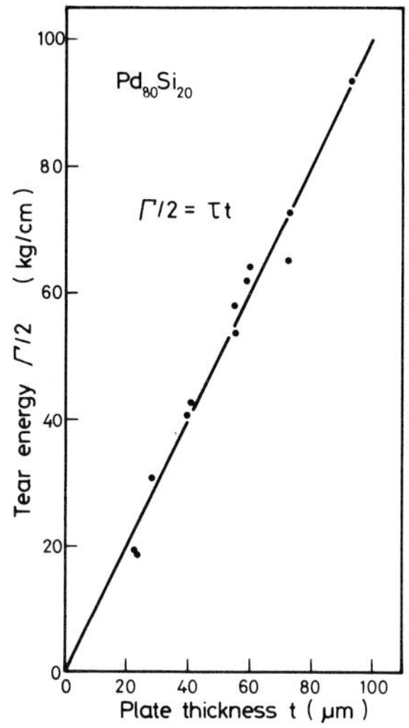

Figure 12.20 Tear energy for amorphous $Pd_{80}Si_{20}$ as a function of plate thickness

Figure 12.20 plots the tear energy for amorphous $Pd_{80}Si_{20}$ against plate thickness[23]. Tear energy is found to be proportional to plate thickness, as equation 12.10 predicts. The visual appearance of the plastic zone is a 'simple' shear band, extending on the slip plane of the maximum shear stress, which is coplanar with and avoids a branching of a crack. Consequently, the value of $100\,kg/mm^2$ for the critical shear stress for amorphous $Pd_{80}Si_{20}$ is derived from the slope of *Figure 12.20* by use of equation 12.10. The uniaxial stress of $173\,kg/mm^2$ is further calculated by assuming von Mises criterion; this value agrees fairly well with the observed value of $165\,kg/mm^2$ for $Pd_{80}Si_{20}$ alloy using a hard machine[25]. The near-equality of the value for 'microscopic' critical shear stress acting on a single shear band and for macroscopic yield stress proves that the inhomogeneity of the flow is of an elastic perfectly plastic nature.

Intersection and multiplication of slip bands

The plastic dèformation in a multiple axial stress state, e.g. in drawing, rolling and V-notch bending, proceeds by gross yielding exhibiting an intersection and multiplication of slip bands, whereas the plastic deformation in uniaxial stress occurs by highly localized shear (i.e., along a single slip plane) wherein the general yielding stress is adequate as a variable for the flow. *Figure 12.21* shows stress–strain curves of undrawn and drawn $Pd_{77.5}Cu_6Si_{16.5}$ amorphous wires[26], (*a*) area reduction $(R_0) = 0\%$, (*b*) $R_0 = 26\%$, and (*c*) $R_0 = 44\%$. The fracture stresses of drawn wires increase slightly (7 per cent) from that of undrawn wire, and macroscopic elongation to fracture increases considerably after drawing. Consider the mechanics of gross yielding and 'apparent' strain hardening. Intrinsic (physical) strain hardening is not responsible for these events, since an individual slip is of an ideal-plastic nature. The intersecting of slip bands inhibits further sliding of pre-existing slip bands and further deformation promotes creation and propagation of new slip bands ahead of terminal slip bands through undeformed material; an increase in the density of mobile slip bands can be created only by raising the applied stress. The pre-existing slip planes tend to slide at a smaller applied stress below macroscopic general yielding under stress concentration (i.e., local yielding); an increase in the density of such pre-existing slip planes accounts for an increase in the elongation of drawn wire.

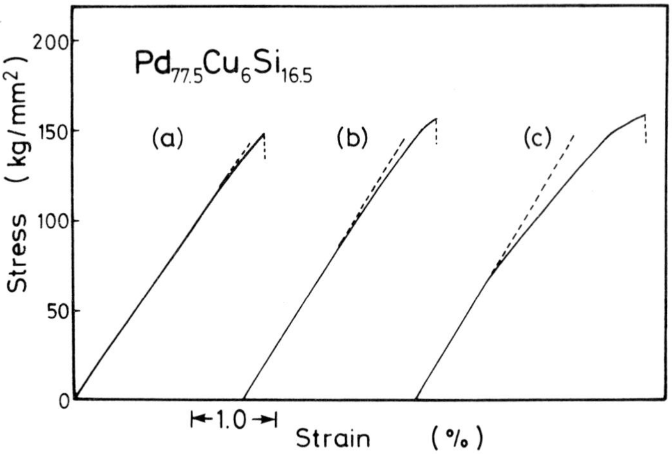

Figure 12.21 The stress–strain curves of undrawn and drawn $Pd_{77.5}Cu_6Si_{16.5}$ amorphous wire: (*a*) area reduction (R_0) = 0%, (*b*) $R_0 = 26\%$ and (*c*) $R_0 = 44\%$

12.2.4 Static Yield Strength — Overview

Elasticity and anelasticity

In the tensile stress–strain curve[7] for amorphous $Pd_{80}Si_{20}$ as shown in *Figure 12.22*, Young's modulus is characterized by: (1) the static value obtained from the slope of the linear region is 40 per cent smaller than that of the crystallized $Pd_{80}Si_{20}$, (2) the relative large anelasticity appears as a deviation from a linearity, including both recoverable strain and irrecoverable strain as can be seen in the loop and in the permanent strain in the unloading stress–strain curve. A deviation from a proportional limit is confirmed to be due to anelasticity, not micro-yielding, by Takayama and Maddin[27] who measured a change of the electrical resistivity in various levels of applied tensile stress for amorphous $Ni_{63}Pd_{17}P_{20}$.

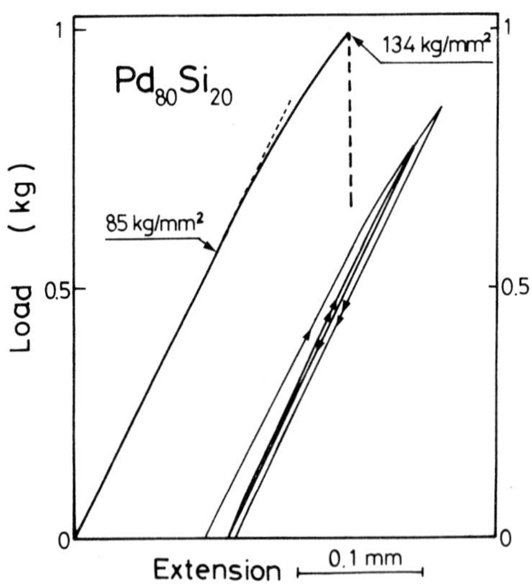

Figure 12.22 The tensile stress–strain curve for an amorphous $Pd_{80}Si_{20}$ alloy

Factors determining strength

The yield stress of materials can be described in terms of variables in a phenomenological equation:

$$\sigma = \sigma(\text{structure}, \dot{\varepsilon}, T, P \text{ and } t) \qquad (12.12)$$

We describe the effect of structure on the 'static' yield strength of amorphous alloys.

Compositional effect *Table 12.1* summarizes the yield strength, hardness and Young's modulus for typical iron, cobalt and nickel based amorphous alloys, and other amorphous alloys. The yield strengths shown in *Table 12.1* are obtained from the fracture strength σ_F in an antiplane strain mode, and so it could be lower than the general yield stress. Amorphous alloys are higher strength materials, especially the iron-based system, which exhibit very high fracture stress, in excess of 350 kg/mm^2. A collection of the yield stresses for various amorphous alloys are given in ref. 6. Hardness can be used to estimate the yield strength of amorphous alloys, since a correlation exists between hardness and derived shear stress for alloys which obey Hill's prediction for a rigid perfectly plastic solid. The small pressure coefficient of yield stress for amorphous alloys does not much affect the hardness which is a yielding under a triaxial stress state. The yield strength of amorphous alloys is strongly dependent on the transition metals (T), which are a major constituent; the yield strength tends to increase with increasing number of both group and period in the periodic table. *Figure 12.23* shows the hardness for Fe–B amorphous alloys as a function of the metalloid (M) content of boron[28]. The hardness in this Fe–B alloy system strongly increases with an increase in the ratio of boron to iron, demonstrating an important role of metalloid–transition metal interaction in the level of yield strength of amorphous

TABLE 12.1 Yield strength, Young's modulus, hardness and the ratio of yield strength to Young's modulus

Alloy composition	*Hardness* H_v/DPN	*Fracture strength* σ_F/kg mm^{-2}	*Young's modulus* E/kg mm^{-2}	σ_F/E	H_v/σ_F
$Fe_{80}P_{20}$	700	—	—	—	—
$Fe_{80}B_{20}$	1080	350	17 × 10^3	0.020	3.1
$Fe_{90}Zr_{10}$	640	220	—	—	2.9
$Fe_{80}P_{13}C_7$	760	310	12.4 × 10^3	0.025	2.5
$F_{78}B_{10}Si_{12}$	910	340	12 × 10^3	0.028	2.7
$Fe_{62}Mo_{20}C_{18}$	970	390	—	—	2.9
$Fe_{62}Cr_{12}Mo_8C_{18}$	900	330	—	—	2.7
$Fe_{46}Cr_{16}Mo_{20}C_{18}$	1130	400	—	—	2.8
$Co_{90}Zr_{10}$	600	190	—	—	3.2
$Co_{78}Si_{15}B_{12}$	910	306	9 × 10^3	0.034	3.0
$Co_{56}Cr_{26}C_{18}$	890	330	—	—	2.7
$Co_{44}Mo_{36}C_{20}$	1190	390	—	—	3.1
$Co_{34}Cr_{28}Mo_{20}C_{18}$	1400	410	—	—	3.4
$Ni_{90}Zr_{10}$	550	180	—	—	3.1
$Ni_{78}Si_{10}B_{12}$	860	250	8 × 10^3	0.034	3.4
$Ni_{34}Cr_{24}Mo_{24}C_{18}$	1060	350	—	—	3.0
$Pd_{80}Si_{20}$	325	136	6.8 × 10^3	0.020	2.4
$Cu_{80}Zr_{20}$	410	190	—	—	2.7
$Nb_{50}Ni_{50}$	893	—	13.2 × 10^3	—	—
$Ti_{50}Cu_{50}$	610	—	10.0 × 10^3	—	—

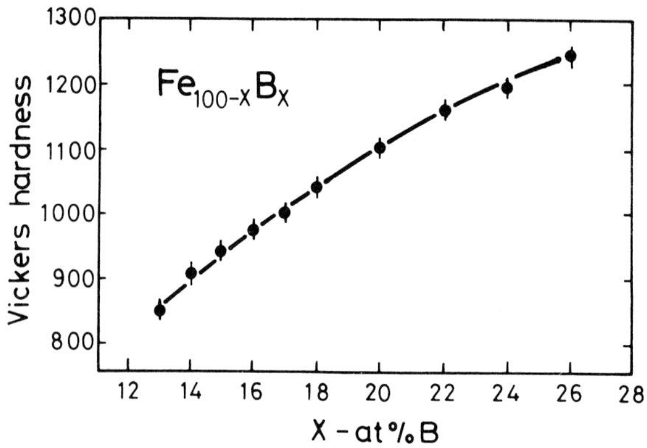

Figure 12.23 Hardness of Fe–B amorphous alloy as a function of the content of boron

alloys. The hardness in the binary system of iron-based alloy is shown to increase in the order $Fe_{80}B_{20} > Fe_{80}C_{20} > Fe_{80}Si_{20} > Fe_{80}P_{20} > Fe_{80}Ge_{20}$ as reported recently by Naka *et al.*[29].

It is of interest to correlate empirically the yield strength (hardness) of amorphous alloys with their elastic moduli, since plastic flow of these alloys is statically characterized as that of an elastic perfectly-plastic solid. *Figure 12.24* shows the correlation of Vickers' hardness (H/kg mm^{-2}) and Young's modulus (E/kg mm^{-2}) for a variety of amorphous alloys[30]. The upper curve displays the correlation for amorphous alloys approximating the $T_{80}M_{20}$ composition, and the lower curve is for the metal–metal amorphous alloys of roughly equiatomic composition. The variation of the hardness for various compositions in the two systems are found to be in good correlation with the dependence of their Young's moduli:

$$\sigma_Y/E = C_0 \qquad (12.13)$$

The ratio (C_0) (i.e., plastic strain) is almost constant; it lies between 1/60 and 1/40

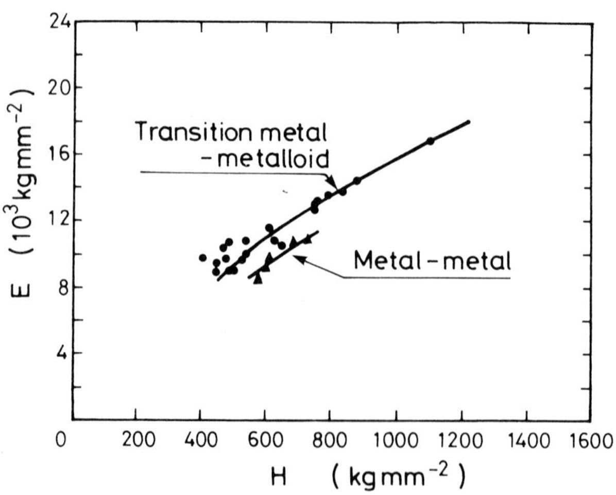

Figure 12.24 Correlation of the hardness (H) and Young's modulus (E) for transition metal–metalloid (●) and metal–metal (▲) amorphous alloys

within the experimental observation[1,4]. This finding indicates that the basic flow mechanism might be similar in all amorphous structures at least in each system as described by Davis *et al.* On the other hand, Chen demonstrated that yield strain (σ_Y/E) is related to Poisson's ratio[31]. The average outer-electron concentration of the transition metals has also a good correlation with the hardness (yield stress); this treatment[32,33] could be recognized as a logical extension of theory underlying equation 12.13. It is natural that yield strength correlates fairly well with elastic moduli, since an amorphous structure may prove to be of an isotropic nature for flow; consider, e.g., composition-independent slip band angle.

Particle dispersion effect Dispersion hardening is a well known mechanism to strengthen crystalline materials. Partial crystallization of amorphous alloys improves the yield strength relative to the as-quenched alloy, but concomitantly the alloy tends to lose its ductility[7,10,34–36]. Dispersion hardening with second phase hard particles is a simple mechanical case of particle hardening[37]. The yield stress for the ductile WC-dispersed amorphous $Ni_{78}Si_{10}B_{12}$ (σ_{YWC}) with volume fraction (V_f) appears to increase according to a mixture rule (upper bound rule) of Young's moduli; therefore, according to Kimura, Cunningham and Ast[37]:

$$\sigma_{YWC} = \sigma_Y\left[1 + V_f\left\{\left(\frac{E_p}{E_m}\right) - 1\right\}\right] \tag{12.14}$$

where E_m and E_p are the Young's modulus of the matrix and second phase particle respectively. If equation 12.14 indeed applies, a considerable improvement in the strength of the material can be obtained with a small fraction of hard particles.

Cold deformation effect A total area reduction in drawing was 93 per cent for a wire specimen of amorphous $Pd_{77.5}Cu_6Si_{16.5}$ after multiple passes[26]; a thickness reduction in rolling[1] was 40 per cent for amorphous $Pd_{80}Si_{20}$; and in these processes, the materials flow without fracturing. There are two features of the cold deformation effect on the strength of amorphous alloys; work hardening and work softening. A strain hardening in cold drawing was confirmed by Takayama who analysed the increase in the stress by drawing under different applied back-pull with reduction by continuum plastic theory[38]. This increase can be accounted for by mechanical intersection of slip bands as already discussed in section 12.2.3. On the other hand, the hardness of amorphous $Pd_{80}Si_{20}$ decreases with reduction in cold rolling; 30 per cent softening of the flow stress is derived for in-plane strain compression (under high pressure) for amorphous $Fe_{40}Ni_{40}Si_{14}B_6$ (*see Figure 12.18*). This softening would not be expected from the mechanics of ideal plastic flow and may be related to structural changes inside shear bands (*see* Section 12.24).

Radiation and structural relaxation effect Amorphous alloys possess good resistance to irradiation. Neutron irradiation of 5×10^{20} n/cm^2 causes a reduction of 10 per cent in the Young's modulus for amorphous $Pd_{80}Si_{20}$ relative to that for unirradiated samples as reported by Kayano *et al.*[39]. This reduction could be the result of a structural change of the amorphous matrix; and may cause a decrease in the fracture strength according to equation 12.13. (The fracture strength of irradiated samples reported by Kayano *et al.* was slightly increased, but this result may be explained by the improvement of the notch-sensitivity to the fracture strength (plastic instability) in a tearing mode, and not to an increase in general yielding stress.) The decrease in Young's modulus agrees with the swelling observed by Chang and Li[40] in the irradiation of $Fe_{40}Ni_{40}P_{14}B_6$ alloy by Ni^{5+}. It is noteworthy that irradiation of amorphous $Mo_{49.2}Ru_{32.8}B_{18}$ improves its ductility. The specimen after irradiation can be bent back without fracture, whereas this unirradiated sample is brittle[41]. The recovery of ductility may be accounted for by an increase in the micro-fracture stress (σ_f) of the irradiated specimen (*see* section 12.4.1 for a definition of σ_f).

On the other hand, the microhardness of $Pd_{77.5}Cu_6Si_{16.5}$ first decreases and then rises with annealing temperature, corresponding to two structural relaxations as reported by Chen and Lo[42]. Amorphous $Fe_{40}Ni_{40}P_{14}B_6$ after structural relaxation undergoes a 30 per cent decrease of yield stress for as-quenched material as observed by Ast and Krenitsky[43].

Implication to atomic flow

Deviation from an ideal plastic flow law The non-specificity of characterization for the flow of an amorphous alloy as being that of an elastic perfect solid makes it difficult to derive a detailed micro-mechanism. Important deviations from an ideal plastic flow law definitely exist, even if they are small, which include compositional dependence of yield strength, pressure dependence of shear stress and strain softening. Some similar relations to equation 12.13 are sought for various implications to the micromechanism of plastic flow. The yield stress depends on shear modulus (μ) for Li's dislocation model[44]; on $B/[1 + (3B/4\mu)]$ for Gilman's dislocation model[45]; and on B for the free volume model of Spaepen and Turnbull[46]. Here B is the bulk modulus.

A pressure-dependent shear stress can be accounted for by the physical processes; i.e. a pressure dependent elastic modulus, an activation volume in activated flow process, and dilatational component of yield stress. A pressure coefficient (ξ) of shear stress for amorphous alloys ($\xi = 0.0039$ for $Fe_{40}Ni_{40}P_{14}B_6$ and $\xi = 0.005\,\text{kbar}^{-1}$ for $Pd_{77.5}Cu_6Si_{16.5}$) are comparable with that of Young's modulus and shear modulus of crystalline materials[20]. The models which analyse the flow of amorphous alloy in terms of dislocations are therefore compatible with the derived pressure coefficient. Flow models based on free volume theory predict that flow stress should vary with pressure as B'/B, where B is the bulk modulus, and B' is a dimensionless pressure derivative[20] of B. The approach to the formulation of yield strength and/or micromechanism, using elastic modulus, could be itself self-consistent, as long as one wants to couple it with a static flow micromechanism. The firm conclusion on flow micromechanism should be drawn, following the exact measurement of yield stress and elastic modulus and their pressure dependence.

Small dilatation and softening are a minor effect on the formula for tearing and slip-band angle and so on, based on an ideal plastic flow. A detailed, exact measurement may permit a derivation of a flow law which describes a detailed model.

The springback in bending can be expressed in terms of bend-back angle (α) and separation distance (D) between the plates[43,47];

$$\frac{1-(\alpha/180)}{D'} = -\frac{12}{Et^3}\int_0^{t/2} \sigma_P(Z)\cdot Z\,dZ \tag{12.15}$$

where σ_P is the longitudinal stress distribution, measured from the neutral axis, and t is plate thickness. For an ideal plastic solid, the equation becomes[43]

$$\frac{1-(\alpha/180)}{D'} = -\frac{1}{2}\left(\frac{\sigma_Y}{Et}\right)^3 D'^2 + \frac{3}{2}\left(\frac{\sigma_Y}{Et}\right) \tag{12.15a}$$

where $D' = D - t$. The first measurement of a springback test for an amorphous alloy was performed by Ast and Krenitsky[43] who derived the yield stress and non-strain hardening property of $Fe_{40}Ni_{40}P_{14}B_6$. The angle α is very sensitive to a small change of strain. A more careful and exact measurement of the same alloy can present two kinds of deviation from a linearity in the plot of $[1 - (\alpha/180)]/D'$ against D'^2 expected from [48] equation 12.15*a*. The necessary condition to account for such deviations is the presence of a plastic stress concentration around an elastic–plastic boundary (*see* section 12.4.1) and softening in the plastic region near a specimen surface. An individual slip band could be recognized as an equivalent crack in fracture mechanics concept, resulting in the physical interaction between 'highly' localized slip bands in amorphous alloys.

A detailed examination of the fracture surface orientation for $Ni_{49}Fe_{29}P_{14}B_6Si_2$ glassy samples over a temperature range from 77 K to 568 K was recently performed by Davis and Yeow[49] who reported the slip band angle averaged 37 degrees; only a few degrees larger than the expected value. They suggest that this small variation may derive from the dynamic process of yielding. The equation presented by Argon[50], who analyses a difference between angles in the loaded and unloaded states, is used to calculate the expected value of slip band angle:

$$(90 - \theta) = \cos^{-1}\left\{\frac{1}{\sqrt{3}}\left[1 - \frac{\sigma_Y}{E(1+\nu)}\right]\right\} \tag{12.16}$$

where ν is Poisson's ratio.

Deformation-induced structural change A structural change due to inhomogeneous plastic flow has been observed in various measurements. The first peak position of rolled $Pd_{80}Si_{20}$ shifts to lower angles and the width of the peak increases with thickness reduction as observed by Masumoto and Koiwa[1]. Pampillo reported that shear bands which appeared in bending are susceptible to etching[51]. Kulik and Matyja have recently reported that plastic deformation in compression causes a definite decrease in the crystallization temperature[52]. The shear stress under plane strain compression has a 30 per cent smaller value than pure shear as reported by Kimura *et al.*, who tested amorphous $Fe_{40}Ni_{40}P_{14}B_6$ specimens in a diamond anvil[20]. These results indicate a more disordered atomic structure inside shear bands than the structure of the as-quenched alloys. The disordering inside shear bands is discussed in the perspective of local dilatation as suggested by Gilman[3,45] and the compositional short range disordering as proposed by Polk and Turnbull[53]. Transmission electron microscopy studies also describe the structure and formation of slip bands[54,55]. Deformation-induced structural change and a change of flow property due to structural relaxation seems to depend on alloy composition.

12.3 Dynamics of inhomogeneous plastic flow

12.3.1 Model mechanics of dynamical flow

Serrated flow is prominent in amorphous alloys as a manifestation of the dynamical nature of their inhomogeneous flow; the mechanics of serrated flow could provide a route to reveal the dynamics underlying the actual yield strength.

Yielding in a perfectly rigid machine

A model of the mechanics was developed from which it is possible to account for serrated flow in a conventional machine by Kimura and Masumoto[56]. The flow law assumes that the level of the observable stress (σ_0) applied to the specimen undergoing serrated flow is composed of two 'independent' physical processes,

$$\sigma_0 = \sigma_{eff} + \Delta\sigma \tag{12.17}$$

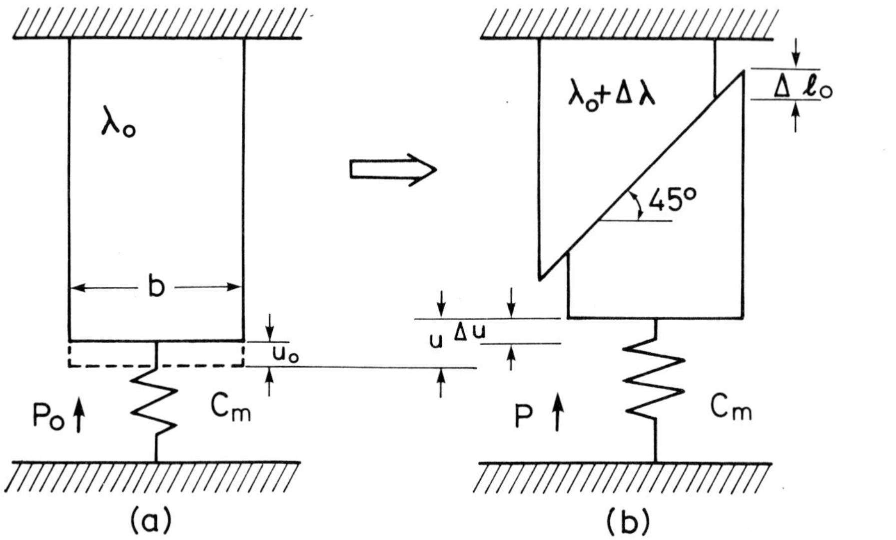

Figure 12.25 A model mechanics of serrated flow occurring on a single slip plane in an amorphous alloy, showing the machine–specimen system during the loading phase and the shear deformation phase

where σ_{eff} is the effective flow stress of the sliding of the slip band, and $\Delta\sigma$ is the serration amplitude. The second flow law assumes that the serration amplitude ($\Delta\sigma$) is at levels above the effective flow stress, and is expressed explicitly as a power law of the aging time (t):

$$\Delta\sigma = C_1 t^n \tag{12.18}$$

Figure 12.25 shows the schematic diagram of the machine–specimen system, (*a*) at the begining of a yield drop caused by serrated flow, and (*b*) after undergoing plastic sliding along the single shearing plane at 45 degrees to the compressive axis, to which the specimen accommodates by one serration. The two flow laws given by equations 12.17 and 12.18 being assumed, the changes in the load and deformation of the loaded specimen during quasi-static slipping, under the configuration of machine–specimen system of *Figure 12.25*, are mechanically characterized as shown in *Figure 12.26*. As

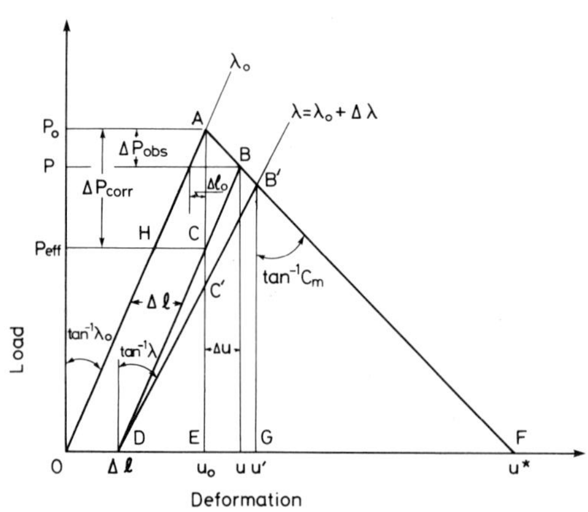

Figure 12.26 The load–deformation curve displaying a yield drop accompanying serrated flow based on a model mechanics

seen in the geometrical relations in *Figure 12.26*, the corrected serration amplitude ($\Delta\sigma_{corr}$) (i.e., the 'true' serration amplitude) that a perfectly rigid machine would produce is, from the observable serration amplitude in a conventional testing machine, given by:

$$\Delta\sigma_{corr} = \frac{\lambda_0 + C_m}{\lambda_0} \Delta\sigma_{obs} \tag{12.19}$$

Equation 12.19 can apply to a variety of machine–specimen systems for which the compliance of the machine (C_m) and the specimen (λ_0) are known. In the case of static aging in *Figure 12.26*, the applied extension rate at a certain level of $\Delta\sigma_{corr}$ in a conventional machine is given by

$$\dot{u} = (\Delta\sigma_{corr})^{n-1/n} \cdot (C_1)^{1/n} \frac{1}{1 - \left(\dfrac{1}{\lambda_0 + C_m}\right)^{1/n}} \cdot \frac{h}{E} \tag{12.20}$$

where E is Young's modulus and h, is the specimen height. Equation 12.20 can be used to derive C_1 and n in equation 12.18. The concept of static aging described by equations 12.17 and 12.18 has proved to apply generally to serrated flow in an amorphous alloy but static model mechanics described in equations 12.19 and 12.20 would be valid[56] only for a hard machine of compliance below about 3×10^{-4} mm/kg.

Analysis of the rate dependence of flow stress

Figure 12.27 shows a typical change in the compressive stress–strain curves for amorphous $Pd_{78}Cu_6Si_{16}$ specimens under different applied extension rates at room temperature for a machine compliance of 2.1×10^{-4} mm/kg, and a specimen's compliance of 1.4×10^{-4} mm/kg (after Kimura and Masumoto[56]). The observed serration amplitude decreases with the increasing applied extension rate, and disappears (at available resolutions) above the extension rate of 1 mm/min. *Figure 12.28* shows the compressive yield stress (σ_0) and the derived effective flow stress (σ_{eff}), from equation 12.19, of amorphous $Pd_{78}Cu_6Si_{16}$ at room temperature as a function of applied extension rate. The yield stress (σ_0) for amorphous $Pd_{78}Cu_6Si_{16}$ is independent of applied extension rate within this observation range. One can see that the dependence

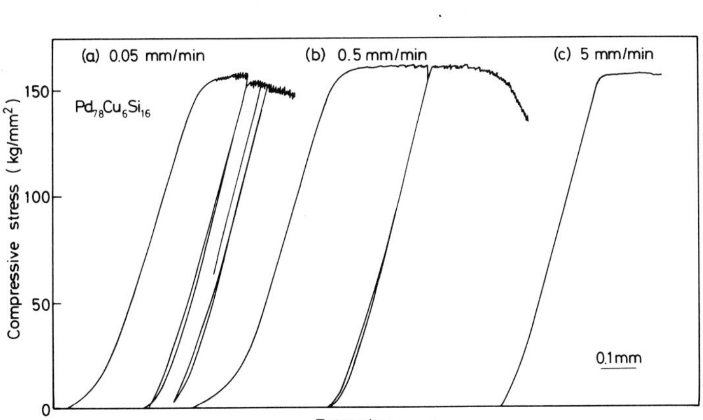

Figure 12.27 Compressive stress–extension curves for amorphous $Pd_{78}Cu_6Si_{16}$ at different applied extension rates

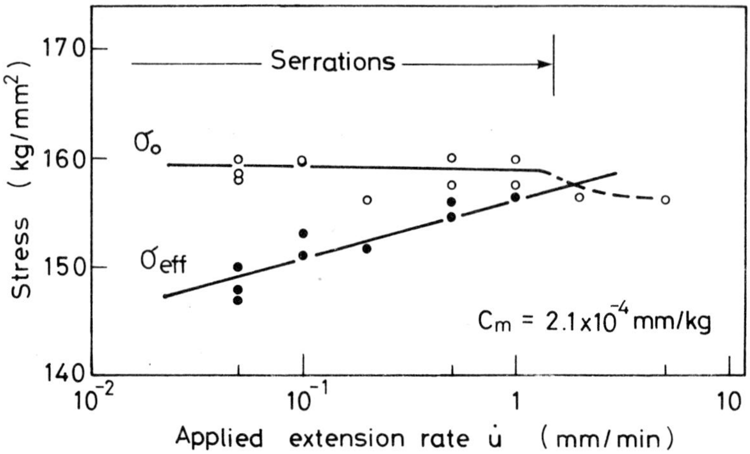

Figure 12.28 The applied extension rate dependence of compressive yield stress σ_0 and effective flow stress σ_{eff} for amorphous $Pd_{78}Cu_6Si_{16}$, for machine compliances of 2.1×10^{-4} mm/kg

of the effective flow stress on the applied extension rate is positive. This dependence is conveniently characterized by the introduction of the strain rate sensitivity (m):

$$m = \left(\frac{\partial \log \dot{u}}{\partial \log \sigma_{eff}}\right)_T \tag{12.21}$$

The strain-rate sensitivity of the effective flow stress for amorphous $Pd_{78}Cu_6Si_{16}$ is about 59 for the extension rates ranging[56] from 0.05 to 1.5 mm/min.

A power law for serrated flow

Figure 12.29 shows the corrected serration amplitude ($\Delta\sigma_{corr}$) plotted against aging time during one serration, observed on the record charts for a machine with a

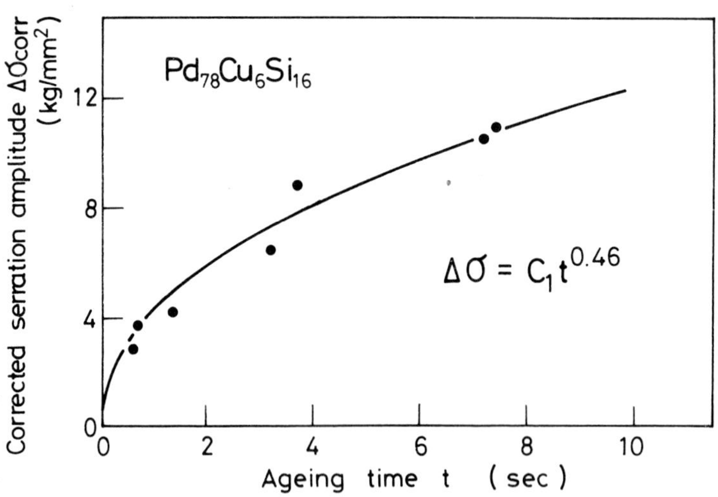

Figure 12.29 Corrected serration amplitude plotted against ageing time for a machine with $C_m = 2.1 \times 10^{-4}$mm/kg. The curve line is the prediction of a power law, $\Delta\sigma = C_1 t^{0.46}$, for serrated flow

compliance[56] of 2.1×10^{-4} mm/kg. It can be seen that the experimental data for amorphous $Pd_{78}Cu_6Si_{16}$ agree fairly well with the predicted curve using a power law for serrated flow with $n = 0.46$.

The $t^{0.46}$ law derived for an amorphous alloy is different from the $t^{2/3}$ law, predicted by Cottrell and Bilby who describe the atomic size effect on the solute–dislocation interaction[57]. Our results are nearly equal to the $t^{1/2}$ law for the shear modulus effect[58]. The empirical $t^{0.28}$ law for Snoek–Schoeck ordering[59] and the $t^{0.3\sim0.39}$ law for electrostatic effects[60] have smaller values of n than our $t^{0.46}$ law. From this comparison, one suggests that the mechanism of the serrated flow in an amorphous alloy does not involve a long-range diffusion process that is vacancy controlled such as the Cottrell effect, but can be related to a short range interaction of a 'defect' in a glassy structure. In other words, the factor which dominates the serrated flow in amorphous alloys is not the atomic size of solute elements, but its 'shear modulus' or chemical short-range ordering around a defect.

12.3.2 Thermally activated slip motion

Mobility of single shear band

A method was developed, which uses a tear test, from which one can derive the mobility (v) of the front of a slip band, by Kimura and Masumoto[61]. This is the fundamental flow unit in amorphous metals at available resolutions. The rate (v) of shear sliding at the slip band front can be given by the relation 12.22 between the rate of separation of grips (extension rate, dl/dt) and the rate of crack propagation (dc/dt) in steady tearing, by assuming v to be equivalent to dc/dt in an ideal plastic solid:

$$2\frac{dc}{dt}(=v) = \frac{dl}{dt} \qquad (12.22)$$

With the tear test providing measures of both variables, shear stress in equation 12.10 and the rate of shear-band front in equation 12.22, one can deal with the various thermodynamical phenomena of inhomogeneous slip, for an amorphous alloy.

Figure 12.30 shows the mobility of a shear-band front as a function of testing

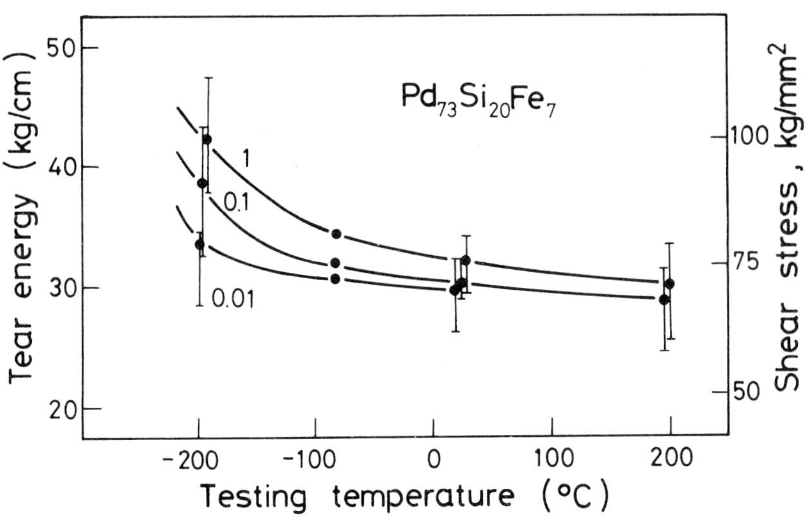

Figure 12.30 Mobility of shear-band front as a function of testing temperature for an amorphous $Pd_{73}Si_{20}Fe_7$

temperature for an amorphous $Pd_{73}Si_{20}Fe_7$ alloy[61]. The stress required to maintain a constant rate of motion at the shear-band front increases rapidly with decreasing temperature and increasing rate of motion of the shear-band front; the stress is almost constant above room temperature where serrated flow predominates. Using the relation $H_0 = kT^2(\partial \ln v/\partial \ln T)$, we can derive the activation energy for inhomogeneous plastic flow as 0.05 eV at a shear stress (τ) of 86 kg/mm^2 in the lower temperature range where serrated flow disappears. This value for inhomogeneous flow is a smaller value than 0.5 eV found[62] in homogeneous flow for amorphous $Pd_{80}Si_{20}$, indicating different mechanisms between both deformation modes.

Activation energy for serrated flow

Figure 12.31 is a simple Arrhenius plot of the critical extension rates for onset of serration (v_c)* in tearing, compression and V-notch bending for amorphous $Pd_{78}Cu_6Si_{16}$. The temperature dependence of the critical extension rate can clearly be

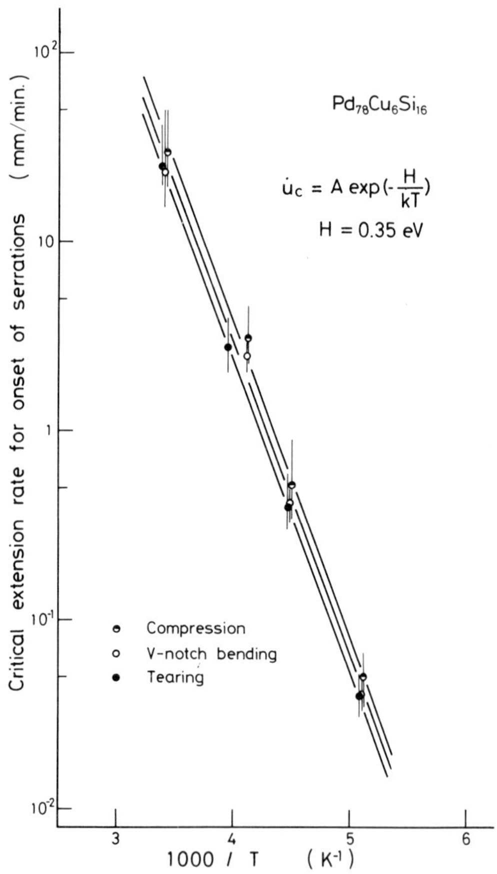

Figure 12.31 Arrhenius plot of the critical extension rate for onset of serrations for amorphous $Pd_{78}Cu_6Si_{16}$ in tearing, compression and V-notch bending

*The critical extension rate for onset of serration is derived by assuming a linear relationship between extension rates and serration amplitude in the lower stress region, and then extrapolating the experimental data to the load fluctuations below a certain amplitude (where the stress–extension curves become practically smooth).

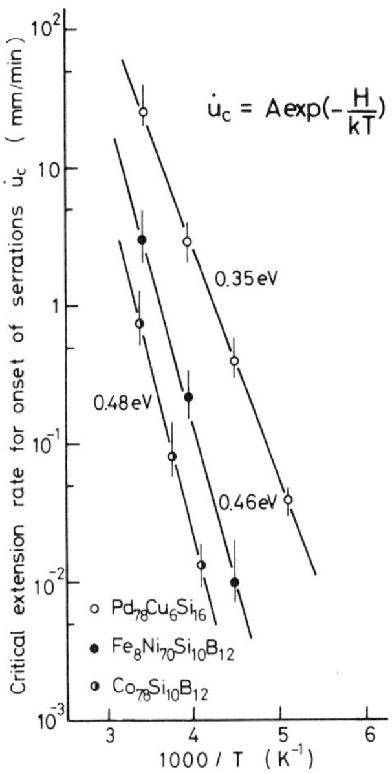

Figure 12.32 Arrhenius relations in serrated flow for various amorphous alloys

represented by

$$v_c = A\exp(-H/kT) \tag{12.23}$$

where H is the activation energy for serrated flow, k the Boltzmann constant, and A the pre-exponential factor. The activation energy for amorphous $Pd_{78}Cu_6Si_{16}$ is thus 0.35 eV and this value is the same for tearing, compression and V-notch bending as shown in *Figure 12.31*. This near-equality of the values for activation energy obtained by various testing methods suggest a small value for the activation volume, when the negative or positive hydrostatic pressure superimposed in compression and V-notch bending is considered. *Figure 12.32* shows the relations between critical extension rates and the reciprocal of the temperature in serrated flow for amorphous $Fe_8Ni_{70}Si_{10}B_{12}$, $Co_{78}Si_{10}B_{12}$ and $Pd_{78}Cu_6Si_{16}$ alloys[61]. All the alloy compositions used (which are categorized as transition-metal–metalloid systems) show serrated flow; they can thus be represented fairly well by the Arrhenius equation 12.23, suggesting that the basic activating mechanism involved in serrated flow is the unique rate process in all amorphous metals. *Table 12.2* shows, for various compositions, the activation energy for serrated flow (H) and 'critical' temperature for onset of serration (T_s) at the extension rate of 0.1 mm/min which corresponds to that customarily used in tensile tests. Thermodynamic parameters in serration are strongly dependent on alloy composition.

TABLE 12.2 Thermodynamic parameters for serrations of various amorphous alloys

Composition	H/eV	T_s/°C
$Pd_{78}Cu_6Si_{16}$	0.35	−67
$Ni_{70}Fe_8Si_{10}B_{12}$	0.46	−26
$Co_{78}Si_{10}B_{12}$	0.48	−12
$Fe_{48}Ni_{30}Si_{10}B_{12}$	—	0

12.3.3 Dynamic flow — overview

Factors determining flow stress

Temperature and pressure effect *Figure 12.33* shows the ratio between fracture stress and shear modulus for a variety of amorphous alloys as a function of temperature (after Pampillo and Polk[63]). Pampillo and Polk point out that there are two salient features of the temperature dependence of yield stress in the regime of inhomogeneous plastic flow: the occurrence of a plateau and a rapid increase in a low temperature range. The stress in the low temperature region can be attributed to a thermal

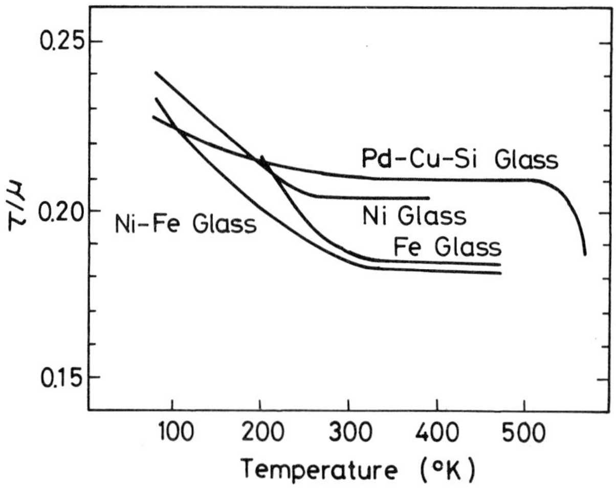

Figure 12.33 The ratio between fracture stress and shear modulus for a variety of amorphous alloys as a function of temperature

component; this idea follows from our analysis of the temperature dependence of the slip bands and other measurements of temperature dependence of fracture stress[49,64], yield stress and general yielding stress[14]. At present, there are two explanations for the occurrence of a plateau in the graph of flow stress against temperature. The stress is ascribed to an athermal character by Pampillo and Polk[59] who refer to Seeger's theory[65] on crystalline alloys. Our model mechanics describe another model based on dynamical flow[61]. *Table 12.3* shows the temperature (T_p) at the beginning of the plateau

TABLE 12.3 Temperature at the beginning of the plateau of the temperature dependence of macroscopic yield stress or fracture stress for various amorphous alloys

Alloy composition	T_p/°C
$Pd_{78}Cu_6Si_{16}$*	−43
$Ni_{72}P_{18}B_7Al_3$†	−14
$Ni_{49}Fe_{29}B_6Al_2$†	13
$Fe_{76}P_{16}C_4Si_2Al_2$†	49

*Kimura and Masumoto (1980).
†Pampillo and Polk (1974).

in the temperature dependence of macroscopic yield stress for various amorphous alloys. T_p was obtained by intersecting the curve of the low-temperature dependence of yield stress with the plateau in *Figure 12.33*. Comparing *Tables 12.2* and *12.3*, one finds a significant correlation between the values of T_s and T_p in the iron-, nickel- and palladium-based alloys. This finding supports an explanation for the occurence of a plateau. The mechanical threshold stress at 0 K continues to decrease with increasing temperature with the aid of thermal activation, but the increase in serration amplitude with increasing temperature balances this decline[14,61].

Where the critical shear stress is exponentially pressure dependent as expressed by $\tau = \tau_0 \exp(\xi P)$, the activation volume (ΔV) and ξ are related by $\Delta V = \xi kT$, where k and T have their usual meanings. The pressure dependence of the flow stress of *Figure 12.18* satisfactorily permit the derivation[20] of the pressure independent activation volume of 0.16 Å^3 for amorphous $Fe_{40}Ni_{40}P_{14}B_6$. This value is incompatible with a simple free volume model.

Composition effect It has been demonstrated that the yield strength of amorphous alloy is composed of two independent flow processes ($\sigma_0 = \sigma_{eff} + \Delta\sigma$) dominated by thermal activation. The serration amplitude of an amorphous alloy, from the combination of equations 12.18, 12.20 and 12.23, is empirically given[56] by:

$$\Delta\sigma = C_2 t^n \exp\left(\frac{n}{1-n} \cdot \frac{H}{kT}\right) \tag{12.24}$$

Equation 12.24 says that equations 12.18 and 12.23 are unique for various amorphous alloys; the exponent n has the constant value of 0.46, and H and the pre-exponential term C_2 are material parameters. Note that the corrected serration amplitude attains at maximum at about 11 per cent of the yield stress for amorphous $Pd_{78}Cu_6Si_{16}$ at a commonly used extension rate as shown in *Figure 12.28*. If one discusses the origin and the deformation map of the yield strength underlying inhomogeneous flow around ambient temperatures for various amorphous alloys, the dynamics describing both equations 12.21 and 12.24 must be taken into account. On the other hand, a correlation between yield stress and elastic modulus such as equation 12.13 should be examined with use of the mechanical threshold stress at 0 K.

Theoretical constitutive law of the flow

A successful phenomenological description of the flow, which includes a prediction for the observable flow, is developed, based on the transition state theory. Spaepen presents a theory for inhomogeneous flow which is based on a dynamic equilibrium between stress-driven creation and diffusional annihilation of structural disorder, and predicts the boundary between the inhomogeneous and homogeneous flow[66]:

$$\frac{\tau}{\mu} = \frac{4kT}{\mu\Omega} \operatorname{arcsin}\left(\frac{\alpha kT}{S\Omega}\right)^{-1/2} \tag{12.25}$$

where $S = (2\mu/3)(1 + v)(1 - v)$; Ω, atomic volume; α, a dimensionless constant dependent on the structure of the glass and the statics of atomic movement; and k, T, μ and v have their usual meanings. Argon advances a theory which is based on thermally activated shear transformation initiated around free volume regions under an applied shear stress, and predicts a transition between homogeneous and inhomogeneous flow[67]:

$$\frac{T}{T_g} = A\left\{\beta - \left(\frac{\hat{\tau}}{\mu}\right)\left(\frac{\sigma}{\hat{\tau}}\right)_{crit}\right\}\frac{1}{\ln(\dot{\gamma}_G/\dot{\gamma})} \tag{12.26}$$

where T_g is the glass temperature; $\hat{\tau}$, inherent shear resistance; γ_G, pre-exponential factor;

$$\beta = \frac{7 - 5\nu}{30(1 - \nu)}\gamma_0 + \frac{\hat{\tau}}{\mu}$$

$$A = \frac{\mu\gamma_0\Omega_f}{kT_g}$$

γ_0 the transformation shear strain and Ω_f is a volume of roughly spherical shape consisting of a flow unit. Argon's theory describes the temperature dependence of the inhomogeneous flow stress at constant strain rate[67,68]

$$\frac{\sigma}{\hat{\tau}} = 1 - \sqrt{(0.22)}\left(\frac{kT}{\hat{\tau}\Omega_f}\right)\ln\left(\frac{\alpha_0\gamma_0\nu_G}{\dot{\gamma}}\right) \tag{12.27}$$

where ν_G is the normal mode frequency of the transforming complex along the activation path, α_o is a numerical constant and the steady state volume fraction of flow units contributing to plastic flow, and γ is the strain rate.

The boundary between inhomogeneous and homogeneous flow is experimentally presented by Maddin and Masumoto[62] who examined the effect of strain rate on both fracture stress and mode of amorphous $Pd_{80}Si_{20}$ over a temperature range of 0 °C—300 °C. Patterson and Jones[69] show that the experimental boundary between homogeneous and inhomogeneous flow for amorphous $Fe_{40}Ni_{40}P_{14}B_6$ agrees with Spaepen's prediction of equation 12.25 with a proper value of the fitting parameter for the atomic volume $\Omega = 9\,\text{Å}^3$. On the other hand, Megusar *et al.* demonstrate[68] that

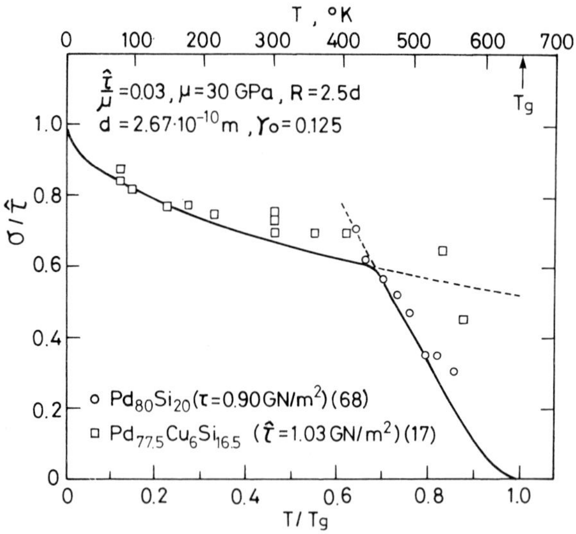

Figure 12.34 Dependence of flow stress on temperature computed from Argon's prediction[67] of equation 12.27, and compared with experimental results of Pampillo and Chen[17]. The experimental results of Megusar *et al.*[68] for homogeneous flow are included.

Argon's prediction of equation 12.26, by choosing the proper values of some fitting parameters, is satisfactorily in accord with the boundary between inhomogeneous and homogeneous deformation for amorphous $Pd_{80}Si_{20}$, although there is some discrepancy between theory and experiment.

Argon compares his prediction of equation 12.27 with the temperature dependence of the compressive flow stress for amorphous $Pd_{77.5}Cu_6Si_{16.5}$ obtained by Pampillo and Chen as shown in *Figure 12.34* and finds good agreement between the theory and the experiment assuming the value of parameters in equation 12.27. The concept of an activated flow underlying Argon's theory for inhomogeneous flow of an amorphous alloy is basically compatible with the thermally activated motion of a macroscopic shear band deduced from the model mechanics as proposed in section 12.3.1. Our model mechanics of the serrated flow may result in the better understanding of an apparent deviation in the plateau region of the temperature dependence of the observed yield stress from the prediction of equation 12.27 (*Figure 12.34*).

In a phenomenological description of the inhomogeneous flow for amorphous alloys, some self-consistent treatments between theory and experiment could be completed, taking into account the proposed mechanics for the flow.

12.4 Mechanics of fracture

12.4.1 Ductility

Most amorphous alloys can be bent completely back on to themselves without fracturing, but the tensile plastic strain is very small because a specimen undergoes highly localized shear deformation. The measure of ductility, the bend-back angle of 180 degrees, indicates that an amorphous alloy is very ductile, but the latter quantity, the small fracture energy given by the stress–strain curve, shows that this alloy is less ductile. On the other hand, the fracture toughness (K_{IC}, K_{IIIC}) of some iron-based amorphous alloys is comparable with that of high strength materials, such as a marageing steel[70–72]. The linear fracture mechanics is, needless to say, useful to describe the resistance of materials to crack propagation. However, most as-quenched amorphous alloys do not undergo crack formation below general yielding in bending, avoiding tensile plastic instability[73,74], so this fracturing is not a valid condition for measuring fracture toughness. Furthermore, the values of K_{IC} for iron-based alloys and its thickness dependence strongly depends on quenching conditions[75]. This fact indicates that, at present, it appears to be difficult to offer fracture toughness value (K_{IC}) as an accepted parameter of toughness for a variety of amorphous compositions, and to formulate a rule which is based on any fracture toughness concepts. Here, we are going to provide fracture parameters with which it is possible to describe ductility of amorphous alloys, and to gain physical understanding of the structure sensitivity of ductility, giving special attention to the actual (local) fracture process.

Perfect ductility and notch bar fracture mechanics

The plastic constraint factor (L) of a notched specimen is a convenient parameter to estimate how much an amorphous alloy is ductile, since L is determined through constraint general yielding, dependent on notch angle. In notch bar fracture mechanics a well-known elastic–plastic distribution ahead of the notch can be used to analyse local

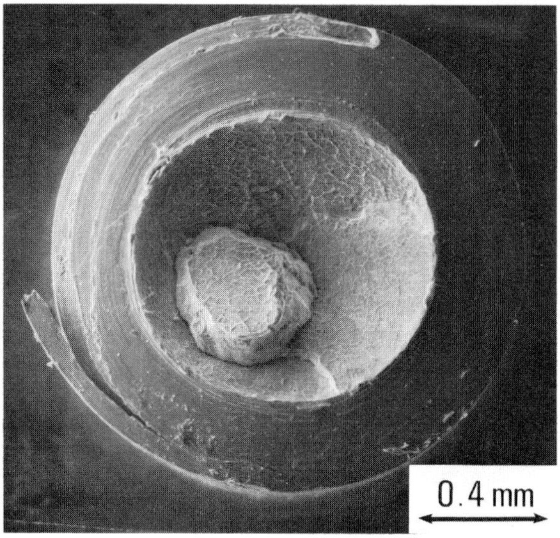

Figure 12.35 Fracture surface of a circumferentially 45 degree notched round specimen of amorphous $Pd_{78}Cu_6Si_{16}$

fracture. The plastic constraint factor of a double-notched tensile specimen is given by[76]

$$L = \frac{P_{GY}}{\sigma_Y(W - 2a)t}$$

where P_{GY} is the general yielding load; W the specimen width; a the notch depth; and t the specimen thickness. *Figure 12.35* shows the fracture surface of a 45 degree notched

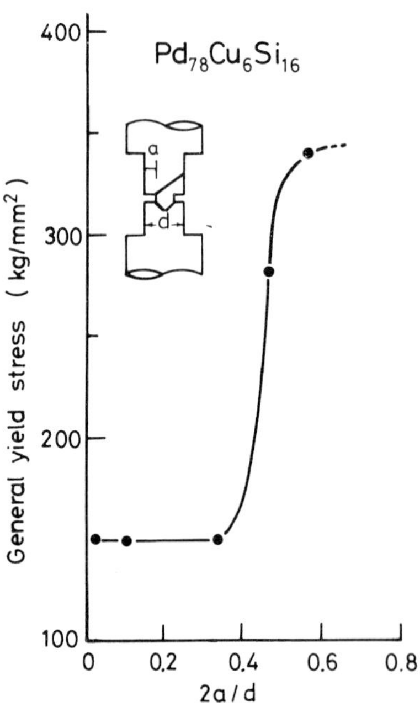

Figure 12.36 Variation of the general yield stress for an amorphous $Pd_{78}Cu_6Si_{16}$ round specimen containing slit notches with the notch depth ratio (2*a*/*d*)

round specimen[11] of amorphous $Pd_{78}Cu_6Si_{16}$. The fracture surface is composed of a circular-smooth region ahead of the notch due to the constrained slipping, and a corresponding plastic elongation in load–extension curve and a characteristic morphology, known as a vein pattern[9,79,80] due to tensile plastic instability. *Figure 12.36* shows a variation of the general yield stress for amorphous $Pd_{78}Cu_6Si_{16}$ round specimens containing exterior slit notches (i.e., notch angle of 0 degrees) with the ratio of notch depth (a) to specimen's diameter (d). Specimens with shallow notches yield from one side of the notch and then over its gross cross-section. The amorphous $Pd_{78}Cu_6Si_{16}$ round specimen with a deep slit notch undergoes constrained yielding through the notch width; its plastic constraint factor attains the value of 2.4, which is nearly equal to the upper limit of 2.96 for a deep crack predicted by slip-line fields, assuming von Mises' criterion. With this information, one can conclude that an amorphous $Pd_{78}Cu_6Si_{16}$ alloy is a 'perfect' ductile material.

Mechanics of ductile–brittle transition

Local tensile criterion for fracture We adopt for our hypothesis the critical tensile criterion to set up a mechanics of ductile–brittle transition for an amorphous alloy[74]. This criterion can be generally expressed in a simple form[76]:

$$\sigma_f = \sigma_{yy,max} = K_{e,}K_p\sigma \tag{12.28}$$

where σ_f is micro-fracture stress, $\sigma_{yy,max}$ is the maximum of longitudinal stress, σ is the applied stress where a crack occurs, and K_e and K_p are the elastic stress concentration factor and the plastic intensification factor. Here, we describe annealing induced embrittlement appearing in V-notched bend specimens of amorphous $Pd_{78}Cu_6Si_{16}$ to test equation 12.28. This alloy, with its high constraint factor of 2.4(*Figure 12.36*), loses ductility gradually. V-Notch specimens enable us to calculate K_p in terms of slip-line fields; it therefore lends itself to our purpose. *Figure 12.37* shows deformation and fracture patterns observed at the surface of the plane-strain specimens by optical

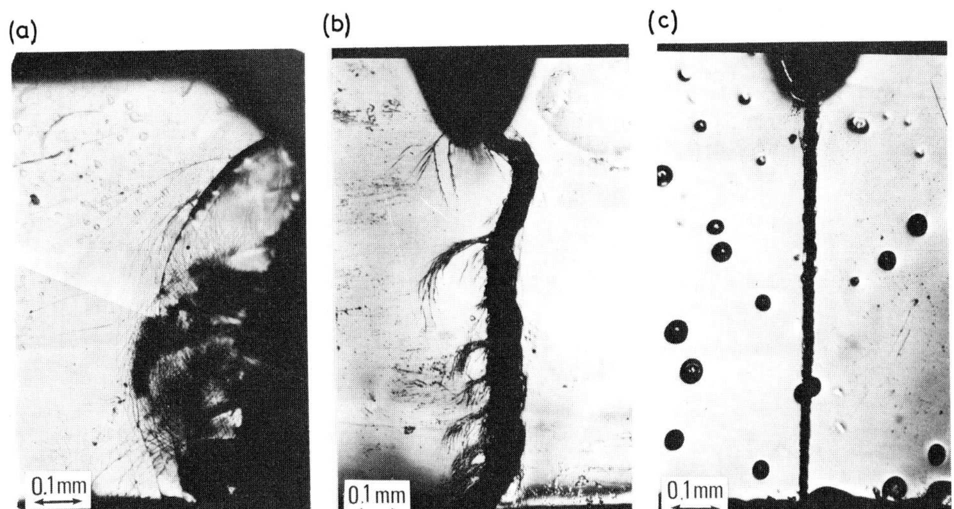

Figure 12.37 Deformation and fracture patterns observed at the surface of the plane–strain specimens of amorphous $Pd_{78}Cu_6Si_{16}$ alloys for (*a*) 260 min. (Region II), (*b*) 330 min. (Region III), and (*c*) 720 min. (Region IV)

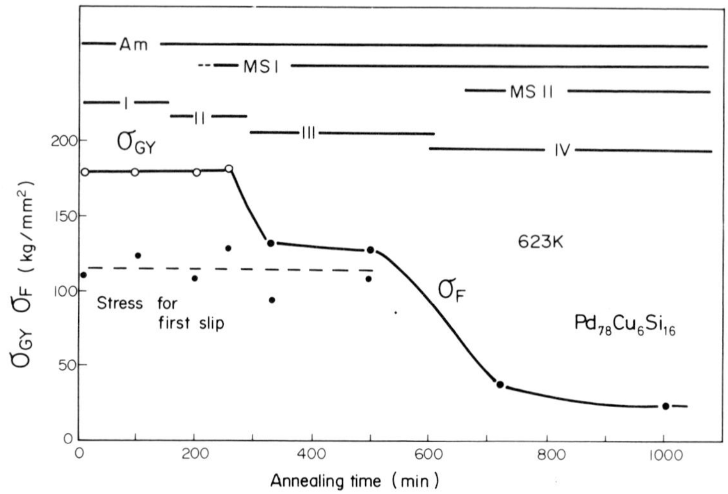

Figure 12.38 General yield stress, fracture stress and stress for a first slip for amorphous $Pd_{78}Cu_6Si_{16}$ alloys in plane strain as a function of annealing time at 623 K

microscopy of amorphous $Pd_{78}Cu_6Si_{16}$ alloys annealed for (*a*) 260 min (Region II), (*b*) 330 min (Region III), and (*c*) 720 min (Region IV). These display various characteristic fracture events[74]. Regions I, II, III and IV are defined, in a plane strain specimen, as sliding off after general yielding, crack initiation and unstable fracture after general yielding, brittle fracture accompanied by plastic flow, and brittle fracture, respectively. There are two characteristics of the morphology of the fractured surface of the

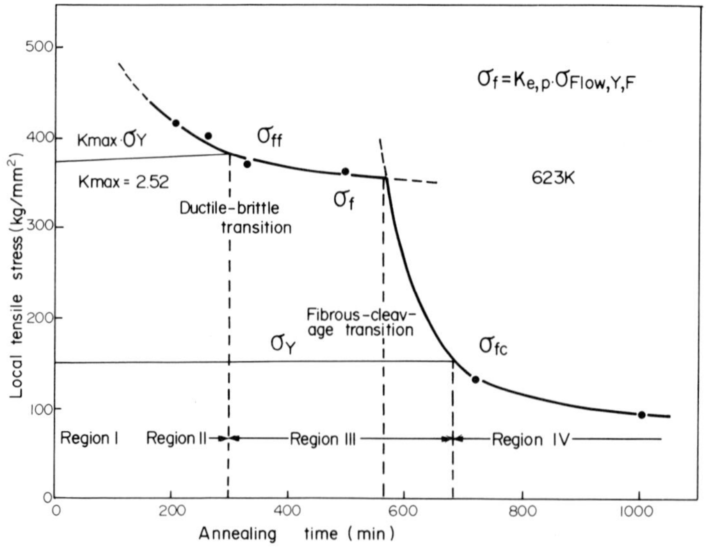

Figure 12.39 Calculated microfracture stress (σ_f), maximum yield stress beneath notch root ($K_{max}\sigma_Y$) and uniaxial yield stress (σ_Y) at notch root as a function of annealing time at 623 K in plane strain $Pd_{78}Cu_6Si_{16}$ amorphous alloys

specimen tested in plane strain: the crack starts well below the tip; and its shape is straight and normal to the tensile axis. These characteristics meet the test conditions of microfracture stress (σ_f) defined in the critical tensile criterion. *Figure 12.38* shows the general yield stress (σ_{GY}), fracture stress (σ_F), and stress for a first slip (σ_{first}) for amorphous $Pd_{78}Cu_6Si_{16}$ in plane strain as a function of annealing time at 623 K, showing the abrupt ductile–brittle transition in terms of macroscopic variables. *Figure 12.39* shows the calculated microfracture stress (σ_f), maximum yield stress beneath notch root ($K_{max}\sigma_Y$) and uniaxial yield stress (σ_Y) at the notch root as a function of annealing time at 623 K in plane strain for $Pd_{78}Cu_6Si_{16}$ amorphous alloys. $\sigma_f = K_{e,p}\sigma_{flow,Y,F}$. The loss of ductility can be explained as a decrease in the microfracture stress, not an inhibiting of the slip process. The critical tensile criterion can explain the ductile–brittle transition in plane strain as shown in *Figure 12.38* as the result of the decrease (between Region II and III) in microfracture stress below the maximum longitudinal yield stress.

The microfracture stress is accepted as another material parameter of fracture for amorphous alloys as well as the plastic constraint factor, L. The microfracture stress of as-quenched amorphous $Pd_{78}Cu_6Si_{16}$ is derived as more than 420 kg/mm^2, which gives the high ratio of $E/20$ for the Young modulus (E) of this alloy ($E = 8970$ kg/mm^2), approaching the value of $E/10$ for the ideal fracture strength. It is concluded that ductility of an amorphous alloy simply originated from possessing a higher microfracture stress in contrast to oxide glasses.

Note that the microfracture stress is a structure-sensitive material parameter for an amorphous structure, unlike flow.

Derivation of microfracture stress A mechanics is presented for the fracture of thin specimens of amorphous alloys in bending. The model permits a derivation of the microfracture stress for a variety of amorphous alloys (Kimura and Ast[81,82]). *Figure 12.40* shows the fracture surface[82] of a thin bent specimen of amorphous

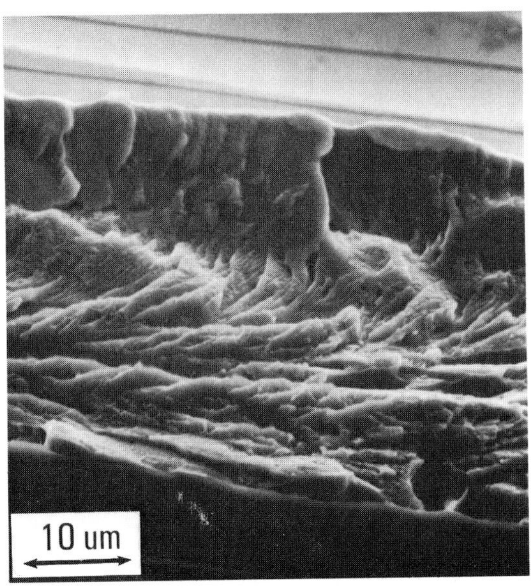

Figure 12.40 Fracture surface of thin bent specimen of amorphous $Fe_{40}Ni_{40}P_{14}B_6$ annealed for 15 minutes at 413 K, fractured beyond the surface plastic strain

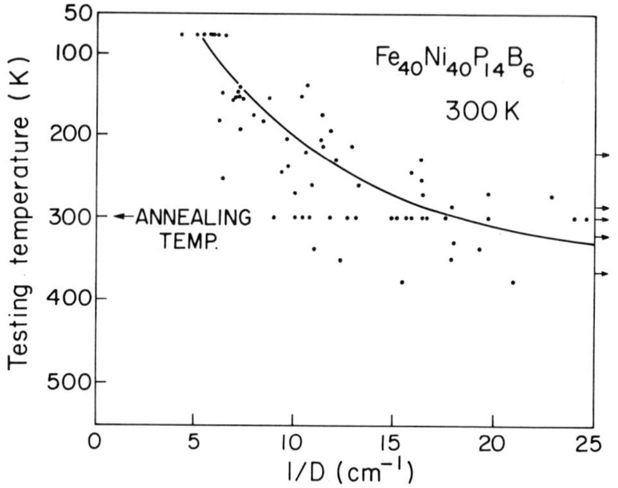

Figure 12.41 The reciprocal fracture separation ($1/D$) of amorphous $Fe_{40}Ni_{40}P_{14}B_6$ as a function of testing temperature. The annealing temperature was 300 K

$Fe_{40}Ni_{40}P_{14}B_6$ annealed for 15 min at 413 K, fractured beyond the surface plastic strain of 1.5×10^{-2}. A line of voids can be easily seen above the neutral axis inside the specimen in plane stress, while slip bands are observed at the specimen surface. Chen *et al.* observed[73] a similar distribution of voids through the cross-section area of a bent specimen, following plastic flow of amorphous $Fe_{75}P_{16}C_9$. These findings strongly point out that the fracture initiates inside the plane-stress specimen, and the presence of plastic stress concentration, followed by slip bands exists. We now define the plastic concentration factor (Q). The critical tensile criterion for fracture for a thin bend specimen of an amorphous alloy is expressed as:

$$\sigma_f = Q\sigma \tag{12.29}$$

An analysis[81,82] of the temperature dependence of the bend ductility (the reciprocal fracture separation, $1/D$) of amorphous $Fe_{40}Ni_{40}P_{14}B_6$ of Krenitsky and Ast[64] is given in *Figure 12.41. Figure 12.42* shows the yield stress (at testing temperature) of the specimen, fractured at various testing temperatures in *Figure 12.41*. The smaller scattering of the experimental data of *Figure 12.42* than of *Figure 12.41*

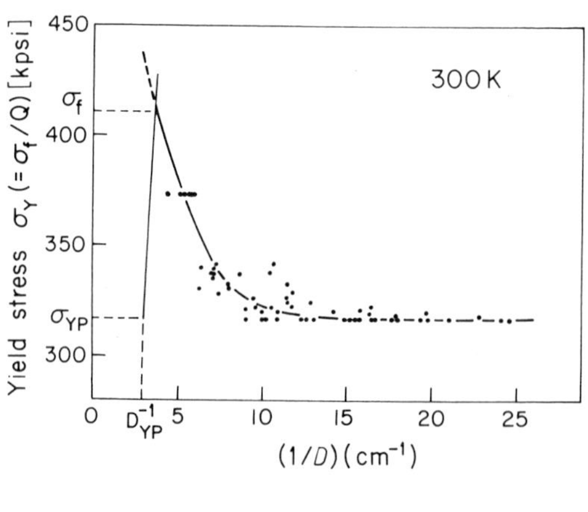

Figure 12.42 The applied yield stress of the specimen described in *Figure 12.41*, fractured at various testing temperatures as a function of $1/D$

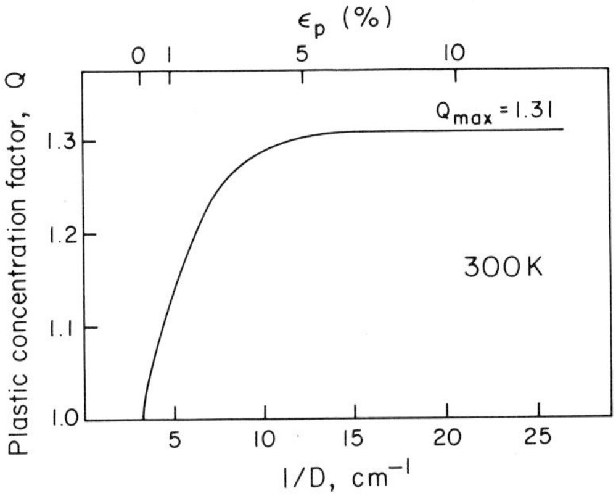

Figure 12.43 The plastic concentration factor described in *Figure 12.41* as a function of $1/D$

may validate this analysis. The microfracture stress can be deduced from an intersection between the line of the surface yield strain of the bent specimen (D_{YP}^{-1}) and the transformed curved line of the yield stress, where a fracture occurs at the specimen surface ($Q = 1$); i.e., the surface applied stress exceeds the surface yield stress. The microfracture stress (σ_f) of as-received amorphous $Fe_{40}Ni_{40}P_{14}B_6$ is thus derived as 411 kpsi (288 kg/mm^2). This value is smaller than 420 kg/mm^2 obtained for amorphous $Pd_{78}Cu_6Si_{16}$.

The critical tensile criterion describes the temperature independent microfracture stress as a parameter of ductility while the reciprocal separation distance (i.e., bend ductility) describes the temperature dependence as shown in *Figure 12.41*. Based on equation 12.28, the plastic concentration factor of a bent amorphous $Fe_{40}Ni_{40}P_{14}B_6$ specimen as a function of $1/D$ can be derived as shown in *Figure 12.43*. One can use this function to calculate the microfracture stress of annealed $Fe_{40}Ni_{40}P_{14}B_6$, by knowing the bend strain at fracture. The maximum plastic concentration factor could be a composition dependent parameter.

12.4.2 Embrittlement — overview

We apply the critical tensile criterion to various embrittlements of amorphous alloys.

Structural relaxation

Figure 12.44 shows the surface bend strain at fracture (ε_f) of amorphous $Fe_{40}Ni_{40}P_{14}B_6$ as a function of annealing temperature (Kimura and Ast[81]). In agreement with the results of Luborsky and Walter[83] and of Davis[84], the annealing-induced embrittlement shows two well-defined stages when surface strain at fracture is taken as a parameter of fracture. Stage I, where fracture follows plastic flow; i.e. ε_f exceeds ε_Y, the yield surface strain (1.5×10^{-2}); Stage II, where fracture occurs below ε_Y. *Figure 12.45* shows the decay of the total enthalpy of Peak I in the DSC curve of structural relaxation of $Fe_{40}Ni_{40}P_{14}B_6$, as a function of increasing pre-annealing temperatures (T_a), together with the decay of the microfracture stress (σ_f) deduced from the fracture strain (ε_f) of *Figure 12.44* by use of the function of the plastic concentration factor[81] in *Figure 12.43*. The two quantities track quite well. The critical transition of the ductile–brittle behaviour in the plot of ε_f against T_a as shown in *Figure 12.44* is not ascribable to the rapid change of microstructure, but to the mechanical factor; i.e., saturation of the plastic concentration factor in a large bend strain. Note that the

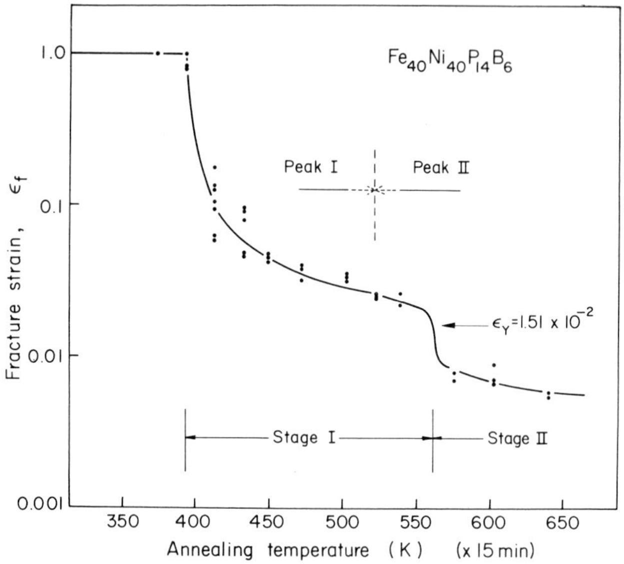

Figure 12.44 Relation of surface bend strain at fracture (ε_f) of amorphous $Fe_{40}Ni_{40}P_{14}B_6$ to annealing temperature

annealing time and/or temperature of the ductile–brittle transition against the fracture strain strongly depends on the maximum level of the plastic concentration factor (Q_{max}) of the materials.

The structural relaxation of $Fe_{40}Ni_{40}P_{14}B_6$, when monitored through the irreversible enthalpy of relaxation in a differential scanning calorimetry, shows a two-peak behaviour (Peak I and Peak II[81,85]). Two activation energies for structural relaxation (H_s) are found, 0.94 eV associated with Peak I and 1.8 eV associated with

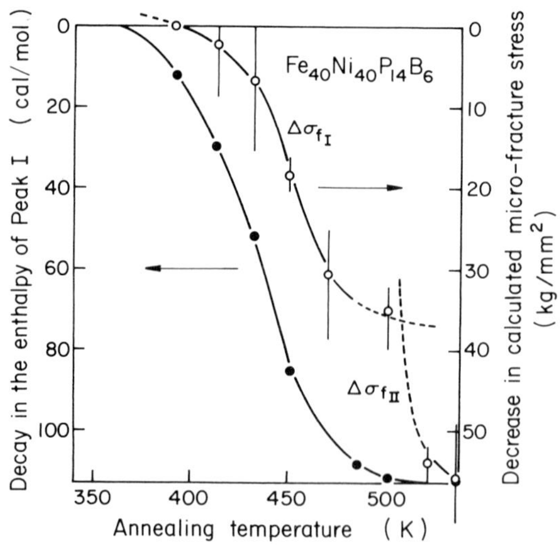

Figure 12.45 Decay of the total enthalpy in Peak I (left vertical axis) and decay of the local microfracture stress in Stage I in $Fe_{40}Ni_{40}P_{14}B_6$ isochronally annealed for 15 minutes

Peak II[81]. *Figure 12.46* shows Arrhenius plots of the annealing time required to reach various fracture strains[81]. The two activation energies of $Fe_{40}Ni_{40}P_{14}B_6$ (H_E) obtained are 0.94 eV for embrittlement of Stage I and 2.47 eV for Stage II. The activation energy 0.94 eV of H_E for Stage I correlates well with the 0.94 eV of H_s for structural relaxation of Peak I. With this information, the structural relaxation of Peak I can be coupled with the decay of microfracture stress (σ_f) in Stage I embrittlement. The activation energy of 0.94 eV is too low to be compatible with long range diffusion. A local ordering process is thought to be responsible for the embrittlement of Stage I.

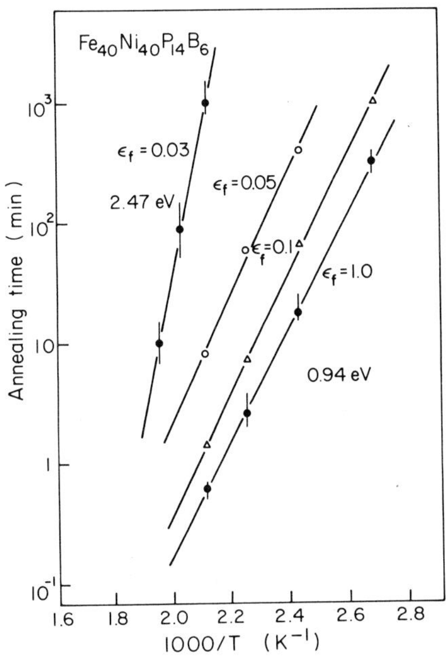

Figure 12.46 Arrhenius plots of the annealing times required to reach various fracture strains in $Fe_{40}Ni_{40}P_{14}B_6$

The value (H_E) of 2.47 eV deduced from embrittlement is somewhat higher than the DSC value of 1.8 eV for Peak II, 1.8 eV deduced by Egami from changes in the X-ray radial distribution function, and 1.95 eV for phosphorus diffusion in the similar testing temperature range 510–588 K. Stage II is still complicated by the possibility that (i) surface segregation of phosphorus causes a brittle surface layer, (ii) irreversible changes upon first heating, and (iii) reversible short range order changes during subsequent thermal cycles. Chi *et al.* demonstrated the drastic influence of quenching rate on the kinetics of the embrittlement of $Fe_{40}Ni_{40}B_{20}$, and found two activation energies, 21 cal/mole (0.91 eV) for the sample produced with a low quenching rate, and 60 cal/mole (2.6 eV) for a high quenching rate sample[86]. The ductile–brittle transition of the high quenching rate sample is postponed to a longer annealing time than that of the low quenching rate sample. Taking into account the two-stage behaviour in fracture mechanics of bending, one can suggest that the low quenching rate sample of $Fe_{40}Ni_{40}B_{20}$ exhibits two-stage behaviour, while the high quenching rate sample shows one-stage behaviour; i.e. only Stage II of embrittlement (Stage I disappears). Therefore, the two activation energies obtained by Chi *et al.* correspond to those for structural relaxation of Peak I and Peak II respectively. Luborsky and Walter reported that the phosphorus-containing amorphous alloy ($Fe_{40}Ni_{40}P_{14}B_6$) embrittles at a lower annealing temperature than the phosphorus-free amorphous alloy ($Fe_{40}Ni_{40}B_{20}$)[83].

They suggested that phosphorus is the element causing embrittlement. The observation of phosphorus segregation by Auger electron spectroscopy supports this idea[87]. In the results of Luborsky and Walter[83], however, the two-stage behaviour of $Fe_{40}Ni_{40}P_{14}B_6$ (the appearance of Stage I resulted from a low quenching rate) as suggested by Chen, is the main cause for its lower ductile–brittle transition temperature than that of $Fe_{40}Ni_{40}B_{20}$, exhibiting one-stage behaviour. The Stage I embrittlement of $Fe_{40}Ni_{40}P_{14}B_6$, strongly coupled with the structural relaxation of Peak I, would occur at a lower testing temperature than segregation of phosphorus[88], and so have a different activation energy (0.94 eV) from phosphorus diffusion of 1.95 eV.

Various investigations have described the effects of composition [89–92] and the effects of the embrittling elements antimony, selenium and tellurium[93] on the annealing-induced embrittlement of amorphous alloys, but their results should be discussed on a micro-mechanism, by knowing how elements and compositions can be related to both processes of structural relaxation and of fracture.

Crystallization

We here consider the physical process implying microfracture stress in our critical tensile criterion. There is a correlation between microfracture stress derived as a structure sensitive parameter and characteristics of crystallization. The appearance of fibrous fracture and cleavage fracture are essential microstructural events. A slow decrease in microfracture stress (σ_{ff}) for fibrous fracture in plane strain as a function of annealing time as shown in *Figure 12.39* is consistent with the nucleation of clusters and the fcc phase (MSI) and also with a slow increase in the quantity of the fcc phase, which has been generally characterized in the crystallization of $Pd_{78}Cu_6Si_{16}$ alloy. Similarly, the specimen in Region IV displays cleavage fracture; its onset and a rapid decrease in the microfracture stress (σ_{fc}) as shown in *Figure 12.39* are ascribed to nucleation of MS II and a rapid growth in its size.

Hydrogen

Hydrogen embrittlement of amorphous alloys is well established[94]. For example, Nagumo and Takahashi reported that amorphous $Fe_{80}P_{13}C_7$ and $Fe_{70}Cr_{10}P_{13}C_7$ lose ductility with the absorption of hydrogen, but the samples recover their ductility when the hydrogen is removed by annealing[95]. They also observed that the hardness of hydrogen-absorbed iron-based amorphous alloys remains the same as that of the as-quenched sample, while brittleness occurs in hydrogen absorbed samples. The critical tensile criterion would be applicable to hydrogen embrittlement, indicating that the ductile–brittle transition is caused by a decrease in the microfracture stress with absorption of hydrogen, and not by inhibition of slip motions. The microfracture stress is also sensitive to absorption of hydrogen.

Low temperature

The local tensile criterion is applicable to low temperature embrittlement as already proved in section 12.4.1. Some iron-based amorphous alloys ($Fe_{76}P_{16}C_4Al_2Si_2$) show a sudden drop in the tensile fracture stress at a low temperature, about 100 K, as reported by Pampillo and Polk[63]. Contrary to this finding, Kimura and Masumoto show that amorphous $Fe_{80}P_{15}C_5$ (which is similar in composition to $Fe_{76}P_{16}C_4Al_2Si_2$) does not become brittle in tension at a similar testing temperature[10]. According to the critical tensile criterion, this embrittlement is likely to occur in iron-based amorphous alloys which possess a low microfracture stress (e.g. 288 kg/mm^2 for $Fe_{40}Ni_{40}P_{14}B_6$)

relative to 420 kg/mm^2 for $Pd_{78}Cu_6Si_{16}$, and a rapid increase of yield stress with decreasing testing temperature (*see Figure 12.33*). The appearance of a low temperature ductile–brittle transition for amorphous $Fe_{76}P_{16}C_4Al_2Si_2$ is simply explained by its lower microfracture stress (which could result from an insufficient quenching rate) than the yield stress at a cryogenic temperature.

12.4.3 Fracture mechanics processes (in an ideal plastic solid)

A model mechanics study, using an amorphous alloy as a model material, is useful

1. To test the continuum plastic theory and the concept of fracture mechanics
2. To formulate an experimental rule for non-linear phenomena of plastic flow and fracture
3. To derive new concepts to treat flow events resulting from the extreme inhomogeneity of plastic flow. This section demonstrates some applications of model mechanics to the general problem of fracture events.

Fatigue crack propagation

Ogura *et al.* studied fatigue crack growth in thin specimens of amorphous $Pd_{80}Si_{20}$ in the tension–tension mode, and measured the fatigue crack propagation rate i.e., increment of crack length (a) per cycle (N), and da/dN[96,97]. The plastic zone size (Z_p) is obtained from the length of slip marking perpendicular to the fatigue crack, and then correlated to the stress intensity factor (K) deduced from the stress amplitude,

$$Z_p = CK^{n_0} \tag{12.30}$$

They found the value of 2 for the exponent n_0; this value accords with the prediction of Rice who calculated the plastic zone size ahead of a fatigue crack in the small scale yielding of Tresca–von Mises material[98].

The fatigue crack propagation rate for an amorphous alloy is given by a power law for cyclic stress intensity factor (ΔK), but there has been a discrepancy between the exponent[97] of 4 for amorphous $Pd_{80}Si_{20}$ and that of 2.25 for amorphous $Ni_{39}Fe_{38}P_{14}B_6Al_3$ as reported by Davis[99]. In order to get a reliable experimental rule with which to test the theoretical model, further studies are needed, to take into account the local fracture process, closure of a crack during unloading, and so on.

Fibrous dimple fracture

Figure 12.47 shows the correlation between critical crack opening displaçement (ϕ_c, COD) and dimple size measured at the fractured surface of amorphous $Pd_{78}Cu_6Si_{16}$ specimens, annealed at 350 °C for various periods, in both plane strain and plane stress. One can see the two experimental correlations: (1) the critical COD is proportional to the dimple size, and (2) for the same dimple size, the critical COD is smaller in plane stress than in plane strain. Based on these relations and the J integral, Kimura and Masumoto formulated an internecking model for the micromechanism of plane-strain fibrous fracture in an ideal plastic solid:

$$J_c = m_0 \sigma_Y \phi \tag{12.31}$$

where m_0 is unity in plane stress and J_c is the critical J integral at fracture. A theoretical consideration in fracture mechanics results in the same function as equation 12.31 which has the potential to describe non-linear fracture events, but its derivation remains

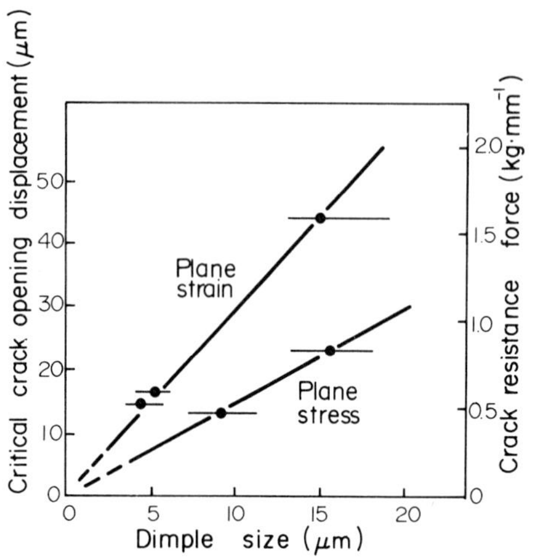

Figure 12.47 Relation between dimple size and crack opening displacement at fracture (ϕ_c) or crack resistance force (G_{IC})

uncertain. Note that a mechanics of the model material easily permits the derivation of equation 12.31 for non-linear fracture.

References

1. Masumoto, T. and Maddin, R. *Mater. Sci. Engng,* **19,** 1 (1975)
2. Pampillo, C. A., *J. mater Sci.,* **10,** 1194 (1975)
3. Gilman, J. J., *J. appl. Phys.,* **46,** 1625 (1975)
4. Davis, L. A., *'Metallic Glasses',* p. 190, American Society of Metals, Metals Park, Ohio (1978)
5. Chen, H. S., *Rep. Prog. Phys.,* **43,** 353 (1980)
6. Li, J. C. M., *'Treatise on Material Science and Technology',* Ed. Herman, H., vol. 20, p. 326, Academic Press, New York (1981)
7. Masumoto, T. and Maddin, R., *Acta metall.,* **19,** 725 (1971)
8. Chen, H. S. and Wang, T. T., *J. appl. Phys.,* **41,** 5338 (1970)
9. Leamy, H. J., Chen, H. S. and Wang, T. T., *Metall. Trans.,* 699 (1972)
10. Masumoto, T. and Kimura, H., *J. Japan Inst. Metals,* **39,** 133 (1975)
11. Kimura, H., Doctoral Thesis, Tôhoku University (1978); in part, Kimura, H. and Masumoto, T., *Metall. Trans.,* in the press
12. Kimura, H. and Masumoto, T., *Mat. Sci. Engng.,* to be published
13. Krenitsky, D. and Ast, D. G., *J. mater. Sci.,* **14,** 275 (1979)
14. Kimura, H. and Masumoto, T., *Acta metall.,* **28,** 1663 (1980)
15. Dugdale, D. S., *J. Mech. Phys. Solids,* **8,** 100 (1960)
16. Davis, K. A., *Scripta Met.,* **9,** 339 (1975)
17. Pampillo, C. A. and Chen, H. S., *Mater. Sci. Engng,* **13,** 181 (1974)
18. Green, A. P., *Q. J. Mech. appl. Math.,* **6,** 223 (1953)
19. Ewing, D. J. F., *J. Mech. Phys. Solids,* **16,** 205 (1968)
20. Kimura, H., Ast, D. G. and Bassett, W. A., *J. appl. Phys.,* **53,** 3523 (1982)
21. Davis, L. A. and Kavesh, S., *J. mater. Sci.,* **10,** 453 (1975)
22. Bowden, P. B. and Jukes, J. A., *J. mater. Sci.,* **7,** 52 (1972)
23. Kimura, H. and Masumoto, T., *Phil. Mag.,* **A44,** 1005 (1981)
24. Rivlin, R. S. and Thomas, A. G., *J. Polym. Sci.,* **10,** 291 (1953)
25. Murata, T., Masumoto, T. and Sakai, M., *Proc. 3rd Int. Conf. on Rapidly Quenched Metals,* Ed. Cantor, B., Vol II, A401, The Metal Society, London (1978)
26. Takayama, S., *Mater. Sci. Engng,* **38,** 41 (1979)
27. Takayama, S. and Maddin, R., *Scripta Met.,* **9,** 343 (1975)
28. Ray, R., Hasegawa, R., Chou, C. P. and Davis, L. A., *Scripta Met.,* **11,** 973 (1977)
29. Naka, M. and Masumoto, T., *J. Phys. (Paris),* **C8,** 839 (1980)

30. Davis, L. A., Chou, C. P., Tanner, L. E. and Ray, R., *Scripta Met.*, **10,** 937 (1976)
31. Chen, H. S., *J. appl. Phys.*, **49,** 462 (1978)
32. Naka, M., Tomizawa, S., Masumoto, T. and Watanabe, T., *'Rapidly Quenched Metals II'*, Eds. Grant, N. J. and Giessen, B. C., vol. I, p. 273, Massachusetts Institute of Technology Press, Boston (1976)
33. Donald, I. W. and Davis, H. A., *Phil. Mag.*, **A42,** 277 (1980)
34. Takayama, S. and Maddin, R., *J. mater. Sci.*, **11,** 22 (1976)
35. Chou, C.-P. P. and Spaepen, F., *Acta metall.*, **23,** 609 (1975)
36. Freed, R. L. and Vandersande, J. B., *Acta metall.*, **28,** 103 (1980)
37. Kimura, H., Cunningham, B. and Ast, D. G., *Proc. 4th Int. Conf. on Rapidly Quenched Metals,* Eds. Masumoto, T. and Suzuki, K., vol II, p. 1385, Japan Institute of Metals, Sendai (1982)
38. Takayama, S., *J. mater. Sci.*, **15,** 649 (1980)
39. Kayano, H., Masumoto, T., Tomizawa, S. and Yajima, S., *Sci. Rep. Res. Insts Tôhoku Univ.*, **261,** 240 (1977)
40. Chang, B. T. A. and Li, J. C. M., *Scripta Met.*, **11,** 933 (1977)
41. Kramer, E. A., Johnson, W. L. and Cline, C., *Appl. Phys. Lett.*, **35,** 815 (1979)
42. Chen, H. S. and Lo. C. C., ref. 32, vol. I, p. 413
43. Ast, D. G. and Krenitsky, D. J., *Mater. Sci. Engng,* **43,** 241 (1980)
44. Li, J. C. M., *'Frontiers in Material Science',* Ed. Murr, L. E. and Stein, C., p. 527, Marcel Dekker, New York (1976)
45. Gilman, J. J., *J. appl. Phys.*, **44,** 675 (1973)
46. Spaepen, F. and Turnbull, D., *Scripta Met.*, **8,** 563 (1974)
47. For example, Dieter, G. E., *'Mechanical Metallurgy',* p. 680, McGraw-Hill, New York (1976)
48. Kimura, H. and Ast, D. G., to be published
49. Davis, L. A. and Yeow, Y. T., *J. mater. Sci.*, **15,** 230 (1980)
50. Argon, A. S., *'The Inhomogeneity of Plastic Deformation',* p. 161, American Society of Metals, Metals Park, Ohio (1973)
51. Pampillo, C. A., *Scripta Met.*, **6,** 915 (1972)
52. Kulik, T. and Matyja, H., *J. mater. Sci.*, **15,** 3169 (1980)
53. Polk, D. E. and Turnbull, D., *Acta metall.*, **20,** 493 (1972)
54. Sethi, V. K., Gibala, R. and Heuer, A. H., *Scripta Met.*, **12,** 207 (1978)
55. Donovan, P. E. and Stobbs, W. M., *Acta metall.*, **29,** 1419 (1981)
56. Kimura, H. and Masumoto, T., *Acta metall.*, **31** (1983)
57. Cottrell, A. H. and Billby, B. A., *Proc. Phys. Soc.*, **A62,** 49 (1949)
58. Fleisher, R. L., *Acta metall.*, **9,** 996 (1961)
59. Mannan, S. L. and Rodriguez, P., *Phil. Mag.*, **25,** 673 (1972)
60. Blanchin, M. G. and Frontaine, G., *Phil. Mag.*, **A41,** 261 (1980)
61. Kimura, H. and Masumoto, T., *Phil. Mag.*, **A44,** 1021 (1981)
62. Maddin, R. and Masumoto, T., *Mater. Sci. Engng,* **9,** 153 (1972)
63. Pampillo, C. A. and Polk, D. E., *Acta metall.*, **22,** 741 (1974)
64. Krenitsky, D. J. and Ast, D. G., *J. mater. Sci.*, **14,** 275 (1979)
65. Seeger, A., *Z. Naturf.*, **9a,** 758, 819, 856 (1954); *'Defects in Crystalline Solids',* The Physical Society, London (1955)
66. Spaepen, F., *Acta metall.*, **25,** 407 (1977)
67. Argon, A. S., *Acta metall.*, **27,** 47 (1979)
68. Megusar, J., Argon, A. S. and Grant, N. J., *Mater. Sci. Engng,* **38,** 63 (1979)
69. Patterson, J. P. and Jones, D. R. H., *Acta metall.*, **28,** 675 (1980)
70. Kimura, H. and Masumoto, T., *Scripta Met.*, **9,** 211 (1975)
71. Davis, L. A., *J. mater. Sci.*, **10,** 1557 (1975)
72. Ast. D. G. and Krenitsky, D., *Mater. Sci. Engng,* **23,** 241 (1976)
73. Chen, H. S., Leamy, H. J. and O'Brien, M. J., *Script Met.*, **7,** 415 (1973)
74. Kimura, H. and Masumoto, T., *Acta metall.*, **28,** 1677 (1980)
75. Davis, L. A., *Metals Trans.*, **10A,** 235 (1979)
76. For example, Knott, J. F., *'Fundamentals of Fracture Mechanics',* p. 77, Butterworths, London (1974)
77. Pampillo, C. A. and Reimschuessel, A. C., *J. mater. Sci.*, **9,** 718 (1974)
78. Takayama, S. and Maddin, R., *Phil. Mag.*, **32,** 457 (1975)
79. Spaepen, F., *Acta metall.*, **23,** 615 (1975)
80. Argon, A. S. and Salama, M., *Mater. Sci. Engng,* **23,** 219 (1976)
81. Kimura, H. and Ast. D. G., vol II, p. 457, in ref. 37
82. Kimura, H. and Ast, D. G., to be published
83. Luborsky, F. E. and Walter, J. L., *J. appl. Phys.*, **47,** 3638 (1976)
84. Davis, L. A., Ray, R., Chou, C.-P. and O'Handley, R. C., *Scripta Met.*, **10,** 541 (1976)

85. Chen, H. S., *Scripta Met.*, **11,** 367 (1977)
86. Chi, G. C., Chen, H. S. and Miller, C. E., *J. appl. Phys.*, **49** (3), 1715 (1978)
87. Walter, J. L., Bacon, F. and Luborsky, F. E., *Mater. Sci. Engng*, **24,** 239 (1976)
88. Baer, D. R., Pederson, L. R. and Thomas, M. T., *Mater. Sci. Engng*, **48,** 283 (1981)
89. Chen, H. S., *Mater. Sci. Engng*, **26,** 79 (1976)
90. Pampillo, C. A. and Polk, D. E., *Mater. Sci. Engng*, **33,** 275 (1978)
91. Prattern, N. A. and Scott, M. G., *Proc. 3rd Int. Conf. on Rapidly Quenched Metals*, Ed. Cantor, B., vol I, p. 387, The Metal Society, London (1978)
92. Nagumo, M., Takahashi, T., Arai, T. and Hasegawa, T., *'Proc. 3rd Int. Conf. on Rapidly Quenched Metals'*, Ed. Cantor, B., vol. 1, p. 383, The Metals Society, London (1978)
93. Lieberman, H. H. and Luborsky, F. E., *Acta metall.*, **29,** 1413 (1981)
94. Hashimoto, K. and Masumoto, T., ref. 6, vol. 20, p. 291
95. Nagumo, M. and Takahashi, T., *Mater. Sci. Engng*, **23,** 257 (1976)
96. Ogura, T., Fukushima, K. and Masumoto, T., *Scripta Met.*, **9,** 979 (1975)
97. Ogura, T., Fukushima, K. and Masumoto, T., *Mater. Sci. Engng*, **23,** 231 (1976)
98. Rice, J. R., *'Fatigue Crack Propagation'*, ASTM Spec. Publ., 415, 247 (1967)
99. Davis, L. A., *J. mater. Sci.*, **10,** 1557 (1975)

Chapter 13

Flow and fracture

F. Spaepen
Division of Engineering and Applied Physics, Harvard University, Cambridge, Massachusetts, USA

A. I. Taub
General Electric Company, Schenectady, New York, USA

13.1 Introduction

The mechanical response of a material can be categorized according to reversibility and time dependence. The elastic response is instantaneous and reversible, the anelastic response is reversible and time-dependent, and the plastic response is irreversible and time-dependent. In metallic glasses, the plastic response occurs in two distinct regimes:

1. Homogeneous flow, in which each volume element of the material contributes to the strain, resulting in uniform deformation for a uniformly stressed specimen, and
2. Inhomogeneous flow, where the deformation is localized in discrete, thin shear bands, leaving the rest of the material plastically undeformed.

These two regimes are illustrated on a schematic deformation mechanism map (*see Figure 13.1*), introduced by Ashby[1] and developed by Frost and Ashby[2] for crystalline materials. On this map, contours of constant *steady state* strain rate, $\dot{\gamma}$, are plotted as a function of stress, τ, and temperature, normalized with respect to the shear modulus, μ, and the melting point, T_m, respectively. For amorphous metals, the glass transition temperature, T_g, is a more natural normalization temperature. Regions in which each of the deformation mechanisms dominates are indicated. The constitutive laws $\dot{\gamma}(\tau, T)$ for each mechanism can be tested by comparing calculated and experimental maps.

The steady state nature of the flow described by the maps requires that the atomic structure of the material, especially its defect content and configuration, be either a function of stress and temperature, or, at least, constant and well-characterized for the whole map. For crystalline materials this requires taking into account relatively well-defined microstructural parameters such as the grain size. Metallic glasses, on the other hand, are in an *unstable* thermodynamic state. As a result, they continuously change their structure to lower free energy states. This *structural relaxation* process affects all the physical properties, but most strongly the atomic transport properties. This is a most important consideration in homogeneous flow, where special care must be taken that all the data on a particular map are *isoconfigurational*, i.e. correspond to the same state of annealing. Inhomogeneous flow is much less sensitive to structural relaxation

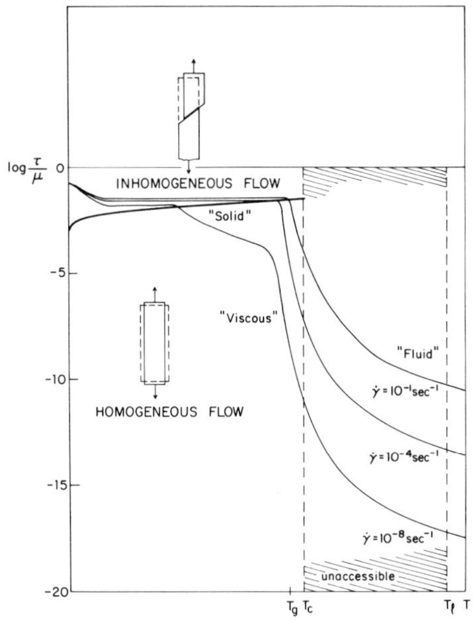

Figure 13.1 Schematic deformation mechanism map for an amorphous metal (in a particular state of annealing), illustrating the temperature and stress regions for homogeneous and inhomogeneous plastic flow (after Spaepen[41])

since, as will be discussed below, the atomic structure in the shear band is a function of stress and temperature, and, at steady state, independent of the initial structure.

Homogeneous flow occurs at low stresses ($\tau < \mu/100$) and is measured in creep or stress relaxation tests. At sufficiently low stresses, the flow is Newtonian viscous ($\dot{\gamma} \propto \tau$), which allows determination of the shear viscosity ($\eta \equiv \tau/\dot{\gamma}$). The results of these measurements for a variety of metallic glasses will be reviewed in this chapter, with special attention to the kinetics of the continuous viscosity rise due to structural relaxation, and to the non-Newtonian flow behaviour and its atomistic interpretation. The scaling relation between the viscosity and the diffusivity is particularly helpful in this respect.

Inhomogeneous flow occurs at high stresses ($\tau > \mu/50$) and low temperatures ($T < 0.7T_g$) and is observed in tensile tests, hardness tests and cold working processes such as rolling and wire drawing. The flow is strain rate insensitive, exhibits no work hardening (i.e. is almost ideally plastic) and is only very weakly temperature dependent. Since the phenomenology of this type of flow is reviewed in Chapter 12, we will mainly discuss the various microscopic mechanisms, together with the macroscopic conditions for flow localization in the shear bands.

Although the local plastic strain in a shear band is very large, the overall strain is determined by the number of shear bands, which in turn depends on the geometrical constraints of the deformation process. For example, in rolling or wire drawing, the reductions can be 50 per cent. In a tensile test, however, only a few shear bands are active and fracture occurs soon after initiation of the flow. The total plastic strain in this case is only a few tenths of a per cent, which makes the stress–strain curve resemble that of a brittle material. However, we will still refer to the fracture process that terminates the flow as a 'ductile' one because it occurs *along* a shear band (not normal to the tensile axis) and forms a 'vein' morphology on the fracture surface, which involves extensive plastic flow. It will be shown that this fracture process is a special case of a general set of instability phenomena of the type first described by G. I. Taylor.

Some metallic glasses, especially the iron-based ones, at low temperatures or after annealing, fail by a mechanism of brittle fracture, which occurs normal to the tensile axis and forms a featureless fracture surface. Its understanding on the atomistic level is far from complete, but we will discuss how some of the ideas on homogeneous and inhomogeneous flow can give insight on the brittle fracture problem.

13.2 Homogeneous flow

13.2.1 Structural state definitions

A recurring theme throughout this book is the effect of structural relaxation with material properties. Almost every intrinsic property of amorphous alloys has been shown to vary as the structure changes[3], with perhaps the greatest effect being the change in the homogeneous flow resistance. Structural relaxation has been shown to increase the viscosity, a direct measure of the flow resistance, by more than five orders of magnitude[4].

The origin of this relaxation effect can be explained with the aid of *Figure 13.2*, which illustrates the variation of the viscosity of a glass-forming alloy with temperature. Consider the alloy at a temperature above the melting temperature T_m. The atomic structure and physical properties of the alloy in this regime are determined by the equilibrium state. If the alloy is cooled below T_m, its atomic structure continuously adopts configurations dictated by equilibrium, until the region of the glass transition temperature T_g is reached. At some temperature near T_g, the atomic configuration begins to 'freeze'. That is, the resistance to atomic motion becomes large enough to restrict structural rearrangement, resulting in the time required for the atoms to adopt their equilibrium configurations becoming greater than the time allowed by the cooling

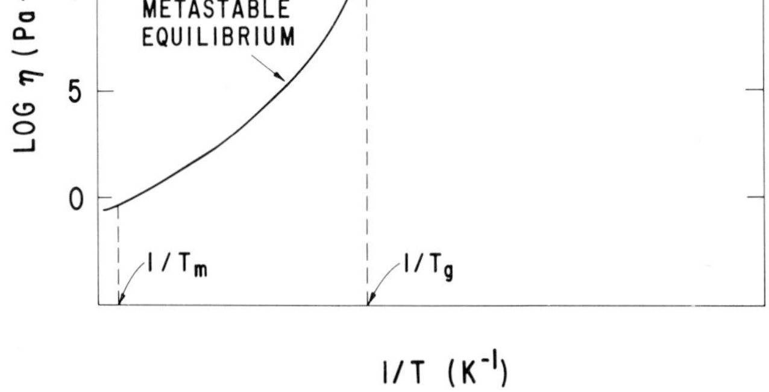

Figure 13.2 Schematic plot of the temperature dependence of the viscosity in a glass forming alloy during quenching and reheating. Curve *a* represents the deviation of the viscosity from the equilibrium curve during the quench. Curves *b* and *c* illustrate the shift in the isoconfigurational viscosity towards the equilibrium curve during annealing

process. The atomic structure then begins to deviate from the configuration required by equilibrium. As the temperature continues to decrease, the resistance to atomic motion increases even more and the time required for structural rearrangements becomes larger. Eventually, significant structural change on the time scale of the cooling process ceases and the atomic configuration is 'frozen in'. Curve *a* in *Figure 13.2* represents the isoconfigurational (i.e. constant structure) viscosity of the alloy in the as-quenched condition.

Subsequent heat treatment of the alloy at temperatures below T_g allows the atomic structure to approach its equilibrium configuration before crystallization commences. This results in a shift of the isoconfigurational curves towards the equilibrium curve, as shown schematically in *Figure 13.2*.

The viscosity of a glass former is therefore not a unique function of temperature in the range $T < T_m$. One must distinguish between the viscosities of the metastable equilibrium liquid, the isoconfigurational glass and the crystallized alloy. In this chapter, care will be taken to identify the structural state of the alloy.

13.2.2 Experimental techniques

No single method of measuring homogeneous flow is applicable over the entire range of viscosities that are of interest in amorphous alloys. Tests in the low viscosity regime ($\eta < 100\,\mathrm{N\,s\,m^{-2}}$) require a dynamic measurement technique where a solid interface creates a controlled velocity gradient in the material. The viscosity is determined from the ratio of the stress required to shear the material to the observed rate of shear. The oscillating cup viscometer is the instrument used most frequently for testing liquid metals[5]. In this technique, the fluid motion is caused by an oscillating stress that is imposed on the liquid test material. Direct rotational viscometry[6] and capillary flow[7] techniques have also been applied to amorphous alloys.

At higher viscosities ($\eta > 10^9\,\mathrm{N\,s\,m^{-2}}$), creep and stress relaxation experiments have been employed. In creep tests, a constant pure bending or uniaxial tensile stress σ is applied to an alloy specimen. The resulting strain ε, is monitored as a function of time and the strain rate determined. The viscosity is the ratio $\eta = (1/3)\sigma/\dot{\varepsilon}$. In stress relaxation tests, the sample is elastically strained to some value ε_0, and allowed to relax by homogeneous flow. The flow produces a decrease in the elastic stress. The viscosity is computed from the strain rate, which in turn is determined from the rate of elastic stress decay $\eta = (1/3)\sigma/\dot{\varepsilon} = (1/3)\sigma E/\dot{\sigma}$.

A method of determining the viscosity from thermal measurements has been developed by Chen[8]. It is based on the observation that the activation energies for the glass transition, crystallization and viscosity are similar. The data for two palladium-based alloys are shown in *Figure 13.3*. The solid lines are experimental data from creep tests, fitted to the Fulcher–Vogel relation (equation 13.1). The symbols are the rates of transformation as measured by differential scanning calorimetry. The transformation rate is determined from the time to the peak of crystallization during isothermal annealing. The rate can also be computed from measurements at constant heating rate α, by the relation T^2/α, where T is respectively the glass transition temperature (○), the temperature at the onset of crystallization (□), and the peak crystallization temperature (△). The measured values have been shifted to coincide with the viscosity data by applying time shift constants A_t and A_α. The need to shift the data in this manner limits this method to a determination of the temperature dependence only, not the actual value of the exact viscosity. However, it has proved to be very useful at temperatures above T_g, where direct flow measurements of the amorphous state are difficult because of the onset of crystallization.

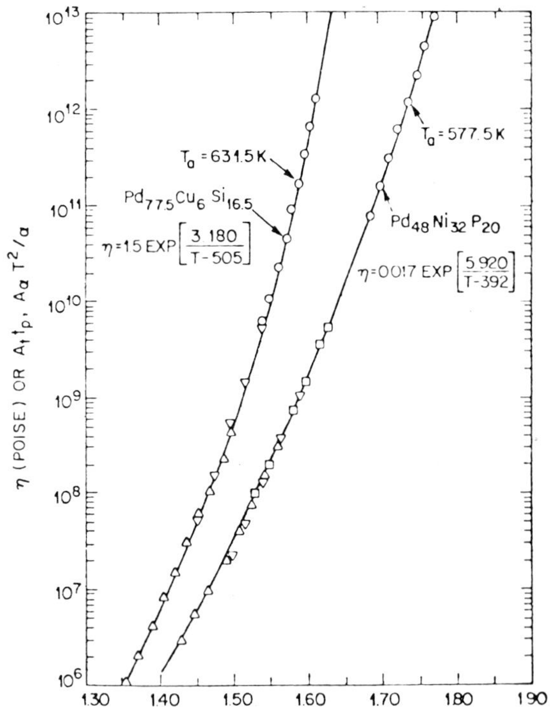

Figure 13.3 Comparison between the reported creep viscosities and the rates of transformations; t_p (▽), T_g (○), T_x (□) and T_p(Δ), A_t and A_2 are time shift constants. Solid lines are experimental viscosity data and dash-dot lines are extrapolated viscosities described by Fulcher's expression. T_a is the glass transition temperature at a scanning rate of 10 K/min. For $Pd_{77.5}Cu_6Si_{16.5}$, $T_a = 631.5$ K and $\eta = 1.5 \exp[3.180/(T - 505)]$ and for $Pd_{48}Ni_{32}P_{20}$, $T_a = 577.5$ K and $\eta = 0.017 \exp[5.920/(T - 392)]$ (after Chen[8])

13.2.3 Empirical flow laws

Equilibrium

As illustrated schematically in *Figure 13.2*, the viscosity of an equilibrium liquid exhibits an increasing activation energy with decreasing temperature. This behaviour has classically been described by the Fulcher–Vogel equation[9,10]

$$\eta = A \exp\left[\frac{B}{T - T_0}\right] \qquad (13.1)$$

However, Berry and Fox[11] have presented a compilation of data on organic and inorganic glasses, from which they concluded that the Fulcher–Vogel relation cannot represent the data on any single liquid over a wide temperature range. Therefore, we will distinguish between a high temperature ($T > T_m$) and a low temperature $T_m < T < T_g$ regime.

In the high temperature regime the equilibrium phase is the liquid state. Since the viscosity of glass-forming alloys in this state is low ($<10\,\mathrm{N\,s\,m^{-2}}$), a dynamic measurement of the viscosity is required. The most recent measurements of this type are those of Steinberg *et al.*[5]. They employed an oscillating cup viscometer to measure the viscosities of molten $Fe_{40}Ni_{40}P_{14}B_6$, $Fe_{40}Ni_{40}B_{20}$ and $Pd_{82}Si_{18}$. The results are shown in *Figure 13.4*, together with the previously reported data[6,7,12]. The iron-based alloys appear to exhibit very similar behaviour when the temperature is normalized by the melting temperature T_m. The viscosities obey an Arrhenius relation with $\eta(T_m) = (1\text{—}2) \times 10^{-2}\,\mathrm{N\,s\,m^{-2}}$ and with an activation energy of approximately 40 kJ/mole. The viscosities measured for $Au_{77.8}Ge_{13.8}Si_{8.4}$ are similar in magnitude to those of the iron-based glasses. In fact, most alloy melts have similar viscosities[13,14] near T_m. The palladium-based alloys appear to be exceptional in this regard. They exhibit a large spread in viscosity with average values more than two orders of magnitude greater than those of the iron-based glass formers. The easy glass forming ability of the palladium-based alloys has been associated with this higher viscosity.

Between the melting temperature T_m and the glass transition temperature T_g, the phase of interest is the metastable liquid. Direct measurement of the viscosity in this state is difficult because of rapid crystallization. The compiled data together with some theoretical estimates are shown in *Figure 13.5*. Davies *et al.*[15] extrapolated the viscosity data of pure nickel. Naka *et al.*[6] extrapolated the data from the equilibrium melt of the alloy, assuming $\eta(T_g) = 10^{12}\,\mathrm{N\,s\,m^{-2}}$. Chen[8] computed the viscosity from the rates of thermal transformations as described previously. Ramachandrarao *et al.*[16] estimated the viscosity from free volume calculations. The convergence of the calculated curves at $10^{12}\,\mathrm{N\,s\,m^{-2}}$ (10^{13} poise) at T_g is a consequence of the assumptions made in the

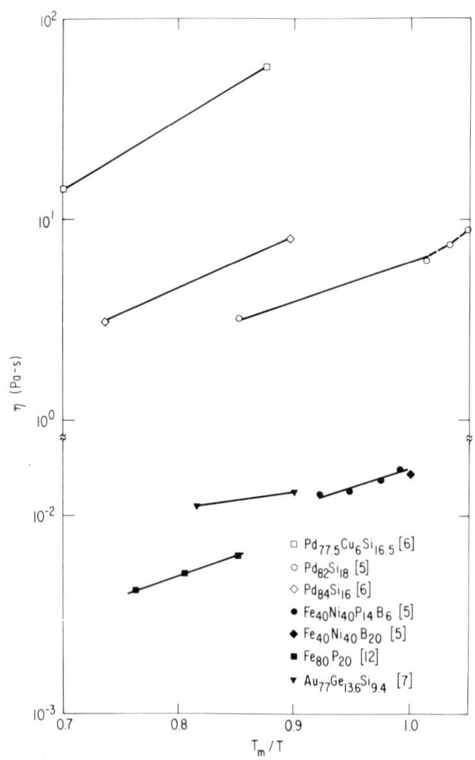

Figure 13.4 Viscosities of several glass formers in the equilibrium liquid state.

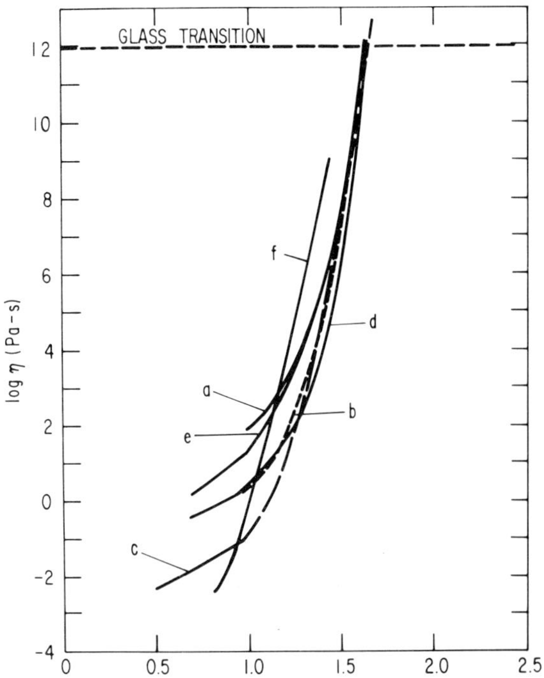

Figure 13.5 Viscosities of several glass forming alloys in the metastable liquid state between T_m and T_g. The alloy compositions are $Pd_{77.5}Cu_6Si_{16.5}$: *a* from ref. 8, *b* from ref. 16, *c* from ref. 15 and *e* from ref. 6; $Pd_{84}Si_{16}$ from ref. 6 (*d*); and $Fe_{40}Ni_{40}P_{14}B_6$ from ref. 96 (*f*)

calculations. The results of the different methods are in fact quite disparate, being almost three orders of magnitude apart at T_m.

In tests near the glass transition temperature T_g, the viscosity is low enough to perform accurate creep measurements in the short time available before the onset of crystallization. Care must be taken, however, to ensure that the metastable equilibrium state has been reached. The only reported direct viscosity measurement of equilibrium flow in this region are those of Chen and Turnbull[17] and Chen and Goldstein[18]. The data are described well by the Fulcher–Vogel relation (equation 13.1). For the three palladium based alloys tested, $A = 0.152$ and $B = 3.180$ and T_0 was in the range of 510 to 515 K. For $Au_{77}Ge_{13.6}Si_{9.4}$, $A = 0.52$, $B = 1360$ and $T_0 = 241$ K.

Non-equilibrium

Equilibrium flow at high temperatures ($T > T_g - 20$), as just described, is characterized by a unique viscosity at all temperatures. When the temperature is raised to some value T_1, the viscosity decreases until equilibrium is attained $\eta = \eta(T_1)$. If the temperature is instead decreased from some higher value to T_1, the viscosity will increase until the *same* equilibrium viscosity $\eta(T_1)$ is attained. Flow at low temperatures ($T < T_g - 20$), on the other hand, has never been observed to equilibrate in this manner. Taub and Spaepen[19] tested $Pd_{82}Si_{18}$ in a creep experiment at constant load and monitored the viscosity as a function of time in the range $0.66 < T/T_g < 0.84$. The viscosity was observed to increase with time at all temperatures with no evidence of asymptotic behaviour, even after a month of isothermal annealing. Similar results have

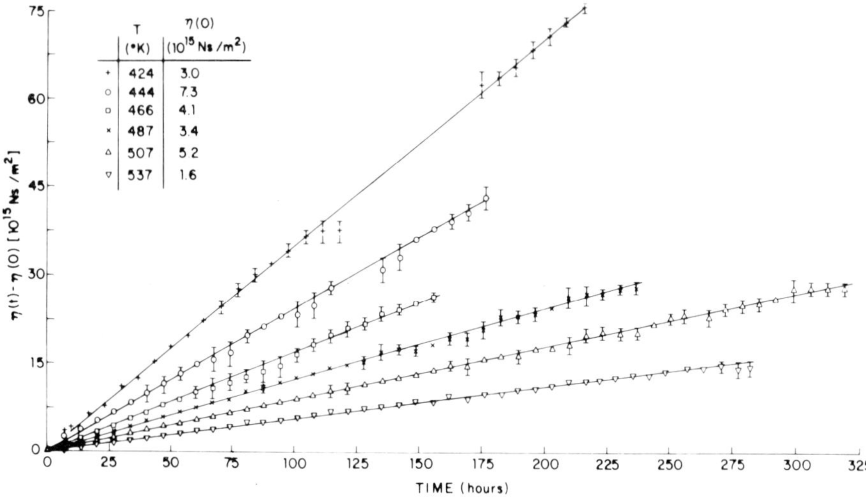

Figure 13.6 Viscosity of glassy $Pd_{82}Si_{18}$ as a function of annealing time. *Table 13.1* lists the initial viscosity η_0 for each of the annealing temperatures (after Taub and Spaepen[19])

been observed for $Pd_{77.5}Cu_6Si_{16.5}$ [20–22] and for $Fe_{40}Ni_{40}P_{14}B_6$ [22–26] even at temperatures near T_g.

In *Figure 13.6* the change in the viscosity of $Pd_{82}Si_{18}$ during isothermal annealing is shown. A linear increase $[n = \mathrm{d}(\ln \eta)/\mathrm{d}(\ln t) = 1]$ is observed at all temperatures. This linear behaviour appears to be a universal property of amorphous alloys in the range $T < T_g - 20$ (PdCuSi, $n = 1.0 \pm 0.1$ [16–18]; FeNiPB, $n = 0.9 \pm 0.1$ [22–26]; NiZr,

The rate of viscosity increase $\dot{\eta}$ varies with temperature, as shown in *Figure 13.7* for FeNiPB. The available data for other alloy systems are similar[21,22]. The temperature dependence can generally be described as Arrhenius-type over limited temperature ranges, with an activation energy $Q_{\dot{\eta}}$ that increases as the temperature approaches T_g.

$$\eta = [\eta_0(T, \mathrm{M}) + \dot{\eta}_0 \mathrm{e}^{Q\dot{\eta}(T,\mathrm{M})/kT} t^n] \tag{13.2}$$
$$\mathrm{M} = \text{material}$$

The data indicate that the measured values of $\dot{\eta}$ are independent of testing method[22,26], prior thermal treatment of the alloy[19,21,22,28], and the magnitude of the testing stress[22,28,29], except at very low stresses, where the relaxation rate appears to be lower[30].

TABLE 13.1 Data used for *Figure 13.6*

Symbol	T/K	$\eta_0/(10^{15}\ \mathrm{N\ s\ m^{-2}})$
+	424	3.0
○	444	7.3
□	466	4.1
×	487	3.4
△	507	5.2
▽	537	1.6

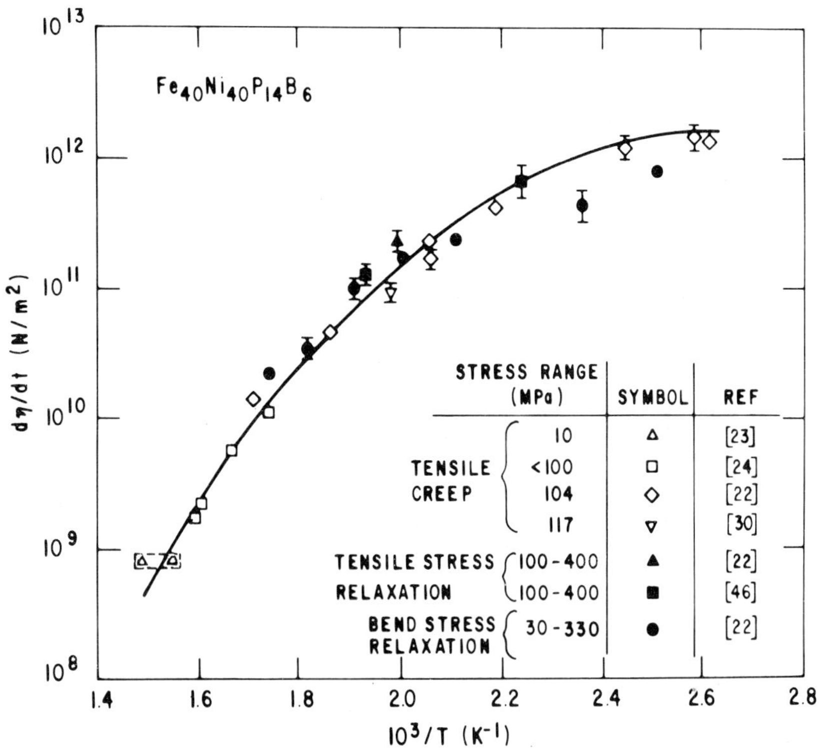

Figure 13.7 Rate of change of viscosity with time as a function of temperature, as measured in tensile creep, tensile stress relaxation and bend stress relaxation tests. The testing stress ranges are indicated (from Taub[30])

In a few of the tests conducted near T_g, two stages of viscosity increase have been reported[23,24]. At the start of the tests the typical linear increase with time was observed. However, Anderson and Lord[23] have shown that this is followed by a second stage of rapid viscosity increase. The onset of the second stage was shown to coincide with the onset of crystallization in the alloy. Patterson and Jones[25] conducted a study of the second stage, comparing the kinetics of viscosity increase with the crystallization kinetics of the alloy. The Einstein[31] formulation for the viscosity of a medium containing a known volume fraction V of non-flowing particles was used: $\eta = \eta_m(1 + 2.5\,V)$, where η_m is the viscosity of the amorphous matrix material. The data are shown in *Figure 13.8*. The volume fraction of crystals was determined by calorimetry. The agreement between the Einstein model and the data is good, except at large V where the assumption of non-interacting crystals breaks down.

Equation 13.2 expresses the time dependence of the viscosity or, equivalently, the structural dependence of the flow rate. With reference to *Figure 13.2*, the equation accounts for the shift of the isoconfigurational curves with structural relaxation. We now examine the temperature dependence of the viscosity at constant structure or, equivalently, the shape of the isoconfigurational curves in *Figure 13.2*.

The data from the early attempts[17,32,33] at measuring the isoconfigurational atomic mobility of amorphous alloys were compiled by Spaepen and Turnbull[34,35].

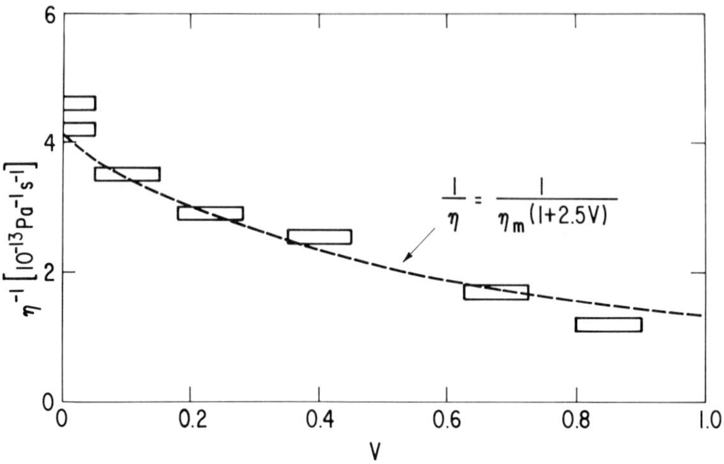

Figure 13.8 Variation of viscosity with volume fraction of crystals for a specimen creeping during crystallization at 646 K under a stress of 150 MPa. The dashed line is the prediction of the Einstein model $1/\eta = 1/[\eta_m(1 + 2.5\,V)]$ (after Patterson and Jones[25])

They showed that the tests were not truly isoconfigurational and attributed the scatter in the data to structural relaxation during the tests. That analysis motivated a study by Taub and Spaepen of amorphous $Pd_{82}Si_{18}$ in which care was taken to stabilize the structure prior to isoconfigurational testing[4,19]. The activation energies obtained in that study were higher than those reported previously, confirming the lack of structural stability in the early tests.

It is important to define what is meant by a 'stable structure'. We know that the structure relaxes and that therefore the viscosity changes continuously during isothermal annealing. However, the fractional rate of increase in viscosity $(1/\eta)(d\eta/dt) = d\ln\eta/dt$ decreases as the test proceeds. If a specimen is annealed until the fractional rate of viscosity change becomes less than 10 per cent per day, then one can lower the temperature, measure the viscosity and then raise the temperature back to the anneal temperature and find that the viscosity has not changed from its original value. The viscosity can also be reproduced after higher temperature cycling if the measurements are made in a short time. This process is shown schematically in *Figure 13.9*. The return to the original viscosity value at the end of the test implies that negligible structural change occurred during the temperature cycle. This ability to reproduce viscosity measurements during temperature cycling is the operational definition of isoconfigurational measurements. The tests are not truly at constant structure, rather the irreversible structural change during the test is negligible.

Figure 13.10 shows the isoconfigurational viscosity data for a $Pd_{82}Si_{18}$ specimen that was structurally stabilized in the manner described above, at successively higher temperatures. As expected, the isoconfigurational curves are seen to shift in the direction of the equilibrium curve with increasing annealing temperature. A constant activation energy was obtained $Q_{\text{isoconfigurational}} = 192 \pm 17$ kJ/mole. Further studies have shown that this activation energy is unaffected by the magnitude of the applied stress during both the stabilizing preanneal and the isoconfigurational testing[22,38]. Mulder *et al.*[36] have structurally stabilized $Ni_{36}Fe_{32}C_{12}P_{14}B_6$ and $Fe_{40}Ni_{40}B_{20}$ at 548 K and measured $Q_{\text{iso}} = 250 \pm 15$ kJ/mole for both alloys. Taub and Spaepen[20] measured $Q_{\text{iso}} = 240 \pm 25$ kJ/mole for PdCuSi preannealed at 534 K.

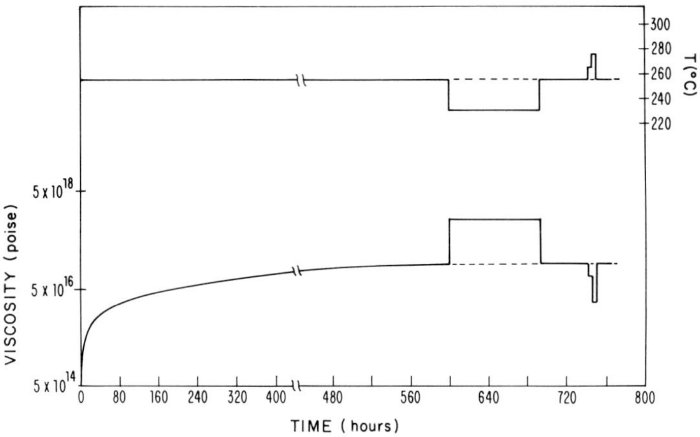

Figure 13.9 Typical annealing and temperature change program for an isoconfigurational creep test on $Pd_{77.5}Cu_6Si_{16.5}$. Annealing must be continued until log η changes slowly enough with t so that the viscosity can be reproduced when the temperature is cycled (S. S. Tsao, personal communication)

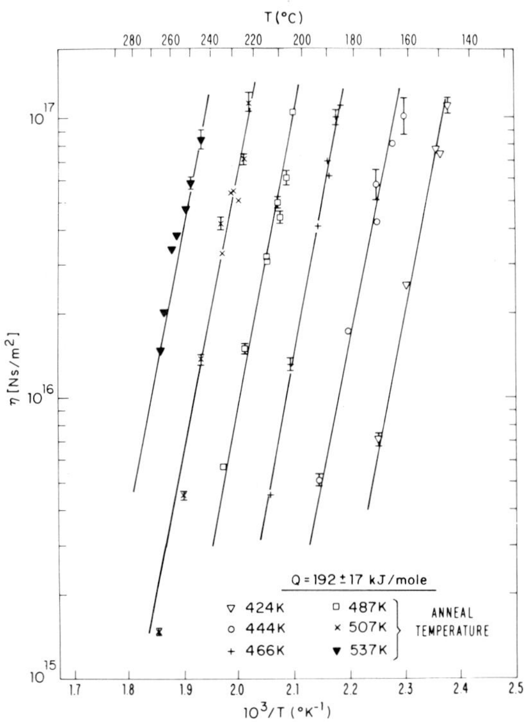

Figure 13.10 Isoconfigurational viscosities of glassy $Pd_{82}Si_{18}$ obtained between the annealing treatments of *Figure 13.6* (from ref. 19).

It has recently been shown that the activation energy for isoconfigurational flow does not remain constant over the entire temperature range but that it increases with temperature (*see Figure 13.11*)[21]. The change in activation energy is small at low temperatures, but it increases rapidly as the glass transition temperature is approached. This trend is similar to that found previously for $Q_{\dot{\eta}}$.

The stress dependence of homogeneous flow in amorphous alloys has been the subject of numerous experimental investigations. Of particular concern has been the determination of the stress exponent $m = \mathrm{d}(\ln \dot{\gamma})/\mathrm{d}(\ln \tau)$. The early work[4,17,18,33,37–40] contained large differences in the reported values of this parameter ($1 < m < 12$). The origin of these discrepancies was later clarified by applying transition state theory to the flow process, which predicts[41–43] a stress-strain rate relation of the form:

$$\dot{\gamma} = \gamma_0 \sinh \left[\frac{\tau \gamma_0 \Omega_f}{2kT} \right] \tag{13.3}$$

Equation 13.3 predicts a strong relation between the stress exponent and the applied stress. In fact, when the widely varying literature data were compiled and tabulated against the applied stress, a consistent trend was observed[25,44]. The stress exponent was found to be unity at low stresses and to increase rapidly at higher stresses, as expected. Furthermore, this behaviour was exhibited in both creep and stress relaxation testing[22,26].

An important parameter in this relation is the volume strain element $\gamma_0 \Omega_f$, where γ_0 is the measure of the local shear strain that an atomic scale volume Ω_f undergoes when contributing to flow. The experimental determinations[45,46] of this parameter are shown in *Figure 13.12*. A strong temperature and composition dependence is indicated.

A recent study of the stress–strain rate relation showed that equation 13.3 breaks down at low stresses[26,46]. In *Figure 13.13*, stress–strain rate data from that study is shown. The stress exponent is seen to be unity only at intermediate stress levels, becoming greater than unity at both very high and very low stresses. Equation 13.3

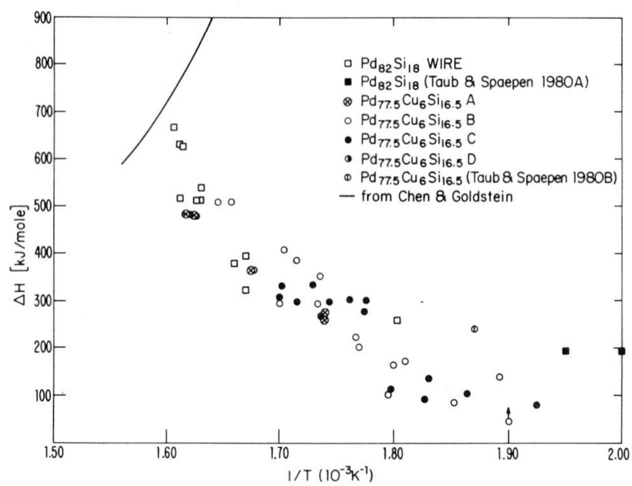

Figure 13.11 Apparent activation energies for a number of isoconfigurational tests on palladium-based glasses. The solid line represents the equilibrium values for Pd–Cu–Si alloys measured by Chen and Goldstein (ref. 18); □, $Pd_{82}Si_{18}$ wire; ■, $Pd_{82}Si_{18}$, Taub and Spaepen (ref. 19); ⊗, $Pd_{77.5}Cu_6Si_{16.5}$ (A); ○, $Pd_{77.5}Cu_6Si_{16.5}$ (B); ●, $Pd_{77.5}Cu_6Si_{16.5}$ (C); ○, $Pd_{77.5}Cu_6Si_{16.5}$ (D); ⦶, $Pd_{77.5}Cu_6Si_{16.5}$, Taub and Spaepen (ref. 20)

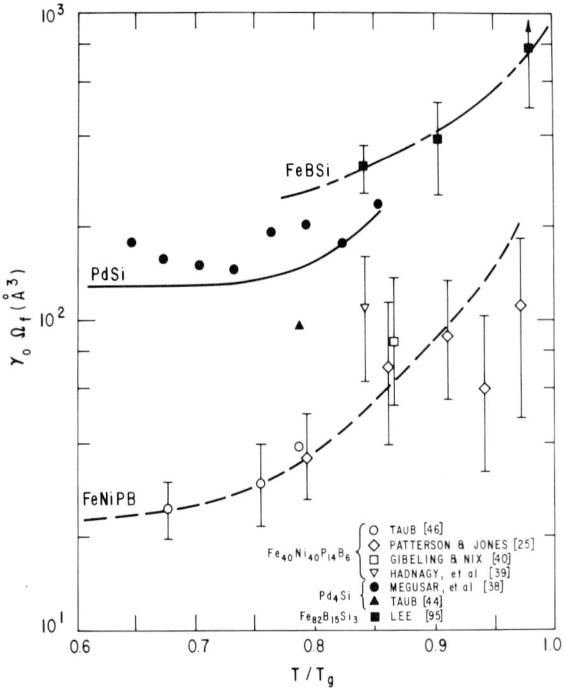

Figure 13.12 Available values of the volume strain element as a function of testing temperature. The temperatures have been normalized by the glass transition temperature in order to make the comparison between different alloys more meaningful (from ref. 46). For $Fe_{40}Ni_{40}P_{14}B_6$: ○, Taub (ref. 46); ◊, Patterson and Jones (ref. 25); □, Gibeling and Nix (ref. 40); ▽, Hadnagy *et al.* (ref. 39); for Pd_4Si: ●, Megusar *et al.* (ref. 38); ▲, Taub (ref. 44); for $Fe_{82}B_{15}Si_3$: ■, Lee (ref. 95)

predicts only the high stress deviation from linear behaviour. However, the equation is easily modified to account for the low stress behaviour by introducing[47–49] a threshold stress τ_0:

$$\dot{\gamma} = \dot{\gamma}_0 \sinh\left[\frac{(\tau - \tau_0)\gamma_0\Omega_f}{2kT}\right] \tag{13.4}$$

The improved fit to the data is shown in *Figure 13.13*.

τ_0 is the stress below which the amorphous alloy will not flow. This threshold stress appears to be dependent on both temperature and composition. For $Fe_{76}B_{18}Si_6$, $\tau_0 \sim 120$ MPa at 498 K[50]. For FeNiPB, the available data[46] fall in the range 10–70 MPa for $448 < T < 520$ K. A weak dependence on the structural state of the alloy is indicated as well.

A complete empirical flow law can now be formulated as follows:

$$\dot{\gamma} = \left[\frac{\sinh\dfrac{(\tau - \tau_0)\gamma_0\Omega_f}{2kT}}{\eta_0 + \dot{\eta}_0(\mathrm{M})(\exp Q_{\dot{\eta}}/kT)t^n}\right](1 + 2.5V)(\exp Q_{iso}/kT) \tag{13.5}$$

$Q_{\dot{\eta}}$, Q_{iso}, T_0 and $\gamma_0\Omega_f$ all depend on temperature and composition.

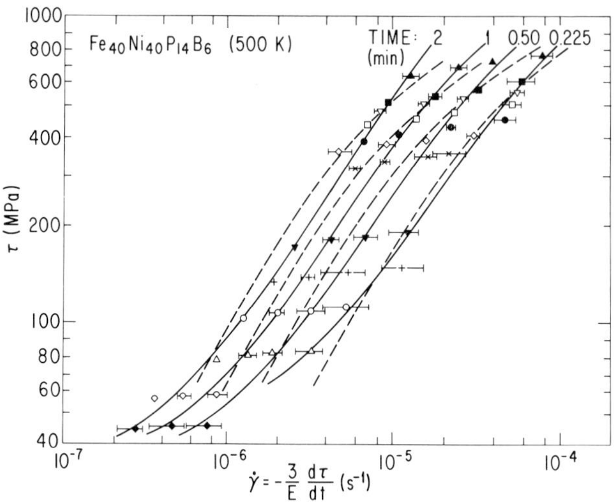

Figure 13.13 Stress–strain rate data for $Fe_{40}Ni_{40}P_{14}B_6$ stress relaxation tested at 500 K. Each isochronal data set corresponds to a different isoconfigurational state of the alloy. The dashed lines are the least squares fits of the original sinh law (equation 13.3) to the data. The solid lines are the fits of the threshold modified sinh law (equation 13.4) (from Taub[46])

13.2.4 Models

Equilibrium

Two models that have been widely used to explain atomic transport in organic and inorganic glasses are the free volume and the configurational entropy models.

Free Volume Model Batschinski[51] first introduced the concept that the resistance to flow in a liquid should depend upon the relative free volume per molecule. However, it was not until the early 1950's that Doolittle[52] provided satisfactory experimental evidence for this proposal. Doolittle's measurements on liquid normal paraffins disclosed a logarithmic relation between the viscosity and free volume.

$$\eta = A \exp \frac{\beta v_0}{v_f} \tag{13.6}$$

where A and β are constants, the value of β being about unity; $v_f(T)$ is the average free volume per molecule; and v_0 is the molecular volume at 0 K.

The free volume is defined as the excess volume arising from the total thermal expansion of the liquid without change of phase. Defining $T = T_0$ as the temperature at which the free volume goes to zero (i.e. $\eta \rightarrow \infty$ in the Fulcher–Vogel relation) the free volume can be expressed as:

$$v_f = \Delta\alpha(T - T_0) \tag{13.7}$$

Substituting equation 13.7 into equation 13.6 we obtain:

$$\eta = A \exp \frac{\beta v_0}{\Delta\alpha(T - T_0)} \tag{13.8}$$

Equation 13.8 is of the form of the Fulcher–Vogel relation where $B = \beta v_0/\Delta\alpha$.

A model to explain this free volume correlation was later developed by Cohen and Turnbull[53–55]. They assumed that each atom or molecule is confined to a cage delineated by its Voronoi nearest neighbours. The free volume is that part of the cage in which an atom can move without an energy change. The molecular motion within this cage is limited to oscillations unless there is a fluctuation in density which opens up a hole large enough to permit a diffusive-type displacement into a neighbouring cage. The authors computed the probability of occurrence of a fluctuation with volume greater than the minimum volume v^* necessary for such a jump as $\exp(-\gamma v^*/v_f)$. Since flow occurs uniformly throughout the material, this probability is proportional to the flow rate. The viscosity can then be expressed as:

$$\eta = C\exp(\gamma v^*/v_f) \tag{13.9}$$

γ is a geometrical factor approximately equal to unity. Equation 13.9 is exactly the Doolittle equation.

Configurational Entropy Model The configurational entropy model, attributed to Adam and Gibbs[56], deals with the flow behaviour of glass-forming liquids in terms of cooperative molecular rearrangements. Whereas the free volume model accounts for the temperature dependence of flow by examining changes in free volume in the equilibrium liquid, the configurational entropy model explains the temperature dependence by considering the changes in the size of the rearranging region.

A cooperatively rearranging region of z molecules is defined as a subsystem of the sample whose energy can fluctuate sufficiently to allow rearrangement into another configuration. If both mechanical and thermal contact between the subsystems are assumed and only small departures from the equilibrium distribution are permitted, the probability of a cooperative transition of a region is $\exp(-z\Delta\mu/kT)$. $z\Delta\mu$ is the activation barrier for rearrangement. As in the free volume model, if one assumes a critical lower limit z^* for rearrangement, then the average transition probability can be computed in terms of z^*. Since flow occurs uniformly in the material, this transition probability is proportional to the flow rate and therefore inversely proportional to the viscosity.

$$\eta = \bar{A}\exp(z^*\Delta\mu/kT) \tag{13.10}$$

A complete expression for the viscosity requires evaluation of the parameter z^*, which Adam and Gibbs determined as a function of the configurational entropy of the system S_c. The result of their analysis is:

$$\eta = \bar{A}\exp(\Delta G^*/TS_c) \tag{13.11}$$

where ΔG^* is the activation energy barrier for rearrangement of the region of critical size z^*.

Defining $T = T_0$ as the temperature at which the configurational entropy goes to zero, the configurational entropy can be expressed as:

$$TS_c = \Delta C_p(T - T_0) \tag{13.12}$$

ΔC_p is the difference in specific heat between the equilibrium liquid and the glass. Substituting equation 13.12 into equation 13.11 we obtain:

$$\eta = \bar{A}\exp[\Delta G^*/\Delta C_p(T - T_0)] \tag{13.13}$$

Equation 13.13 is of the form of the Fulcher–Vogel relation where $B = \Delta G^*/\Delta C_p$.

Non-equilibrium

Nature of the Microscopic Shear Transformation In the glass, as in the equilibrium liquid state, macroscopic plastic strain is the sum of the strain contributions, γ_0, from a large number of localized microscopic shear transformations or 'flow defects' of volume Ω_f. In order to elucidate the nature of these flow defects, it is very useful to consider the scaling relation between the viscosity and the diffusivity:

$$\eta D = \frac{kT}{L} \tag{13.14}$$

where L is a characteristic length related to the dimensions of the defects governing diffusion and flow[42]. In the liquid state, the well-known Stokes–Einstein (S–E) relation is known to hold for many metallic systems, to within ± 50 per cent, with $L = 6\pi r$, where r is the average ionic radius[57]. Chen *et al.*[58] have confirmed the validity of the S–E relation around T_g to within 70 per cent and recent evidence seems to indicate that it is valid in the glassy state as well. This evidence is based on comparisons between the results from interdiffusion measurements on palladium-containing compositionally modulated amorphous films[59,60] with the results of the flow measurements on palladium-based glasses, discussed above:

1. The isoconfigurational activation energies for both sets of measurements are very similar: 195 kJ/mole in diffusion[60] compared with 192 kJ/mole in flow[19].
2. D and η in samples with a similar thermal history scale according to the S–E relation, to within 80 per cent.
3. Although the magnitudes of D and η are very dependent on thermal history, the relaxation *rates*, discussed above, are not. It is observed that $(1/D)/\mathrm{d}t$ and $\mathrm{d}\eta/\mathrm{d}t$ scale according to the S–E relation, to within 80 per cent.

The continuity of the scaling relation, all the way from the liquid to the glassy state, indicates that the flow and diffusion defects are similar in nature, and the magnitude of the scaling length, $L = 6\pi r$, indicates that their size is of the order of an atomic diameter. This seems plausible for a diffusional rearrangement, where only a change of nearest neighbours is required. It is also consistent with the measurements of $\gamma_0\Omega_f \approx 2$–$8\Omega$ (Ω = atomic volume), which can be interpreted as either a small number of atoms undergoing an extensive shear transformation ($\gamma_D \approx 1$) or a larger number of atoms undergoing a small shear, as in Argon's high temperature, low stress model[61]. It seems, however, that dislocation models (*see* below), for which $\gamma_0 \approx 1$, are less appropriate here, since the operating segments would only be a few atoms long.

The physical interpretation of the threshold stress τ_0 is related to the driving potential for flow. Transition state theory gives the sinh-dependence on the stress as a result of biasing of the forward flux across the activation barrier over the backward flux. The biasing is the result of the forward state having a lower energy of $\tau\gamma_0\Omega_f$, the work done by the applied stress, as a result of one microscopic rearrangement[46]. If, as a result of the shear transformation, the energy of the system increases (strain energy, topological or chemical disorder), this increase can be written *formally* as $\tau_0\gamma_0\Omega_f$. The driving energy for the transformation is then decreased by this amount and becomes[46] $(\tau - \tau_0)\gamma_0\Omega_f$. Since the total strain in homogeneous flow is small, the effect of strain disordering on the structure is negligible. However, as will be discussed below, it is a crucial element in strain-induced softening during inhomogeneous flow.

Structural Relaxation The viscosity increase as a result of structural relaxation can most easily be interpreted as an increase in structural order, and hence a decrease in the flow defect concentration, n_f. The linear increase of η with t has also been observed in other amorphous systems such as covalent glasses[62,63] and polymers[64]. In covalent networks, where the flow defect is a broken bond, the linear kinetics are easily explained as a result of the bimolecular annihilation reaction

$$\dot{n}_f \propto -n_f^2 \tag{13.15}$$

since it takes two broken bonds to annihilate one another. It is easily seen that this integrates to

$$n_f^{-1}(t) - n_f^{-1}(0) \propto t \tag{13.16}$$

or since, $\eta \propto n_f^{-1}$:

$$\eta(t) - \eta(0) \propto t \tag{13.17}$$

The extension of this bimolecular mechanism to metallic glasses remains speculative. In highly ordered binary alloys, such as the metal (M)–metalloid (X) alloys in which ideally each X atom is surrounded by only M atoms, one might postulate that the flow defect consists of X–X and complementary excess M–M wrong bonds. Alternatively, topological defects, such as density fluctuations in the free volume theory, require an appropriate annihilation site, such as a stress concentration, analogously to a crystal vacancy needing an edge dislocation core site as a sink. Egami, Vitek and their coworkers[65,66] have analysed the internal stresses in model clusters and have proposed that structural relaxation is governed by annihilation of dilated and compressed regions ('*p*' and '*n*' defects) in the system. It remains to be shown that these regions can move, produce flow and have the right lifetime.

In order to obtain bimolecular kinetics it is then necessary to postulate that, for structural reasons, the concentration of annihilation sites is proportional to[21,67] the concentration of flow sites $n_r = Cn_f$. Then

$$\dot{n}_f = k_r n_f n_r = -k_r C n_f^2 \tag{13.18}$$

where k_r is the jump frequency for annihilation. This integrates to

$$\eta(t) - \eta(0) = \frac{k_r C}{k_f'} \frac{\Omega k T}{(\gamma_0 \Omega_f)^2} t \tag{13.19}$$

where k_f' is the unbiased jump frequency for a flow arrangement. From measurements of $\eta(t)$ and $\gamma_0\Omega_f$ it is found that the factor $k_r C/k_f'$ in equation 13.19 equals about 500 in a typical creep experiment. The proportionality constant C can be estimated from the observation of Chen *et al.*[58] that in their diffusion experiments each atom could make at least a few hundred jumps without the system reaching equilibrium. This means that the concentration of annihilation sites, n_r, must be much less than the concentration of diffusion defects, and hence, as discussed in the previous section, also much less than n_f, so that $C \ll 1$. Consequently, the jump frequency for annihilation must be much greater than the frequency for flow rearrangement: ($k_r \gg k_f'$). This result is reasonable, given the much higher driving force for annihilation.

All the bimolecular models discussed above are 'non-redistributive', meaning that the defects are assumed to keep their identity, at least statistically. Taub and Spaepen[19] have proposed a unimolecular relaxation model, in which free volume is annihilated by collapse of free volume fluctuations above a certain size v_a^*. This model is 'redistributive', meaning that after each annihilation event it is assumed that the system immediately redistributes the left-over free volume to form a new constrained equilibrium distribution. It can be shown easily[67] that this model, although unimolecular in the free volume reaction, is formally bimolecular in the defect concentration if $(v_a^* - v^*)$ is small.

Isoconfigurational Flow The interpretation of the temperature dependence of the viscosity measured under operationally 'isoconfigurational' conditions is not straightforward.

Since the temperature dependence of η is dominated by that of the product $n_f k_f'$, Spaepen and Turnbull[34] originally proposed that, since n_f was thought to be constant under isoconfigurational conditions, the observed activation energy reflected the temperature dependence of the jump frequency k_f'. However, this leads to a pre-exponential that is unphysically low[21]:

$$\eta_{iso}(T) = \eta_0 \exp(Q_{iso}/kT) \tag{13.20}$$

The data of Taub and Spaepen on $Pd_{82}Si_{18}$, for example[19], give $\eta_0 = 1.5 \times 10^{-7}\,N\,s\,m^{-2}$, which is far below the minimum value of $10^{-3}\,N\,s\,m^{-2}$, corresponding to the viscosity of a liquid at high temperature.

For this reason, the defect concentration n_f must change under operationally isoconfigurational conditions. Since the crucial element in such experiments is the *reproducibility* of the viscosity with cycling of the temperature, the changes in n_f must be *reversible* and *fast* on the time scale of the temperature changes. This is in contrast with the changes in n_f as a result of structural relaxation, which are *irreversible* and much slower.

Atomistically, the reversible changes in n_f can, for example, be thought of as governed by slight local adjustments in the atom positions, without change in the average free volume. Simple model calculations have shown that, if the volume change due to thermal expansion is not distributed at all, but simply leads to a rescaling of the interatomic distances, the temperature dependence of n_f would be very weak. A full redistribution, to a new constrained equilibrium free volume distribution, however, leads to a strong temperature dependence. Recent data, taken near equilibrium[21], fall between these two extremes, indicating a partial redistribution of free volume.

13.3 Inhomogeneous flow

The phenomenology of this type of flow is reviewed in detail in Chapter 12. For the discussion of atomistic mechanisms the following points are worth summarizing:

1. Plastic flow is confined to shear bands, whose number and orientation are determined by the geometry of the deformation; the rest of the specimen deforms only elastically.
2. The shear bands are very thin[68,69], form very quickly when the yield stress is reached[70], and have sharp slip line-like marks at the specimen surface.
3. The strain inside the band is very large, and is accompanied by chemical disordering (observed by etching[71]) and dilatation (observed by direct density measurements[72]).
4. Over a large temperature range, the flow stress is only very weakly temperature dependent[74], and is very strain rate insensitive[41,74].

The macroscopic resemblance of the sharp surface offsets produced by the shear bands to the slip lines observed after plastic deformation of single crystals has led a number of authors to propose the motion of localized, atomic size, crystal-like dislocations to explain the phenomenon of inhomogeneous flow[75,76].

The topology of forming localized dislocations in amorphous systems has been investigated by Chaudhari *et al.*[77,78] using static computer relaxation of monatomic Lennard-Jones clusters. The dislocations were introduced by applying the procedure for creating Volterra dislocations in a continuum: a planar cut was made halfway down the cluster, the two sides of the cut were displaced with respect to one another by a distance on the order of an atomic diameter (if necessary with removal of material), and reconnected. After relaxation, the stress field in the cluster was investigated. The early results[77] showed that a stress field characteristic of a screw dislocation could be found in the relaxed cluster. The edge dislocation turned out to be more difficult to stabilize[77], and although more recent results[78] show a remaining edge dislocation-like stress field, some questions remain about its asymmetry and irregularity. Although it is plausible that the length of the Burgers vector of these dislocations should be of the order of the atomic diameter (otherwise, the system would simply move back to its original configuration upon relaxation), it is not clear yet how large the Burgers vector must be to make separation into 'partials' energetically favourable. There always remains a possibility that the static relaxation procedure produces configurations that are locally metastable, but does not allow for transition to lower energetic minima. Molecular dynamics experiments on some of the dislocated clusters have not been conclusive yet: a dislocation stress field could still be observed at low temperatures but disappeared at the glass transition temperature.

The main question about these dislocation modelling experiments, however, is whether the stress fields that have been identified are in fact governing inhomogeneous flow: can they move, do they remain stable upon moving, and how is the structure affected by their passage? Computer[79,80] and bubble raft[81,82] modelling of deformation at high stresses show that flow is indeed governed by a variety of localized shear rearrangements, but these cannot be identified with long-lived extended line defects such as dislocations. On the macroscopic level, the dislocation models have so far not accounted for the localization of flow into a small number of shear bonds, nor for the subsequent fracture *along* the shear band resulting in a very characteristic 'vein'-like fracture morphology.

The latter phenomena can be explained by any model that leads to a strong softening, i.e. lowering of the viscosity, in the shear band. Softening concentrates the deformation into the band and weakens the band against fracture by an instability that produces the vein pattern (*see* section 13.4.1). The lowered viscosity also explains the decrease in chemical order or, as expected from a free volume model, the decrease in density.

In Argon's model[43] inhomogeneous flow is governed by localized shear transformations, analogous to the formation of dislocation loops, that are different from those used to describe homogeneous flow. The parameters of both mechanisms, assumed to operate at steady state, can be adjusted to account for the temperature and stress regimes of the two types of flow. The dilatation is introduced as an independent free volume creation parameter accompanying the shear transformation. By assuming that the threshold stress for this latter transformation is lowered by an increase in free volume, the required softening is achieved and localization of the flow can be demonstrated.

Spaepen and his collaborators have treated the same problem in a more unified way by extending the free volume model discussed above for homogeneous flow. Spaepen and Turnbull[83] have pointed out that the dilatation at a stress concentrator, such as a microcrack, is large enough to cause a large drop in viscosity. This mechanism is thought to be important for the *initiation* of inhomogeneous flow, and also, as discussed below, for the circumvention of brittle fracture by 'blunting' of susceptible cracks. However, after initiation of a shear band, the softening is *sustained* by a process of strain-induced disordering: the deformation process continuously disorders the material in the shear band, at a faster rate than the structural relaxation processes can restore it. This idea was introduced by Polk and Turnbull[7] and made quantitative by Spaepen[41]. In this model, disordering is assumed to be a shear-induced free volume creation process, as a result of atoms pushing their neighbours aside at high stress levels. This process is balanced by an ordering process during which free volume is annihilated. In the original paper[41] it was assumed that the annihilation process consisted simply of a series of diffusional jumps. Later work on homogeneous flow, discussed above, has shown that the annihilation process is one of structural relaxation, with kinetics that can be quite different from those of the purely diffusional processes, at least in the high-viscosity regime. Nevertheless, the original model has had some success by assuming that the creation and annihilation processes establish a *steady state*, in which the free volume, and hence the viscosity, is a strong function of the stress. This is contrasted with homogeneous flow, where the viscosity is determined by the thermal history (below T_g) or the equilibrium structure (above T_g). The model could account for the strain rate insensitivity, and for the homogeneous–inhomogeneous flow transition with only a single adjustable parameter.

Most recently, Steif *et al.*[84], using the free volume creation and annihilation models from ref. 41, have analysed the establishment of inhomogeneous flow as a transient, i.e. time-dependent, problem, together with the requirements for localization of the plastic strain. They concluded that the free volume creation model indeed leads to catastrophic softening and formation of a shear band. It also became apparent, in agreement with earlier observations, that the band forms so quickly that the steady state is probably never reached. In that case, the annihilation processes can be neglected and their precise nature does not have to be identified to describe inhomogeneous flow. The time-dependent analysis can account for the experimentally observed stress for onset of the strain localization and its temperature dependence.

13.4 Fracture

An overview of the experimental work on fracture of metallic glasses is given in Chapter 12. We will therefore emphasize again the micro-mechanisms of fracture, especially how they relate to those governing homogeneous and inhomogeneous flow.

13.4.1 Ductile fracture

We term fracture of metallic glasses 'ductile' if it occurs as a result of, at least locally, large plastic strain. Specifically, this applies to the fracture process that terminates inhomogeneous flow. In uniaxial tension, it occurs along the plane of the shear band, not along the plane normal to the tensile axis, which has the largest normal stress. *Figure 13.14* shows the characteristic 'vein' or 'river' pattern of the resulting fracture morphology. *Figure 13.15* shows how the direction of the veins is related to the shear band and the direction of the fracture process. The similarity between the pattern of *Figure 13.14* and that obtained by separating two solid surfaces containing a fluid layer, such as vaseline or grease, is obvious[83] and supports the softening mechanism for inhomogeneous flow discussed above.

The morphology is the result of an instability that occurs during the motion of the interface between the fluid and the outside air. This is a special case of general type of instability, first identified by Taylor[85], and applied to this fracture problem by Spaepen[86] and Argon and Salama[87]. Pitts and Greiller[88] give an instructive

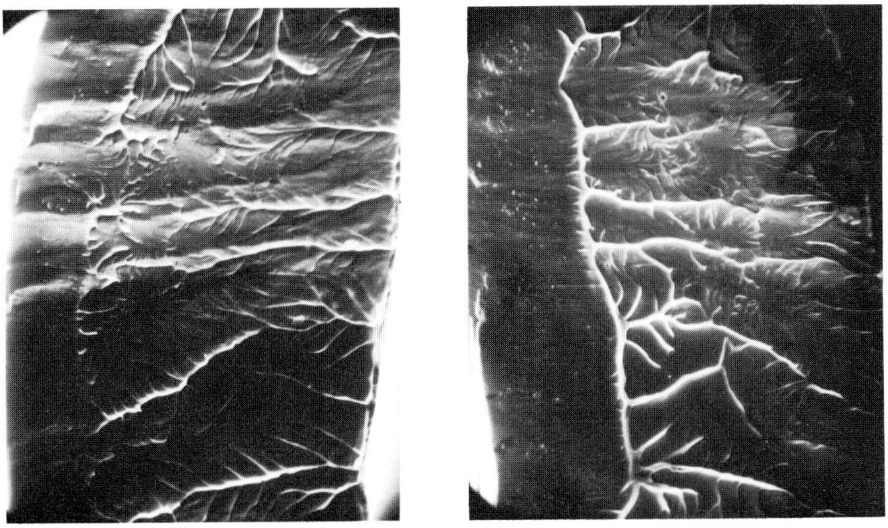

Figure 13.14 Matching fracture surface of an amorphous metal ribbon 100 μm thick, broken in the ductile fracture mode (SEM micrograph from Spaepen[86])

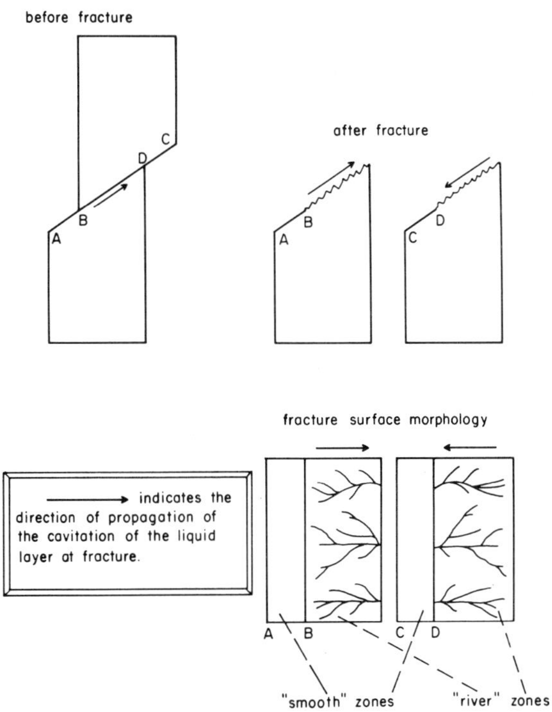

Figure 13.15 Diagram explaining the origin of the features in *Figure 13.14* and their relation to the shear band. The arrows indicate the direction of propagation of the cavitation of the liquid layer at fracture (after Spaepen[42])

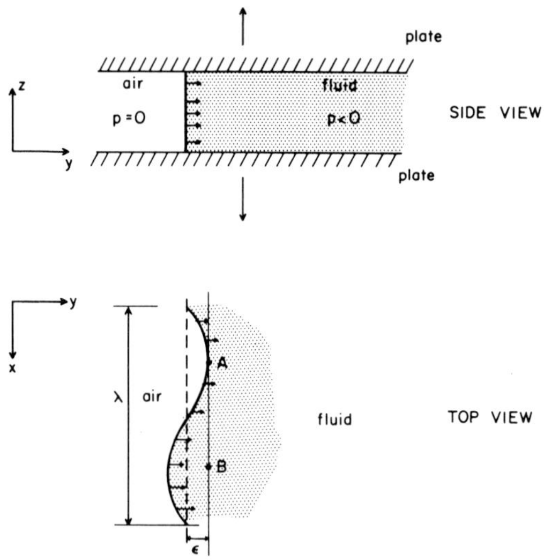

Figure 13.16 Diagram illustrating the stability of a perturbation in a moving fluid–air interface (*see* text for calculation)

approximate analysis of the conditions for instability, which is illustrated in *Figure 13.16*. The fluid–air interface lies in the xy-plane and is driven in the y-direction by an externally applied negative pressure gradient $(\partial p/\partial y)$. An infinitesimal sinusoidal perturbation

$$y = \varepsilon \sin\left(\frac{2\pi x}{\lambda}\right)$$

causes the pressure at a point A, just inside the fluid, to be determined by the surface tension γ, and the local curvature κ

$$P_A = \gamma\kappa = -\gamma\varepsilon\left(\frac{2\pi}{\lambda}\right)^2$$

At point B, the pressure effect due to capillarity has the reverse sign and the effect of the applied pressure gradient over a distance 2ε has to be taken into account as well:

$$P_B = \gamma\kappa + 2\varepsilon\frac{\partial p}{\partial y}$$

If $p_A < p_B$, matter will flow from B to A and the perturbation will die out. If $p_A > p_B$, the reverse will happen and the perturbation will grow as a protrusion into the fluid layer. These protrusions then run into one another on an increasingly larger scale, resulting in the overall 'vein'-pattern effect. The critical wavelength that can still grow then corresponds to $p_A = p_B$, or:

$$\lambda_c = 2\pi\sqrt{\frac{-\gamma}{\partial p/\partial y}}$$

The scale of the instability is seen to be the result of the balance between the pressure gradient driving the interface motion, and the surface tension resisting growth of perturbations. This is similar to the dendritic instability in crystal growth, which is governed by the balance between a temperature or concentration driving gradient and the crystal–melt surface tension. In fact, the Mullins and Sekerka theory for this phenomenon[89] can be adapted easily to the fracture problem[86]. It is found that the wavelength of the fastest growing perturbation is $\lambda_m = 3\lambda_c$. Insertion of appropriate values of the pressure gradient and surface tension gives a value for λ_m in good agreement with the finest vein spacings in *Figure 13.14*.

13.4.2 Brittle fracture

The principle causes of embrittlement, as reviewed in Chapter 12, are: annealing, low temperature, and crystallization. The following points about annealing-induced embrittlement are worth emphasizing:

1. It is *very* composition-dependent: iron-based glasses, especially with large metalloid additions[90,91], embrittle very easily; palladium-based alloys, on the other hand, have been annealed for very long times without embrittling, if no crystallization occurs.
2. It correlates with structural relaxation[92].
3. It can be reversed by neutron irradiation[93].

Although the fundamental understanding of embrittlement is far from complete, it is worth pointing out how some of the insights gained in the study of homogeneous and inhomogeneous flow can be applied to the brittle fracture problem.

Brittle fracture is a cleavage process that can most simply be said to occur if the tensile stress at a stress concentrator, such as a microcrack, reaches the theoretical fracture stress of the material, σ_{th}, before shear-induced plastic flow can blunt the crack and relieve the stress. It seems plausible that the large viscosity increase resulting from structural relaxation (up to five orders of magnitude, as discussed in the section on homogeneous flow), or the viscosity decrease associated with the density decrease upon irradiation[93], should have an effect on this crack blunting process.

Steif[94] has recently attempted to model this by calculating, as a function of time, the stress field around an elliptical hole in a strip subjected to a constant uniaxial tensile strain rate. The material was assumed to be non-linear viscoelastic, with a viscosity that depended on the local dilatation, via the free volume formulation. The results showed that, if local free volume created by the dilatation was allowed to redistribute itself *instantaneously* according to a new constrained equilibrium distribution, the tensile stress around the hole never became large enough for cleavage, i.e. the stress relief process was very effective.

This seems plausible in view of Spaepen and Turnbull's original argument[83] about the initiation of flow at stress concentrators. They pointed out that the temperature change necessary to give dilatation equivalent to that just preceding fracture is

$$\Delta T \simeq \frac{\sigma_{th}}{K\alpha}$$

and can be estimated from $\sigma_{th} \approx E/10$ (E = Young's modulus), the compressibility $K \approx E$ and the volume thermal expansion coefficient $\alpha = 10^{-4}\,K^{-1}$, to give $\Delta T \approx 10^3$ K. This corresponds to a drastic decrease in viscosity, more than enough to relieve the stress at the singularity.

In order for the elliptical hole model to explain brittle behaviour it was necessary to assume that free volume created by the dilatation is *not* instantaneously redistributed, but that a certain amount of time is required to reach the new equilibrium distribution. By introducing a time constant for the redistribution process, Steif was able to show that the theoretical fracture stress can be reached if the time constant and the initial viscosity of the material are large enough. His model suggests that structural relaxation, therefore, not only increases the viscosity, as observed in the creep experiments but also increases the time constant for free volume redistribution, and that differences in embrittlement kinetics between different metallic glasses, such as iron- and palladium-based ones, may be attributed to differences in the redistribution kinetics. The free volume redistribution process discussed here is qualitatively similar to the one discussed for isoconfigurational flow where a volume change caused by thermal expansion is redistributed. However, the values of the time constants involved are quite different (the present one is faster), which makes experimental verification of this conjecture difficult.

Acknowledgements

F.S. acknowledges research support from the Office of Naval Research, Contract # N00014-77-C-0002.

References

1. Ashby, M. F., *Acta metall.*, **20,** 887 (1972)
2. Frost, H. J. and Ashby, M. F., *'Proc. John E. Dorn Symposium'*, Eds. Li, J. C. M. and Mukherjee, A. K., p. 70 American Society of Metals, Cleveland, Ohio (1975)
3. Egami, T., *'Proceedings of Fourth International Conference on Rapidly Quenched Metals'*, Eds. Masumoto, T. and Susuki, K., Japan Institute of Metals, Sendai (1982)
4. Taub, A. I. and Spaepen, F., *Scripta Met.*, **13,** 195 (1979)
5. Steinberg, J., Tyagi, S. and Lord, A. E. (jun.), *Acta metall.*, **29,** 1309 (1981)
6. Naka, M., Nishi, Y. and Masumoto, T., *Proceedings of the Third International Conference on Rapidly Quenched Metals'*, Brighton, England, Ed. Cantor, B., vol. I, p. 231, The Metals Society, London (1978)
7. Polk, D. E. and Turnbull, D., *Acta metall.*, **20,** 493 (1972)
8. Chen, H. S., *J. Non-cryst. Solids*, **27,** 257 (1978)
9. Vogel, H., *Z. Phys.*, **22,** 645 (1921)
10. Fulcher, G., *J. Amer. Ceram. Soc.*, **6,** 339 (1925)
11. Berry, C. and Fox, T., *Adv. Polymer Sci.*, **5,** 261 (1968)
12. Dragomir, I., *'The Properties of Liquid Metals'*, Ed. Takeuchi, S., Taylor and Francis, London (1973)
13. Wilson, J. R., *Metall. Rev.*, **10,** 381 (1965)
14. Jones, L. V., Ofte, D., Rohr, W. G. and Wittenberg, L. J., *Trans. Am. Soc. Metals*, **55,** 819 (1962)
15. Davies, H. A., Aucote, J. and Hall, J. B., *Scripta Met.*, **8,** 1179 (1974)
16. Ramachandrarao, P., Cantor, B. and Cahn, R. W., *J. Non-cryst. Solids*, **24,** 109 (1977)
17. Chen, H. S. and Turnbull, D., *J. chem. Phys.*, **48,** 2560 (1978)
18. Chen, H. S. and Goldstein, M., *J. appl. Phys.*, **43,** 1642 (1972)
19. Taub, A. I. and Spaepen, F., *Acta metall.*, **28,** 1781 (1980)
20. Taub, A. I. and Spaepen, F., *Scripta Met.*, **14,** 1197 (1980)
21. Tsao S. and Spaepen, F., ref. 3, p. 463
22. Taub, A. I. and Luborsky, F. E., *Acta. metall.*, **29,** 1939 (1981)
23. Anderson, P. M. (III) and Lord, A. E., *Mater. Sci. Engng*, **44,** 279 (1980)
24. Takemori, T., Mizoguchi, T. and McGuire, T. R., *Mater. Res. Bull.*, **15,** 81 (1980)
25. Patterson, J. P. and Jones, D. R. H., *Acta metall.*, **28,** 675 (1980)
26. Taub, A. I. and Luborsky, ref. 3, p. 1341
27. Taub, A. I., unpublished results
28. Taub, A. I. and Spaepen, F., *J. mater. Sci.*, **16,** 3087 (1981)
29. Taub, A. I. and Spaepen, F., *Scripta Met.*, **13,** 883 (1979)
30. Taub, A. I., *Acta metall.*, **30,** 2129 (1982)
31. Einstein, A., *Annln Phys.*, **13,** 289 (1906)
32. Maddin, R. and Masumoto, T., *Mater. Sci. Engng*, **9,** 153 (1972)
33. Logan, J. and Ashby, M. F., *Acta metall.*, **22,** 1047 (1974)

34. Spaepen, F. and Turnbull, D., *'Metallic Glasses'*, Eds. Gilman, J. J. and Leamy, H. J., p. 114, American Society of Metals, Cleveland, Ohio (1978)
35. Spaepen, F., *J. Non-cryst. Solids*, **31**, 207 (1978)
36. Mulder, A. L., Drijver, J. W. and Radelaar, S., *J. Phys. (Paris)*, **41**, C8, 843 (1980); ref. 3, p. 1345
37. Murata, T., Kimura, H. and Masumoto, T., *Scripta, Met.*, **10**, 705 (1976)
38. Megusar, J. Argon, A. S. and Grant, N. J., ref. 6, p. 329
39. Hadnagy, T. D., Krenitsky, D. J., Ast, D. G. and Chi-Yu Li, *Scripta Met.*, **12**, 45 (1978)
40. Gibeling, J. C. and Nix, W. D., *Scripta Met.*, **13**, 919 (1978)
41. Spaepen, F. *Acta metall.*, **25**, 407 (1977)
42. Spaepen, F., in *'Physics of Defects'*, Eds. Balian, R., Kleman, M., and Poirier, J.-P., Les Houches Lectures XXXV, p. 133, North-Holland, Amsterdam (1981)
43. Argon, A. S., *Acta metall.*, **27**, 47 (1979)
44. Taub, A. I., *Acta metall.*, **28**, 633 (1980)
45. Li, J. C. M., ref. 3, p. 1335
46. Taub, A. I., *Acta metall.*, **30**, 2117 (1982)
47. Jonas, J. J., *Acta metall.*, **17**, 397 (1969)
48. Kocks, U. F., Argon, A. S. and Ashby, M. F., *Prog. mater. Sci.*, **19**, 291 (1975)
49. Gibeling, J. C. and Nix, W. D., *Mater. Sci. Engng*, **45**, 123 (1980)
50. Taub, A. I., unpublished results
51. Batschinski, A., *Z. phys. Chem.*, **84**, 644 (1913)
52. Doolittle, A., *J. appl. Phys.*, **22**, 1471 (1951)
53. Cohen, M. H. and Turnbull, D., *J. chem. Phys.*, **31**, 1164 (1959)
54. Turnbull, D. and Cohen, M. H., *J. chem. Phys.*, **34**, 120 (1961)
55. Turnbull, D. and Cohen, M. H., *J. chem. Phys.*, **52**, 3088 (1970)
56. Adam, G. and Gibbs, J., *J. chem. Phys.*, **43**, 139 (1965)
57. Nachtrieb, N. H., in *'Liquid Metals and Solidification'*, p. 49, American Society of Metals, Cleveland, Ohio (1958)
58. Chen, H. S., Kimerling, L. C., Poate, J. M. and Brown, W. L., *Appl. Phys. Lett.*, **32**, 461 (1978)
59. Rosenblum, M. P., Spaepen, F. and Turnbull, D., *Appl. Phys. Lett.*, **37**, 184 (1980)
60. Greer, A. L., Lin, C. J. and Spaepen, F., ref. 3, p. 567
61. Argon, A. S., *Acta metall.*, **27**, 47 (1979)
62. Roberts, G. J. and Roberts, J. P., *'Proc. 7th Int. Conf. on Glass'*, p. 31 (1965)
63. Lillie, H. R., *J. Am. Ceram. Soc.*, **16**, 619 (1933)
64. Struik, L. C. E., *Pol. Sci. Engng*, **17**, 2 (1977)
65. Egami, T., Maeda, K. and Vitek, V., *Phil. Mag.*, **A41**, 883 (1980)
66. Srolovitz, D., Egami, T. and Vitek, V., ref. 3
67. Greer, A. L. and Spaepen, F., *Ann. N.Y. Acad. Sci.*, **371**, 218 (1981)
68. Sethi, K. V., Gibala, R. and Heuer, A. H., *Scripta Met.*, **12**, 207 (1978)
69. Donovan, P. E. and Stobbs, W. M., *Acta metall.* **29**, 1419 (1981)
70. Neuhäuser, H., *Scripta Met.*, **12**, 471 (1978)
71. Pampillo, C. A., *Scripta Met.*, **6**, 915 (1972)
72. Pratten, N. and Cahn, R. W., personal communication
73. Pampillo, C. A. and Polk, D. E., *Acta metall.*, **22**, 741 (1974)
74. Murata, T., Masumoto, T. and Sakai, M. ref. 6, vol. II, p. 401
75. Gilman, J. J., *J. appl. Phys.*, **44**, 675 (1973)
76. Li, J. C. M., in *'Metallic Glasses'*, Eds. Gilman, J. J. and Leamy, H. J., p. 224, American Society of Metals, Cleveland, Ohio (1976)
77. Chaudhari, P., Levi, A. and Steinhardt, P., *Phys. Rev. Lett.*, **43**, 1517 (1979)
78. Chaudhari, P., Spaepen, F. and Steinhardt, P., in 'Glassy Metals II', Ed. Güntherodt, H.-J., Springer-Verlag, Berlin (1982)
79. Kobayashi, S., Maeda, K. and Takeuchi, S., *Acta metall.*, **28**, 1641 (1980)
80. Vitek, V., Srolovitz, D. and Egami, T., ref. 3, p. 241
81. Argon, A. S. and Kuo, H. Y., ref. 6, p. 269
82. Argon, A. S. and Kuo, H. Y., *Mater. Sci. Engng*, **39**, 101 (1979)
83. Spaepen, F. and Turnbull, D., *Scripta Met.*, **8**, 563 (1974)
84. Steif, P. S., Spaepen, F. and Hutchinson, J. W., *Acta metall.*, **30**, 447 (1982)
85. Saffman, P. G. and Taylor, G. I., *Proc. R. Soc.*, **A245**, 312 (1958)
86. Spaepen, F., *Acta metall.*, **23**, 615 (1975)
87. Argon, A. S. and Salama, M., *Mater. Sci. Engng*, **23**, 219 (1976)
88. Pitts, E. and Greiller, J., *J. Fluid Met.*, **11**, 33 (1961)
89. Mullins, W. W. and Sekerka, R. F., *J. appl. Phys.*, **35**, 444 (1964)
90. Luborsky, F. E. and Walter, J. L., *J. appl. Phys.*, **47**, 3648 (1976)

91. Liebermann, H. H. and Luborsky, F. E., *Acta metall.,* **29,** 1413 (1981)
92. Kimura, H. and Ast, D. G., ref. 3, p. 475
93. Kramer, E. A., Johnson, W. L. and Cline, C., *Appl. Phys. Lett.,* **35,** 815 (1979)
94. Steif, P. S., Thesis, Harvard University (1982)
95. Lee, D., *Metall. Trans.,* **12,** 419 (1981)
96. Anderson, P. M. and Lord, A. E., *J. Non-cryst. Solids,* **37,** 219 (1980)

Chapter 14

Fundamental magnetic properties

R. C. O'Handley

Massachusetts Institute of Technology, Cambridge, USA

14.1 Introduction

A great volume of data exists, in many scattered sources, with much redundancy. In this review an attempt will be made to tie the data together, put it in a coherent framework, and offer a reasonable explanation for some of the interesting trends observed. Obviously, much must be omitted. We will focus on what have been called readily glass-forming systems, namely those fabricable as glasses using quench rates typical of melt spinning ($10^6\,K\,s^{-1}$). We exclude much useful and interesting information on amorphous systems which can be made only by sputtering or electrodeposition. The scope excludes explicit treatment of Mössbauer, FMR, spin-glass, and percolation phenomena in metallic glasses, as well as much of the data on amorphous rare-earth–transition metal (RE–T) thin films with their rich variety of magnetic structures.

Several reviews of fundamental magnetic properties of metallic glasses can be found in the literature. Graham and Egami[1] summarize and comment on most of the data (magnetic moments, Curie temperatures, anisotropy, and magnetostriction) extant to 1977. Luborsky[2] has compared fundamental as well as technical data from a variety of sources. Masumoto's review[3] covers more recent developments, notably the interesting zirconium- and hafnium-containing glasses. Mizoguchi[4] reviews a variety of fundamental magnetic phenomena and Wohlfarth[5] focuses on properties described by itinerant electron theory (*see* Chapter 15). Alben, Budnick and Cargill[6] have made no effort to summarize the data in the literature. They have focused instead on a few fundamental issues of general relevance to magnetism in amorphous metallic alloys. Consequently their review is well worth returning to, despite its earlier date than the others. Kronmüller[7] reviews some fundamental magnetic fluctuation phenomena within the framework of micromagnetics. Chapters in forthcoming books edited by Hasegawa[8] and Güntherodt[9] should be noted. What I shall try to do here is address some of the problems of fundamental significance that have emerged from the data and, hopefully, move toward their resolution.

Symbols and abbreviations frequently used throughout the text are defined here for convenience.

D	= Spin-wave stiffness constant.
DOS	= Density of states.

J	= Phenomenological exchange constant.
K	= Intrinsic magnetic anisotropy, K_u = induced uniaxial K.
M	= a metalloid such as boron, silicon, carbon, or phosphorus.
n_B	= magneton number, i.e. magnetic moment expressed in Bohr magnetons. Usually reserved for alloy with single magnetic species.
RE	= Rare-earth metal.
SRO	= Short-range order.
T	= a transition metal, TE = an early T metal (e.g., chromium, niobium, hafnium), TL = a late T metal (e.g., iron, cobalt, nickel).
T_C	= Curie temperature.
T_x	= Crystallization temperature.
VBS	= Virtual-bound state.
Z	= Coordination number.
ΔZ	= difference in atomic number between two alloy constituents.
λ	= saturation magnetostriction.
σ	= specific magnetization (emu/g).
μ	= mean magnetic moment (in emu/g or Bohr magnetons) of an alloy containing more than one magnetic species.

14.2 Saturation moments and Curie temperatures: compositional dependence

14.2.1 Transition metal variations

The first complete set of published data showing the variation of the average magnetic moment μ and Curie temperature T_C for amorphous alloys as a function of

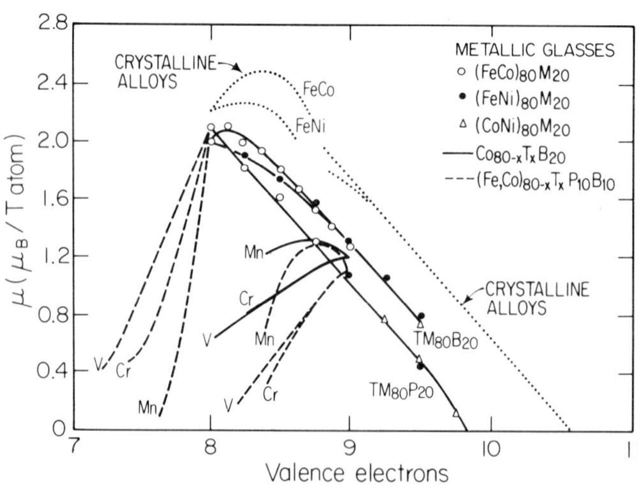

Figure 14.1 Dependence of mean saturation moment per transition metal atom on average transition metal valence electron concentration for several metallic glasses. Solid lines for melt-quenched amorphous ribbons: $(FeCoNi)_{80}M_{20}$ from ref. 12 and $Co_{80-x}TE_xB_{20}$ from ref. 13. Dashed lines for evaporated amorphous thin films $TL_{80-x}TE_xP_{10}B_{10}$: TL = Fe from ref. 10, TL = Co from ref. 4. Moments for $(FeCoNi)_{80}P_{10}B_{10}$ thin films fall just above the line for the $T_{80}P_{20}$ glasses

transition metal content is that of Mizoguchi *et al.*[10]. They plotted μ and T_C as functions of the number of valence electrons for amorphous $T_{80}B_{10}P_{10}$ thin films. To interpret the magnetic moment behaviour, a parallel was drawn with the Slater–Pauling curve for pure transition metal alloys (no metalloids). The difference between the crystalline and amorphous data along the right hand side of the curve (strong ferromagnets) was later described in terms of charge transfer from the metalloid elements to the transition metal *d* band[11](*see* p. 267). The different effects on magnetic moment of boron and phosphorus were clarified when single-metalloid glassy alloys could be fabricated. The magnetic moments of $T_{80}B_{20}$ and $T_{80}P_{20}$ amorphous alloys[12] shown in *Figure 14.1*, were interpreted to suggest a metalloid effect equivalent to a charge transfer of 1.6 and 2.4 electrons per boron and per phosphorus, respectively, to the transition metal *d* band. The effects of transition metals lighter than iron are described on p. 264.

Local moments

Knowledge of the moment associated with each magnetic species in an alloy is important in understanding the metalloid–metal interaction in metallic glasses as well as the local nature of the magnetic moments in transition metal alloys in general. Polarized neutron scattering, so useful in studying crystalline magnets, does not yield atomic moments for amorphous alloys. However, with a few assumptions, it can be shown that the behaviour of the local moments in T–M glasses is probably similar to that observed in comparable crystalline alloys[14–16].

In alloys based on phosphorus, nickel is paramagnetic, i.e. $\mu_{Ni}(0) = 0$ (zero in parentheses means no other transition metal is present in the alloy). The linear increase in μ $[\mu = x\mu_{Fe} + (1 - x)\mu_{Ni}]$ with x in $(Fe_xNi_{1-x})_{80}P_{20}$ (*Figure 14.1*) can be accounted for by assuming that μ_{Ni} remains unchanged (zero) over the linear range of μ. This is justified by noting the alloy's strong ferromagnetic nature and hence the insensitivity of its

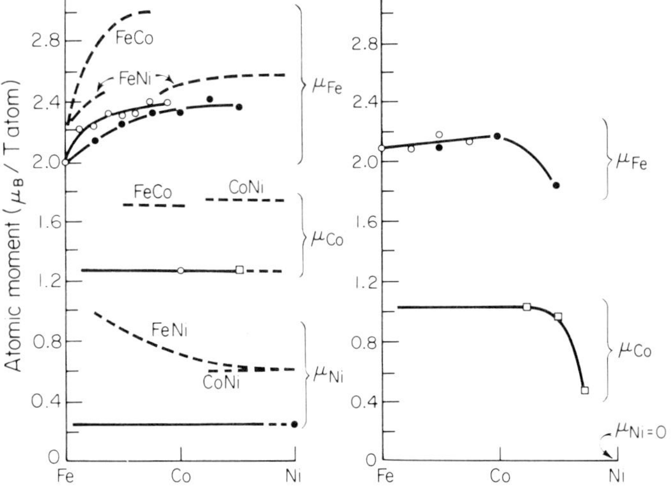

Figure 14.2 Local moments associated with specific transition metal T species for $T_{80}M_{20}$ glasses (solid lines and data points) from ref. 15 by analysis of average moment data (*see* text). Crystalline data from polarized neutron diffraction measurements are from ref. 17

moment to small changes in exchange splitting. Then $\partial\mu/\partial x$ determines μ_{Fe} to be 2.1 μ_B (*Figure 14.2*)[15]. Similarly for $T_{80}B_{20}$ glasses, we assume $\mu_{Ni} = \mu_{Ni}(0)$ which by extrapolation is approximately 0.2 μ_B and, therefore, $\mu_{Fe} = 2.35\,\mu_B$. However, in this $(Fe_xNi_{1-x})_{80}B_{20}$ series, as x approaches 1, μ decreases and, if μ_{Ni} remains constant, $\mu_{Fe}(0) = 2.0\,\mu_B$. The point at which μ_{Fe} begins to drop from its nickel-rich-environment value of 2.35 μ_B is associated with the transition to weak ferromagnetism, i.e. the appearance of iron majority-spin holes. *Figure 14.2* compares the results of such an analysis for these glasses with polarized neutron diffraction results for crystalline transition metal–metal alloys[17]. If one views the similarity of the two sets of data with favour, then self-consistency can be taken to justify the original assumption that effectively begged the question, namely, that the atomic moments of strongly ferromagnetic species are invariant with composition.

The apparent discrepancy between the local moment values reported by different groups is due to the way the metalloids are treated. For example, in amorphous $T_{80}B_{20}$ alloys, $\mu_{Ni} = 0.2\,\mu_B$ applies to the nickel moment *as perturbed by boron*[15] whereas $\mu_{Ni} = 0.6\,\mu_B$ is the nickel moment with an assumed metalloid perturbation removed[14,16].

Curie temperatures

The Curie temperatures for amorphous alloys are generally lower than those of pure transition metal alloys of the same composition. Also there is no evidence of phase transformation effects on T_C in the glassy alloys as observed in crystalline alloys, e.g. at the bcc–fcc transition in crystalline Fe–Ni alloys (*Figure 14.3*).

The variation of T_C with x in the $Fe_{75-x}Ni_xP_{15}C_{10}$ series can be fitted with an expression from CPA (short dashed line, *Figure 14.4*)[18] giving the exchange constants $J_{Fe\text{–}Fe} = 1$, $J_{Fe\text{–}Ni} = 1.2$ and $J_{Ni\text{–}Ni} = 0$. With boron as the only metalloid, the higher

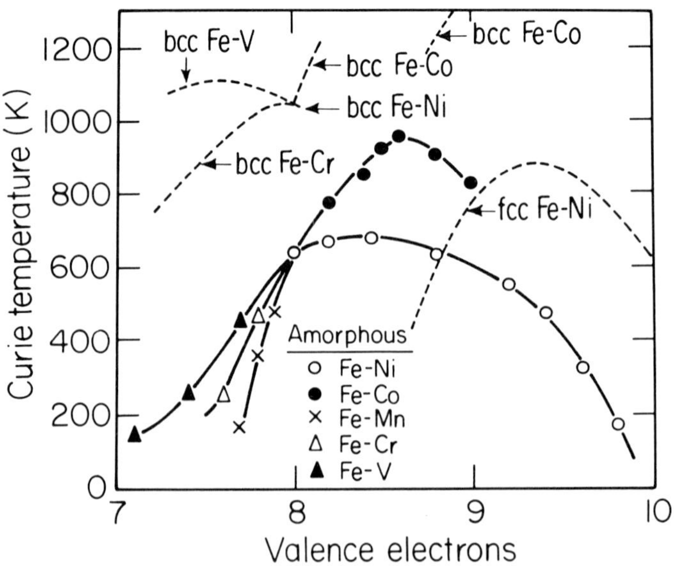

Figure 14.3 Curie temperatures for various $T_{80}P_{10}B_{10}$ amorphous thin films as a function of average transition metal valence electron concentration. Dashed line is data for crystalline alloys (after Mizoguchi *et al.*[10])

Curie temperatures can be fitted with the parameters $J_{Fe\text{—}Fe} = 1$, $J_{Fe\text{—}Ni} = 1.5$, and $J_{Ni\text{—}Ni} = 0.2$ (solid line connecting open datum points in *Figure 14.4*)[15]. This suggests that in the $T_{80}B_{20}$ system nickel is ferromagnetic whereas it is non-magnetic when appreciable phosphorus is present. However, T_C in the $(FeNi)_{80}B_{20}$ system can also be fitted[14] with $J_{Ni\text{—}Ni} < 0$. Luborsky has summarized such results for a number of glassy alloy systems[20]. From *Figure 14.4* it is evident that boron has a much weaker effect[12,15,21] than phosphorus in suppressing T_C.

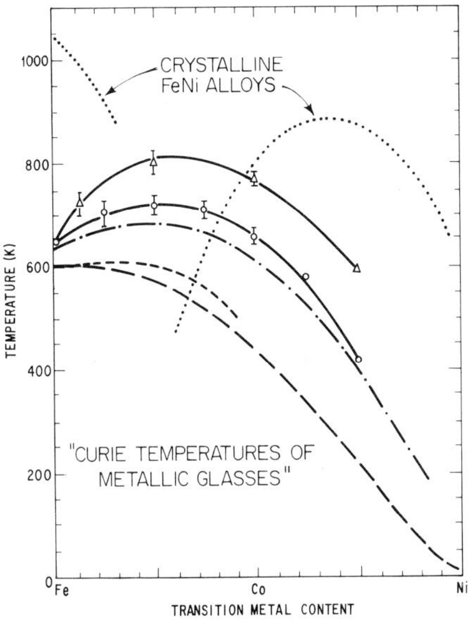

Figure 14.4 Curie temperatures as a function of transition metal content for a variety of metallic glasses with differing metalloid content. $T_{80}B_{20}$ glasses: △ corresponds to T = Fe–Co and Co–Ni and ○ to T = Fe–Ni. Solid lines are a CPA fit to date (ref. 15). Dot-dashed line: amorphous $(FeNi)P_{10}B_{10}$ thin films from Figure 3 of ref. 10. Short dashed line: $(FeNi)_{75}P_{15}C_{10}$ data fit with CPA expression (ref. 18). Long dashed line: $(FeNi)_{75}P_{16}B_6Al_3$ melt-quenched ribbons (ref. 19)

There exists no widely accepted, first-principle theory of Curie temperatures despite recent progress in this area[22]. Perhaps metallic glasses offer a unique opportunity for theoretical comparison because the Curie temperature varies smoothly with composition owing to the absence of structural phase transformations in amorphous materials.

14.2.2 Metalloid variations

The effects of metalloid variations (including carbon and silicon) on magnetic properties of iron-based glasses have been studied[21]. More recent results, including the effects of germanium[23,24], are neatly depicted in *Figure 14.5(a)*. The data suggest that the metalloids can be described as either substitutional (P, Si, Ge) which lower the mass density appreciably, or interstitial (B, C) which lower the density weakly. The members of the interstitial group suppress n_B and T_C more than do those of the substitutional group of the same valence. Within either group of metalloids, the magnetic moment and Curie temperature increase with increasing metalloid radius. The trend observed in T_C is interpreted in terms of the spatial dependence of the exchange interaction and implies weaker ferromagnetic exchange and, ultimately, antiferromagnetic exchange, as iron atoms move closer together.

It is interesting that more recent data[25] on the $Co_{75}B_{25-x}M_x$ glasses show that T_C decreases for all metalloids with $r_M > r_B$, and T_C increases for all metalloids with $r_M < r_B$ [*Figure 14.5(b)*]. This suggests that exchange between cobalt atoms is enhanced at smaller r in contrast to the behaviour of iron atoms.

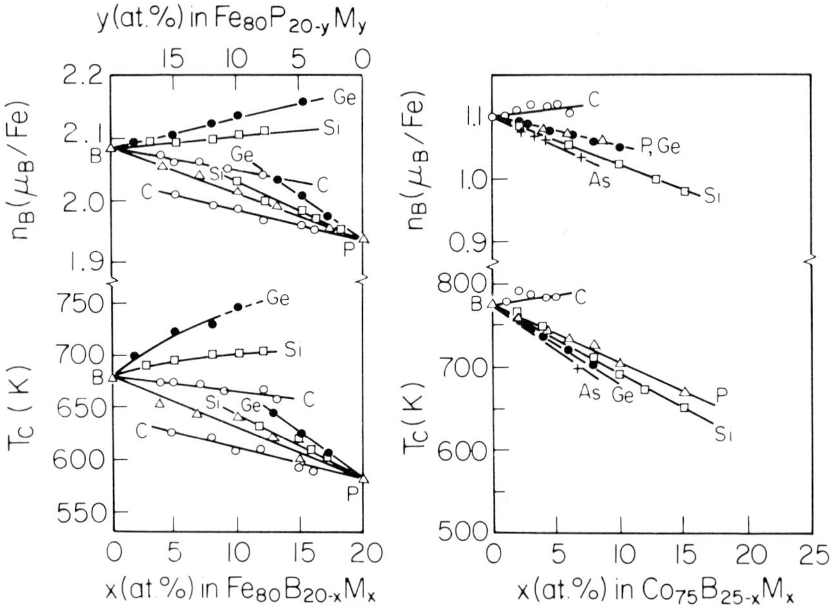

Figure 14.5 Effect of metalloid variations on saturation moment and on Curie temperature of $Fe_{80}M_{20}$ glasses (refs. 23 and 24) and of $Co_{75}B_{25-x}M_x$ glasses (ref. 25)

14.2.3 Varying T/M ratio

Interesting effects are observed when the metal/metalloid ratio is varied. Early studies[21,26–28] were limited to mixed metalloid systems, $Fe_{100-a-x}A_aB_x$. More recent results for the single-metalloid systems are easier to interpret. Some representative data are depicted in *Figures 14.6* and *14.7*.

In $Co_{80-x}M_x$ glasses (M = P, B) the magnetic moment increases toward that of crystalline cobalt as metalloid content decreases. T_C also increases with decreasing x, extrapolating to a large value but one well below that of pure cobalt (*Figure 14.6*). This behaviour is consistent with *Figure 14.5(b)*.

The $Fe_{100-x}M_x$ system (*Figure 14.7*)[21,23,32–36] shows the magnetic moment to increase toward that of α-Fe as metalloid content decreases. However, the observed behaviour of T_C is unexpected; it *decreases* with decreasing metalloid content, extrapolating to a value near or below room temperature at zero metalloid content. Thus, Fe–Fe exchange again appears to be weakened by decreasing the iron atom separation r. This behaviour is discussed below (p. 269).

Chien and Unruh[36] have recently made a series of amorphous FeB thin films over a much wider composition range than could be achieved by liquid quenching. Their results, shown in *Figure 14.8*, show lower T_C's than for liquid quenched glasses, but otherwise confirm earlier data[21,23] of *Figure 14.7*.

Fukamichi *et al.*[34] report that the magnetic moments of FeB glasses increase for boron concentrations down to 14 atom% but then decrease down to the limit of glass formability, 11 atom% (*Figure 14.7*). They interpret this as a transition to weak ferromagnetism as in the Invar alloys. Such a transition would not be expected for this series [assuming weaker exchange splitting (based on decreasing T_C) with decreasing boron content] because $Fe_{80}B_{20}$ is already in a weak magnetic state[12,15] (*see* p. 260 Other authors[23,32,33] see no decrease in moment down to 12 atom% of boron.

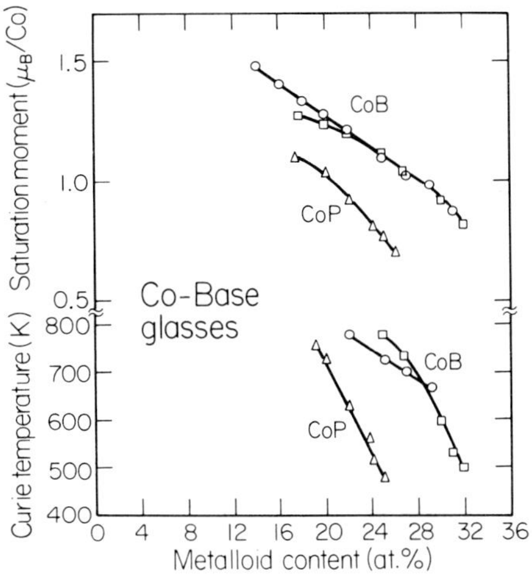

Figure 14.6 Variation of magnetic moment per cobalt atom and T_C for $Co_{100-x}M_x$ glasses as a function of x. M = B, open circles from ref. 29, squares from ref. 23; M = P, triangles from refs. 30 and 31

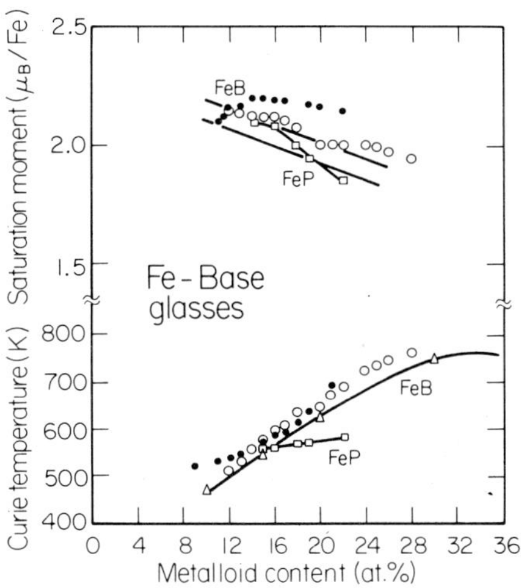

Figure 14.7 Variation of magnetic moment per iron atom and T_C for $Fe_{100-x}M_x$ glasses as a function of x. Moments, M = B, open circles from ref. 33, closed datum points from ref. 34, upper solid line from ref. 32. Moments, M = P, squares from refs. 23 and 34, lower solid line from ref. 32. Curie temperatures: same symbols as for moments with the exception that here some of the data from ref. 36 (triangle and connecting line) are shown instead of those from ref. 32

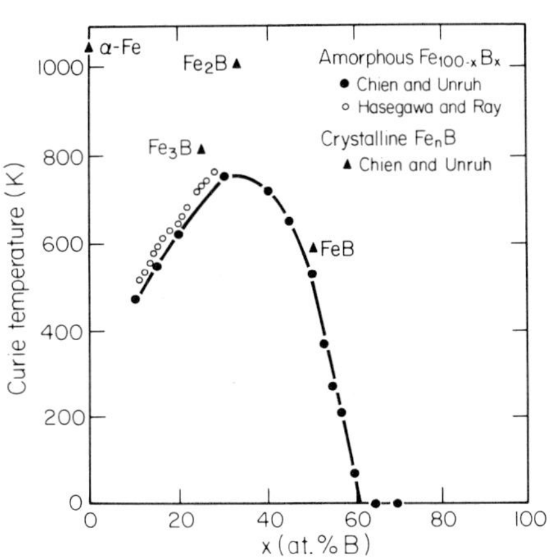

Figure 14.8 Curie temperatures of $Fe_{100-x}B_x$ as a function of x for sputtered amorphous thin films (Chien and Unrue, ref. 36) and melt spun ribbons (open datum points from Hasegawa and Ray, ref. 33). Values of T_C for α–Fe and three crystalline iron boron compounds are also shown for comparison (triangles)

14.2.4 Special alloy systems

Iron–boron–beryllium

In 1981 Hasegawa[37] and Severin *et al.*[38] independently reported the effects of additions of beryllium to FeB glasses. One question of interest is whether the beryllium replaces iron or boron in the amorphous structure. Hasegawa concluded that up to approximately 3 atom %, the beryllium resides in iron sites and beyond that, a transition in moment behaviour suggests that it occupies boron sites. The variation of the magnetic moment and Curie temperature in these systems is generally consistent with that for $Fe_{100-x}M_x$ glasses (beryllium being regarded as a metalloid). That is, increasing beryllium content at the expense of iron lowers n_B and increases T_C. An added feature is that T_C increases initially as beryllium replaces boron in the series $Fe_{80}Be_xB_{20-x}$. Again the suggestion is of stronger Fe–Fe exchange at larger iron separation.

Iron–early transition metal–metalloid

The effects of transition metal additions lighter than iron on the saturation moment[10] are shown in *Figure 14.1*. The sharp decrease in moment with additions of manganese, chromium or vanadium is generally attributed to an antiferromagnetic coupling of the moments of the lighter transition metal species with that of iron, much as is the case in crystalline alloys.

Cobalt–early transition metal–metalloid

More varied effects are observed with the substitution of lighter transition metal species in a cobalt-based glass (*Figure 14.1*). The magnetic moment of $Co_{80-x}Mn_xB_{20}$ glasses increase up to 6 atom % of manganese, then decrease[4,13]. Because it is well established that CoB glasses are strong ferromagnets (minority-spin holes only) this transition at 6 atom % of manganese can be ascribed to the onset of weak ferromagnetism, i.e. the depopulation of majority-spin states.

A detailed model of the density of states of TE–TL systems has been proposed[13,39]. In a cobalt-based glass (TL = Co) with TE = Fe, majority-spin 3*d* states remain below E_F and the moment increases by 1 Bohr magneton per added iron atom

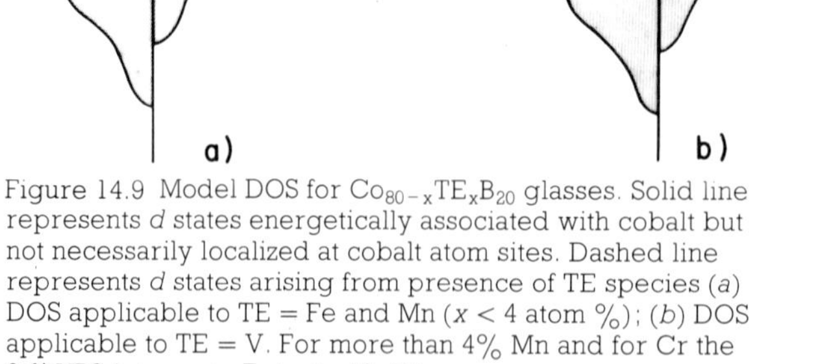

Figure 14.9 Model DOS for $Co_{80-x}TE_xB_{20}$ glasses. Solid line represents *d* states energetically associated with cobalt but not necessarily localized at cobalt atom sites. Dashed line represents *d* states arising from presence of TE species (*a*) DOS applicable to TE = Fe and Mn ($x < 4$ atom %); (*b*) DOS applicable to TE = V. For more than 4% Mn and for Cr the $3d\uparrow$ VBS intersects E_F (refs. 13, 39)

(*Figure 14.9(a)*). The situation is similar for the case TE = Mn up to $x_{Mn} = 4$ atom % but now $\partial\mu/\partial c_{Mn} = 2\,\mu_B$/Mn. Above 4 atom % Mn the majority-spin Mn 3*d* states begin to push above E_F (*Figure 14.9(b)*). Electrons are therefore transferred from spin-up to spin-down states, causing the magnetic moment variation to level off then decrease with further increases in manganese content. For T = Cr or V, the majority-spin 3*d* states intersect E_F and the magnetic moment decreases even at the lowest concentrations (*Figure 14.9(b)*). The magnetic moment variations in these systems have also been explained with the local-environment model[40]. However, this requires several assumptions, perhaps the most severe of which is that the moments of the constituents are independent of composition. Cable and Child[41] have shown this not to be the case in the similarly-behaving crystalline system NiMn.

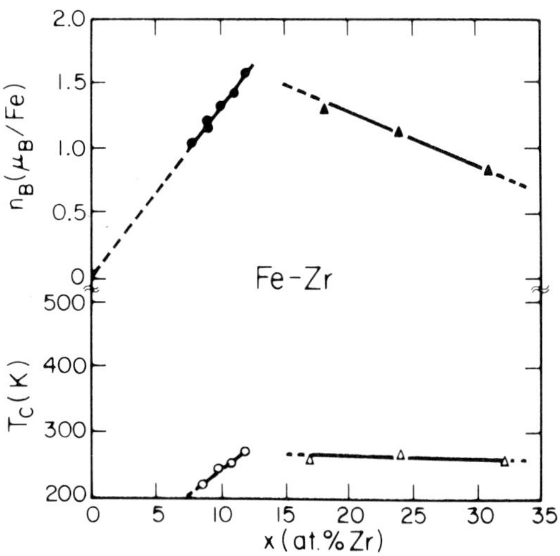

Figure 14.10 Variation with Zr content of T_C and μ per iron atom in FeZr glasses (ref. 46)

Observed trends in magnetostriction are consistent with this model and suggest new features concerning the minority-spin DOS (*see* p. 278).

When TE = niobium, the moment suppression is not the same as for TE = vanadium, even though Friedel's VBS model makes no distinction between TE solutes of the same column. Specifically, $\partial\mu/\partial c_V = -4\mu_B$/V atom in $Co_{80-x}V_xB_{20}$ [13,39] and $\partial\mu/\partial c_{Nb} = -4.5$, -5.5 and $-7\,\mu_B$/Nb atom in $Co_{100-x}Nb_x$[42], $Co_{94-x}Nb_xB_6$[43,44] and $Co_{78-x}Nb_xB_{22}$[45] glasses, respectively. The latter values are closer than the former to the VBS prediction of -6, suggesting that the Nb *d* states are more completely above E_F than are those of vanadium. This is in fact observed in band structure calculations[44] comparing Co_3V and Co_3Nb. Analysis of the data for these glasses is difficult because of the large effect of boron on the magnetic moment: $\partial\mu/\partial c_B = -3.3\,\mu_B$/B atom for $Co_{100-x}B_x$ (*Figure 14.6*).

Early transition element–late transition element

The magnetic moment variation in the $Fe_{100-x}Zr_x$ series[46] shows an increasing moment as *x* decreases to 14 atom %. Below this limit, the moment begins to decrease [*Figure 14.10*]. The authors interpret this in terms of the appearance below $x = 14$% of two magnetic states of iron, Fe-I resembling ferromagnetic, bcc Fe and Fe-II resembling antiferromagnetic, fcc Fe.

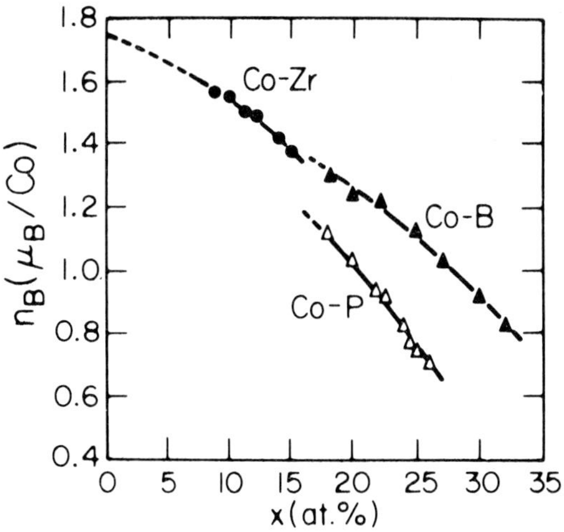

Figure 14.11 Variation with Zr content of μ per cobalt atom for Co–Zr glasses from ref. 46, compared with some data from *Figure 14.6*

The decreasing T_C's observed in Fe–M systems with decreasing M content are also seen in FeZr with decreasing zirconium content, but only below about 14 atom % of zirconium. Thus in the FeZr system, both T_C and n_B increase as x_{Zr} decreases to 14 atom %; below this limit, they both decrease. This parallel behaviour of T_C and n_B is not seen in FeM systems. It is tempting to identify the decrease in moment at low zirconium content in the Fe–Zr system with that observed by Fukamichi in Fe–B for low boron content (*Figure 14.7*)[34].

In $Co_{100-x}Zr_x$, the moment increases as x_{Zr} decreases, extrapolating to approximately 1.7 μ_B at pure Co (*Figure 14.11*)[46]. Curie temperatures are above crystallization in this system.

Very interesting results are now becoming available on a variety of TE–TL thin films[47]. Trends from column to column for TE addition are described to first order by Friedel's theory. That is $\partial\mu/\partial c_{TE} = -(10 + \Delta Z)\mu_B$. However, for different TE substituents from the same column (where Friedel's model makes no distinction), one needs to consider more subtle effects such as the spatial extent of *d* orbitals[47] and hybridization[44].

14.2.5 Size (disorder) effect in fine powders

For the most part, magnetic moments reported by different groups using different melt spinning or splat quenching techniques (presumably with slightly different quench rates) have been in fair agreement for alloys of the same composition (*Figure 14.1*). This agreement even extends to data taken on thin films[4,10] where similar compositions are compared; moment variations are typically less than 5 per cent. The observed Curie temperature differences are generally less than those shown in *Figure 14.8* between melt-spun and sputtered amorphous FeB ($\Delta T_C < 40$ K and $\Delta T_C/T_C < 6\%$).

It is interesting to note, therefore, that significant property variations in $Fe_{75}Si_{15}B_{10}$ spark-eroded powders of different sizes are linked, through their different quench rates, to varying degrees of chemical disorder[48]. The magnetic moments and Curie temperatures of the finer particles (produced at higher quench rates and, therefore, possessing greater chemical disorder) are lower than those of melt-spun ribbons by as much as 20 per cent (*Figure 14.12*). X-Ray diffraction and Mössbauer

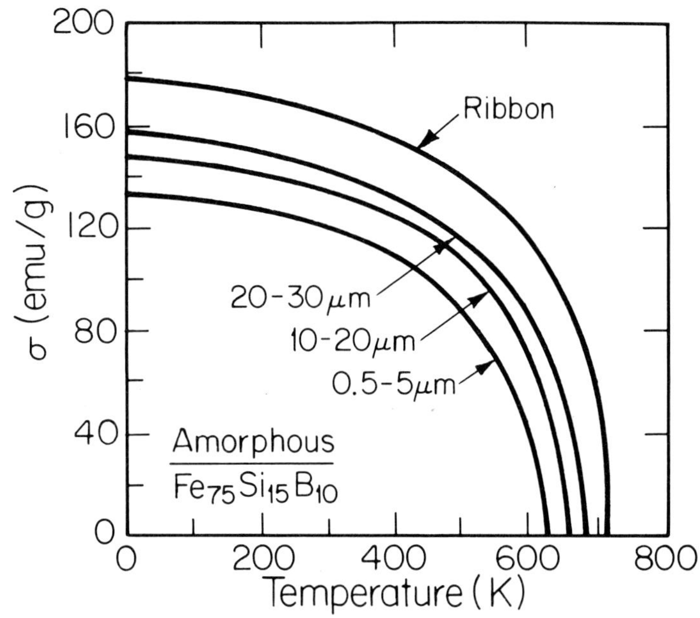

Figure 14.12 Magnetization as a function of temperature for melt-quenched ribbon and spark-eroded powders of amorphous $Fe_{75}Si_{15}B_{10}$ (after Berkowitz *et al.*[48])

spectra were used to verify the absence of any difference in the amount of chemical contamination among the samples of different sizes. The Mössbauer spectra indicate broader hyperfine field distributions nn the finer powders supporting the model of increased chemical disorder in the more rapidly quenched particles. More research in this interesting field is expected as fine powders are produced by a greater variety of rapid solidification processes (*see* Chapter 26).

14.3 Discussion

14.3.1 Charge transfer or bonding?

Problem

The moment variations reported by Mizoguchi *et al.*[10] (p. 258) were later interpreted in terms of charge transfer from the metalloid atoms ($P_{10}B_{10}$) to the transition metal *d* band[11]. It was assumed that boron and phosphorus transfer 1 and 3 electrons, respectively, to the *d*-band. The different roles of these two metalloids were clarified with measurements on single-metalloid systems[12]. These data suggested that 1.6 and 2.4 electrons per boron atom and phosphorus atom, respectively, transfer to the *d* band.

The use of the charge transfer concept resurrected an old controversy that had been actively studied in a variety of crystalline alloys, most notably transition metal borides (*see* footnotes 26 and 31 in reference 15). At issue is not only the magnitude of the charge transfer required to fit the data but also the direction in which charge would be expected to flow based on electronegativity and energy level considerations.

Alben *et al.*[6] came out strongly against the use of the rigid-band model and the substantial charge transfer thereby required to explain the suppressed magnetic moments of metallic glasses. They suggested instead that the moment suppression is a result of a loss of *d* character by the magnetic states because of their participation in bonding with the metalloid species (*p–d* hybridization).

The *p–d* bonding model was developed further by Allen *et al.*[49] who offered thermodynamic evidence for the *p–d* hybridization which stabilizes eutectic glasses. They cite neutron scattering studies of the magnetic form factor of manganese pnictides which suggest that as many as two *d* electrons are lost to non-polarized, delocalized *p–d* bonding and antibonding states. The situation should be similar in metallic glasses, thus obviating the need to invoke charge transfers of ionic magnitude.

Negligible net charge transfer

Watson and Bennett[50] note that the Fermi level drops (increases) by as much as 14 eV upon removal (addition) of one *d* electron from a transition metal. (They compare d^n with d^{n-1} or d^{n+1} configuration energies.) Conversely, to bring E_F to a common value in a TE–TL alloy (where Fermi levels vary by 1 eV or less from column to column) should require charge transfers less than 0.1 e/atom. However, the situation is not quite this simple.

In summarizing the extensive theoretical work on charge transfer in alloys, Watson and Bennett formulate a model that is consistent with the most direct experimental evidence available, viz. Mössbauer isomer shifts. In simplest terms, three steps are involved. First hybridization (*d–d* in their examples) occurs between compatible valence levels drawing to the TE site some *d* states formerly localized at TL sites. Secondly, these new, low-lying *d–d* hybrid states dictate the extent and direction of *d*-electron transfer Δn_d. (The new TE *d* states are populated, adding to the *d*-like charge at the TE sites.) Finally, the conduction electrons are redistributed, Δn_c, to screen the *d* transfer effects. It is found that $\Delta n_c/\Delta n_d < -1.0$, i.e. the conduction electrons move in the direction opposite to that of the *d* electrons (because of the more attractive core potential the *d* electrons leave behind), overcompensating for Δn_d. Thus, the *net* charge transfer is small and it is in the direction of conduction electron transfer. Δn_c is consistent with the relative electronegativities even though the *d* charge moves in the direction opposite to that expected based on relative electronegativities.

A similar picture is expected to apply to transition metal–metalloid alloys although detailed experimental and theoretical results are more scarce than for TE–TL alloys. *d* charge is expected to flow from the T sites into the T–M bonding hybrid states which are not easily polarized. It is the loss of magnetic *d* character at T (and not the amorphous state *per se*) that accounts for the reduced moments and Curie temperatures in glassy alloys[51].

Theoretical support for bonding

The concept that transition metal–metalloid bonding reduces the magnetic moment by tying up *d* electrons in weakly polarizable *p–d* bonds received quantitative confirmation from the cluster calculations of Messmer[52]. By comparison of eigenstates for a tetrahedral Ni_2Fe_2 cluster with those of a similar cluster containing a boron atom at its centre (*Figure 14.13*), it was shown that the *d* band becomes broadened (owing to hybridization with the B 2*p* orbital) and that the number of effective ferromagnetic *d* electrons and the moment arising from their net imbalance are reduced. Furthermore, any measureable net charge transfer (seen more easily with phosphorus than with boron as the metalloid) is found to be in the direction of the metalloid site.

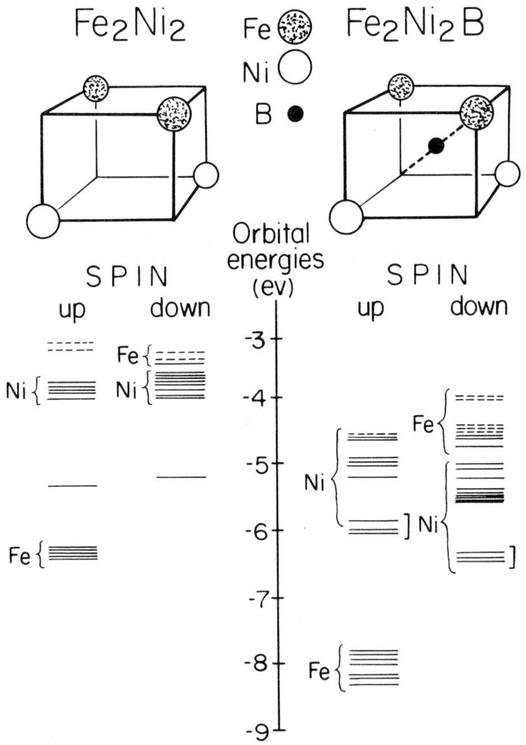

Figure 14.13 Energy levels calculated for the clusters Fe_2Ni_2 (left panel) and Fe_2Ni_2B (right panel). Occupied orbitals are represented by solid lines and unoccupied by dashed lines. *d* orbitals localized at particular transition metal sites are so identified. Square brackets adjacent to certain levels in the right panel indicate *p–d* bonding orbitals stabilized by hybridization which are largely responsible for formation of metastable glassy state and for reduction in polarizable *d* character (after Messmer[52])

The consistency of the shifts in moment data for the boron-containing glassy alloys (*Figure 14.1*) with published results for crystalline TB and T_2B alloys (Cadville and Daniel[53]) has frequently been cited in support of the concept of charge transfer[5,12]. We see it now rather as an indicator of the extent to which boron forms *p–d* bonds with transition metals. This hybridization increases quite linearly with boron content whether the alloys are crystalline or amorphous.

14.3.2 Curie temperature of amorphous iron

Spatial dependence of exchange

The decrease in T_C as $Fe_{100-x}M_x$ glasses approach the amorphous Fe limit ($x = 0$, *Figures 14.7* and *14.8*) is interesting. Chen[26] attributes this behaviour to a decrease in Fe–Fe spacing with decreasing boron content. This leads to lower T_C's according to the Bethe–Slater curve if iron lies in the region of positive $\partial J(r)/\partial r$. This model, though satisfying the data at hand, may be criticized because it neglects the oscillations in $J(r)$ as well as the contribution of second nearest neighbours, which may be significant.

Support for this model is found in *Figure 14.5*. With the exception of phosphorus, large metalloid atoms increase T_C of FeM alloys [*Figure 14.5(a)*] whereas large M atoms decrease T_C of CoM alloys [*Figure 14.5(b)*]. The behaviour of the Fe–B–Be glasses is also consistent with this model.

Coordination numbers

Hasegawa and Ray[33] interpret their T_C data in terms of a mean field theory to suggest that the Fe–Fe coordination decreases from approximately 12 to approximately 7 as boron content decreases from 28 to 12 atom %. (This suggests that over this composition range, amorphous FeB is going from a densely-packed to loosely-packed random structure.) These authors arrive at a similar conclusion from their density measurements on these alloys. In the low-metalloid region, Hasegawa and Ray[54] find that the measured density for the glassy $Fe_{88}B_{12}$ is 7.45 gm/cm^3, only 1 per cent below that of the super-saturated bcc solid solution of the same composition, 7.52 gm/cm^3. They take this to suggest a local atomic arrangement in this low-metalloid glass similar to that in the bcc structure. Assuming a bcc-like SRO for the 88–12 glass leads to the contradiction that a glass with bcc-like local order ($Fe_{88}B_{12}$) exhibits a lower T_C than those with densely-packed, fcc-like local order (e.g. $Fe_{80}B_{20}$). It also leaves unexplained the very large difference between the Curie temperatures of the two phases of $Fe_{88}B_{12}$, 509 K (glassy) and 878 K (crystalline), despite their similar magnetic moments.

Resolution

The observed density of the 88–12 glass could also be explained by a change in SRO from the bcc-like structure of the metastable solid solution, toward one like that of low-T_C, fcc iron with $Z = 12$ and an excess free volume. The observed variation in T_C with x_B would then be linked more to $J(r)$ than to Z and, according to the Bethe–Slater interpretation, would be expected to decrease with decreasing x_B. Chien *et al.*[35] favour a decreasing exchange interaction $J(r)$ rather than a decreasing coordination number Z to explain their data (*Figure 14.8*) at low boron content. Again they point to the weak exchange in γ-Fe.

It is difficult to get independent structural information on the effect of T/M ratio on mean atom spacing or coordination number to clarify the situation. The first peak of the RDF of $Fe_{100-x}B_x$ is observed to shift toward smaller r as x decreases[55a,55b]. This is consistent with Chen's model. Two methods of integrating the first RDF peak have been used to get T–T coordination numbers for the FeB glasses[55a]. This coordination either remains constant or decreases for decreasing metalloid content below approximately 20 atom % of boron depending on how the first peak is defined. Results of Waseda and Chen[56] are not inconsistent with this. This tentative result is not in conflict with the conclusions of Hasegawa and Ray.

It seems likely that both Chen's model, based on the r dependence of the exchange interaction, and the model of Hasegawa and Ray, which emphasizes the coordination number of the magnetic species, are incomplete. The true determinant of T_C is the short-range order[51] which includes both of these ingredients although $J(r)$ alone seems to be able to account for more data than Z alone. For glasses near the stoichiometric composition $Fe_{75}B_{25}$ ($T_C = 735$ K) the SRO is closest to that of Fe_3B ($T_C = 820$ K). At lower boron content the SRO is frustrated from approaching that of Fe_3B and is more random in its dense packing. A lower T_C results because the local order approaches that of γ-Fe rather than that of either Fe_3B or α-Fe. Recent interpretation of structural relaxation in FeB glasses as measured by changes in T_C supports a model in which the extent of Fe_3B-like SRO is responsible for the magnitude of the Curie temperature[57].

A two-site model analogous to that invoked to explain the Invar anomalies as γ-FeNi is enriched with iron, has been proposed to interpret the low T_C of amorphous iron[27]. This model can be thought of as belonging to the category above in which the magnetic properties are dictated by SRO. However, I see no need, nor any evidence, for a dual or phase-separated local atomic arrangement in these glasses.

14.4 Temperature dependence of magnetization

14.4.1 Low-temperature spin waves

An excellent treatment of the problem of magnons in amorphous metals can be found in the review of Alben *et al.*[6].

Early measurements of low temperature $T^{3/2}$ demagnetization behaviour[30,58], $\sigma(T)/\sigma(0) = (1 - BT^{3/2} + \ldots)$, suggested the existence of spin-wave excitations in amorphous alloys. The coefficient B was found to be larger in the amorphous alloys than in comparable crystalline materials[59]. This is understood to be a consequence of the chemical disorder (owing to the presence of metalloids). Inelastic neutron scattering[60] confirmed the existence of well defined spin-wave excitations[61] obeying a quadratic dispersion law $E = E_0 + Dq^2 + \ldots\ldots$

In many amorphous systems, it is observed that D determined from neutron scattering measurements is larger than that calculated from the demagnetization coefficient B. This implies that the neutron measurements do not detect some low-q spin excitations that contribute to the statically measured demagnetization. Although the measurements of Birgeneau *et al.*[62] show spin waves that accounted fully for the demagnetization without need to invoke other low energy excitations, more recent measurements on the $Fe_{100-x}B_x$ glasses[63,64] continue to show that discrepancy.

Luborsky *et al.*[65] show clearly how D scales linearly with T_C for a given series of alloys. Three alloy familes, FeBSi, FeNiPB, and CoBSi, show similar slopes but different intercepts. Although D/T_C is known to be related to the range of the exchange interaction, it is still not clear how to intepret these data other than to make the general statement that the range of the exchange interaction in disordered alloys appears to be attenuated relative to that in ordered alloys.

14.4.2 Corresponding states curve

The generally flatter magnetization curves obtained by plotting $\sigma(T)/\sigma(0)$ against T/T_C (corresponding states curves) of amorphous metallic alloys have been attributed to exchange fluctuations[66]. Tsuei and Lillienthal[67] made a careful Mössbauer study of the temperature dependence of magnetization in amorphous $Fe_{75}P_{15}C_{10}$ and compared the reduced magnetization $\sigma(T/T_C)/\sigma(0)$ of the glass with that of Fe_3P and α-Fe. They attributed the flatter curve of the amorphous material to exchange fluctuations ($\delta = 0.3$ in Handrich's model) arising from the local disorder. They pointed out that the absence of long-range order seems to have no effect on the fundamental magnetic properties.

Alben *et al.*[6] point out that the same flattening effect is observed in disordered crystalline alloys[68]. Thus the exchange fluctuations of Handrich's model may be due to local chemical disorder (in Tsuei's case due possibly to the mixed metalloid phosphorus and carbon) rather than to local structural disorder.

14.5 Anisotropy and magnetostriction

14.5.1 Anisotropy

Alben *et al.*[6] present a simple, quantitative and illuminating argument concerning the range L over which the magnetization is able to follow the local crystal-field magnetic anisotropy which fluctuates over a characteristic distance d in an amorphous material. d is typically of the order 10 Ångströms and it is expected that $L > d$. The competing interactions in determining the orientation of the local magnetization are its exchange interaction J with its neighbouring moments and its coupling D to the local crystal field anisotropy. For $D/J > 1$ (the case for most RE–T alloys) L approaches d, i.e. the magnetization tends to follow the local anisotropy and interesting dispersed magnetic structures (speromagnetic, sperimagnetic, asperomagnetic) are observed[69]. For $D/J < 1$ (the case for most T–M alloys), $L \gg d$, i.e. the magnetization tends to follow a fixed direction over a range much greater than the distance over which the local anisotropy direction changes. Thus, on a macroscopic scale, the local anisotropy is averaged out; $\langle K \rangle = 0$. The magnetization direction in the absence of an applied magnetic field is determined by stress (as long as magnetostriction is non-zero), exchange, and magnetostatic energies. The importance of internal stress is evident from pictures of magnetic domain structures in amorphous materials[70].

Temperature dependence

The weak macroscopic magnetic anisotropy in amorphous transition metal–metalloid alloys is largely responsible for their attractive soft magnetic properties (*see* Chapter 16). This weak anisotropy has also made fundamental investigations of the origin of the anisotropy in these materials difficult. Several authors have measured the temperature dependence of the intrinsic macroscopic anisotropy of metallic glasses and attempted to express it as a power of the magnetization. Some observers find a σ^2 dependence for the intrinsic magnetic anisotropy[71,72] while others[73] find a $\sigma^{2.6}$ dependence. Further analysis by Takahashi and coworkers[74] has shown that the anisotropy of amorphous $Fe_{80}P_{13}C_7$ results largely from a small volume fraction (0.1 to 0.3 per cent) of tetragonal Fe_3P crystal nuclei. Shape anisotropy is responsible for their observed σ^2 dependence for $K(T)$. They rule out magnetoelastic and quenched-in pair-ordering mechanisms for the anisotropy in their samples.

Field-induced anisotropy

The susceptibility of amorphous magnetic materials to field annealing was demonstrated by Berry and Pritchet[75] and graphically illustrated in the scanning electron micrographs of the Bell group[76]. The anisotropy field induced in $Fe_{80-x}Ni_xM_{20}$ glassy alloys shows a maximum near $x = 40$ suggesting that the origin of this anisotropy is mainly Fe–Ni pair ordering[77]. A more general mechanism, allowing for metal–metalloid ordering (in order to account for the induced anisotropy observed in FeB glasses) has been proposed[78,79]. The temperature dependence of this induced anisotropy has been measured to be like[73,79] σ^2 and $\sigma^{3.7}$. An exponent of 2 would be expected for pair-ordering anisotropy.

The order of magnitude of the field-induced anisotropy in metallic glasses is comparable with that observed in crystalline alloys[1]. Furthermore, the magnitude of the anisotropy induced in FeB glasses increases as the composition approaches that of an ordered crystalline phase[80]. This observation suggests that field-induced anisotropy may be the result of (1) the development in the glass of a more ordered local atomic arrangement as during relaxation[57] (Chapter 11), but now with a preferred orientation to that local order, or (2) the growth (again with a preferred orientation) of microcrystal nuclei present in the glass[74] in greater concentrations closer to the composition of the ordered phase.

14.5.2 Magnetostriction

Magnetoelastic effects in metallic glasses have been reviewed (*a*) from the point of view of the variety of isotropic and anisotropic effects on sample dimensions and elastic constants[81], (*b*) with more emphasis on forced and spontaneous volume magnetostriction[82], and (*c*) most recently, in a more mathematical treatment[83]. Here I treat only the anisotropic saturation magnetostriction.

Compositional dependence

The compositional dependence of magnetostriction in $(FeCoNi)_{80}B_{20}$ glasses is shown in *Figures 14.14* and *14.15*. In the iron-rich glasses there is little resemblance to the behaviour of comparable crystalline alloys. In the nickel-rich glasses there appears to be a trend toward zero magnetostriction but at a composition richer in nickel than occurs for crystalline Fe–Ni. In cobalt-rich glasses, the magnetostriction closely follows that of the crystalline alloys, passing from positive to negative values as cobalt content exceeds approximately 93 per cent of transition metal content, regardless of metalloid type[84,85]. The greater disparity between the magnetostrictions of crystalline and amorphous phases of iron-rich alloys is probably related to the greater difference in SRO between these two phases than between those of cobalt- or nickel-rich alloys. This difference in SRO is also reflected in T_C (p. 269).

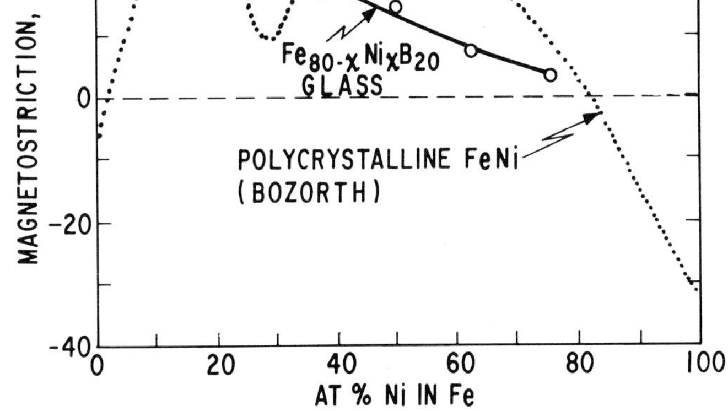

Figure 14.14 Compositional dependence of magnetostriction in Fe–Ni–B glasses $Fe_{80-x}Ni_xB_{20}$ (open datum points and full line) and Fe–Ni polycrystalline alloys (Bozorth) (dotted line) (after O'Handley[84,85])

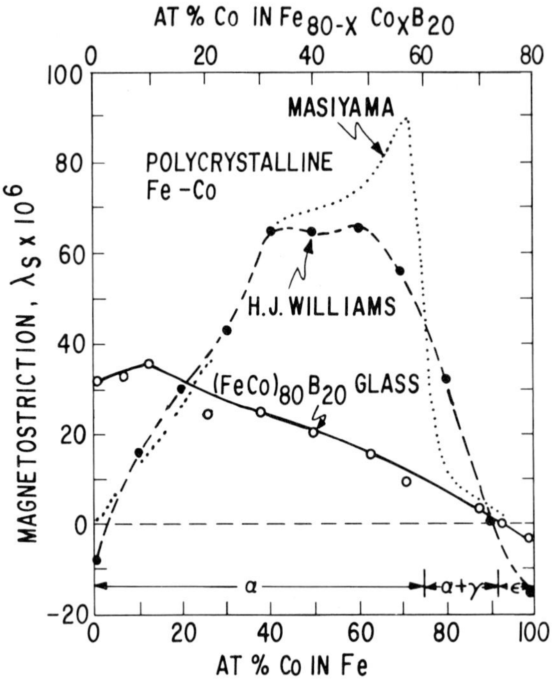

Figure 14.15 Compositional dependence of magnetostriction in Fe–Co–B glasses $Fe_{80-x}Co_xB_{20}$ (open datum points and full line) and in polycrystalline Fe–Co alloys (dotted line from Masiyama and dashed line from H. J. Williams) (after O'Handley[84,85])

It has been noted that at room temperature, the magnetostriction varies with composition as the square of the magnetization, except in cobalt-rich glasses[85]. This should not be construed to imply that a two-ion mechanism is responsible for the magnetoelastic effects because[86] in Fe–Ni based glasses room temperature exceeds $0.6T_C$, the limit above which even single-ion magnetostriction varies as σ^2.

The ternary diagram (*Figure 14.16*) showing magnetostriction for $(FeCoNi)_{80}B_{20}$ glasses[85] fairly well describes the results of studies in various other (FeCoNi)M glasses[19,82,87]. This is because the magnetostriction is not strongly affected by either metalloid type[84,88] or T/M ratio[89]. A large part of the variation that is observed for different metalloids or various T/M ratios can be ascribed either to changes in elastic constants (mechanically softer materials show more strain in response to the same magnetoelastic stress)[90,91] or to changes in saturation magnetization ($\lambda \propto \sigma^2$).

Magnetostriction arises from the strain derivative of the magnetic anisotropy energy and hence reflects the same atomic-scale mechanisms that give rise to magnetic anisotropy. The magnitude of λ (unlike that of the magnetic anisotropy) is not significantly reduced in metallic glasses relative to its values in related crystalline alloys (*Figures 14.14* and *14.15*). It may, therefore, offer an alternative to direct measurements of anisotropy when one is interested in the atomic-scale origin of magnetic anisotropy. However, as noted above, the σ^2 compositional dependence of λ is not significant in this regard. The temperature dependence of λ must be studied in order to probe the microscopic origins of anisotropic magnetic interactions.

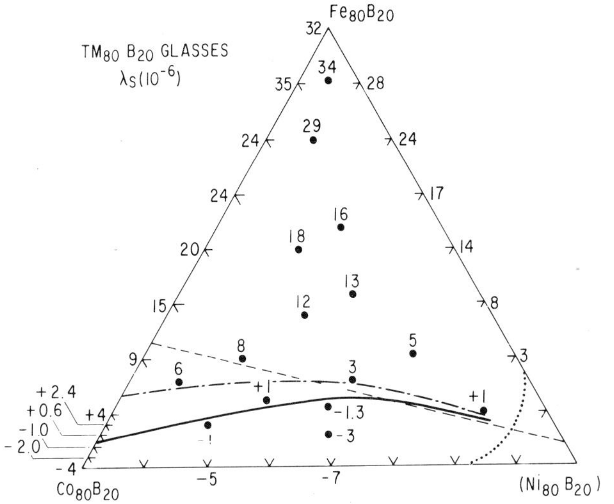

Figure 14.16 Compositional dependence of room temperature magnetostriction ($\lambda_S/10^{-6}$) in (Fe, Co, Ni)$_{80}$B$_{20}$ glasses. Dotted curve delineates a region in nickel-rich corner where $TM_{80}B_{20}$ glasses are difficult to fabricate. Solid line shows observed variation of $\lambda = 0$ glasses ($\lambda > 0$ above this line) (after O'Handley[86])

Temperature dependence

Determination of the temperature dependence of λ can, in many cases, identify the origin of the local magnetic anisotropy as being either single-ion, $\lambda^{\mathrm{I}} \propto \langle (S_i^z)^2 \rangle$, or two-ion $\lambda^{\mathrm{II}} \propto \langle S_i^z S_j^z \rangle$, and of a given symmetry[92].

The temperature dependence of several $TL_{80}B_{20}$ glasses have been measured and shown to follow well the predictions of Callen and Callen's theory for positive single-ion anisotropy of uniaxial symmetry[86,90,93] (*Figure 14.17*). The most significant deviation from one-ion behaviour was observed in glasses containing cobalt. The

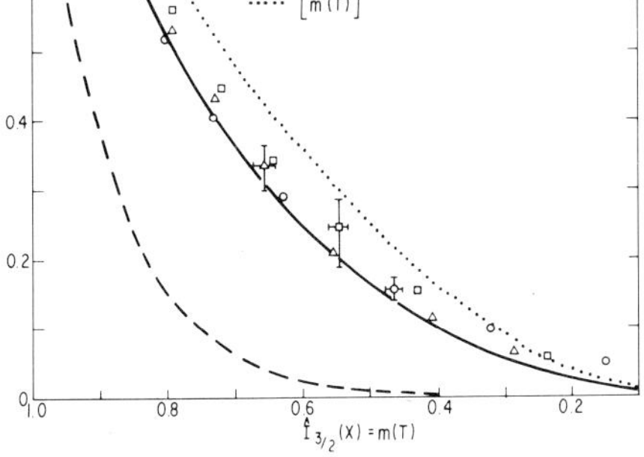

Figure 14.17 Variations of reduced magnetization $\lambda(T)/\lambda(0)$ with temperature expressed in terms of reduced magnetization $m = \sigma(T)/\sigma(0)$ for three representative metallic glasses. Theoretical expressions: solid line, uniaxial one-ion anisotropy $\hat{I}_{5/2}(X)$; dashed line, cubic one-ion anisotropy $\hat{I}_{9/2}(X)$; dotted line, two-ion anisotropy $[m(T)]^2$. ○, $Fe_{80}B_{20}$ glass; △, $Ni_{50}Fe_{30}B_{20}$ glass; □, $Co_{40}Ni_{40}B_{20}$ glass (after O'Handley[86])

deviation is such as to suggest the presence of a significant two-ion component in cobalt-rich glasses. When the temperature dependence of the cobalt-rich glasses was examined in more detail, sign changes were observed[13,86] which suggest that the single-ion and two-ion components were of opposite sign and competing with each other with varying outcomes depending on temperature (*Figure 14.18*). The relative magnitudes of

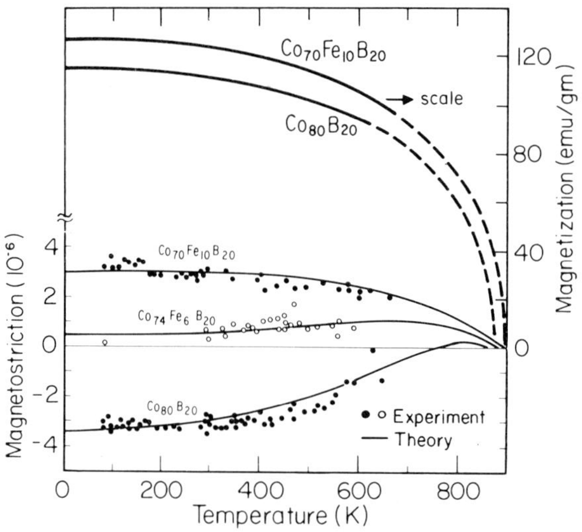

Figure 14.18 Temperature dependence of magnetization and magnetostriction in some cobalt-rich amorphous alloys. The magnetization data (solid lines) are extrapolated above T_x (dashed line) with a best fit to a Brillouin function. The magnetostriction data (dots) are accompanied by theoretical curves fitted to the $Co_{80}B_{20}$ data below T_x. The theoretical curves combine one-ion and two-ion contributions of opposite sign which are responsible for the magnetostriction compensation temperature suggested by the $Co_{80}B_{20}$ data (after O'Handley[86])

λ^{I} and λ^{II} assigned to the cobalt and iron species on the basis of the temperature dependence of λ in $Co_{80}B_{20}$ and $Fe_{80}B_{20}$ describe very well the different temperature dependences of other cobalt-rich compositions when combined proportionately (theoretical curves of *Figure 14.18*)[86]. In summary, we can write for iron-rich (and possibly nickel-rich) glasses:

$$\lambda^{I} > 0, \lambda^{II} \approx 0$$

whereas for cobalt-rich glasses;

$$\lambda^{I} < 0 < \lambda^{II}$$

It is interesting to note that the temperature at which λ goes positive in $Co_{80}B_{20}$ glass (*Figure 14.18*) is close to that for which λ goes positive in polycrystalline cobalt and that this temperature is approximately that[84] of the phase transition $\varepsilon \rightarrow \gamma$. Thus the temperature and compositional dependences of λ in a cobalt-rich glass track the temperature and compositional variations observed for λ in polycrystalline cobalt and CoFe as the order changes from hcp to fcc. This suggests that the mechanism that drives the transition from the hcp phase in crystalline cobalt (either by the addition of iron or by an increase in temperature) also operates in cobalt-rich glasses to produce a change in SRO.

Magnetostriction zeros and the split-band model

Figure 14.16 shows on a ternary diagram the compositional dependence of magnetostriction (295 K) in $(Fe–Co–Ni)_{80}B_{20}$ glasses[86]. Measurements of $\lambda_s(T)$ down to 80 K (e.g. *Figure 14.18*) indicate that the course of the $\lambda_s = 0$ line, shown by the full curve in *Figure 14.16*, is essentially unchanged below room temperature (*see Figure 14.18*). The compositions in the nickel-rich corner beyond the dotted curve were not able to be fabricated by the techniques then available. Extrapolation of observed magnetostrictions to the $Ni_{80-x}Fe_xB_{20}$ limit suggests that $\lambda_s = 0$ would be observed for $6 < x < 12$ atom %. Apparently the presence of boron in the metallic glass shifts the non-magnetostrictive compositions toward the nickel corner from 18 per cent Fe in crystalline Fe–Ni. The observed behaviour of magnetostriction in *Figure 14.16* is much simpler than that reported for Fe–Co–Ni crystalline alloys[94].

The split-band model is a convenient, pictorial representation of the density of states (DOS) in certain alloy systems where the atomic number or valence electron concentration of the constituents differ from each other by at least two[95,96]. The parameters needed to understand the conditions under which the *d* bands of an alloy can be considered to be split on an energy scale ($\varepsilon_A - \varepsilon_B > 2w$ for $\Delta Z > 2$) are shown in *Figure 14.19* (which is essentially a simplification of the state densities shown in *Figure 14.9* for the split-band $(CoTE)_{80}B_{20}$ system). A zero in magnetostriction should occur when the Fermi energy coincides with the boundary between the split bands. The magnetostriction can, therefore, vanish while the magnetic moment remains non-zero.

The $\lambda_s = 0$ line predicted by the split-band model and shown as the dashed line in *Figure 14.16* describes well the slope and intercept of the observed $\lambda_s = 0$ line on the Fe–Ni side of the diagram where $\Delta Z = 2$. The observed shift of the non-magnetostrictive composition from 18 per cent Fe in crystalline Fe–Ni to approximately 7 per cent Fe in the non-crystalline alloys is well described by the split-band model only when the bonding effect of the metalloid boron is taken into account, and done so consistently with the consequent effect on the magnetic moment[86,96]. Increasing the metalloid content increases the shift toward the nickel-rich corner[89]. The observed shift is approximately half the magnitude of that which would be predicted by a rigid-band charge transfer argument.

The observed zero-magnetostriction line departs from the split-band prediction in cobalt-rich glasses because the model is not applicable there owing to the appreciable energy overlap of Fe and Co 3*d* bands ($\Delta Z = 1$).

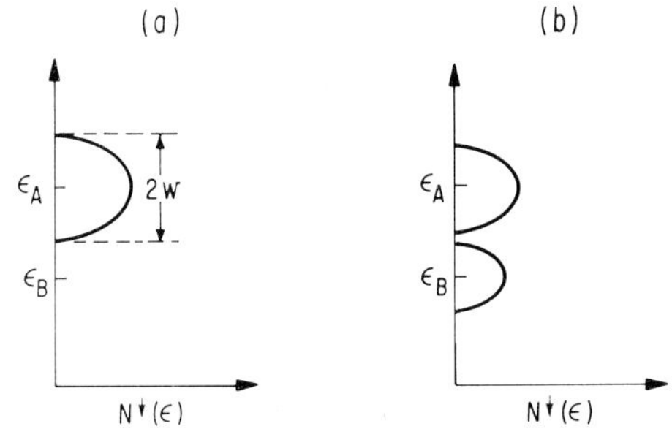

Figure 14.19 Schematic minority-spin *d*-state densities for split-band alloy A_xB_{1-x}: (*a*) $x = 1$, (*b*) $x = 0.5$. Bandwidth for $x = 0$ or 1 is $2w$; ε_A and ε_B give *d*-band centres of gravity (*see* O'Handley and Berger[96] for sources)

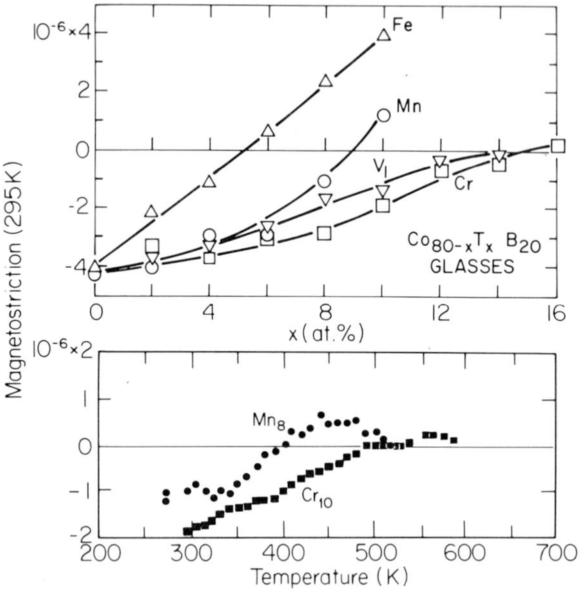

Figure 14.20 Top panel: compositional dependence of room temperature magnetostriction in $Co_{80-x}T_xB_{20}$ glasses. Bottom panel: temperature dependence of magnetostriction for two of the compositions showing magnetostriction compensation at elevated temperatures (after O'Handley and Sullivan[13])

The split-band model has been tested in the cobalt-rich regime by designing appropriate split-band alloys $Co_{80-x}TE_xB_{20}$ (TE = Mn, Cr, V) with $\Delta Z = 2$, 3, and 4 respectively (*Figure 14.20*). The sequence, relative spacings and absolute compositions at which $\lambda = 0$ are predicted by the split-band model with modest success[13,39]. The temperature compensation in the magnetostriction of cobalt-rich glasses inferred in amorphous $Co_{80-x}Fe_xB_{20}$ (*Figure 14.18*), where $T_C > T_x$, is now directly observed in these cobalt-rich glasses where $T_C < T_x$.

Predictions of the split-band model

The split-band model predicts surfaces of zero magnetostriction in Co–Fe–TE–B space[97] as shown in *Figure 14.21*. Here it is assumed that the majority spin VBS of the TE is entirely above, and that of cobalt is entirely below, the Fermi level. (Iron and less than 4 atom% of manganese introduce no majority-spin states at or above the Fermi level (p. 264). (In *Figure 14.16* the Fe–Co–B $\lambda = 0$ composition was calculated assuming Fe $3d\uparrow$ states below E_F). The region in the cobalt-rich corner of *Figure 14.21* is of negative magnetostriction. The predictions of this figure are fairly well born out not only along the $Co_{80-x}TE_xB_{20}$ line as discussed above but also within a constant metalloid plane where Ohnuma *et al.*[98] have measured several low-magnetostriction glasses based on $B_{10}Si_6$. Their data show $\lambda_s = 0$ lines closely following the intersections of the $\lambda = 0$ planes with the B_{20} plane of *Figure 14.21*. Also Aboaf and Klokholm[99] as well as Narita *et al.*[100] have measured magnetostriction in Co–B thin films. The results are again consistent with *Figure 14.21*, showing $\lambda_s < 0$ over the measured range $12 < x_B < 30$ atom% with an extrapolated crossover to $\lambda_s = 0$ near $x_B = 40$–50 atom% close to the split-band model prediction. In short the predictions of *Figure*

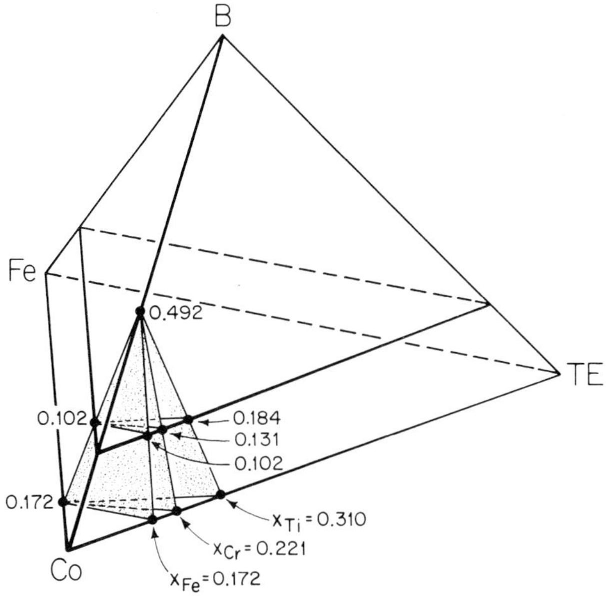

Figure 14.21 Quaternary diagram for magnetostriction in $Co_{1-x-y-z}TE_xFe_yB_z$ alloys. The 20 atom% boron plane is indicated parallel to base of tetrahedron. Three nearly vertical planes and their intercepts describe split-band model predictions for $\lambda_s^1 = 0$ in TE = Fe, Cr and Ti alloys. (Manganese and vanadium $\lambda_s^1 = 0$ planes lie between those of iron and chromium and between those of chromium and titanium respectively.) The $\lambda^1 = 0$ plane for nickel intercepts the B_{20} plane as shown in *Figure 14.16* (dashed line) and does not follow the sequence in *Figure 14.21* for TE substitutes because the nickel majority spin *d* states remain filled unlike those of the TE species shown in this figure (after O'Handley[97])

14.21 are consistent with the limited data extant. Details of the split-band model can be expected to be refined as it is applied to more of these TE–TL systems.

14.6 Conclusions

Fundamental magnetic properties of metallic glasses generally conform to those of related crystalline materials when proper account is taken of the bonding effects of the metalloids or other glass-forming elements. Exceptions occur when the SRO of the glassy phase (generally a highly coordinated, dense, randomly packed one) differs significantly from that of the crystal as in the case of bcc iron. This difference in SRO between α-Fe and iron-rich glasses appears to be responsible for the anomalies observed when comparing T_C and λ between the two phases.

The difference in local order between α-Fe and iron-rich glasses has no clear effect on the saturation moment. Apparently the eight-fold coordination of bcc-Fe is sufficient to establish a maximum moment/Fe atom and an increase in iron nearest-neighbour coordination has no further moment-enhancing effect.

A considerable body of data suggests that $\partial J(r)/\partial r > 0$ in iron-based glasses:

Curie temperatures of $Fe_{100-x}B_x$, of $Fe_{80}B_{20-x}M_x$, and of Fe–B–Be decrease as the Fe–Fe distance decreases. This behaviour clearly contrasts with the increasing T_C's of cobalt-rich glasses as the Co–Co distance decreases. We are, of course, making assumptions about the T–T atom spacing based mostly on assumed neutral atom sizes.

The cobalt-rich glasses are unique in their magnetoelastic properties for several reasons:

1. Cobalt-rich glasses do not show $\lambda \propto \sigma^2$.
2. Only in the cobalt-rich glasses has an appreciable two-ion component to magnetic anisotropy been confirmed.
3. Magnetostriction of $Co_{80}B_{20}$ glass goes positive at elevated temperatures or with the addition of iron just as does crystalline cobalt (at the hcp–fcc transformation).

Available data also suggest $\lambda = 0$ in nickel-rich glasses as in crystalline NiFe. Iron-rich glasses stand out as showing no parallel to their crystalline counterparts. We can understand this on the basis of the more highly coordinated local order of the glasses compared to that of α-Fe.

Thus the fundamental magnetic properties are determined mainly by short range order. The absence of long-range order has little effect on μ, T_C, λ, and K_{local}. (When $D/J \ll 1$, $K_{macro} \approx 0$ and low-field properties are sensitive to the long-range order because of the width of the domain walls.)

Attempted explanations of the complicated effects of metalloid substitutions on magnetic properties (*Figure 14.5*) on the basis of the size and valence of the constituent atoms are incomplete. Efforts to understand the reduced Curie temperatures and SRO of iron-rich glasses (*Figures 14.7* and *14.8*) on the basis of the Bethe–Slater curve and coordination numbers are inadequate. Because of the importance of chemical SRO and bonding in determining fundamental magnetic properties, their basic understanding in amorphous metallic alloys can only come from careful comparison with band structure or cluster calculations.

Acknowledgement

I acknowledge the Army Research Office (contract No. DAAG-29-80-K-0088) and the National Science Foundation (grant No. 78-24185 DMR) for support. I am also grateful to Professor N. J. Grant for encouragement.

References

1. Graham, C. D. (jun.) and Egami, T., in *'Annual Reviews of Materials Science'*, Ed. Huggins, R. A., Bube, R. H. and Roberts, R. W., vol. 8, p. 423, Annual Reviews Inc., Palo Alto, CA (1978)
2. Luborsky, F. E., *J. Magnetism magn. Mater.*, **7**, 143 (1978); Luborsky, F. E., in *'Ferromagnetic Materials'*, Ed. Wohlfarth, E. P., p. 451, North Holland Publ. Co., Amsterdam (1980); Luborsky, F. E. in *'Kirk-Othmer Encyclopedia of Chemical Technology'*, 3rd edn., vol. II, p. 537, Wiley, New York (1978)
3. Masumoto, T., in *'Proc. Conf. on Metallic Glasses: Science and Technology'*, Eds. Hargitai, C., Bakonyi, I. and Kemeny, T., vol. I, p. 121, Kultura, Budapest (1980)
4. Mizoguchi, T., American Institute of Physics Conf. Proc. No. 34, p. 286 (1976)
5. Wohlfarth, E. P., *IEEE Trans. Magn.*, **MAG-14,** 933 (1978)
6. Alben, R. A., Budnick, J. I. and Cargill, G. S. (III), in *'Metallic Glasses'*, Eds. Gilman, J. J. and Leamy, H. J., p. 304, American Society of Metals, Metals Park, Ohio (1978)
7. Kronmuller, H., *J. appl. Phys.*, **52,** 1859 (1981)
8. Hasegawa, R., to be published by CRC Press
9. Guntherodt, H.-J., *'Glassy Metals II'*, Springer Verlag, Berlin (1978)

10. Mizoguchi, T. Yamauchi, K. and Miyajima, H. in *'Amorphous Magnetism'*, Eds. Hooper, H. O. and de Graaf, A. M., p. 325, Plenum Press, New York (1973)
11. Yamauchi, K. and Mizoguchi, T., *J. phys. Soc. Japan*, **39,** 541 (1975)
12. O'Handley, R. C., Hasegawa, R., Ray, R. and Chou, C.-P., *Appl. Phys. Lett.*, **29,** 330 (1976), O'Handley, R. C., Hasegawa, R., Ray, R. and Chou, C.-P., *J. appl. Phys.*, **48,** 2095 (1977)
13. O'Handley, R. C. and Sullivan, M. O., *J. appl. Phys.*, **52,** 1841 (1981)
14. Becker, J. J., Luborsky, F. E. and Walter, J. L., *IEEE Trans Magn.*, **MAG-13,** 988 (1977)
15. O'Handley, R. C. and Boudreaux, D. S., *Phys. Status Solidi* (*a*) **45,** 607 (1978)
16. Ward, K. D., Crangle, J. and Davies, H. A., *'Rapidly Quenched Metals IV'*, Eds. Masumoto, T. and Susuki, K., p. 1141, Japan Institute of Metals, Sendai (1982)
17. Collins, M. F. and Wheeler, D. A. *Proc. R. Soc.*, **82,** 633 (1963) and Collins, M. F. and Forsyth, J. B., *Phil. Mag.*, **8,** 401 (1963)
18. Hasegawa, R. and Dermon, J. A., *Phys. Lett.*, **42A,** 407 (1973)
19. Sherwood, R. C., Gyorgy, E. M., Chen, H. S., Ferris, S. D., Norman, G. and Leamy, H. J., American Institute of Physics Conf. Proc. No. 24, p. 745 (1975)
20. Luborsky, F. E., *J. appl. Phys.*, **51,** 2808 (1980)
21. Chen, H. S., Sherwood, R. C. and Gyorgy, E. M., *IEEE Trans. Magn.*, **MAG-13,** 1538 (1977)
22. Korenman, V., Murray, J. L. and Prange, R. E., *Phys. Rev.*, **B16,** 4032 (1977); Hubbard, J., *Phys. Rev.*, **B19,** 2626 (1979); Cappellman, H., *Z. Phys.* **B34,** 29 (1979) and *Solid St. Communs,* **30,** 7 (1979)
23. Mitera, M., Naka, M., Masumoto, T., Kazama, M. and Watanabe, K., *Phys. Status Solidi,* (*a*), **49,** K163 (1978)
24. Kazama, N. S., Mitera, M. and Masumoto, T., in *'Proc. 3rd Int. Conf. on Rapidly Quenched Metals'*, Ed. Cantor, B., vol. II, p. 164, The Metals Society, London (1978)
25. Kazama, N. S., Masumoto, T. and Mitera, M., *J. Magnetism magn. Mater.* **15—18,** 1331 (1980)
26. Chen, H. S., *Phys. Status Solidi,* (*a*), **17,** 561 (1973)
27. Durand, J., *IEEE Trans. Magn.*, **MAG-12,** 945 (1976)
28. Narita, K., Fukunaga, H. and Yamasaki, J., *Jap. J. appl. Phys.*, **16,** 2063 (1977)
29. Hasegawa, R. and Ray, R., *J. appl. Phys.*, **50,** 1586 (1979)
30. Cochrane, R. W. and Cargill, G. S., *Phys. Rev. Lett.*, **32,** 476 (1974)
31. Pan, D. and Turnbull, D., *J. appl. Phys.*, **45,** 1406 (174)
32. Durand, J. and Yung, M. in *'Amorphous Magnetism II'*, Eds. Hasegawa, R. and Levy, R., p. 275, Plenum Press, New York (1977)
33. Hasegawa, R. and Ray, R., *J. appl. Phys.*, **49,** 4174 (1978)
34. Fukamichi, K., Kikuchi, M., Arakawa, S. and Masumoto, T., *Solid St. Communs,* **23,** 955 (1977); Fukamichi, K., Kikuchi, M., Hiroyoshi, H. and Masumoto, T., ref. 24, vol. II, p. 117 (1978); Fukamichi, K., Hiroyoshi, H., Kikuchi, M., and Masumoto, T., *J. Magnetism magn. Mat.*, **10,** 294 (1979)
35. Chien, C-L, Musser, D, Gyorgy, E. M., Sherwood, R. C., Chen, H. S., Luborsky, F. E. and Walter, J. L., *Phys. Rev.*, **B20,** 283 (1979)
36. Chien, C-L, and Unruh, K. M., *Phys. Rev.*, **B24,** 1556 (1981)
37. Hasegawa, R., *J. appl. Phys.*, **52,** 1847 (1981)
38. Severin, C. S., Chen, C. W., Belova, A. J., and Lin, M. C., *J. appl. Phys.*, **52,** 1850 (1981)
39. O'Handley, R. C., *Solid St. Communs,* **38,** 703 (1981)
40. Obi, Y, Morita, H. and Fujimori, H., *IEEE Trans. Magn.*, **MAG-16,** 1129 (1980)
41. Cable, J. W. and Child, H. R., *J. Phys. (Paris)* **32,** CI-67, (1967)
42. Naka, M., Kazama, N., Fujimori, H. and Masumoto, T., ref. 16, p. 919
43. O'Handley, R. C., Corb, B., Hara, Y., Grant, N. J. and Hines, W. A., *J. appl. Phys.*, **53,** 7753 (1982)
44. Corb, B., O'Handley, R. C., Grant, N. J. and Moruzzi, V., *J. Mag. magn. Mater.*, in the press
45. Ohnuma, S., Kanehira, J., Shirakawa, K., Egami, T. and Masumoto, T., ref. 16, p. 1044
46. Ohnuma, S., Shirakawa, K., Nose, M. and Masumoto, T., *IEEE Trans. Magn.*, **MAG-16,** 1129 (1980)
47. Fukamichi, K. and Gambino, R. J., *IEEE Trans. Magn.*, **MAG-17,** 3059 (1981)
48. Berkowitz, A. E., Walter, J. L. and Wall, K. F. *Phys. Rev. Lett.*, **46,** 1484 (1981)
49. Allen, J. W., Wright, A. C. and Connell, G. A. N., *J. Non-cryst. Solids,* **40,** 509 (1980)
50. Watson, R. E. and Bennett, L. H., in *'Theory of Alloy Phase Formation'*, Ed. Bennett, L. H., p. 245 The Metallurgical Society of AIME, New York (1980)
51. Corb, B. W., O'Handley, R. C. and Grant, N. J., *J. appl. Phys.*, **53,** 7728 (1982)
52. Messmer, R. P., *Phys Rev.*, **B23,** 1616 (1981)
53. Cadeville, M. C. and Daniel, E., *J. Phys.*, **27,** 29 (1966)
54. Hasegawa, R. and Ray, R. *Solid St. Communs,* **27,** 471 (1978)
55a. Fukunaga, T., Misawa, M., Fukamichi, K., Masumoto, T. and Suzuki, K., ref. 24, vol. II, p. 325
55b. Waseda, Y., *Prog. mater. Sci.*, **26,** 1 (1981)

56. Waseda, Y. and Chen, H. S., *Phys. Status Solidi, (a)* **49,** 387 (1978)
57. So, F. F., SM Thesis, Massachusetts Institute of Technology, 1982; So, F. F., Kaplow, R. and O'Handley, R. C., submitted for publication
58. Kazama, N. S., Masumoto, T. and Watanabe, H., *J. phys. Soc. Japan,* **37,** 1171 (1974)
59. Kazama, N. S. and Kameda, M., American Institute of Physics Conf. Proc. No. 34, p. 307 (1976)
60. Axe, J. D., Passell, L. and Tsuei, C. C. American Institute of Physics Conf. Proc. No. 24, p. 119 (1975), Mook, H. A., Wakabayashi, N. and Pan, D., *Phys. Rev. Lett.,* **34,** 1029 (1975)
61. Well defined, standing spin waves with relatively little damping have also been observed by FMR in thin films [Gambino, R. J. and Suran, G., *J. appl. Phys.,* **50,** 7671 (1979)]
62. Birgeneau, R. J., Tarvin, J. A., Shirane, G., Gyorgy, E. M., Sherwood, R. C., Chen, H. S. and Chien, C-L., *Phys. Rev.,* **B18,** 2192 (1978)
63. Rhyne, J. J., Fish, G. E. and Lynn, J. W., *J. appl. Phys.,* **53,** 2316 (1982)
64. Hasegawa, R. and Ray, R., *Phys. Rev.,* **B20,** 211 (1979)
65. Luborsky, F. E., Walter, J. L., Liebermann, H. H., and Wohlfarth, E. P., *J. Magnetism magn. Mater.,* **15—18,** 1351 (1980)
66. Handrich, K., *Phys. Status Solidi,* **32,** K55 (1969)
67. Tsuei, C. C. and Lillienthal, H., *Phys. Rev.,* **B13,** 4899 (1976)
68. Crangle, J. and Hallam, G. C., *Proc. R. Soc.,* **A272,** 119 (1963)
69. Coey, J. D. M., *J. appl. Phys.,* **49,** 1646 (1978)
70. Leamy, H. J., Ferris, S. D., Norman, G. Joy, D. C., Sherwood, R. C., Gyorgy, E. M. and Chen, H. S., *Appl. Phys. Lett.,* **26,** 259 (1975); Hasegawa, R., O'Handley, R. C., Tanner, L. E., Ray, R. and Kavesh, S., *Appl. Phys. Lett.,* **29,** 219 (1976); Livingston, J. and Morris, W. G., *IEEE Trans. Magn.,* **MAG-17,** 2624 (1981)
71 Hasegawa, R., American Institute of Physics Conf. Proc. No. 29, p. 216 (1976)
72. Takahashi, M. and Kim, C. O., *Jap. J. appl. Phys.,* **16,** 2061 (1977)
73. Egami, T. and Flanders, P. J., ref. 71, p. 220
74. Takahashi, M. and Miyazaki, T., *Jap. J. appl. Phys.,* **18,** 743 (1979)
75. Berry, B. S. and Pritchet, W. C., *Phys. Rev. Lett.,* **34,** 1022 (1975)
76. Chen, H. S., Ferris, S. D., Gyorgy, E. M., Leamy, H. J. and Sherwood, R. C., *Appl. Phys. Lett.,* **26,** 405 (1975)
77. Luborsky, F. E. and Walter, J. L., *IEEE Trans. Magn.,* **MAG-13,** 1635 (1977)
78. Allia, P. and Vinai, F., *IEEE Trans. Magn.,* **MAG-14,** 1050 (1978); Becker, J. J., *IEEE Trans. Magn.,* **MAG-14,** 938 (1978)
79. Luborsky, F. E., ref. 71, p. 209
80. Nesbitt, E. A., Hasegawa, R., O'Handley, R. C. and Sullivan, M. O., *J. appl. Phys.,* **51,** 3328 (1980)
81. O'Handley, R. C. and Chou, C-P., *J. appl. Phys.,* **49,** 1659 (1978)
82. Jagielinski, T., *IEEE Trans. Magn.,* **MAG-17,** 2825 (1981)
83. Lacheisserie, E., du Tremolet de, *J. Magnetism magn. Mater.,* **25,** 251 (1982)
84. O'Handley, R. C., in *'Amorphous Metals II',* Eds. Levy, R. and Hasegawa, R., p. 379, Plenum Press, New York (1977)
85. O'Handley, R. C., *Solid St. Communs,* **21,** 1119 (1977)
86. O'Handley, R. C., *Phys. Rev.,* **B18,** 930 (1978)
87. Ohnuma, S., Watanabe, K. and Masumoto, T., *Phys. Status Solidi, (a),* **44,** K151 (1977)
88. Luborsky, F. E., Flanders, P. J., Liebermann, H. H. and Walter, J. L., *IEEE Trans. Magn.,* **MAG-15,** 1961 (1979)
89. Kohmoto, O., Ohya, K., Yamaguchi, N., Fujishima, H. and Ojuma, T., *J. appl. Phys.,* **51,** 4342 (1980)
90. O'Handley, R. C., Narasimhan, M. C. and Sullivan, M. O., *J. appl. Phys.,* **50,** 1633 (1979)
91. Dmowski, W., Jagielinski, T. and Matyja, H., ref. 3, p. 21
92. Callen, E. R. and Callen, H. B., *Phys. Rev.* **129,** 578 (1963); *Phys. Rev.,* **A139,** 455 (1965)
93. Berry, B. S. and Pritchet, W. C., *Solid St. Communs,* **26,** 827 (1978)
94. Sirota, N. N. and Mazovka, A. V., *Dokl. Akad. Nauk SSSR,* **16,** 596 (1972)
95. Ashworth, H., Sengupta, D., Schnakenberg, G., Shapiro, L. and Berger, L., *Phys Rev.,* **185** 792 (1969); Berger, L., ref. 4, p. 355 (1976); Berger, L., *Physica,* **B91,** 31 (1977)
96. O'Handley, R. C. and Berger, L., in *'Physics of Transition Metals',* Institute of Physics Conf. Series, No. 39, p. 477 (1978)
97. O'Handley, R. C. unpublished calculations
98. Ohnuma, S. and Masumoto, T., *J. appl. Phys.,* **50,** 7957 (1979)
99. Aboaf, J. and Klokholm, E., *J. Magnetism magn. Mater.,* **15—18,** 1385 (1980)
100. Narita, K., Yamasaki, J. and Fukunaga, H., *J. appl. Phys.,* **50,** 7591 (1979)

Chapter 15

Itinerant electron model of magnetic properties

E. P. Wohlfarth

Dept of Mathematics, Imperial College, London, UK

15.1 Introduction to the itinerant electron model

There are two theoretical approaches to metallic magnetism, namely the localized and the itinerant electron approaches. The former is concerned with localized magnetic moments in direct exchange interaction. It is easily visualized and thus has great pedagogical advantages. In its simplest form it applies to non-metallic magnets. For metals and alloys the conceptually more difficult itinerant electron model is more appropriate. In applying this model to amorphous alloys many 'localized' features must also be considered, as we shall see. Unfortunately, the itinerant electron model is difficult to apply even to the pure metals such as iron and nickel. Hence the application to complicated alloys, which the amorphous magnets certainly are, requires a judicious blend of theory and empiricism of a type which is not everywhere fully appreciated. This chapter nevertheless attempts to explain the matter using this philosophy, in the hope that it will thus be of some help to the experimental and technological community concerned with this fascinating subject.

The conceptual inputs of the model are the following:

15.1.1 Electron states

In a metal the individual atomic states are broadened into energy bands whose features are given by the interaction of the electrons with the crystal lattice. For iron-based amorphous alloys the atomic $3d$-states give rise to the $3d$-band, and the vacant states in this band, or holes, are the essential carriers of their magnetism. The s and p electrons of the iron atoms and of the glass formers (B,P,Si,...) are, however, also important, as we shall see. A simple description of the energy band structure is the tight binding approximation[1]. The wave functions of the electrons are of the general form

$$\psi_{\mathbf{k}}(\mathbf{r}) = \textstyle\sum_l \phi(\mathbf{r} - \mathbf{R}_l)e^{i\mathbf{k}.\mathbf{R}_l} \tag{15.1}$$

where ϕ is an atomic wave function at a position $\mathbf{r}$ near to a lattice site $\mathbf{R}_l$, $\mathbf{k}$ is the wave vector describing the hopping motion of the electron through the lattice and the summation is over all lattice sites. This hopping from site to site makes the electrons itinerant. For the $3d$ electrons equation 15.1 is actually more complicated due to the five-fold degeneracy[1]. The energy of the itinerant electrons depends on their wave vector and is for this non-degenerate case given approximately by

$$\varepsilon_{\mathbf{k}} = E_0 + A + B\sum e^{i\mathbf{k}\cdot\mathbf{R}_1} \tag{15.2}$$

where

$$A = \int |\phi(\mathbf{r})|^2 \Delta V \, d\mathbf{r} \tag{15.3a}$$

$$B = \int \phi(\mathbf{r})\phi(\mathbf{r} - \mathbf{R}_1)\, \Delta V \, d\mathbf{r} \tag{15.3b}$$

Here $\Delta V(<0)$ is the interaction energy of the itinerant electrons with the lattice, $\mathbf{R}_1$ is a nearest neighbour distance and the summation in equation 15.2 is taken over nearest neighbours. The so-called hopping integral B (for degenerate *d* electrons there are more) determines the features of the band structure, in particular the energy density of electron states $N(\varepsilon)$ and the width W of the energy band which also depends on the local environment of an atom, e.g. an iron atom, due to the summation in equation 15.2. On this basis a very crude description is possible of the magnetic state of amorphous alloys (*see* section 15.4.2). The energy position of the band is given by $E_0 + A$, where E_0 is the atomic energy of an isolated atom.

15.1.2 Charge transfer*

As we shall see, the magnetization of amorphous alloys varies in a characteristic manner with the alloying content, both transition metal and metalloid. The simplest explanation is in terms of a transfer of charge between these two species of atoms (e.g. $B \rightarrow Fe$). This transfer is, however, not a simple movement of electrons in space but is related to the hybridization of the *d* electrons on the transition metal atoms with the s–p electrons on the metalloids. As an example, a paramagnetic band calculation[2] on crystalline Fe_3Si shows that there is, apart from a full low lying band containing two states per formula unit, a broad band of higher density of states holding precisely 30 states of which only 26 are occupied. If this band were taken to be localized on iron atoms we might imagine that each of these atoms contributed eight electrons to the *d*-band and a further two electrons were transferred spatially from Si to Fe *d*-states. However, the calculation shows that the wave functions corresponding to states in this band have components not only on iron atoms but also on silicon atoms. Hence there is no need to assume that there is spatial charge transfer of any magnitude at all. More generally, then, a 3*d* band containing ten states per transition metal atom can be regarded as being gradually filled as the concentration of the metalloids increases. These states are not localized on the transition metal atoms but their wave functions also have a considerable weight on the metalloids. This concept replaces that of spatial charge transfer but has the same final effect.

15.1.3 Electron interactions

Ferromagnetism is essentially a many body phenomenon in that the itinerant electrons interact with each other. In the simplest approximation this interaction is between two electrons on the same lattice site of the form

$$I_0 = \int \frac{e^2}{r_{12}} |(\mathbf{r}_1)|^2 |\varphi(\mathbf{r}_2)|^2 \, d\mathbf{r}_1 \, d\mathbf{r}_2 \tag{15.4}$$

This interaction is by itself too large since it is based on an inadequate approximation (Hartree–Fock) to this problem. However, an improvement to include correlation

*I am grateful to R. L. Jacobs for this account.

effections, due to Kanamori[3] and Hubbard[4], leads to a smaller value I given approximately by

$$I = I_0/(1 + \gamma I_0/W) \qquad (15.5)$$

where γ is a constant of order unity and W the energy band width determined by the hopping integrals B of the form 15.3(b). This formula has been used to discuss the magnetovolume anomalies of many amorphous ferromagnets (*see* p. 292), since the bandwidth W of the $3d$ electrons varies[5] approximately as R_1^{-5}. It is an amazing result: for narrow bands the interaction is effectively limited by W!

Having determined $N(\varepsilon_F)$, where the density of states is taken at the Fermi energy ε_F up to which states are occupied, and the effective interaction I it is possible to use the Stoner criterion[6]

$$\bar{I} = IN(\varepsilon_F) > 1 \qquad (15.6)$$

to determine the condition for ferromagnetism and to calculate other observables. Temperature effects are included in this model via the appropriate statistical mechanics, which will now be discussed.

15.1.4 Statistical mechanics

Electrons in metals at finite temperatures obey the formulae appropriate to the Pauli exclusion principle, i.e. Fermi–Dirac statistics. The corresponding energy distribution function is

$$f(\varepsilon, T) = [e^{(\varepsilon - \mu)/kT} + 1]^{-1} \qquad (15.7)$$

where μ is the chemical potential determined by the density of states per spin $N(\varepsilon)$ and the particle numbers n as

$$\tfrac{1}{2}n = \int_0^\infty N(\varepsilon) f(\varepsilon)\, d\varepsilon \qquad (15.8)$$

Using expansions of equation 15.7 it is frequently possible to obtain a number of simple formulae for such temperature effects as are observable. Some simple examples are:

1. The Curie temperature T_C, given by

$$I \int_0^\infty N(\varepsilon) \left| \frac{\partial f(\varepsilon, T_C)}{\partial \varepsilon} \right| d\varepsilon = 1 \qquad (15.9)$$

2. The temperature dependence of the magnetization $M(T)$, given for the case, which frequently occurs, $IN(\varepsilon_F) - 1 \ll 1$, by

$$\{M(T)/M(0)\}^2 = 1 - T^2/T_C^2 \qquad (15.10)$$

3. The magnetic field dependence of the magnetization, given by

$$\left\{\frac{M(H,T)}{M(0,0)}\right\}^3 - \frac{M(H,T)}{M(0,0)}\left(1 - \frac{T^2}{T_C^2}\right) = \frac{2\chi_0 H}{M(0,0)} \qquad (15.11)$$

where

$$\chi_0 = \mu_B^2 N(\varepsilon_F)/(\bar{I} - 1) \qquad (15.12)$$

These and many other formulae of this model may be tested by experiment on amorphous ferromagnets, as we shall see. Where there are deviations these may arise for a number of reasons. Thus deviations from equation 15.10 may arise owing to the neglected influence of spin wave effects, which will be referred to below, and other spin fluctuation effects[24]. Again, deviations from equation 15.11 may arise owing to metallurgical effects such as atomic scale concentration fluctuation effects. Finally, there are some materials like metallic iron for which the Curie temperature given by equation 15.9 is a lot larger than that observed. This is partly due to the on-site model used being insufficient for this metal. Much work on extensions of this model is therefore currently in progress.

15.1.5 Spin waves

If an itinerant electron of wave vector **k** is excited to a state of wave vector $\mathbf{k} + \mathbf{q}$ and its spin is reversed, it can form with the hole left behind an electron-hole pair with energy $\hbar\omega$ where, for small q,

$$\hbar\omega = Dq^2 \tag{15.13}$$

These elementary excitations are called spin waves and the constant D the spin wave stiffness. This may be observed by neutron scattering or from the temperature dependence of the magnetization arising from *this* origin (additionally to equation 15.10):

$$M(T)/M(0) = 1 - c\,(T/D)^{3/2} \tag{15.14}$$

where c is a constant. The value of D at 0 K may be calculated from the formula[7]

$$D = \frac{IR_1^2}{3(\Delta E)^2}\int_{\varepsilon_-}^{\varepsilon_+} \varepsilon(\mu - \varepsilon)N(\varepsilon)\,\mathrm{d}\varepsilon \tag{15.15}$$

where ΔE is the exchange splitting between the up and down spin energies ε_+ and ε_- and μ is the chemical potential at 0 K; equation 15.15 applies in the case of the tight binding approximation and a non-degenerate band. The spin wave stiffness D must be regarded as equally characteristic a property of a ferromagnet as is the saturation magnetization and the Curie temperature.

The itinerant electron model of ferromagnetism is a complicated many body theory which it is intended to apply to amorphous materials. This brief outline of some of the basic concepts of this model should suffice to prepare for some of the more specific applications which are described in the following sections.

15.2 Some itinerant aspects of amorphous ferromagnetism

15.2.1 Magnetization and Curie temperature

The saturation magnetization of magnetic alloys is traditionally represented by the Slater–Pauling curve, giving the variation of this property with the electron concentration. For crystalline alloys this curve is shown in *Figure 15.1* (*see also* Chapter 14), which also includes the results for the crystalline borides[8] T_2B and TB. The curves are seen to have been shifted systematically as a result of the charge transfer effect discussed above. A similar Slater–Pauling curve for amorphous transition metal–metalloid systems has been constructed by Mizoguchi *et al.* (*see* Figure 20 of Luborsky[9]; *also* Chapter 14). From the analysis for a wide variety of amorphous as well as crystalline alloys Luborsky[9] concludes that, very roughly, boron atoms transfer a charge of about one electron each to the transition metal atoms and phosphorus about two. The effect is to be understood in the sense of the *sp–d* hybridization discussed on

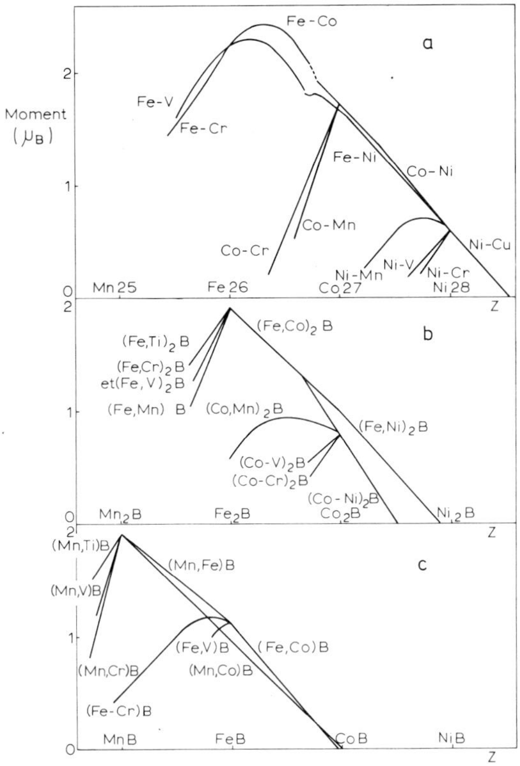

Figure 15.1 Slater–Pauling curves of crystalline alloys and borides (after Cadeville[8])

p. 284. The overall evidence is also that it makes little difference if this concept is applied to crystalline or amorphous alloys.

A more explicit discussion of the influence of metalloids on the saturation magnetization has been given by Luborsky *et al.*[10] for amorphous Fe–X, where X is one of, or a combination of, B, Al, Ga, C, Si, Ge, P. The saturation magnetization as a function of the *total* metalloid content is shown in *Figure 15.2.* The results are complicated but are compatible with a charge transfer determined on the whole by just the effective valency of the added metalloid, i.e. 1 for boron, 2 for carbon, silicon and germanium and 3 for phosphorus. It would require a much more sophisticated alloy model to go beyond these crude concepts.

The same authors[10] also gave data for the Curie temperature of Fe–B–X and these are shown in part in *Figure 15.3.* Apart from Fe–B–Ge the data of T_C against the *total* metalloid (B + X) content tend to lie on a single non-linear curve which rises to maximum values about 800 K. These results imply that the Curie temperatures are much less sensitive to the charge transfer effect than is the saturation magnetization. A very qualitative discussion of this result was given[10] in terms of size and local environment effects. The first of these was related in this discussion to the magnetovolume effects characteristic of these alloys (section 15.3) and the second to a crude tight binding model of the amorphous state (section 15.4). The weak effect of charge transfer on T_C had been noted earlier for crystalline Fe–Ni–C alloys[11], and seems reasonable on general grounds, using equation 15.9. That there is some effect is shown in ref. 43 by reference to Fe–B alloys.

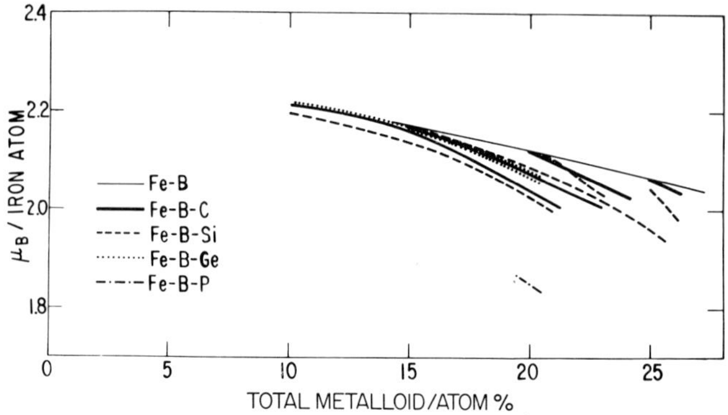

Figure 15.2 Saturation magnetizations of some amorphous alloys (after Luborsky[10])

In fact, the maximum T_C values shown in *Figure 15.3* have been observed more markedly for binary Fe–B alloys[12], where the maximum occurs at about 30 atom% of boron. The addition of further boron atoms leads to a rapid reduction of T_C which vanishes at about 60 atom% of boron.

As seen in *Figures 15.2* and *15.3* as well as the work by Chien and Unruh[12] there are ranges of metalloid content where the saturation magnetization $M(0)$ and T_C vary in the same sense and others where this sense is opposite. Now it follows that for weak itinerant ferromagnets ($\bar{I} - 1 \ll 1$) a variation in the *same* sense is expected since, in fact, from equation 15.9 (*see also* equation 15.17 and 15.18):

$$M(0) \sim T_C \sim (\bar{I} - 1)^{1/2} \tag{15.16}$$

Hence we may use this criterion to establish for Fe–B the phase diagram of *Figure 15.4*, showing schematically the full curves as representing observations and the broken curves as imaginative extrapolations into alloy regimes where the amorphous state becomes unstable. The most controversial point is that amorphous iron is here proposed to be non-magnetic! A test of this proposal would add immeasurably to our understanding of the magnetism of iron (bcc = ferro.; fcc = antiferro.; hcp = para.; amorph. = para?). It has also been discussed (inconclusively) by consideration of the

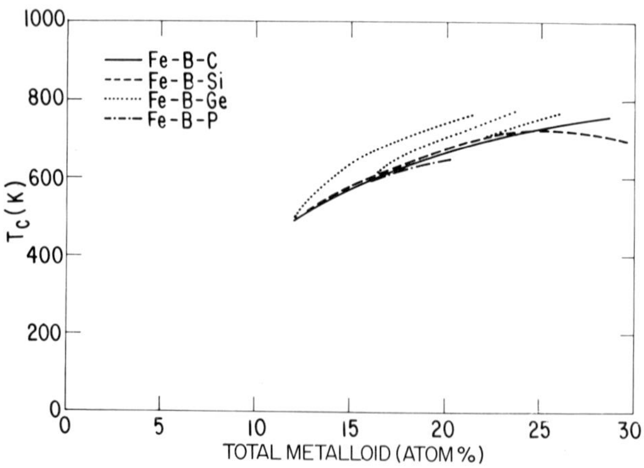

Figure 15.3 Curie temperatures of some amorphous alloys (after Luborsky[10])

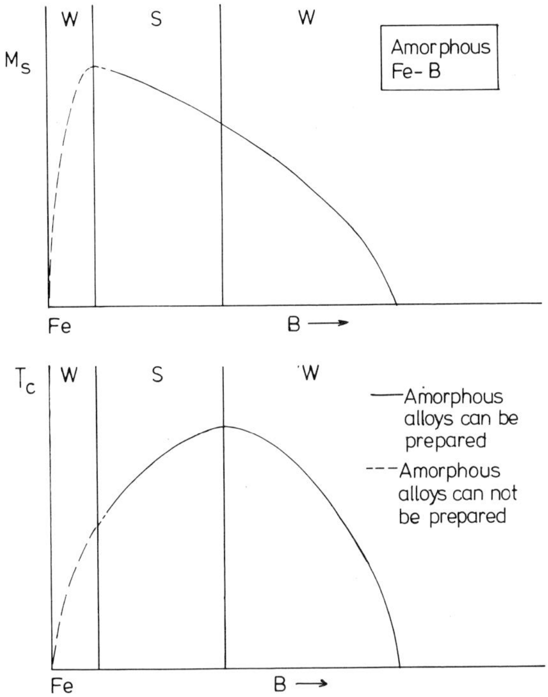

Figure 15.4 Imaginative phase diagram for Fe–B; S, strong itinerant; W, weak itinerant. Full lines represent amorphous alloys that can be prepared, and broken lines amorphous alloys that cannot be prepared

hyperfine fields in the alloys[12a]. Perhaps amorphous Fe should be treated as a spin glass?

15.2.2 Rhodes–Wohlfarth plot[13]

This is a plot of the ratio q_c/q_s versus T_C, where q_s is the number of magnetic carriers from the saturation magnetization $M(0)$ and q_c that from the Curie–Weiss law for the susceptibility above T_C. The itinerant model in the form given in section 15.1 gives[13] this ratio $\sim T_C^{-1}$, but, as noted in this chapter, although crystalline alloys tend to follow this trend rather well, the data for amorphous alloys lie systematically higher. This effect may be explained using the work described in section 15.4 in terms of the effects of concentration fluctuations characteristic of the metallurgical state of amorphous alloys. The anomalous increase of the ratio is shown by use of the approach in section 15.4 to be proportional to the relative difference between the paramagnetic and ferromagnetic Curie temperatures which arises from the same fluctuations. Measurements to test this result were made on amorphous alloys with relatively low Curie temperatures[14], $(Fe,Ni)_{77}Si_{10}B_{13}$ giving reasonable agreement.

15.2.3 Mathon plot

It follows from equation 15.16 that for an alloy system having a critical concentration c_0 where ferromagnetism disappears the saturation magnetization and Curie temperature should follow the parabolic dependences on the concentration c

$$M(0)^2 \sim T_C^2 \sim |c - c_0| \tag{15.17}$$

Mathon plots are of these squares against the concentration. For many crystalline alloys with magnetic homogeneity (*see* section 15.4), such as Ni–Pt, this formula is well obeyed. For a number of amorphous alloy systems such as Ni–Y [15] and Ni–P [16] the Mathon plots are, however, not linear in the sense that this linear variation is approximately observed for $M(0)$ and T_C and not their squares. It must be concluded from these results that the alloys are magnetically heterogeneous as will be discussed in section 15.4.

15.2.4 Arrott plots

A more widely used method of assessing these alloy structure effects is one which is based on equation 15.11. This leads to the result that plots of $M(H,T)^2$ against $H/M(H,T)$ are the straight lines, with slopes independent of T and intercepts $M(0,T) \equiv M(T)$, given by equation 15.10. For homogeneous alloys like Ni–Pt these straight line Arrott plots have in fact been observed (Type 1 of *Figure 15.5*). Again, deviations give an indication that the materials are not homogeneous. There are several measurements for amorphous alloys. Although these have not been done very systematically we show in *Table 15.1* a brief summary of the results (for another useful

TABLE 15.1 Arrott plot characteristics (*see Figure 15.5*)

Alloy	*Reference*	*Type*
Ni–P	16	2
Ni–B	41	1
$Y_{9.5}Ni_{90.5}$	15	2—3
$Y_{13}Ni_{87}$	17	2
(Fe, Ni)$_{80}B_{10}P_{10}$	18	3
Fe–Ge	19	2—3
Fe–Zr	42	3
(Fe, Mn)P–C	20	4
(Fe, Ni)B–Si	14	3, 4
Tb–Fe	21	3
Y–Fe	40	3

account of Arrott plots, *see* Schneider and Zaveta[18]). The types of plot noted are those shown schematically in *Figure 15.5*. There are several ways of interpreting these results. Reference is made in section 15.4 to Landau–Ginsburg type calculations which lead to Type 2 and a mild version of Type 3. Type 4 occurs more rarely as *Table 15.1* shows; it has been obtained theoretically by Acker and Huguenin[22] using a superparamagnetic model and also by Herzer *et al.*[23] using a variant of Landau–Ginzburg theory. None of these theories is all that satisfactory owing to the theoretical difficulties inherent in the problem (i.e. the magnetization depends on temperature and field as well as on position). It is thus desirable to have a more systematic experimental investigation.

15.2.5 T^2 law of the magnetization $M(0,T)$

For several homogeneous crystalline alloys the relation 15.10 has been observed to hold over a wide range of temperatures. Systematic measurements for amorphous alloys have been carried out by Liénard *et al.*[15,17] for Y_xNi_{100-x} and the results are shown in *Figure 15.6*. Relation 15.10 is obeyed surprisingly well up to 7 atom % of Y with marked deviations for the alloys with 9.5 and 12.8 atom % of Y. Liénard *et al.* ascribe these deviations to spin fluctuations[24]. We are, however, of the opinion, based on the present imperfect systematics of crystalline and amorphous alloys, that the

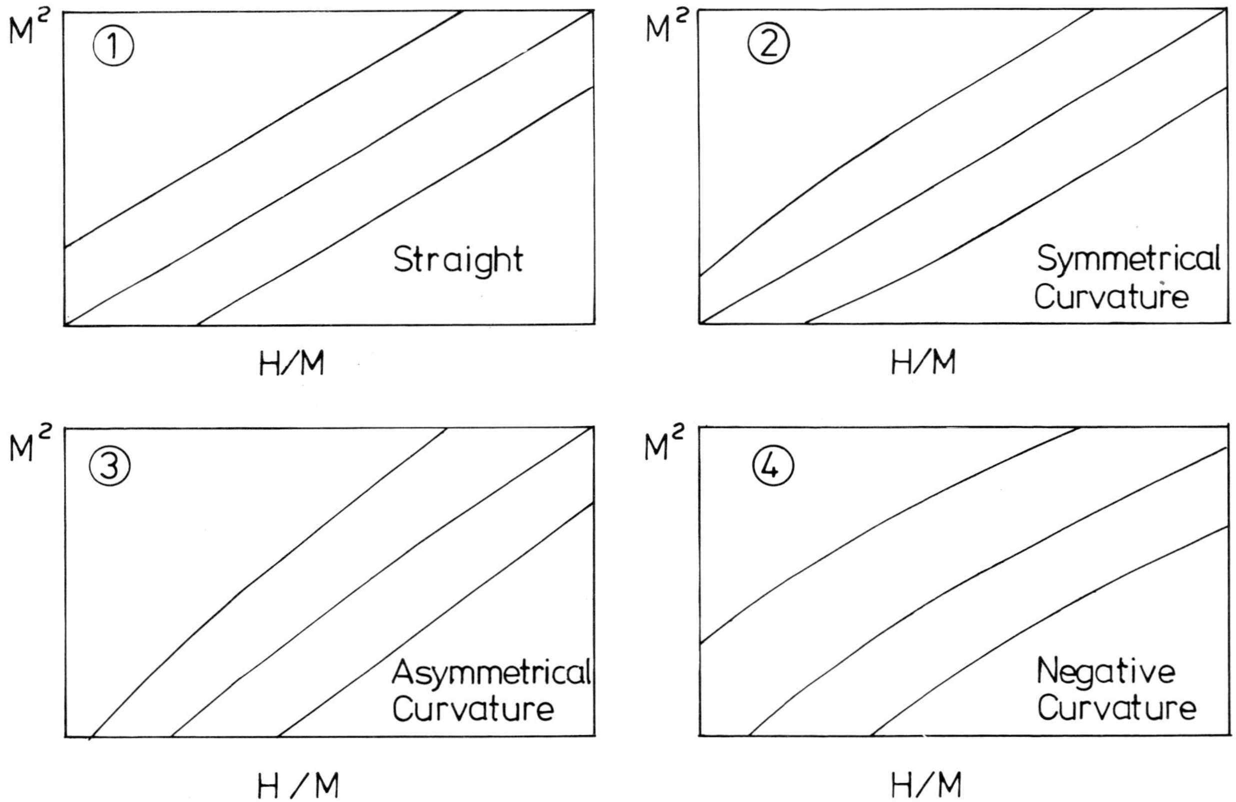

Figure 15.5 Types of Arrott plots (*see Table 15.1*)

$B(x) = B\{R_1(x)\}$ where R_1 is the nearest neighbour distance; as on p. 293 $B \sim R_1^{-5}$. Hence the spatial fluctuations of M are determined by the spatial fluctuations of the transition metal concentration $[1 - c(x)]$ and of the nearest neighbour distance between the T atoms, $R_1(x)$. It seems rather clear on physical grounds that both these fluctuations are on the atomic scale. Wohlfarth[34] contrasts this scale of distances to that corresponding to the exchange and anisotropy conditioned Bloch walls which is much larger for these T–M alloys. The physical measurements required to assess these fluctuations might include Mössbauer, neutron, EXAFS, etc. The aim is to correlate such physical measurements with the results of the magnetic measurements discussed in sections 15.2 and 15.3. The electronic structure of amorphous materials is discussed in Chapters 8 and 9. A good survey is also given by Güntherodt *et al.*[35].

15.4.3 Landau–Ginzburg model

How do these fluctuation effects enter the itinerant electron model and how do they therefore help to explain the magnetic data of sections 15.2 and 15.3? Apart from some other viable models (*see*, for example Acker and Huguenin[22]) we have developed the application to this problem of the Landau–Ginzburg model. Here the thermodynamic potential 15.24 ($\omega = 0$ for present purposes) is amended such that the Landau coefficients A and B are now functions of position x. The resulting fluctuations of $M(x)$ are limited by adding to 15.24 the gradient term

$$\tfrac{1}{2}C'|\nabla M(x)|^2 \tag{15.37}$$

In equation 15.37 the coefficient C' determines the effective correlation length ξ of the spatial fluctuations of $M(x)$ such that

$$\xi^{-2} = |A|/C' \tag{15.38}$$

From the discussion of p. 297 it follows that for the simple non-degenerate tight binding model used there

$$\xi^{-2} = \xi_{R1}^{-2} + \xi_c^{-2} \tag{15.39}$$

where
$$\xi_{R1}^{-2} = (5\alpha)^2\overline{(R_1^{-1}\mathrm{d}R_1/\mathrm{d}x)^2} \tag{15.40}$$

$$\xi_c^{-2} = (2\alpha)^2\overline{[(1-c)\mathrm{d}c/\mathrm{d}x]^2} \tag{15.41}$$

where
$$\alpha = [\bar{I}/2(\bar{I}-1)](I/I_0) \sim 1 \tag{15.42}$$

Shtrikman and Wohlfarth[36] expressed the Landau coefficient A as a function of $c = c(x)$ and the Arrott plots were obtained on this basis by minimizing 15.24 ($\omega = 0$) and in terms of the second moment of the spectrum of the concentration fluctuations $\overline{\Delta c^2}$. This first calculation was fitted to several features of the itinerant electron model and finally led to Arrott plots of Type 2 of *Figure 15.5*. The differences between Types 1 and 2 were thus determined by $\overline{\Delta c^2}$. In later work[37,38] this calculation was extended and applied in several ways but it was only found possible, in all this work, to have rather mild indications of Type 3 of *Figure 15.5*. A more pronounced asymmetry of Arrott plots was obtained in a different solution of the Landau–Ginzburg problem involving some localization of magnetic moments, as in crystalline Ni–Pd alloys[39]. Other discussions of the plots of *Figure 15.5* were given by Acker and Huguenin[22] and Herzer *et al.*[23]. The work was extended to the magnetovolume problems by Wagner and Wohlfarth[27] with the essential results already discussed (15.3.1), including *Table 15.2*. Applications to Rhodes–Wohlfarth plots, differences between Curie temperatures, etc. were given by Wagner and Wohlfarth[38].

Using this approach to discuss spatially varying alloy concentrations in amorphous alloys it has thus been found possible to give some interpretation of the various amorphicity dependent properties described in sections 15.2 and 15.3. Owing to the inherently great difficulties of the physical situation involving itinerant electrons in disturbed lattices this theoretical approach, though already found to be valuable, should only be regarded as a stepping stone to further theoretical work but also as an indication as to where further experimental work would be valuable.

References

1. Fletcher, G. C., *'Electron Band Theory of Solids'*, North Holland, Amsterdam (1971)
2. Switendick, A. C., *Solid St. Communs*, **19,** 511 (1976)
3. Kanamori, J., *Prog. theor. Phys., Osaka*, **30,** 275 (1963)
4. Hubbard, J., *Proc. phys. Soc.*, **84,** 455 (1964)
5. Heine, V., *Phys. Rev.*, **153,** 673 (1967)
6. Stoner, E. C., *Proc. R. Soc.*, **A165,** 372 (1938)
7. Edwards, D. M., *Phys. Lett.*, **24A,** 350 (1967).
8. Cadeville, M. C., Thesis, Strasbourg (1965)
9. Luborsky, F. E., *'Handbook on Ferromagnetic Materials*, vol. I, p. 451, North Holland, Amsterdam (1980)
10. Luborsky, F. E., Walter, J. L. and Wohlfarth, E. P., *J. Phys. F*, **10,** 959 (1980)
11. Cadeville, M. C. and Wohlfarth, E. P., *Phys. Status Solidi, (a)*, **26,** K157 (1974)
12. Chien, C. L. and Unruh, K. M., *Phys. Rev.*, **B24,** 1556 (1981); 12a. Takács, L. and Hargitai, C. *J. Phys. F* (1983)
13. Wohlfarth, E. P., *J. Magnetism magn. Mater.*, **7,** 113 (1978)
14. Hilscher, G., *et. al.*, *J. Phys. F*, **11,** 2429 (1981); Kaul, S. N. and Rosenberg, M., *Phil. Mag.*, **B44,** 357 (1981)
15. Liénard, A. and Rebouillat, J. P., *J. appl. Phys.*, **49,** 1680 (1978)
16. Berrada, A., *et al.*, *Solid St. Communs*, **21,** 671 (1977)
17. Liénard, A., Thesis, Grenoble (1981)
18. Schneider, J. and Zaveta, K., *Proc. Conf. on High Purity Materials, Dresden*, vol. III, p. 18 (1980); *'Proc. 4th Int. Conf. on Rapidly Quenched Metals'*, Japan Institute of Metals, Sendai (1982)
19. Endoh, Y., *et al.*, *Solid St. Communs*, **18,** 735 (1976)
20. Wieser, E. *et al.*, *Phys. Status Solidi, (a)*, **66,** 607 (1981)
21. Alperin, H. A., Cullen, J. R. and Clark, A. E., American Institute of Physics Conf. Proc. No. 29, p. 186 (1976)
22. Acker, F. and Huguenin, R., *J. Magnetism magn. Mater.*, **12,** 58 (1979)
23. Herzer, G., *et al.*, *Phys. Status Solidi, (b)*, **101,** 713 (1980)
24. Moriya, T., *J. Magnetism magn. Mater.*, **14,** 1 (1979)
25. Luborsky, F. E., *et al.*, *J. Magnetism magn. Mater.*, **15—18,** (1980); 25a. Babić, E., Marohnic, Z. and Wohlfarth, E. P., to be published
26. Wohlfarth, E. P., Honda Memorial Series (Maruzen) **3,** 327 (1978); *J. Magnetism magn. Mater.*, **10,** 120 (1979); *'Proc. Int. Symp. on Physics at High Pressures'*, p. 175, North Holland, Amsterdam (1981)
27. Wagner, D. and Wohlfarth, E. P., *J. Phys. F*, **11,** 2417 (1981)
28. Fukamichi, K., Masumoto, T. and Kikuchi, M., *IEEE Trans. Magn.*, **15,** 1404 (1979)
29. Shirakawa, K., *et al.*, *J. appl. Phys.*, **52,** 1829 (1981)
30. Shirakawa, K. *et al.*, *IEEE Trans. Magn.*, **16,** 910 (1980)
31. Steinemann, S. G., *J. Magnetism magn. Mater.*, **7,** 84 (1978); **12,** 191 (1979)
32. Fletcher, G. C., *Proc. phys. Soc.*, **68,** 1066 (1955); **78,** 145 (1961); *see also* Swirkowicz, R. and Sukiennicki, A., *Acta Phys. pol.*, **A57,** 517 (1980)
33. Briane, M., *C.r. hebd. Séanc Acad. Sci., Paris*, **276,** 139 (1973); **277,** 695 (1973)
34. Wohlfarth, E. P., *Phys. Lett.*, **69A,** 222 (1978)
35. Güntherodt, H. J., *et al.*, Institute of Physics Conf. Series, No. 55, p. 619 (1981)
36. Shtrikman, S. and Wohlfarth, E. P., *Physica*, **60,** 427 (1972)
37. Yamada, H. and Wohlfarth, E. P., *Phys. Lett.*, **51A,** 65 (1975)
38. Wagner, D. and Wohlfarth, E. P., *J. Magnetism magn. Mater.*, **15—18,** 1345 (1980); *Physica*, **112,** 1 (1982)
39. Edwards, D. M., Mathon, J. and Wohlfarth, E. P., *J. Phys. F*, **5,** 1619 (1975)
40. Coey, J. M. D. *et al.*, *J. Phys. F*, **11,** 2707 (1981); Croat, J. J. and Herbst, J. F., *J. Appl. Phys.*, **53,** 2294 (1982)
41. Kaul, S. N. and Rosenberg, M., *Phys. Rev.*, **B25,** 5863 (1982)
42. Hiroyishi, H., *et al.*, *Int. Symp. High Field Mag.*, Osaka, to be published
43. Khanna, S. N. and Wohlfarth, E. P., to be published

Chapter 16

Magnetic anisotropy

H. Fujimori

The Research Institute for Iron, Steel and Other Metals, Tohoku University, Sendai, Japan

16.1 Introduction

The structure of amorphous metals consists of a random array of atoms, and is normally considered to be isotropic on a long range scale. As a result, the ferromagnetic magnetization vector is supposed to be free from any special anisotropy. This concept provided the strong motive for the development of soft magnetic materials, and in fact excellent soft magnetic properties were realized in various amorphous ferromagnets based on iron, cobalt and nickel.

However, if one goes into the details, it is seen that most amorphous ferromagnets behave anisotropically. Relatively hard magnetization beyond the hysteresis loop range, sharp 180 degree type or complex maze type magnetic domains and two-fold symmetrical magnetic torque, which can all be observed in most melt-quenched amorphous ribbons, are the typical properties reflecting magnetic anisotropies. For example, in the ideal case of a completely isotropic magnet, domain walls do not appear and the spin moments must be smoothly rotated inside the whole specimen (circular magnetic domains). But real amorphous materials have sharp 180 degree walls by which the permeability and the coercivity or Barkhausen noise in regard to magnetization hysteresis are strongly governed.

In addition, it is well known that magnetic field cooling, cold rolling and stress annealing of the amorphous materials can induce additional magnetic anisotropies superposed on the as-received anisotropy. These induced anisotropies are also characterized by two-fold symmetry and respond reversibly to reannealing.

The characteristics of these anisotropies will now be described, and the physical origins discussed. The amorphous materials concerned here are limited to melt-quenched ribbons; sputter-deposited materials are excluded.

16.2 As-received magnetic anisotropies

16.2.1 Local anisotropy

The hysteresis loop of the melt-quenched amorphous ribbons, measured along the ribbon axis, is fairly rectangular, exhibiting low coercivities of the order of 0.01—0.1 Oe

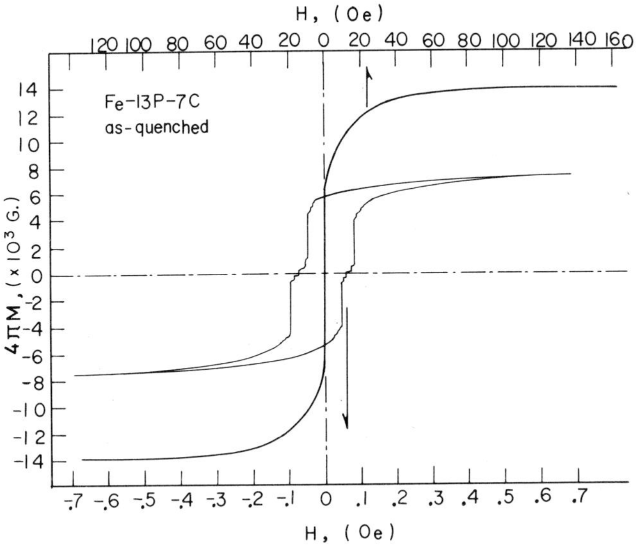

Figure 16.1 Magnetization curves in the hysteresis range and in the approach-to-saturation range of $Fe_{80}P_{13}C_7$ amorphous ribbon in the as-received state. The measurements were done along the ribbon long axis (after Fujimori *et al.*[1])

but with rather small remanences. *Figure 16.1* shows a typical example ($Fe_{80}P_{13}C_7$ in the as-received state)[1]. Closely related to the small remanence, the magnetization beyond the coercive field range increases slowly so that saturation does not take place until a field corresponding to about 1000 times the coercive field is reached. The magnetic domain observations have revealed that the existence of maze domains is responsible for this hard magnetization. *Figure 16.2* shows the Bitter figures[2] of the maze domains observed in the sample used in *Figure 16.1*. Such complex domain patterns appear irregularly and locally on the ribbon surface. By applying a perpendicular field, the zigzag stripes becomes much clearer, and the black and white stripes are exchanged if the direction of perpendicular field is reversed (see *b* and *c* of the *Figure 16.2*). From these properties, it is said that the maze domains are attributed to a perpendicular magnetic anisotropy. A realistic model has been proposed[3–5]. The effects of composition, heat-treatment and applied stress on anisotropy have been investigated extensively and as a result, it is believed that the origin of the perpendicular anisotropy is due to the magnetostriction–stress coupling effect. The following are the experimental evidence and the mechanism involved.

By adopting the ratio of remanence magnetization (M_r)/saturation magnetization (M_s), Luborsky *et al.*[6] have examined how the hardness in magnetization is affected by annealing. *Figure 16.3* shows their results obtained on the Fe–Ni–P–B amorphous alloys. The ratio M_r/M_s increases gradually with increasing annealing temperature (T_a), reaching almost the saturation value $M_r/M_s = 1$ around $T_a = 350$—$400\,°C$. On the other hand, they have measured the stress relief of the same ribbon samples, where the stress relief has been estimated from the change in r/r' on annealing. r is the diameter of the ring bobbin on which the ribbon sample is wound tightly and r' is the effective relaxed diameter of the sample that is removed from the bobbin after the annealing. The results obtained for r/r' are also shown in *Figure 16.3*. The significant point is that M_r/M_s and r/r' change on annealing in the same manner. This coincidence implies that the hard magnetization in the as-received ribbon is caused by internal stress.

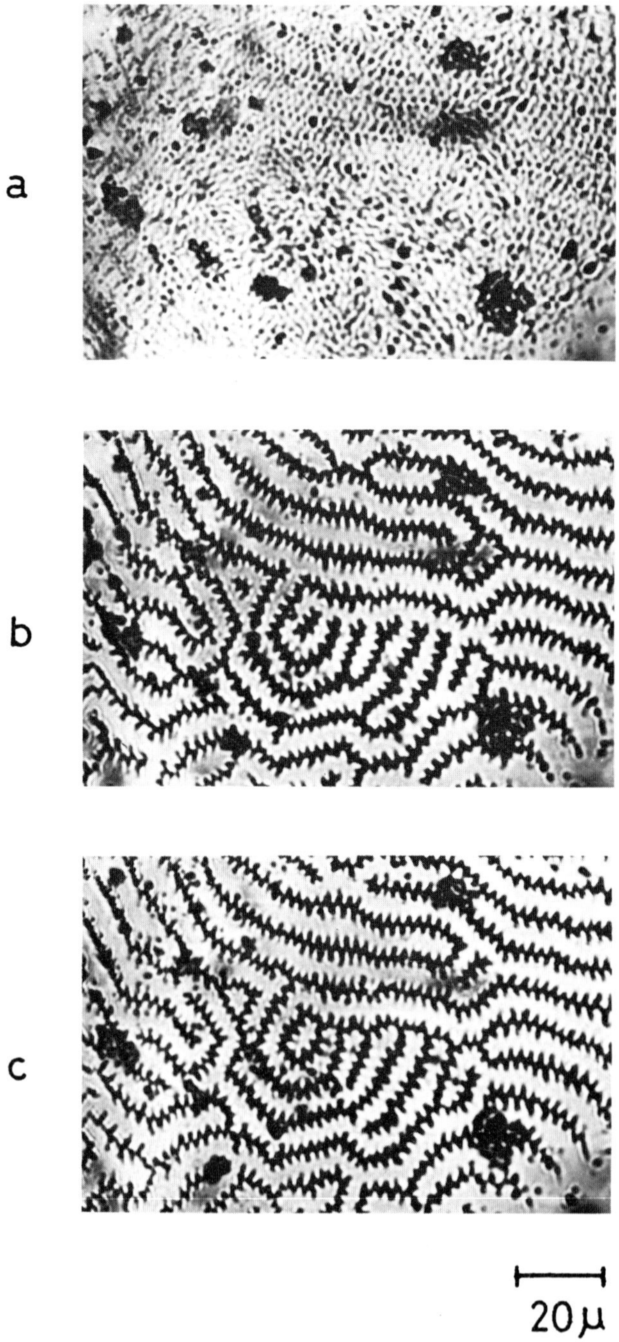

Figure 16.2 Magnetic domains by means of Bitter technique. The figures shown here are the so-called maze domains appearing on the same ribbon surfaces as in *Figure 16.1*: (*a*) without external field; (*b*) with a field 100 Oe applied perpendicularly to the ribbon surface; (*c*) with the same field but applied opposite to the direction for (*b*) (after Obi *et al.*[2])

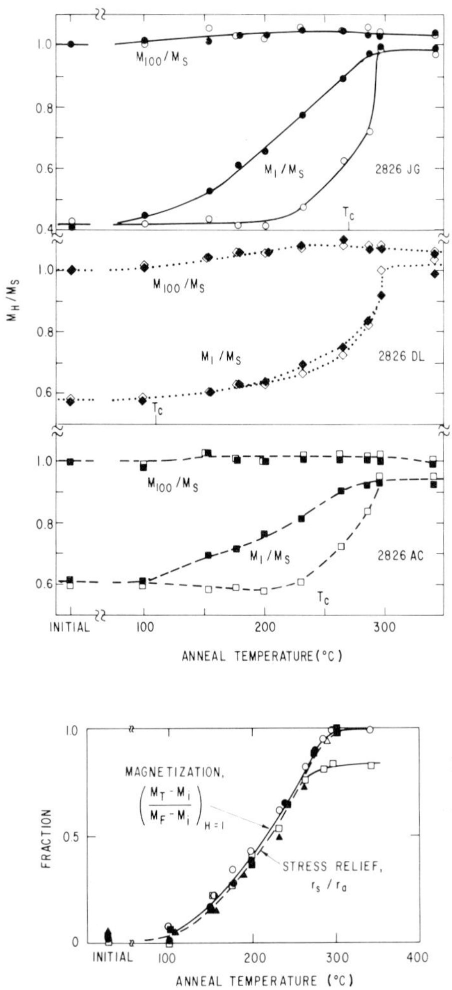

Figure 16.3 Ratio of magnetization at 1 Oe (nearly equal to remanence value) and 100 Oe to saturation magnetization M_s as a function of annealing temperature T_a for the METGLAS 2826 $Ni_{40}Fe_{40}P_{14}B_6$. Closed symbols: 21 Oe annealing field. Open symbols: without annealing field. The bottom figure shows the fraction of the magnetization change on annealing, together with the fraction of the stress relief r/r' (after Luborsky *et al.*[6])

Instead of M_r/M_s, the magnetization energy defined as

$$E_a = \int_0^{H_a} (M_s - M)\,dH \tag{16.1}$$

where M is the magnetization, H the field, M_s the saturation magnetization and H_a the field necessary to bring M to M_s, can measure the magnetizing energy more quantitatively. O'Handley *et al.*[7] have shown that the compositional dependence of E_a for Fe–Ni–B amorphous alloy is linearly related to the saturation magnetostriction (λ_s). Further, Egami *et al.*[8] have shown that E_a decreases when an external tension is applied along the ribbon axis, and that the temperature dependence of E_a is exactly the same as the dependence of magnetostriction on temperature. Fujimori *et al.*[3] have, on the other hand, examined the relation between magnetostriction and magnetic domain structure in the $(Fe_{1-x}Co_x)_{80}P_{13}C_7$ and $(Fe_{1-x}Co_x)_{75}Si_{10}B_{15}$ amorphous alloys. In *Figure 16.4*, magnetostriction λ_s, anisotropy field H_a and field necessary to make the maze domains *Hm* disappear are shown as a function of concentration x. Note that λ_s of these systems changes from large positive values in the iron-rich region to small negative values in the

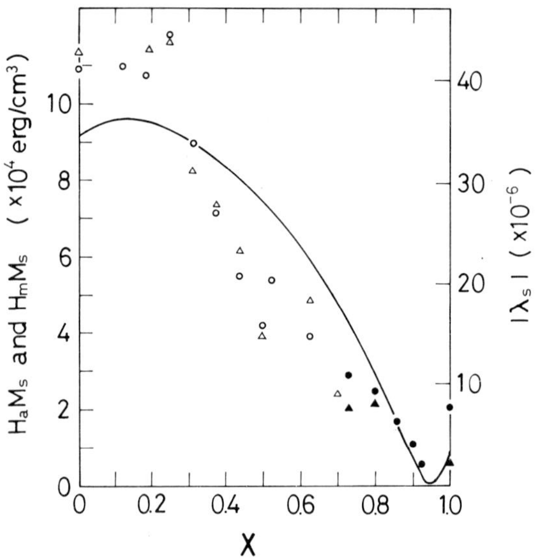

Figure 16.4 Comparison between H_aM_s (○, ●) or H_mM_s (Δ, ▲) and magnetostriction λ_s (solid line) as a function of concentration x for $(Fe_{1-x}Co_x)_{80}P_{13}C_7$ (○, Δ) and $(Fe_{1-x}Co_x)_{75}Si_{15}B_{10}$ (●, ▲) amorphous alloys. H_a is the field necessary to obtain M_s, H_m the field along the ribbon axis necessary to make the maze domains disappear. Thus, H_aM_s or H_mM_s can measure the perpendicular magnetic anisotropy energy (after Fujimori *et al.*[3])

cobalt-rich region, passing through zero[9] in the vicinity $x = 0.094$. (It is well known that the cobalt-rich Co–Fe based amorphous alloys with zero magnetostriction are superior high permeability materials[10–12]. As *Figure 16.4* shows, the concentration dependence of H_a and H_m is nicely coincident with the absolute value of λ_s. Let us discuss this correlation in terms of the anisotropic stress–magnetostriction coupling effect. The magnetoelastic energy arising from this coupling, E_σ, can be expressed as

$$E_\sigma = -\tfrac{3}{2}\lambda_s\sigma\cos^2\theta \qquad (16.2)$$

where σ is the local stress component assumed to be normal to the ribbon surface and θ is the angle between the stress axis and magnetization vector. Owing to E_σ, magnetization along the ribbon axis is easy or hard depending on the sign of λ_s as well as on whether the stress is tensile or compressive with respect to the ribbon surface. In the case of tensile stress, the magnetization is hard when $\lambda_s > 0$, but easy when $\lambda_s < 0$ (contrary to the case of compressive stress). Thus, if a uniform stress is assumed in order to explain the relation of *Figure 16.4*, the stress in the positive λ_s alloys must be tensile, while compressive in the negative λ_s alloys. However, the structure of the stresses to obtain the observed magnetic domain structure is different from the above simple speculation. It is deduced from the domain observation (*Figure 16.5*) that the as-received ribbon involves internal stresses in the form of a coexistence of tensile and compressive stresses distributed locally. In other words, the maze domains that can be seen in both the positive and negative λ_s alloys (note that the maze domains do not appear in the zero λ_s alloy) can be explained as follows: The maze domains for $\lambda_s > 0$

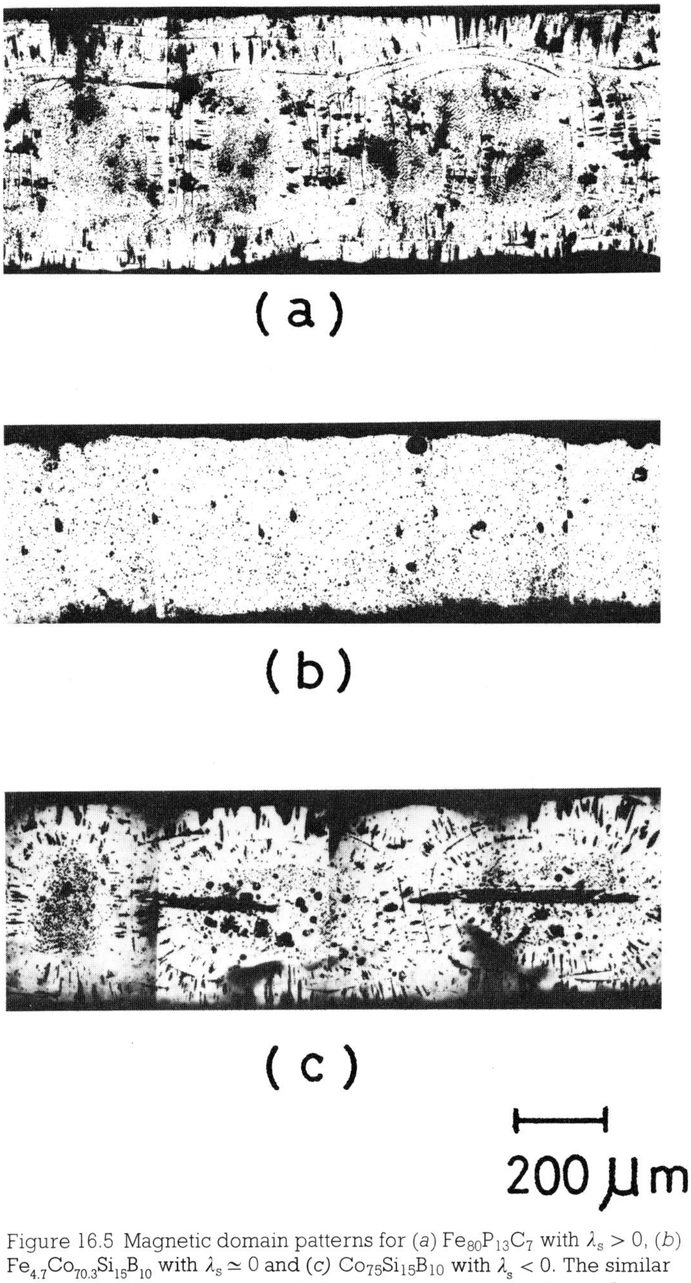

Figure 16.5 Magnetic domain patterns for (*a*) $Fe_{80}P_{13}C_7$ with $\lambda_s > 0$, (*b*) $Fe_{4.7}Co_{70.3}Si_{15}B_{10}$ with $\lambda_s \simeq 0$ and (*c*) $Co_{75}Si_{15}B_{10}$ with $\lambda_s < 0$. The similar complex domains consisting of 180 degree and maze walls can be seen in the alloys of $\lambda_s > 0$ and < 0, while they do not appear in alloys with $\lambda_s = 0$ (after Fujimori *et al.*[3])

appear by reflecting the tensile regions while those for $\lambda_s < 0$ reflect the compressive regions.

Thus, we have seen that the melt-quenched amorphous ribbons are magnetically anisotropic, being characterized by the spontaneous magnetization deviating from the plane of the ribbon. This anisotropy is not uniform. Internal stress fluctuation–magnetostriction coupling is thought to be the origin of the local fluctuations in the anisotropy. The stress fluctuations may arise from the fluctuations in the adherence of the ribbon to the melt-quenching wheel during its formation. The Mössbauer experiments have also confirmed the existence of the perpendicular magnetization component in the melt-quenched ribbons[13,14].

16.2.2 Two-fold in-plane anisotropy

In addition to the local perpendicular magnetic anisotropy, it is known that the melt-quenched amorphous ribbon is also anisotropic in its in-plane magnetization. Typical experimental evidence for this can be seen in the magnetic torque curve of a disc sample made from the wide ribbon[15–17]. The in-plane magnetic torque has clearly revealed that the anisotropy is characterized by a two-fold symmetry as shown[15] in *Figure 16.6*. Therefore, the in-plane anisotropy energy E_u can be expressed as

$$E_u = -K_u \cos^2\theta \qquad (16.3)$$

where K_u is the anisotropy constant and θ the angle between magnetization vector and its easy direction. Then, K_u can be calculated from the torque $L = -dE_u/d\theta$ as

$$K_u = \tfrac{1}{2}L_{max} \qquad (16.4)$$

where L_{max} is the amplitude of the torque curve.

In order to understand the origin for the in-plane anisotropy, Morita *et al.*[18] have examined the compositional dependence of K_u in the $(Fe_{1-x}Co_x)_{78}Si_{10}B_{12}$ amorphous alloys, and they have found that neither magnetostriction–stress coupling nor internal shape effect can explain the result, since the curve of K_u against x is quite different from

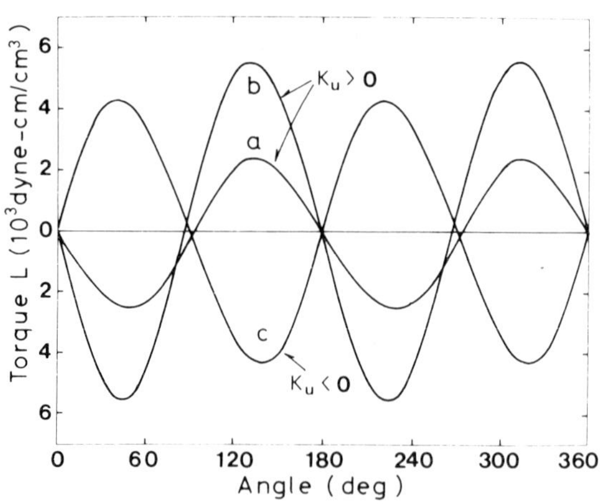

Figure 16.6 Torque curves of the in-plane magnetic anisotropy for $Fe_{4.7}Co_{70.3}Si_{15}B_{10}$ amorphous alloy ($\lambda_s \simeq 0$). *a*, as-received; *b*, cold-rolled along the ribbon width; *c*, cold-rolled along the ribbon axis (after Morita *et al.*[15])

the relations of λ_s to x and M_s^2 to x. In addition, the anisotropy has been found not to be much altered by heat treatment, even at high temperatures where the internal stress relief was completed. Surprisingly, almost the same anisotropy remained after the crystallization by a higher temperature or prolonged annealing. Takahashi *et al.*[19] have interpreted this peculiar behaviour in terms of the microcrystalline effect. They assumed that a very weak orientation of microcrystallites, which is possible in the amorphous state, gives rise to the magnetic anisotropy. If so, it is likely that the crystallization occurs in such a way as to fit the crystal axis with those of the initially existing microcrystallites in the amorphous state. Then the crystallized sample has an anisotropy similarly oriented to the direction of energy minimum in the amorphous state. Particularly, they emphasized that the anisotropy observed in Fe–P amorphous alloys can be interpreted by this criterion. The reason is that the curve of K_u against measuring-temperature for the $Fe_{82}P_{18}$ amorphous alloy shows a peak at the low temperature region, similarly to its crystallized state, as shown in *Figure 16.7*. On the

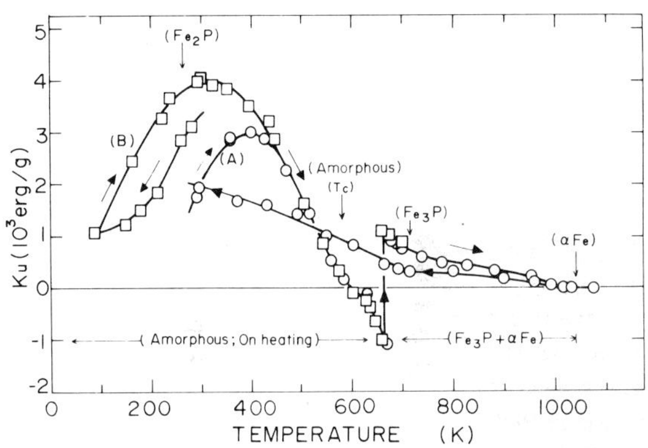

Figure 16.7 Dependence of measuring temperature on the uniaxial anisotropy K_u for $Fe_{82}P_{18}$ amorphous alloy. The arrows above the curves indicate the Curie point for Fe_2P, Fe_3P, α-Fe and the amorphous $Fe_{82}P_{18}$ alloy, respectively. Note that the curve of K_u against T_a shows a peak at the low temperature region in both the amorphous and crystalline states (after Takahashi *et al.*[19])

other hand, Fujimori and his co-workers[18,20] considered the effect of atomic pair ordering. Their idea arose from the observation of roll induced magnetic anisotropy[15]. The essential point is that in the process of melt-quenching, the molten alloy or supercooled viscous liquid close to the surface of the quenching wheel may suffer from a shear stress, and it may be deformed similarly to the plastic deformation which occurs by cold-rolling. Such a deformation seems to induce a two-fold anisotropic array of atoms or directional ordering of atomic pairs in the frozen amorphous solid. It is thought that the main driving force for the directional ordering is the different chemical bonding force between metal–metal and metal–metalloid interactions. The above two arguments are, however, still hypothetical, and we need further detailed investigations.

16.3 Induced magnetic anisotropies

16.3.1 Field induced anisotropy

As is well known, many ferromagnetic crystalline materials can respond to field cooling, by which a two-fold magnetic anisotropy is induced. Directional atomic pair ordering in disordered alloys[21,22], preferred crystalline orientation due to martensitic transformation in pure cobalt[23], elongated precipitates in precipitation type permanent magnets[24], and directional ion–ion pair ordering in ferrites[25] are the main origins for the anisotropy. It has been known that amorphous alloys also exhibit uniaxial magnetic anisotropy by field cooling. The observed magnitudes of the anisotropy constant range from 10^2 to $10^4\,\mathrm{erg/cm^3}$ for most amorphous alloys, which are the same order of magnitude as for crystalline disordered alloys.

So far, Fe–P–C (Berry *et al.*[26]), Fe–Ni–P–B and Fe–Ni–B (Luborsky *et al.*[6,27–29]), Fe–P–C, Fe–Co–Si–B and Co–M–Si–B (Fujimori *et al.*[3,30,31]), Fe–Co–Si–B (Miyazaki *et al.*[36]) and Fe–B (Allia *et al.*[17]) have been examined in detail. To summarize, the following are the characteristic properties of the field induced magnetic anisotropy:

1. The development of the anisotropy is governed by a thermal activation process.
2. The driving force is the magnetic interaction.
3. The coexistence of two or more different kinds of metallic element is much more effective in developing stronger anisotropies than those of amorphous alloys including only one metallic element.

Figure 16.8 shows the development of induced anisotropy for $(Fe_{0.2}Co_{0.8})_{70}Si_{2.5}B_{27.5}$ amorphous alloy as a function of isothermal annealing time[31,32]. Here, the anisotropy constant K_u^f was obtained as

$$K_u^f = \int_0^{H_a} (M_{\parallel} - M_{\perp})\,\mathrm{d}H \tag{16.5}$$

where $M_{\parallel}$ is the magnetization of a long ribbon annealed in a field of about 100 Oe directed along the ribbon axis, $M_{\perp}$ the magnetization of the same ribbon but reannealed in a field directed across the ribbon width, and H_a the magnetizing field necessary for saturation magnetization. K_u^f thus defined can also be obtained by torque measurements according to equations 16.3 and 16.4. *Figure 16.8* shows that K_u^f changes with time and annealing temperature, showing a rapid saturation at higher temperatures. The almost saturated value, K_{us}^f, increases from the Curie temperature with decreasing annealing temperature[33] as shown in *Figure 16.9*. On the other hand, the characteristic time or relaxation time τ necessary to bring K_u^f to half of K_{us}^f at each annealing temperature follows approximately the Arrhenius plot, as represented[33] in *Figure 16.10*. The result gives the activation energy of about 1.4 eV for the presently used amorphous alloy. Similar estimations for other magnetic amorphous alloys yield activation energies from 0.7 to 1.5 eV which are smaller than those for the crystalline solid-solution-type alloys.

Concerning the origin of K_u^f, the most provable model is based on the Néel–Taniguchi theory[21,22] of the directional atomic pair ordering in crystalline alloys. The physical concept is as follows: Pseudo-dipole–dipole interactions are different in the different kinds of atomic pair existing in a magnetic alloy. Thus, if the system is thermally activated in the presence of an external magnetic field, the atomic pairs tend to align so as to minimize the total magnetic energy. Then, when the system is cooled to

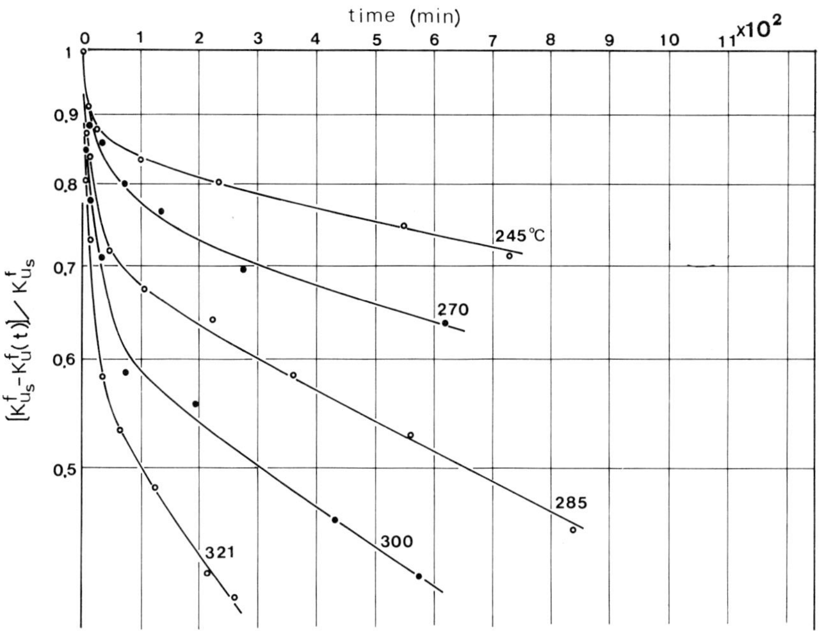

Figure 16.8 Relaxation process of the development of field-induced uniaxial magnetic anisotropy K^f_u for $(Fe_{0.2}Co_{0.8})_{70}Si_{2.5}B_{27.5}$ amorphous alloy with $\lambda_s \simeq 0$. K^f_{us} used is the saturation value of K^f_u, and it was obtained by the successive long annealings from 450 °C to 200 °C (after Fujimori *et al.*[32])

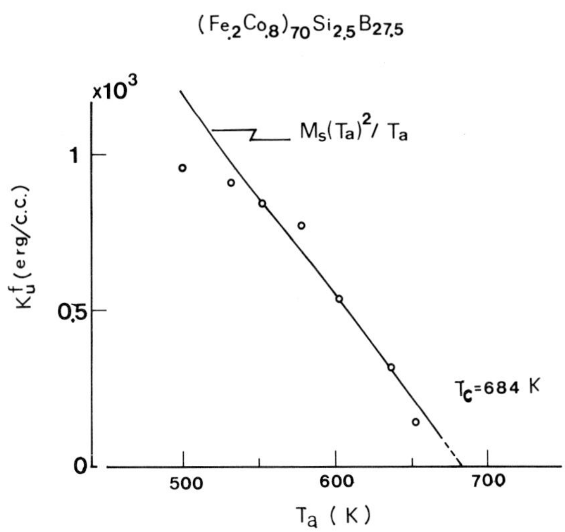

Figure 16.9 K^f_{us} as a function of annealing-temperature T_a for the same alloy as in *Figure 16.8*. The solid line is the theoretical dependence according to equation 16.6 (after Fujimori *et al.*[33])

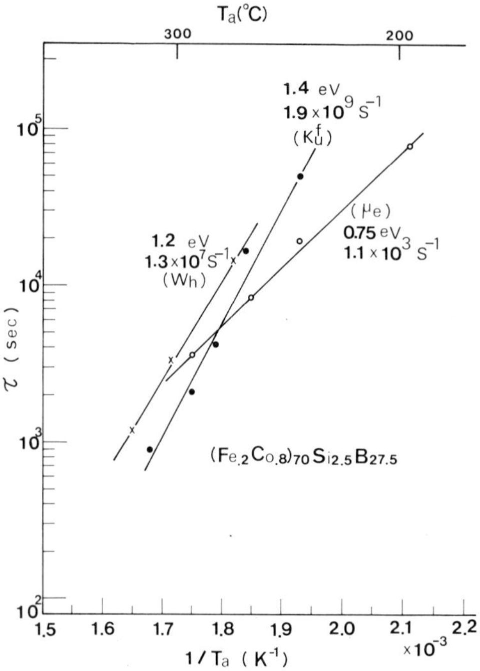

Figure 16.10 Arrhenius plots of the relaxation time for K_u^f obtained in *Figure 16.8*. The same plots for the increase in magnetization hysteresis loss W_h and the decrease in AC permeability μ_e, both of which take place by ageing in the absence of external field, are also shown for comparison. As seen here, these ageing effects are related to the development of K_u^f, being due to the so-called domain wall fixing[20,33,35] (after Fujimori *et al.*[33]). Activation energies and frequency factors are shown on the curves

low temperatures, where atomic diffusion is suppressed, the directionally ordered pairs are frozen. As the result, the system has a uniaxial magnetic anisotropy. The directional pair ordering in the crystalline lattice occurs by an interexchange diffusion between two different kinds of atom (*see* the schematic picture after Graham[34] for example). However, the corresponding directionality in the amorphous alloy may occur by a short-range diffusion and in the form of a local collective mode in atomic arrangement, since the well defined lattice sites do not exist and the interexchange diffusion between atoms is not likely to happen within the amorphous state. *Figure 16.11* is the artifical two-dimensional picture to explain a possible directional collective arrangement of atoms in the amorphous state. Since the thermal equilibrium distribution of directional atomic pairs is decided by the Boltzmann distribution function, K'_u, *which is the* anisotropy constant defined by the Néel–Taniguchi model, increases in proportion to $M_s(T_a)^2/T_a$. M_s is the saturation magnetization at the annealing temperature T_a, and K'_u depends on concentration x. For ideal solutions $K'_u \propto x^2(1-x)^2$. The explicit expression of the energy E'_u is given as

$$E'_u = -K'_u \cos^2\theta$$

$$K'_u = -\frac{A}{kT_a}[M_s(T_a)/M_s(0)]^2[M_s(T)/M_s(0)]^2 x^2(1-x)^2 \qquad (16.6)$$

where k is the Boltzmann factor, $M_s(T)$ the saturation magnetization at room temperature, θ the angle between the magnetizing field and the direction of the annealing field and A the numerical constant including the coefficients of dipole interactions.

Figures 16.9 and *16.12* compare the experimental K_u^f with the theoretical expression 16.6. As seen in *Figure 16.9*, K_{us}^f increases from the Curie point in good agreement with the theoretically expected curve of $M_s(T_a)T_a$ against T_a, although a

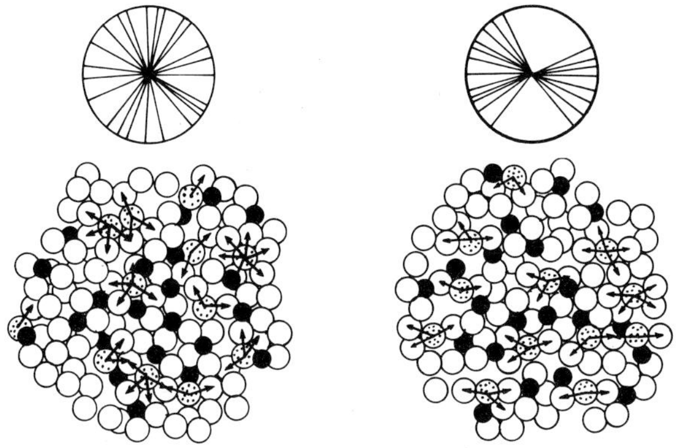

Figure 16.11 Schematic two-dimentional pictures of the atomic pair correlation in the amorphous alloy; (a) isotropic correlation in the amorphous alloy; (b) collective directional ordering in pairs represented by empty circles and dot-filled circles in the nearest neighbour state. These two types of circle represent the different kinds of metal atoms and the filled circles represent the metalloids. The basic model structure of the arrangement of empty and filled (black) circles is after Fukunaga and Suzuki[38]

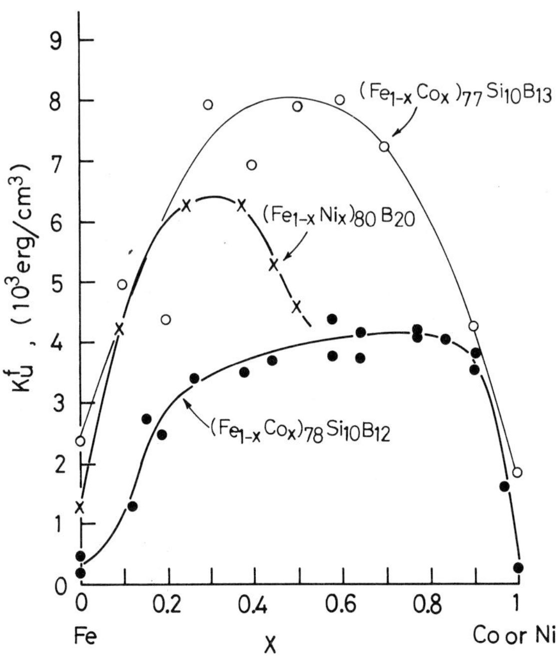

Figure 16.12 Concentration dependence of K_u^f at room temperature. Open circles (after Miyazaki *et al.*[36]) for $(Fe_{1-x}Co_x)_{77}Si_{10}B_{13}$ and crosses (after Luborksy *et al.*[27]) for $(Fe_{1-x}Ni_x)_{80}B_{20}$ are the nearly saturated values obtained for the field-annealing temperatures of 300°C and 226°C, respectively. Filled circles (after Fujimori *et al.*[31]) for $(Fe_{1-x}Co_x)_{78}Si_{10}B_{12}$ are the unsaturated values because of the simple field-cooling

slight deviation of experimental points can be seen particularly in the low temperature side. This deviation seems to be caused by the lack of saturation of K_u^f. Concerning the concentration dependence of K_{us}^f shown in *Figure 16.12*, the significant point to note is the tendency for a peak in K_{us}^f as a function of x in all the alloy cases examined. According to equation 16.6

$$\frac{K_{us}^f \times T_a}{[M_s(T_a)/M_s(0)]^2[M_s(T)/M_s(0)]^2} \propto x^2(1-x)^2 \tag{16.7}$$

The plots of the relation of the left hand side of equation 16.7 (labelled Q for convenience) are represented in *Figure 16.13* for the Fe–Ni–B system[27] and compared with the theoretical relation $x^2(1-x)^2$ calculated by normalizing to give the best fit at $x = 0.5$. Note that in *Figure 16.13* the non-zero value of K_{us}^f at $x = 0$ is subtracted from the values at the other concentrations, because the pair ordering theory accounts only for a simple binary alloy system in which the anisotropy is zero at $x = 0$ or 1 owing to the absence of different kinds of atomic pair. The non-zero values at $x = 0$ and 1 in the amorphous alloys (*Figure 16.12*) are obviously due to the directional pair ordering between metal and metalloid atoms, but these anisotropies are smaller than 10^3 erg/cm^3. *Figure 16.13* shows that the experimental values thus obtained are far above $x^2(1-x)^2$ in the intermediate region of x. Luborsky *et al.*[27] and later on Miyazaki *et al.*[36] have improved the theory by taking into account precipitation type

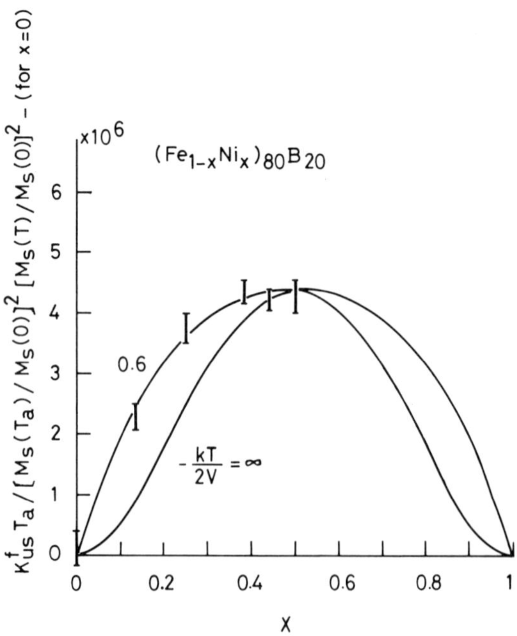

Figure 16.13 Comparison of experiment and theory on K_{us}^f for amorphous $(Fe_{1-x}Ni_x)_{80}B_{20}$ alloys. Data points are the K_{us}^f corrected by annealing temperature and magnetization according to equation 16.7. Solid lines with $-kT/2V = \infty$ and 0.6 are the theoretical relations for an ideal solid solution or $x^2(1-x)^2$ and for a precipitation type solid or $V < 0$, respectively (after Luborsky *et al.*[27])

atomic pair interactions. According to Iwata,[37] the Néel and Taniguchi theory, for the case of ideal solid solution, must be modified to account for the effect of superlattice and precipitation type interactions. The upper solid line in *Figure 16.13* represents Iwata's relation in which $-kT/2V$ was taken to be 0.6 for the best fit for Fe–Ni–B system. V used here is the atomic pair potential, being positive for superlattice type, negative for precipitation type and zero for ideal solid solutions. Thus, the presently found negative value of V suggests that atomic cohesive interactions among the constituent atoms in the amorphous alloys are stronger between the same kinds of atoms rather than for the different kinds of atom.

We have seen that the field induced anisotropy of amorphous alloys can be interpreted within the framework of the directional atomic pair ordering. The pairs between the different kinds of metal–metal atom are much more important than the metal–metalloid pairs, so that the results of *Figure 16.12* can be satisfactorily explained. Finally, we should note that Becker[39] proposed an alternative model based on a single atom mechanism. The composition dependence predicted agrees well with the results of the Fe–Ni–B alloys[27] without any modification for tendency towards precipitation type clustering or compound formation and it accounts for magnetic annealing effects in iron-only alloys.

16.3.2 Roll-induced anisotropy

Roll induced magnetic anisotropy has been investigated for $(Fe_{1-x}Co_x)_{78}Si_{10}B_{12}$ amorphous alloys by torque measurement (Morita *et al.*[15,40]), and the related extrinsic magnetic properties have been examined for the $Fe_{40}Ni_{40}P_{14}B_6$ amorphous alloy (Luborsky *et al.*[41]). For these alloys, cold rolling is possible up to about 30—40 per cent reduction in thickness, by using a conventional roll technique for thin sheets. It has been found that deformation of amorphous ribbons occurs, on a macroscopic level, in a slip-band-like mode[42], but the amorphous structure, as confirmed by the X-ray diffraction, is not altered.

Roll-induced anisotropy is again uniaxial in the plane of the rolled ribbon, being described with a two-fold symmetry energy E_u^r equivalent to equation 16.3. An example of the torque curves is shown in *Figure 16.6*. Cold rolling will increase the internal strain in the ribbon, but the anisotropy does not seem to be correlated with a magnetostriction–stress coupling effect, because, as seen in *Figure 16.6*, the anisotropy can appear even in the zero magnetostriction alloy. The interesting behaviour of E_u^r to note is the direction of the energy minimum. As seen in *Figure 16.6*, the direction of easy magnetization is always perpendicular to the roll direction, irrespective of the roll direction, i.e., rolling along the long axis or along the width axis. Therefore, the longitudinal reduction results in a hard magnetization along the ribbon axis. The tendency for the decrease in M_1/M_s by rolling found by Luborsky *et al.*[41] for the $Fe_{40}Ni_{40}P_{14}B_6$ amorphous ribbon (*Figure 16.14*) is in reasonable agreement with the result of the torque measurement, where M_1 is the magnetization at 1 Oe (and consequently close to the remanence value) and M_s is the saturation magnetization. This relation between the direction of easy magnetization and the roll direction is attributed to the mechanism by which the slip bands are formed perpendicular to the roll direction. Morita *et al.*[15,40] considered that atomic pair directional ordering may be developed in association with the slip-bands and be responsible for the uniaxial induced anisotropy. On the other hand, Luborsky *et al.*[41] emphasized the importance of a magnetostriction–strain effect, by showing that the decrease in M_1/M_s by rolling can be recovered by a long annealing around 350 °C.

At the present, therefore, the origin for E_u^r is still not clear.

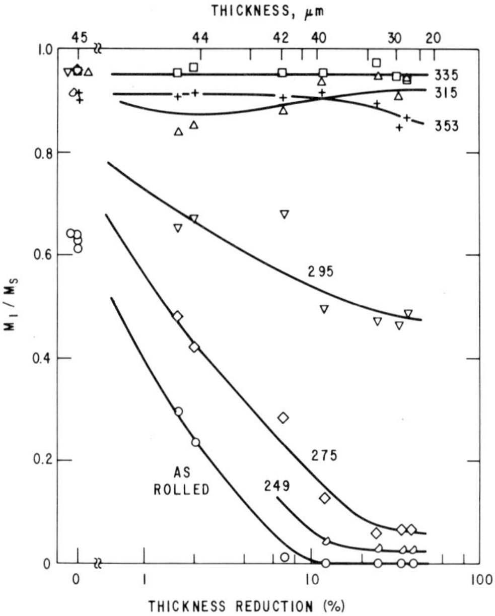

Figure 16.14 Change in ratio of magnetization at 1 Oe/saturation magnetization for METGLAS 2826 as a function of thickness reduction by cold rolling. The subsequent changes on annealing at 249–335 °C are shown with the numbers (after Luborsky *et al*[41])

16.3.3 Strain-anneal induced anisotropy

In the early work on soft magnetic properties[3,30,43] it was found that an application of tensile strain along the ribbon axis makes the magnetization easy, in the case of positive magnetostrictive amorphous alloys. This is due to the simple magnetoelastic effect and can be explained in terms of equation 16.2. Thus, the strain-induced anisotropy is caused to vanish reversibly by removal of the external strain. Contrary to this, strain-anneal induced anisotropy was recently found[44,45,46]. This anisotropy can remain in the strain-free sample after the annealing. Such a strain-annealing effect was first shown by Luborsky *et al.*[47] during a systematic study of the peculiar behaviour of the winding stresses. During annealing of toroidal samples the coercivity and the core losses in iron-rich amorphous alloys remained high, being quite different from the remarkable reduction in losses in the annealed straight ribbon sample (pointed out also by Fujimori *et al.*[49]).

According to the results of Nielsen *et al.*[44,45] and Hilzinger,[46] after annealing under tensile stress, amorphous alloys showed nearly the equivalent magnetic anisotropy perpendicular to the ribbon axis, irrespective of the sign of magnetostriction. *Figure 16.15* shows the magnetization curves[46] of the $Co_{66}Fe_4Si_{16}B_{12}$ amorphous alloy with $\lambda \simeq 0$. The value of the anisotropy constant, K_u^σ, estimated from *Figure 16.15*, was found to increase linearly with increasing tensile stress σ applied during the annealing. K_u^σ is about $160\,J/m^3$ for $\sigma = 500\,MPa$. K_u^σ behaves reversibly in a cyclic annealing with and without tensile stress; K_u^σ rises by stress annealing and disappears completely in annealing free from stress. Furthermore, it is emphasized that K_u^σ can be induced even at higher temperatures than the Curie point, suggesting that the K_u^σ is attributed to the different mechanism from that for the field anneal anisotropy. Nielsen *et al.*[44,45] and Hilzinger[46] considered that K_u^σ may be interpreted as a transient creep effect which has been observed by Kimura *et al.*[48]. Nielsen *et al.* predicted the coupling effect between magnetostriction and compressive strain that is frozen in during the creep process, while, in order to explain the finite K_u^σ observed in the zero

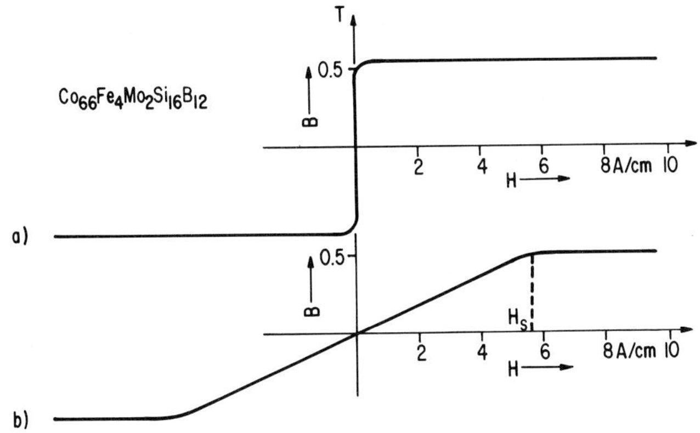

Figure 16.15 Change in magnetization curve due to stress-anneal induced anisotropy for a $Co_{66}Fe_4Mo_2Si_{16}B_{12}$ amorphous alloy: (*a*) annealed at 400 °C for 1 h without stress, (*b*) annealed under tensile stress $\sigma = 272$ MPa (after Hilzinger[46])

magnetostrictive alloy, Hilzinger postulated that a structural rearrangement occurred without any internal stress in the strain-annealed amorphous samples.

The mechanism of K_u^σ is still questionable too. It is of interest that the direction of the induced energy minimum for K_u^σ is perpendicular to the direction of applied tensile stress. This relation is just the same as in the case of cold-roll-induced anisotropy, as already discussed.

References

1. Fujimori, H., Masumoto, T., Obi, Y. and Kikuchi, M., *Jap. J. appl. Phys.*, **13,** 1889 (1974)
2. Obi, Y., Fujimori, H. and Saito, H., *Jap. J. appl. Phys.*, **15,** 611 (1976)
3. Fujimori, H., Obi, Y., Masumoto, T. and Saito, H., *Mater. Sci. Engng*, **23,** 281 (1976)
4. Takahashi, M., Suzuki, T. and Miyazaki, T., *Jap. J. Appl. Phys.*, **16,** 521 (1977)
5. Becker, J. J., American Institute of Physics Conf. Proc. No. 29, p. 204 (1976)
6. Luborsky, F. E., Becker, J. J. and McCary, R. O., *IEEE Trans. Magn.*, **MAG-11** 1644 (1975)
7. O'Handley, R. C., *IEEE Trans. Magn.*, **MAG-11,** 206 (1975)
8. Egami, T. and Flanders, P. J., *IEEE Trans. Magn.*, **MAG-11,** 220 (1976)
9. Fujimori, H., Arai, K. I., Shirae, H., Saito, H., Masumoto, T. and Tsuya, N., *Jap. appl. Phys.*, **15,** 705 (1976)
10. Kikuchi, M., Fujimori, H., Obi, Y. and Masumoto, T., *Jap. J. appl. Phys.*, **14,** 1077, 1975)
11. Fujimori, H., Kikuchi, M., Obi, Y. and Masumoto, T., *Sci. Rep. Res. Insts Tôhoku Univ.*, **26A,** 36 (1976)
12. Sherwood, R. C., Gyorgy, E. M. and Leamy, H. J., American Institute of Physics Conf. Proc. No. 24, p. 745 (1975)
13. Ura, M., Thesis, Osaka University (1974)
14. Ito, A., Torikai, E., Morimoto, S., Shiiki, K. and Kudo, M., *'Proc. 4th Int. Conf. on Rapidly Quenched Metals'*, Eds. Masumoto, T. and Suzuki, K., vol. II, p. 1101 Japan Institute of Metals, Sendai (1982)
15. Morita, H., Fujimori, H. and Obi, Y., *Jap. J. appl. Phys.*, **18,** 683 (1979)
16. Takahashi, M., Ono, F. and Takakura, K., *Jap. J. appl. Phys.*, **15,** 183, 1821 (1976)
17. Allia, P., Luborsky, F. E., Soardo, G. P. and Vinai, F., *J. appl. Phys.*, **52,** 3553 (1981)
18. Morita, H., Fujimori, H. and Obi, Y., *Jap. J. appl. Phys.*, **20,** 126 (1979)
19. Takahashi, M. and Ishio, S., *Jap. J. appl. Phys.*, **16,** 2273 (1977)
20. Fujimori, H. *'Proc 5th Int. Symp. on High Purity Materials in Science and Technology'*, Dresden, vol. III, p. 198 (1980)
21. Néel, L., *C.r. hebd. Séanc Acad. Sci., Paris*, **237,** 1613 (1953)
22. Taniguchi, S., *Sci. Rep. Res. Insts Tôhoku Univ.*, **A7,** 269 (1955)

23. Takahashi, M. and Kono, T., *J. phys. Soc. Japan*, **15,** 936 (1960); Sambogi, T. and Mitsui, T., *J. phys. Soc. Japan*, **16,** 1478 (1961)
24. Kittel, C., Nesbitt, E. A. and Shockley, W., *Phys. Rev.*, **77,** 839 (1950)
25. *See* Chikazumi, S., *'Physics of Magnetism'*, p. 369, John Wiley & Sons, Inc., New York (1964)
26. Berry, B. S. and Pritchet, W. C., *Phys. Rev. Letters*, **34,** 1022 (1975)
27. Luborsky, F. E. and Walter, J. L., *IEEE Trans. Magn.*, **MAG-13,** 953 (1977)
28. Luborsky, F. E. and Walter, J. L., *IEEE Trans. Magn.*, **MAG-13,** 1635 (1977)
29. Luborsky, F. E. and Walter, J. L., *Mater. Sci. Engng*, **28,** 77 (1977)
30. Fujimori, H. and Masumoto, T., *Trans. Japan Inst. Metals*, **17,** 175 (1976)
31. Fujimori, H., Morita, H., Obi, Y. and Ohta, S., *'Amorphous Magnetism II'*, Eds. Levy, R. A. and Hasegawa, R., p. 393, Plenum Press, New York (1977)
32. Fujimori, H., unpublished work
33. Fujimori, H., Ohta, S., Masumoto, T. and Nakamoto, K., *'Proc. 3rd Int. Conf. on Rapidly Quenched Metals'*, Ed. Cantor, B., vol. II, p. 232, The Metals Society, London (1978)
34. Graham, C. D. (jun.), *'Magnetic Properties of Metals and Alloys'*, pp. 288—329, American Society for Metals, Metals Park, Ohio (1959)
35. Fujimori, H., Yoshimoto, H., Masumoto, T. and Mitera, T., *J. appl. Phys.*, **52,** 1893 (1981)
36. Miyazaki, T. and Takahashi, M., *Jap. J. appl. Phys.*, **17,** 1755 (1978)
37. Iwata, T. *Sci. Rep. Res. Insts Tôhoku Univ.*, **13,** 356 (1961)
38. Fukunaga, T. and Suzuki, K., *Sci. Rep. Res. Insts. Tôhoku Univ.*, **A29,** 153 (1981)
39. Becker, J. J., *IEEE Trans. Magn.*, **MAG-14,** 938 (1978)
40. Morita, H., Fujimori, H. and Obi, Y., *J. Magnetism magn. Mater.*, **15—18,** 1359 (1980)
41. Luborsky, F. E., Walter, J. L. and LeGrand, D. G., *IEEE Trans. Magn.*, **MAG-12,** 930 (1976)
42. Chen, H. S. and Polk, D. E., *J. Non-cryst. Solids*, **15,** 174 (1974)
43. Egami, T., Flanders, P. J. and Graham, C. D. (jun.), American Institute of Physics Conf. Proc. No. 24, p. 697 (1975)
44. Nielsen, O. V. and Nielsen, H. J. V., *Solid St. Communs*, **35,** 281 (1980)
45. Nielsen, O. V. and Nielsen, H. J. V., *J. Magnetism magn. Mater.*, **22,** 21 (1980)
46. Hilzinger, H. R., ref. 14, p. 791
47. Luborsky, F. E. and Becker, J. J. *IEEE Trans. Magn.*, **MAG-15,** 1939 (1979)
48. Kimura, H., Murata, T. and Masumoto, T., *Sci. Rep. Res. Insts Tôhoku Univ.*, **A26,** 270 (1977)
49. Fujimori, H., Kato, T., Masumoto, T. and Morita, H., ref. 33, p. 240

Chapter 17

Magnetovolume effects in amorphous alloys

K. Fukamichi

The Research Institute for Iron, Steel and Other Metals, Tohoku University, Sendai, Japan

17.1 Introduction

Invar alloys have practically zero thermal expansion coefficient around room temperature, which is quite useful for engineering applications such as bimetals and cavity resonators. Fe–Ni crystalline alloys in the Invar region of composition exhibit many remarkable behaviours, particularly the drastic deviation of saturation magnetization from the Slater–Pauling curve and the disappearance of ferromagnetism associated with anomalies in various physical properties[1]. These anomalous properties, that is, the Invar effects, have been extensively investigated because they provide a key for understanding the mechanism of ferromagnetism of the iron-group transition metals[2]. Therefore, the Invar effects seem to be of great interest not only for metallurgists but also for physicists. The thermal expansion anomaly below the Curie temperature, the forced volume magnetostriction and the effects of pressure on magnetization and the Curie temperature are called the magnetovolume effects because they have their origin in the interdependence of the magnetic interaction and the specific volume.

Hitherto, several kinds of crystalline Invar alloy have been reported. In these alloys, the Fe–Ni alloy system has been investigated in great detail. It should be noted that the Invar effects of this alloy system are observed in the composition near the boundary of an $\gamma \rightleftarrows \alpha$ phase transformation, accompanying the following various peculiar phenomena:[1]

1. Decrease in the magnetic moment and the Curie temperature at low nickel concentration ranges.
2. Small thermal expansion or large positive spontaneous volume magnetostriction.
3. $\Delta\alpha$ effect; the thermal expansion behaviour is modified by cold-rolling.
4. Large high-field susceptibility and forced volume magnetostriction.
5. Remarkable pressure dependence of the Curie temperature and the magnetization.
6. Deviation of curve of magnetization against temperature from the standard Brillouin functions.
7. Small spin wave stiffness constant.
8. Wide dispersion of internal magnetic field in the Mössbauer spectra.

9. Large γ-value in the low temperature specific heat.
10. Large value of the electrical resistivity.
11. Anomalous ΔE effect; the softening of Young's modulus even in the saturated magnetic field.

A number of models have been proposed in order to explain the anomalous behaviour described above, but there is as yet no complete and final theory.

In the amorphous alloys containing metalloids, iron-based and cobalt-based alloys are ferromagnetic and almost all nickel-based alloys are paramagnetic around room temperature. Some iron-based alloys exhibit the remarkable magnetovolume effects[3]. Especially, the Fe–B amorphous alloys show Invar characteristics over a wide temperature range[4], and their magnetic properties are very similar to those of Fe–Ni crystalline Invar alloys[5,6]. The comparison of these anomalies between these crystalline and amorphous alloys would throw light on the main causes of these phenomena. In this chapter, the magnetovolume effects and other physical properties of iron-based amorphous alloys containing metalloids such as boron and phosphorus are discussed and compared with those of the Fe–Ni crystalline Invar alloys.

17.2 The Curie temperature and magnetic moment

Iron-based amorphous alloys are regarded as weak ferromagnets (i.e. holes of both spin type exist) and cobalt-based amorphous alloys are strong ferromagnets (only minority-spin holes are present)[7]. *Figure 17.1* shows the concentration dependences of the Curie temperature T_C and the magnetic moment μ of Fe–B, Fe–P, Co–B and Co–P amorphous alloys[3,6,8–10]. Both values of cobalt-based amorphous alloys increase with decreasing boron or phosphorus content. On the other hand, the Curie temperature of iron-based amorphous alloys decreases with decreasing boron or phosphorus content. The magnetic moment of these alloys exhibit a similar tendency to that of cobalt-based

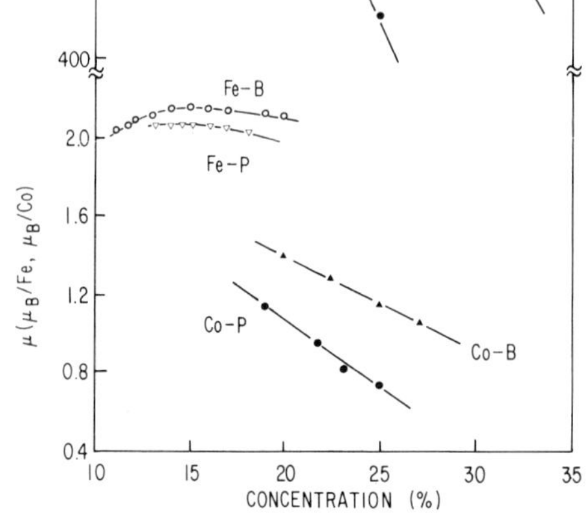

Figure 17.1 Concentration dependences of the Curie temperature and the magnetic moment of Fe–B, Fe–P, Co–B and Co–P amorphous alloys

amorphous alloys except in the low metalloid concentration range, and the value of amorphous pure iron is linearly extrapolated to be about 2.2 μ_B from the high boron and phosphorus concentration range, which is equal to that of crystalline pure iron. However, as shown in *Figure 17.1*, the variation of the magnetic moment with the composition below 15 per cent is not so simple, that is, the value of Fe–B decreases and that of Fe–P hardly varies with decreasing boron and phosphorus content, respectively. From *Figure 17.1* it is concluded that the concentration dependences of the Curie temperature and the magnetic moment of Fe–B amorphous alloys are similar to those of Fe–Ni crystalline Invar alloys. The decrease in the magnetic moment at low concentration ranges seems to be a common phenomenon among iron-based amorphous alloys because many other iron-based amorphous alloys such as Fe–Au[11], Fe–Si, Fe–O, Fe–Ge[12], Fe–Zr[13,14] and Fe–Hf[15] show a similar tendency. These phenomena would be explained by the models proposed in order to explain the Invar effects of Fe–Ni crystalline alloys.

17.3 Thermal expansion anomaly

Figure 17.2 shows the thermal expansion curves of some iron-based and cobalt-based amorphous alloys[3,6]. All samples exhibit a drastic shrinkage at the crystallization temperature T_x. It is worth noting that no thermal expansion anomaly is observed below the Curie temperature T_C in cobalt-based amorphous alloys, but the large anomaly due to the spontaneous volume magnetostriction ω_s is observed below T_C in iron-based amorphous alloys. The spontaneous volume magnetostriction ω_s is caused as a result of the sum of thermal, elastic and magnetic energies becoming a minimum. The thermal expansion anomaly, that is, the magnitude of ω_s of amorphous alloys, is sensitively effected by alloying elements. Usually, amorphous alloys contain two or three kinds of metalloid in order to stabilize the amorphous phase and the thermal expansion anomaly of Fe–Si–B amorphous alloys becomes more remarkable with

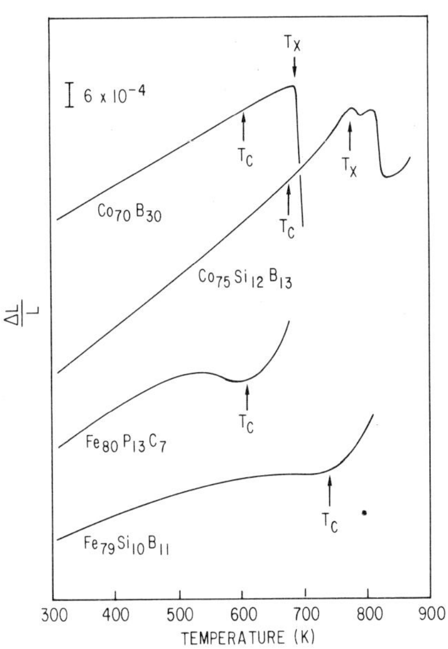

Figure 17.2 Thermal expansion curves of some iron-based and cobalt-based amorphous alloys

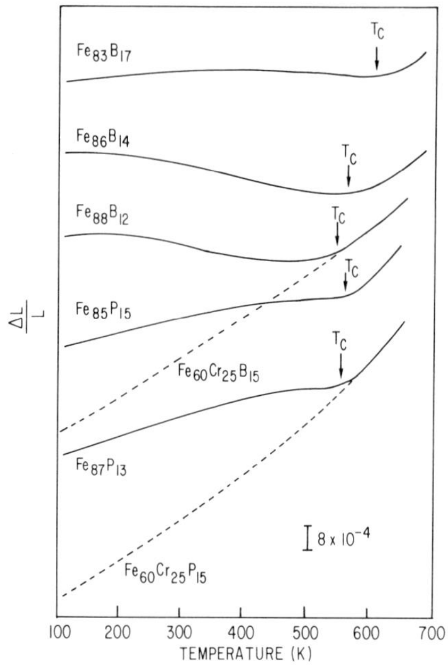

Figure 17.3 Thermal expansion curves of Fe–B and Fe–P amorphous alloys. Dashed lines are for a paramagnetic state

decreasing silicon or boron content[17]. In order to understand the magnetovolume effects in more detail, it is desirable to investigate an alloy system which should be as simple as possible. It is known that Fe–B and Fe–P binary alloy systems can be made amorphous over a relatively wide composition range. The thermal expansion curves of these alloys[6] are shown in *Figure 17.3*. These curves exhibit a remarkable anomaly below the Curie temperature T_C. Especially in the case of $Fe_{83}B_{17}$ alloy, the spontaneous volume magnetostriction ω_s is comparable with the usual thermal expansion due to anharmonicity, resulting in almost zero thermal expansion around room temperature.

As is well known, the magnetic properties of 3*d* transition metal alloys and the Invar effects are discussed by using a localized or itinerant electron model. The judgement between the localized state and the itinerant one can be made by considering the Rhodes–Wohlfarth plot[18]. As described later, Fe–B and Fe–P amorphous alloys seem to be itinerant ferromagnets. We will discuss the spontaneous volume magnetostriction by using Wohlfarth's itinerant electron model. According to his theory, the spontaneous volume magnetostriction at temperature T, $\omega_s(T)$, is given by the equation[19]

$$\omega_s(T) = \omega_s(0) - \kappa C M^2(0,0)\left(\frac{T}{T_C}\right)^2 \tag{17.1}$$

where $\omega_s(0)$ is the spontaneous volume magnetostriction at 0 K, κ the compressibility, C the magnetoelastic coupling constant, and $M(0,0)$ is the magnetization at 0 K and zero magnetic field. Kalinin *et al.* have derived the same equation by using Stoner's formula for magnetization[20]. The value of $\omega_s(T)$ may also be given by

$$\omega_s(T) = 3\int(\alpha_p - \alpha)\,dT \tag{17.2}$$

where α_p is the thermal expansion coefficient if the alloy is assumed to be paramagnetic and α is the measured thermal expansion coefficient. In *Figure 17.3*, the thermal

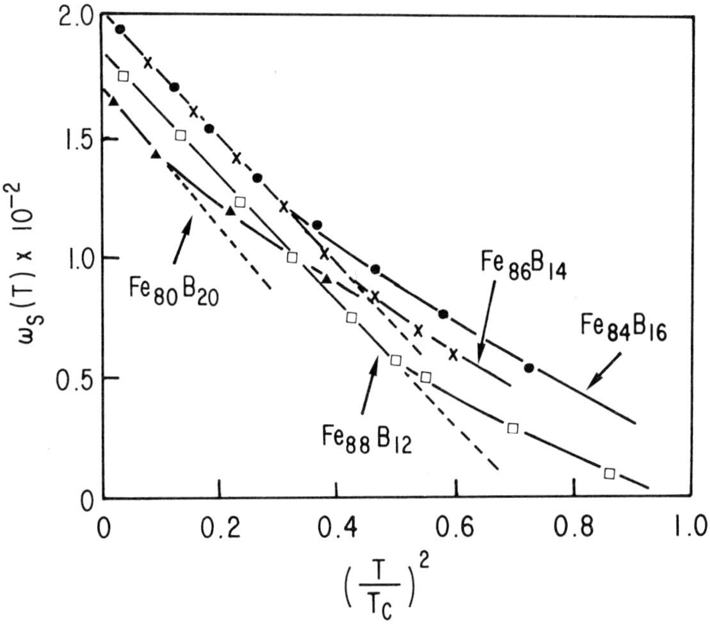

Figure 17.4 Spontaneous volume magnetostriction of Fe–B amorphous alloys as a function of $(T/T_C)^2$

expansion curve of $Fe_{60}Cr_{25}B_{15}$, which is paramagnetic above 100 K, is also shown by the broken line. Therefore, the value of $\omega_s(T)$ is equal to three times the difference of the thermal expansion curves between the solid and broken lines. *Figure 17.4* shows the spontaneous volume magnetostriction $\omega_s(T)$ for Fe–B amorphous alloys as a function of $(T/T_C)^2$. The temperature range where a linear relationship holds becomes wider with decreasing boron content[6], and by extrapolating to 0 K, we get the value of $\omega_s(0)$. The concentration dependence of $\omega_s(0)$ of Fe–B and Fe–P amorphous alloys is given in *Figure 17.5*, together with that of Fe–Ni crystalline Invar alloys[1]. The Curie temperatures of these alloys are not so different from one another, but the thermal expansion anomaly of Fe–B amorphous alloys is more remarkable accompanying a negative thermal expansion coefficient at low boron content. The value of $\omega_s(0)$ for Fe–B amorphous alloys is much larger than that of Fe–P amorphous and Fe–Ni crystalline alloys.

Shiga[21] has found a relationship between the lattice constant and the atomic moment of 3*d*-transition metal crystalline alloys expressed by the equation

$$a(x) = a_A(1 - x) + a_B x + c\langle|\mu|\rangle \qquad (17.3)$$

where x is the atomic fraction of element B, $\langle|\mu|\rangle$ is the site average of the magnitude of atomic moment, and a_A, a_B and c are adjustable parameters. For simple ferromagnetic alloys, $\langle|\mu|\rangle$ is equal to the spontaneous atomic moment at 0 K. From the thermal expansion curves and the temperature dependence of the spontaneous volume magnetostriction shown in *Figures 17.3 and 17.4*, we can estimate the nearest neighbour distance at 0 K of Fe–B amorphous alloys because the nearest neighbour distance at room temperature has been reported[22]. The deduced value at 0 K decreases monotonically with decreasing boron content, indicating a similar trend as the room temperature results[22]. Therefore, Shiga's rule is not valid in Fe–B amorphous alloys

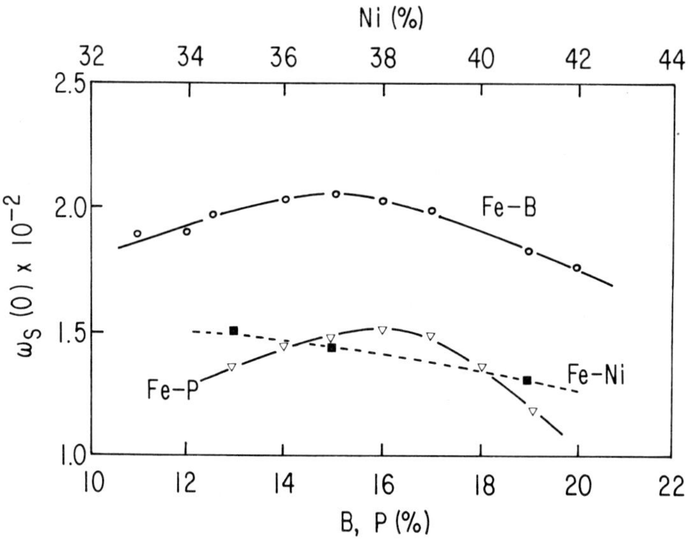

Figure 17.5 Concentration dependence of the spontaneous volume magnetostriction at 0 K of Fe–B and Fe–P amorphous alloys, together with that of Fe–Ni crystalline alloys

because the concentration dependence of the magnetic moment does not show the monotonic behaviour shown in *Figure 17.1*.

The Invar effects of Fe–Ni crystalline Invar alloys are remarkably affected by environment[1]. For example, the thermal expansion characteristics are modified by cold-rolling, and the thermal expansion coefficient is different depending on the rolling direction, which is termed the $\Delta\alpha$ effect[23]. This behaviour is accounted for by the change in the number of Fe–Fe pairs by cold-rolling as well as the roll magnetic

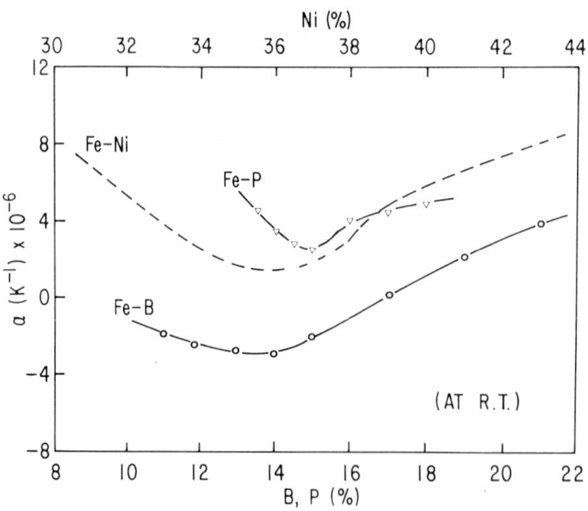

Figure 17.6 Concentration dependence of the room temperature thermal expansion coefficient of Fe–B and Fe–P amorphous alloys, together with that of Fe–Ni crystalline alloys

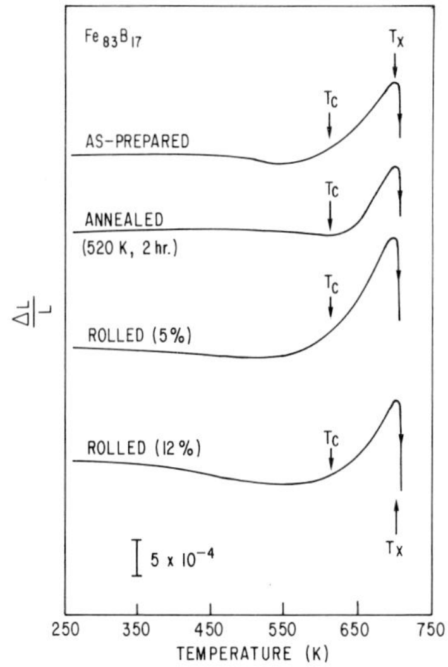

Figure 17.7 Thermal expansion curves of the annealed and cold-rolled specimens of $Fe_{83}B_{17}$ amorphous alloy

anisotropy[24]. The room temperature thermal expansion coefficient of Fe–B and Fe–P amorphous alloys[25] is shown in *Figure 17.6*, together with that of Fe–Ni crystalline Invar alloys. These three curves have a broad minimum. The effect of cold-rolling on the thermal expansion measured in the rolling direction of $Fe_{83}B_{17}$ amorphous alloy[26] is shown in *Figure 17.7*. In this figure, the rolling ratio is defined by the change of the length. With increasing rolling ratio, the thermal expansion anomaly below the Curie temperature is enhanced, indicating a negative thermal expansion coefficient around room temperature. Therefore, the thermal expansion curves of cold-rolled specimens apparently correspond to those of the amorphous alloys with lower boron content (*see Figure 17.3*). In other words, the number of Fe–Fe pairs seems to increase in the roll direction. Such a presumption is confirmed by the following results on the $Fe_{86}B_{14}$ amorphous alloy. *Figure 17.8* shows the room temperature thermal expansion coefficient α of $Fe_{83}B_{17}$ and $Fe_{86}B_{14}$ amorphous alloys as a function of the rolling ratio[27]. With increasing rolling ratio, the value of α for the $Fe_{83}B_{17}$ alloy decreases but α for the $Fe_{86}B_{14}$ alloy increases. Then, these values correspond to the values of the alloys with lower boron content as seen from *Figure 17.6*. In order to discuss in more detail the $\Delta\alpha$ effect, it is necessary to investigate the magnetic anisotropy and the thermal expansion in the direction perpendicular to the rolling direction. However, the

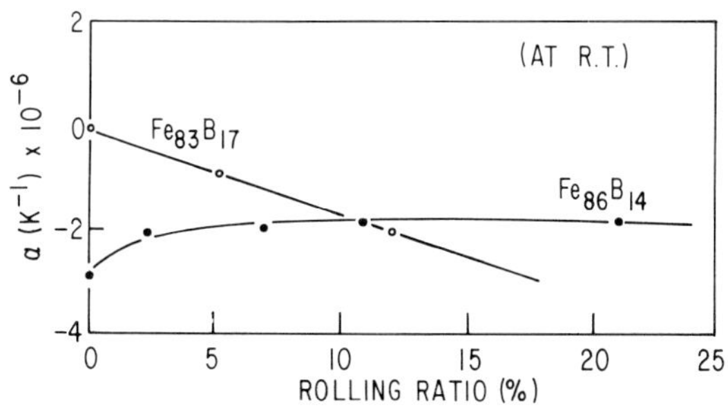

Figure 17.8 Room temperature thermal expansion coefficient of $Fe_{83}B_{17}$ and $Fe_{86}B_{14}$ amorphous alloys as a function of the cold-rolling ratio

amorphous ribbon samples are originally very thin and their thickness is not uniform, that is, the centre is thicker than the edge. At the present stage of work, the cold-rolling was also very non-uniform[27].

17.4 High-field susceptibility

The high-field susceptibility is obtained from the slope of the linear portion of magnetization curves at higher magnetic fields. The concentration dependence of the high-field susceptibility χ_{hf} at 4.2 K of Fe–B and Fe–P amorphous alloys[5,28] is shown in *Figure 17.9*, together with that of Fe–Ni crystalline Invar alloys. The magnitude of χ_{hf} for these three alloys is almost the same and these values increase with increasing iron content. The high-field susceptibility χ_{hf} may be caused by several factors. Assuming the Pauli paramagnetism of 3*d* electrons mainly contributes to χ_{hf}, the value of χ_{hf} depends on the density of states at the Fermi level and the following equation is given in the itinerant electron model[29,30];

$$\chi_{hf} \approx \chi_p = \left[\frac{1}{4\mu_B^2}\left(\frac{1}{N_+(\varepsilon_F)}\right) + \frac{1}{N_-(\varepsilon_F)} - \phi\right]^{-1} \tag{17.4}$$

where $N_+(\varepsilon_F)$, $N_-(\varepsilon_F)$ are the density of states of positive spin and negative spin bands at the Fermi level, respectively, ϕ is the molecular field coefficient, and μ_B is the Bohr magneton. Therefore, such a large high-field susceptibility shown in *Figure 17.9* is attributed to the high density of state at the Fermi level. On the other hand, such a large high-field susceptibility χ_{hf} would be also explained in terms of the localized electron model. According to this model, this large value of χ_{hf} is caused by the flipping of weakly coupled antiferromagnetic spin under high magnetic fields[1].

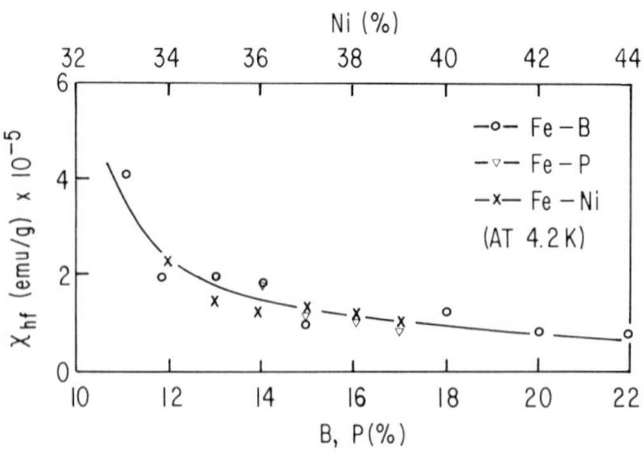

Figure 17.9 Concentration dependence of the high-field susceptibility at 4.2 K of Fe–B and Fe–P amorphous alloys, together with that of Fe–Ni crystalline Invar alloys

17.5 Forced volume magnetostriction

The forced volume magnetostricten $\partial\omega/\partial H$ is obtained from the slope of the linear portion of the magnetostriction as a function of the magnetic field, and is given by the equation;

$$\partial\omega/\partial H = h_{\parallel} + 2h_{\perp} \tag{17.5}$$

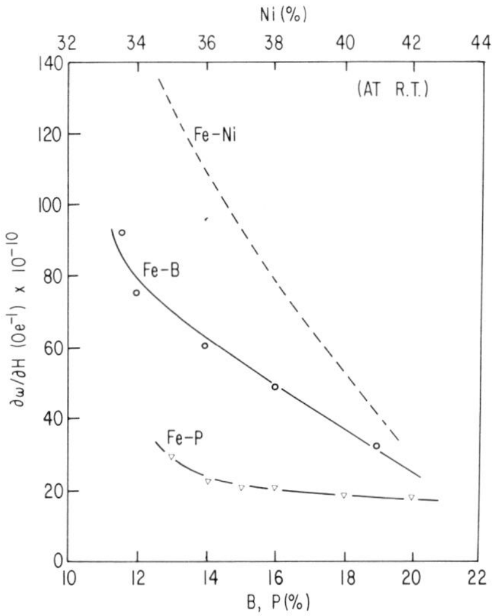

Figure 17.10 Concentration dependence of the room temperature forced volume magnetostriction of Fe–B and Fe–P amorphous alloys, together with that of Fe–Ni crystalline Invar alloys

where $h_{\parallel}$ and $h_{\perp}$ are the slopes of magnetostriction curves in the parallel and perpendicular directions to the magnetic field, respectively. The concentration dependence of the room temperature values of $\partial\omega/\partial H$ of Fe–B and Fe–P amorphous alloys[31,32] is shown in *Figure 17.10*, together with that of Fe–Ni crystalline Invar alloys. In the case of Fe–B amorphous alloys, $\partial\omega/\partial H$ increases remarkably with decreasing boron content, whereas that of Fe–P amorphous alloys only slightly increases with decreasing phosphorus content. The relationship between the spontaneous volume magnetostriction ω_s and the spontaneous magnetization M_s is given by

$$\omega_s = \kappa C M_s^2 \tag{17.6}$$

where κ is the compressibility and C is the magnetoelastic coupling constant. By differentiating this equation with respect to the magnetic field H, we obtain

$$\partial\omega/\partial H = 2\kappa C M_s \chi_{hf} \tag{17.7}$$

Therefore, the value of $\partial\omega/\partial H$ at 0 K, $\partial\omega(0)/\partial H$, is calculated from $\omega_s(0)$, $M_s(0)$ and χ_{hf} as obtained from *Figures 17.1, 17.5,* and *17.9*. For example, the value of $\partial\omega(0)/\partial H$ of $Fe_{88}B_{12}$ amorphous alloy is about half the room temperature value. Such a tendency of the temperature dependence is also observed in Fe–Ni crystalline Invar alloys. On the other hand, the value of $\partial\omega(0)/\partial H$ of $Fe_{86}P_{14}$ is almost the same as the room temperature value.

Thermodynamically, the forced volume magnetostriction is related to the saturation magnetization by the equation

$$(\partial\omega/\partial H)_{P,T} = -\rho(\partial M_s/\partial P)_{H,T} \tag{17.8}$$

where ρ is the density and M_s is the saturation magnetization per unit volume. Anderson *et al.*[33] obtained $\partial\omega/\partial H$ of $Fe_{29}Ni_{49}P_{14}B_6Si_2$ amorphous alloy from the pressure dependence of the room temperature value of M_s. Although this alloy exhibits no Invar effect, a relatively large value is estimated because the Curie temperature is very close to room temperature and usually the value of $\partial\omega/\partial H$ drastically changes around the Curie temperature.

17.6 Pressure effect on the Curie temperature

Both itinerant and localized electron models predict that the alloys with a large spontaneous volume magnetostriction exhibit a large pressure effect on the Curie temperature as given[1] by the equations;

$$\omega_s(0) = -\frac{M_s^2(0,0)}{2\chi_{hf}}\frac{1}{T_C}\frac{\partial T_C}{\partial P} \tag{17.9}$$

and

$$\omega_s(0) = -\frac{3}{2}k_B N \frac{S^2}{S(S+1)}\frac{\partial T_C}{\partial P} \tag{17.10}$$

where k_B is Boltzmann's constant, S the spin and N is the atomic number. In fact, the values of $\partial T_C/\partial P$ of Fe–Ni crystalline Invar alloys are very large[34], being $-(7 \sim 3)$ K/kbar. As shown in *Figure 17.5*, Fe–B and Fe–P amorphous alloys have a large spontaneous volume magnetostriction $\omega_s(0)$. Thus we expect a large value of $\partial T_C/\partial P$ in these alloys. *Figure 17.11* shows the value of $\partial T_C/\partial P$ as a function of the Curie temperature of some amorphous alloys[35–38], together with that of Fe–Ni crystalline alloys. It is noteworthy that the values for Fe–B and Fe–Cr–B amorphous alloys are very large and comparable with those of Fe–Ni crystalline Invar alloys. Kamarád *et al.*[35] discussed the results by using Wohlfarth's theory on the itinerant ferromagnets. According to this model, the pressure coefficient of the Curie temperature $\partial T_C/\partial P$ is given by

$$\frac{\partial T_C}{\partial P} = -\frac{A}{T_C} \tag{17.11}$$

where A is a constant containing some factors such as the compressibility, the density of state and so on. The volume V and the Curie temperature T_C are connected by the equation

$$\Gamma = \frac{\partial \ln T_C}{\partial \ln V} = \frac{A}{\kappa T_C^2} \tag{17.12}$$

where κ is the compressibility and Γ is called Cochrane and Graham's Grüneisen parameter[39]. Kamarád *et al.* have given the result $\Gamma = (-5/6) + 4.3 \times 10^{-3} T_C$ for (NiFe)PB amorphous alloys. Lie *et al.* have obtained the value $\partial T_C/\partial P$ from Mössbauer spectra under high pressures[33]. In the case of $Fe_{32}Ni_{36}Cr_{14}P_{12}B_6$, the value of $\partial T_C/\partial P$ is about -0.5 K/kbar. This value is much smaller than that for Fe–B amorphous alloys, because this alloy exhibits no Invar characteristics. Schneider *et al.* have pointed out that the pressure coefficient for many crystalline and amorphous alloys can be fitted with the expression[37]

$$\partial T_C/\partial P = \eta T_C - \lambda T_C^2 \tag{17.13}$$

where η and λ are constants.

Figure 17.12 shows the pressure dependence of the nearest neighbour distance and the lattice constant of $Fe_{83}B_{17}$ amorphous alloy and crystalline pure iron[40]. The inflection point may indicate the magnetic phase transition. From this result, the value of $\partial T_C/\partial P$ is estimated to be -2.7 K/kbar and corresponds to the value measured directly as seen from *Figure 17.11*. The slope is 1/3rd the compressibility κ. The value of κ for $Fe_{83}B_{17}$ amorphous alloy in the ferromagnetic range is extremely large being about 14.3×10^{-4}/kbar, and its value in the paramagnetic range is small, being comparable

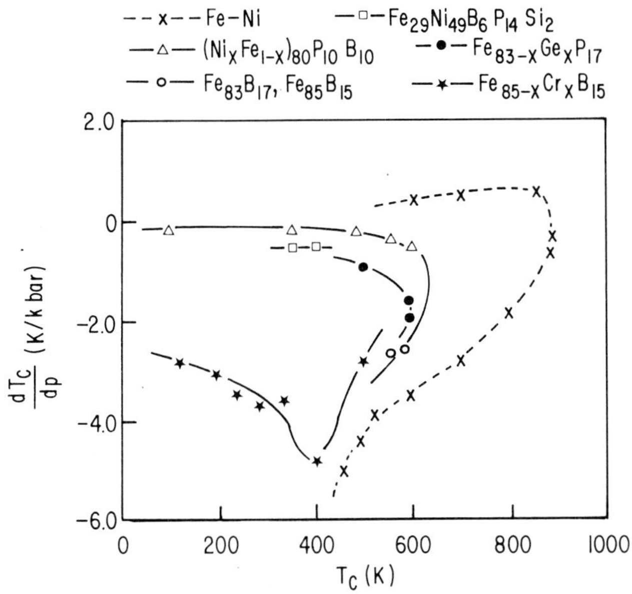

Figure 17.11 Pressure coefficient of the Curie temperature of some amorphous alloys, together with that of Fe–Ni crystalline Invar alloys

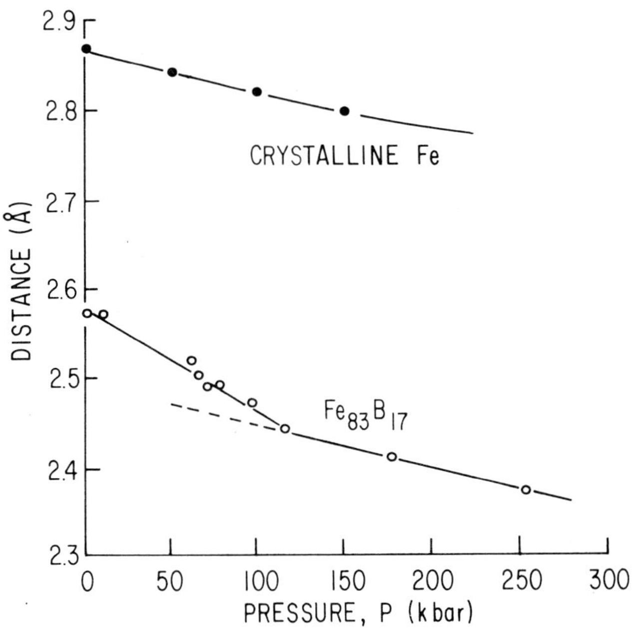

Figure 17.12 Pressure dependence of the nearest neighbour distance and the lattice constant of $Fe_{83}B_{17}$ amorphous alloy and crystalline pure iron

TABLE 17.1 The Curie temperature, T_C, high-field magnetic susceptibility, χ_{hf}, forced volume magnetostriction, $\partial\omega/\partial H$, compressibility, κ, and the pressure coefficients of the Curie temperature, $\partial T_C/\partial P$, for some amorphous and crystalline materials

Material	T_C/K	$10^3\chi_{hf}$ emu mol^{-1}	$10^9\partial\omega/\partial H$ Oe^{-1}	$10^4\kappa$ kbar	$\partial T_C/\partial P$ K kbar^{-1}
Amorphous $Fe_{83}B_{17}$	600	0.9	4.5	14.3	−2.6
Amorphous $Fe_{80}B_{20}$	650	0.7	2.1	7.1	—
Amorphous $Fe_{90}Zr_{10}$	238	—	28.5	—	−5.6
Crystalline Fe	1043	0.34	0.45	5.9	~0
Crystalline Ni	630	0.13	0.12	4.6	+0.4
Crystalline Fe–35%Ni	495	1.1	5.5	9.3	−3.5
Crystalline Fe–28%Pt	448	0.8	1.8	7.7	−4.0

with that of crystalline iron. This means that the magnetic effect on the compressibility for the amorphous Invar alloys is significant. In *Table 17.1*, the compressibility κ measured at room temperature is presented, together with the Curie temperature T_C and its pressure coefficient $\partial T_C/\partial P$, the high-field susceptibility χ_{hf} and the forced volume magnetostriction $\partial\omega/\partial H$ at about 0 K[40–46]. In the case of Fe–Ni crystalline alloys, κ drastically increases and Poisson's ratio v decreases at the composition range where the Invar effects are observed[43]. As shown in *Table 17.1* κ of $Fe_{83}B_{17}$ amorphous alloy is much larger than that of $Fe_{80}B_{20}$ amorphous alloy. This indicates the same tendency as the spontaneous volume magnetostriction (*see Figure 17.5*). It would be expected that Poisson's ratio becomes smaller with decreasing boron content in a similar manner to Fe–Ni crystalline Invar alloys.

17.7 Reduced magnetization curves

The reduced magnetization–temperature curves of some materials are shown in *Figure 17.13.* As is well known, the curves of Fe–Ni crystalline Invar alloys are much flatter than that of crystalline pure iron, deviating from the standard Brillouin functions. This behaviour has been explained by the models of latent antiferromagnetism, two-electron configuration of the iron atom or very weak itinerant electron ferromagnetism[1]. Generally, the curves of amorphous alloys are also flatter than that of crystalline pure iron. Handrich[47] attributed this behaviour to the fluctuation of the exchange integral, and explained it by using the parameter

$$\delta = \langle \Delta J^2 \rangle^{1/2} / \langle J \rangle \tag{17.14}$$

The parameter of many amorphous alloys which show no Invar effect such as $Co_{70}B_{20}P_{10}$ is about 0.3 as seen from *Figure 17.13.* However, the curves of iron-based amorphous alloys are much flatter than those of $Co_{70}B_{20}P_{10}$ amorphous alloy. For example, the curve of $Fe_{83}B_{17}$ amorphous alloy almost overlaps with that of Fe–36%Ni crystalline Invar alloy, and the curve of $Fe_{88}B_{12}$ is flatter[4] than that of $Fe_{83}B_{17}$. If we take into consideration Handrich's model, the fluctuation of the exchange integral of iron-based amorphous alloys should be much larger than that of cobalt-based ones. On the other hand, since iron-based amorphous alloys exhibit the Invar effects as shown in *Figure 17.2,* such a flat reduced magnetization–temperature curve should be explained by taking into account the Invar effects as well, as for Fe–Ni crystalline Invar alloys.

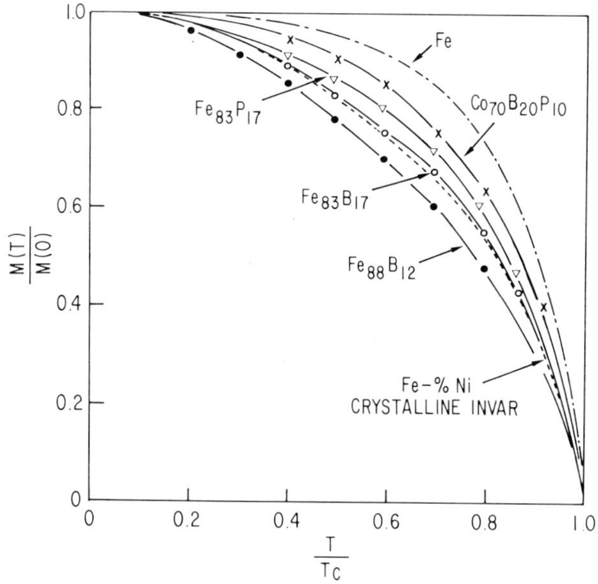

Figure 17.13 Reduced magnetization–temperature curves of some amorphous and crystalline materials

17.8 Mössbauer effect

From the flat reduced magnetization–temperature curves of Fe-base amorphous alloys, it is expected that these alloys have a large fluctuation of the exchange integral which should be reflected in the internal magnetic field distribution. On the other hand, the Mössbauer spectra of Fe–Ni crystalline Invar alloys give a broad distribution of the internal magnetic field suggesting weak ferromagnetic and/or antiferromagnetic components depending on the alloy composition.

Figure 17.14 shows the internal magnetic field distribution of Fe–B amorphous and Fe–Ni crystalline alloys[48,49]. These results have been obtained by the same calculation because the distribution is sensitively affected by the analytical method. It is

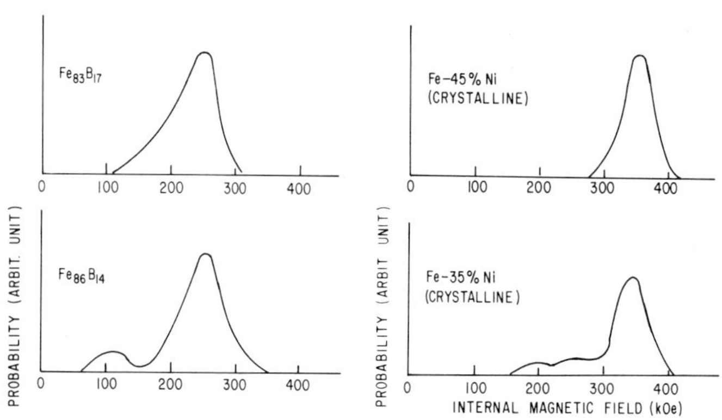

Figure 17.14 Internal magnetic field distribution of Fe–B amorphous and Fe–Ni crystalline alloys

noted that Fe–45%Ni non-Invar alloy exhibits a relatively narrow symmetrical distribution. On the contrary, the distribution of Fe–B amorphous alloys is broad and the low field components increase with decreasing boron content, being similar to that of Fe–35% Ni crystalline Invar alloy. These results suggest that Fe–B amorphous alloys have a large fluctuation of the exchange integral and also have weak ferromagnetic and/or antiferromagnetic components depending on the boron content. The internal magnetic field distribution of Fe–P amorphous alloys is also relatively broad, although it hardly depends on phosphorus content[50].

17.9 Rhodes–Wohlfarth plot

The ratio of the effective magnetic moment per magnetic atom q_c and the saturation magnetization per magnetic atom q_s when graphed against the Curie temperature T_C is called the Rhodes–Wohlfarth plot[18]. Here, q_c is obtained from the Curie–Weiss constant given by the equation:

$$C_A = (\mu_B^2/3R)q_c(q_c + 2) \tag{17.15}$$

where R is the gas constant per mole. The value of q_s is obtained from the low temperature magnetization and is given by

$$q_s = M_s/\mu_B \tag{17.16}$$

If the localized state occurs, q_c/q_s should be unity, while for the itinerant state with very weak ferromagnetism, q_s is close to zero or T_C goes to zero, and then the Rhodes–Wohlfarth plot diverges with decreasing T_C because the Curie–Weiss constant C_A is independent of q_s and T_C as given by equation 17.15. Therefore, we can define the localized state or the itinerant one from the Rhodes–Wohlfarth plot. It has been pointed out from this plot that various amorphous alloys are itinerant ferromagnets[51,52]. *Figure 17.15* shows the Rhodes–Wohlfarth plots of various kinds of materials[53]. In the case of crystalline pure iron, the Curie temperature is too high to define whether it is a localized state or an itinerant one, because these two lines almost coincide with each other above about 600 K. On the other hand, the ratio of q_c/q_s of crystalline pure gadolinium is unity,

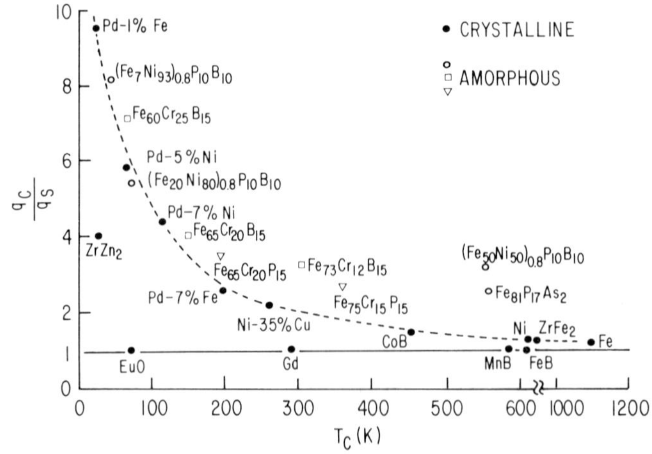

Figure 17.15 Rhodes-Wohlfarth plots of various crystalline and amorphous materials

indicating a localized state. In the case of many amorphous alloys, the Curie temperature is close to the crystallization temperature and then q_c is not defined precisely, as is the case for Fe–B and Fe–P amorphous alloys. However, by addition of chromium to Fe–B and Fe–P alloys, the Curie temperature T_C decreases and the crystallization temperature T_x increases, keeping the Invar effects[54,55]. Roughly speaking, itinerant crystalline ferromagnets are on the dotted line, but the plots of FeCrB and FeCrP amorphous alloys are above the line[55,56] as shown in *Figure 17.15*. One of the reasons for this behaviour may be attributed to the fact that the magnetization curves of Invar alloys are not easily saturated. This underestimates the value of q_s resulting in a large ratio of q_c/q_s. In any event, Fe–B and Fe–P alloy systems are regarded as itinerant ferromagnets. Therefore, the spontaneous volume magnetostriction has been discussed by using Wohlfarth's theory on itinerant ferromagnets in *Figure 17.3*.

17.10 Spin wave stiffness constant

The low temperature magnetization gives information on the spin waves. The change in magnetization with temperature is given by

$$M = M(0,0)(1 - BT^{3/2} - CT^{5/2}) \tag{17.17}$$

with

$$B = 2.612\left(\frac{g\mu_B}{M(0,0)}\right)\left(\frac{k_B}{4\pi D}\right)^{3/2} \tag{17.18}$$

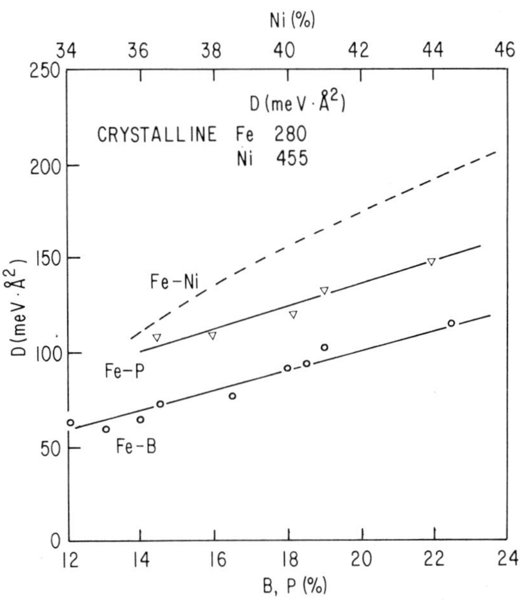

Figure 17.16 Concentration dependence of the spin stiffness constant of Fe–B and Fe–P amorphous alloys, together with that of Fe–Ni crystalline Invar alloys. *D* for crystalline iron is 280 and for crystalline nickel 455 MeV Å^{-2} respectively

where $M(0, 0)$ is the magnetization at 0 K and zero magnetic field and D is the spin wave stiffness constant. The concentration dependence of D for Fe–B and Fe–P amorphous and Fe–Ni crystalline alloys[1,57] is shown in *Figure 17.16*. The small D value is characteristic of the Invar alloys, reflecting the spin wave instability. The values of Fe–B and Fe–P amorphous alloys are remarkably small and they become much smaller with decreasing boron or phosphorus content. The coefficient of D is obtained from the neutron diffraction data by use of the relation

$$\hbar\omega = Dq^2 \tag{17.19}$$

where ω is the frequency, $\hbar\omega$ the energy and q is the wave vector. The value of D obtained from the neutron diffraction is usually larger by about 10—20 per cent than that obtained from the low temperature magnetization. This discrepancy has been accounted for by taking into consideration the temperature dependence of D. However, the spin stiffness constant $D(T)$ of $Fe_{86}B_{14}$ amorphous alloy varies with temperature according to the following equation[59] as predicted by the two magnon theory for a Heisenberg system, contrary to the itinerant electron model:

$$D(T) = D(0)\left(1 - a\left(\frac{T}{T_C}\right)^{5/2}\right) \tag{17.20}$$

The value at 0 K, $D(0)$, thus determined is almost twice as large as the value deduced from the magnetization data (*see Figure 17.16*). Such a discrepancy is also observed in the Brillouin scattering data[60]. The coefficient D is correlated with the equation[61]

$$\omega = \varepsilon(H + 2\pi M_s + \xi D k_y^2) \tag{17.21}$$

where ω is the angular frequency, ε the gyromagnetic ratio, M_s the magnetization, H the applied field in the sample plane, k_y the wave vector in the sample plane and perpendicular to H, and ξ is a constant. The value for $Fe_{80}B_{20}$ amorphous alloy obtained from equation 17.21 is about twice the value obtained from equation 17.19. At present, such a discrepancy, especially in the Invar alloys, is not understood.

17.11 Low temperature specific heat

The low temperature specific heat of ferromagnetic amorphous alloys C_p can be analysed in terms of the following three principal contributions[62];

$$C_p = \gamma T + \beta T^3 + \delta T^{3/2} \tag{17.22}$$

with

$$\gamma = \frac{1}{3}\pi^2 k_B{}^2 N(\varepsilon_F) \tag{17.23}$$

$$\beta = \frac{12}{5}\pi^4 N k_B \left(\frac{1}{\theta_D}\right)^3 \tag{17.24}$$

$$\delta = 0.113 k_B \left(\frac{k_B}{D}\right)^{3/2} \tag{17.25}$$

where $N(\varepsilon_F)$ is the density of states at the Fermi level, N the number of atoms per unit volume, θ_D the Debye temperature, D the spin wave stiffness constant and k_B is Boltzmann's constant. *Figure 17.17* shows the concentration dependence of the γ-value of Fe–B and Fe–P amorphous alloys[63], together with that of Fe–Ni crystalline Invar

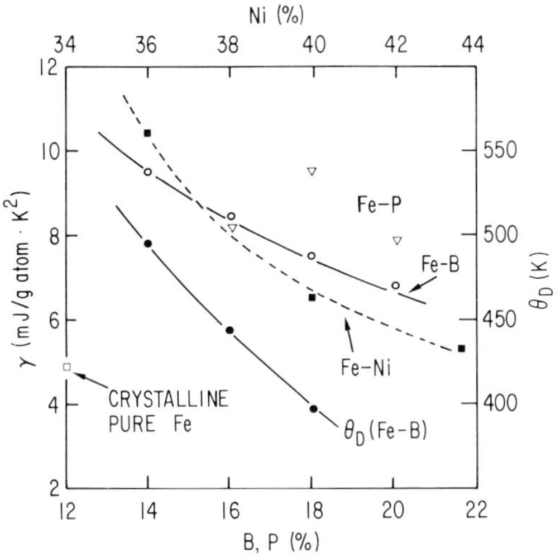

Figure 17.17 Concentration dependence of the γ-value of amorphous Fe–B and Fe–P and crystalline Fe–Ni alloys, together with the Debye temperature of Fe–B amorphous alloys

alloys[1]. With decreasing metalloid contents, γ remarkably increases and becomes larger than that of crystalline pure iron. Equation 17.23 shows that γ is proportional to the density of states, that is

$$\gamma \propto [N_+(\varepsilon_F) + N_-(\varepsilon_F)] \tag{17.26}$$

where $N_+(\varepsilon_F)$ and $N_-(\varepsilon_F)$ are the density of states of up and down spins, respectively. Therefore, the concentration dependence of the γ-value should be similar to that of the high-field susceptibility χ_{hf}. This interpretation is supported by the values of χ_{hf} which increase with decreasing boron content as seen from *Figure 17.9*. On the other hand, such a large γ-value would also be explained by the localized model which assumes the coexistence of different spin states such as ferromagnetic and antiferromagnetic states[64]. Such large γ-values have also been observed in other iron-based amorphous alloys[65,66]. The concentration dependence of the Debye temperature θ_D of Fe–B amorphous alloys[67] is also shown in *Figure 17.17*. The value of θ_D remarkably increases with decreasing boron content. It should be noted that the concentration dependence and the magnitude of θ_D of Fe–Ni crystalline Invar alloys are very similar to those of Fe–B amorphous alloys[1].

17.12 Electrical resistivity and galvanomagnetic effects

The resistivity of Fe–Ni crystalline alloys exhibits an enormous increase around the Invar composition. This anomaly in the resistivity has been considered to be one of the striking characteristics of the Invar alloys. This behaviour has been attributed to the scattering of the conduction electrons by the inhomogeneity of magnetic moment or as due to the transition from strong to weak ferromagnetism[1].

Figure 17.18 shows the concentration dependence of the electrical resistivity measured at 4.2 K, $\rho_{4.2}$, of some iron-based amorphous alloys, together with that of Fe–Ni crystalline alloys. The values of Fe–B and Fe–P amorphous alloys decrease with decreasing boron of phosphorus content[68]. On the other hand, the values of Fe–Au and Fe–Zr amorphous alloys increase at low solute contents where the magnetic moments decrease with decreasing contents[69,70]. As is well known the resistivity of amorphous alloys is very high, compared with that of crystalline alloys. Then, a drastic increase in

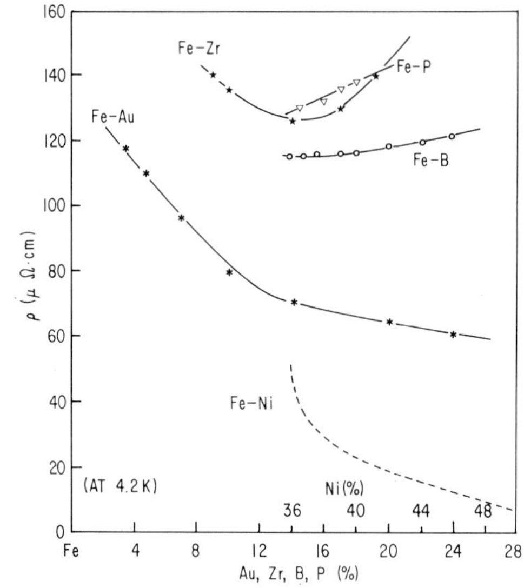

Figure 17.18 Concentration dependence of the electrical resistivity at 4.2 K of some iron-based amorphous and Fe–Ni crystalline alloys

the resistivity such as in Fe–Ni crystalline alloys is not observed at high iron concentration ranges. The decrement of ρ of Fe–B amorphous alloys becomes sluggish in the lower boron concentration range, suggesting that ρ increases at much lower concentration ranges in the same manner as Fe–Au and Fe–Zr amorphous alloys.

The forced magnetoresistivity $(1/\rho)(\partial\rho/\partial H)$ as a function of boron content of Fe–B amorphous and Fe–Ni crystalline alloys[71] is shown in *Figure 17.19*. The value of $(1/\rho)(\partial\rho/\partial H)$ increases with decreasing boron content. This behaviour would be explained by taking into account the contribution of the large forced volume magnetostriction $\partial\omega/\partial H$. If it is assumed that the forced magnetoresistivity is a function

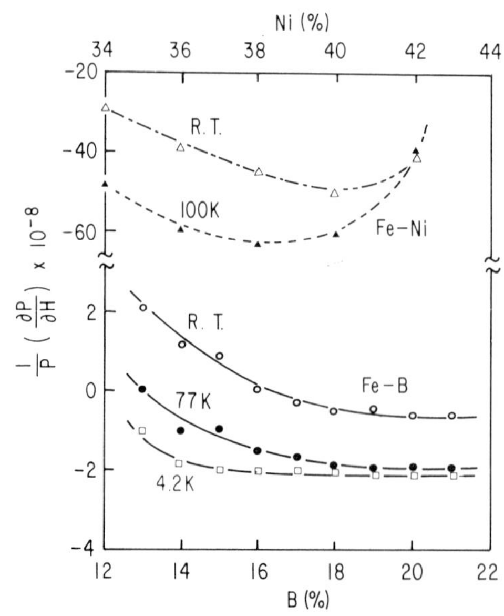

Figure 17.19 Forced magnetoresistivity at room temperature (R.T.), 77 K and 4.2 K as a function of boron content of Fe–B amorphous alloys, together with that at R.T. and 100 K of Fe–Ni crystalline alloys

of volume V and magnetic field H, the following expression is given at constant pressure[72]:

$$\frac{1}{\rho}\left(\frac{\partial \rho}{\partial H}\right)_P = \left(\frac{\partial \ln \rho}{\partial \ln V}\right)_H \left(\frac{\partial \omega}{\partial H}\right)_P + \frac{1}{\rho}\left(\frac{\partial \rho}{\partial H}\right)_V \tag{17.27}$$

Generally, the first term is negligibly small, but the Invar alloys have a large value of $\partial\omega/\partial H$ as seen from *Figure 17.10*. Therefore, the contribution due to $\partial\omega/\partial H$ should be taken into consideration as well as the spin disorder scattering. Since the value of $\partial\omega/\partial H$ becomes larger with decreasing boron content (*see Figure 17.10*), the first term in equation 17.27 would become larger. Consequently, the concentration dependence of $(1/\rho)(\partial\omega/\partial\rho)$ could be explained by taking into account the large forced volume magnetostriction. It is worth noting that Fe–Ni crystalline alloys exhibit a similar concentration dependence[72].

The concentration dependence of the spontaneous Hall coefficient R_s of Fe–B amorphous alloys shows a similar tendency as that of Fe–Ni crystalline Invar alloys, that is, R_s decreases remarkably with decreasing solute elements[73]. Almost all reported data on the ordinary Hall coefficient of amorphous alloys include the contributions from the spontaneous Hall coefficient R_s and the high-field susceptibility χ_{hf}. The apparent ordinary Hall coefficient R_0^* is given by

$$R_0^* \equiv \frac{\partial \rho_H}{\partial H} = R_0 + R_s 4\pi\chi_{hf} \tag{17.28}$$

where ρ_H is the Hall resistivity, H the applied magnetic field and R_0 is the ordinary Hall coefficient. As seen in *Figure 17.9* the values of χ_{hf} of Fe–B amorphous alloys are remarkably large. Therefore, in such a case the last term in equation 17.28 cannot be ignored.

17.13 Elastic properties

The magnetic contribution to the elastic properties of ferromagnetic alloys is closely related to the magnetostriction. Fe–Ni crystalline Invar alloys exhibit various elastic anomalies such as the large compressibility and the softening in Young's modulus[1]. As shown in *Table 17.1*, Fe–B amorphous alloys also have a large compressibility. However, these alloys do not show a softening in Young's modulus. It has been known that iron-based amorphous alloys exhibit a large ΔE effect, accompanying a large linear magnetostriction[6]. Berry and Pritchet have revealed that a thin flexible reed apparently exhibits a large ΔE effect because of superposition of the so-called pole effect[74]. Török and Hausch have given the following equation for this phenomenon[75]:

$$\frac{E - E_D}{E_D} = \frac{M \cdot H}{(2\pi f)^2 \rho l^2} \cdot K \tag{17.29}$$

where E and E_D are Young's moduli in a magnetic field and zero field, respectively. M is the magnetization, H the applied magnetic field, f the frequency, ρ the density, l the length and K is the constant depending on the overtone of frequencies. From equation 17.29, the pole effect is reduced when the measurement is carried out at high frequencies. *Figure 17.20* shows the temperature dependence of Young's modulus of $Fe_{85}B_{15}$ and $Fe_{81}Cr_4B_{15}$ amorphous alloys[76], together with that of Fe–35%Ni crystalline Invar alloy[77]. At zero field, Young's modulus is almost independent of the temperature below

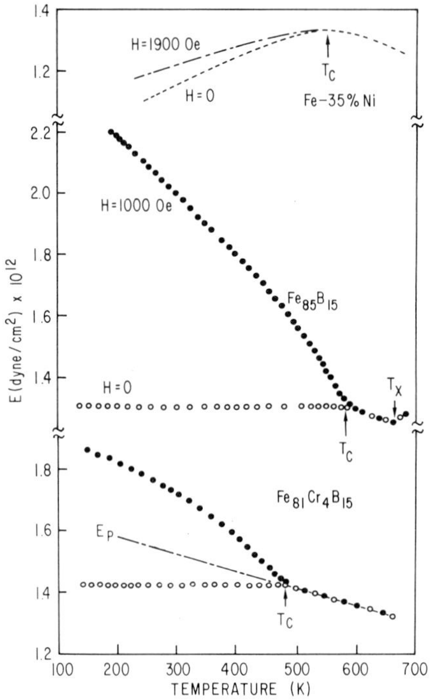

Figure 17.20 Temperature dependence of Young's modulus of $Fe_{85}B_{15}$ and $Fe_{81}Cr_4B_{15}$ amorphous alloys, together with that of Fe–35% Ni crystalline Invar alloy

the Curie temperature T_C, indicating Elinvar characteristics. By applying a magnetic field, these alloys exhibit a large ΔE effect accompanying the stiffening which is not in connection with the pole effect. Because of this stiffening, Young's modulus E_s in saturated magnetic fields becomes larger than the paramagnetic Young's modulus E_p. Such a peculiar phenomenon is connected with neither the phase transformation nor the amorphous structure, because Fe–B amorphous alloys, needless to say, do not have a phase transformation in an amorphous state and other alloys such as cobalt-based amorphous alloys do not show such a stiffening[78]. On the other hand, as *Figure 17.20* shows, the softening in Young's modulus of Fe–35%Ni crystalline Invar alloy is observed even in the saturated magnetic field[77]. This behaviour may be connected with the pre-martensitic transformation.

From the practical viewpoint, Fe–B amorphous alloys are promising as delay lines. The temperature coefficient of delay time t is given by [6,17]

$$t = -\tfrac{1}{2}(\alpha + e) \tag{17.30}$$

where α and e are the thermal expansion coefficient and the temperature coefficient of Young's modulus, respectively. As shown in *Figures 17.3* and *17.20* Fe–B amorphous alloys simultaneously exhibit the Invar and Elinvar characteristics, consequently the value of t in equation 17.30 becomes very small.

17.14 Recent results: metal–metal alloy systems

Recently, Fe–Zr, Fe–Hf and Fe–Ti amorphous alloys have been prepared[13–17]. In these alloy systems, the concentration dependences of the magnetic moment and the Curie temperature are very similar to those of Fe–Ni crystalline and Fe–B amorphous

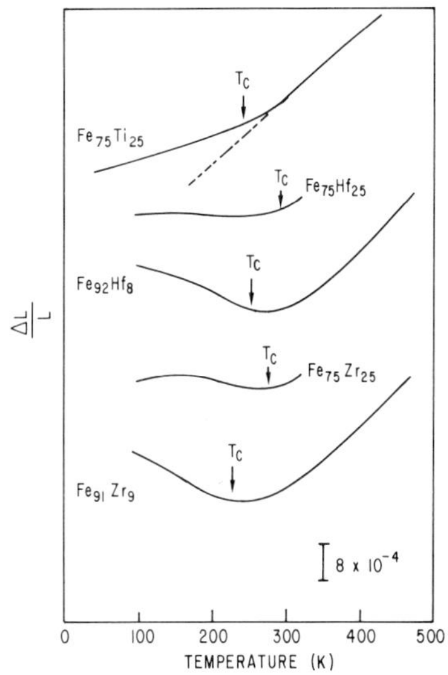

Figure 17.21 Thermal expansion curves of Fe–Zr, Fe–Hf and Fe–Ti amorphous alloys

alloys. They also exhibit remarkable magnetovolume effects such as large spontaneous and forced volume magnetostrictions, high-field susceptibility and the pressure coefficient of the Curie temperature[79,80]. *Figure 17.21* shows the thermal expansion curves of some Fe–Zr, Fe–Hf and Fe–Ti amorphous alloys[79]. A large thermal expansion anomaly is observed. It should be noted that the large anomaly in Fe–Zr and Fe–Hf amorphous alloys is observed even above the Curie temperature. This peculiar behaviour may originate from the magnetic inhomogeneity or the persistence of the localized magnetic moment above the Curie temperature. A recent theoretical study predicts that the spontaneous volume magnetostriction is proportional to the square of localized moment instead of the bulk magnetization[81]. Fe–Zr amorphous alloys also reveal mictomagnetic or spin-glass like behaviours such as a remarkable relaxation of remanence and an exchange anisotropy at low temperature[82,83], suggesting the coexistence of ferromagnetic and antiferromagnetic states. In addition, the Mössbauer spectra of these alloys also imply that there exist magnetic inhomogeneities[84].

17.15 Summary and remarks

In this chapter it has been emphasized that all physical properties except for the temperature dependence of Young's modulus of Fe–B amorphous alloys are very similar to those of Fe–Ni crystalline Invar alloys. It is clear that the remarkable magnetovolume effects are common to iron-based amorphous alloys not only for the alloys with metalloids but also for alloys with metals. We have discussed the various magnetovolume effects and other related phenomena by using the itinerant or localized electron model, but there may be some criticism against these interpretations. The argument as to whether the itinerant model or the localized model gives a better understanding of 3*d*-transition metal alloys is very old. Nevertheless, complete comprehension is a future problem even in crystalline states.

In order to make the situation of iron-based binary Invar alloys more visible, the schematic *Figure 17.22* is presented, following the precedent set by the crystalline Invar alloys[85]. The horizontal line stands for the iron concentration of some kinds of amorphous alloy. On going to the left-hand side, the ferromagnetism disappears. The

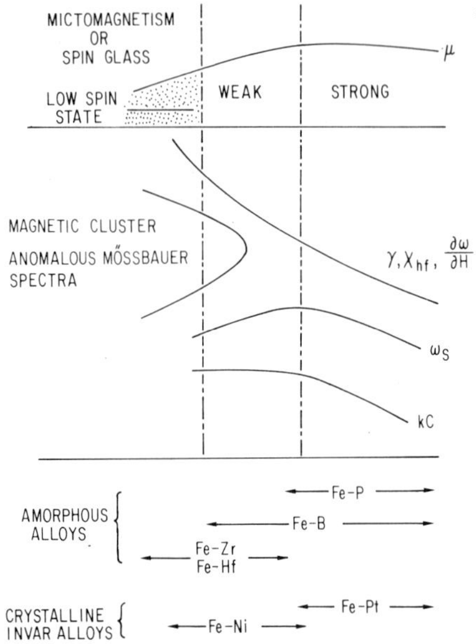

Figure 17.22 Schematic diagram which shows the relation between the magnetism and the occurrence of the magnetovolume effects in some iron-based amorphous and crystalline alloys

high-field susceptibility χ_{hf}, the forced volume magnetostriction $\partial\omega/\partial H$, the γ-value of low temperature specific heat and the magnetovolume coupling constant κC, increase with going to the left. The spontaneous volume magnetostriction ω_s takes a maximum value at the composition of the maximum magnetization. There are several different views on how the ferromagnetism collapses with increasing iron content. One view is that localized moments persist and couple each other antiferromagnetically. Another is that the iron atom transforms to a low spin state at a certain critical concentration[1]. In the case of Fe–Ni crystalline Invar alloys, the phase instability due to the martensitic phase transformation makes it difficult to offer a definite conclusion. Fortunately, iron-based amorphous alloys have no such phase transformation. Therefore, these amorphous alloys are suitable candidates for study of the magnetovolume effects. In fact, peculiar phenomena such as mictomagnetic or spin-glass like behaviours have been observed in Fe–Zr amorphous alloys.

Acknowledgements

Many of the results presented are due to the fruitful cooperation of Drs M. Kikuchi and H. Hiroyoshi. The author thanks Prof. T. Masumoto for his encouragement and interest.

References

1. Saito, H., *'Physics and Applications of Invar alloys', Maruzen Co. Ltd., Tokyo, Japan (1978)*
2. Freeman, A. J. and Shimizu, M., *J. Magnetism magn. Mater.*, **10,** 113 (1979)
3. Fukamichi, K., Kikuchi, M. and Masumoto, T., *Sci. Rep. Res. Insts Tôhoku Univ.*, **A26,** 225 (1977)
4. Fukamichi, K., Kikuchi, M., Arakawa, S. and Masumoto, T., *Solid St. Communs,* **23,** 955 (1977)
5. Fukamichi, K., Hiroyoshi, H., Kikuchi, M. and Masumoto, T., *J. Magnetism magn. Mater.*, **10,** 294 (1979)
6. Fukamichi, K., Masumoto, T. and Kikuchi, M., *IEEE Trans. Magn.*, **MAG-15,** 1404 (1979)

7. O'Handley, R. C. and Boudreaux, D. S., *Phys. Status Solidi, (a)*, **45,** 607 (1978)
8. Watanabe, H., Morita, H. and Yamauchi, H., *IEEE Trans. Magn.*, **MAG-14,** 944 (1978)
9. Pan, D. and Turnbull, D., *J. appl. Phys.*, **45,** 1406 (1974)
10. Cargill, G. S. (III) and Cochrane, R. W., *J. Phys. (Paris)*, **35,** C4, 269 (1974)
11. Felsch, W., *Z. Phys.*, **219,** 280 (1969)
12. Felsch, W., *Z. angew. Phys.*, **29,** 217 (1970)
13. Ohnuma, S., Shirakawa, K., Nose, M. and Masumoto, T., *IEEE Trans. Magn.*, **MAG-16,** 1129 (1980)
14. Fukamichi, K. and Gambino, R. J., *IEEE Trans. Magn.*, **MAG-17,** 3059 (1981)
15. Fukamichi, K. and Hiroyoshi, H., to be published
16. Fukamichi, K., Kikuchi, M., Hiroyoshi, H. and Masumoto, T., *Sci. Rep. Res. Insts Tôhoku Univ., A.* Suppl. 199 (1978)
17. Kikuchi, M., Fukamichi, K. and Masumoto, T., *Sci. Rep. Res. Insts Tôhoku Univ.*, **A26,** 232 (1977)
18. Rhodes, P. and Wohlfarth, E. P., *Proc. R. Soc.*, **A273,** 247 (1963)
19. Wohlfarth, E. P., *Phys. Lett.*, **28A,** 569 (1969)
20. Kalinin, V. M., Beskachko, V. P. and Khomenko, O. A., *Phys. Metals Metallogr.*, **37,** 184 (1974)
21. Shiga, M., American Institute of Physics Conf. Proc. No. 18, p. 463 (1973)
22. Fukunaga, T., Misawa, M., Fukamichi, K., Masumoto, T. and Suzuki, K., *'Proc. 3rd Int. Conf. on Rapidly Quenched Metals'*, Ed. Cantor, B., vol. II, p. 325, The Metals Society, London (1978)
23. Schulze, R., *Z. angew. Phys.*, **7,** 57 (1955)
24. Kagawa, H. and Chikazumi, S., *J. phys. Soc. Japan*, **43,** 1097 (1977)
25. Fukamichi, K., Kikuchi, M., Hiroyoshi, H. and Masumoto, T., ref. 22, p. 117
26. Fukamichi, K., Kimura, H. M., Kikuchi, M. and Masumoto, T., *IEEE Trans. Magn.*, **MAG-17,** 2701 (1981)
27. Fukamichi, K., Kimura, H. M., Kikuchi, M. and Masumoto, T., *Proc. 7th Int. Symp. Thermal Expansion*, Ed. Larsen, D. G., p. 163, Plenum Press, New York (1982)
28. Hiroyoshi, H., Fukamichi, K, Kikuchi, M., Hoshi, A. and Masumoto, T., *Phys. Lett.*, **65A,** 163 (1978)
29. Wohlfarth, E. P., *Phys. Lett.*, **3,** 17 (1962)
30. Gersdorf, R., *J. Phys. Radium*, **23,** 726 (1962)
31. Fukamichi, K., Kikuchi, M., Arakawa, S., Masumoto, T., Jagielinski, T., Arai, K. I. and Tsuya, N., *Solid St. Communs*, **27,** 405 (1978)
32. Kikuchi, M., Fukamichi, K., Satoh, T., Masumoto, T., Ohmori, K. and Tsuya, N., *J. Phys. F*, **12,** 2427 (1982)
33. Liu, C. M., Ingalls, R., Whitmore, J. E., Rao, K. V. and Bhagt, S. M., *J. appl. Phys.*, **50,** 1577 (1979)
34. Dubovka, G. T. and Ponyatovskiy, Ye. G., *Phys. Metals Metallogr.*, **33,** 179 (1972)
35. Kamarád, J., Arnold, Z., Schneider, J. and Krupicka, S., *J. Magnetism magn. Mater.*, **15—18,** 1409 (1980)
36. Kamarád, J. and Arnold, Z., *J. Magnetism magn. Mater.*, **23,** 69 (1981)
37. Schneider, J., Arnold, Z., Kamarád, J. and Handstein, A., *Phys. Status Solidi, (a)*, **64,** K133 (1981)
38. Fukamichi, K., Shirakawa, K., Kaneko, T. and Masumoto, T., *J. appl. Phys.*, **53,** 2246 (1982)
39. Cochrane, R. W. and Graham, G. M., *Can. J. Phys.*, **48,** 264 (1970)
40. Fukamichi, K., Kikegawa, S. and Iwasaki, H., to be published
41. Chou, C.-P., Davis, L. A. and Hasegawa, R., *J. appl. Phys.*, **50,** 3334 (1979)
42. Shimizu, M., *J. Magnetism magn. Mater.*, **20,** 47 (1980)
43. Tanji, Y., Shirakawa, Y. and Moriya, H., *Sci. Rep. Res. Insts Tôhoku Univ.*, **A22,** 93 (1970)
44. Jagielinski, T., Wegrzyn, A., Ohnuma, A. and Masumoto, T., *Solid st. Communs*, **44,** 255 (1982)
45. Shirakawa, K., Fukamichi, K., Kaneko, T. and Masumoto, T., to be published
46. Hausch, G., *J. phys. Soc. Japan*, **37,** 819 (1974)
47. Handrich, K., *Phys. Status Solidi*, **32,** K55 (1969)
48. Fujita, F. E. and Sohmura, T., *J. Phys. (Paris)*, **37,** C6 397 (1976)
49. Oshima, R. and Fujita, F. E., *Jap. J. appl. Phys.*, **20,** 1 (1981)
50. Takahashi, M. and Koshimura, M., *Jap. J. appl. Phys.*, **18,** 685 (1979)
51. Schneider, J., Handstein, A., Hesske, R. and Zaveta, K., *Physica*, **86—88B,** 301 (1977)
52. Sinha, A. K., *J. appl. Phys.*, **42,** 338 (1971)
53. Fukamichi, K., *J. Magn. Soc. Japan*, **4,** 62 (1980) (in Japanese)
54. Kikuchi, M., Fukamichi, K. and Masumoto, T., *Sci. Rep. Res. Insts Tôhoku Univ.*, **A27,** 210 (1979)
55. Fukamichi, K., Kikuchi, M., Satoh, T., Masumoto, T., Ohmori, K. and Tsuya, N., *'Proc. 4th Int. Conf. Rapidly Quenched Metals*, Eds. Masumoto, T. and Suzuki, K., p. 832, Japan Institute of Metals, Sendai (1982)
56. Fukamichi, K., Kikuchi, M. and Masumoto, T., to be published
57. Kazama, N., Mitera, M. and Masumoto, T., ref. 22, vol. II, p. 164

58. Birgeneau, R. J., Tarvin, J. A., Shirane, G., Gyorgy, E. M., Sherwood, R. C., Chen, H. S. and Chien, C. L., *Phys. Rev.*, **B18,** 2192 (1978)
59. Ishikawa, Y., Yamada, K., Tajima, K. and Fukamichi, K., *J. phys. Soc. Japan,* **50,** 1958 (1981)
60. Grimsditch, M., Malozemoff, A. P. and Brunsch, A., *Phys. Rev. Lett.*, **43,** 711 (1979)
61. Damon, R. W. and Eshbach, J. R., *J. Phys. Chem. Solids,* **19,** 308 (1961)
62. Onn, D. G., *J. appl. Phys.*, **52,** 1788 (1981)
63. Matsuura, M. and Mizutani, U., ref. 55, p. 1291
64. Gupta, K. P., Gheng, C. H. and Beck, P. A., *J. Phys. Chem. Solids,* **25,** 73 (1964)
65. Onn, D. G., Johnson, W. D., Gleeson, P. E., Donnelly, T. A., Egami, T. and Liebermann, H. H., *J. Phys. C,* **10,** L639 (1977)
66. Long, T. C., Donnelly, T. A., Durand, J. and Onn, D. G., ref. 22, p. 69
67. Matsuura, M., Mizutani, U. and Yazawa, Y., *J. Phys. F,* **11,** 1393 (1981)
68. Fukamichi, K., *Metals,* **49,** No. 9, 21 (1979) (in Japanese)
69. Bergmann, G. and Marquardt, P., *Phys. Rev.*, **B18,** 326 (1978)
70. Fukamichi, K., Gambino, R. J. and McGuire, T. R., *J. appl. Phys.*, **53,** 2310 (1982)
71. Fukamichi, K., Gambino, R. J. and McGuire, T. R., ref. 55, p. 835
72. Soumura, T., *J. phys. Soc. Japan,* **42,** 826 (1977)
73. Fukamichi, K., Gambino, R. J. and McGuire, T. R., *J. appl. Phys.*, **52,** 2199 (1981)
74. Berry, B. S. and Pritchet, W. C., American Institute of Physics Conf. Proc. No. 34, p. 292 (1977)
75. Török, E. and Hausch, G., *J. Magnetism magn. Mater.*, **10,** 303 (1979)
76. Kikuchi, M., Fukamichi, K., Masumoto, T., Jagielinski, T., Arai, K. I. and Tsuya, N., *Phys. Status. Solidi,* (*a*), **48,** 175 (1978)
77. Maeda, T., *J. phys. Soc. Japan,* **30,** 375 (1971)
78. Kikuchi, M., Fukamichi, K. and Masumoto, T., *Sci. Rep. Res. Insts Tôhoku Univ.*, **A28,** 242 (1980)
79. Hiroyoshi, H. and Fukamichi, K., to be published
80. Shirakawa, K., Fukamichi, K., Kaneko, T. and Masumoto, T., to be published
81. Shiga, M., *J. phys. Soc. Japan,* **50,** 2573 (1981)
82. Hiroyoshi, H. and Fukamichi, K., *Phys. Lett.*, **85A,** 242 (1981)
83. Hiroyoshi, H. and Fukamichi, K., *J. appl. Phys.*, **53,** 2226 (1982)
84. Fujita, F. E., Kiritani, M., Oshima, R. and Nasu, S., to be published
85. Shiga, M., Muraoka, Y. and Nakamura, Y., *J. Magnetism magn. Mater.*, **10,** 280 (1979)

Chapter 18

Magnetic after-effects and the hysteresis loop

H. Kronmüller and N. Moser

Max Planck Institut für Metallforschung, Institut für Physik, Stuttgart, FDR

18.1 Introduction

In recent years amorphous ferromagnetic alloys fabricated by the spin-quenching technique have become attractive, soft magnetic materials expanding and improving the field of possible technical applications[1–4]. In particular a large elastic flexibility guarantees excellent insensitivity with respect to plastic deformations and a small electrical conductivity reduces the eddy-current losses.

A prerequisite for the application of amorphous alloys is their structural and magnetic stability over long times. The coercive field as well as the permeability and the energy losses are expected not to become worse after some time of applications. The guarantee of these properties requires a full understanding of the phenomena of magnetic after-effects (MAEs) observed in ferromagnetic amorphous alloys. Previous investigations have shown that MAEs may change the following characteristic properties of the hysteresis loop by more than 50 per cent:

1. Reduction of the initial and the maximum permeability
2. Increase of the coercive field
3. Increase of the hysteresis loop
4. Increase of the anisotropy energy

The present chapter gives a review on our present knowledge of MAEs in the as-quenched materials as well as after various pretreatments, e.g., annealing above room temperature, magnetic field annealing or plastic deformation. A further interesting aspect of MAE measurements is their relation to the diffusional properties of amorphous alloys. It will be shown that MAEs are directly related to the presence of the free volume which supports microstructural reordering of atom pairs within domain walls. The presence of MAEs[5–12] has been demonstrated in a multitude of amorphous alloys. In this chapter the common aspects of these MAE measurements are outlined and a series of very recent results will be presented.

18.2 Experimental techniques for the study of magnetic after-effects

18.2.1 General remarks

Several techniques may be used for the investigation of MAEs in ferromagnetic materials. First we have to distinguish between so-called irreversible and reversible types of after-effect. Irreversible types of relaxation are those connected with thermally activated microscopic Barkhausen jumps of domain walls whereas reversible MAEs are due to a local deepening of the domain wall potential resulting from microstructural atomic rearrangements within the domain walls. The first type of these MAEs is of importance for the case of permanent magnets with high remanent magnetizations whereas the reversible type of MAEs influences mainly the relevant properties of soft magnetic alloys. A theoretical description of both types of MAE has been given by several authors[13–18].

As is well known from previous experiments all characteristic properties of the hysteresis loop of amorphous alloys change with time. In particular the reversible susceptibility, χ, and the coercive field, H_c, are time-dependent properties. Consequently, without special pretreatments ferromagnetic materials show instabilities seriously influencing their quality and the range of applications.

18.2.2 Isothermal relaxation curves

For the study of MAEs the method of disaccommodation experiments has been found very useful. In this type of experiment the reversible susceptibility is measured by an AC technique[19] as a function of time, t, after demagnetization at a certain temperature T. As shown schematically in *Figure 18.1* the reversible susceptibility decreases from its initial value χ_0 as a function of time approaching for large times a limiting value χ_∞. In simple cases this time dependence can be described by exponential relaxation functions characterized by a limited number of relaxation times τ_i:

$$\chi(t) = \chi_0 - \sum_i \Delta\chi_i(1 - e^{-t/\tau_i}) \tag{18.1}$$

Here $\Delta\chi_i$ denotes the relaxation amplitude of the ith relaxation mode, and χ_0 the

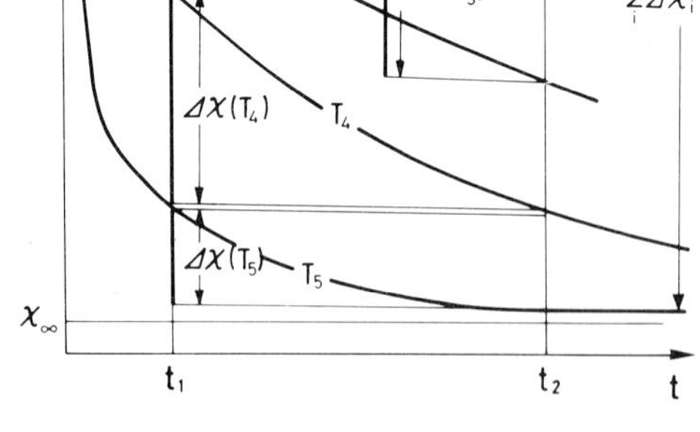

Figure 18.1 Isothermal relaxation curves for different measuring temperatures. The relaxation amplitudes $\Delta\chi$ as measured between the times t_1 and t_2 are indicated by vertical bars

susceptibility at $t = 0$. In the case of thermally activated processes the relaxation times, τ_i, are given by Arrhenius equations

$$\tau_i = \tau_{0,i} \exp [Q_i/kT] \tag{18.2}$$

with the pre-exponential factors $\tau_{0,i}$ and the activation energies Q_i. In the case of quantum mechanical tunnelling processes the relaxation times τ_i are temperature-independent.

18.2.3 Isochronal relaxation curves

The construction of isochronal relaxation curves from isotherms has been proposed previously[17,18,20,21] for a study of the temperature dependence of the MAEs. This type of representation is especially suitable for a separation of overlapping relaxation processes. Furthermore, rapidly varying thermally activated processes can be separated easily from 'background relaxations' which vary slowly with temperature. The method of construction of isochronal relaxation curves is shown in *Figure 18.2* for the susceptibility. The relaxation amplitudes of the susceptibility measured between t_1 and t_2

$$\Delta\chi(t_1, t_2, T) = \chi(t_1, T) - \chi(t_2, T) \tag{18.3}$$

are plotted as a function of T. An alternative representation uses the reluctivity $r(t, T)$ defined as the reciprocal susceptibility $r(t, T) = 1/\chi(t, T)$. If a relaxation process exists in the temperature range investigated a relaxation maximum is observed as shown in *Figure 18.2*. The characteristic parameters of the relaxation peak are the temperature, T_{max}, of the relaxation peak, the half-width, ΔT_H, and the relaxation amplitude $\Delta\chi_{max}$ at $T = T_{max}$.

In the case where we are dealing with a simple exponential time dependence of the type 18.1, the relaxation time τ at the temperature T_{max} of the relaxation peak is given by

$$\tau(T_{max}) = \frac{t_2 - t_1}{\ln (t_2/t_1)} \tag{18.4}$$

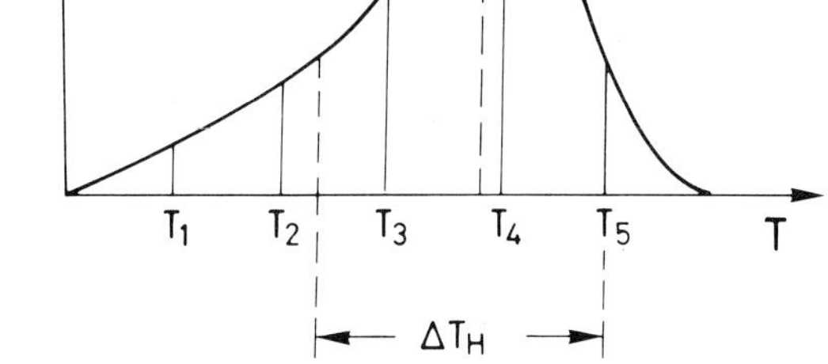

Figure 18.2 Isochronal relaxation curves obtained by plotting $\Delta\chi(t_1, t_2, T)$ against temperature T

According to equation 18.4 the temperature dependence of τ can be determined either by a variation of t_2 for fixed t_1 or by a variation of t_1 for fixed t_2.

18.2.4 Spectrum of activation energies

Because of the non-periodicity of the lattice potential in amorphous alloys, we cannot expect to find discrete activation parameters τ_0 and Q for the description of the MAEs. Owing to the randomness of the atomic distribution function it is assumed that thermally activated atomic rearrangements must be described by a spectrum of activation energies. As proposed by Richter[13] we may use a box-type distribution function for the activation energy Q within the intervals $Q_1 \leqslant Q \leqslant Q_2$. The time dependence of the reluctivity then is given by

$$r(t) = r_\infty + \Delta r[E_1(t/\tau_1) - E_1(t/\tau_2)]\frac{1}{\ln(\tau_2/\tau_1)} \tag{18.5}$$

where $r_\infty = 1/\chi(\infty)$; $\Delta r = r_0 - r_\infty$, $\tau_1 = \tau_0 \exp[Q_1/kT]$, $\tau_2 = \tau_0 \exp[Q_2/kT]$. $E_1(t/\tau)$ denotes the exponential integral

$$E_1(t/\tau) = \int_{t/\tau}^{\infty} \frac{e^u}{u}\,du \tag{18.6}$$

By introducing a spectrum of activation energies we have the possibility to describe the behaviour of amorphous structures in a first approximation.

18.3 Magnetic after-effect spectra of amorphous alloys

18.3.1 $Fe_{40}Ni_{40}P_{14}B_6$, METGLAS R 2826

Figure 18.3 shows the isochronal relaxation spectra of $Fe_{40}Ni_{40}P_{14}B_6$ as measured for the as-quenched state and after an annealing treatment of 4 h at 400 K. Both relaxation spectra show the same qualitative behaviour where we may distinguish three characteristic temperature ranges:

1. *Temperature range A:* $4.2\,K \leqslant T \leqslant 300\,K$. This relaxation is nearly temperature-independent and is ascribed to tunnelling-controlled reversible, microstructural rearrangements in the amorphous state which may be described by the two-level Anderson model.
2. *Temperature range B:* $300\,K \leqslant T \leqslant 450\,K$. Relaxation B corresponds to a wide relaxation peak with its position depending sensitively on the type of pretreatment.
3. *Temperature range C:* $450\,K \leqslant T \leqslant T_C$. In this temperature range just below the Curie temperature, T_C, an additional relaxation peak is observed corresponding to the so-called Hopkinson effect which results from the magnetic phase transition.

Whereas relaxation A is rather insensitive to annealing treatments relaxation B

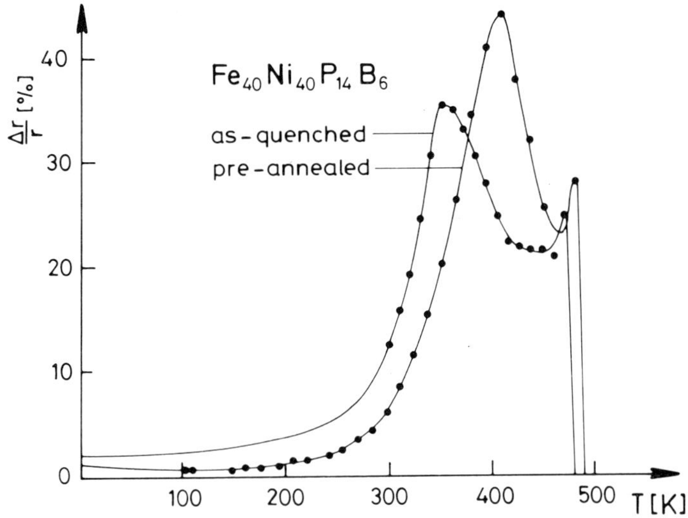

Figure 18.3 MAE spectrum of $Fe_{40}Ni_{40}P_{14}B_6$ as measured for the as-quenched state and after an annealing treatment of 4 h at 400 K ($t_2 = 180$ s)

turns out to be extremely sensitive on the annealing and cooling procedure. In order to display this characteristic feature the relative relaxation amplitude $\Delta r/r(t_1)$ has been measured at $T = 390$ K as a function of the annealing temperature, T_a, and after different cooling conditions. The annealing time t_a for each annealing temperature was $t_a = 4$ h. After each annealing treatment one of the specimens was slowly cooled to room temperature within about 2.5 h, whereas the other one was quenched to liquid nitrogen temperature. There is a significant effect of the cooling rate on the annealing curves of both types of specimen. In rapidly quenched FeNiPB the relaxation amplitude anneals in one single annealing stage at about 480 K. In the slowly cooled specimen annealing is observed also in the temperature range between 300 K and 480 K followed by a steep annealing stage as in the quenched specimen.

From these results we may conclude that slowly cooled specimens in general show a smaller relaxation amplitude. An optimum elimination of the relaxation is obtained for an annealing treatment above the Curie temperature of $T_C \sim 510$ K.

18.3.2 Iron–nickel–boron alloys

Figure 18.4 shows the relaxation spectra of FeNiB-alloys $Fe_{80-x}Ni_xB_{20}$. As in the case of $Fe_{40}Ni_{40}P_{14}B_6$ the relaxation spectra of these alloys show in general the relaxation types A, B and C as defined in *Figure 18.3*. The Hopkinson peak is only observed for $Fe_{20}Ni_{60}B_{20}$ because the Curie temperature of the other alloys lies about 500 K. The relaxation spectra for a measuring field amplitude of $\mu_0\hat{H} = 2.5\,\mu T = 25$ mG ($t_1 = 1$ s; $t_2 = 180$ s) are increased approximately by a factor of 2 in comparison to the results obtained for $\mu_0\hat{H} = 0.25\,\mu T$. This behaviour is attributed to the fact that in the case of larger field amplitudes domain walls move partially by irreversible Barkhausen jumps which avoid a full stabilization of the domain walls. It is of special interest that the FeNiB alloys qualitatively show the same relaxation modes as FeNiPB alloys.

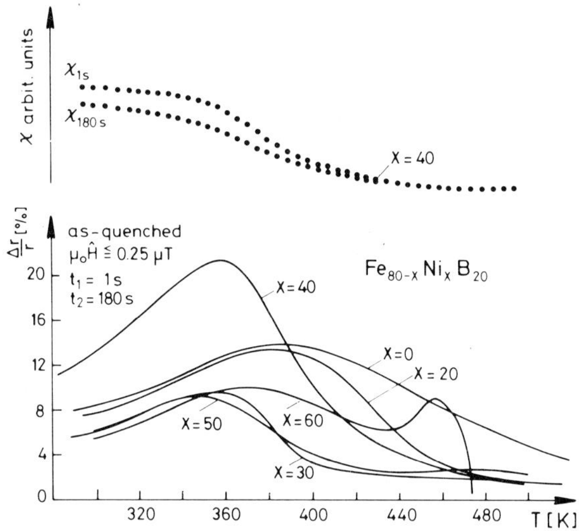

Figure 18.4 MAE spectra of as-quenched $Fe_{80-x}Ni_xB_{20}$ alloys. Amplitude of the A.C. measuring field $\mu_0\hat{H} = 0.25\,\mu T$. The upper part of the figure shows the temperature dependence of χ for $t_2 = 1\,s$ and $t_2 = 180\,s$

18.3.3 $Fe_{80}B_{20}$

The two-component alloy $Fe_{80}B_{20}$ is of special interest concerning the MAE because in this alloy no atom pair composed of two different transition metal atoms is present. Relaxation phenomena in this alloy therefore should be due to some kind of atomic defects similar to vacancies or interstitial atoms in crystalline material. *Figure 18.5* shows the relaxation spectrum for the as-quenched alloy and that obtained after an annealing treatment for 4 h at 400 K. As in other amorphous alloys a nearly temperature independent relaxation, A, is observed below room temperature whereas above 300 K a relaxation peak appears at about 430 K corresponding to relaxation B.

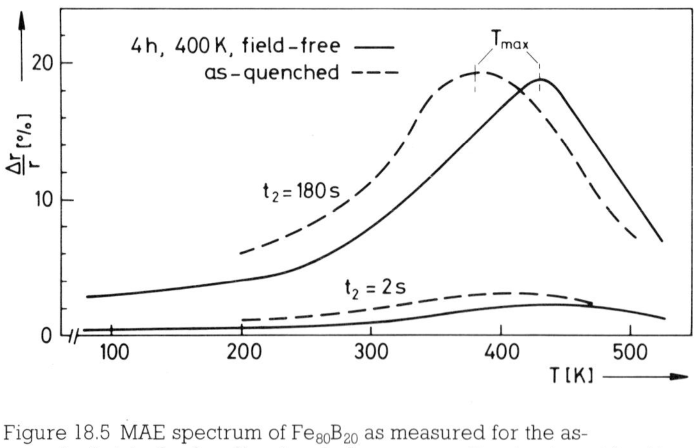

Figure 18.5 MAE spectrum of $Fe_{80}B_{20}$ as measured for the as-quenched state (broken lines) and after an annealing treatment for 4 h at 400 K (full lines). The isochronal relaxation curves are given for $t_1 = 2\,s$ and $t_2 = 180\,s$

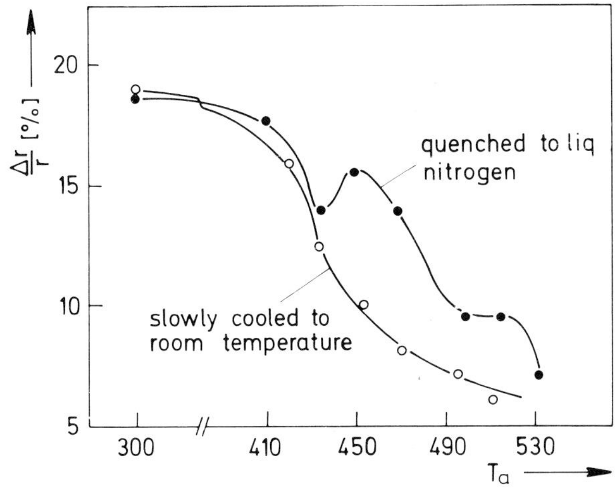

Figure 18.6 Annealing curves of the relative relaxation amplitude of $Fe_{80}B_{20}$ as measured at 400 K for slowly cooled samples (open circles) and those quenched to liquid nitrogen temperature (filled circles)

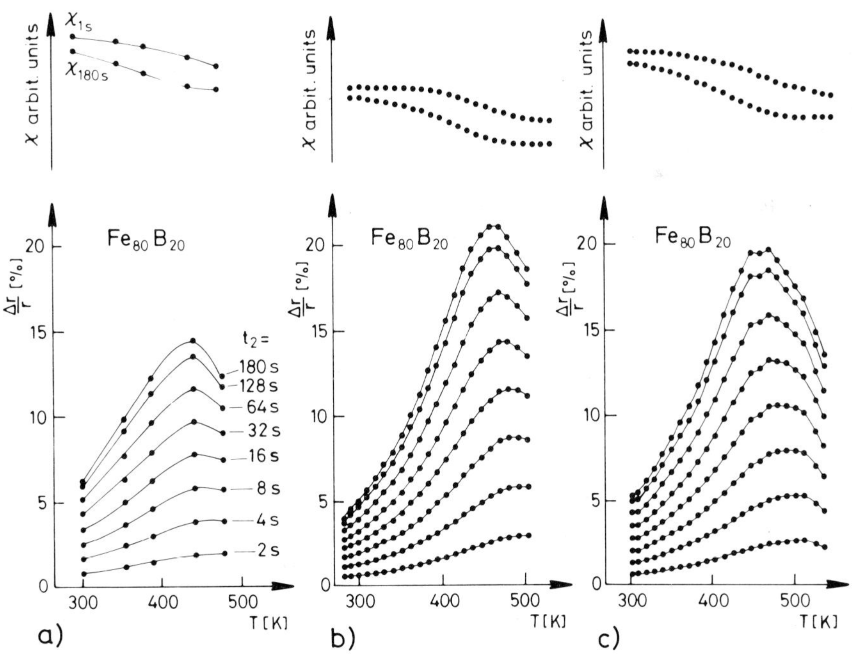

Figure 18.7 MAE spectrum of $Fe_{80}B_{20}$ after different pretreatments. The upper part of the Figure shows the temperature dependence of the susceptibility for $t_1 = 1$ s and $t_2 = 180$ s; (*a*) pre-annealed at 520 K for 30 min; (*b*) after plastic bending; (*c*) after annealing treatment for 4 h at 540 K

The Hopkinson maximum is not observable here because the Curie temperature lies above the temperature range investigated.

In order to investigate the possible sources of the MAEs in $Fe_{80}B_{20}$ different pretreatments were applied. *Figure 18.6* shows the annealing curves of the relative relaxation amplitude measured at 400 K for a slowly cooled specimen and a specimen quenched into liquid nitrogen. Whereas in the slowly cooled specimen the annealing takes place in one single stage centred at 430 K, in the case of the rapidly quenched specimen an additional annealing stage is observed around 475 K. This indicates clearly that the diffusion processes may be influenced by pretreatments. In a series of relaxation spectra measured for increasing and decreasing temperature for the as-quenched material, and after annealing treatments at 480 K in a magnetic field of 750 Oe, and a successive quenching to liquid nitrogen temperature, the shape of the spectrum remains unchanged by the annealing treatment but the peak temperature T_{max} shifts to larger temperatures and the relaxation amplitude increases. It is of interest that the relaxation can be produced repeatedly just by quenching to liquid nitrogen temperature.

Figure 18.7 shows another interesting result obtained after a weak plastic deformation by bending. The relaxation peak at 430 K increases from 13 per cent to 21 per cent and shifts to larger temperatures. After an annealing treatment of 4 h at 540 K the relaxation amplitude decreases only slightly to 20 per cent. The isochronal relaxation curves obtained for various measuring times t_2 reveal a shift of T_{max} to larger temperatures with decreasing t_2. According to equation 18.4 this is expected for a thermally activated relaxation process. If the relaxation times, $\tau(T_{max})$, as determined according to equation 18.4, are plotted in an Arrhenius plot, $\ln \tau(T_{max})$ against $1/T_{max}$, a straight line is obtained giving an activation energy of $Q = 1.66\,\text{eV}$ and a pre-exponential factor of $\tau_0 = 2.2 \times 10^{-17}\,\text{s}$.

18.3.4 Non-magnetostrictive amorphous alloys

Amorphous alloys with a vanishing magnetostriction are of particular interest for applications because these materials are insensitive to external as well as internal stresses of defect structures. The main pinning effects for domain walls in this case are due to relaxation processes and surface irregularities[22,23]. The elimination of relaxation processes therefore seems to be a necessary pre-condition for an improvement of these alloys. *Figure 18.8* shows the relaxation spectra of the nearly non-magnetostrictive alloys $Co_{58}Fe_5Ni_{10}Si_{11}B_{16}$, $Co_{70}Fe_5Si_7B_{18}$ and $Co_{70.4}Fe_{4.6}Si_{15}B_{10}$. All three alloys show a relaxation A below 300 K, however, with rather small amplitudes $\Delta r/r \leqslant$ about 1 per cent. Relaxation B above 300 K only in the case of $Co_{58}Fe_5Ni_{10}Si_{11}B_{16}$ shows a well-developed, however small, maximum. After an annealing treatment for 4 h at 470 K and 570 K relaxation B further develops as demonstrated in *Figure 18.9*. The

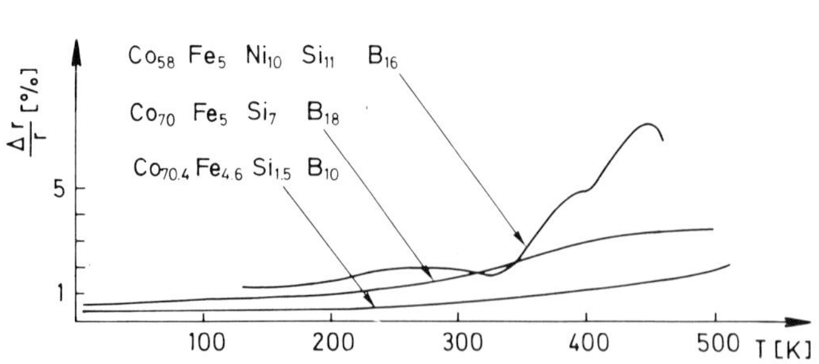

Figure 18.8 MAE spectra of several as-quenched non-magnetostrictive alloys for $t_2 = 180\,\text{s}$

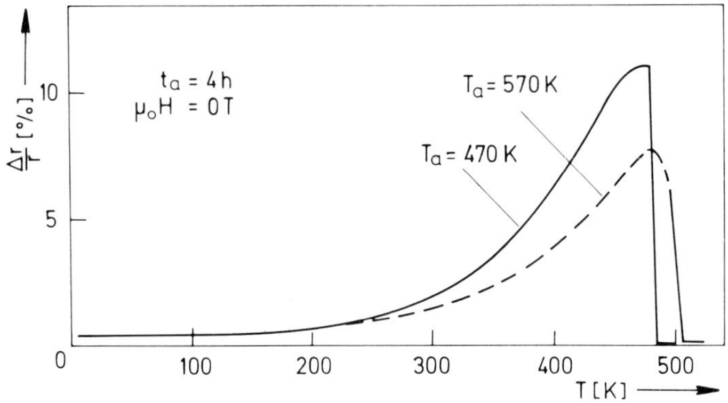

Figure 18.9 MAE spectra of $Co_{58}Fe_5Ni_{10}Si_{11}B_{16}$ after annealing treatments for 4 h at 470 K and 570 K ($t_2 = 180$ s); $\mu_0 H = 0$ T

effect of a longitudinal or a perpendicular magnetic field applied during the annealing treatment is shown in *Figure 18.10*. Obviously, a longitudinal field has a significant enhancing effect on the MAEs in the whole temperature range. In addition to the increased relaxations A and B, the Hopkinson peak is observed at 480 K and a new relaxation peak appears at 200 K. Whereas the longitudinal field annealing increases relaxation A by a factor of 10 a transversal field annealing even reduces relaxation A to an amplitude of about 0.2 per cent.

Also the MAE of the amorphous alloy $Co_{70}Fe_5Si_7B_{18}$ is extremely sensitive to longitudinal magnetic field annealing. The relaxation amplitude increases up to 36 per cent after an annealing treatment at 570 K in a longitudinal field of 700 Oe.

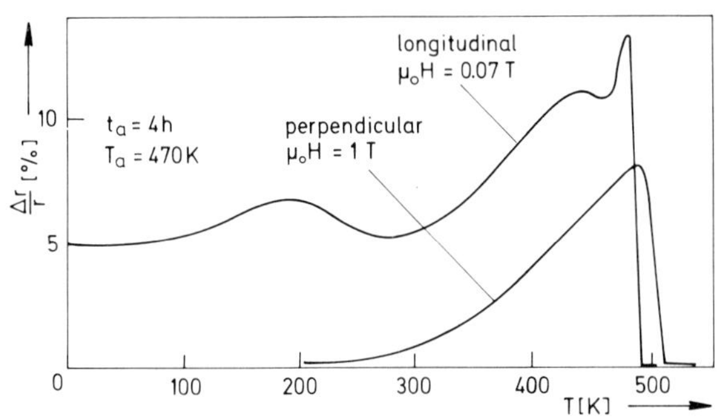

Figure 18.10 MAE spectra of $Co_{58}Fe_5Ni_{10}Si_{11}B_{16}$ after annealing treatment for 4 h at 470 K in a longitudinal (parallel to the ribbon's plane) magnetic field of 0.07 T and a perpendicular magnetic field of $\mu_0 H = 1$ T

18.4 Magnetic after-effects of hydrogen-charged amorphous alloys

18.4.1 $Co_{75}Si_{15}B_{10}$

As is well known from crystalline ferromagnetic alloys hydrogen isotopes give rise to MAEs below room temperature. A similar behaviour is expected for amorphous alloys because it is suggested that hydrogen atoms occupy some kind of 'interstitial sites' in amorphous alloys. Rearrangements of the hydrogen atoms between these 'interstitial sites' give rise to MAEs as well as internal friction phenomena[24–26]. Investigations of this kind are of interest for several points of view:

1. Energy storage by hydrogen.
2. Mechanical embrittlement.
3. Diffusional properties of light atoms.

In particular MAE measurements are suitable for a study of diffusional properties of hydrogen in amorphous alloys. In *Figure 18.11* the MAE spectra of the as-quenched $Co_{75}Si_{15}B_{10}$ alloy is compared with that obtained after electrolytic hydrogen charging. Whereas in the as-quenched specimen relaxation A and B are observed in the hydrogen-charged specimen three additional relaxation processes are observed. The generation and the annealing behaviour of these different MAEs are represented in *Figures 18.12*

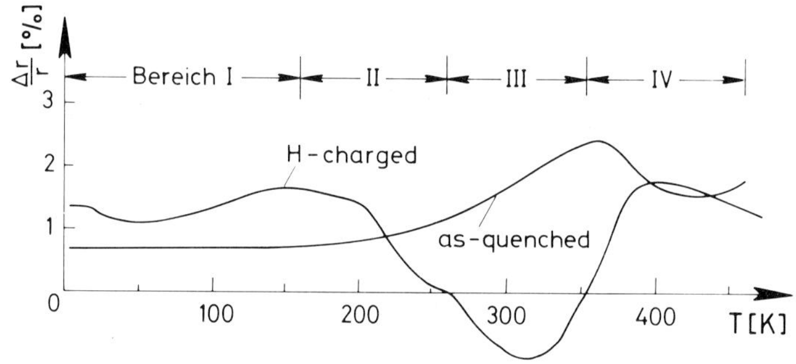

Figure 18.11 MAE spectra of as-quenched $Co_{75}Si_{15}B_{10}$ and after electrolytic hydrogen charging

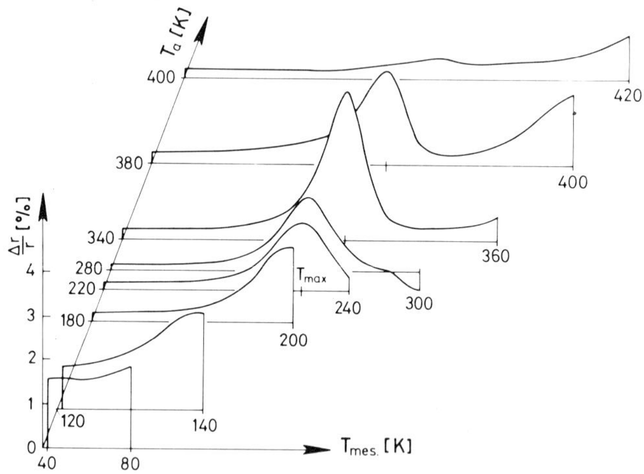

Figure 18.12 MAE spectra of hydrogen charged $Co_{75}Si_{15}B_{10}$ as measured successively after heating to the temperatures indicated on the T_{mes} axis

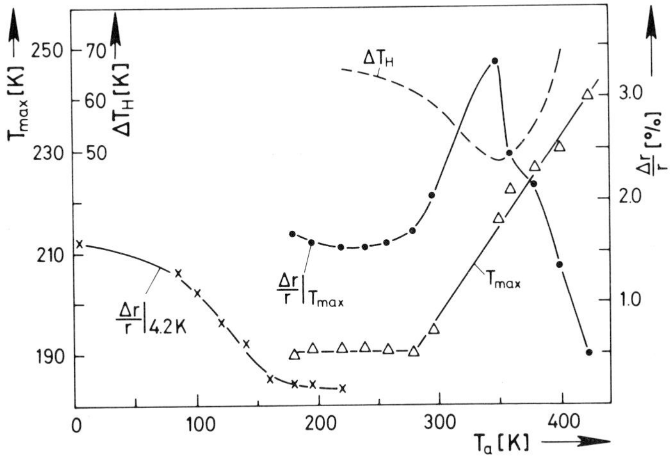

Figure 18.13 Annealing behaviour of the characteristic properties of the low temperature relaxation and of the relaxation maximum of $Co_{75}Si_{15}B_{10}$

and *18.13*. From the properties of these MAEs we may draw the following conclusions:

1. Below 100 K a nearly temperature independent relaxation process is observed annealing in one stage at 120 K. This relaxation is attributed to a tunnelling motion of a metastable intersitial configuration of hydrogen atoms.
2. The relaxation maximum near 200 K corresponds to a thermally activated relaxation process of a rather stable interstitial configuration of hydrogen atoms. In this configuration the hydrogen atoms probably are trapped by silicon atoms, thus forming SiH pairs where the hydrogen atoms perform hopping processes around the silicon atom.
3. The negative relaxation process observed between 260 K and 355 K is attributed to the dissolution of larger hydride complexes which act as trapping centres for domain walls. This assumption also is supported by investigations of the coercive field of hydrogen charged $Co_{75}Si_{15}B_{10}$ alloys[27].

An analysis of the time and temperature dependence of the 200 K peak shows that this MAE should be described by a spectrum of activation energies. Using equation 18.5 we obtain from the isochronal relaxation curves a pre-exponential factor $\tau_0 = 7.51 \times 10^{-15}$ s and for the lower and upper boundaries of the spectrum of activation energies $Q_1 = 0.46$ eV, $Q_2 = 0.67$ eV. These values for the activation parameters of hydrogen are compatible with those determined for face centred cubic alloys[28].

18.5 Interpretations of magnetic after-effects in amorphous alloys

18.5.1 Atom pair reorientation after-effect

From crystalline binary alloys, AB, it is well known that magnetic annealing induces an anisotropy energy owing to the alignment of atom pairs parallel to the spontaneous magnetization M_s. In amorphous alloys containing two types of transition metal similar atom pairs of type AA, BB and AB exist. We may characterize these atom pairs by their interaction energy, ε_i with M_s and the orientation, $\mathbf{P}_i$, of the pair axis. The index, i, denotes the atom pair position $\mathbf{r}_i$. Assuming for the pairs a uniaxial anisotropy we may write for the interaction energy

$$\varepsilon_i = \varepsilon_{0,i} \cos^2 \theta_i(\mathbf{r}_i) \tag{18.7}$$

where $\varepsilon_{0,1}$ corresponds to an interaction constant and $\theta_i(\mathbf{r}_i)$ to the angle between the local anisotropy axis, $\mathbf{P}_i$, and the spontaneous magnetization. This interaction energy probably results from the spin orbit coupling and may be either of magnetocrystalline or magnetostrictive origin.

In general the atom pairs have to be considered as immobile. However, in the neighbourhood of free volumes, i.e., vacancy type empty spaces between the atoms, the atom pairs may reorient their anisotropy axis as illustrated in *Figure 18.14*. Since all neighbouring atoms are involved in such a rearrangement the whole reorientation process corresponds to a collective phenomenon of high activation entropy. In the following we approximate the reorientation process by a 'two level system' similar to Anderson's[29] model developed for the excitations in the glassy state. We assume that each atom pair in the neighbourhood of a free volume may occupy two energetically different orientations $\mathbf{P}_1$ and $\mathbf{P}_2$ as illustrated in *Figure 18.15*. In each orientation the interaction energy ε_i is different. In thermal equilibrium the local occupation numbers f_i^∞ are given by

$$f_{1,2}^\infty = \frac{\exp(\pm\Delta\varepsilon/kT)}{2\cosh(\Delta\varepsilon/kT)} \tag{18.8}$$

where

$$2\Delta = \varepsilon_2 - \varepsilon_1 \tag{18.9}$$

corresponds to the energy difference of the two configurations P_1 and P_2. At high temperatures the rearrangement between the two energy levels occurs by thermally activated processes whereas at low temperatures this process may occur by quantum mechanical tunnelling. The high temperature rearrangements are characterized by a relaxation time $\tau = \tau_0 \exp(Q/kT)$ where Q corresponds to the activation energy and the pre-exponential factor

$$\tau_0 = \frac{1}{2\nu_0} \exp[-S_m/k] \tag{18.10}$$

contains the activation entropy, S_m, and an atomic attempt frequency ν_0 of the order of magnitude of the Debye frequency. If the occupation numbers at $t = 0$ are $f_1 = f_2 = 1/2$

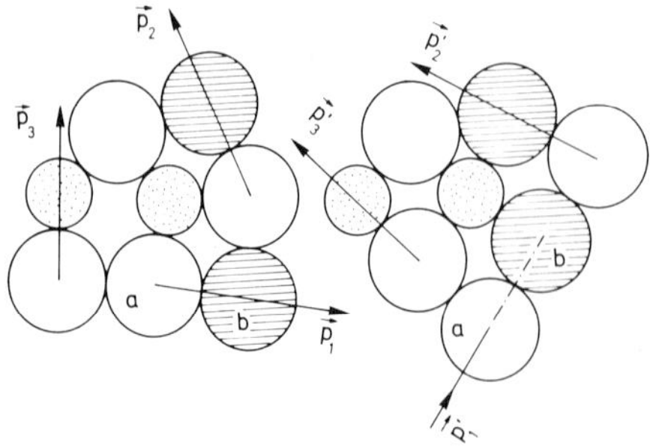

Figure 18.14 Reorientation of pairs of metal atoms in the neighbourhood of free volumes (big empty circle = iron atoms, big shaded circle = nickel atoms, small circles = metalloid atoms). The transitions of the local anisotropy axes from $\mathbf{P}_i$ to $\mathbf{P}_j$ are indicated

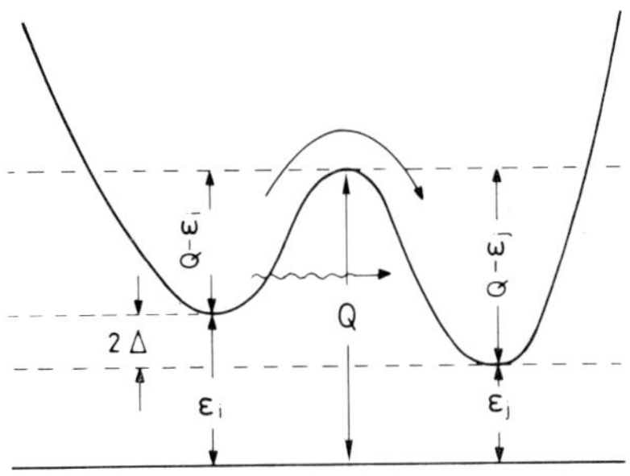

Figure 18.15 Two-level model for the local reorientation of atom pairs. The arrows indicate thermally activated and tunnelling transitions into energetically favourable positions ($\varepsilon_j < \varepsilon_i$)

we obtain for the occupation numbers at time t

$$f_j(t) = \frac{1}{2} - \frac{1}{2kT}(\varepsilon_j - \bar{\varepsilon})(1 - e^{-t/\tau}) \qquad (18.11)$$

where $\bar{\varepsilon} = (\varepsilon_1 + \varepsilon_2)/2$ corresponds to the average interaction energy.

In the present state of our knowledge on the microstructural properties of amorphous alloys we do not have detailed information on the concentration, c_p, of mobile atom pairs. We may, however, assume that c_p is proportional to the concentration, c_f, of free volumes, thus giving

$$c_p = \alpha c_f \qquad (18.12)$$

where α corresponds to a structural factor of the order of magnitude unity.

The relaxation process defined by equation 18.11 leads to a decrease of the total interaction energy of a domain wall with the atom pairs. This energy is denoted as the stabilization energy, W_s, of a domain wall because the rearrangement of atom pairs leads to an additional pinning potential of the domain wall. The local density of the stabilization energy is given by

$$W_s'(r, t) = -\frac{c_p}{2kT} \sum_{j=1,2} \{\varepsilon_j - \bar{\varepsilon}\}\varepsilon_j(1 - e^{-t/\tau}) \qquad (18.13)$$

In order to determine the total stabilization potential we assume that at time t the domain wall is displaced by a distance, U, from its original position which gives us

$$W_s(t, U) = -\frac{c_p}{2kT} \sum_{j=1,2} \int_{-\infty}^{\infty} \langle\{[\varepsilon_j(z) - \bar{\varepsilon}(z)]\varepsilon_j(z - U)\}\rangle(1 - e^{-t/\tau})\, dz \qquad (18.14)$$

where the integration extends over the z-coordinate which is assumed to be parallel to the domain wall normal. The angle brackets $\langle\rangle$ of equation 18.14 indicate an averaging over the statistical variables $\varepsilon_{0,i}$ and θ_i as defined in equation 18.7. Performing the summation in expression 18.14 we obtain

$$W_s(t, U) = -\frac{c_p}{4kT}(\langle F_1(U)\rangle - \langle \bar{F}_1(U)\rangle(1 - e^{-t/\tau}) \qquad (18.15)$$

With substitutions $z = Z + U/2$; $z - U = Z - U/2$; $\langle F_1\rangle$ and $\langle \bar{F}_1\rangle$ are defined as

$$\langle F_1\rangle = \sum_{j=1,2} \left\langle \int_{-\infty}^{\infty} \varepsilon_j\left(Z - \frac{U}{2}\right)\varepsilon_j\left(Z + \frac{U}{2}\right)\right\rangle dZ \qquad (18.16)$$

$$\langle \bar{F}_1\rangle = 2\left\langle \int_{-\infty}^{\infty} \bar{\varepsilon}\left(Z - \frac{U}{2}\right)\bar{\varepsilon}\left(Z + \frac{U}{2}\right)\right\rangle dZ \qquad (18.17)$$

For a quantitative calculation of expressions 18.16 and 18.17 we have to insert equation 18.7 into these equations with the angle θ_i referred to the coordinate system of the domain wall under consideration. In the following we consider a 180 degree domain wall with the wall normal parallel to the z-axis and the plane of the wall parallel to the (x, y) plane. Within the domain wall the angle of rotation of M_s with respect to the x-axis is given by[23]

$$\sin\zeta_0 = \frac{1}{\cosh(\pi z/\delta_B)} \tag{18.18}$$

where the domain wall thickness is defined as

$$\delta_B = \pi\sqrt{\langle A\rangle/\langle K\rangle} = \pi\delta_0 \tag{18.19}$$

Here $\langle A\rangle$ corresponds to the average exchange constant and $\langle K\rangle$ to the average anisotropy energy[30–32]. If we denote the polar and the azimuthal angle of the local anisotropy axis at position r_i by ϑ_i and ζ_i we find

$$\cos\theta_i = \sin\vartheta_i \sin(\zeta_{0,i} - \zeta_i) \tag{18.20}$$

where $\zeta_{0,i}$ denotes the rotation angle of M_s within the domain wall as given by (18.18). The evaluation of $\langle F_1\rangle$ and $\langle\bar{F}_1\rangle$ is drastically simplified if we assume that the interaction constants $\varepsilon_{0,1}$ and $\varepsilon_{0,2}$ are statistically fluctuating parameters with a symmetrical distribution of positive and negative values. In this case the averages $\langle\varepsilon_{0,1}\,\varepsilon_{0,2}\rangle$ vanish and

$$\langle\varepsilon_{0,1}^2\rangle = \langle\varepsilon_{0,2}^2\rangle = \langle\varepsilon_0^2\rangle;\ \langle\sin^4\vartheta_1\rangle = \langle\sin^4\vartheta_2\rangle = \frac{8}{15} \tag{18.21}$$

This gives us finally

$$\langle F_1(U)\rangle - \langle\bar{F}_1(U)\rangle = -\frac{8}{15}\langle\varepsilon_0^2\rangle\left\langle\int_{-\infty}^{\infty}\sin^2\left(\zeta_0\left(Z - \frac{U}{2}\right) - \zeta_1\right)\right. \times \left.\sin^2\left(\zeta_0\left(Z + \frac{U}{2}\right) - \zeta_1\right)\right\rangle dZ \tag{18.22}$$

Here we have taken into account that the averaging over ζ_1 and ζ_2 gives the same result.

Neglecting such terms which are independent of the displacement U we obtain for the stabilization energy

$$W_s(t, U) = \frac{2}{15}\langle\varepsilon_0^2\rangle\frac{c_p}{kT}\left\{I_1(U) + \frac{1}{4}I_2(U)\right\}\cdot(1 - e^{-t/\tau}) \tag{18.23}$$

where

$$\begin{aligned} I_1(U) &= -\int_{-\infty}^{\infty}\sin^2\left[\zeta_0\left(Z - \frac{U}{2}\right)\right]\sin^2\left[\zeta_0\left(Z + \frac{U}{2}\right)\right] dZ \\ &= 4\delta_0\frac{1 - (U/\delta_0)\,\text{cotanh}\,(U/\delta_0)}{\sinh^2(U/\delta_0)} \end{aligned} \tag{18.24}$$

$$\begin{aligned} I_2(U) &= -\int_{-\infty}^{\infty}\sin\left[2\zeta_0\left(Z - \frac{U}{2}\right)\right]\sin\left[2\zeta_0\left(Z + \frac{U}{2}\right)\right] dZ \\ &= 4I_1\cosh\left(\frac{U}{\delta_0}\right) - 8\delta_0\frac{(U/\delta_0)}{\sinh(U/\delta_0)} \end{aligned} \tag{18.25}$$

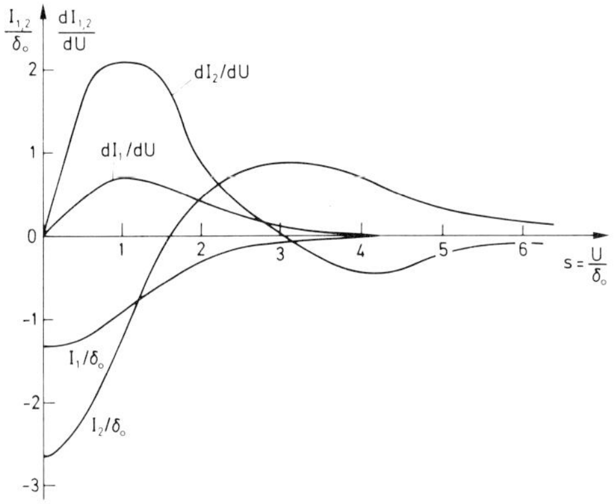

Figure 18.16 The functions I_1 and I_2 and their derivatives $dI_{1,2}/dU$

The numerical results for the integrals I_1 and I_2 are represented in *Figure 18.16*. Both functions show a parabolic behaviour for small displacement and a point of inflection at $U \sim \delta_0$. The derivations $dI_{1,2}/dU$ shown in *Figure 18.16* correspond to the force exerted by the stabilization potential on the domain wall.

According to our present understanding the MAE is due to a decrease of the total interaction energy between mobile atom pairs and the domain walls. Equation 18.23 gives a full description of the time and temperature dependence of the so-called stabilization potential and all relevant properties of the MAEs can be derived from the functions $I_1(U)$ and $I_2(U)$. The relaxation amplitude $\Delta r(t_1, t_2, T)$ introduced in section 18.2.3 is related to the curvature of the stabilization energy at $U = 0$ in a simple manner:

$$\Delta r(t_1, t_2, T) = \frac{1}{4\cos^2\phi_0 M_s^2 S}\left[\frac{d^2W_s(U, t_2)}{dU^2} - \frac{d^2W_s(U, t_1)}{dU^2}\right]_{U=0} \tag{18.26}$$

where ϕ_0 corresponds to the angle between the applied magnetic field and the spontaneous magnetization within the domain, and S to the total domain wall surface per unit volume. With the analytical results of equations 18.24 and 18.25 we find

$$\left(\frac{d^2W_s(U, t)}{dU^2}\right)_{U=0} = \frac{4}{15}\langle\varepsilon_0^2\rangle\frac{c_p}{kT}\frac{1}{\delta_0}(1 - e^{t/\tau}) \tag{18.27}$$

The relaxation amplitude now may be written as

$$\Delta r(t_1, t_2, T) = \frac{1}{15 M_s^2 S \cos^2\phi_0}\langle\varepsilon_0^2\rangle\frac{c_p}{kT}\frac{1}{\delta_0}(e^{-t_1/\tau} - e^{-t_2/\tau}) \tag{18.28}$$

There is no restriction to modify equation 18.28 by introducing a spectrum of relaxation times as proposed in section 18.2.4. In this case the exponential functions are replaced by exponential integrals.

18.5.2 Snoek-type magnetic relaxations of metalloid atoms

Metalloid atoms in amorphous alloys in general are assumed to occupy the centres of the Bernal polyhedra[33]. These positions of the metalloid atoms are rather stable ones and it seems to be unlikely that these configurations give rise to significant MAEs. As discussed previously[10,34] there exist 'anomalous' sites for the metalloid atoms corresponding to positions between two next nearest neighbour transition metal atoms. Transitions between these types of 'interstitial sites' may give rise to a Snoek-type MAE because the orientation of the local anisotropy axis changes. The quantitative description of this type of MAE is analogous to that of the atom pair relaxation model of section 18.5.1.

Figure 18.17 gives a schematic model for the two configurations of metalloid atoms. The occupation of 'anomalous' metalloid sites may be enhanced by small over-concentrations of metalloid atoms or by the presence of free volumes. From a thermodynamical point of view it is however quite reasonable that in thermal

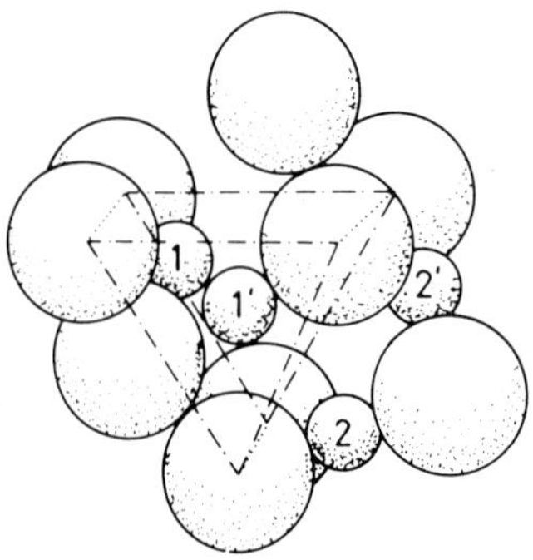

Figure 18.17 Bernal polyhedron showing 'normal' metalloid sites 1 and 1′ as well as 'anomalous' metalloid sites 2 and 2′; big spheres correspond to transition metal atoms

equilibrium always a certain number of 'anomalous' sites are occupied. In the as-quenched materials it is self-suggesting that the distribution of metalloid atoms on 'normal' and 'anomalous' sites does not correspond to a thermal equilibrium. Annealing treatments therefore give rise to drastic recovery effects as has been shown in section 18.3. Similarly, the MAEs will depend sensitively on the quenching conditions[36]. The present model may well describe the special behaviour of $Fe_{80}B_{20}$ where it seems unlikely that atom pairs produce the MAE. The magnetic relaxations observed around 430 K therefore are attributed to the relaxation of boron atoms. This interpretation is also supported by the agreement between the activation energy of $Q = 1.66\,eV$ determined after plastic deformation for relaxation B, and the activation energy of $Q \simeq 1.87\,eV$ derived by Köster *et al.*[36] from the crystallization which is governed by boron diffusion.

18.6 Influence of magnetic after-effects on magnetic properties

18.6.1 Coercive field

As discussed in section 18.5.1 the atomic rearrangements within a domain wall lead to the formation of a stabilization potential which acts as pinning potential. In order to push a domain wall out of its potential a finite magnetic field is required, usually called the coercive field. In general static pinning potentials will be superimposed on the stabilization potential. In the case of non-magnetostrictive alloys where the static pinning potential is expected to be rather small the stabilization potential is the main source of a finite coercive field. This coercive field H_c^{rel} may be determined from the maxima of the derivatives of the functions I_1 and I_2 plotted in

Figure 18.16. The coercive field due to MAEs is defined as

$$H_c^{rel} = \frac{1}{2M_s \cos\phi_0} \left.\frac{dW_s(U,t)}{dU}\right|_{max} \tag{18.29}$$

From *Figure 18.16* and equation 18.18 we then obtain

$$H_c^{rel}(t) = \frac{4}{15}\frac{1}{2M_s \cos\phi_0}\langle\varepsilon_0^2\rangle\frac{c_p}{kT}(1 - e^{-t/\tau}) \tag{18.30}$$

At large times we expect a saturation of H_c^{rel} given by

$$H_c^{rel}(\infty) = \frac{4}{15}\frac{1}{2M_s \cos\phi_0}\langle\varepsilon_0^2\rangle\frac{c_p}{kT} \tag{18.31}$$

A time dependence of H_c in fact has been observed recently in the non-magnetostrictive alloy $Co_{58}Ni_{10}Fe_5Si_{11}B_{16}$. *Figure 18.18* presents relaxation curves of $H_c(t)$ as obtained after different pre-treatments[37]. In the as-quenched state $H_c^{rel}(\infty)$ contributes about 50 per cent to the total coercive field.

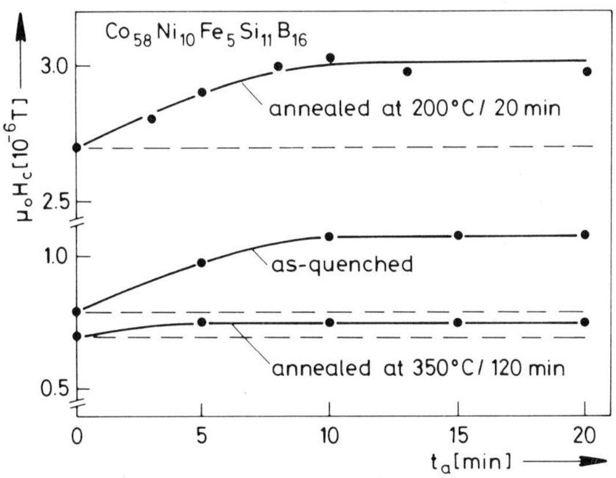

Figure 18.18 Magnetic after-effect of the coercive field of $Co_{58}Fe_5Ni_{10}Si_{11}B_{16}$ after various annealing treatments. H_c is measured after different waiting times, t_a, in the remanent state

18.6.2 Initial susceptibility

Similarly to the coercive field, the reversible susceptibility is also influenced by MAEs. If we take into account that a domain wall is pinned by a static potential, $W_0(U)$, as well as the stabilization potential $W_s(U, t)$, we find for the initial reluctivity

$$r(t) = \frac{1}{\chi(t)} = \frac{1}{\chi_0} + \frac{1}{15 M_s^2 S \cos^2\phi_0}\langle\varepsilon_0^2\rangle\frac{c_p}{kT}\frac{1}{\delta_0}(1 - e^{-t/\tau}) \tag{18.32}$$

Here $1/\chi_0$ corresponds to the static initial susceptibility measured for $t = 0$ immediately after demagnetization. For an infinite initial static susceptibility the second term of equation 18.32 defines the maximum initial susceptibility attainable in soft magnetic alloys.

18.6.3 Domain wall width

In amorphous alloys the domain wall widths are rather large because the effective magnetic anisotropy energies are a factor of 10—100 smaller than in crystalline materials[23,31,32]. As discussed elsewhere[23,31] in amorphous alloys long-range internal stresses and induced anisotropies determine the wall width $\delta_B = \pi(\langle A\rangle/\langle K\rangle)^{1/2}$.

The anisotropy energies in magnetostrictive amorphous alloys were found to be 10^3 erg/cm^3—10^4 erg/cm^3, giving δ_B values of 5×10^2—5×10^3 Å. In non-magnetostrictive alloys anisotropy energies as small as 10^2 erg/cm^3 are expected which would lead to δ_B's of 10^4 Å. Therefore all other anisotropy energies will have a significant effect on δ_B. From equations 18.15 and 18.22 we may derive the locally induced stabilization anisotropy energy which is given by

$$W_s'(z, U, t) = -\frac{1}{30}\langle\varepsilon_0^2\rangle\frac{c_p}{kT}\cos 2[\zeta_0(z - U) - \zeta_0(z)](1 - e^{-t/\tau}) \qquad (18.33)$$

If the domain wall is kept fixed, i.e., $U = 0$, the induced anisotropy is constant all over the domain wall and as a consequence the spin structure of the wall remains unchanged. Similarly, for large elongations $U > \delta_B$ the domain wall is not influenced by the MAE. However, for intermediate elongations, $U \leqslant \delta_B$, the stabilization energy, W_s', may influence the domain wall structure, the effective anisotropy constant being now

$$\langle K_{\text{eff}}\rangle = \langle K\rangle + \frac{1}{30}\langle\varepsilon_0^2\rangle\frac{c_p}{kT}(1 - e^{-t/\tau}) \qquad (18.34)$$

An estimate of the second term in equation 18.34 gives for $\varepsilon_0 = 10^{-3}$ eV, $c_p = 10^{21}$/cm^3, $kT = (1/40)$ eV a contribution to the anisotropy constant of 2×10^3 erg/cm^3 which may be even larger than the unrelaxed anisotropy energy. It is thus evident that the MAE produces a shrinkage of the domain wall width if the domain wall moves within the stabilization potential.

References

1. Luborsky, F. E. and Walter, J. L., *IEEE Trans. Magn.*, **MAG-13**, 1635 (1977)
2. Hilzinger, H. R., Mager, A. and Warlimont, H., *J. Magnetism magn. Mater.*, **9**, 191 (1978)
3. Luborsky, F. E. and Johnson, L. A., *J. Phys. (Paris) Colloq.*, **41**, C8, 820 (1980)
4. Graham, C. D. and Egami, T., *A. Rev. mater. Sci.*, **8**, 423 (1978)
5. Boucher, B. and Barbara, B., *J. Phys. F*, **9**, 151 (1970)
6. Fujimori, S., Ohta, S., Masumoto, T. and Nakamoto, K., '*Proc. 3rd Int. Conf. on Rapidly Quenched Metals*', Ed. Cantor, B., vol. II, p. 232, The Metals Society, London (1978)
7. Allia, P., Mazzetti, P. and Vinai, F., *J. Magnetism magn. Mater.*, **19**, 281 (1980)
8. Cecchetti, A., Masoli, F., Poppi, M. and Soardo, G. P., *J. Magnetism magn. Mater.*, **19**, 284 (1980)
9. Kisdi-Koszó, E., Vojtanik, P. and Potocky, L., *J. Magnetism magn. Mater.*, **19**, 159 (1980)
10. Moser, N. and Kronmüller, H., *J. Magnetism magn. Mater.*, **19**, 275 (1980)
11. Miyazaki, T., Takahashi, M., Lee, Y. H., Hisatake, K. and Yamamoto, T., '*Proc. 4th Int. Conf. on Rapidly Quenched Metals*', Eds. Masumoto, T. and Suzuki, K., Japan Institute of Metals, Sendai, Japan (1981)
12. Miyazaki, T. and Takahashi, M., *Sci. Rep. Res. Insts Tôhoku Univ., Suppl.*, **A28**, 311 (1980)
13. Richter, G., *Annln. Phys.*, **29**, 605 (1937)
14. Jordan, H., *Z. tech. Phys.*, **11**, 2 (1930)
15. Néel, L., *J. Phys. Radium*, **12**, 339 (1951)
16. Néel, L., *J. Phys. Radium*, **13**, 249 (1951)
17. Kronmüller, H., '*Nachwirkung in Ferromagnetika*', Springer-Verlag, Berlin, Heidelberg, New York (1968)
18. Kronmüller, H., in '*Vacancies and Interstitials in Metals*', Eds. Seeger, A., Schumacher, D., Schilling, W. and Diehl, J., p. 667, North-Holland, Amsterdam (1969)

19. Walz, F., *Phys. Status Solidi, (a)*, **8,** 125 (1971)
20. Kronmüller, H., Schaefer, H.-E. and Rieger, H., *Phys. Status Solidi*, **9,** 863 (1965)
21. Seeger, A., Kronmüller, H., and Rieger, H., *Z. angew. Phys.*, **18,** 377 (1965)
22. Kronmüller, H., *J. Magnetism magn. Mater.*, **24,** 159 (1981)
23. Kronmüller, H., Gröger, B., *J. Phys. (Paris)*, **42,** 1285 (1981)
24. Berry, B. S. and Pritchet, W. C., *Scripta Met.*, **15,** 637 (1981)
25. Berry, B. S. and Pritchet, W. C., *J. appl. Phys.*, **52,** 1865 (1981)
26. Künzi, H. W. and Agyeman, K., *'Proc. Conf. on Internal Friction and Ultrasonic Attenuation in Solids'*, Ed. C. C. Smith, p. 371, Pergamon Press, Oxford (1980)
27. Wang, Y. Z. and Kronmüller, H., *Phys. Status Solidi, (a)*, **68,** 499 (1981)
28. Hohler, B. and Kronmüller, H., *Phil. Mag.*, **A43,** 1189 (1981)
29. Anderson, P. W., 'Lectures on Amorphous Systems', in Physics of Ill–Condensed Matter, Eds. Balian, R., Maynard, R. and Toulouse, G., North-Holland, Amsterdam (1979)
30. Luborsky, F., Walter, J. L., Liebermann, H. H. and Wohlfarth, E. P., *J. Magnetism magn. Mater.*, **15–18,** 1351 (1980)
31. Kronmüller, H., *Atomic Energy Review, Suppl.* **1,** 255 (1981)
32. Domann, M., Grimm, H. and Kronmüller, H., *J. Magnetism magn. Mater.*, **13,** 81 (1979)
33. Bernal, J. D., *Nature*, **185,** 68 (1960)
34. Moser, N., Kronmüller, H. and Frank, W., Proc. Conf. on Internal Friction and Ultrasonic Attenuation in Solids, Lausanne (1981), *J. Phys. (Paris)*, **10,** 647 (1981)
35. Köster, U., Herold, U., Hillenbrand, H.-G. and Denis, J., *J. mater. Sci.*, **15,** 2155 (1980)
36, Allia, P. Luborsky, F. E., Turtelli, R. S., Soardo, G. P. and Vinai, F., *IEEE Trans. Magn.*, **MAG-17,** 2615 (1981)
37. Gröger, B., Beck, W., Dong, X.-Z., Moser, N. and Kronmüller, H. *J. Magnetism magn. Mater.*, **26,** 264 (1982)

Chapter 19

Applications-oriented magnetic properties

F. E. Luborsky

General Electric Company, Schenectady, New York, USA

19.1 Introduction

In this chapter we will discuss various magnetic properties, other than the so-called fundamental properties discussed in Chapters 13—16 such as Curie temperature, saturation moment, spin wave constant, anisotropy constants, magnetostriction and magnetovolume effects. The properties to be discussed include coercivity, remanence-to-saturation ratio, losses, permeability, exciting volt-amperes and the change of these properties with temperature and time. All of these can be discussed from a fundamental view and many of the characteristics are critical parameters influencing the performance of the device. For example, the losses and exciting power are critical criteria in the design of large transformers, while the initial permeability is a critical parameter in the design of small ground-fault detector transformers. The coercivity, on the other hand is a controlling factor in the losses.

19.2 Coercivity

The coercivity of amorphous ribbons has been interpreted in the same terms as used for conventional crystalline ribbons. We will use the discussions of Gyorgy[1] on the effect of surface imperfections, of Goto *et al.*[2] on the effect of fluctuations in composition or internal stress, of Evetts *et al.*[3] and Alben *et al.*[4] on the effect of local anisotropy fluctuations, and of Kronmüller and Groger[5] and of Kronmüller[6,7] on the effect of many possible contributions to the coercivity.

At least five pinning effects have been identified and discussed by Kronmüller[6,7] as contributing to the total coercivity. In order of increasing importance in amorphous alloys these are:

1. Intrinsic fluctuations of exchange energies and local anisotropies (10^{-3}—1 mOe), H_c(i).
2. Clusters of chemical short range ordered regions (<1 mOe), H_c(SO).
3. Surface irregularities (<5 mOe), H_c(surf)
4. Relaxation effects due to local structural rearrangements (0.1—10 mOe), H_c(rel).
5. Volume pinning of domain walls by defect structures in magnetostrictive alloys (10—100 mOe), $H_c(\sigma)$.

Within the framework of the statistical potential theory[8,9], on the basis of which the coercivity has to be determined, the contributions of the different statistically distributed pinning centres must be added quadratically. However, the effect of H_c(rel) has to be added linearly because each domain wall sees a stabilizing energy due to the relaxation with a wavelength of $2\delta_0$ where δ_0 is related to the domain width. This wavelength, λ, is of the same order of magnitude as the λ of the potential due to statistical fluctuations. Thus, the resultant total coercivity

$$H_c(\text{total}) = [H_c^2(\sigma) + H_c^2(\text{surf}) + H_c^2(\text{SO}) + H_c^2(\text{i})]^{1/2} + H_c(\text{rel}) \tag{19.1}$$

In the special case where the contribution of the pinning due to surface irregularities is larger than all other contributions the various terms add linearly, i.e.,

$$H_c(\text{total}) = H_c(\sigma) + H_c(\text{rel}) + H_c(\text{surf}) + H_c(\text{SO}) + H_c(\text{i}) \tag{19.2}$$

This is so because the wavelength of the surface pinning is much larger ($\sim 10\,\mu$m) than that of the intrinsic fluctuations ($< 0.5\,\mu$m).

We will now discuss each of these contributions in more detail.

19.2.1 H_c due to intrinsic fluctuations

The lower bound to the coercivity H_c(i) has been defined[5] as due to intrinsic fluctuations of the exchange and of the local anisotropy. These act as pinning centres for the domain walls. According to the statistical potential theory[8,9]

$$H_c(\text{i}) = \frac{\rho_M^{1/2}}{M_s(2F_B)^{1/2}}\left[\int_{-\infty}^{\infty} \langle P_{a,i}^2(Z) + P_{ex,i}^2\rangle\, dz\right]^{1/2} \cdot \left[\ln\frac{L_3}{2\delta_0}\right]^{1/2} \tag{19.3}$$

where ρ_M is the density of magnetic ions and is equal to the density of pinning centres with spacing of L_3 corresponding to the average domain width. The parameter δ_0 is related to the domain width $\delta_B = \pi\delta_0 = \pi(A/K_0)^{1/2}$ where A is the exchange constant and K_0 is the anisotropy constant. The domain wall area is F_B. The local pinning forces due to anisotropy and exchange fluctuations are

$$P_a = -\frac{d}{dz}(\delta\phi_a) \qquad P_{ex} = -\frac{d}{dz}(\delta\phi_{ex}) \tag{19.4}$$

Integrating and structure-averaging equation 19.3, we obtain

$$H_c(\text{i}) = \frac{\rho_M^{1/2}}{M_s(2F_B\delta_0)^{1/2}}\left[\frac{8}{15}\langle\xi_{a,i}^2\rangle + \frac{11}{15}\langle\xi_{ex,i}^2\rangle\left(\frac{1}{\delta_0^4}\right)\right]^{1/2}\left[\ln\frac{L_3}{2\delta_0}\right]^{1/2} \tag{19.5}$$

For soft materials, to obtain the minimum coercivity, we neglect the anisotropy fluctuations since they are small compared with the exchange fluctuations. Then $\langle\xi_{ex,i}^2\rangle$ is calculated on the basis of a nearest neighbour model for *nn* exchange interactions between iron atoms only

$$\langle\xi_{ex,i}^2\rangle = \frac{1}{9}r_0^4\sum w_n n^2 J_{FeFe}^2 = \frac{13}{3}r_0^4 J_{FeFe}^2 \tag{19.6}$$

where $\xi_{ex,i} = r_0^2 J_i/3$ with r_0 corresponding to the Fe–Fe distance. Substitution of equation 19.6 into 19.5 gives

$$H_c(\text{i}) = \frac{\rho_M^{1/2}}{3M_s(2F_B)^{1/2}}\frac{J_{FeFe}r_0^2}{\delta_0^{5/2}}\left[\frac{39\cdot 11}{15}\right]^{1/2} \cdot \left[\ln\frac{L_3}{2\delta_0}\right]^{1/2} \tag{19.7}$$

The value of $H_c(\text{min})$ is estimated to be about 2×10^{-6} Oe using $F_B = 6 \times 10^{-5}\,\text{cm}^2$, $\delta_0 = 10^{-5}$ cm, $M_s = 1030$ G, $\rho_M = 0.5/\langle\Omega\rangle = 4.5 \times 10^{22}\,\text{cm}^{-3}$, $\langle\Omega\rangle = 1.1 \times 10^{-23}\,\text{cm}^3$, $J_{\text{FeFe}} = kT_c/8$, $r_0 = 2.5 \times 10^{-8}$ cm and $L_3 = 100\,\mu$m. This analysis is similar to the analysis given by Evetts *et al.*[3] and by Alben *et al.*[4]. This value of H_c is about two orders of magnitude lower than the lowest coercivity observed in amorphous alloys and thus does not represent a significant contribution. Even for materials with finite K_0 the dispersion in K, σ_K, can be taken as comparable with the value for crystalline iron ($\sim 10^6$ erg/cm^3). Then $H_c(\text{i}) = \sim 0.3$ mOe.

19.2.2 H_c due to short range order

A contribution to the coercivity may also arise from clustering of atoms due to short range ordering. Experimentally it is believed that the peak in H_c obtained on annealing may be attributed to this effect as shown in *Figure 19.1*. These results are very similar to the original results reported by Luborsky *et al.*[10]. Local, chemical short-range ordering without the loss of the topological amorphous structure results in an increase in the fluctuations as illustrated by the development of an induced anisotropy[11,12]. In

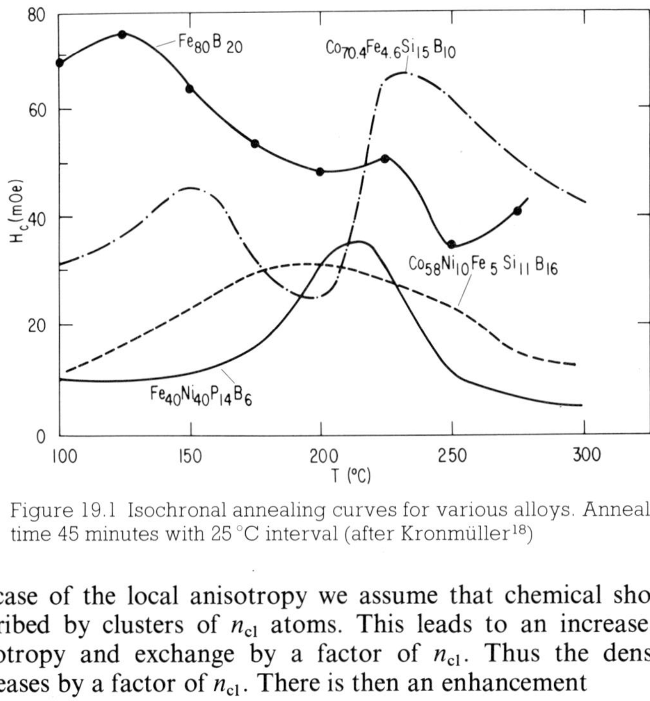

Figure 19.1 Isochronal annealing curves for various alloys. Anneal time 45 minutes with 25 °C interval (after Kronmüller[18])

the case of the local anisotropy we assume that chemical short range order may be described by clusters of n_{cl} atoms. This leads to an increase of the fluctuations of anisotropy and exchange by a factor of n_{cl}. Thus the density of pinning centres decreases by a factor of n_{cl}. There is then an enhancement

$$H_c(\text{cl}) = H_c(\text{min})\, n_{cl}^{1/2} \tag{19.8}$$

Assuming that the clusters are composed of about 12 atoms the coercivity of the alloy would increase from 0.3 mOe calculated for $H_c(\text{min})$ to 1.1 mOe.

19.2.3 H_c due to surface irregularities

There is considerable evidence that the surface roughness of the amorphous metal ribbons is also a large factor in controlling the coercivity[1,6,7,13,14]. The two sides of the ribbon show different surface features. The side in contact with the wheel reflects the roughness and scratches in the wheel and in addition shows varying sizes, shapes, and orientation of air pockets as controlled by the casting conditions[12–14]. The upper surface of the ribbon shows a long-wavelength undulation but a relatively smooth background. If we define a roughness ratio $\Delta T/T$ where ΔT corresponds to the amplitude of the local surface irregularity and T is the average thickness of ribbon, then $\Delta T/T$ is found to be about 0.03—0.2 for the underside and about 0.01 for the top side of the ribbon. The wide variation can result from changes in the casting parameters[12–14]. The force exerted on the domain wall due to the surface irregularities is given by

$$P_{\mathrm{surf}} = -\gamma_{\mathrm{B}} L_2 \,\mathrm{d}\langle T(z)\rangle/\mathrm{d}z \qquad (19.9)$$

where γ_{B} is the specific wall energy, L_2 is the domain wall length and $\langle T(z)\rangle$ is the average thickness. If we assume that $T(z)$ varies sinusoidally with z, i.e.

$$\langle T(z)\rangle = \langle T\rangle + \Delta T \sin(2\pi z/\gamma) \qquad (19.10)$$

then from equation 19.3, modified to consider only a two-dimensional distribution of pinning centres with density ρ_{s},

$$H_{\mathrm{c}}(\mathrm{surf}) = \frac{\rho_{\mathrm{s}}^{1/2}}{2M_{\mathrm{s}}L_2^{1/2}\langle T\rangle}\left[\int P_{\mathrm{surf}}^2(t)\,\mathrm{d}z\right]^{1/2}\left[\ln\frac{L_3}{2\delta_0}\right]^{1/2} \qquad (19.11)$$

Substituting equation 19.10 into 19.11 and assuming that an upper bound to the density of pinning centres is $\rho_{\mathrm{s}} = 1/\lambda^2$, we get

$$H_{\mathrm{c}}(\mathrm{surf}) = \frac{\pi\Delta T\gamma_{\mathrm{b}}}{M_{\mathrm{s}}\langle T\rangle\lambda}\left(\frac{L_2}{2\lambda}\right)^{1/2}\left[\ln\frac{L_3}{2\delta_0}\right]^{1/2} \qquad (19.12)$$

For an FeNiBP alloy we obtain H_{c}(surf) about 6 mOe using $\Delta T/\gamma = 0.1$, $\Delta T/\langle T\rangle = 0.05$, $\gamma_{\mathrm{B}} = 0.1\,\mathrm{erg/cm^2}$, $\lambda = \sim\langle T\rangle = 50\,\mu\mathrm{m}$, $L_2 = L_3 = 100\,\mu\mathrm{m}$ and $M_{\mathrm{s}} = 1000$ G. This is an upper bound since we used an upper bound for ρ_{s}. This value is also in the range of values obtained for the coercivity of ribbons and thus must represent a significant factor. In fact, different means of thinning ribbons all produce a linear dependence of H_{c} on $1/T$ but with slopes which increase with increasing degree of roughness as developed by thinning[6,7] (*see Figure 19.2*).

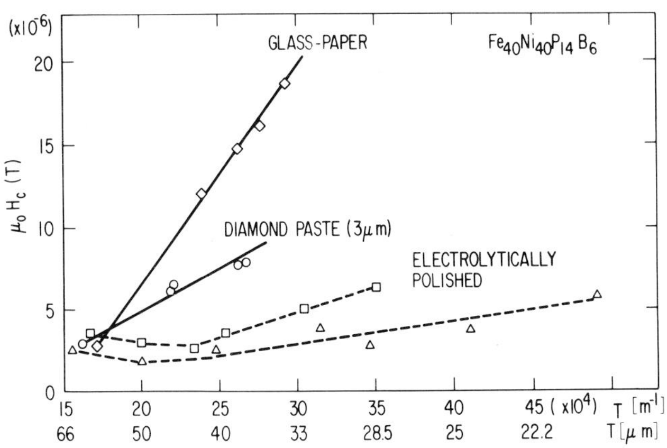

Figure 19.2 The effect of surface treatment on the thickness dependence of coercivity (after Kronmüller[6]) for $Fe_{40}Ni_{40}P_{14}B_6$

19.2.4 H_c due to relaxation effects

Domain wall mobility will also be affected by the rearrangement of interstitial type impurities, i.e., due to magnetic after-effect. This rearrangement has been shown to occur in Fe–B and Fe–B–P alloys[11,12,15,16], resulting in an induced anisotropy. In amorphous alloys this rearrangement can occur from a small displacement of the free volume resulting in a reorientation of the axis of an atom pair. These reorientations are thermally activated and thus the mobility of the domain walls become time-dependent. The decrease of the interaction energy between the spontaneous magnetization and the atom pair axis results in a stabilization of the domain wall. The additional field required to overcome this statistical force field is the coercivity caused by the relaxation H_c(rel). This is time-dependent with

$$H_c(\mathrm{rel})(t) = \mathrm{const}\,\frac{\xi_p^2 \rho_p}{kT}(1 - e^{-t/\tau}) \tag{19.13}$$

where ρ_p is the atom-pair density reorienting with a relaxation time τ, and ξ_p is the interaction energy of atom pairs with the spontaneous magnetization. In this case

$$\tau = \tau_0 \exp\left(\frac{\Delta E}{kT}\right) \tag{19.14}$$

Equation 19.13 holds for a single relaxation time. If a spectrum of τ's exist, it must be modified as recently discussed[15,17] for amorphous alloys. For the case when the displacement of the domain walls can be described by the statistical potential theory[8,9] the contribution H_c(rel) can be calculated from

$$H_c(t)\chi(t) \sim \frac{M_s \delta_B}{L_3} \tag{19.15}$$

According to this a relaxation $\Delta\chi$ of the initial susceptibility χ_0 is related to an increase of the coercivity ΔH_c as given by

$$H_c(\mathrm{rel}, t) = \Delta H_c(t) = \frac{\Delta\chi(t)}{\chi_0} H_c(0) \tag{19.16}$$

where $H_c(0)$ is the unrelaxed coercivity. Experimentally the relative relaxation $-\Delta\chi/\chi_0$ is found[15,17] to vary from -0.05 to 1.0. Thus the contribution of H_c(rel) may vary from an insignificant amount to a significant amount.

19.2.5 H_c due to volume pinning by defects

The coercivity due to interactions of walls with defect structures $H_c(\sigma)$ occurs because in magnetostrictive materials defects exert elastic stresses. This assertion is supported by the generally increasing H_c with increase in magnetostriction in a variety of amorphous alloy systems[6,7] (*see Figure 19.3*). It was shown[9] that quasi-dislocation dipoles which exert stress fields, proportional to $1/r^2$, are the main sources of elastic stress. It was suggested that these dipoles are created by an agglomeration of the free volume during the rapid quenching of the melt. These quasi-dislocation dipoles are characterized by a dipole width, D, a length, L_3, and an effective Burgers vector, b. Calculating the interaction force between the dipole and the domain wall and integrating equation 19.1, we obtain

$$H_c(\sigma) = \frac{12G\lambda_s \Delta V \rho_{\mathrm{dip}}^{1/2}}{M_s(30F_B)^{1/2}\delta_0^{1/2}}\left[\ln\frac{L_2}{2\delta_0}\right]^{1/2} \tag{19.17}$$

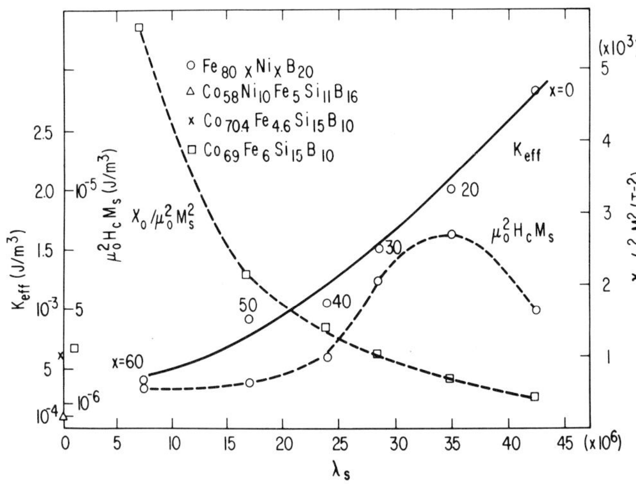

Figure 19.3 Magnetization work, K_{eff}, coercive field (H_cM_s) and susceptibility (χ_0/M_s) for $Fe_{80-x}Ni_xB_{20}$ alloys as a function of λ_s. For nearly non-magnetostrictive cobalt-alloys the results for H_cM_s are shown in the lower left side, (after Kronmüller[6])

where ρ_{dip} is the density of quasi-dislocation dipoles, $\Delta V = DL_3b$ is the volume contraction due to the dipole, and G is the elastic modulus. Values of $H_c(\sigma)$ are calculated from 10 to 100 mOe, using $G = 8 \times 10^{11}$ erg/cm, $\lambda_s = 1.5 \times 10^{-6}$, $\Delta V = 10^{-19}$ cm³, $\rho_{dip} = 2 \times 10^{17}$ cm^{-3}, $M_s = 640$ G, $F_B = 4 \times 10^{-5}$ cm², $\delta_0 = 1000$ Å $L_2 = 100\,\mu$m. This range of values calculated for $H_c(\sigma)$ are of the same order of magnitude as found experimentally for H_c.

The question now is how can we minimize the coercivity and thus the losses based on the results of the above discussion. Defect structures which give rise to the maximum contribution to the coercivity cannot be suppressed. These defect structures are generated during the rapid quenching process by an agglomeration of the free volume. This free volume is stable with respect to annealing and thus cannot be removed completely. Thus, the interaction of domain walls with defects can only be reduced significantly by using non-magnetostrictive alloys. Then short-range order effects and relaxation effects become important. These latter effects may be minimized by a suitable anneal treatment. The intrinsic coercivity may only be detectable in ribbons with extremely smooth surfaces, with $\lambda = 0$, and after an anneal arranged to avoid relaxation and short range order effects.

19.3 Remanence-to-saturation ratio and the hysteresis curves

The shape of the hysteresis loop, in particular its squareness, is important both theoretically and technologically. From the theoretical viewpoint it is important because its value is determined by the magnitudes and directions of the various anisotropies present in the material. In amorphous alloys of the transition metal–metalloid type and the transition metal–metal type the anisotropies in the as-quenched ribbons are dominated by the strain-magnetostriction anisotropy. The magnitude and direction of this strain varies with the details of the quenching, e.g. with the melt temperature, wheel temperature, wheel speed, sticking distance on the wheel, melt purity, ribbon thickness and the crucible-to-wheel spacing as well as perturbations of these variables with time[18]. Thus, the squareness ratio, M_r/M_s, will vary also with changes in the quenching details. In particular, the effect of melt temperature has been studied by Luborsky and Liebermann[19], the effect of ribbon thickness has been studied by Luborsky *et al.*[20] and by Huang *et al.*[21]. The effect of various surface features on the ribbon has been studied by Luborsky *et al.*[14] and the effect of impurities has been reported by Fiedler *et al.*[22]. These studies all show changes in M_r/M_s or changes in the hysteresis behaviour which are attributed to changes in the strain in the ribbons.

After annealing these ribbons, to stress-relieve them without producing any crystallization, the stress–magnetostriction anisotropy disappears[23]. The remaining anisotropy has been ascribed to directional ordering, either magnetically induced[24,25] or strain induced[24,25]. The shape of the hysteresis curve, and the value of M_r/M_s, is then determined by the resultant direction and magnitude of this directional order anisotropy. Since the stresses produced by the quenching during casting are not necessarily along the length of the ribbon, so the anisotropies resulting from these stresses are also not along the ribbon length. The superposition of a magnetization induced anisotropy will change the resultant magnitude and direction of the final anisotropy. This magnetization-induced anisotropy may be random in orientation if annealed after demagnetization or unidirectional if annealed after magnetization to saturation if the remanent state is still saturated or if annealed under the influence of a saturating field. Values of M_r/M_s less than 0.1 have been obtained on annealing Fe–Ni–B alloys in transverse fields, while values greater than 0.9 have been obtained on annealing the same alloys in fields parallel to the measuring direction[26].

Hysteresis curves have been calculated for a dense random packing model of an amorphous material containing random uniaxial anisotropy[27] arising from single-ion anisotropy. The spins were taken as classical unit vectors located at the atomic sites of a dense random packing model and coupled by nearest neighbour exchange interactions. An easy axis direction is then chosen at random for each site. For large anisotropy, the coercivity is very large and the hysteresis curve is square, resembling that found experimentally for $TbFe_2$ at low temperature. As the anisotropy is reduced, there is a fairly abrupt change in behaviour of the model indicating that sufficiently weak random anisotropy is effectively averaged away.

19.4 Losses

Power loss in any magnetic material is represented by the area enclosed by the hysteresis loop. Mathematically, this may be expressed as

$$W = \int H \mathrm{d}B/\mathrm{d}t \qquad (19.18)$$

Classically, this total power loss may be divided into three parts; namely, static hysteresis loss, classical eddy current loss and excess eddy current loss (formerly called anomalous loss). The frequency dependence of these three constituents of the total power loss are shown in *Figure 19.4*. The total loss per cycle is obtained by measuring the loss over a frequency range, e.g., down to 10 to 20 Hz and extrapolating to $f = 0$ to obtain the static hysteresis loss, P_h. The classical eddy current loss is then calculated (*see* Bozorth[28]). For example for sheets subjected to a sinusoidally varying field and when flux penetration is complete, the power loss is

$$P_e = (\pi d f B_m)^2/6\rho \qquad (19.19)$$

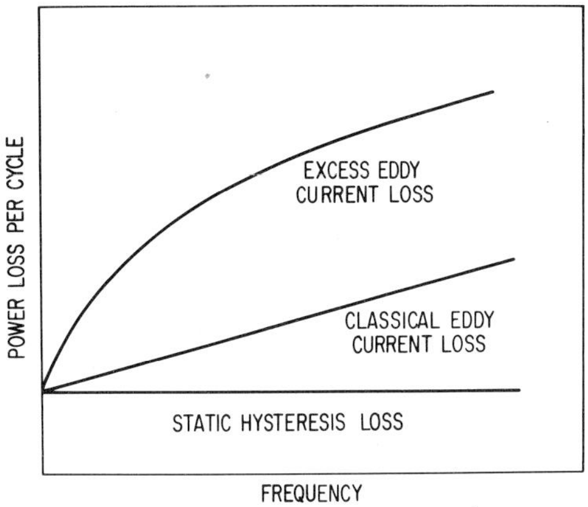

Figure 19.4 Classical separation of power loss as a function of frequency into its components

where d is the sheet thickness, f is the frequency, B_m is the peak flux density and ρ is the resistivity. For other geometries and for incomplete flux penetration this expression changes[28]. For example, at high frequencies, when flux penetration is very small, then

$$P_e = H_m^2(\mu\rho f)^{1/2}/4\pi \qquad (19.20)$$

where H_m is the amplitude of the field and μ is the permeability.

The excess eddy current loss is then the difference between the total loss and $P_e + P_h$. From experimental measurements it has been shown[29] that the excess eddy current loss is responsible for 90—99 per cent of the total power loss of amorphous materials at power frequencies. This should be compared with the 30—90 per cent range for grain-oriented Fe—3.2%Si. The origins of this excess eddy current loss has been attributed to various domain effects[30]:

1. Occurence of domain walls and wall angles
2. Non-sinusoidal, non-uniform and non-repetitive domain wall motion
3. Lack of flux penetration and domain wall bowing
4. Non-sinusoidal flux density and localized variation of flux density
5. Interaction between grains, grain size, grain orientation, and specimen thickness effects
6. Nucleation and annihilation of domain walls.

It has been concluded[30] that the principal cause of the excess eddy current loss in amorphous ribbons is the non-sinusoidal, non-uniform and non-repetitive domain wall motion which is reflected by the non-sinusoidal and localized variation of the flux density.

The total loss, at intermediate and high inductions but at frequencies less than about 100 Hz, can still be expressed by the same equation used for conventional crystalline alloys as the sum of hysteresis and eddy current losses, namely,

$$P = \eta B^{1.6} f + eB^2 f^2 \qquad (19.21)$$

At low inductions the hysteresis loss term approaches $B^3 f$. At higher frequencies the losses become controlled by eddy currents as is given, for example, by equation 19.19. Thus, the anneal used to optimize the properties for low frequency power application is not the same as used to optimize the properties for high frequency applications. Thus, it has been reported[31,32] that the precipitation of less than one volume percent of particles greatly improves the high frequency losses and permeability. They[31,32] have examined

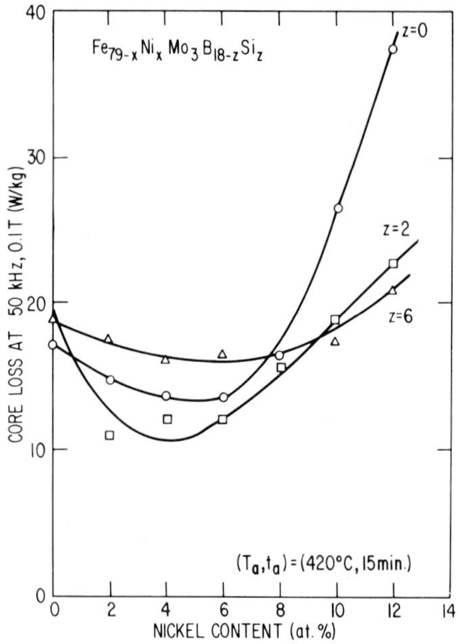

Figure 19.5 Effects of the nickel and silicon contents on AC core loss measured at $f = 50\,\text{kHz}$ and $B_m = 0.1\,\text{T}$ for the glassy $Fe_{79-x}Ni_xMo_3B_{18-z}Si_z$ alloys annealed at 420 °C for 15 minutes (after Hasegawa, Fish and Ramanan[32])

the effect of nickel content, silicon content and anneal temperature on the losses at 50 kHz. These results, shown in *Figures 19.5* and *19.6*, show that an alloy of $Fe_{75}Ni_4Mo_3B_{16}Si_2$ exhibits minimum core loss, after a suitable anneal at temperatures without an applied field, between about 400 and 420 °C. At these temperatures transmission microscopy revealed small star shaped dendritic bcc iron crystals occupying less than 1 volume percent of the sample. This small percentage is sufficient to reduce eddy current losses by preventing wall motion by decreasing the domain size.

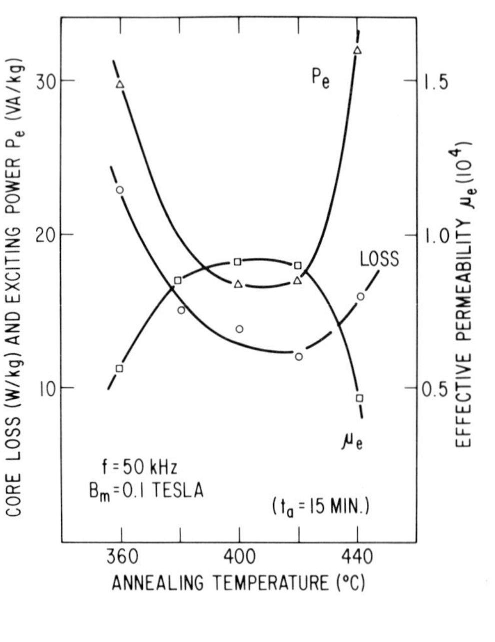

Figure 19.6 AC core loss (L), exciting power (P_e) and effective permeability (μ_e) at $f = 10\,\text{kHz}$ and $B_m = 0.1\,\text{T}$ as functions of annealing temperature T_a. Annealing time was 15 minutes (after Hasegawa *et al.*[32])

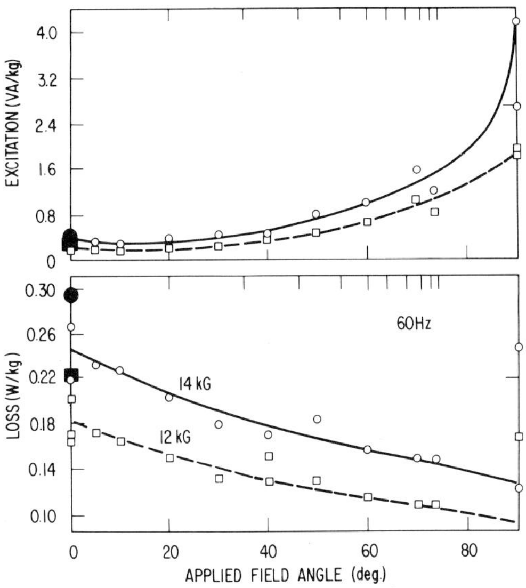

Figure 19.7 Effect of field angle on the excitation losses measured at $B = 12$ kG and 14 kG at 60 Hz (after Luborsky and Livingston[35])

Another method of reducing core loss is to induce an easy axis at some angle to the measuring direction. This modification of the domain structure by misorientation of the easy axis, to produce lower losses, has been recognized to be effective in Fe–3.2 %Si as reported by Shilling *et al.*[33] and Nozawa *et al.*[34]. It has recently been applied to amorphous alloys by annealing in a field oriented at some angle to the measuring direction[35,36]. The results, shown in *Figure 19.7*, show a steady decrease in loss with misorientation. In addition, the exciting volt-amperes exhibits a slight decrease before increasing. Thus, the losses can be decreased by about 20 per cent without any increase in exciting volt-amperes.

A further method of decreasing core loss, particularly applicable to amorphous alloys, because of the wide range of properties available through the wide range of alloys possible, is to decrease the magnetostriction to zero. This effect is due to the decrease in magnetostrictive-strain interaction anisotropy with decrease in magnetostriction; the strain arising from the magnetization. This produces a phase shift between

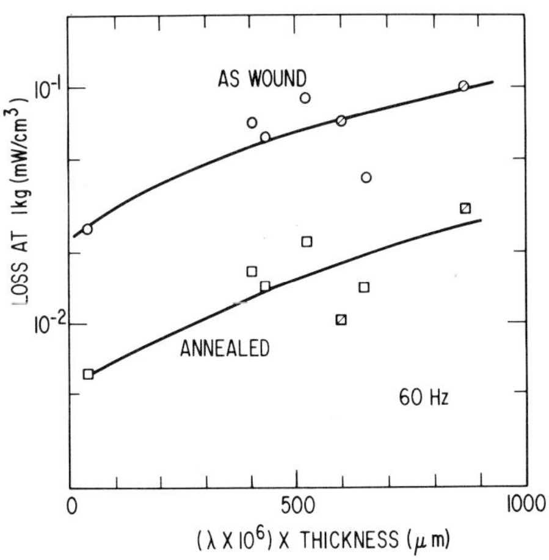

Figure 19.8 Loss as a function of the product of magnetostriction and thickness for as-wound and annealed toroids at 60 Hz (after Luborsky *et al.*[26])

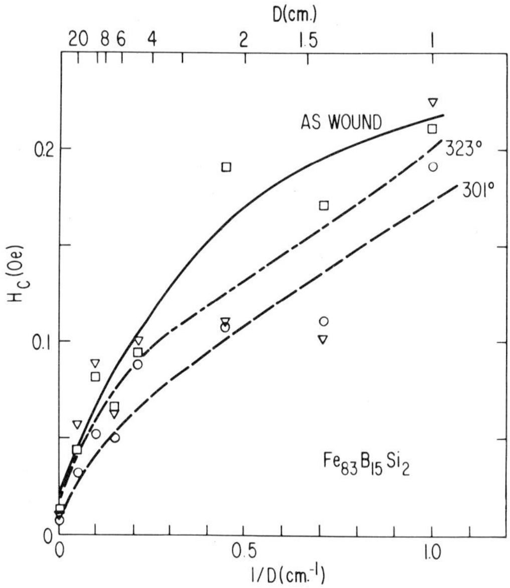

Figure 19.9 Coercive field as a function of diameter of the toroid for samples prepared from ~ 1 mm wide ribbons of $Fe_{83}B_{15}Si_2$, as-wound and then annealed for 2 h at 301 °C and 323 °C in a circumferential field = $20/D$ (after Luborsky and Becker[39])

the magnetic excitation and the resultant magnetization in the material. With decreasing magnetostriction the phase shift, and thus the exciting power and losses, decrease. This was recognized by Luborsky[37,38] as shown by the plot of losses against magnetostriction in *Figure 19.8* for a variety of amorphous alloys with magnetostrictions from about 35×10^{-6} to near zero.

Another method of minimizing core loss is to minimize the deterimental effects of stress-induced anisotropies. One way of doing this was demonstrated by Luborsky and Becker[39] where toroids were fabricated with increasing diameters. It was found that

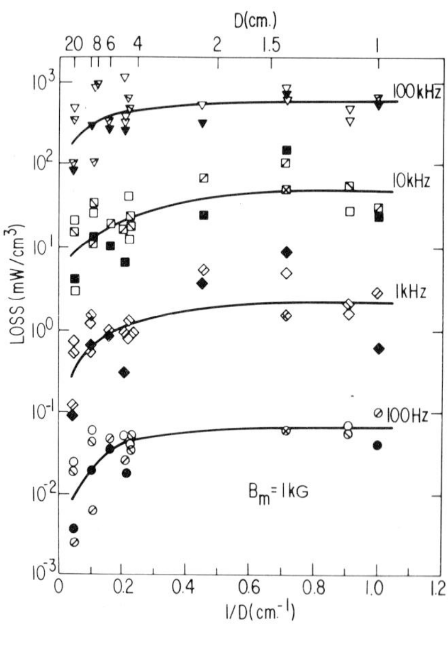

Figure 19.10 Loss as a function of toroid diameter measured at various frequencies with $B_m = 1$ kG. ⊗, taken from Luborsky[38] for $Fe_{80}B_{20}$ and $Fe_{82}B_{18}$; ○, ⦸, $Fe_{82}B_{15}Si_3$ 2.5 cm wide after annealing at 300 °C and 323 °C; ⊘, ●, $Fe_{83}B_{15}Si_2$ 1 mm wide after annealing at 300 °C and 323 °C; all annealed in a circumferential field

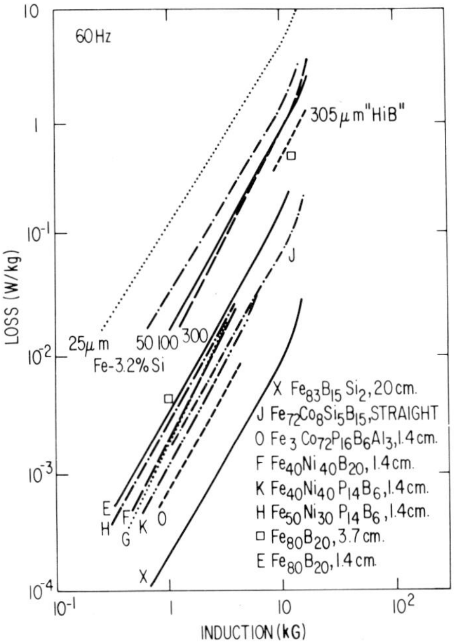

Figure 19.11 Core loss at 60 Hz as a function of maximum measuring induction for various amorphous alloys compared with Fe–3.2% Si of various thicknesses. □, O'Handley *et al.*, *J. Appl. Phys.* **47,** 4660 (1976) (after Luborsky[38])

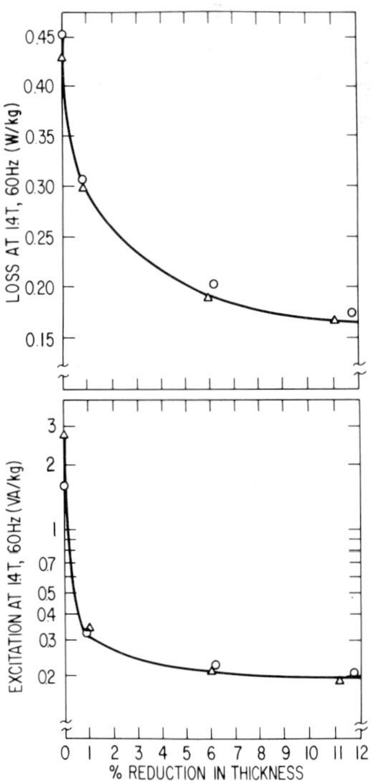

Figure 19.12 Effect of etching on the magnetic properties of ribbon, excited at 1.4 T at 60 Hz (after Fiedler *et al.*[40])

straight samples of high magnetostriction alloy, $Fe_{83}B_{15}Si_2$, had the lowest coercivity and core loss both as-wound as well as after annealing at the highest possible temperature. Typical results are shown in *Figures 19.9* and *19.10*. The effect of toroid diameter on properties was attributed to the stresses introduced; on the inner half of each ribbon a compressive stress is produced which rotates the magnetization out of the easy direction of magnetization. On annealing, this stress results in a stress induced anisotropy in the same poor direction. These effects result in the increase in H_c and losses with decreasing diameter, i.e., increasing stress. It was also demonstrated that this effect of toroid diameter on properties decreased, and approached zero, as the magnetostriction was decreased towards zero. As a result, measurements on a 20 cm diameter toroid of this same Fe–B–Si alloy produced the lowest losses yet reported (*see Figure 19.11*). The reported losses are over a factor of ten below the previously reported values for high magnetostriction Fe–B and Fe–B–Si alloys; four times lower than the zero magnetostriction alloy; and a factor of two below the zero magnetostriction Supermalloy measured at 50 kHz.

Another source of detrimental stress induced anisotropy is due to the presence of small amounts of aluminium in the alloy[22,40]. This aluminium is normally introduced from carbon-reduced ferroboron and has been shown to be extremely detrimental to the magnetic properties of planar flow cast wide ribbons as shown in *Table 19.1* by the increase in loss and exciting power with increase in aluminium. This degradation of properties is attributed to the presence of a thin layer of crystalline material on the top surface of the as-cast ribbon which introduces a biaxial stress into the bulk of the ribbon. Etching away even only a few percent of the ribbon restores the properties as shown in *Figure 19.12*.

TABLE 19.1 Effect of aluminium on magnetic properties[22] of alloy ribbons measured at 60 Hz with excitation of 1.4 T

Ribbon	*Weight % aluminium* Added	Acid soluble	*Thickness* μm	*Loss* W kg^{-1}	*Excitation* VA kg^{-1}
A	0	<0.002	30.9	0.196	0.24
B	0.01	<0.005	33.5	0.227	3.1
C	0.02	0.015	30.6	0.379	2.3
D	0.04	0.026	33.0	0.383	5.3

Properties are averages of at least three samples. Samples all annealed for 2 h at 345 °C in a field.

19.5 Susceptibility, permeability and exciting volt-amperes

We will consider the response of a demagnetized amorphous alloy ribbon (i.e. $B = 0$) to a small oscillating magnetic field. This response, or low field susceptibility $x = M/H$, can also be calculated by use of the model used to calculate the coercivity as previously described. We assume that 180 degree walls are pinned at the ribbon surfaces and have a wall spacing S and a ribbon thickness t. The equation for wall motion by bowing is then:

$$m\ddot{x} + \beta\dot{x} + kx = 2M_0H \qquad (19.22)$$

This applies particularly to alloys with zero magnetostriction or to a stress-free amorphous alloy which has been field-annealed along the ribbon length. At low values of applied field the walls are rigidly pinned at the surfaces. The term that contains the wall mass, m, is neglected for the low frequencies considered here. The damping constant, β, arises only from eddy currents and the spring constant, k, is due to the added energy required to bow the walls. At sufficiently low frequencies where the damping term, $\beta\dot{x}$, can also be neglected we obtain from equation 19.22:

$$\chi = M/H = \frac{2M_0}{S}\left(\frac{x}{H}\right) = \frac{4M_0^2}{SK} \tag{19.23}$$

Equating the magnetic energy gained by the wall distortion, $E_w = -\bar{H}_a \cdot \bar{M}$, to the added wall energy we obtain

$$K = 18\frac{\sigma_w}{t^2} \tag{19.24}$$

Substitution of equation 19.24 and $\sigma_w = 4(AK_u)^{1\cdot 2}$ into equation 19.23 gives

$$\chi = \frac{M_0^2 t^2}{18S(AK_u)^{1/2}}$$

Thus, we see that increasing ribbon thickness t will increase χ. This has been observed in amorphous alloys[20]. Further, increasing the applied magnetic field increases χ also because the domain spacing S decreases as more walls are nucleated.

The 'initial' effective permeability μ_e has been measured for a variety of alloys prepared under a variety of conditions. The value of μ_e reflects the anisotropies developed; namely magnetostriction–strain anisotropy and field or stress induced anisotropy as has already been discussed. The contributions of these sources of anisotropy to μ_e is clearly seen in two different annealing experiments. In the first the composition of iron, cobalt and nickel is adjusted so that the Curie temperature is about 300 °C but the magnetostriction changes from 12×10^{-6} for $(Fe_{0.4}Ni_{0.6})_{78}Si_8B_{14}$ to -4×10^{-6} for $(Co_{0.65}Ni_{0.35})_{78}Si_8B_{14}$. The permeability in the as-quenched state is low except for the $\lambda \sim 0$ composition, as shown in *Figure 19.13*, indicating that the magnetostrictive anisotropy is the main factor in determining the value of μ_e. The magnetostrictive anisotropy can be largely removed by annealing at

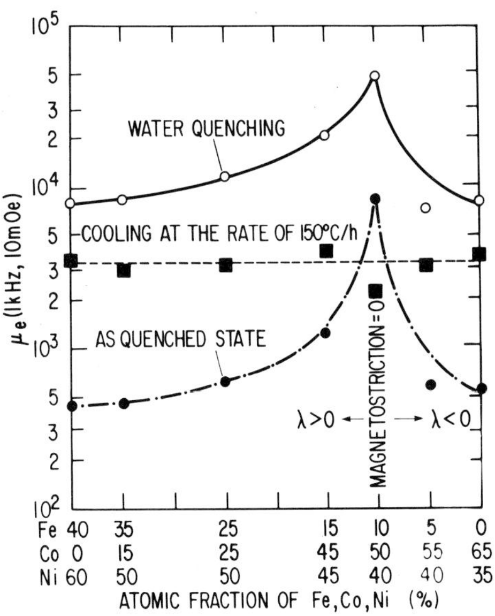

Figure 19.13 Permeability measured at 1 kHz and 10 mOe of various (Fe, Co, Ni)$_{78}$Si$_8$B$_{14}$ alloys with T_c near 300 °C, ●, as quenched; ○, water quenched from 450 °C or ■, slowly cooled (150 °C h^{-1}) from 450 °C (after Hasegawa and O'Handley[41])

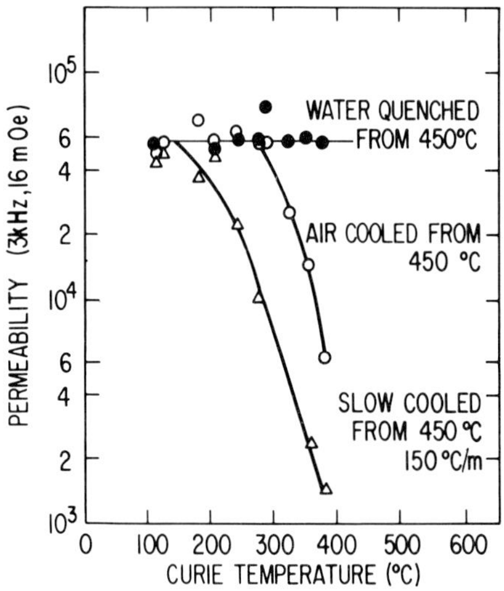

Figure 19.14 Permeability measured at 3 kHz and 16 mOe of $(Fe, Ni, Co)_{74}Si_{10}B_{16}$ zero-magnetostriction ($\lambda_s = 0$) alloys, as a function of T_c

temperatures 50—80 K below T_x and, if the alloy is water quenched so that there is little chance for the induced anisotropy to develop, the value of μ_e is increased by an order of magnitude. However, if the alloy is slowly cooled through T_C, an induced anisotropy develops and μ_e is significantly reduced. In this particular case, μ_e is independent of the metal composition and is certainly unrelated to the magnetostriction. In the second experiment (*Figure 19.14*) the magnetostriction was kept at nearly zero but the Curie temperature was changed by adjusting the composition. If the alloys are water quenched from 450 °C, μ_e is a strong function of T_C for the reasons discussed above. Since K_u is induced only below T_C, if T_C is too low for diffusion to occur, the reduction in μ_e does not happen even when the sample is cooled slowly.

The behaviour of μ_e described above is qualitatively understandable in terms of magnetostrictive and induced anisotropies. The value of μ_e, however, depends on composition in a more indirect way. *Figure 19.15* shows the value of μ_e for

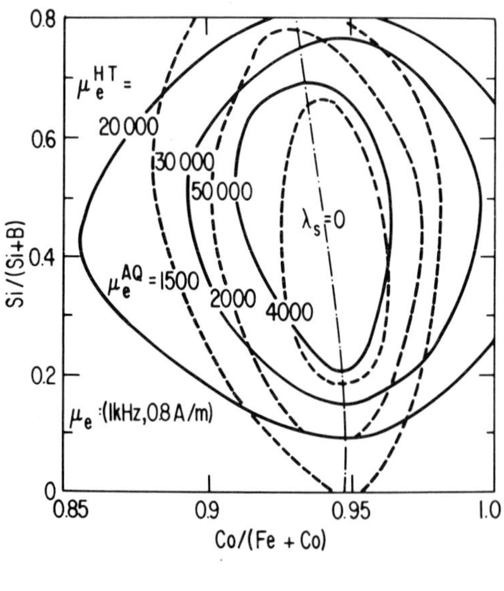

Figure 19.15 Permeability measured at 1 kHz and 0.8 A/m of $(Co, Fe)_{70}(Si, B)_{30}$ alloys, as-quenched (μ_e^{AQ}) or heat-treated at 450 °C and water-quenched (μ_e^{HT}) in a combined pseudobinary diagram (after Kisdi-Koszo *et al.*[43])

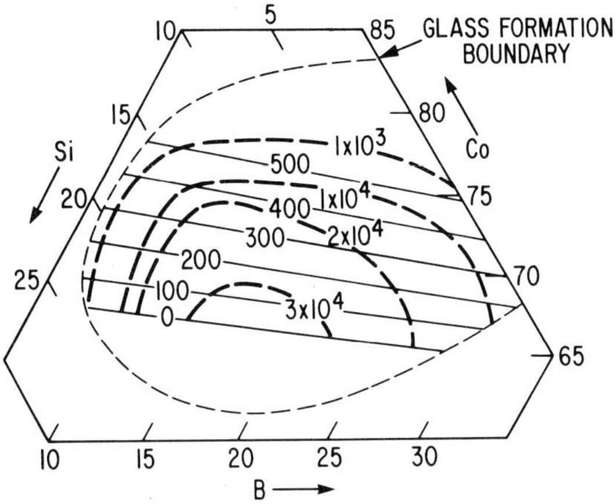

Figure 19.16 Permeability of the Co–B–Si system after heat treatment to maximize μ_e (heavy broken lines) shown together with T_C/°C (full lines) (1 kHz; 10 mOe) (after Hasegawa and O'Handley[41])

$(Co, Fe)_{70}(B, Si)_{30}$ alloys both as-quenched and after a heat treatment. As the ratio of cobalt to iron is changed, the alloy goes through a zero-magnetostriction line and μ_e goes through a maximum. The value of μ_e, however, is not constant along the zero-magnetostriction composition line but is dependent on the ratio of silicon to boron. Since the Curie temperature and magnetization stay more or less constant along this line, the origin of this variation in μ_e must be sought in the degree of structural inhomogeneity in the alloy. Replacing boron by silicon drastically increases T_x and simultaneously decreases the critical cooling rate, making it easier to produce the alloy. It is most likely that this increase in the glass-forming tendency directly causes μ_e to be a maximum near the composition $Si_{0.4}B_{0.6}$. Furthermore, the ratio between the

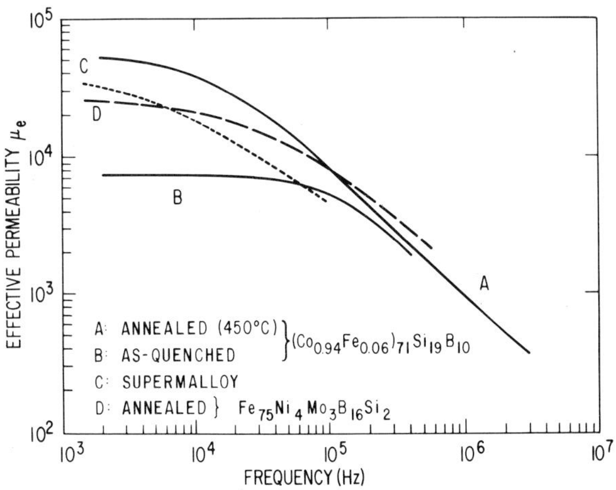

Figure 19.17 Effective permeability as a function of frequency for $(Co_{0.94}F_{0.06})_{71}Si_{19}B_{10}$ *A*, annealed at 450 °C; *B*, as quenched (Blundell and Overshot[29]), compared with *C*, Supermalloy (Luborsky and Becker[39]) and *D*, annealed $Fe_{75}Ni_4Mo_3B_{16}Si_2$, 32 μm thick (Luborsky and Walter[11])

permeability after heat treatment and the as-quenched permeability is less than 10 for $(Co, Fe)_{70}B_{30}$alloys but is more than 20 for $(Co, Fe)_{70}(Si_{0.4}B_{0.6})_{30}$ alloys, indicating that the increased glass-forming tendency is also concomitant with the better response to annealing. Similar results for the Co–B–Si system are given in *Figure 19.16*. In the central portion of the glass-forming region, μ_e rapidly decreases, accompanied by a decrease in T_x.

It is well known that the permeability is independent of frequency for low frequencies while at very high frequencies μ approaches unity. The frequency at which μ starts to fall off depends on the occurrence of various types of resonance; e.g., dimensional resonance due to the size of the core or magnetic resonance due to the rotation of the magnetization against the action of the anisotropy field or due to the resonant oscillation of domain walls. Thus, it has been shown that the drop in μ shifts towards higher frequencies with an increase in μ_e although μ decreases with the increase in μ_e. Typical curves of μ_e as a function of frequency are shown in *Figure 19.17* and are compared with those of some conventional crystalline alloys[32,38,41,42]. The best amorphous alloy is roughly equivalent to the best soft magnetic alloy, namely Supermalloy.

The exciting volt amperes, VA, commonly used in engineering practice to measure the VA necessary to achieve a given level of flux, is another measure of the susceptibility. As amorphous alloys begin to be designed into transformers and other magnetic devices, these engineering oriented measurements are beginning to appear. Of particular interest is the comparison of the VA for an amorphous alloy and oriented Fe–3.2 %Si sheet steel. As seen in *Figure 19.18* the VA required to reach the knee of the curve for the amorphous alloy is about a factor of five smaller than for the Fe–3.2 %Si sheet steel[42]. Another interesting trend is shown by the VA obtained as a function of ribbon thickness[21] shown in *Figure 19.19*. This trend with ribbon thickness follows the previously discussed trend of coercivity with thickness and the minimum occurs for the same reasons; namely, ribbons thinner than the minimum are all amorphous and, as just discussed, the surface pinning becomes less important as the thickness increases leading to a decrease in VA, i.e. an increase in χ. This downward trend in VA would continue if crystallization of the top surface of the ribbon did not start as the thickness of the ribbon increased. As the thickness increases, the percentage of the sample which is crystalline increases leading to an increase in VA.

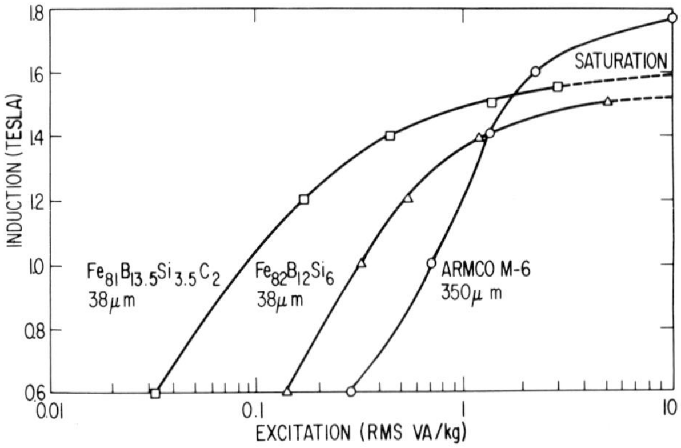

Figure 19.18 Exciting power of $Fe_{81}B_{13.5}Si_{3.5}C_2$ (38 μm thick, 2605SC); $Fe_{82}B_{12}Si_6$ (38 μm thick, 2605S); and Armco M-6 (350 μm thick) at 60 Hz

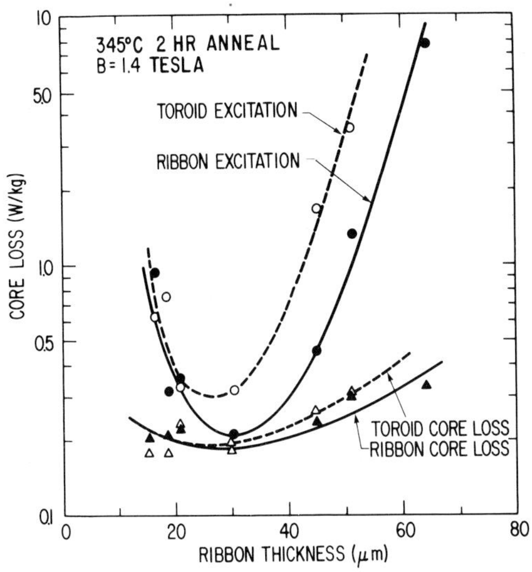

Figure 19.19 The ribbon thickness dependence of the core loss and excitation for straight ribbons and 7.5 cm toroids, annealed for 2 h at 345 °C; $B = 1.4\ T$

19.6 The effects of temperature and time

19.6.1 Temperature dependence of H_c and χ

The temperature dependence of H_c can be calculated from the temperature dependence of the terms in equation 19.1. If we assume that we have H_c controlled by walls pinned owing to defects in the sample volume, i.e. the largest of the five possible contributions to H_c, and rewrite equation 19.17 to include only the temperature dependent terms:

$$H_c(\sigma) \propto \frac{\lambda_s K_0^{1/4}}{M_s} \tag{19.25}$$

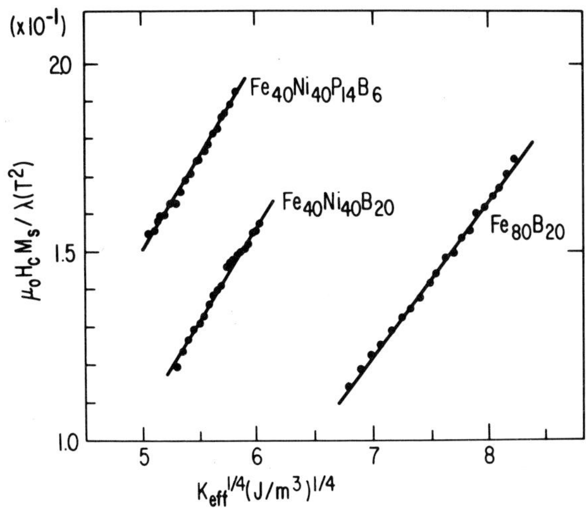

Figure 19.20 Temperature dependence of $H_c M_s / \lambda_s$ on $K_{eff}^{1/4}$ for FeNiPB alloys (after Kronmüller[7])

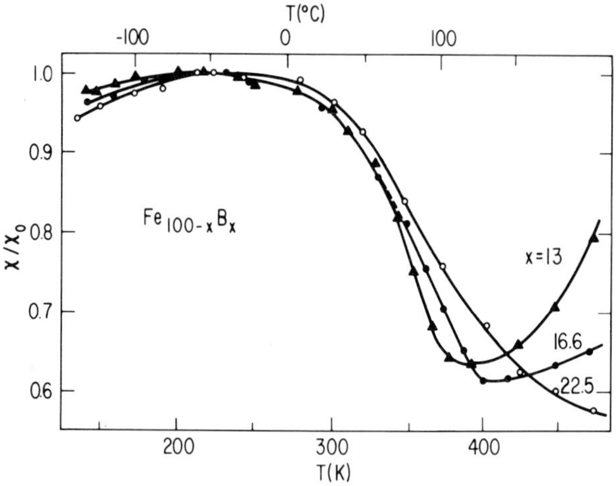

Figure 19.21 Temperature dependence of relative initial susceptibility for three amorphous Fe–B alloys $Fe_{100-x}B_x$ with various boron contents x

where $\delta_0 = \pi(AK_0)^{1/2}$. Thus plotting $H_c M_s/\lambda_s$ against $K^{1/4}$ gives straight lines as illustrated in *Figure 19.20* from Kronmüller[7].

The temperature dependence of χ has also been reported. It is almost independent of composition in the series of Fe_xB_{100-x} alloys as shown in *Figure 19.21* from Kisdi-Koszo *et al.*[43].

19.6.2 Time and temperature dependence of μ_0

There have been many studies of the time and temperature dependence of the change in permeability. There is obvious significance to these measurements for

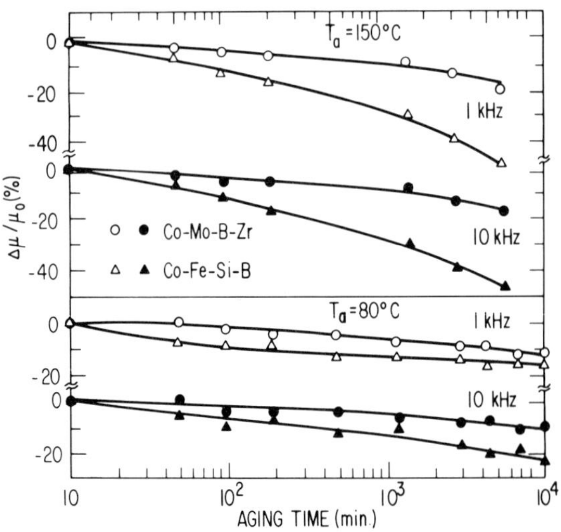

Figure 19.22 Effect of aging temperature and measuring frequency on the change in the permeability

application of amorphous alloys since it is important not only to achieve high permeability but also to achieve stable permeability. This magnetic after-effect or disaccommodation DA is the change in permeability with time, i.e.

$$DA = 100(\mu_0 - \mu_t)/\mu_0 \tag{19.26}$$

Some typical results are shown in *Figure 19.22* for two different alloys measured at two different temperatures. Note the relatively large changes at these relatively low temperatures. These changes are attributed to the stabilization of the domain structure owing to the development or rearrangement of the magnetically induced anisotropy, K_u. This rearrangement of K_u has been shown to occur at these low temperatures by Luborsky and Walter[11,12,44,45]. The spectrum of relaxation times for disaccommodation has been found[15] to be very broad, indicating a wide range of atomic environments to be involved in the rearrangements. Additions of various elements have been found[46] to minimize the disaccommodation and processing parameters have been found[47] to influence drastically the time dependence of χ shown in *Figure 19.23* as well as the kinetics of the stress relaxation[44,45].

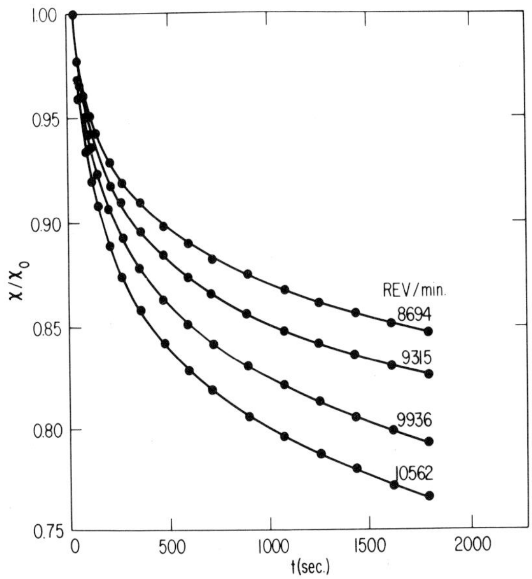

Figure 19.23 Time dependence of relative susceptibility measured on as-quenched $Fe_{83.6}B_{16.6}$ samples prepared with various cooling rates: $T_{melt} = 1570\,K$, $T_{meas} = 600\,K$. The figures on the curves are revolutions per minute

References

1. Gyorgy, E. M., *'Metallic Glasses'*, p. 275, American Society for Metals, Metals Park, Ohio (1978)
2. Goto, M. Tange, H. and Hamatake, T., *J. appl. Phys.*, **52**, 1914 (1981)
3. Evetts, J. E., Hawarth, W. and Gibbs, M. R. J., *'Proc. 3rd Int. Conf. on Rapidly Quenched Metals'*, Ed. Cantor, B., vol. II, p. 127, The Metals Society, London (1978)
4. Alben, R., Becker, J. J. and Chi, M. C., *J. appl. Phys.*, **49**, 1653 (1978)
5. Kronmüller, H. and Groger, B., *J. Phys. (Paris)*, **42**, 1285 (1981)
6. Kronmüller, H., *J. Magnetism magn. Mater.*, **24**, 159 (1981)
7. Kronmüller, H., *J. appl. Phys.*, **52**, 1859 (1981)
8. Kronmüller, H., *Z. angew, Phys.*, **30**, 9 (1970)
9. Kronmüller, H., Fahnle, M., Domann, M., Grimm, H., Grimm, R. and Groger, B., *J. Magnetism magn. Mater.*, **13**, 53 (1979)
10. Luborsky, F. E., Becker, J. J. and McCary, R. O., *IEEE Trans. Magn.*, **MAG-11**, 1644 (1975)

11. Luborsky, F. E. and Walter, J. L., *IEEE Trans. Magn.*, **MAG-13,** 1635 (1977)
12. Luborsky, F. E. and Walter, J. L., *IEEE Trans. Magn.*, **MAG-13,** 953 (1977)
13. Huang, S. C. and Fiedler, H. C., *Metals Trans.*, **12A,** 1107 (1981)
14. Luborsky, F. E., Huang, S. C. and Fiedler, H. C., *IEEE Trans. Magn.*, **MAG-17,** 3463 (1981)
15. Allia, P., Mazzetti, P. and Vinai, F., *J. Magnetism magn. Mater.*, **19,** 281 (1980)
16. Allia, P., Luborsky, F. E., Soardo, G. P. and Vinai, F., *J. appl. Phys.*, **52,** 3553 (1981)
17. Moser, N. and Kronmüller, H., *J. Magnetism magn. Mater.*, **19,** 21 (1980)
18. Kronmüller, H., *'Proc. 4th Int. Conf. on Rapidly Quenched Metals'*, Eds. Masumoto, T. and Suzuki, K., Japan Institute of Metals, Sendai (1982)
19. Luborsky, F. E. and Liebermann, H. H., *Mater. Sci. Engng.*, **49,** 257 (1981)
20. Luborsky, F. E., Liebermann, H. H. and Walter, J. L., *'Proc. Conf. Metallic Glasses: Science and Technology'*, Eds. Hargitai, C., Bakanyi, I. and Kemeny, T., vol. I, p. 203, Central Research Institute for Physics, Budapest (1980)
21. Huang, S. C., Frischmann, P. G., Luborsky, F. E., Livingston, J. D. and Mogro-Campero, A., Proc. Materials Res. Conf., Boston (1981) (to be published)
22. Fiedler, H. C., Livingston, J. D. and Huang, S. C., *J. Magnetism magn. Mater.*, **26,** 157 (1982)
23. Chambron, W. and Chamberod, A., *Solid St. Communs*, **35,** 61 (1980)
24. Hilzinger, H. R., ref. 18
25. Nielsen, O. V. and Nielsen, H. J. V., *J. Magnetism magn. Mater.*, **22,** 21 (1980)
26. Luborsky, F. E., McCary, R. O. and Becker, J. J., *'Rapidly Quenched Metals'*, Section 1, Eds. Grant, N. J. and Giessen, B. C., p. 467, Massachusetts Institute of Technology Press, Cambridge, Mass. (1977)
27. Chi, M. C. and Alben, R., *J. appl. Phys.*, **48,** 2987 (1977)
28. Bozorth, R. M. *'Ferromagnetism'*, chap. 17, Van Nostrand Co., New York
29. Blundell, M. G. and Overshott, K. J., *IEEE Trans. Magn.*, **MAG-16,** 1224 (1980)
30. Overshott, K. J., *IEEE Trans. Magn.*, **MAG-17,** 2698 (1981)
31. Nathasingh, D. and Smith, C. H., Proc. Conf. on Power Conversion 7, Paper B2, San Diego (1980), to be published
32. Hasegawa, R., Fish, G. E. and Ramanan, V. R. V., ref. 18
33. Shilling, J. W., Morris, W. G., Osborn, M. L. and Rao, P., *IEEE Trans. Magn.*, **MAG-14,** 104 (1978)
34. Nozawa, T., Yamamoto, T. Matsuo, Y. and Ohya, T., *IEEE Trans. Magn.*, **MAG-14,** 252 (1978)
35. Luborsky, F. E. and Livingston, J. D., *IEEE Trans. Magn.*, **MAG-18,** 908 (1982)
36. Fujimori, H., Yoshimoto, H. and Morita, H., *IEEE Trans. Magn.*, **MAG-16,** 1227 (1980)
37. Luborsky, F. E., *'Amorphous Magnetism II'*, Eds. Levy, R. A. and Hasegawa, R., p. 345, Plenum Press, New York (1977)
38. Luborsky, F. E., in *'Ferromagnetic Materials'*, Ed. Wohlfarth, E. W., vol. I, p. 451, North-Holland, Amsterdam (1980)
39. Luborsky, F. E. and Becker, J. J., *IEEE Trans. Magn.*, **MAG-15,** 1939 (1979)
40. Fiedler, H. C., *IEEE Trans. Magn.*, **MAG-18,** 1388 (1982)
41. Hasegawa, R. and O'Handley, R. C., *J. appl. Phys.*, **50,** 1551 (1979)
42. Luborsky, F. E., Frischmann, P. G. and Johnson, L. A., *J. Magnetism magn. Mater.*, **19,** 130 (1980)
43. Kisdi-Koszo, E., Vojtanik, P. and Potocky, L., *J. Magnetism magn. Mater.*, **19,** 159 (1980)
44. Luborsky, F. E. and Walter, J. L., *Mater. Sci. Engng.*, **28,** 77 (1977)
45. Luborsky, F. E., American Institute of Physics Conf. Proc. No. 29, p. 209 (1976)
46. Inomata, K., Hasegawa, M. and Shimanuki, S., *Jap. J. appl. Phys.*, **18,** 937 (1979)
47. Vojtanik, P., Kisdi-Koszo, E., Lovas, A. and Potocky, L., ref. 20, p. 247

Chapter 20

Applications of amorphous metals: progress and prospects

D. Raskin and C. H. Smith

Allied Corporation, Metglas Products, 6 Eastmans Road, Parsippany, New Jersey 07054, USA

20.1 Introduction

The world-wide interest in amorphous metals has been sustained to a great degree by the clear benefits seen in the use of these materials in a number of application areas[1]. Not unexpectedly, active programmes of scientific research into the properties and creation of these unusual alloys were under way long before their practical applications were fully defined (*see* Chapter 1). At the present time, however, the scale of production investment and application development has far exceeded that of laboratory investigations into a curious class of materials; such present investment is directly attributable to a world-wide appreciation for the energy conservation and performance potential of the metallic glasses.

It is ironic to note, in view of the great commercial interest in ferromagnetism in amorphous metals, that the initial application thrust for the metallic glasses was in the mechanical and structural areas. This early interest resulted from the discovery of alloys with extraordinary strength[2]. As our knowledge of metallic glass materials has evolved, the initial promise of the high strength alloys is not yet close to realization. This shortfall is due in part to the relatively low resistance to cyclic fatigue under tension in the amorphous metals[3] and in part to the difficulty of producing such extreme high strength characteristics on a commercial scale.

It was the discovery of ferromagnetism in the amorphous metals (a phenomenon originally though to be impossible in non-crystalline materials) that began the large-scale scientific and engineering effort that has made possible the production of high quality amorphous metal strip in tonnage quantities. In order to appreciate the basis for this statement, it should be noted that the full-scale substitution of ferromagnetic metallic glasses for conventional electrical steels in a single application (transformer cores for electric utility distribution) has the potential for saving two-thirds of the more than 3×10^{10} kWh of electrical energy annually wasted by these cores in the United States, alone, as estimated by Werner[4]. Widespread use of these materials in other transformers, in electric motors, and in power conditioning equipment could reduce energy use by another similar amount. The conviction that cost-effective methods are being found for producing these alloys and for incorporating them into such devices has spurred application work in the Americas, the Far East, the Middle East and Europe.

In view of the inherent advantages of producing amorphous metals in large volume by continuous casting — low capital investment, high throughput, low energy costs and relatively low alloy costs — these large volume applications have been pulling down the prices of these alloys, thus opening up ever widening areas of cost-effective application. For example, when the authors' company first offered for sale kilogram lots of ferromagnetic amorphous metals (1979), the price was \$300/kg; at the present time (1982) tonnage quantities are available for sale and are priced at \$20/kg for certain alloys. Substantial further price decreases are expected as the penetration of these materials into electrical power applications becomes increasingly widespread.

The electric power and electronic applications of these materials at both low frequencies (50—60 Hz) and high (up to approximately 200 kHz) will be discussed in detail in this chapter, with both the underlying justification for such use and the application technology being described. Following that will be descriptions and discussions of other applications where the amorphous metals offer significant performance or economic advantages to the user. In addition, a variety of amorphous-metal-based devices will be described, such as transducers, magnetic tape heads and security tags.

20.2 Soft magnetic applications

AC power distribution (50–60 Hz)

Distribution transformers are used by electrical utilities in the final voltage step-down in the network that supplies power to residences, stores, offices and small industries. The principal characteristic of distribution transformers, distinguishing them from essentially all other transformers, is that they are put into service for periods of 25—40 years and that during that time, the primary coil is continuously energized by the utility supply (except for brief maintenance interruptions and rare failure periods).

The significance of this duty pattern is that the magnetic core material linking the primary and secondary coils is thus cycled around its *B–H* loop at the line frequency continually for up to 40 years. Since magnetic materials will incur losses owing to hysteresis and eddy currents when cycled, there will be heat generated in the core and this energy must be supplied by the utility. As an example to quantify this loss, a certain commercial distribution transformer with a grain-oriented silicon iron core (M-4), which is representative of *high efficiency* designs and of an average rating size (25 kVA), operates at 98.7 per cent efficiency *at full load*, but generates 85 W of core loss, independent of loading. Over a 35-year period, this loss amounts to more than 26 000 kWh, which can be considered as costing \$1300 for electricity valued at 0.05 \$/k Wh.

In actuality the monetary evaluation of losses by the utilities is a great deal more complicated, since it must account for a multitude of considerations, including local fuel mix, finance rates, cost escalation rates, etc. Nonetheless, the figures given in the first column in *Table 20.1* indicate that the *core* losses, as evaluated by one Northeastern United States utility, carry a loss evaluation penalty that is approximately equal to the initial purchase price of this transformer. (Notice that the per unit full-load *coil* losses are valued well below the core losses. This is typical of such loss evaluation formulas, since the transformer operates significantly below full load, on average; 30—40 per cent of full load is a typical diurnal average loading.)

The metallic glasses offer the opportunity to decrease transformer core losses by as much as 75 per cent, a possibility that is offered *only* by such materials. In fact it has been pointed out that only amorphous-metal-cored transformers can be designed to

TABLE 20.1 Loss comparison of two 25 kVA distribution transformers

	Commercial high efficiency	*Prototype toroidal*
Core material	Grain-oriented silicon iron (M-4)	METGLAS 2605S-2
Core loss (W)	85	16
Coil loss (W)	240	235
Loss evaluation*		
Core ($7.462/W)	$634.27	$119.39
Coil ($1.538/W)	369.12	361.43
	$1003.38	$480.82
Core Weight (kg)	65	77
Total Weight (kg)	182	164

*Loss evaluation rates used by Public Service Electric and Gas, Newark, NJ (1981).

achieve greater than 99 per cent efficiency[5]. Efforts to improve the efficiency beyond that point in silicon–iron transformers by, for example, increasing core area to reduce induction levels (hence to reduce core losses) has the unwanted effect of raising coil losses, owing to the increase in mean turn length.

Also important to the designer of distribution transformers is the low exciting power of amorphous metals (approximately one-third that of conventional silicon iron). The saturation induction of the iron-based amorphous alloys, while not equalling that of 3 per cent-silicon iron (1.6 T compared with 2.0 T, respectively), appears to be adequate for this application. This is especially true in view of the marked decrease in the design induction level to 1.2 T in recent commercial loss-evaluated silicon iron distribution transformers[6]. Furthermore, amorphous alloys have been developed[7] with core loss and exciting power characteristics optimized for operation at 100—125 °C, which is the typical service temperature range for distribution transformers. Additionally, the inherent high strength and flexibility of the metallic glasses facilitate certain aspects of core forming.

On the other hand, there are characteristics of the amorphous metals that have slowed their introduction into this field and which are presently challenging the ingenuity of a multitude of designers and manufacturing experts. Among these detriments are thinness of gauge[8], post-anneal brittleness and stress sensitivity. The gauge thinness (25 to 40 μm, generally) affects the construction of cores in two ways: first, in requiring large numbers of layers of material in order to build up a given thickness of winding or laminations and, secondly, in reducing the effective space factor (or packing fraction) to the range of 75—85 per cent, as compared with 95 per cent or better in the 300 μm thick conventional silicon irons. Thinness affects space factor because it amplifies the role of surface irregularities in reducing stackability. Additionally, as will be discussed below, many present core construction methods involve the interleaving of laminations; this process is made more costly by an increase in the number of layers.

The brittleness and stress sensitivity of annealed amorphous metal material also reduce the number of practicable design and manufacturing choices. As has been mentioned in previous chapters, the square-loop properties of the amorphous metal materials are optimized during a magnetic field anneal at temperatures of approximately 400 °C for times of the order of two hours. This anneal cycle is both shorter in time and lower in temperature than that which is given to crystalline electrical steels and it produces cores with very low acoustical noise[8], owing to the creation of long stripe domains with 180 degree walls parallel to the direction of the operating excitation. However, one has very little ability to deform the annealed core for the purpose of aiding the installation of electrical coils, both because the material is brittle and because the stresses generated by unrelieved deformations tend to distort the magnetic domain structure, thus increasing core losses and audible noise.

In summary then, an extremely low level of magnetic losses is achievable with the use of amorphous metals in distribution transformers. Conventional design and manufacturing techniques are not readily applicable, however, if these magnetic performance levels are to be attained. Faced with this opportunity and challenge, different approaches are being developed by groups of engineers and scientists. Public information on their work indicates two basic thrusts: variants of the round-wound or cruciform construction[8–10] and attempts to find cost-effective fabrication methods for toroidal-type designs[11]. These designs will be discussed in detail in this chapter. In the United States this effort to commercialize the manufacture of amorphous-metal-cored power distribution transformers has been encouraged and supported by the utilities through their research arm, the Electric Power Research Institute (EPRI).

The techniques for constructing conventional distribution transformers have evolved over the years to the point where nearly all units are presently built with completed electrical coils being slipped on to butt-jointed, wound magnetic cores (*Figure 20.1*). This construction requires a method for opening the magnetic circuit, of course, but the simple straight-through cutting of a magnetic core (into two 'U's') would result in excessive exciting power in the finished transformer. The general solution to this requirement is to divide the core into packs of layers, each several laminations thick. All the individual laminations in a pack have a cut in approximately (but not exactly) the same location. For the annealing of the core material, all of the packs are assembled together, one inside the other, as they will be in the finished transformer. After anneal, the packs are separated from one another and set aside in size order. Then the packs are opened at their joints and inserted through the coils one-by-one in size order, until the full core is completed in the coils. This procedure, though labour-intensive, has generally been found to be the most effective method for making distribution transformers with conventional electrical steels.

A successful adaptation of this method to amorphous metals has not yet been found, which is not surprising in view of the thinness and brittleness of these materials. The most straightforward commercial approach to the use of amorphous metals is to construct what is called a 'wound-round' or cruciform-core transformer. In that design, an uncut core is wound using a progression of ribbon widths so that the cross-section approximates a circle (*Figure 20.2*). After annealing of the core, a split bobbin is placed

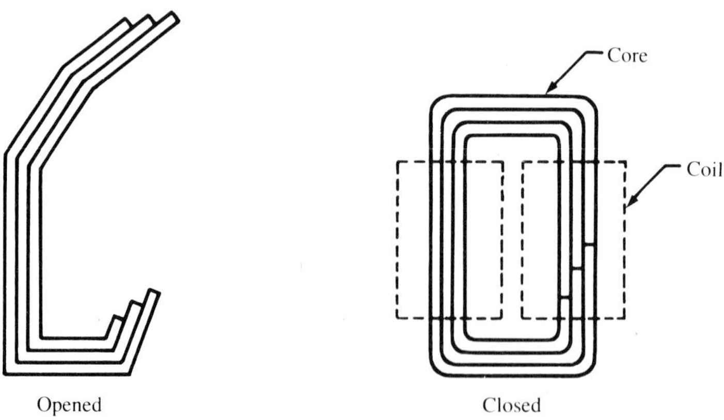

Figure 20.1 Distribution transformer with stepped-joint core

Figure 20.2 Cruciform core cross-section

around each of the long core legs and a coil is wound by rotating the bobbin by its geared flange (*Figure 20.3*). This technique is familiar to distribution transformer manufacturers, since it was commonly used by them until the 1940's (and is still being used by certain manufacturers). Over the years most manufacturers changed over to the stepped-joint construction described in the previous paragraph, because the new method allowed them to wind coils in multiple, independent of the core. The cruciform technique required more time for core and for coil construction. (However, it did not require the laborious disassembly and reassembly process for inserting the core into the coil.)

A variation on the stepped cross-section round-wound core has been developed[10], which uses only one width of ribbon and produces an octagonal approximation to the circle [*Figure 20.4*(*a*)]. In this method, the core-making begins with bias slitting of the appropriate length of ribbon [*Figure 20.4*(*b*)] after which one of the slit pieces is wound with its narrow end against the core-forming mandrel. When this piece has been wound up, an appropriate length of full-width (rectangular) ribbon is added to the core, followed by the second slit piece (this piece being wound in the direction of decreasing width). This technique, which can be made to be highly automated, produces cores that

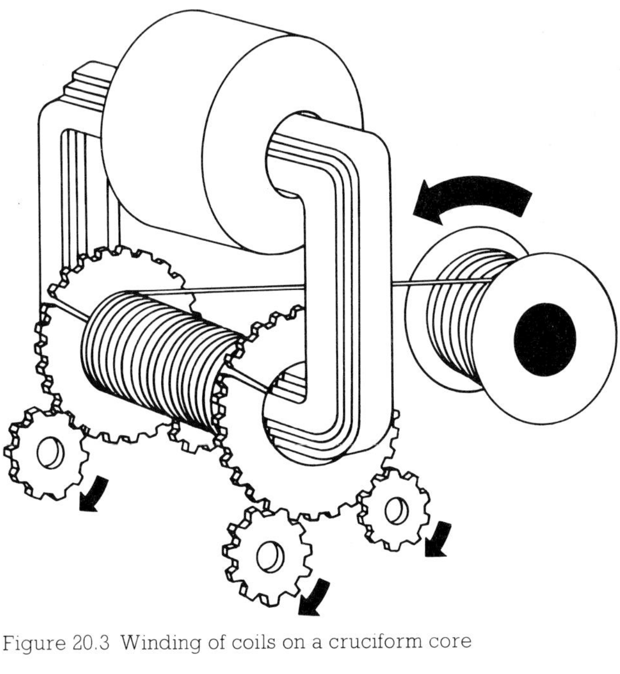

Figure 20.3 Winding of coils on a cruciform core

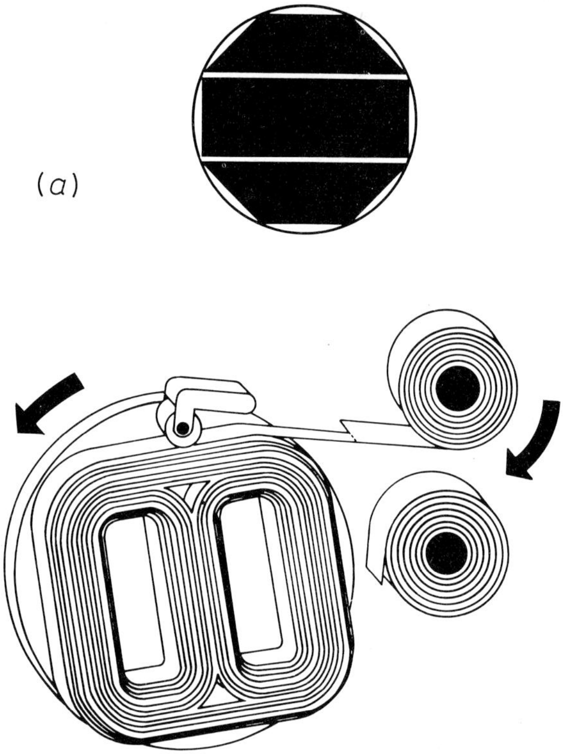

(*a*)

(*b*)

Figure 20.4 Octagonal core: (*a*) cross section, (*b*) core winding

fill 90 per cent of the area of the circumscribed circle (before correction for the non-unity stacking factor of the ribbon). The stepped-width method produces 90.8 per cent circle filling with the use of five ribbon widths and 88.6 per cent filling with four ribbon widths (again before stacking factor correction).

The round-wound design is not remarkably efficient in its use of materials, however. This is because its requirement for a circular cross-section has the effect of reducing the design degrees-of-freedom by one (the core cross-section area is characterized by only one dimension). Since minimum weight designs of silicon–iron cores with *rectangular* cross-sections tend to have dimensions approximately the ratio 2:1 (the core leg is twice as 'deep' as it is 'wide'), it is clear that purely from the materials usage point of view the circular cross section is non-optimal.

In terms of minimizing use of both magnetic and electrical materials, the optimal transformer type is a toroid, since it allows the entire core to be encircled by coils. While this statement is true for all core materials, this geometry is particularly compatible with the use of amorphous metals, in view of the relative ease of winding amorphous metal cores of this type. The shortcoming of toroidal designs is, of course, the difficulty of applying the electrical coils. These coils must be constructed in-place on to the core and, in view of the heavy gauge of the secondary conductors, conventional toroidal winding machinery is unsuitable.

One novel method for constructing coils for toroidal distribution transformers has been under development by Mas at Allied Corporation[11]. The method involves the use of insulated aluminium or copper ribbon for the primary winding along with a fabricated secondary winding. The primary is wound with a toroidal winder into segments arrayed around the core (*Figure 20.5*). A 36 inch diameter winder has a shuttle large enough to hold a quantity of insulated conductor ribbon that is sufficient to wind the entire primary of a 25 kVA unit. Surrounding the core and primary is a secondary composed of heavy conductor 'U' sections that are interconnected by straps. Unlike the stepped-joint core transformer, this entire process has possibilities for a high level of automation, but the actual manufacturing costs have not yet been quantified.

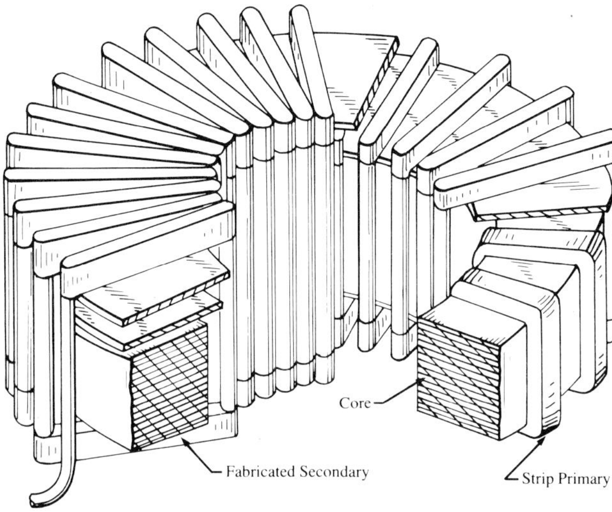

Figure 20.5 Mas toroidal transformer

20.2.2 400 Hz power transformers

In 400 Hz transformers, which are primarily used in airborne and military applications, the low loss characteristic of the metallic glasses can in a sense be 'traded off' against volume or weight, since the latter qualities are often more critically important than direct energy consumption in such uses. With a better than 5:1 advantage in core losses when compared with thin (0.1 mm) silicon iron at 400 Hz and 1.0 T, it is clear that a transformer utilizing a metallic glass core can be designed to permit much higher copper losses without exceeding a given maximum allowable temperature rise. Thus the winding volume can be decreased by using lighter gauge wire (which has the secondary effect of allowing a somewhat smaller core).

Boll and Warlimont[12] have presented this feature of 400 Hz applications of metallic glasses in a somewhat different form. They measured the maximum power capability of toroidal cores of like dimensions (90 mm OD, 50 mm ID and 25 mm height) made of METGLAS® * alloy 2605SC and of grain oriented silicon iron in two thicknesses (0.1 mm and 0.3 mm) and constrained to a maximum temperature rise of 75 K above a 40 °C ambient. Their results, as shown in *Figure 20.6*, indicate 20 per cent and 60 per cent higher output power for the amorphous metal core as compared with the silicon iron cores, respectively.

* METGLAS is Allied Corporation's registered trademark for amorphous alloys of metals.

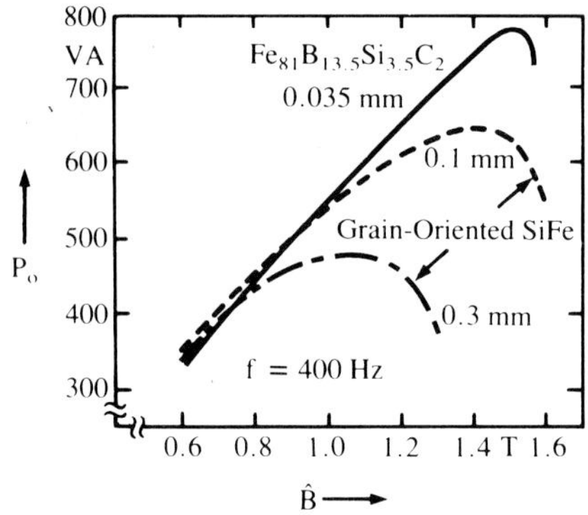

Figure 20.6 Output power P_0 of 400 Hz transformers as a function of B for toroidal strip wound cores (from Boll and Warlimont[12])

The relatively small size of such transformers reduces the difficulties of both toroidal winding and core cutting. Hence the application of metallic glasses in airborne and military power circuits is delayed only by the need for appropriate qualification trials.

20.2.3 Motor applications

In utility transformers, the value of metallic glasses results from their 65—75 per cent lower core losses when compared with grain oriented silicon steel. Grain oriented steels have been used in such transformers because the direction of the applied flux is uniaxial and the magnetic anisotropy of these steels can be aligned with that directionality. In *motors* such uniformity of orientation does not exist, as is shown in *Figure 20.7* for the example of a simple three-phase motor. As can be seen, grain oriented steels cannot be used to advantage for such flux paths, whether the steel is utilized in a laminated construction (in the plane of the diagram) or in a wound configuration. Thus non-oriented steels are used (except in extremely large machines where oriented strips can be used as chordal segments) with their concomitant higher magnetic losses.

Amorphous metals, on the other hand, can provide extremely high magnetic efficiency in motors because their anisotropy can be 'built-in' during a final anneal to develop low reluctance paths in the exact configuration required by the motor[13]. Thus in motors, the use of amorphous metals can reduce core losses by as much as 90 per cent. Such losses are particularly onerous in industrial AC motors that drive pumps, blowers, conveyors, etc., since these motors are of significant size and are energized continuously. As solid state controllers are increasingly widely applied to these motors, thereby reducing the load losses, the core losses will become a more and more significant energy drain.

In AC motors, the core losses in the stator are incurred at the line excitation frequency and are thus more important than the rotor core losses, which occur at the slip frequency (a small fraction of the line frequency). In DC motors, it is the rotor that incurs the major share of core losses, by a similar argument. In either case, the application of metallic glasses will require the solution to some very challenging manufacturing problems. These result largely from the difficulty of fabricating motor laminations from such thin and hard material. Novel motor constructions are being investigated in an effort to capitalize on the enormous energy efficiency potential of these materials in this application.

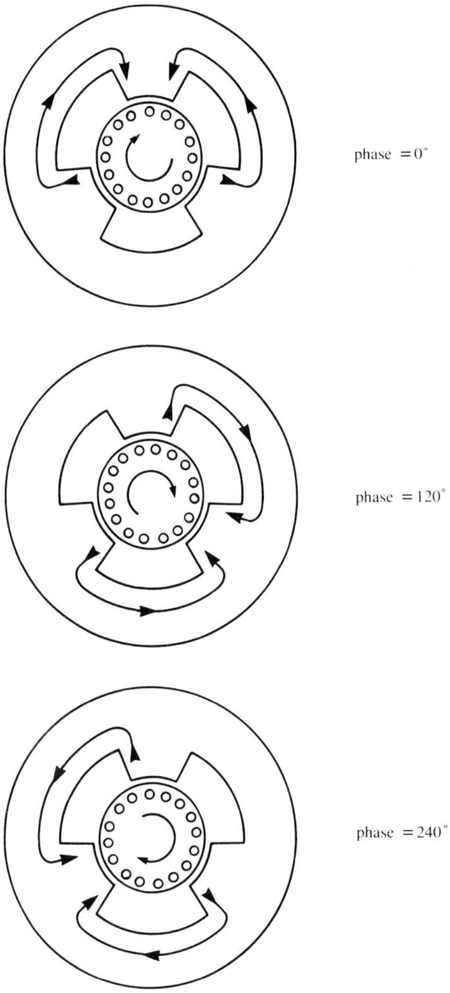

Figure 20.7 Stator flux paths in a simplified three-phase motor

20.2.4 Lamp ballasts

Close behind the magnitude of core losses in electrical distribution and in motors, core losses in lamp ballasts in the US accounted for an estimated 1.6×10^{10} kWh in 1977[4]. Such ballasts are used in fluorescent and metal-vapour arc lamps, whose discharge has a negative resistance characteristic. Thus a choke or ballast is required in series with each lamp. These ballasts have cores that are typically fabricated from non-oriented M45 silicon steel laminations. Core losses in such ballasts could be reduced to one tenth their value by using amorphous alloys, but this would require a substantial device redesign. Today's typical 40 W dual-lamp ballast is usually optimized for initial cost rather than low losses, and economic forces have not yet driven a significant fraction of the ballast market even to thinner, oriented silicon steel laminations, which would result in an approximately two-thirds savings in core losses. It is considered by many to be more advantageous to jump to the next generation (high frequency lamp ballasts) rather than to attempt material substitutions at 50—60 Hz. However, designs have been suggested in which a spiral of amorphous ribbon functions as an integral capacitor and core for a lamp ballast[14].

High frequency electronic lamp ballasts operating at frequencies of 20 kHz to above 100 kHz offer several advantages over ballasts operating from line frequency[15].

1. *Reduction in size and weight.* High frequency magnetics are smaller and lighter than 60 Hz magnetics, even when the extra solid state components are added.
2. *Efficiency.* Electronic ballasts are more efficient than traditional magnetic ballasts, which saturate to achieve constant current. Also the arc lamps themselves operate more efficiently with high frequency and/or square wave voltage excitation.
3. *Elimination of flicker and hum.* High frequency electronic ballasts do not hum and the lamps do not flicker at twice line frequency as do conventional ballasts.
4. *Wider range of voltage sources.* High frequency electronic ballasts can be operated from varying frequency or even DC power sources.

The frequency range of electronic ballasts is very similar to the range of frequencies for the switched-mode power supplies discussed below, and the advantages of metallic glasses as core materials are very similar. Additionally, for compact screw-in units used as direct replacements of incandescent bulbs, ballast temperatures due to the close proximity between lamp and ballast require higher Curie temperatures and low core losses at elevated temperatures. These advantages of metallic glasses are especially important for designs in which the magnetic core directly excites the arc[16] and for outdoor lighting applications, where the temperature insensitivity of these materials makes them superior to the ferrites.

The present higher purchase costs of electronic ballasts limit their use to applications to which their advantages uniquely suit them. With energy costs rising in the future, the efficiency of electronic ballasts, especially ones made with amorphous magnetic cores, is expected to become increasingly attractive.

20.2.5 Switched-mode power supplies

Switched-mode power supplies are rapidly replacing the less efficient linear power supplies in many electronic applications. This growing market requires magnetic components operating at 10 kHz to 200 kHz. These components, shown schematically in *Figure 20.8*, include transformers, inductors and even magnetic switches used in output regulators. The most important magnetic characteristics in these applications are high saturation induction, together with the low losses and high Curie temperatures that allow the transformer designer to utilize the available induction. In addition, filter inductor applications require low effective permeability, especially when a DC component of the current is present, and switching applications require high remanence and low 'saturated' permeability together with low losses.

A variety of amorphous alloys have properties which make them attractive for various components in switched-mode power supplies. High frequency losses in ferromagnetic ribbons are dominated by eddy current losses and those losses in turn are determined by the ribbon thickness, the resistivity of the material and the domain structure. The latter is very important since inhomogeneous eddy currents caused by the localized nature of $\mathrm{d}B/\mathrm{d}t$ in the vicinity of a domain wall can result in losses much larger than those predicted by classical eddy current losses, which assume uniform changes in magnetization[17].

The resistivities of metallic glasses are typically 2—3 times those of crystalline ferromagnetic materials used at high frequencies (130 μohm cm compared with 45—65

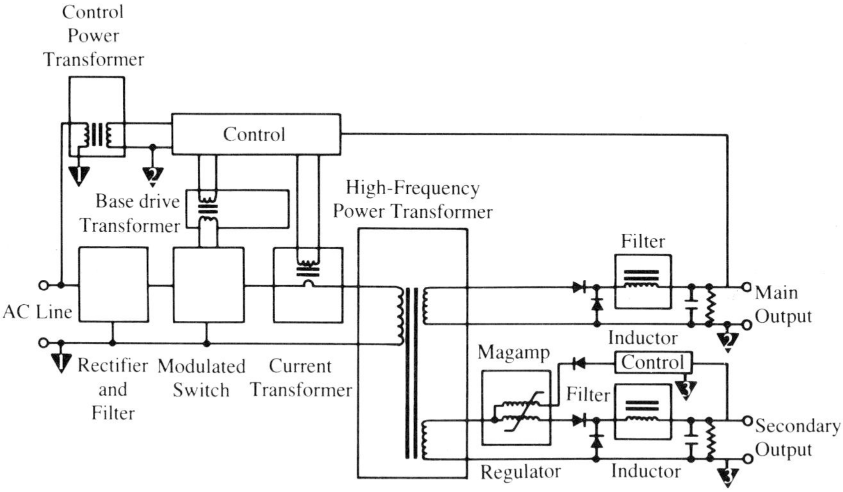

Figure 20.8 Uses of soft magnetic material in a multi-output switched-mode power supply. Note the isolation of the input and both outputs

for nickel–iron tapes). The thickness of amorphous ribbon is currently 20–40 μm, very similar to the thinner gauges of crystalline nickel–iron tapes. These two factors alone make amorphous alloys potentially interesting at high frequencies, and several early articles have discussed this potential[18].

The discovery that reductions in high-frequency losses could be achieved in an iron-based amorphous alloy[19] and the realization that the low losses were achieved by precipitating α-iron crystals and refining the domain structure[20] have resulted in potentially low-cost high-frequency materials with losses that are superior to the nickel–iron tapes and ferrites, even when compared on a volumetric basis, as shown in *Figure 20.9*. Chen has compared the losses of some amorphous materials with nickel–iron tapes and ferrites and found that the crossover frequency (the frequency below which amorphous alloys have lower losses) increases with increasing induction[21]. This effect is shown in *Figure 20.10*.

Transformers are used in switched-mode power supplies for several purposes. Power may be furnished for control circuits from a small 60 Hz transformer off the mains. Currents may be sensed for control purposes by current transformers. The base drive for switching transistors is often provided via a transformer. By far, the largest magnetic component is the power transformer that provides isolation of one or more outputs along with voltage step-up or step-down, as required. Although amorphous metals cannot compete commercially with small ferrite pot cores in low-power applications, high power supplies can economically utilize the lower losses and much higher saturation inductions of metallic glasses in compact toroidal transformer designs. Switched-mode power supplies of up to 1 kW output have been constructed utilizing metallic glass cores[22]. Kunz and Gratzer have calculated that the capability of amorphous alloys in terms of maximum transferable power per transformer volume exceeds that of MnZn ferrites at 20 and 50 kHz and is similar to that of the more expensive 0.03 mm permalloy[23].

To date, the use of amorphous alloy cores has been limited to toroidal (ungapped) configurations, which require symmetrical voltage excitation or some system to block any DC component of the primary current. At present, cut cores are becoming commercially available and many more applications in transformers and inductors will

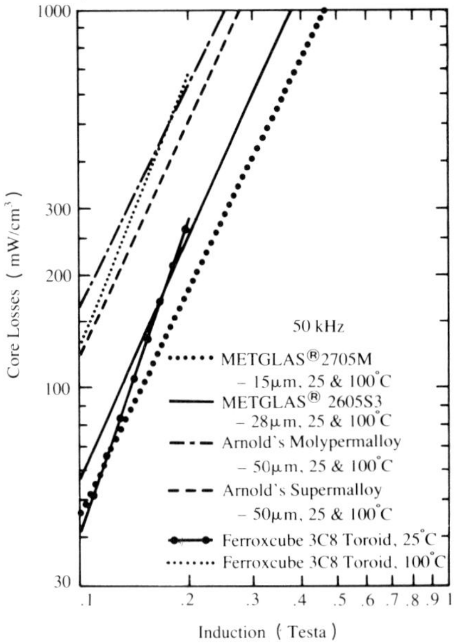

Figure 20.9 Comparison of volumetric core losses as a function of induction for various magnetic materials at 50 kHz, sinusoidal voltage drive

become feasible, both because of the lower permeability available in a gapped cut core and of the easier coil manufacturing techniques for cut-core transformers.

An additional use of toroidal amorphous alloy cores in switched-mode power supplies is found in magamp output regulators. The rugged, reliable magnetic amplifier can be utilized at high frequencies. The inductors, made with square-loop amorphous alloy cores, act as magnetic switches to obtain regulation through pulse-width modulation of the applied square-wave voltage. This use is discussed in the following section with other saturating applications.

The properties of amorphous metal alloys which make them attractive in switched-mode power supplies — low losses, high saturation induction and high Curie

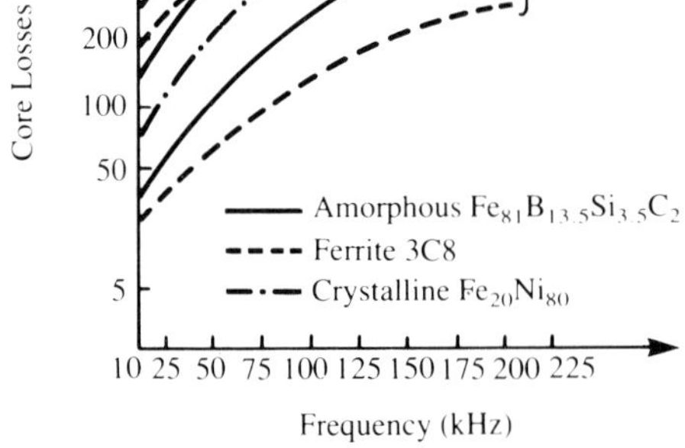

Figure 20.10 Core loss as a function of frequency at 45 °C. Ribbon thickness 50 μm, sinusoidal voltage drive (from Chen[21])

temperature — are also necessary for magnetic components in high frequency inverters and in high-frequency power supplies for welding. Push-pull type circuits can utilize toroidal cores, but forward converters require low remanence, which is usually obtained by gapping the core.

20.2.6 Saturating applications

Most magnetic applications carefully avoid the saturation of the core material, but one class consciously makes use of the large change in permeability between the saturated and the unsaturated state. These saturating applications utilize a magnetic core as a magnetic switch that supports a large voltage until saturation and then provides a low impedance path after saturation.

The important properties for core materials in saturating devices are high squareness ratio (B_r/B_1), high saturation induction, low losses, and low saturated permeability. Such characteristics provide high gain and low power usage. In order to achieve high gain, toroidal cores are used in these saturating applications, since the introduction of an air gap would reduce the squareness ratio. The high resistivity, high saturation induction and thin ribbon form, together with the high squareness ratio that is obtainable by magnetic annealing of amorphous alloys, all suggest the suitability of these alloys in saturating applications[24], such as magnetic amplifiers, active power filters, ferroresonant transformers, and magnetic modulators.

Magnetic amplifiers

The principles of magnetic switches have been used for years in magnetic amplifiers. These devices, usually operating in the 50—400 Hz frequency range with sinusoidal excitation, have been recognized as rugged and reliable units capable of handling large power levels[25]. Recently interest has been expressed in the application of magnetic amplifiers at frequencies between 10 kHz and 100 kHz for output regulators on multi-output switched-mode power supplies[26]. Magnetic amplifiers with square-wave excitation act as magnetic switches, effectively pulse-width modulating the outputs and offering the advantages of low component count, high reliability, high efficiency and inherently low EMI generation. An example of a magamp output regulator is shown in *Figure 20.11.* Either half-wave or full-wave units can be designed,

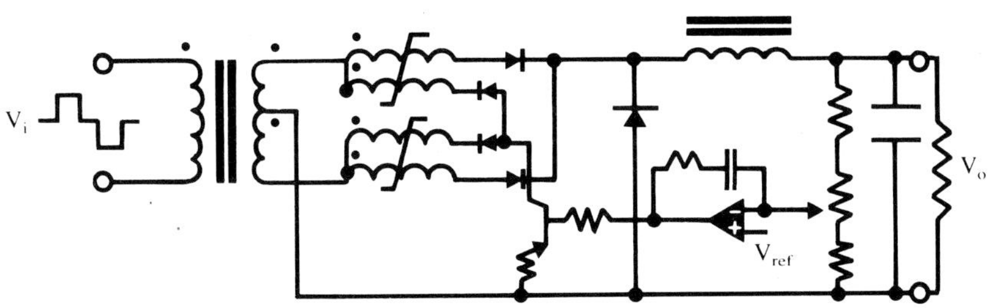

Figure 20.11 Magamp output regulator for switched-mode power supply

and the same principles can be utilized to control high frequency AC power as shown in *Figure 20.12.* A magamp can serve as a DC high-current transducer by using the unknown current as the control current, thus causing the output of the magamp to be proportional to the DC current.

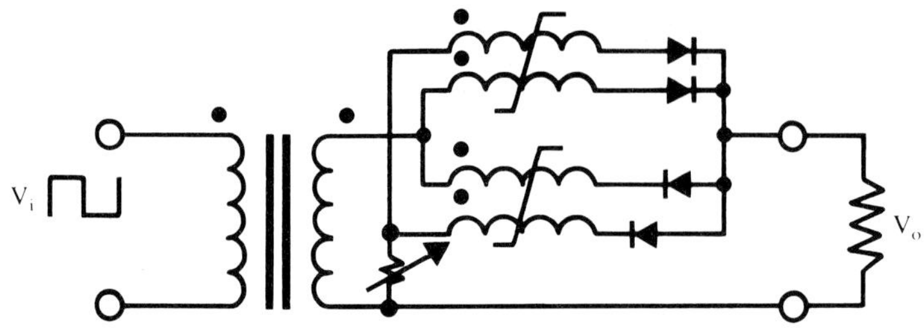

Figure 20.12 High-frequency variable AC output magamp

Active power filter

An AC inverter usually requires a relatively large low-pass filter to produce a sine wave from the chopped DC. Active filters, by using pulse width modulation, can eliminate the lower harmonics of the waveform and allow the use of higher frequency low-pass filters. A smaller and more efficient unit is thereby achieved. By utilizing magnetic switches rather than semiconductor components to accomplish this pulse-width modulation, lower component count and higher efficiency are possible. An active filter utilizing amorphous saturable cores was recently reported which achieved 90 per cent efficiency at 400 Hz[27].

Ferroresonant transformers

Ferroresonant or constant voltage transformers utilize core saturation to provide constant output voltage over a range of input voltages. High-frequency inverter technology provides the means for overcoming their shortcoming of large size and weight. Harada and Murata used a square-loop amorphous alloy in a 20 kHz ferroresonant inverter[28]. The low saturated inductance of the amorphous core provided much better regulation than a Supermalloy core over an input voltage range of 50 to 150 volts. The efficiency of the circuit with the amorphous core (88 per cent) was also higher than with the Supermalloy core (80 per cent).

Magnetic modulators

The techniques of compressing voltage or current pulses by using saturable inductors have been known since the 1950s and were extensively developed for use in radar. Recently, the need for reliable, repetitive, extremely-high-power pulsed sources for lasers, accelerators, and fusion drivers has caused renewed interest in magnetic modulators and magnetic switching techniques[29,30]. The properties of materials necessary for these applications are high saturation induction, low saturated permeability, high resistivity and thin ribbon geometry. In addition, high remanence is valuable when pulsed reset of the core is used. In fast-pulse applications the behaviour of tape-wound toroids seems to be well described by saturation wave-theory, in which a single domain wall rapidly collapses to the centre of each layer of ribbon. This theory predicts that switching time will scale as $(t^2/\rho)^{1/2}$ and eddy current losses as (t^2/ρ) where t and ρ are the ribbon thickness and resistivity, respectively.

The high resistivity of amorphous alloys and their high saturation induction, together with ribbon thickness as low as 18 μm are ideal for this application. Sandia Laboratory, Albuquerque, NM has constructed a pulse compression system using amorphous cores in a coaxial water transmission line which delivered a 60 ns (FWHM) 100 kV pulse into 1 ohm (10 GW peak power)[31]. Lawrence Livermore National

Laboratories, Livermore, CA, has made several magnetic switches for the ETA accelerator utilizing amorphous alloys in toroidal cores. One magnetic modulator using an FeBSiC amorphous alloy in the final core achieved a 10 ns rise time, 40 kA pulse into 10 ohms (16 GW peak power)[30].

High repetition rate, ultra high-power pulse systems for accelerators in the future may well be designed to use amorphous alloys[32]. Also ultra-reliable pulse power sources for lasers in satellites will probably require the reliability of magnetic switches made of efficient amorphous metals.

20.2.7 Linear accelerator applications

Linear accelerators for fusion drivers and for producing intense charged particle beams require not only the pulse power sources and pulse forming lines discussed above but also large quantities of induction materials which, however, are generally not operated into saturation. Linear induction accelerators (linacs) utilize large toroids of ferromagnetic materials essentially as 1:1 transformers surrounding the beam tube. A pulse forming network supplies the required short, high-energy pulse to magnetize the core from minus remanence to plus saturation. The secondary voltage which appears across an accelerating gap in the centre of the core serves to accelerate the beam of charged particles (*Figure 20.13*).

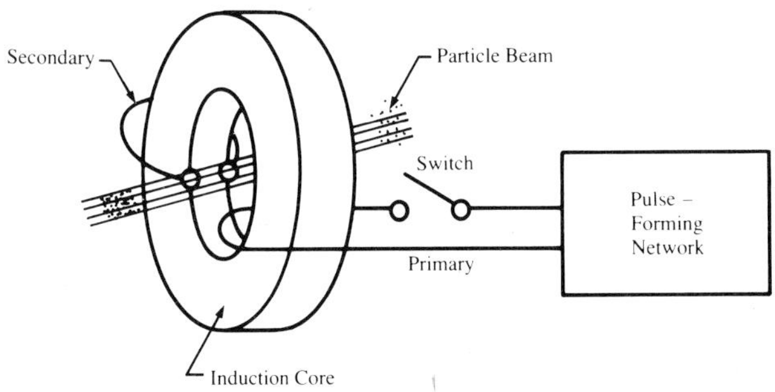

Figure 20.13 Operating principle of a linear induction accelerator

The materials characteristics desirable for this application include a large ΔB swing ($B_r + B_s$) to minimize the amount of materials needed, low excitation current at the required pulse width, and relatively low cost because of the large quantities involved. A reasonably sized accelerator could have of the order of 100 toroids, each 50 to 500 kg. The ferromagnetic materials used in the past have been silicon steel, 50 per cent nickel–iron, and ferrites. The choice of material depends upon the pulse width requirement and cost.

Lawrence Berkeley Laboratory, when studying ferromagnetic materials for a proposed heavy-ion linac for inertial confinement fusion, found that iron-based amorphous alloys offered significant advantages in losses over silicon steel and nickel–iron in the 1 to 10 μs pulse duration range[33]. Much of the accelerator length operates in this range, with only the submicrosecond stages requiring high resistivity ferrites. The noteworthy point here is that the potential loss advantage of ferrites, due to their extremely high resistivities, is outweighed by their volume disadvantage, due to lower ΔB swing.

More recent studies for the ATA accelerator now under construction at Lawrence

Livermore National Laboratories have indicated that thin amorphous ribbons, of the order of 18 μm, would be suitable even for the faster (50 ns) pulses used in this linear induction electron accelerator[32]. By then, however, the design had already been committed to the use of ferrite cores.

20.2.8 Magnetic shielding applications

The combination of high permeability and high yield strength allows amorphous alloys to be used in flexible magnetic shields[34]. One of the early METGLAS alloy products offered by Allied Chemical Corporation was a metre-wide fabric woven from narrow (3 mm) $Fe_{40}Ni_{40}P_{14}B_6$ ribbon. The material could be handled with little or no degradation in properties because of its high yield strength[35]. The strength and flexibility also led to the fabrication of flexible braided shields of amorphous ribbon[36,37].

Although the shielding ratio of one layer of fabric at 100 to 200 A/m is of the order of 10—20, the low *initial* permeability of most amorphous alloys results in relatively poor shielding properties at very low fields. The cobalt-rich, near-zero magnetostriction alloys, however, can be fabricated[37] into shields with good properties below 10 A/m. Such alloys are especially attractive in applications that require small radius bends, such as in shielding for small cable. In other amorphous alloys the bending stresses act via the magnetostriction to reduce the effective permeability.

The combination of high yield strength, high Young's modulus and high permeability can be utilized in applications requiring a shield that also serves as a spring. No other soft-magnetic materials have spring properties which approach those of amorphous alloys. (In fact, Yost[38] has suggested that certain amorphous alloys compare favourably with the *best* mechanical spring materials.) Magnetic tape cassettes require both a magnetic shield to protect the tape head from stray fields and, in the same vicinity, a spring to hold the tape against the head. A combination spring and shield, fabricated from an amorphous alloy, has been suggested[39].

20.3 Mechanical applications

20.3.1 Brazing filler materials

High strength, high temperature metal joints are often produced by brazing, which uses foils of eutectic compositions to flow into the joint at a temperature that is as low as possible, but still well above the service temperature of the joint. Such crystalline compositions are generally either noble metal foils or are brittle powders held in an organic binder. However, since these compositions are at or near eutectic, they are potential glass formers, when cast via the technology of extremely rapid quenching[40]. The amorphous metal brazing ribbon so formed is strong and flexible, so it conforms well to curved surfaces and provides an optimal, controlled gap. Because they contain no binder, the metallic glass foils are 100 per cent dense, non-contaminating and infinitely long-lived on the shelf. In addition, noble-metal-based crystalline foils are being replaced by amorphous compositions of relatively lower-cost elements (e.g., the replacement of gold-based foils by amorphous $Ni_{bal}Co_{13}Fe_{4.2}Si_{4.5}B_{2.8}$), in applications where the higher brazing temperature of the nickel-based foil can be tolerated (2100 °F compared with 1825 °F for gold).

The fact that the amorphous foil becomes crystalline during brazing is of no detriment, since the alloy is in fact combining with the joint materials at that time.

Comparative joint strength tests have shown the amorphous metal foil providing superior joints[41]

20.3.2 Composites and reinforcement

Because of their high strengths and because of their isotropic two-dimensional form, amorphous metals have held promise for planar composites. Indeed high quality plates have been made of half-inch METGLAS 2826MB ribbons in 0°–90° cross ply with an epoxy matrix[42]. These specimens have shown a remarkable degree of in-plane isotropy and strength.

However, the use of such composites in cylindrical pressure vessels has been impeded by the inability, in a sense, to close off the ends of the chamber. The use of amorphous metals in composite flywheels, either of the flat disk type or the tape-wound type, has been essentially discarded because of the relatively low life under cyclic tensional loads of these materials. While in many ways the amorphous metals were ideally suited for flywheels, especially as applied to electric vehicles where energy stored per unit volume is particularly important, flywheel applications are, per force, high cycle uses.

Amorphous metal ribbon fibres have been evaluated in cement reinforcement[43]. In that application, the surfaces of the METGLAS 2826B fibres were roughened by sandblasting to improve adhesion to the mortar, and superior reinforcement performance — an increase of specific fracture work by 2000-fold, with only a 1 per cent volume fraction of fibre — was noted.

20.4 Device-oriented applications

20.4.1 Transducers

In a previous section it was mentioned that power transformers utilizing amorphous metal cores are acoustically quiet because the absence of crystalline anisotropy allows the core maker to anneal-in a longitudinal stripe domain pattern. Under such circumstances there are no domain rotations during the excitation of the core, thus there is no magnetostrain[44]. If desired, however, one could just as well anneal-in a domain structure that is transverse to the ribbon length (again because of the low inherent anisotropy of the metallic glasses) by applying a transverse field during the annealing cycle. This would have the effect of maximizing the rotation of domains and hence the magnetostrain during any subsequent longitudinal excitation.

By such cross-field annealing techniques, it has been shown[45] that amorphous metal specimens can be made to have magnetomechanical coupling coefficients, k, as high as 0.96, which implies efficiencies greater than 92 per cent for converting magnetic energy into mechanical or vice versa. It is important to note in addition that under certain casting conditions for certain alloys, k-values of 0.4 or higher are produced in *as-cast* ribbons. This is noteworthy because this value compares with the best achievable coupling coefficient for *annealed* nickel (*Table 20.2*). Thus, the amorphous metals hold great promise for the construction of rugged and inexpensive sensors that provide, at the same time, high levels of sensitivity.

Mohri and Sudoh[46] have described the design and performance of force, pressure and displacement transducers based on the change of permeability in various segments of a toroidal core that is compressed along a diameter. A simple multivibrator circuit converts these changes into a linear output voltage signal. A tuned vibration sensor for

TABLE 20.2 Comparison of stress-sensitive materials

	Coupling coefficient, k	*Efficiency, k^2*
Magnetomechanical materials		
Annealed		
Amorphous $Fe_{81.5}B_{13}Si_{3.5}C_2$	0.95	0.90
RE–Fe	0.60	0.36
Ni–Co	0.51	0.26
Fe–Ni	0.40	0.16
Ni	0.31	0.10
Unannealed (as-cast)		
Amorphous $Fe_{66}Co_{18}B_{15}Si_1$	0.40	0.16
Piezoelectric materials		
PZT-4 (ceramic)	0.70	0.49
PVF_2 (polymer film)	0.19	0.04

use, for example, in detecting automotive engine knock or incipient machine bearing failure has recently been commercially announced by Allied Corporation. Vibratory signals that are within the resonant bandwidth of the mechanical system of the device cause strains in a METGLAS alloy strip, which in turn generate a voltage in a surrounding electrical coil (a permanent magnet provides bias to the ribbon). Such a device has been shown to have a readily adjustable resonant frequency and bandwidth.

Again because of the remarkable combination of ruggedness with sensitivity in the amorphous metals, these materials are being applied in underwater sound uses. The '*gd* product', or material figure-of-merit, for certain annealed amorphous alloys is the order of $80 \times 10^{-12}\,m^2/N$, as compared with $9.7 \times 10^{-12}\,m^2/N$ for piezoelectric polymer (PVF_2) and $7.8 \times 10^{-12}\,m^2/N$ for piezoceramic (PZT-4)[47]. This sensitivity advantage of the metallic glasses is of particular use in sonar *receivers*. In battery-powered sonar *projectors*, where efficient power transfer is critical, it is the tunability of the Young's modulus of the amorphous metal material via the ΔE effect, that is of importance in coupling the mechanical system to the power supply.

20.4.2 High-fidelity communications components

Record/playback heads for audio magnetic tape systems have been offered for sale by at least three companies in Japan (TDK, Matsushita and Sony). In addition, amorphous alloys suitable for video tape heads have been reported[48]. The characteristics of note for amorphous metals in this application are high saturation induction, high permeability, low attenuation, and high surface hardness (for abrasion resistance).

In addition, Sony has offered commercially a stereo phonograph cartridge using a four-pole structure made of two tiny C-cores made of amorphous metal arrayed around a moving magnet mounted on the stylus support arm.

20.4.3 Security systems

The high magnetic permeability of the ferromagnetic amorphous metals makes them good generators of harmonics when they are placed in suitable AC and DC fields. This characteristic makes them useful as antitheft tags for protecting store merchandise. A typical system consists of field-generating coils arrayed around a passageway and a detector of high order harmonics ($n \sim 10$—20). Since commonly-occurring ferromagnetic materials do not generate such high order harmonics, 'false alarm' signals are rare. Other high quality ferromagnetic materials like square permalloy can serve in such targets, as well, but are less reliable in practice than the metallic glasses owing to their lower mechanical strength (hence the degradation of their permeability during manufacture, dispensing and handling). In addition, of course, the decreasing price of the metallic glasses has made them economically attractive when compared with the permalloys.

References

1. Gilman, J. J., *Physics today,* **28,** No. 5, 46 (1975); Luborsky, F. E., *IEEE Trans. Magn.,* **MAG-14,** 1008 (1978); Warlimont, H., *Phys. in Technol.,* **11,** 28 (1980)
2. Polk, D. E. and Pampillo, C. A., *Scripta Met.,* **7,** 1161 (1973)
3. Davis, L. A., *J. mater. Sci.,* **11,** 711 (1976)
4. Werner, F. E., in *'Energy Efficient Steels',* Eds. Marder, A. R. and Stephenson, E. T., p. 1, TSM–AIME, Pittsburgh, PA (1981)
5. Mas, J. A., personal communication
6. Douglas, J. and Norton, E., *EPRI J.,* **7,** No. 1, 14 (1982)
7. DeCristofaro, N., Datta, A., Davis, L. A. and Hasegawa, R., *'Proc. 4th Int. Conf. on Rapidly Quenched Metals.,* Japan Institute of Metals, Sendai (1982)
8. Johnson, L. A., Cornell, E. P., Bailey, D. J. and Hegyi, S. M., *IEEE PES* 1981 Transmission and Distribution Conf., Minneapolis, MN, September 1981, paper 81 TD 641-0
9. Osaka Transformer Co., Ltd., Cat. No. 8051H (1981)
10. Walker Power, Inc., Worcester, MA, USA
11. Toroidal Transformer exhibited at National Inventors Day Exhibit Washington, DC, February 1982. Patent applied for
12. Boll, R. and Warlimont, H., *IEEE Trans. Magn.,* **MAG-17,** 3053 (1981)
13. Mischler, W. R., Rosenberry, G. M., Frischmann, P. G. and Tompkins, R. E., IEEE PES Winter Meeting, Atlanta, GA, February 1981, paper 81 WM 189—0
14. Alley, R. P., Haller, T. R. and Tompkins, R. E., U.S. Pat., 4211 1957 (1980)
15. Paul, J. D., Proceedings of 3rd International Power Conversion Conf., Munich, W. Germany, September 1981, p. 467; Hirshmann, W., *ibid.,* p. 451; Alling, W. R., Lawrence Berkeley Laboratory Publication LBL-11845UC-95d., September 1980
16. Luborsky, F., *U.S. Pat.* 4227 120 (1980)
17. Pry, R. H. and Bean, C. P., *J. appl. Phys.,* **29,** 532 (1958); Charap, S. H., *J. appl. Phys.,* **50,** 7701 (1979); Bertotti, G., Mazzetti, P. and Soardo, G. P., Soft Magnetic Materials 5 Conf., Grenoble, France, September 1981 *J. Magnetism magn. Mater.,* (1982)
18. Egami, T., Flanders, P. J. and Graham, C. D. (jun.), 20th Annual Conf. on Magnetism and Magnetic Materials, San Francisco, CA, December 1979, Published in conference proceedings; F. E. Luborsky, in *'Amorphous Magnetism II',* Eds. Levy, R. A. and Hasegawa, R., p. 345, Plenum Press, New York (1977); Chen, D., Proceedings of IEEE PESC Conference, San Diego, CA, June 1979, p. 321
19. Nathasingh, D. and Smith, C. H., *POWERCON 7,* San Diego, CA, March 1980, p. B2—1
20. Davis, L. A., DeCristofaro, N. and Smith, C. H., in *'Metallic Glasses: Science and Technology',* Budapest, Hungary, July 1980, in press; Datta, A., DeCristofaro, N. J. and Davis, L. A., ref. 7, Hasegawa, R., Fish, G. E. and Ramanan, V. R. V., in ref. 7
21. Chen, D. Y., *Proc. IEEE,* **69,** 853 (1981)
22. Torre, J. J., Smith, C. H. and Rosen, M., Proceedings of 4th Int. Power Conversion Conf., San Francisco, CA, March 1982, p. 278
23. Kunz, W. and Gratzer, D., *J. Magnetism magn. Mater.,* **19,** 183 (1980)
24. Smith, C. H. and Rosen, M., ref. 15, p. 13
25. Dornhoefer, W. J. and Krummenacher, V. H., Collected reprints from *'Electrical Manufacturing',* The Gage Publishing Co., New York (1951); Storm, H. F., *'Magnetic Amplifiers',* Wiley, New York (1963)
26. Smith, S. and Mullett, C., Professional Advancement Seminar presented during POWERCON 7, San Diego, CA, March 1979, Published by Power Concepts, Inc., Ventura, CA; Gratzer, D., *IEEE Trans. Magn.,* **MAG-16,** 922 (1980); Hiramatsu, R., Havada, K. and Ninomiya, T., *Power Conversion Int.,* **6,** No. 2, 75 (1980)
27. Harada, K. and Murata, K., Proc. IEEE PESC Conf. San Diego, CA, June 1979, p. 206

28. Harada, K. and Murata, K., Proceedings of 2nd International Power Conversion Conf., Munich, W. Germany, September 1980, p. 3A.7
29. Aaland, K., 2nd International Conf. on Energy Storage, Compression and Switching, Venice, Italy, December 1978
30. Birx, D. L., Lauer, E. J., Reginato, L. L., Rogers, D. (jun.), Smith, M. W. and Zimmerman, T., Proc. 3rd IEEE Conf. on Pulsed Power, Albuquerque, NM, June 1981, p. 262; VanDevender, J. P. and Reber, R. A., *ibid.*, p. 256; Nunnally, W. C., *ibid.*, p. 210
31. Stockton, M., Neau, E. L. and VanDevender, J. P., 27th Conf. on Magnetism and Magnetic Materials, Atlanta, GA, November 1981; *J. appl. Phys., (1982)*
32. Schwarzschild, B. M., *Physics today*, **35,** No. 2, 20 (1982)
33. Heavy Ion Fusion Staff, University of California, Lawrence Berkeley Laboratory, PUB 5031, September 1979
34. Mendelsohn, L. I., Nesbitt, E. A. and Bretts, G. R., *IEEE Trans. Magn.*, **MAG-12,** 924 (1976)
35. Sellers, G. J., Proceedings of 1977 IEEE International Symp. on Electromagnetic Compatibility, Seattle, WA, August 1977, p. 129
36. Sellers, G. J., Dismukes, J. P., Shiau, Y., Krstansky, J. J., Sharp, R. E. and Kincaid, J. W. (jun.), Proceedings of 1978 IEEE Int. Symp. on Electromagnetic Compatibility, Atlanta, GA, June 1978, p. 152; Shiau, Y., Bridges, J. and Sellers, G. J., *ibid.*, p. 155
37. Warlimont, H. and Boll, R., Soft Magnetics Materials 5 Conference, Grenoble, France, September 1981; *J. Magnetism magn. Mater.*, (1982); Borek, L., *Elektronik* **43,** 26 (1982)
38. Yost, F. G., *J. mater. Sci.*, **16,** 3039 (1981)
39. Sellers, G. J., Durand, K. J. and Bretts, G. R., *U.S. Pat.* 4 300 180 (1981)
40. DeCristofaro, N. and Henschel, C., *Weld. J.*, **57,** No. 7, 33 (1978)
41. Bose, D., Datta, A. and DeCristofaro, N., *Weld. J.*, **60,** No. 10, 29 (1981)
42. Yeow, Y. T., *Jnl. Compos. Mater.*, Supplement, **14,** (1980)
43. Argon, A. S., Hawkins, G. W. and Kuo, H. Y., *J. mater. Sci.*, **14,** 1707 (1979)
44. Cullity, B. D., *'Introduction to Magnetic Materials'*, p. 248, Addison Wesley, Redding, MA (1972)
45. Modzelewski, C., Savage, H. T., Kabacoff, L. T. and Clark, A. E., *IEEE Trans. Magn.*, **MAG-17,** 2837 (1981)
46. Mohri, K. and Sudoh, E., *IEEE Trans. Magn.*, **MAG-17,** 1317 (1981)
47. Powers, J. M., Proceedings of IEEE Electronics and Aerospace Systems Conf., IEEE Publ. 79CH 1476—1 AES, p. 517
48. Shiiki, K., Otomo, S. and Kudo, M., *J. appl. Phys.*, **52,** 2483 (1981)

Chapter 21

Electrical transport properties

K. V. Rao

Central Research Laboratories, MMM Company, St Paul, Minneapolis, USA

21.1 Introduction

Amorphous metallic alloys are fascinating from both theoretical and experimental points of view. For the experimentalist they provide an opportunity to study systematically the electronic structure of a disordered alloy series in a 'single phase' over the whole composition range without having to be unduly concerned about stoichiometry, and other metallurgical problems attendant upon crystallinity. As with their crystalline analogues, rapidly quenched alloys exhibit a full spectrum of electronic behaviour which ranges from a metallic conductor, superconductor, semiconductor to even that of an insulator. The interpretation of the properties of metallic glasses imposes a particular challenge, especially so, since most of the properties of crystalline solids have been explained in terms of theories based on crystalline periodicity. No such basic theory has yet been developed for the disordered state. The influence of the lack of periodicity on the electronic states, the density of states (which is primary information needed to explain properties like magnetism, superconductivity etc.), and the possible relation between band structure and glass formability of amorphous alloys are therefore some of the fundamental questions of interest.

Much of the information on the electronic transport properties of a system comes from the measurements of electrical resistivity, thermopower, and the Hall coefficient. By now many experimental data, especially that of electrical resistivity, exist. These indicate that the high temperature ($T > \theta_D$) properties of these glasses extrapolate in many cases very close to those of their liquid state. Hence, although because of the lack of translational invariance in the atomic arrangement the conduction electrons can no longer be described by Bloch-waves, they can still be described by plane wave states, i.e. essentially by a 'nearly-free-electron-model'. The amorphous material may thus be considered as a frozen liquid with a fairly sharp and spherical Fermi surface, by which one means the energy boundary between the occupied and unoccupied 'one-electron' states. This follows from symmetry arguments.

Nevertheless, it is also well known that there are some thermal, elastic and transport properties of these rapidly quenched alloys which are clearly a characteristic of the glassy state alone. For example, at low temperatures the specific heat has an additional linear temperature dependent term, the thermal conductivity has a term which varies as the square of the temperature, and the sound velocity and attenuation have peculiar anomalies. The origin of these anomalies, which are also seen in insulating glasses, have been attributed to the disordered state itself[1].

Another interesting aspect of the studies of amorphous materials arises from the fact that, because of the high disorder in these systems, the electron mean free path is short ($\sim$3—5 Å), and of the order of atomic distances. In such a situation, the Boltzmann transport equation may not be valid. A viable microscopic theory that deals with electron transport for systems with such a short mean free path does not yet exist. In principle, for systems with high disorder, the possibility of electron localization exists. Recently Mooij[2], from an analysis of a large number of transport data in transition metal alloys, both crystalline and disordered, established the correlation that the temperature coefficient of the resistivity of alloys, with resistivity $> 150\,\mu\Omega\,\text{cm}$ is as a rule *negative*. This feature is believed to be a consequence of the approach of the magnitude of the electron mean free path towards that of the interatomic spacing. This represents an important theoretical approach to the electron transport problem.

In order to test the applicability of many of the approaches mentioned above and to distinguish between them, it is imperative to look into the ability of these theories to describe the experimentally observed facts, e.g. the magnitude and temperature dependence of the electrical resistivity, Hall coefficient, magnetoresistance, and thermopower. This chapter is written with such questions in mind.

21.2 Electrical resistivity

Transport properties, electrical resistivity in particular, provide a very sensitive probe with which to understand the various scattering processes that occur in a given material. However, it is by no means trivial to interpret the observed magnitude and temperature dependence of the resistivity. This is because, in addition to the dependence on the intrinsic mechanisms involved, the electrical resistance is sensitive to metallurgical factors like disorder, stress relaxation, possible differences in the free volume in an amorphous ribbon as a consequence of different speeds at which the melt has been rapidly quenched (the so-called size effect)[3], structural relaxation[4], changes in local atomic arrangements[5], and remnant crystallinity[6] to name a few. Most of the resistivity data reported, in particular the so-called systematic studies covering a wide concentration range of any chosen binary system for example, are on 'as-quenched' alloys. The possibility of errors arising from effects mentioned above thus render meaningful only a qualitative comparison of the data for different alloys.

In general the resistivity of an amorphous metal is dominated by the disorder scattering (100–$300\,\mu\Omega\,\text{cm}$). It has a very small temperature dependence (TCR) which could be positive, negative, or even zero over a part or whole temperature range up to crystallization. For some alloys this TCR can be changed continuously by changing the alloy composition. The overall change in resistance from the lowest temperature to the crystallization temperature is usually less than 10 per cent. Because of the high residual resistance even a weak temperature dependent contribution to the resistivity can therefore be seen conspicuously in a plot of resistance against temperature. There are four classes of amorphous metallic alloys that have been studied extensively because of their technological interest:

1. Transition metal–metalloid alloys of the type Fe–Ni–P–B.
2. Early and late transition metal-based binary alloys of the type Nb–Ni, Zr–Cu etc.
3. Alloys containing non-*s*-state rare earth ions of the type RE–T (where RE = Gd, Tb, Dy, Ho etc and T can be a transition or noble metal).

4. Alloys of high valency (*s–p* type) elements such as gallium, tin and lead with noble metals copper, gold and silver.

In order to give a comprehensive picture of some of the general characteristic features observed in the electrical resistivity of glassy alloys examples from each of the above four categories will be discussed. Theories that provide a plausible explanation for the observed electrical resistivity behaviour will then be critically examined.

21.3 Characteristic features of the resistivity of amorphous metallic alloys

Although amorphous metallic alloys are characterized by a relatively high concentration of conduction electrons ($10^{22}/cm^3$), the behaviour of the electrical resistivity differs fundamentally from that of its crystalline state. *Figures 21.1—21.5* illustrate the general characteristic features of the resistivity of amorphous metallic alloys.

Figure 21.1 shows the electrical resistivity of a metallic glass $Fe_{32}Ni_{36}Cr_{14}P_{12}B_6$ (commercially available as METGLAS 2826A from Allied Chemicals) measured[7] both in the glassy state as well as after crystallization in the crystalline state. The striking sharp drops in resistivity around 625 K, 690 K, and possibly 765 K indicate that there are at least three major crystallization stages from the amorphous state before the magnitude and temperature dependence of the resistivity becomes identical with that of the crystalline states. Resistivity is thus the simplest, and perhaps best, technique to detect the onset of crystallization. Transformation kinetics for a material can therefore be studied by resistance measurements. However, quantitative determination of the amount of the metastable phase present during a phase transition may not be possible because of the difficulty in relating the changes to the residual. There is a large difference in the room temperature resistivity between the crystalline and amorphous state $[\rho(300\,K)_{am}/\rho(300\,K)_{crys}] \simeq 1.78$. This probably reflects the importance of periodicity of the lattice for electron transport, in addition to the effect of changes in the chemical short range order on crystallization. In the crystalline phase the resistivity looks typical for disordered alloys, increasing monotonically to about 900 K with a relatively large positive temperature coefficient. In the amorphous state the total change in resistivity over the 100 to 600 K temperature range is less than 0.4 per cent. Yet, the weak

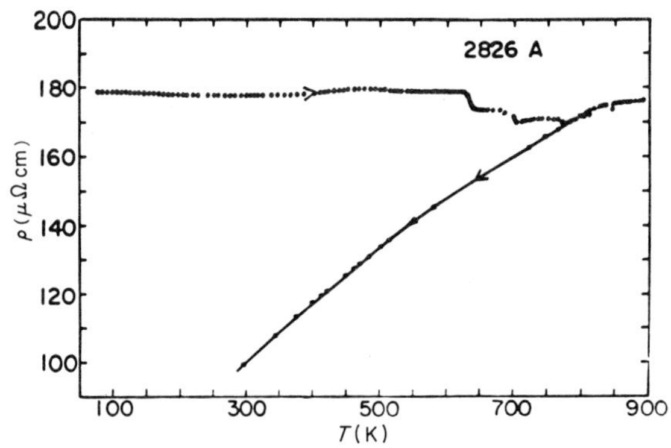

Figure 21.1 Electrical resistivity of METGLAS 2826A (from Teoh, Teoh and Arajs[7])

temperature dependences in the resistivity clearly indicate many an underlying scattering mechanism that must be present in the glassy state. Below around 250 K, which happens to be the ferromagnetic transition temperature for this alloy[8], the resistivity is found to have a negative temperature coefficient (TCR). Above 250 K initially the resistivity is found to have a positive quadratic temperature dependence. The origin of this resistance minimum phenomenon observed in metallic glasses has been a topic of considerable research activity of late both from a theoretical and experimental point of view.

Figure 21.2 is from a series of systematic resistivity studies on amorphous $(A_xB_{1-x})_{75}G_{25}$ (where A, B ≡ Fe, Co, Ni, Cr, Mn etc, and G is the glass former

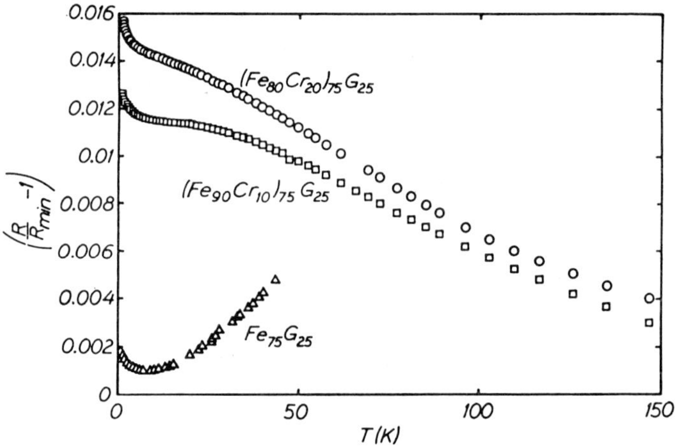

Figure 21.2 Electrical resistivity of some $(Fe_{1-x}Cr_x)_{75}P_{16}B_6Al_3$ alloys: ○, $x = 10$; □, $x = 10$; △, $x = 0$ (from Rao *et al.*[9])

$P_{16}B_6Al_3$) carried out by the Stockholm group[9]. In these studies the effect of changing the composition as well as the combination of the transition elements A and B on the resistance minimum has been carried out over the whole concentration range while maintaining the same glass former. The resistivity of the matrix $Fe_{75}G_{25}$ (triangles) is found to exhibit a minimum around 8 K, with an initial positive quadratic temperature dependence above 8 K which eventually reduces to almost linear behaviour above 250 K. Replacing iron with chromium shifts this minimum to temperatures higher than 150 K. For dilute concentrations of chromium, one can even see an additional shallow minimum around 10 K below which the resistance increases rather sharply. The main point of *Figure 21.2* is to demonstrate that in glassy systems

1. The observed resistance minimum is very sensitive to the nature of the local impurity in the alloy.
2. This can result in a negative TCR over wide range of temperatures.

It is useful to point out that all the alloys referred to in *Figure 21.2* have a ferromagnetic transition above room temperature. The occurrence of a resistance minimum well below a ferromagnetic transition is unusual and has not been seen in crystalline materials. Clearly, in the glassy state where the mean free path is of the order of interatomic distances, electrical resistivity appears to probe the details of the local environment in a matrix preferentially over other long range properties. Such resistivity behaviour as described here has also been reported for amorphous Ni–Cr–Pd–B,

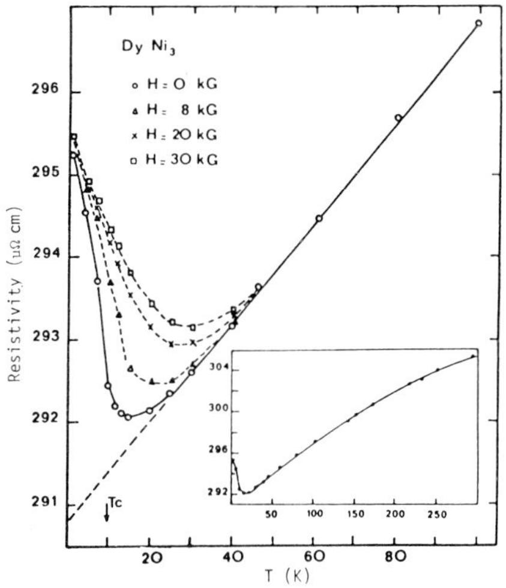

Figure 21.3 Electrical resistivity of $DyNi_3$ binary alloys in several magnetic fields H: ○, 0 kG; △, 8 kG; ~, 20 kG; □, 30 kG (after Asomoza *et al.*[12])

Cr–Pd–Si, and Mn–Pd–Si alloys[10]. Further discussions on the resistance phenomena will be presented later.

Electrical resistivity of rare-earth based binary alloys (Ce–Au, Ni–RE, Ag–RE, RE–Co, Gd–La–Au)[11–15] reveal new aspects of the influence of the amorphous structure. In particular, as shown in *Figure 21.3*, the temperature dependence of the resistivity[12] is characterized by a minimum at a temperature close to the magnetic ordering temperature T_C. On cooling the material from room temperature the resistivity continuously decreases to about T_C, and then increases sharply at lower temperatures as shown in the insert to *Figure 21.3*. Such a positive contribution to the resistivity from magnetic ordering is contrary to what is observed in crystalline alloys for which magnetic ordering generally results in a decrease of the resistivity. The unusual behaviour described above has also been reported for amorphous binary alloys of the type Fe–Zr and Fe–Hf which do not contain rare earth elements[16,17]. Also shown in *Figure 21.3* is the effect of the external magnetic field on the resistivity. The observed magnetoresistivity which in this alloy is *positive* has its maximum value close to T_C (*Figure 21.4*). The magnetic origin of this anomalous resistivity behaviour has been explained with a model based on a modification of the Ziman theory of liquids and amorphous alloys, taking into account the magnetization and the nearest-neighbour spin correlations[12]. This 'coherent exchange scattering' model involving spin–spin correlations, to be discussed later, might well be the general approach needed to explain the resistance minimum phenomenon observed in transition metal-based amorphous alloys[18] as well.

Almost all amorphous binary alloys involving early and late transition metals [Nb–Ni; Zr–(Cu, Pd, Fe, Ni, Co etc)][19–22] exhibit a negative TCR in their electrical resistivity up to the crystallization temperature. Such a resistivity behaviour has also been reported as a function of alloy concentration in ternary transition metal based amorphous alloys which will be referred to while discussing plausible theories for this phenomenon. A typical example of the resistivity behaviour of these alloy systems is shown in *Figure 21.5* for 'as quenched' Zr_{70}(Fe, Ni, Pd and Co)$_{30}$ amorphous ribbons. In this figure deviations from a negative TCR close to 600 K are due to the effects of structural relaxation and possible onset of crystallization. The resistivity continues to increase at low temperatures below 100 K. However, strong deviations from a negative TCR are observed below 30 K with an apparent shallow maximum in the resistivity data

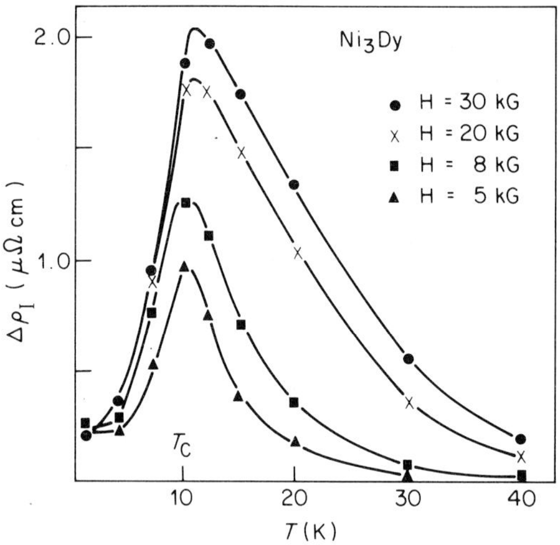

Figure 21.4 Magnetoresistance of $DyNi_3$ in several magnetic fields H/kG, shown on the curves (after Asomoza *et al.*[12])

taken above 4 K. This is because almost all of these alloys become superconductors[23] below 4 K (*see* insert in *Figure 21.5*). The occurrence of negative TCR over such a large temperature range, often extrapolating to its value in the liquid state[24], is a rather uncommon feature in crystalline metallic systems. This fact has played a key role in modifying the Ziman theory of resistivity for liquid metals and extending it to amorphous metallic alloys.

In summary, the main features in the resistivity of amorphous metallic alloys that need to be explained are:

1. The negative TCR observed both as a function of alloy concentration and the temperature range, which can often extend up to the crystallization temperature.
2. The *positive* quadratic temperature dependence at low temperatures above the resistance minimum, which eventually changes to an almost linear behaviour at higher temperatures.
3. Resistance minimum observed in dilute alloys.

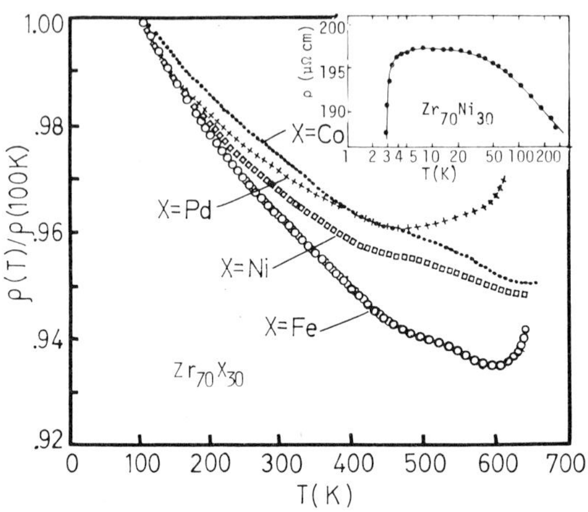

Figure 21.5 Electrical resistance of some zirconium-based alloys $Zr_{70}X_{30}$ with (inset) resistivity of $Zr_{70}Ni_{30}$ (after Rao *et al.*[20])

4. Resistance minimum found in concentrated alloys in which long range magnetic order is known to exist at temperatures above the minimum.

21.4 Theoretical approaches to the electron transport in amorphous alloys

There are at least three major theoretical approaches to explain the experimentally observed resistivity behaviour in amorphous metallic alloys: The natural extension of Ziman theory for the resistivity of simple liquid metals (the so-called 'diffraction' model), the structural analogue of Kondo phenomena, and the 'coherent exchange scattering' model. While discussing these we shall also touch on other relevant theoretical approaches of interest.

21.4.1 The diffraction model

In view of the strong similarities of the ionic and electronic properties of the liquid and glassy states in many metallic systems, extension of the Ziman theory to metallic glasses was a natural starting point to take in order to understand the resistivity of amorphous metallic alloys. The original Ziman theory[25] deals with the potential scattering of conduction electrons by a disordered set of scattering centres. In simple single element metallic liquids these scattering centres are represented by psuedopotentials. In the frame work of the Ziman approach the temperature dependence of the resistivity is governed by that of the interference function or the structure factor $s(k)$. Hence, the general name 'diffraction model' is rather appropriate for a Ziman theory modified to incorporate the role of partial structure factors in multielement-based disordered alloys. In such a model, the magnitude as well as the temperature coefficient of the resistivity of an alloy would be determined by the relative position of $2k_F$, where k_F is the Fermi-momentum vector, with respect to k_p, the position of the main peak in the structure factor $s(k)$. In this picture then the electronic details contribute to the overall magnitude of the resistivity but not to the TCR directly. The dynamic effects appear through the Debye–Waller factor that describes the temperature dependence of $s(k)$.

The basic Ziman theory was extended by Evans *et al.*[26] to transition metal liquid alloys by replacing the pseudopotential matrix element with a t-matrix in order to incorporate scattering phase shifts for non-overlapping potentials. The diffraction model expression for the resistivity of a pure liquid metal is

$$\rho = \frac{12\pi\Omega_0}{e^2\hbar v_F}\int_0^1 s(k)|t(k)|^2\left(\frac{k}{k_F}\right)^3 d\left(\frac{k}{k_F}\right) \tag{21.1}$$

where the symbols have the usual meaning. Since the k^3 term in the integral heavily weighs on the integrand close to $k = k_F$, it follows from equation 21.1 that the temperature dependence of the resistivity is primarily determined by $s(k = 2k_F)$. A general formalism for the temperature dependence of $s(k)$ for an amorphous solid for $T < \theta_D$ was developed by Nagel[27] using a Debye spectrum. For the temperature dependence of the static structure factor Nagel obtains the expression

$$s_T(k) = 1 + [s_0(k) - 1]e^{-2[W(T)-W(0)]} \tag{21.2}$$

where $s_0(k)$ is the static structure factor at $T = 0$ K and $e^{-W(T)}$ is the Debye–Waller factor at a temperature T. Note that from equation 21.2

$$\frac{1}{\rho}\frac{d\rho}{dT} = \frac{1}{s(k)}\frac{ds(k)}{dt} \tag{21.3}$$

Thus, the static structure factor plays a central role in determining the transport properties of amorphous metals. In order to account for contributions from both the elastic and inelastic (arising from electron–phonon interactions) scattering Meisel and Cote[28] expand the resistivity structure factor as

$$s^{\rho}(k) = s^{\rho}_0(k) + s^{\rho}_1(k) + s^{\rho}_2(k) + \ldots \tag{21.4}$$

where $s^{\rho}_n(k)$ is an n-phonon term. Then the elastic term is related to the Debye–Waller factor as

$$s^{\rho}_0(k) = a(k)\, e^{-2W(k)} \tag{21.5}$$

Here, $a(k)$ is the geometric structure factor. Evaluating this for transition metal based alloys in which the largest contribution to resistivity arises from the d-wave phase shifts, and using an approximation to one-phonon resistivity structure factor term, Meisel and Cote[28] evaluate the temperature dependence of the averaged structure factors and demonstrate that at low T, i.e. $T < \theta_D/2$

$$s^{\rho}(k) \sim 1 + \frac{b}{\theta_D^3} T^2 \tag{21.6}$$

where b is *always positive*, and independent of temperature, while at high T, i.e. $T > \theta_D/2$

$$s^{\rho}(k) \sim \pm \frac{c}{\theta_D^2} T \tag{21.7}$$

The negative sign occurs when $2k_F = k_p$, the position of the main peak in the geometric structure factor $a(k)$. This same result as in equation 21.7 has also been obtained by Nagel[27]. Similar results to those obtained by Meisel and Cote have also been independently deduced by Froböse and Jäckle[29] who used Einstein's phonon spectrum.

From equations 21.6 and 21.7 it is easily seen that the 'diffraction model' essentially predicts most of the resistivity behaviour observed in many amorphous alloys. As an example, consider the resistivity behaviour[30] of $(Pd_{50}Ni_{50})_{100-x}P_x$ alloys (*Figure 21.6*) in the temperature range 1.2 K to 450 K measured for increasing concentrations of phosphorus from 15 to 27.5 atom %. The small upturn in the

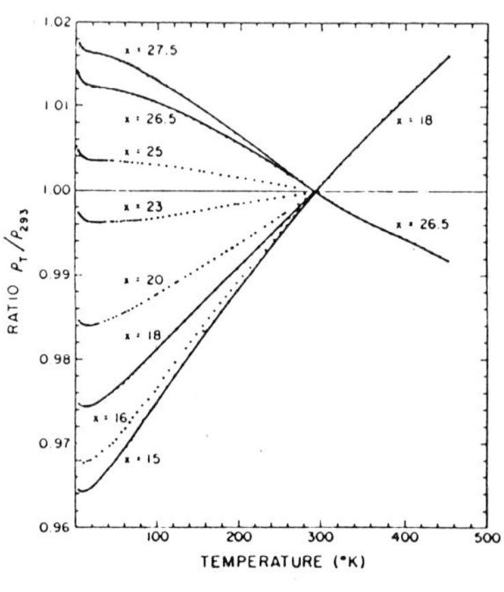

Figure 21.6 Resistivity of $(Pd_{50}Ni_{50})_{100-x}P_x$ alloys for the values of x shown on the curves (from Boucher[30])

resistivity at very low temperatures being neglected for the present discussion, it is clear from *Figure 21.6* that

1. All the curves initially increase almost quadratically with temperature as expected from equation 21.6.
2. When x is less than about 23 atom%, TCR is positive at high temperatures, thus exhibiting an S-shaped relation of resistivity with temperature.
3. When x exceeds 23 atom%, TCR is negative which, along with the low temperature behaviour produces a shallow maximum around $\theta_D/2$ where $k = k_p$.

The resistivity behaviour discussed here has also been observed in $(Pt_{1-x}Ni_x)_{75}P_{25}$ by Sinha[31] who first suggested the possible applicability of the Ziman theory to transport properties of amorphous metals. Resistivity studies[32,33] on $(Pd_{1-x}Cu_x)_{80}P_{20}$ and $(Pd_{1-x}Cu_x)_{80}Ge_{20}$ also conform to the Ziman picture.

At this stage, it may be instructive to consider in more detail the conditions for the applicability of the diffraction model and look into other interesting manifestations of this model that must be observable experimentally. *Figure 21.7* is a schematic (not drawn to scale and somewhat exaggerated for clarity) of a structure factor at a temperature T_1 plotted as a function of the k vector. Quantitatively this may be obtained from the X-ray diffraction pattern using the formula $k = 2\pi \sin(\theta/\lambda)$, where 2θ is the Bragg angle corresponding to the structure $s(k)$. For our purposes consider only the region around k_p the major peak in $s(k)$. As the temperature is increased to say T_2, this peak broadens and all the values near k_p are reduced. For an alloy with $2k_F$ in the vicinity of k_p then a *negative* TCR is expected. On the other hand if $2k$ is far away from k_p, all the values of the interference function increase, and so a *positive* TCR is predicted. Thus, for example, for alloying a monovalent with a multivalent element a positive TCR for electron concentrations less than 1.5, and a negative TCR for 1.5 to 2 effective conduction electrons per ion should be observable. The largest TCR will be when $2k_F = k_p$. This simple picture has been quite successful in explaining the resistivity behaviour of many amorphous alloys. However, it is important to remember that, while

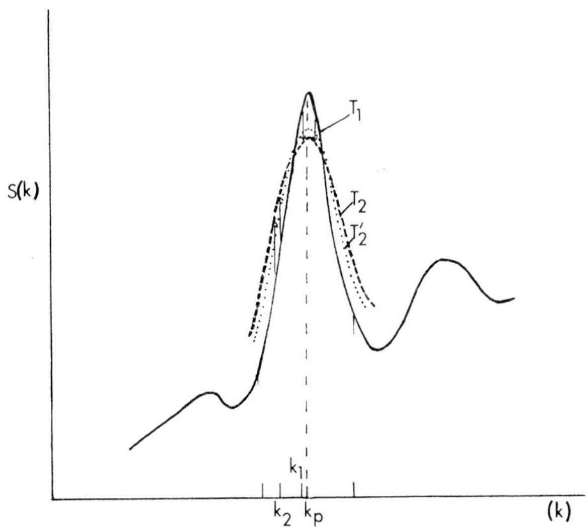

Figure 21.7 A general schematic diagram of a structure factor for an amorphous alloy. For a non-magnetic alloy the small first peak would not exist

$s(k)$ can be experimentally determined for an amorphous material, at present there is no simple way to determine $2k_F$ directly. k_F can be calculated by use of the free electron theory, where $k_F = (3\pi^2 NdZ/A)^{1/3}$. Here N is the Avogadro number, d the density, and Z the effective number of conduction electrons of the material of atomic weight A. Quite often simple extrapolation of the number of valence electrons from the values for pure components can give $2k_F$ values totally inconsistent with the experimentally observed TCR. Such apparent inconsistencies have been found for a number of amorphous metallic systems like Zr based alloys[21,22], alloys of gallium, tin and lead with copper, silver and gold[34], lanthanum-based alloys (La–Au in particular)[35], etc. Usually this discrepancy is explained away by assuming charge transfer[22] between the constituent elements in the alloy which tends to fill up the d-bands of the transition metal. In order to be consistent with the diffraction model, Meisel and Cote[22] had to assume a valence 5 state for phosphorus in NiP, which may not be true[37] because it is well known that phosphorus fills the d-state rendering nickel almost diamagnetic in NiP. Recently, Waseda and Chen[38] computed the resistivity and TCR of amorphous $Cu_{60}Zr_{40}$ using partial structure factors and obtained reasonable agreement with experiment. These results probably indicate that the diffraction model contains much of the physics of transport at least in some of the amorphous systems. We shall return to this point.

There is an interesting feature of the resistivity in amorphous systems for which the Ziman model provides adequate explanation. *Figure 21.8* shows the electrical and Hall resistivities of amorphous $Be_{40}Ti_{50}Zr_{10}$ (commercially available as METGLAS 2204 from Allied Corporation) measured[38] almost up to its glass transition temperature which for this alloy is about 600 K. The electrical resistivity of this alloy has a negative TCR in the entire temperature range. After warming the sample to a high temperature, say 550 K, the resistance measured while cooling the sample is found to be larger than the one observed during the warming run at the same temperature. However, no change in TCR is observed. It is as if the whole resistivity curve is laterally shifted to the high temperature at which the sample was cooled. At these temperatures amorphous materials structurally relax. Egami *et al.*[39] have found that when a metallic glass is structurally relaxed the first peak of $s(k)$ was raised and sharpened by relaxation annealing (curve marked T_2' in *Figure 21.7*). The change thus produced is qualitatively similar to the result if the sample had been cooled to a lower temperature. A consequence

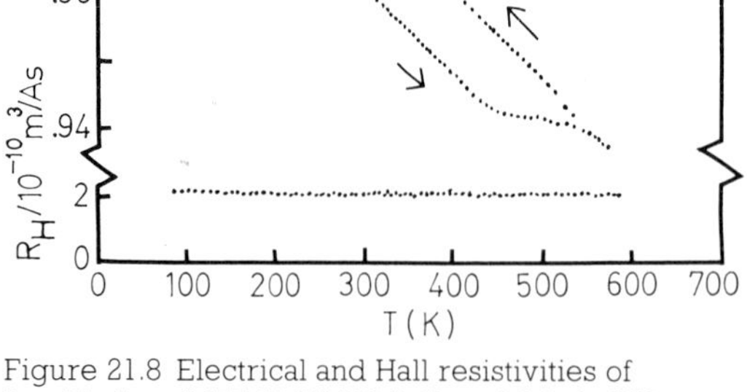

Figure 21.8 Electrical and Hall resistivities of $Be_{40}Ti_{50}Zr_{10}$ (METGLAS 2204) (from Malmhäll[38])

of such a relaxation then is to raise or lower the value of the resistivity (but not its TCR) depending on how close $2k_F$ is to k_p, for the alloy concerned. Both these effects, i.e. the resistivity shift to higher or a lower value in the relaxed states depending on the relative position of $2k_F$ with respect to k_p have been demonstrated in $Pd_{82-x}V_xSi_{18}$ alloys[4] by varying the vanadium concentration from 0 to 6 atom %. Further remarkable properties of this alloy, which can be explained only on the basis of the Ziman model, will be discussed later.

Another accomplishment of the modified Ziman theory is to provide an explanation as to why many metallic glasses are stable and easily formed around a concentration centred around 80 per cent of the transition metal (alloys like $Au_{81}Si_{19}$, $Pd_{80}Si_{20}$, $Fe_{80}B_{20}$ etc.). To explain the increased stability around this composition a criterion was put forth by Nagel and Tauc[40] based on the electronic properties of metallic glasses. They observe that the glass formability is enhanced when the alloy concentration is such that the diameter of the Fermi surface, $2k_F$, is of the order of the wave-number k_p. Since k_p represents roughly the first shell of a reciprocal lattice of a close-packed structure, a dip or a minimum is expected in the density of states in such a situation. This would ensure the stability of the glass against crystallization. This has been demonstrated via photoemission measurements on Pd–Cu–Si and Nb–Ni alloys[41,19]. Unfortunately, there is no other clear evidence for this local minimum in the density of states. In fact careful specific heat measurements by Mizutani *et al.*[42] on Pd–Si alloys show a high density of states and a maximum at about 1.6 electrons per atom. In the case of $Fe_{100-x}B_x$ alloys in the region of stability $15 < x < 22$, the resistivity does not have a negative TCR as expected from the above stability criterion.

21.4.2 Some relevant models of non-magnetic origin

We have pointed out many successes and yet obvious inconsistencies in applying the extended Ziman theory to specific systems. These results indirectly cast doubts about the interpretation of the negative TCR of resistivity in terms of the extended Ziman theory alone. In view of this, several mechanisms have been invoked or revived in attempting to explain the negative TCR of metallic glasses.

Since almost all metallic glasses are made with at least one transition metal component in them, the high density of *d*-states at the Fermi level and the role of *s–d* mixing must play a role in determining the resistivity of these materials. Owing to the short mean free path in these systems, it can be expected that the influence of atomic properties increases in the amorphous phase. In fact this has been commented upon from the resistivity studies on Pd–Ni–P alloys by Boucher[30] while comparing it with those on Pt–Ni–P by Sinha[31]. For the same concentration of phosphorus, on varying the Pt/Ni concentration ratio it was observed that the TCR changes its sign and becomes *negative* for the platinum-rich alloys. No such change in TCR was observed under a similar situation in Pd–Ni–P. Since palladium and platinum have the same electronic structure, the observed differences must be related to those in the band structure of palladium and platinum. Rather convincing evidence for the role of the density of *d*-states comes from the studies by Korn *et al.*[34,43] on amorphous $Sn_{1-x}Cu_x$ alloys shown in *Figure 21.9*. Increasing copper concentration not only increases the total resistivity by almost a factor of 2 (65 to 112 Ω cm), it also changes the sign of TCR which becomes negative for $x > 0.5$. This suggests that the $3d$ state of copper is responsible both for the magnitude of the total resistivity as well as the TCR. The positive TCR observed at low temperatures in these alloys arises from the paraconductivity occurring close to the superconduction transition. Korn *et al.* explain their results in terms of the scattering from *d*-states in an excited Cu d^9s^2 configuration. It may be worth pointing out

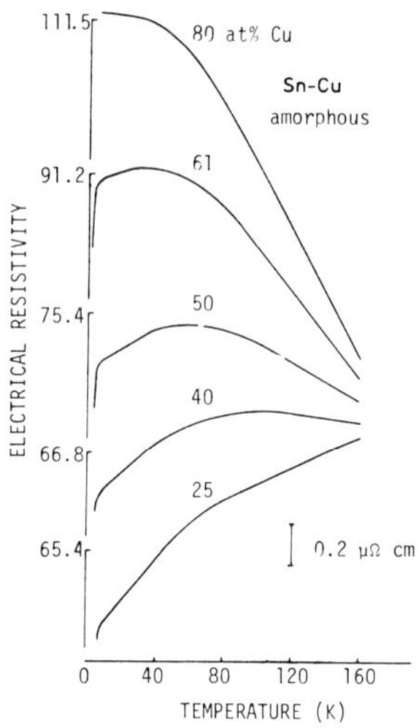

Figure 21.9 Resistivity of amorphous $Sn_{100-x}Cu_x$ alloys for the values of x shown on the curves (from Korn *et al.*[34])

that the Ziman model cannot explain satisfactorily the properties seen above in the extended concentration range of Sn–Cu alloys since no maximum is observed in the resistivity as a function of x.

Mott's s–d scattering model[44] relates negative TCR to the density of states at the Fermi energy. This model assumes two distinct groups of carriers at the Fermi surface. The s or p electrons which carry the current can be scattered from the d holes at the Fermi surface so that the resistivity is proportional to $N_d(E_F)$, the density of d-states at the Fermi-level. As the temperature is raised there is a shift in E_F due to thermal broadening which produces a temperature dependence of the resistivity given by

$$\rho_{s-d}(T) = \rho_0\left[1 - \frac{\pi^2}{6}(kT)^2(E_0 - E_F)^{-2}\right] \tag{21.8}$$

For a nearly filled d-band, ρ_{s-d} decreases with increasing temperature.

The local spin-fluctuation model has sometimes been invoked to explain the negative TCR observed in some amorphous alloys[45]. In this model, originally proposed by Kaiser and Doniach[46] and later developed by Zuckermann[47], for dilute alloys, one considers the localized virtual d-levels and the spin fluctuations associated with them as being responsible for the additional scattering which has a negative TCR. Direct applicability of these ideas to amorphous materials is however questionable in view of the difficulties involved in separating out the spin-fluctuation part from the total resistivity.

In order to identify the appropriate scattering mechanism which describes the transport properties of a given system, it is clear we need to study some other property like thermopower or magnetoresistance for which these theories predict substantially different behaviour. It is helpful first to look into theories which are based on structural disorder alone.

21.4.3 Theories based on structural disorder — 'the tunnelling model'

The widespread occurrence of a minimum in the resistivity of amorphous metallic alloys, irrespective of the magnetic state of the alloy, has led to the pursuit of mechanisms which could be a direct consequence of the non-periodic structure.

For kinetic reasons a glass finds itself trapped in a region of configurational space far different from the ground state which is the crystalline state. Since this region of configurational space is characterized by a relatively high energy there will be numerous other energetically equivalent states in which the glass would be trapped. Of these, roughly equivalent states which can make transitions within an experimental time scale, will contribute to the thermal and transport properties. It is these changes in configuration or rearrangement of small groups of atoms at the 'equivalent' sites that are called 'glassy' excitations or the so-called 'tunnelling levels'. In its simplest form they are the 'two-level-systems' (TLS). Clarly, for these states to occur some short range ordering in conjunction with a large number of atoms participating for a closer energy equivalence, is desirable. Thus it is the competition of accessibility and degeneracy that determines the size of the TLS. In an insulating glass the coupling between TLS and phonons leads to anomalies in specific heat, thermal conductivity, and ultrasonic attenuation. In metallic glasses there is the additional possibility of coupling between TLS and conduction electrons. Cochrane *et al.*[48] suggested that this causes a divergence in the rate of electron scattering as the temperature is lowered just as in the Kondo-effect. In their analyses they consider a potential-well with two minima of equal energy so that the eigenstates are the symmetric and antisymmetric tunnelling states which are split by an energy 2Δ (~ 0.2 MeV). Because the relaxation time of the electron sea ($\sim 10^{-13}$s) is much smaller than the tunnelling time ($\sim 10^{-10}$s) the electrons can distinguish between the environment provided by the tunnelling states. Assuming a Hamiltonian $H = H_0 + H_1$, where H_0 is the unperturbed term, they describe the scattering of electron from the tunnelling states by

$$H_1 = \sum_{\substack{kk' \\ \alpha,\alpha' = \pm}} V_{kk'}^{\alpha\alpha} a_{k\alpha}^{\dagger} a_{k'\alpha'} b_{\alpha}^{\dagger} b_{\alpha'} \qquad (21.9)$$

Here, $a\dagger$, a, $b\dagger$, b are creation and annihilation operators of the electrons and tunnelling states respectively. The resistivity as a function of temperature calculated in this case leads naturally into a Kondo-type expression

$$\rho_{\mathrm{TLS}}(T) = -c \ln (k_{\mathrm{B}}^2 T^2 + \Delta^2) \qquad (21.10)$$

where $\Delta \lesssim 1$ K. The 'tunnelling-model' thus predicts an increase in the resistivity with decreasing temperatures which should eventually saturate to a constant value below about 0.5 K. In addition, no significant or systematic dependence on changing the alloy concentration, the nature of the impurity, or the application of external magnetic fields should be observed. There have been a number of attempts to test this model and the experimental findings are:

1. No saturation in the resistance has been observed in a number of metallic glasses[49] on which measurements have been made down to a few mK.
2. Resistance minimum has never been observed in an amorphous metallic alloy which is totally free from magnetic impurities like iron or chromium (in a few ppm range) (for example PdSi, PdB alloys of ultra-high purity show no resistance minimum[10]).
3. Small additions of chromium or manganese result in large changes in the resistance minimum phenomenon[9,10] as illustrated in *Figure 21.2)*.
4. There are a number of amorphous alloys[34,43] of comparable resistivity that do not contain transition elements as components in which no resistance minima is observed.
5. The sign of the magnetoresistance in many of these alloys at low temperatures has been found to be very dependent on the magnetic state of the material[50].

Recent calculations by Black and Gyorffy[51] for the resistivity due to the two-level systems, using results obtained from ultrasonic experiments and specific heat studies, suggest that the scattering from such a system is at least three orders of magnitude smaller than expected from experimental observation. Furthermore, the contribution from such a mechanism was found to have a *positive* rather than a negative temperature coefficient. Kondo[52] has analysed the lowest order singularity terms of the perturbation theory to look for the divergence response of an electron gas to a localized perturbation with internal dynamics. He found the lowest order term in the elastic scattering rate to be of the *fourth* order given by

$$\tau^{-1} \sim (\rho_{\parallel} \cdot V_{\parallel})^2 (\rho_{\perp} \cdot V_{\perp})^2 [\ln (T/W)]^2 \tag{21.11}$$

where W is the conduction band width and $V_{\parallel}$, and $V_{\perp}$ denote coupling energies for the scattering of the electron between the two level states. However, for the well known Kondo effect in the magnetic spin-flip case the divergence occurs in the *third* order

$$\tau^{-1} \sim -J^3 \ln (T/W) \tag{21.12}$$

Despite the sign of the $[\ln (T/W)]^2$ term, Stewart and Phillips[52] demonstrate that equation 21.11 fits the resistivity data for a number of metallic glasses (*Figure 21.10*), much better than equation 21.10 and also over a wider temperature range. The continuous line fits shown in *Figure 21.10* clearly show that that data do not obey the oft claimed ln T type of behaviour, although in approximation for the resistivity below $T_{\min}$ is often claimed to be the cause although in approximation. The latter observation has also been implied by Rayne and Levy[53].

Recently Tsuei has reinterpreted[54] the Kondo approach taken by Cochrane *et al*, in order to explain the negative TCR observed for the resistivity of a number of both disordered (crystalline) and non-crystalline metallic systems. This included structurally disordered materials like lanthanum-based alloys[35], as well as disordered crystalline A-15 structured superconductors like Nb_3Ge, Ti_2Al etc., all of which have a negative TCR up to very high temperatures. Tsuei points out that in the structural analogue of the Kondo approach the internal degree of freedom need not arise from quantum mechanical tunnelling between structural fluctuations approximated by a two-level system. Any localized internal degree of freedom (such as localized phonon modes or even localized electron states) 'which give rise to excitations that are degenerate on the

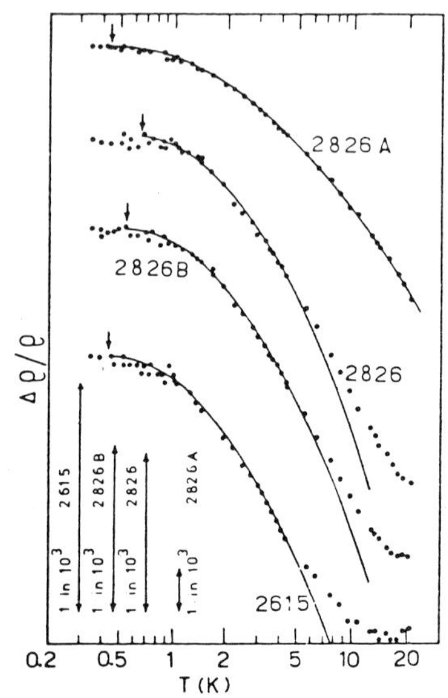

Figure 21.10 Low temperature resistivity of some METGLAS samples. The full lines are a fit to the Kondo structural model of equation 21.11. The arrows in the lower left corner represent the value 1/1000 in $\Delta\rho/\rho$ (from Stewart and Phillips[52])

scale of $k_B T$ ($T \geqslant \theta_D$) can lead to a negative TCR through the structural Kondo-mechanism'. This interpretation would extend the negative TCR anomaly over a wide range of temperature since Δ in equation 21.10 can now acquire a large value almost compatible with θ_D. To illustrate this Kondo-type resistivity anomaly Tsuei plots the resistivity data both on a linear and logarithmic scale. If the data followed the Ziman theory predictions according to equation 21.7, at high temperatures, the resistivity would be linear with temperature, whereas it would be linear on a ln T plot if a Kondo-type model was appropriate. This approach is particularly convincing, at least for an experimentalist, especially in the case of $La_{100-x}M_x$ alloys in which case the additional fact that the negative TCR is independent of the valence of the second element casts doubt on the applicability of the Ziman criteria ($2k_F = k_p$) for this system.

21.5 Thermopower

Study of thermopower is particularly valuable to test theories of electrical resistance since it is the energy derivative of electrical resistivity. Mott's expression for diffusion thermopower is as follows[55]:

$$S = -\frac{\pi^2 k_B T}{3|e|E_F}\left|\frac{\partial \ln \rho(E)}{\partial E}\right|_{E=E_F} \tag{21.13}$$

where $\rho(E)$ is the electrical resistivity. The derivation of equation 21.13 assumes elastic scattering of electrons and that the density of states is a smoothly varying function around the Fermi energy.

In the s–d scattering model equation 21.13 shows that the thermopower will be large and negative for partially filled d-band elements (at least comparable with that in the crystalline state for transition elements). Furthermore, it will be positive or negative depending on whether the d-band is less or more than half filled. Experimental evidence on amorphous metallic alloys contradicts this requirement. The thermopower of zirconium-based alloys[56] for example is found to have the same sign as that of Ni–Pt–P alloys[31].

In crystalline systems the Kondo effect gives rise to large ($\sim 30\,\mu$V/K) thermopower anomalies with a maximum near a characteristic temperature called the Kondo temperature, T_K. In fact, it is this property of Au–Fe alloys that makes them good thermometers at cryogenic temperatures. In an amorphous alloy, since the effects of phonon-drag[55] would not arise, Kondo-type anomalies should be observable although they are expected to be much smaller in magnitude owing to the Gorter–Nordheim rule[55], which says that the contribution to thermopower from a scattering mechanism is weighted in proportion to its contribution to the total resistivity, i.e., $S_i = (\rho_i/\rho_T)S_T$. The same type of anomalies must also be observable in the thermopower of glassy metals if the Kondo-type 'tunnelling-level model' were operative.

In the first thermopower studies made on amorphous metals Sinha[31] found that for Pt–Ni–P alloys the thermopower was linear in temperature ($\sim 2\,\mu$V/K at 300 K), and suggested the possible application of the Ziman model to determine the 'form factor' dependence on the Ni/Pt ratio in the alloy. Recently, Nagel and his coworkers have made extensive studies to correlate thermopower, structure factor and resistivity studies in a number of metallic glasses[56–58]. From the Ziman theory the following expression is obtained[57] for the thermopower:

$$S = -\frac{\pi^2 k_B T}{3|e|E_F}\left[3 - 2q - \frac{1}{2}r\right] \tag{21.14}$$

where

$$q = \frac{s(2k_F)|t(2k_F)|^2}{\int_0^1 |t(k)|^2 s(k)^4 (2k_F)^{-4} k^3 \, dk} \tag{21.15}$$

Here r contains the energy dependence of the t-matrix which is usually small. The sign of S then is mainly determined by the term $(3 - 2q)$ in equation 21.14. Based on the Ziman model the following observations for the temperature dependence of the thermopower of glassy metals can be made:

1. If the Ziman criterion is satisfied, the thermopower should be a linear function of temperature with a small positive slope.
2. If $2k_F < k_p$, S will be negative but again linear in temperature.
3. In a multielement system, because of the partial structure factors involved, S could change slope as a function of composition.

All of these cases have been identified experimentally[54–60]. *Figure 21.11* shows some of the recent thermopower studies by Nagel[56] both on non-magnetic and magnetic metallic glasses. For the non-magnetic glasses the thermopower is found to be remarkably linear over the whole temperature range as expected from the Ziman model. Notice also that the resistivity of CuZr and Be–Ti–Zr (METGLAS 2204) alloys have a negative TCR (*Figures 21.5 and 21.8*). It is interesting to note that the METGLAS 2204 alloy has perhaps the largest negative temperature coefficient for the resistivity of any known amorphous alloy. Thermopower in this case has a slope $5.7 \times 10^{-3}\ \mu V/K^2$. The absolute value of the thermopower is, as expected, small in all cases. A similar behaviour has also been recently reported[60] for an alloy containing only normal metals, $Ca_{0.8}Al_{0.2}$. It is useful to point out that for a Pd–Cu–Si alloy where the resistivity has a

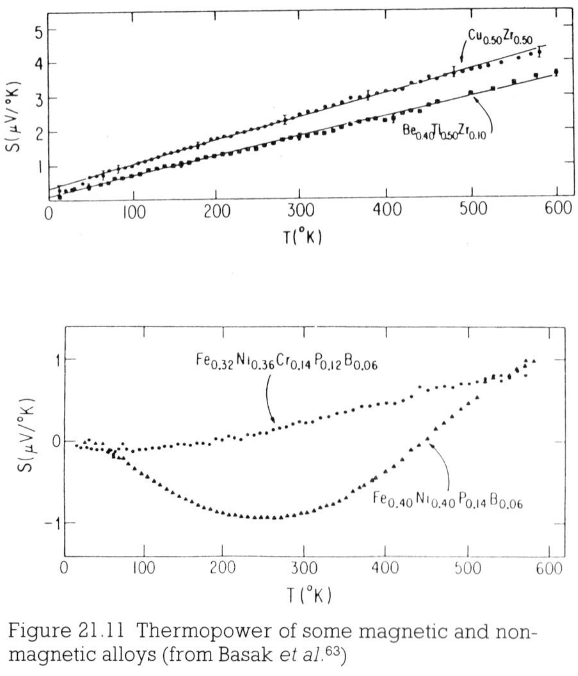

Figure 21.11 Thermopower of some magnetic and non-magnetic alloys (from Basak *et al.*[63])

positive TCR, the measured thermopower is found to be negative[58] which again is in accord with the diffraction model.

It thus appears that thermopower studies clearly help distinguish between models. The linear temperature dependence of the thermopower and the correlation of its slope with that of resistivity seems to be a reasonable way to separate out systems which are consistent with the diffraction model. However, the first clear breakdown of this model in describing the transport properties of a non-magnetic amorphous metal has recently been reported by Armbrüster and Naugle[61] from their resistivity and thermopower studies on $La_{0.82}Ga_{0.18}$ and $La_{0.78}Ga_{0.22}$ alloys. They find that while the resistivity for these two alloys have different signs for the slope α, the thermopower is found to be negative for both cases. Furthermore, the diffraction model predicts $S(\alpha < 0)/T > S(\alpha > 0)/T$, which contradicts the experimental finding. It is useful to recall our earlier reference[35] to the independence of α ($\equiv d\rho/dT$) from valence for the lanthanum-based alloys. Clearly for the lanthanum-based alloys the diffraction model is inadequate.

The temperature dependence of the thermopower of magnetic amorphous alloys (*Figure 21.11*) is clearly more complicated. They do show a minimum, and a behaviour almost similar to the electrical resistivity[58,62,63]; however, a quantitative analysis is not yet possible. Qualitatively these results can be accounted for[63] on the basis of a recent theory by Grest and Nagel[64] which indicates the possible consequences of Kondo-type scattering of electrons by spins situated at the 'low-field' sites in a random system with long range order.

21.5.1 Theories of magnetic origin: Kondo approach

The striking resistance rise at low temperatures and its sensitivity to the nature and concentration of the impurity in dilute transition metal–metalloid-based amorphous alloys has been associated with the Kondo effect which has been studied extensively in crystalline materials. The Kondo effect is associated with the scattering of conduction electrons from a local magnetic impurity in an otherwise non-magnetic matrix. This gives rise to a spin-dependent increase in the resistivity at low temperatures which, coupled with the usual increased scattering by phonons at higher temperatures, produces a minimum in the total resistivity of these alloys. Since this effect is closely related to the local moment problem, studies of resistance minima in various matrices have been an active subject over almost a decade now.

Using a perturbation approach, in the second Born-approximation, Kondo[65] showed that this spin-dependent resistivity is given by

$$R_{\mathrm{spin}} = R_{\mathrm{M}}[1 + Jc\ln(T/T_{\mathrm{K}})] \tag{21.16}$$

where c is the magnetic impurity concentration in the alloy, J the local exchange, and T_K defined by $k_B T_K = D$ the conduction band width is the so-called Kondo temperature. When J is negative one observes the rise in resistance at low temperatures. To summarize, if the local impurity in a metal has a moment or a sufficient long-lived local spin-correlation, then the following will be observed:

1. $R(T)$ has a minimum at a temperature, T_{min}.
2. $T_{\mathrm{min}} \sim (\mathrm{conc})^{1/5}$. (Here the 1/5 power arises from the T^5 dependence of the resistance in normal metals due to the phonon term).
3. The resistance below T_{min} increases as $-Bc\ln(T/T_K)$.

The unphysical logarithmic divergence at low temperature was removed by the concept of 'spin-compensation' where the effective local moment decreases with decreasing temperature to vanish at $T = 0$. Taking this into consideration, Hamann[66] obtained the analytical expression for the resistivity rise as

$$\frac{2\Delta\rho}{\rho} = \left[1 - \frac{\ln(T/T_K)}{[\ln^2(T/T_K) + s(s+1)\pi^2]^{1/2}}\right] \tag{21.17}$$

From the value of T_K estimated from a fit of the resistivity data to equation 21.17 an estimate of the effective *s–d* exchange interaction can be obtained via

$$T_K = D\exp(-n(E_F)|J_{eff}|)^{-1}. \tag{21.18}$$

Extensive experimental studies looking for the manifestations of the Kondo effect in amorphous alloys have been carried out initially by the Caltech group[10,67–70] particularly on Pd–Si containing between 0.5 and 10 atom% of chromium, manganese, iron and cobalt, Ni–Pd–P, and Ni–Pd–B containing chromium. From these studies, fitting the general characteristics of the resistivity to equations of the type 21.16 to 21.18 they obtain a value for the exchange interaction in the range 0.2—0.6 eV. However, there are a number of inconsistencies observed in applying the Kondo approach to these concentrated alloys. In fact, the approach itself may be questionable in view of recent low field magnetic studies which reveal a new 'glassy-magnetic' state in them[71,72]. Furthermore,

1. The resistivity never really saturates even at the lowest of temperatures (few mK) to which they have been measured[49,73]. A logarithmic plot of *R* and against *T* can be misleading in this respect. Hasegawa found on substituting a few atom% of iron or manganese for nickel in a $Ni_{40}Pd_{40}P_{20}$ host matrix, a $T^{-1/2}$ singularity[74] in the resistivity.
2. Rather than the expected (concentration)$^{1/5}$ dependence, T_{min} is found to vary almost linearly with impurity concentration[69]. Since in these alloys the resistivity increases as T^2 above the minimum, one would expect a (conc)$^{1/2}$ dependence of T_{min}. It is also useful to point out that in $Fe_xPd_{80-x}Si_{20}$ alloys it was found[69] that for $x > 3$ atom% of Fe, T_{min} starts to *decrease* with increasing x.
3. Most importantly the Kondo phenomenon is a single impurity spin effect. In crystalline alloys the Kondo regime is probable for impurity concentrations of the order of a few ppm. Also, the Kondo temperature, T_K, should be independent of the spin concentration. In addition, any long range coupling between the spins is detrimental to the Kondo effect. From this point of view most of the amorphous alloys that have been studied are really in the high concentration regime where clusters, and intercluster interactions, could play a role.
4. Recent low field (~3 Oe rms) magnetic susceptibility studies[71,72] show that for alloy concentrations in the range 1 to 10 atom% all the Pd–Si based alloys exhibit a 'spin-glass' type of behaviour, i.e., below a characteristic temperature the impurity spins interact strongly with the local field and get locked into a frozen 'glassy-magnetic' state. This would at least reduce considerably the number of possible isolated impurities that could give rise to a Kondo type of scattering.

Resistivity data on 'well-characterized' alloys made of high purity components and with magnetic impurity concentrations less than 100 ppm are scarce. However, it may be appropriate to apply the Kondo approach to analyse the low temperature rise in the resistivity of a so-called 'pure' amorphous alloy like $Pd_{80}Si_{20}$, because palladium usually will have at least a few ppm of iron as impurity. Pure crystalline palladium with less than 0.1 atom % of iron is known to give rise to a giant moment ($\sim 10\,\mu_B$) at the Fe-site and exhibit a ferromagnetic transition around 4 K. However, with 20 atom % of silicon, amorphous $Pd_{80}Si_{20}$ is almost diamagnetic[10], and so any observed resistance minimum in such an alloy is likely to arise from the iron impurity in a ppm range. We have analysed the resistivity data of one such alloy which was measured down to 50 mK by Dierker *et al.*[75]. These measurements, presented in *Figure 21.12*, were carried out on $Pd_{77.5}Si_{16.5}Cu_6$ well characterized samples both in the form of a rod and a wire. The continuous line on *Figure 21.12* represents the locus of the resistivity data points taken

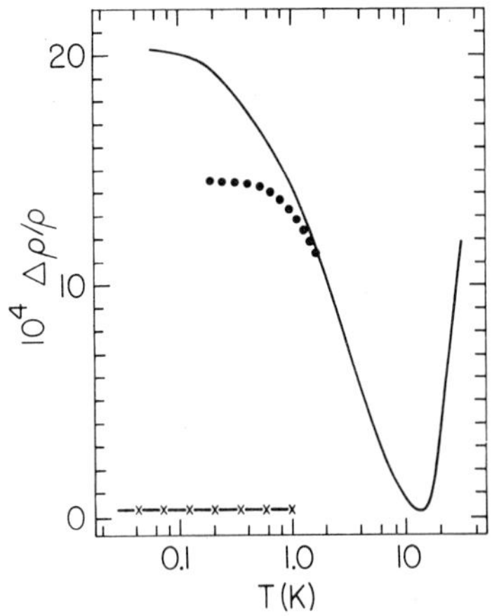

Figure 21.12 Low-temperature electrical resistance of $Pd_{77.5}Si_{16.5}Cu_6$ wire in applied fields $B_a = 0$ and 0.7 T (dotted points). The × data points are for a rod of PdSiCu with a smaller iron impurity content (after Dierker *et al.*[75])

on the wire which when analysed was found to contain about 700 ppm Fe. A resistance minimum is observed in this alloy around 14 K. On applying an external 0.7 T magnetic field, the dotted curve was obtained indicating clearly that the resistance rise observed here is due to magnetic scattering. The resistivity of the rod (shown by crossed points) with less than 50 ppm Fe impurity after a mild heat treatment, but still amorphous, was found to be independent of temperature. We have analysed[73] these resistivity data from the Kondo point of view using the Hamann expression given by equation 21.17, and find that the data can be fitted well in the whole temperature range below T_{min} if we assume that iron is in a spin $\frac{1}{2}$ state. This result is found to be in agreement with independent magnetic low field susceptibility studies on the same alloy. However, we also find that the same data can be fitted, perhaps with better precision, to the exponential temperature dependent behaviour which is predicted by the 'coherent-exchange scattering' model. This model is based on the temperature dependence of the spin–spin correlation which cannot be ruled out in the present case.

21.5.2 Coherent-exchange scattering model

This model was proposed by Asomoza *et al.*[76] in order to explain the unusual resistivity behaviour of rare-earth based amorphous alloys discussed before in describing *Figure 21.3*. It considers a system of magnetic ions of angular momenta J_R, coupled to the conduction electron spin S by an exchange interaction which is written as

$$H_{ex} = -\Gamma \sum_R S \cdot J_R\, \delta(r - R) \tag{21.19}$$

The contribution to the resistivity is then calculated in the nearly-free-electron-model of Ziman for the transport properties of liquid metals and amorphous alloys by making use of the quasi-elastic De Gennes–Friedel[77] approximation. The exchange scattering resistivity is thus obtained from the expression

$$\rho_{\mathrm{m}} = \frac{2\pi\hbar k_{\mathrm{F}}}{ne^2} \int_0^{\pi} \sin\theta(1 - \cos\theta)\frac{\mathrm{d}\sigma}{\mathrm{d}\Omega}\,\mathrm{d}\theta \tag{21.20}$$

where n is the number of conduction electrons per atom, and

$$\frac{\mathrm{d}\sigma}{\mathrm{d}\Omega} = \left(\frac{m}{2\pi\hbar^2}\right)^2 \sum_{\alpha,\beta} W_\alpha \left|\langle\alpha|\int \mathrm{d}r\, \mathrm{e}^{-ik\cdot r} H \mathrm{e}^{-ik'\cdot r}|\beta\rangle\right|^2 \tag{21.21}$$

where $|\alpha\rangle$ and $|\beta\rangle$ are the spin states of the system before and after the scattering and W_α the probability of the state $|\alpha\rangle$. By taking into account the interferences between the scatterings on neighbouring magnetic ions the magnetic contribution to the resistivity is obtained as

$$\rho_{\mathrm{m}} = \rho_{\mathrm{M}}[c + c^2 m(2k_{\mathrm{F}})] \tag{21.22}$$

where

$$\rho_{\mathrm{M}} = \frac{m^2 k_{\mathrm{F}}^2 J(J+1)}{4ne^2\hbar^3\pi} \tag{21.23}$$

and

$$m(2k_{\mathrm{F}}) \sim \sum_{i,j} \langle S_i \cdot S_j\rangle\, \mathrm{e}^{-\mathrm{i}(2k_{\mathrm{F}})(R_i - R_j)} \tag{21.24}$$

is the spin correlation function or the magnetic structure factor evaluated at its dominant contribution to the integral, i.e. $q = 2k_{\mathrm{F}}$.

The first term in equation 21.22 results from the independent scattering by the

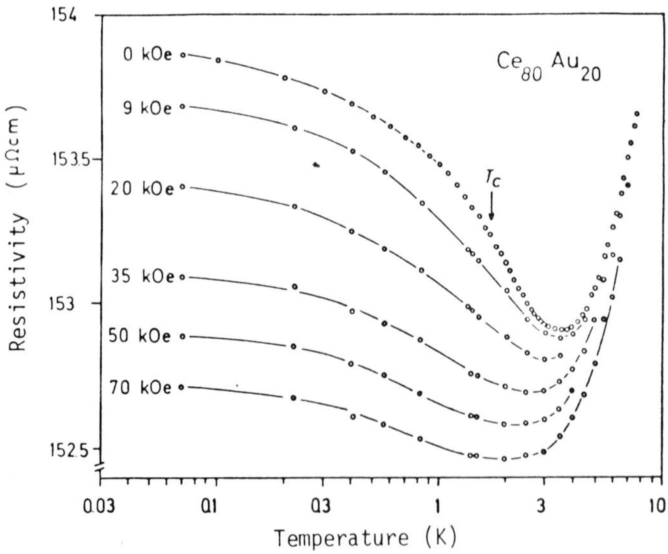

Figure 21.13 Electrical and magnetoresistance of $Ce_{80}Au_{20}$ (after Ernst *et al.*[11])

magnetic ions and does not depend on magnetic ordering. The second term results from coherent exchange scattering by the neighbouring ions. This is non-zero at low temperatures where spins are correlated, and vanishes at high temperatures. Owing to the amorphous structure it is reasonable to expect that spin correlation in glassy alloys depends only on the short range order in the system. To evaluate $m(2k_F)$ for a particular alloy, when the range of the magnetic order is longer than the structural order, we have to consider the partial structure factors which will involve individual magnetic species and the possible mixture of ferromagnetic and antiferromagnetic correlations between them. If the net interferences are constructive, i.e., $m(2k_F) > 1$, then spin ordering is found to result in a resistivity *increase*. If the interferences are destructive, i.e., $m(2k_F) < 1$, spin ordering results in a *decrease* of the resistivity. Bhattacharjee and Coqblin[78] have analysed theoretically the magnetoresistivity behaviour in the framework of this model. Taking into consideration the presence of both ferromagnetic and antiferromagnetic interactions between the rare earth ions, in a two-spin cluster approximation within the random anisotropy model, 'with a reasonable choice of parameters' they are able to account both for the resistivity rise and the positive (*Figure 21.4*) or negative (*Figure 21.13*) magnetoresistivity observed with a maximum around T_C in a number of rare earth based amorphous alloys[12–15,76,79].

21.6 Resistivity of magnetic amorphous alloys: a unified approach

One unusual feature in the resistivity of amorphous magnetic alloys is the existence of the low temperature resistance minimum, whether the alloy be dilute or a concentrated ferromagnet with a high Curie temperature. In crystalline alloys such a Kondo type of resistivity behaviour in a system with long range order is not possible because for Kondo scattering, not only the molecular field on the spin which is the scattering centre must be weak, but the molecular field at the neighbouring spins have also to be weak, so that the orbit of the resonant state of the conduction electron remains reasonably undisturbed. The fact that low-temperature anomalies are observed in almost every amorphous alloy containing an element capable of carrying a moment, and not in alloys that are nonmagnetic, indicates a magnetic origin.

In order to explain the observed resistivity behaviour it is most helpful to discuss it in terms of the magnetic phase diagram of these systems. There are some general features of the magnetic phase diagram obtained for various alloys systems, both crystalline as well as amorphous, which are rather similar and in accordance with the predictions of Sherrington and Kirkpatrick[80]. This universal type of magnetic phase diagram, on a temperature versus concentration plane, shows as a function of increasing concentration of the magnetic component in the alloy, boundaries representing spin-glass, T_f, ferromagnetic T_C, and ferromagnetic to 'spin-glass-like', T_{fg} transitions. Such a phase diagram for a given system can be determined from low field AC-susceptibility measurements.

To illustrate a possible unified approach to the resistivity behaviour of amorphous magnetic alloys, we shall discuss the resistivity observed for an Fe–Ni alloy system. *Figure 21.14* shows the magnetic phase diagram of the amorphous $(Fe–Ni)_{75}P_{16}B_6Al_3$ alloy system which is qualitatively similar to those obtained for a number of amorphous alloys containing cobalt, manganese, chromium and molybdenum in addition to iron for the transition metal combinations[81–84]. *Figure 21.14* concentrates on the region close to the multicritical point, MCP, which corresponds to an iron concentration $x_{MCP} = 12.75$ atom% where the boundary lines of the three

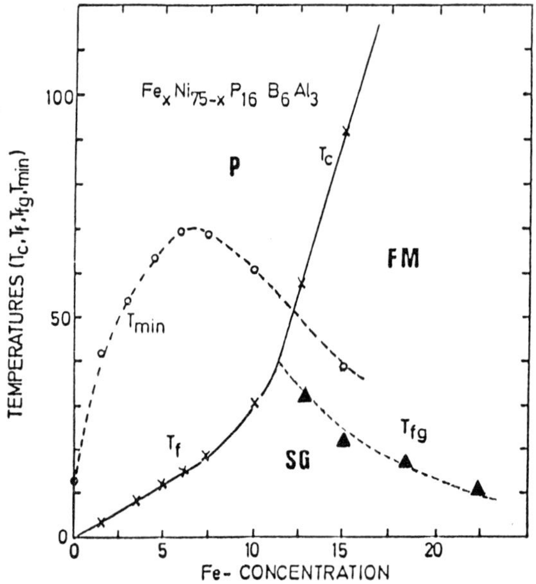

Figure 21.14 Magnetic phase diagram of $Fe_xNi_{75x}P_{16}B_6Al_3$. The value T_{min} correspondes to the temperature where resistance is a minimum (after Rao *et al.*[18])

magnetic states intersect. This is the region of interest which provides further insight into the resistance minimum phenomena. Although all the alloys in the whole concentration range exhibit a minimum in the temperature dependence of the resistivity, T_{min} has a maximum around $x = 6$ atom % of iron, a concentration at which T_f starts deviating from a linear concentration dependence. That this correlation is more general is demonstrated in *Figure 21.15*, where T_{min} for two different Fe–Ni based alloy series[18] is plotted along with their concentration dependence of T_f. The consistently higher values observed for T_{min} for the same iron concentration in the case of alloys containing P_{16} as compared to those for P_{14} alloys suggest that the metalloid concentration has a significant influence on the phenomena. Notice also for the alloys with P_{14}, T_f deviates from linearity around 5 atom % of iron. Such a correlation between T_{min} and linearity of T_f with concentration has also been pointed out by Poon and Durand for Gd–La–Au alloys[83]. The value of T_{min}, however, is determined by the balance between the two components of the resistivity, i.e. the low temperature rise which is of magnetic origin and the T^2 dependence term above T_{min} which in these alloys appears to persist to above 130 K. Since T_{min} is determined by the balance between two independent contributions to the resistivity it does not have a physical significance of its own. It appears that the logarithmic inflection point, T_i defined by

$$\left.\frac{\partial^2 R}{\partial(\ln T)^2}\right|_{T=T_i} = 0 \qquad (21.25)$$

is a better defined quantity to characterize the low temperature component of the resistivity. Indeed, while T_{min} can vary by almost a factor of two for the two alloy series, the value of T_i for both the alloy series agree well with T_f as shown in *Figure 21.15*. It is

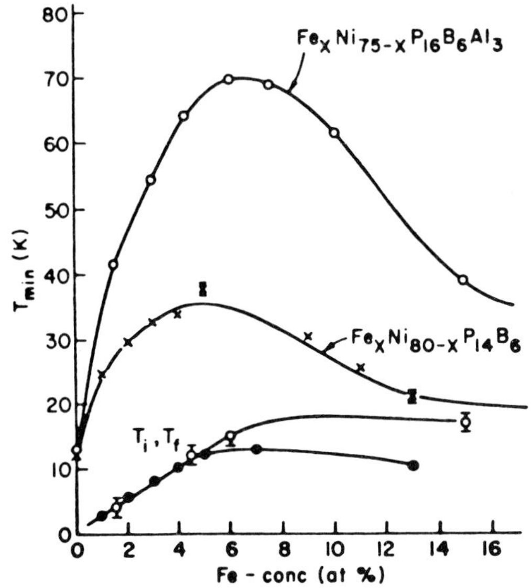

Figure 21.15 Concentration dependence of T_{min} for two Fe–Ni amorphous series. T_f corresponds to the spin-glass transition temperature and T_i is defined in equation 19.26 (after Rao *et al.*[18])

useful to recall that this linear concentration dependence of T_f for spin glasses has been predicted in a mean field approach[85]. Furthermore, Rao *et al.*[18] found that when the relative change in the resistivity

$$\Delta R(T) = \frac{R(T) - R(T_i)}{R(T_i)} \tag{21.26}$$

is scaled by an appropriate factor, which depends on the composition, the reduced resistivity for *both* the alloys coalesced into a *single universal curve* shown in *Figure 21.16*, over a wide composition range. This indicates that a single mechanism dominates at least in the spin-glass regime. Above T_i, this universal curve in *Figure 21.16* can be approximated by

$$g(T) \propto e^{-aT} \tag{21.30}$$

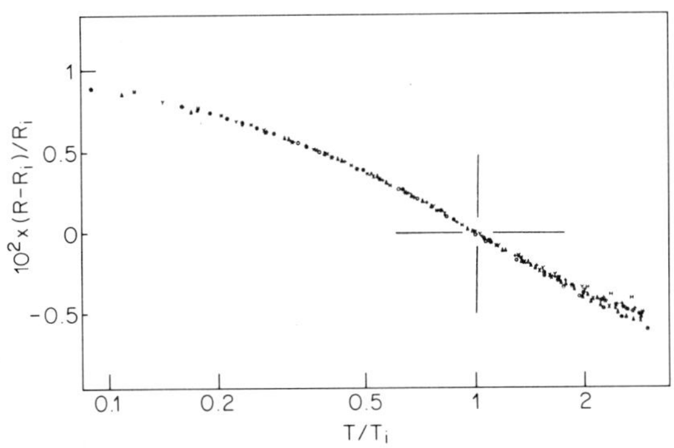

Figure 21.16 The universal curve of $f \cdot \Delta R(T)$ as a function of T/T_i for nine Fe–Ni based alloys from the two series in *Figure 21.15* (after Rao *et al.*[18])

Such an exponential temperature dependence, which is rather similar to the Debye–Waller factor for phonons discussed earlier, suggests that the resistivity anomaly in these alloys is probably related to the magnetic structure factor $m(2k_F)$ discussed by Asomoza *et al.*[76] (equation 21.22) in their coherence-exchange scattering model. It is very likely then that $m(k)$ has a pre-peak, or the antiferromagnetic peak, inside the first peak of the structural interference function $s(k)$ as illustrated in *Figure 21.7*. The positive temperature coefficient of the resistivity above T_{min} indicates that $s(2k_F) < 1$, which is common for the transition metal–metalloid amorphous alloys. Therefore $2k_F$ is most likely situated at or near the antiferromagnetic prepeak, so that $m(2k_F) > 0$ and $(\partial m(2k_F)/\partial T) < 0$ which results in the resistivity minimum behaviour. Recent neutron studies[86] of the magnetic structure of amorphous spin-glass $(MnO)_{0.5}(SiO_2)_{14}(Al_2O_3)_{0.1}$ at $k = 1.36\,Å^{-1}$ found a temperature dependence similar to the temperature dependence of the universal curve (equation 21.30) and the proposed ideas by Asomoza *et al.*[76]. Thus it appears that the low temperature resistivity behaviour of amorphous alloys in the spin-glass regime can be explained on the basis of a model based on spin–spin correlation and its temperature dependence. Clearly, the Kondo approach can be meaningful only in alloys of magnetic impurity concentration in the ppm range.

It now remains to test if the same spin-correlation effects can explain the resistance minimum behaviour observed for alloys with long range order, i.e. for $x > x_{MCP}$ in *Figure 21.14*. A typical resistivity study[87] in this regime for a $(Fe_{0.2}Ni_{0.8})_{75}P_{16}B_6Al_3$ alloy shown in *Figure 21.17* is particularly revealing. In order to correlate with the magnetic state at various temperatures, low field AC-susceptibility data (measured at 3 Oe rms) for samples from the same stock is also shown in *Figure 21.17*. Around the ferromagnetic transition the resistivity exhibits an anomaly very similar to that seen in pure crystalline nickel. The temperature derivative of the resistivity around T_C exhibiting a characteristic cusp has been analysed recently by Källbäck *et al.*[88] to study the contributions from critical fluctuations. They observe that above T_C the cusp yields a critical exponent $\alpha = -0.37 \pm 0.05$, which is found to be compatible via the scaling relationships with other known exponents β and γ obtained from

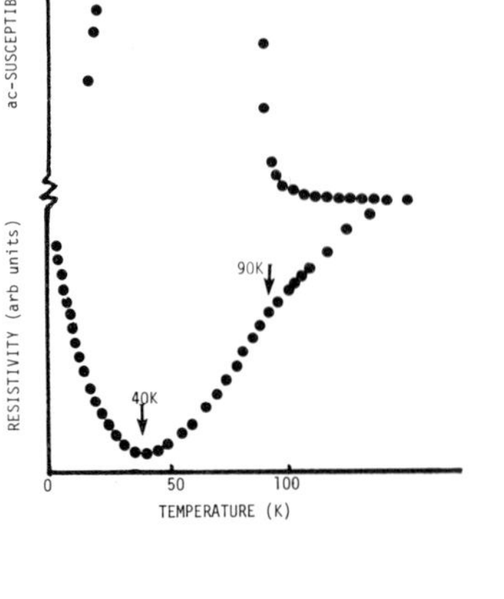

Figure 21.17 Electrical resistivity and AC susceptibility (both in arbitrary units) for $(Fe_{0.2}Ni_{0.8})_{75}P_{16}B_6Al_3$.

magnetic measurements on the same sample[82]. Below T_C down to about 60 K, a T^2 dependence is observed in the resistivity, the magnitude of whose coefficient agrees well with what one expects for scattering of electrons from spin wave excitations. Thus in magnetic systems the T^2 term above T_{min} is of magnetic origin. Recently a $T^{3/2}$ behaviour has been predicted for this term by Richter *et al.*[89]. Mogro-Compero[90] and others have found that the resistivity data can be fitted to this $T^{3/2}$ term if in addition one takes into consideration the possible usual phonon term.

Below 40 K in *Figure 21.17* the AC-susceptibility begins to decrease with decreasing temperatures signalling the onset of the collapse of long range order expected on crossing from a ferromagnetic to 'spin-glass-like' state as seen from the phase diagram in *Figure 21.14*. This is manifested as a large increasing resistivity below T_{min}. It turns out again that the temperature corresponding to the inflection point as defined by equation 21.25 in this part of the resistivity agrees well with T_{fg} obtained by other techniques from magnetic data[82]. This suggests that the resistance minimum observed in magnetic systems with long range order can also be explained on the basis of the temperature dependence of spin–spin correlation in the system. In this case, however, the correlation will be among spins at 'weak-field-sites' one expects in a random system with mixed interactions. Recently, another theoretical formalism has been proposed by Grest and Nagel[64]. They suggest that, owing to the intervening metalloid atoms, the next nearest neighbour atoms interact via superexchange. This then, along with the regular long range order, produces a field distribution which allows a number of sites in the matrix at which the local field would be very weak. Any spin located at such sites would give rise to a Kondo-type scattering. Although this is qualitatively plausible, no quantitative predictions have yet been made. Besides, for reasons stated earlier, the Kondo mechanism and its extensions have many drawbacks. Clearly, neutron studies of the magnetic structure factor would help distinguish between these models. There is another piece of experimental evidence which suggests that the spin-correlation approach may be appropriate. As expected from the spin-correlation model, Bhatacharjee and Coqblin[78] show that the magnetoresistance will have a strong temperature dependence centred around the magnetic transition. Qualitatively the temperature-dependence of the magnetoresistance observed for a

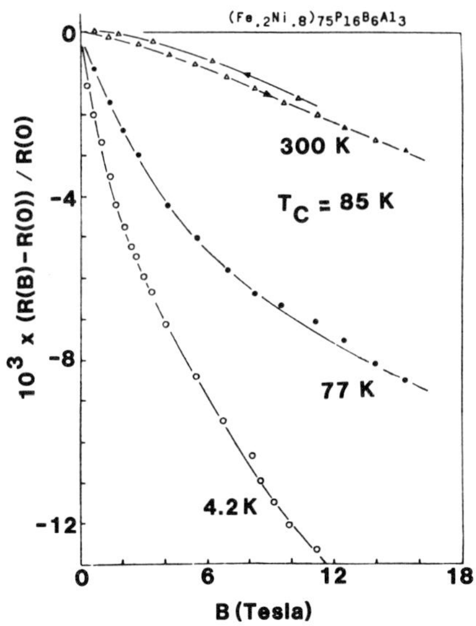

Figure 21.18 Magnetoresistance of $(Fe_{0.2}Ni_{0.8})_{75}P_{16}B_6Al_3$ alloys at high magnetic fields. Temperature are shown on the appropriate curves; T_C is 85 K

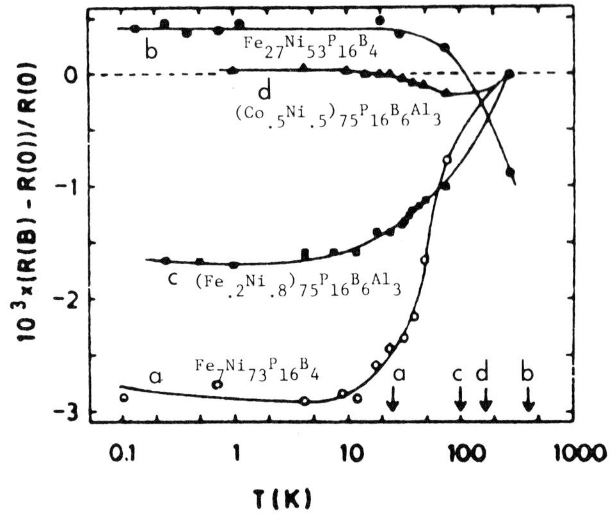

Figure 21.19 Magnetoresistance of some amorphous alloys through their magnetic transitions: *a*, $Fe_7Ni_{73}P_{16}B_4$; *b*, $Fe_{27}Ni_{53}P_{16}B_4$; *c*, $(Fe_{0.2}Ni_{0.8})_{75}P_{16}B_6Al_3$; *d*, $(Co_{0.5}Ni_{0.5})_{75}P_{16}B_6Al_3$
(after Gudmundsson *et al.*[50])

number of alloys[50] fits into the above picture as shown in *Figures 21.18* and *21.19*. These measurements taken in fields up to 180 kOe (*Figure 21.18*), and in some cases down to 100 mK (*Figure 21.19*) clearly show that the magnetoresistance can be large and comparable with that of crystalline alloys, and that it is negative over a wide range of temperature centred around T_C. The persistence of the negative magnetoresistance much above T_C is a strong indication of spin-correlation effects playing a significant part in the amorphous systems. A detailed quantitative analysis of these magnetoresistance data is however not easily approachable at present for disordered systems.

In summary, while resistance minimum is admittedly of magnetic origin the exact interplay between various possible mechanisms is still not tractable quantitatively. However, the coherent-exchange scattering approach does give a single unified picture.

21.7 Hall effect

The close analogy of glassy metals with the liquid state, rather than that of the crystalline state, suggests that the Hall coefficient of simple metallic glasses should be nearly free-electron like. To discuss in most general terms, we shall consider the Hall resistivity of a soft magnetic material. The Hall resistivity measured in an applied field B_a for such a material is given by

$$\rho_H = R_0 B_a + \mu_0 M_s [R_1 - N R_0]$$
$$= R_0 B_a + \mu_0 M R_s \quad (21.31)$$

where M is the internal magnetization, N the demagnetization factor, R_0 the ordinary Hall coefficient, and R_1 (or R_s) the extra-ordinary or (spontaneous) Hall coefficient. In amorphous materials it is found that $R_1 \gg R_0$. For B_a well below technical saturation

$$R_H = [\partial \rho_H / \partial B_a]_{\beta a \to 0} = R_1$$

TABLE 21.1 Room-temperature Hall resistivity data for transition metal alloys

	$\frac{\rho}{\mu\Omega\,m}$	$\frac{\mu_0 M_s}{T}$	$\frac{10^8 R_1}{m^3 A^{-1} s^{-1}}$	$\frac{10^8 R_{HF}}{m^3 A^{-1} s^{-1}}$
Amorphous				
$Fe_{75}G_{25}$*	1.8	1.34	6.3	<0.12
$(Fe_{0.8}Ni_{0.2})_{75}G_{25}$	1.7	1.10	6.5	0.05
$(Fe_{0.5}Ni_{0.5})_{75}G_{25}$	1.8	0.64	6.1	0.06
$(Fe_{0.4}Ni_{0.6})_{75}G_{25}$	2.3	0.33	6.1	0.12
$(Fe_{0.3}Ni_{0.7})_{75}G_{25}$	1.9	—	(1.1)†	0.19
$(Fe_{0.2}Ni_{0.8})_{75}G_{25}$	1.8	—	(0.2)†	0.07
$Ni_{75}G_{25}$	1.3	—	( )†	−0.006
$(Co_{0.2}Ni_{0.8})_{75}G_{25}$	1.6	—	( )†	−0.003
$(Co_{0.6}Ni_{0.4})_{75}G_{25}$	1.6	—	(0.8)†	<0.10
$Co_{75}G_{25}$	1.5	0.66	3.7	<0.01
Polycrystalline				
Ni[a]	0.08	—	−0.06	−0.006
$Fe_{0.68}Ni_{0.32}$[b]	0.83	—	2.34	0.09
$Fe_{0.38}Ni_{0.62}$[b]	0.20	—	0.07	−0.02

*G represents the Glass former of composition $P_{16}B_6Al_3$.
†T_C < 300 K.
[a]Malmhall *et al.*, ref. 96.
[b]Jellinghaus and de Andres, *Ann. Phys.* 7, 187 (1960).

At high fields

$$R_{HF} = [\partial \rho_H / \partial B_a]_{B_a \gg M_s}$$

and R_H should be equal to R_0. This is specially true for amorphous systems in which the high field susceptibility is non-negligible. The absolute value of R_1 is not known to better than 10–20 per cent.

Table 21.1 lists the values of R_1, R_{HF}, and electrical resistivity of a number of amorphous ribbons as well as for a few polycrystalline Fe–Ni alloys. One notices that R_{HF} is positive for all the iron-rich amorphous alloys. As one goes to the nickel rich end R_{HF} at 300 K is found to be negative. *Figure 21.20* shows how ρ_H varies with B_a for some nickel rich amorphous alloys. Neglecting the behaviour at low fields, we see that at high fields all have negative values for R_{HF}. The numerical value -0.6×10^{-10} m^3/A s is found to be close to the value of high temperature R_0 for polycrystalline nickel. The

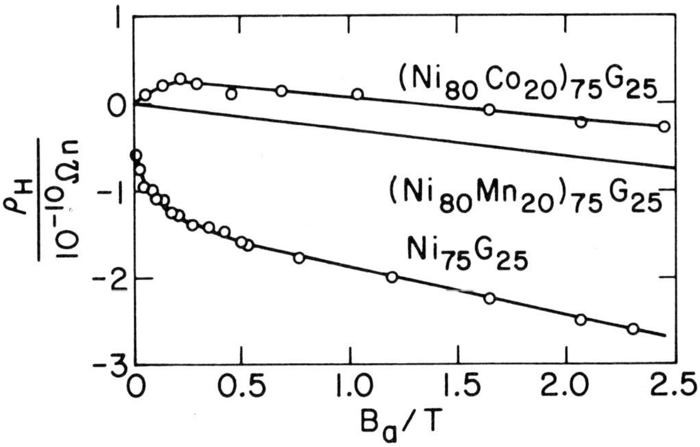

Figure 21.20 Hall resistivity of three nickel-rich amorphous glasses (after Rao *et al.*[20])

positive R_0 of liquids and glassy metals containing transition and rare earth atoms is still an unexplained fact. R_0 data can be useful in comparing other transport data with theory. Of course, the results can be quite embarrassing to live with. We have seen earlier how the resistivity and thermopower of many non-magnetic amorphous alloys can be explained on the basis of the Ziman model which is based on the free-electron theory. Quite often in such systems R_0 is found to be positive as for example in the case of zirconium-based alloys, METGLAS 2204 (*Figure 21.8*), etc. R_0 data for PdSi and PdSiCu alloys[91] are found to give an effective electron ratio of Pd/Si $\simeq 0.38$ and 4 respectively. This result is consistent with the specific heat data which showed no minimum in the density of states but contradicts the condition $2k_F = k_p$ expected on the basis of resistivity and photoemission studies[42]. Similarly, R_0 in amorphous La–Ga is found to be positive[92] which extrapolates to a value observed for liquids rather than for the crystalline phase of the material. There is no satisfactory explanation for these positive Hall coefficients either in the amorphous or liquid state.

The first measurement of the Hall resistivity of a magnetic amorphous alloy was made on $Fe_{80}P_{13}C_7$ by Lin[93]. Recently, extensive Hall effect studies have been carried out mainly by McGuire[94] and Malmhäll *et al.*[38]. In view of comprehensive reviews[94,20] on this topic only a few salient features will be touched upon. The anomalous Hall coefficient R_s is found to be two orders of magnitude larger in amorphous magnetic materials than in crystalline equivalents! Furthermore, below the Curie temperature R_H is found to be almost temperature independent. In contrast, for crystalline systems, like rare earths, with comparable electrical resistivity, R_1 is found to be strongly temperature dependent. The temperature dependence of R_1 thus has been used as an excellent tool to follow the process of crystallization, changes in short-ordering, etc., on heat treatment of the sample. *Figure 21.21* illustrates this point in showing the temperature dependence of R_H for METGLAS 2826 studied by Malmhäll *et al.*[95]. The curves marked 1 and 2a are for the amorphous state showing a sharp drop in R_H above the ferromagnetic transition around 250 K. Curve 2b already displays the onset of crystallization when R_H was measured after *annealing* the sample at 600 K as compared with 2a in which the sample was only thermally cycled to 600 K. Further annealing at

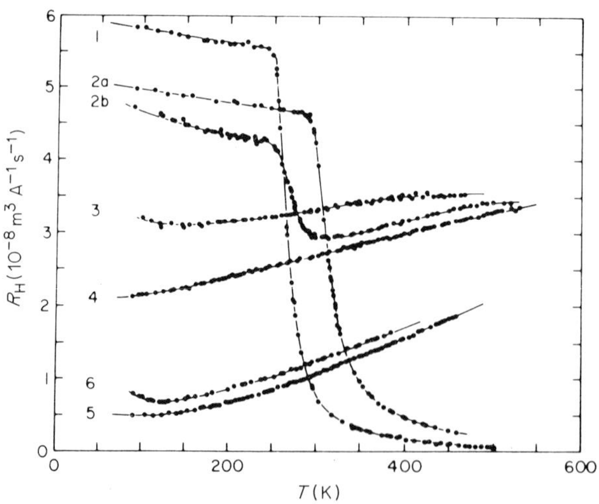

Figure 21.21 Hall resistivity of METGLAS 2826A under various heat treatments described in the text until it was crystallized (after Malmhäll *et al.*[95])

higher temperatures results in dramatic changes in the temperature dependence until the completely crystallized state is reached at which the curve 5 is reproduced. Changes in R_H can be a much more sensitive probe to study thermal effects on amorphous magnetic materials than electrical resistivity. The temperature dependence of R_H above T_C has been used to deduce the magnetic susceptibility via the expression

$$\chi = [R_H - R_0]/[R_1 - R_H],$$

where $R_H(T_C) = R_1$. The susceptibility determined by this technique has been used to determine the critical exponent[38,96] γ which was found to be equal to 1.6. This value is found to be in agreement with independent magnetic studies[82].

Finally, as in crystalline systems, R_s is related to the electrical resistivity via

$$R_s = a\rho + b\rho^2 \qquad 21.32$$

Here, the first term, which is important only at low temperatures, is due to skew-scattering originally proposed by Smit[97], and the second term arises from a 'side-jump' mechanism proposed by Berger[98]. From extensive studies on a number of systems[94,96,38] the magnitude of the side-jump evaluated from the coefficient b is found to be of the same magnitude as in crystalline systems, supporting the view that the magnitude of this side-jump is rather insensitive to the nature of the scattering centre.

Hall resistivity studies have thus been a valuable tool to study amorphous materials.

21.8 Conclusion

The Ziman theory devised for pure liquids in the weak scattering limit can be successfully applied with suitable modifications to a large number of amorphous solids. However, its success depends mostly on the assumption that the Ziman criterion $2k_F = k_p$ is satisfied. This assumption has not been substantiated. Furthermore, inconsistencies in the detailed quantitative interpretation of the transport data strongly suggest the need for further partial structure factor studies using both neutron and X-ray techniques as has been attempted by Clarke *et al.*[99]. As discussed in the text, at present there are a number of amorphous systems (lanthanum-based alloys for example) whose transport properties do not correlate in the framework of the Ziman theory. Thermopower studies have helped distinguish those which fit into the Ziman model.

Transport properties of magnetic amorphous alloys have been approached from many points of view. As suggested in this chapter, in the context of the complicated magnetic phase diagrams of amorphous systems, the coherent-exchange-scattering model appears to give a consistent picture for the observed transport behaviour. Clearly, further neutron studies to explore the magnetic structure of these systems would help resolve alternative explanations. Finally, the Kondo problem in dilute amorphous alloys (ppm range) well characterized and prepared from high purity starting materials is still to be explored.

In amorphous and disordered metals with $\rho \gtrsim 150\,\mu\Omega$, the electronic states are probably close to being localized. Recently a number of theories[100,101] have suggested that incipient Anderson localization could be responsible for the various electron transport anomalies like the Mooij correlation[2]. Existing extensive magneto-resistance and thermopower studies do not support the possible interpretation of the electrical resistivity data along these lines for metallic glasses. Clearly, this interesting approach will be the field of intense research.

References

1. For an excellent review *see:* Black, J. L., *'Metallic Glasses I'*. Eds. Güntherodt, H. J. and Beck, H., pp. 167—189, Springer-Verlag, New York (1980)
2. Mooij, J. H., *Phys. Status Solidi,* **A17,** 521 (1973)
3. Mogro-Compero, A. and Luborsky, F. E., *J. appl. Phys.,* **52,** 515 (1981)
4. Lin, C. H., Bev., K. J. and Turnbull, D., *Solid St. Communs,* **29,** 641 (1979)
5. Balanzat, M., *Scripta Met.,* **14,** 173 (1980)
6. Mogro-Compero, A. and Walter, J. L., *J. Phys. Colloque,* **41,** C8 497 (1980)
7. Teoh, W., Teoh, N. and Arajs, S., *'Amorphous Magnetism II',* Eds. Levy, R. and Hasegawa, R., p. 327, Plenum Press, New York (1977)
8. Malmhäll, R., Bäckström, G., Bhagat, S. M. and Rao, K. V., *J. Non-cryst. Solids,* **28,** 159 (1978)
9. Rao, K. V., Gudmundsson, H., Aström, H. U. and Chen, H. S., *J. appl. Phys.,* **50,** 1592 (1979)
10. Liang, V. K. C. and Tsuei, C. C., *Phys Rev,* **B7,** 3215 (1973); Hasegawa, R. and Tsuei, C. C., *Phys. Rev,* **B2,** 1631 (1970)
11. Ernst, U., Felsch, W. and Samwer, K., *J. Magnetism magn. Mater.* **15—18,** 1375 (1980)
12. Asomoza, R., Campbell, I. A., Fert, A., Liénard, A. and Rebouillat, J. P., *J. Phys. F,* **9,** 349, (1979)
13. Fert, A. and Asomoza, R., *J. appl. Phys,* **50,** 1886 (1979)
14. Berrada, A., Durand, Mizoguchi, T., Budnick, J., Loegel, B., Ousset, J. C., Askenazy, S. and Güntherodt, H. J., *'Proc. 4th Int. Conf. on Rapidly Quenched Metals',* Eds. Masumoto, T. and Suzuki, K., p. 829, Japan Institute of Metals, Sendai (1982)
15. Poon, S. J., Durand, J. and Yung, M., *Solid St. Communs,* **22,** 475 (1977)
16. Fukamichi, K., Gambino, R. J. and McGuire, T. R., *J. appl. Phys.,* **53,** 2310 (1982)
17. Chien, C. L., Unruh, K. M., Levy, A., Liou, S. H., Stokes, J. P., Gambino, R. J. and Fukamichi, K., *J. appl. Phys.,* **53,** 2307 (1982)
18. Rao, K. V., Egami, T., Gudmundsson, H., Åström, H. U., Chen, H. S. and Nägele, W., *J. appl. Phys.,* **52,** 2187 (1981)
19. Nagel, S. R., Vassilov, J., Horn, P. M. and Giessen, B. C., *Phys. Rev.,* **B17,** 462 (1978)
20. Rao, K. V., Malmhäll, R., Bhagat, S. M., Bäckström, G. and Chen, H. S., *IEEE Trans. Magn.,* **MAG-16,** 896 (1980)
21. Szofran, F. R., Gruzalski, G. R., Weymouth, J. W., Sellmyer, D. J. and Giessen, B. C., *Phys. Rev.,* **B14,** 2160 (1976)
22. Buschow, K. H. J. and Beckmans, N. M., *Phys. Rev.,* **B19,** 3843 (1979)
23. *See,* for example, Rapp, Ö., Lindberg, B., Chen, H. S. and Rao, K. V., *J. less-Common Metals,* **62,** 221 (1978)
24. Güntherodt, H. J., *'Metallic Glasses',* Festköperprobleme, *Adv. Solid State Physics,* **17,** 25 (1977)
25. Ziman, J. M., *Phil. Mag.,* **6,** 1014 (1961)
26. Evans, R., Greenwood, D. A. and Lloyd, P., *Phys. Lett.,* **A35,** 57, (1971); **A38,** 51 (1972)
27. Nagel, S. R., *Phys. Rev.,* **B16,** 1694 (1977)
28. Meisel, L. V. and Cote, P. J., *Phys. Rev.,* **B16,** 2978 (1977); *Phys. Rev. Lett.,* **39,** 102 (1977)
29. Froböse, K. and Jäckle, J. J., *J. Phys. F,* **7,** 2331 (1977)
30. Boucher, B. Y., *J. Non-cryst. Solids,* **7,** 277 (1972)
31. Sinha, A. K., *Phys. Rev.,* **B1,** 4541 (1970)
32. Tangonan, G. L., *Phys. Lett.,* **54A,** 307 (1975)
33. Oberle, R., Künzi, H. U., and Güntherodt, *Phys. Lett.,* **58A,** 272 (1976)
34. Korn, D., Pfeifle, H. and Zibold, G., *Z. Phys.,* **270,** 195 (1974)
35. Mueller, R., Agyeman, K. and Tsuei, C. C., *Phys. Rev.,* **B22,** 2665 (1980)
36. Meisel, L. V. and Cote, P. J., *Phys. Rev.,* **B15,** 2970 (1977)
37. Esposito, E., Ehrenrich, H. and Gelatt, C. D. (jun.), *Phys. Rev.,* **B18,** 3913 (1978)
38. Malmhäll, R., PhD Thesis, University of Umea, Sweden; Waseda, Y. and Chen, H. S., *Phys. stat. Solidi,* **87,** 777 (1978)
39. Egami, T. and Ichikawa, T., *Mater. Sci. Engng,* **32,** 293 (1978)
40. Nagel, S. R. and Tauc, J., *Phys. Rev. Lett.,* **35,** 380 (1975)
41. Nagel, S. R., Fisher, G. B., Tauc, J. and Baley, B. G., *Phys. Rev.,* **B13,** 3284 (1976)
42. Mizutani, U., Harting, K. T., Massalski, T. B. and Hooper, R. W., *Phys. Rev. Lett.,* **41,** 661 (1978)
43. Korn, D., Mürer, W. and Zibold, G., *Phys. Lett.,* **47A,** 117 (1974)
44. Mott, N. F., *Phil. Mag.,* **26,** 1249 (1972)
45. Hasegawa, R., *Phys. Lett.,* **38A,** 5 (1972)
46. Kaiser, A. B. and Doniach, S., *Int. J. Magn.,* **1,** 11 (1970)
47. Zuckermann, M. J., *J. Phys. F,* **2,** L25 (1972)
48. Cochrane, R. W., Harris, R., Ström-Olson, J. O. and Zuckermann, M. J., *Phys. Rev. Lett.,* **35,** 670 (1975)

49. Rapp, Ö., Bhagat, S. M., Johannesson, Ch., *Solid St. Communs,* **21,** 83 (1977)
50. Gudmundsson, H., Rao, K. V. and Anderson, A. C., *'Liquids and Amorphous Metals',* Eds. Lüscher, E. and Coufal, H., p. 653, Sijthoff and Noordhoff, Amsterdam (1980)
51. Black, J. L. and Gyorffy, B. L., *Phil. Mag.,* **B40,** 331 (1979)
52. Stewart, A. M. and Phillips, W. A., *Phil. Mag.,* **B37,** 561 (1978); Kondo, J., *Physica,* **84B,** 207 (1976)
53. Rayne, J. A. and Levy, R. A., ref. 7, p. 319
54. Tsuei, C. C., *Solid St. Communs,* **27,** 691 (1978)
55. *See* for example, Barnard, R. D., *'Thermoelectricity in Metals and Alloys',* Taylor and Francis, London (1972)
56. Nagel, S. R., *'Proc. Conf. on Metallic Glasses: Science and Technology',* Eds. Hargitai, C., Bakanyi, I. and Kemeny, T., p. 133, Central Research Institute for Physics, Budapest (1980)
57. Nagel, S. R., *Phys. Rev. Lett.,* **41,** 990 (1978)
58. Carini, J. P., Basak, S. and Nagel, S. R., *J. Phys. Colloque,* **41,** C8, 463 (1980)
59. Baibich, M. N., Muir, W. B., Bellanger, G., Destry, J., Elzinga, H. S., and Schroeder, P. J., *Phys. Lett.,* **73A,** 328 (1979)
60. Carini, J. P., Basak, S., Nagel, S. R., Giessen, B. C. and Tsai, C. L., *Phys. Lett.,* **81A,** 525 (1981)
61. Armbüster, H. and Naugle, D. G., *Solid St. Communs,* **39,** 675 (1981)
62. Bhanu Prasad, B. and Bhatnagar, A. K., *Solid St. Communs,* **41,** 381 (1982)
63. Basak, S., Nagel, S. R. and Giessen, B. C., *Phys. Rev.,* **B121,** 4049 (1980)
64. Grest, G. S. and Nagel, S. R., *Phys. Rev.,* **B19,** 3571 (1979)
65. Kondo, J., *Solid St. Phys.,* **23,** 183 (1967)
66. Hamann, D. R., *Phys. Rev.,* **158,** 570 (1967)
67. Tsuei, C. C. and Hasegawa, R., *Solid St. Communs,* **7,** 1581 (1969)
68. Hasegawa, R., *J. appl. Phys.,* **41,** 4096 (1970)
69. Hasegawa, R. and Tsuei, C. C., *Phys. Rev.,* **B3,** 214 (1971)
70. Hasegawa, R., *J. Phys. Chem. Solids,* **32,** 2487 (1971)
71. Dublon, G., *Phys. Status Solidi (a),* **60,** 287 (1980)
72. Zentko, A., Duhaj, P., Potočky, L., Tima, T. and Bānsky, J., *Phys. Status Solidi, (a),* **31,** K41 (1975)
73. Rao, K. V., Steinback, M., Anderson, A. C. and Chen, H. S., to be published
74. Hasegawa, R., *Phys. Rev. Lett.,* **28,** 1376 (1982)
75. Dierker, H., Gudmundsson, H. and Anderson, A. C., *Solid St. Communs,* **29,** 767 (1979)
76. Asomoza, R., Fert, A., Campbell, I. A. and Meyer, R., *J. Phys. F,* **7,** L327 (1977)
77. De Gennes, P. G. and Friedel, J., *J. Phys. Chem. Solids,* **4,** 71 (1958)
78. Bhattacharjee, A. K. and Coqblin, B., *J. Phys. F,* L221 (1978)
79. Fert, A. and Asomoza, R., *J. appl. Phys.,* **50,** 1886 (1979), and references therein
80. Sherrington, D. and Kirkpatrick, S., *Phys. Rev. Lett.,* **35,** 1792 (1975)
81. Rao, K. V. and Chen, H. S., ref. 14, p. 829
82. Yeshurun, Y., Salamon, M. B., Rao, K. V. and Chen, H. S., *Phys. Rev.,* **B24,** 1536 (1981)
83. Poon, S. J. and Durand, J., ref. 7, p. 245
85. Anderson, P. W., *Mater. Res. Bull.,* **5,** 549 (1970)
86. Nägele, W., Knorr, K., Prandl, W., Convert, P. and Bevoz, J. L., *J. Phys. C,* **11,** 3295 (1978)
87. Gudmundsson, H., Rao, K. V., Åström, H. U. and Chen, H. S., to be published
88. Källbäck, O., Gudmundsson, H., Rao., K. V. and Åström, H. U., *Physica Scripta,* **25,** 755 (1982)
89. Richter, R., Wolf, M. and Goedsche, F., *Phys. Status Solidi, (b),* **95,** 473 (1979)
90. Mogro-Compero, A. and Walter, J. L., G.E. Report No. 81CRD082 (1981)
91. Mizutani, U. and Massalski, R. B., *Phys. Rev.,* **B21,** 3180 (1980)
92. Calter, P. C., Adair, T. W. (III), and Naugle, D. G., *Phys. Rev.,* **B20,** 2959 (1979)
93. Lin, S. V. H., *J. appl. Phys.,* **40,** 2175 (1969)
94. McGuire, T. R., Gambino, R. J., O'Handley, R. C., 'Hall Effect in Amorphous Metals', in *'Hall Effect and its Applications',* Eds. Chien, C. L. and Westgate, C. R., p. 137, Plenum Press, New York (1978)
95. Malmhäll, R., Bäckstrom, G., Bhagat, S. M. and Rao, K. V. *J. Non-cryst. Solids,* **28,** 159 (1978)
96. Malmhäll, R., Bhagat, S. M., Rao, K. V. and Bäckström, G., *Phys. Status Solidi, (a),* **53,** 641 (1979)
97. Smit, J., *Physica,* **24,** 39 (1958)
98. Berger, L., *Phys. Rev.,* **B2,** 4559 (1970)
99. Clarke, R., Nagel, S. R., Hitterman, R. L. and Mueller, M. H., *Solid St. Communs,* **36,** 751 (1980)
100. Altshuler, B. L. and Aronov, A. G. *Soviet Phys. JETP,* **50,** 968 (1979)
101. Y. Imry, *J. appl. Phys.,* **52,** 1817 (1981) and references therein

Chapter 22

Superconducting properties of amorphous metallic alloys

S. J. Poon

Department of Physics, University of Virginia, Charlottesville, Virginia, USA

22.1 Introduction

Almost three decades ago, Buckel and Hilsch[1] discovered the first amorphous superconductors of bismuth and gallium by vapour deposition on to cryogenic substrates. Since then, the superconducting properties of simple metal thin films have been studied rather extensively[2]. It was not until a decade ago that the first systematic study of amorphous transition metal superconductors was undertaken[3]. In 1975, the first series of amorphous superconductors obtained by liquid quenching was reported[4]. The discovery of this new class of bulk superconductor has added a new dimension to the research and industry of metallic glasses.

It is the purpose of this chapter to survey the recent developments in the area of amorphous superconductors. For earlier information, the readers can refer to other reviews[2,5]. Materials which have already been discussed by them will not be covered here. The format of this chapter is as follows. Occurrence of superconductivity in simple-metal and transition-metal alloys will be discussed in section 22.2. Comparison with theories will be made whenever possible. Section 22.3 discusses the effects of inhomogeneities on the superconducting properties of metallic glasses. Emphasis will be placed on paraconductivity and upper critical field. Possible correlation with low angle X-ray scattering will be mentioned. Results on thermal relaxation and high pressure studies will be presented in section 22.4 and discussed, and the effect of quenching rate and substrate temperature on the as-quenched amorphous state will be reported.

22.2 Occurrence of superconductivity in amorphous alloys

Let us first define some parameters to be used. Following the strong-coupling theory of superconductivity, the relevant parameters in determining the transition temperature (T_c) are given by the general moment $\langle\omega^n\rangle$ of the Eliashberg function $\alpha^2F(\omega)$[6,7]

$$\langle\omega^n\rangle = \frac{2}{\lambda}\int_0^\infty \alpha^2F(\omega)\,\omega^{n-1}\,d\omega \qquad (22.1)$$

where

$$\lambda \equiv 2\int_0^\infty \frac{\alpha^2 F(\omega)}{\omega}\,\mathrm{d}\omega = \frac{N(0)\langle I^2\rangle}{M\langle\omega^2\rangle} \tag{22.2}$$

is the dimensionless electron–phonon coupling constant first obtained by McMillan.[8] The symbols have their usual meanings.

22.2.1 Simple metal alloys

Strong-coupling superconductors

In amorphous metals where lattice periodicity is absent, the electronic and phononic states must have very complicated forms. In simple metals where the pseudopotential theory is well established, one can apply the Faber–Ziman liquid metal theory[9] of nearly free electrons (NFE) to amorphous metals. It has been shown that this approach accounts rather well for the transport and optical properties of amorphous and liquid simple metals[2,10]. To obtain the Eliashberg function, one also needs to know the phonon dispersion relation. Jäckle and Froböse[11], and Meisel and Cote[12] obtained the following expression in the pseudopotential (V_k) formalism

$$\alpha^2 F(\omega) = \frac{1}{(2\pi)^3 h^2 n_\mathrm{i} v_\mathrm{F}}\int_0^{2k_\mathrm{F}} kV_\mathrm{k}^2 \frac{S_\mathrm{inel}(k,\omega)}{1+n(\omega)}\,\mathrm{d}k \tag{22.3}$$

where n_i denotes the ionic number density, v_F and k_F the Fermi velocity and wave vector, $S_\mathrm{inel}(k, \omega)$ is the inelastic part of the dynamic structure factor of the amorphous network, and $n(\omega)$ the Bose function. In the low-ω region where the continuum model for acoustic phonons is valid, one can obtain an explicit expression for $\alpha^2F(\omega)$ in terms of material parameters. Earlier, an expression was derived by Bergmann[13] and Poon and Geballe[14] in the form

$$\alpha^2 F(\omega) = \frac{e^2 E_\mathrm{F} k_\mathrm{F}^2 N(0)\rho}{3\pi^3\langle v^3\rangle mM}\,\omega \tag{22.4}$$

where ρ is the resistivity, $\langle v^3\rangle$ is the mean cube sound velocity, and m is the electronic mass. Thus $\alpha^2F(\omega)$ was found to be linear in ω in the low-ω regime. This result compares well with tunnelling data[2]. The general enhancement of the electron–phonon coupling constant λ in amorphous metals over their crystalline counterparts is caused by the significant increase of $\alpha^2F(\omega)$ for low energy phonons. This also explains the tendency towards strong coupling behaviour (i.e. $\lambda > 1$) as the degree of disorder increases in simple metals. The linear dependence of $\alpha^2F(\omega)$ on ω originates from the expanded phase space in the NFE model for electrons scattering off acoustic Umklapp phonons[14]. This scattering process was termed 'pseudo Umklapp process'[13,14] or 'diffuse Umklapp process'[11]. Umklapp phonons in amorphous metal were observed in a neutron scattering experiment on Mg–Zn alloys by Suck *et al.*[15]. Model calculation of vibrational spectrum was due to von Heimendahl[16]. The enhancement in electron–phonon interaction certainly favours superconductivity, but it alone does not necessarily guarantee high transition temperatures. From numerical studies of the Eliashberg equations[6,7], the coupling constant λ is not the only important parameter determining T_c. It is useful to have a general expression relating material parameters for the optimization of T_c. In what follows, we present a plausible expression and compare our analysis with available data.

Optimization of T_c.

For strong-coupling isotropic superconductors, Leavens[7] obtained the upper bound $T_c \leqslant C(\mu^*)\lambda\langle\omega\rangle$ where μ^* is the repulsive Coulomb pseudopotential. When the phonon frequency ω is normalized by the maximum frequency ω_0, $\alpha^2F(\omega)$ becomes $\alpha^2f(x)$ with $x = \omega/\omega_0$. By use of relations 22.1 and 22.2, Leaven's criterion becomes $T_c \leqslant C(\mu^*)N(0)\langle I^2\rangle/k\omega_0 M$ where

$$k = \frac{\int_0^1 x\alpha^2 f(x)\,\mathrm{d}x}{\int_0^1 \alpha^2 f(x)\,\mathrm{d}x}$$

For amorphous simple metals, the Hopfield–McMillan parameter $N(0)\langle I^2\rangle$ was shown[14] to be adequately given in the NFE approximation by the integral

$$\sim N(0)\left(\frac{1}{k_F}\right)^2 \int_0^{2k_F} V_k^2 k^3\,\mathrm{d}k$$

where V_k is the Fourier transform of the atomic pseudopotential[17]. The latter is approximately proportional to $k_F^6 N(0)$. This can further be reduced to

$$N(0)\langle I^2\rangle \propto n^{4/3}z$$

in the NFE theory, n being the density of electrons. Grouping the relevant terms together, one obtains

$$(T_c)_{max} \propto \frac{C(\mu^*)n^{4/3}z}{kM\omega_0} \tag{22.5}$$

Except for pathological shapes of $\alpha^2F(\omega)$ which are unlikely for amorphous metals, k can be treated as a constant. In *Figure 22.1* we have plotted $(T_c)_{max}$ against $n^{4/3}z/M\omega_0$. For materials with unknown ω_0, we take $\omega_0 \propto \theta_D$, the Debye temperature. As pointed out by Leavens, lead is among the few superconductors reaching their optimal T_c. The open squares denote elements in their crystalline state. The three allotropic forms of

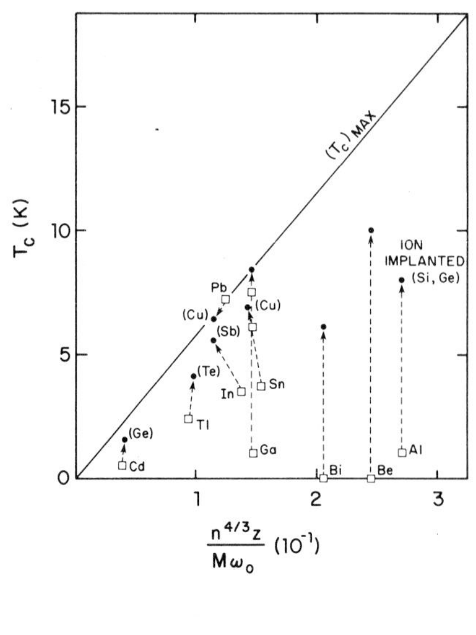

Figure 22.1 Maximum predicted T_c as a function of material parameters. n is the density of electrons, z is the valency, M is the atomic mass, and ω_0 is the maximum phonon energy. Elements in parenthesis stabilize the amorphous or disordered host metals

gallium are also included. Dashed lines show changes in T_c as the elements are stabilized (by a second constituent) in their amorphous or 'disordered' forms. General enhancements of T_c towards the optimal values are observed. Included also are beryllium with an unknown structure and ion-implanted aluminium. For quench condensed bismuth, it was found that the NFE model was inadequate to explain its transport properties[2]. It seems that Al–Be alloys, be they in 'disordered' or amorphous form, have the potential of achieving relatively high T_c. It is noteworthy that the optimization of T_c as given in expression 22.5 does not distinguish between the origins of T_c enhancement, whether it is due to the softening of phonon modes or an increase in $N(0)\langle I^2 \rangle$. The results are direct consequences of the nearly free-electron model and the strong-coupling theory of superconductivity.

22.2.2 Transition metal alloys

Electron–phonon interaction

The electronic states in *d*-band metals are more appropriately described by the tight-binding model. Using the Barisić, Labbé, and Friedel[18] rigid-atom approximation, the author extended the calculation of $\alpha^2F(\omega)$ to the nearly-localized-electrons case[19]. Again, an explicit expression was given only for acoustic phonons. It was found that $\alpha^2F(\omega) \propto \omega^3$. The proportionality constant was given by a double sum over momentum space weighted appropriately by momentum distribution functions of electrons. This finding was in agreement with the tunnelling data of Kimhi and Geballe on amorphous molybdenum and niobium[20]. One can compare the present result to the linear behaviour in simple metals. This contrasting feature is the consequence of the interplay of 'additional' scattering processes and the range of localization of electron–phonon interaction L. In simple metals L is greater than a (where a is the interatomic spacing) while in transition metals L is less than or about equal to a. For long wavelength perturbation with $q^{-1} > L$, electrons couple ineffectively to phonons in the presence of disorder. The latter bears similarity with the Pippard–Ziman condition (as stated by Ziman[21]): 'Phonons whose wavelengths q^{-1} exceeds the electron mean free path Λ are ineffective electron scatterers'. The present result is a direct consequence of the localization of d electrons in the tight-binding approximation. Hence, electron–phonon coupling benefits less from 'additional' scatterings in transition metals than in simple metals. The attenuation of electron–phonon interaction at low ω is consistent with the observation that amorphous transition metal alloys are weak-coupled or intermediate-coupled superconductors (i.e. the ratio $2\Delta/k_BT$ near BCS value)[20,22]. On the other hand, using the Pippard function[23], Meisel and Cote[12] obtained $\alpha^2F(\omega) \propto \omega^2$, a result familiar in crystalline metals. The small reduction in electron–phonon interaction was due to the absence of *d*-band structure from the Pippard function. The latter was derived from the NFE model with a finite electronic mean free path. Therefore, the reduction of electron–phonon interaction in amorphous *d*-band metals might have been underestimated.

Alloys and their T_c

As mentioned earlier, the first systematic study of amorphous transition metal superconductors was performed on quench condensed thin films[3]. Since the discovery of the first liquid quenched superconductors[4], several dozen more have been added to the list of bulk superconductors. References will be given in the following discussion. In general, the binary alloys investigated so far can be categorized into three groups: ones that are made of transition metals only, others made of transition metals and noble metals and transition metals and metalloids. In addition, there have been studies in ternary alloys. A summary of transition temperatures for both vapour-quenched and liquid-quenched binary alloys (containing no metalloid) based on the 4*d* series as a function of group number (*e*/*a* ratio) is reproduced from Poon and Carter[24]. This is shown in *Figure 22.2*. The references in this figure refer to those of reference 24. The T_c

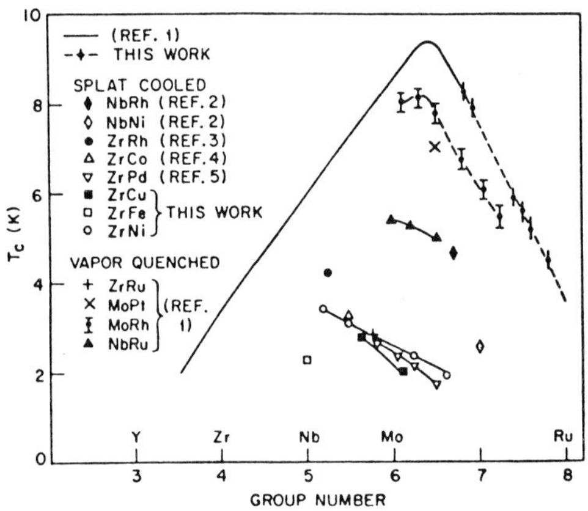

Figure 22.2 Superconducting transition temperature as a function of group number for alloys based on the 4*d* series. References refer to those of Poon and Carter[24]

envelope is the Collver–Hammond curve corrected for Mo–Ru alloys. Alloys of the 5*d* series follow a similar trend. Since then, systematic studies on liquid quenched ternary alloys made of niobium, tantalum, rhodium and iridium have been reported[25]. Recently, the superconducting properties of Zr–Cu and Zr–Ni alloys have been analysed by other authors[26,27]. For the transition metal–metalloid system, both vapour-quenched (e.g. Mo–N) and sputtered refractory metal–metalloid[22,28] (e.g. Mo–Ge) films have been produced and studied. Several years ago, the first liquid-quenched refractory metal–metalloid (e.g. Mo–Ru–P) superconducting glasses were reported[29]. The transition temperature in superconducting metallic glasses was then raised from 4.7 K in Nb–Rh[4] to near 10 K in this new class of alloy. The significant increase in T_c, together with their thermal stability (crystallization temperature near 700 °C), and attractive mechanical properties have greatly enhanced the potential of superconducting glasses for technological applications and basic research.

More recently, additional refractory metal–metalloid glasses have been produced and investigated by other groups[30,31]. Of special interest is the work of Strudthoff *et al.*[31], where it was found that by alloying $Mo_{0.6}Rh_{0.4}$ with 20 per cent of phosphorus, T_c was enhanced from around 3.7 K to 6.5 K. This is in contrast to the usual reductions in T_c when metalloids were incorporated into refractory metal alloys[24,32]. Summarizing, amorphous transition metal alloys with transition temperatures near 8 K are easily attainable. Superconductors with higher T_c (e.g. >10 K) remain to be found.

Comparison with microscopic theories

It is noted in *Figure 22.2* that alloys prepared from constituents (not containing metalloids) with increasing difference in *e*/*a* values tend to have progressively lower T_c than the Collver–Hammond curve. Earlier, it was pointed out[24] that this trend can be understood in terms of the Varma–Dynes empirical relations[33]. The Varma–Dynes result expresses the dimensionless electron–phonon coupling constant in terms of the tight-binding d-band parameters

$$\lambda \propto N(0)W(1 \mp 2S) \tag{22.6}$$

where W is the band width, S is the overlap integral, the upper sign corresponds to the bonding states in the lower half d-band and the lower sign corresponds to the antibonding states in the upper half d-band. The proportionality constant and parameter S can be deduced from figure 1 of reference 33. Meaningful comparison with relation 22.6 can only be made with elements from the same row in the transition metal series so that $\langle I^2 \rangle / M \langle \omega^2 \rangle$ can be taken as constant. Following Poon and Carter[24], we have plotted the ratio $\lambda/(1 \mp 2S)$ for the elements zirconium, niobium and molybdenum and alloy $Mo_{0.6}Ru_{0.4}$ in *Figure 22.3*. Data on λ are taken from specific heat[34,35] and tunnelling measurements[21]. Taking the proportionality constant to be about $(1.4/W)$, one obtains $N(0)$ for these systems to be 0.64, 0.78, 0.9 and 0.75 spin states per eV per atom respectively. Since specific heat data on amorphous elements are not available these results can only be compared with data on amorphous Zr–Pd[34] and Mo–Ru[35] alloys. The values are 0.61 and 0.8 respectively. Recently, a cluster model calculation for transition metal glasses was carried out by Zwicknagl[36]. He configurationally modelled the amorphous system with short-range order by incorporating different topological types of polyhedra filling space. For the band structure, features of the bonding and antibonding lobes remained while an additional maximum in the 'valley' emerged. This is characteristic of the energy level distribution for the 13-atom icosohedral cluster, a change from the short range order of the body centred cubic structure. He obtained $N(0)$ for niobium, molybdenum, and technetium to be 0.6, 0.7, and 0.62 respectively. Qualitatively speaking it can be said that T_c goes hand in hand with $N(0)$ in the $4d$ series.

In binary alloys, the contribution from $N(0)$ of the individual constituents to λ has been emphasized[37,38]. Recent photoemission results indicated that the band structure of amorphous binary alloys can be constructed from the superposition of the d-bands of

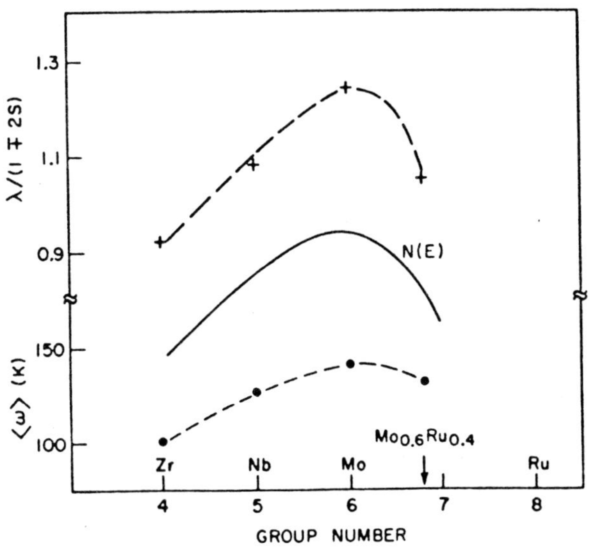

Figure 22.3 Microscopic parameters $\langle\omega\rangle$ and $\lambda/(1 \mp 2S)$ as a function of group number. The density of states curve is schematic

individual constituents[39,40]. In particular, split bands occur for elements separated far apart. By combining these facts and the one-to-one correspondence between T_c and $N(0)$ for the elements, the T_c 'depression' below the envelope by alloying two elements separated by $\Delta(e/a) \gtrsim 2$ can be qualitatively understood. However these results cannot be directly applied to ternary alloys composed of elements from different rows in the periodic table[25]. For ternary alloys containing metalloids, it is not clear what role the latter play in determining the electronic structure of the system. Magnetic susceptibility[29], specific heat and upper critical field[5,27] results, however, also indicate the correlation of T_c with $N(0)$. More amazing is the enhancement (almost double) in T_c in Mo–Rh–P mentioned beforehand[31]. As yet, there are inadequate data for the author to elaborate on this point.

Finally, the occurrence of superconductivity in amorphous alloys containing magnetic constituents ought to be mentioned. Poon[41], Tenhover *et al.*[42], and O. Rapp[43] have studied the systematics of superconductivity in Zr–3*d* metallic glasses. We reproduce in *Figure 22.4* the T_c values of $Zr_{70}X_{30}$ (X = 3*d* element) alloys as a function

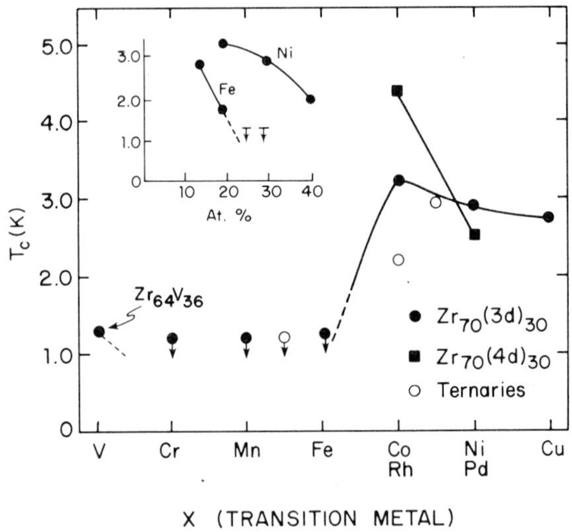

Figure 22.4 Superconducting transition temperatures of amorphous $Zr_{70}(3d)_{30}$ alloys from Tenhover and Johnson[42]. Inset shows T_c as a function of composition for Zr–Fe and Zr–Ni alloys from Poon[41]

of X from Tenhover and Johnson[42]. We have corrected the T_c value for $Zr_{70}Fe_{30}$ (i.e. $T_c < 1.2$ K). The T_c minimum is due to spin fluctuation and formation of localized magnetic moment in these alloys. In *Figure 22.4* we compare T_c results as a function of composition for Zr–Fe and Zr–Ni alloys taken from Poon[41]. A drastic decrease in T_c at relatively low iron concentration is noted. Meanwhile, photoemission data[39,44] and upper critical field results[42] show enhanced density of states around the T_c minimum. In the split band picture, as the 3*d* band moves towards E_F its contribution to $N(E_F)$ increases, thus increasing the $N(E_F)$ of the alloy. This favours the formation of localized moment. The dratic depression of T_c in Zr–Fe alloys can be understood as a precursor to moment formation, that is, spin fluctuation.

22.3 Degree of homogeneity and superconducting properties

In this section we first review and discuss results on paraconductivity and upper critical field measurements. The dependence of these properties on sample homogeneity is pointed out. The nature of quenched-in inhomogeneities and their possible origins are suggested. We also present results of X-ray scattering on as-quenched and irradiated samples. Correlations with superconducting properties are pointed out.

22.3.1 Paraconductivity

Suppression of the Maki–Thompson contribution

The excess electrical conductivity due to the fluctuation of the order parameter consists of two distinct terms; the Aslamazov–Larkin[45] (AL) term contributed from the virtual Cooper pairs created by thermal fluctuation, and the Maki–Thompson [46,47] (MT) term coming from the interaction of normal conduction electrons and the superfluid. Fluctuations near superconducting phase transitions have been reviewed by Skocpol and Tinkham[48]. Data on simple metals are also given there. Here we focus our attention on metallic glasses. The first studies of paraconductivity on bulk amorphous superconductors were carried out by Johnson and Tsuei[49]. They observed a universal behaviour of the excess conductivity as a function of temperature described by a plot of $\ln(\sigma'/\sigma_0)$ against $\ln\varepsilon$, where σ' and σ_0 are the excess conductivity and normal conductivity respectively, and $\varepsilon = (T - T_c)/T_c$ is the normalized reduced temperature. From their analysis, it was concluded that the MT contribution to σ' was absent from this class of bulk superconductor. A short-wavelength cut-off in the fluctuation spectrum of AL contribution was introduced to fit their data. Recently, additional bulk superconductors are measured and analysed[50]. The data are in general agreement with previous findings. From recent tunnelling data[20] and theoretical calculations[12,19], however, it can be argued that the MT term is very likely to exist in amorphous transition metal superconductors and that one has to invoke a short-wavelength cut-off to both the AL and MT contributions in order to explain the data reasonably well.

In the presence of pair breaking, the MT contribution to σ' is given by[46]

$$\sigma'_{MT} = \frac{4\sigma'_{AL}}{1 + \left(\frac{\varepsilon_c}{\varepsilon}\right)^{1/2}} \tag{22.7}$$

where ε_c is the pair breaking parameter, $\sigma'_{MT} \to 0$ as $T \to T_c$. When electron–phonon collision is the dominant contribution to pair breaking, ε_c is given by[51]

$$\varepsilon_c = \frac{\pi^2\hbar}{2k_B T_c}\int_0^\infty \frac{\alpha^2 F(\omega)}{\sinh\left(\frac{\omega}{k_B T_c}\right)}\,d\omega \tag{22.8}$$

For amorphous simple metals, ε_c was found to be in the range 1.5—2.2. The relatively large values of ε_c can be understood from the large enhancement in $\alpha^2F(\omega)$ in the region of interest, that is $\omega \lesssim k_B T_c$. However for transition metals, it was pointed out beforehand that such an enhancement in $\alpha^2F(\omega)$ was absent. From the data on amorphous molybdenum and niobium[20], ε_c is found to be about 0.5. σ'_{MT} is then less than half of σ'_{AL} only when $\varepsilon \lesssim 10^{-2}$ or $\ln\varepsilon \lesssim -4.5$. In the region where a large negative deviation from the AL contribution alone is observed, that is $-2 < \ln\varepsilon < 0$, then $\sigma'_{MT} > \sigma'_{AL}$ from equation 22.7. Even if we use a high value of $\varepsilon_c \simeq 2$, σ'_{MT} cannot be neglected for $\ln \varepsilon \gtrsim -3$. It is reasonable to introduce a short-wavelength cut-off in the fluctuation spectrum of the MT contribution. The following expression can be obtained

$$\sigma'_{MT} = \frac{8\sigma'_{AL}}{\pi}\left[\frac{1}{1+\left(\frac{\varepsilon_c}{\varepsilon}\right)^{1/2}}\tan^{-1}\frac{q_c\xi_0}{\varepsilon_c^{1/2}} + \frac{1}{1-\left(\frac{\varepsilon_c}{\varepsilon}\right)}\left(\tan^{-1}\frac{q_c\xi_0}{\varepsilon_c^{1/2}} - \tan^{-1}\frac{q_c\xi_0}{\varepsilon_c^{1/2}}\right)\right] \tag{22.9}$$

where q_c is the cutoff wave vector and ξ_0 is the zero temperature coherence length. Notice that the second term on the right hand side of equation 22.9 is always negative. One finds that in order to suppress $\sigma' = \sigma'_{MT} + \sigma'_{AL}$ for $-2 < \ln \varepsilon < 0$, q_c has to be around $0.5\,\xi_0^{-1}$. It turns out that σ'_{MT} rather than σ'_{AL} is a more important contribution to σ' in this region. Summarizing, in the high temperature regime ($T \sim 2T_c$), one cannot neglect the Maki–Thompson contribution even in the presence of significant pair breaking ($\varepsilon_c \simeq 0.5$—2). The general suppression of σ' suggests a breakdown of the mean field theory far away from T_c.

Paraconductivity near T_c and sample homogeneity

In *Figure 22.5*, we have plotted $\ln(\sigma'/\sigma_0)$ again $\log \varepsilon$ for as-quenched and annealed amorphous Zr_3Pd samples. For the two as-quenched samples #1 and #2, different behaviours are observed. Sample #2 follows the trend observed in amorphous Zr_3Rh[49],

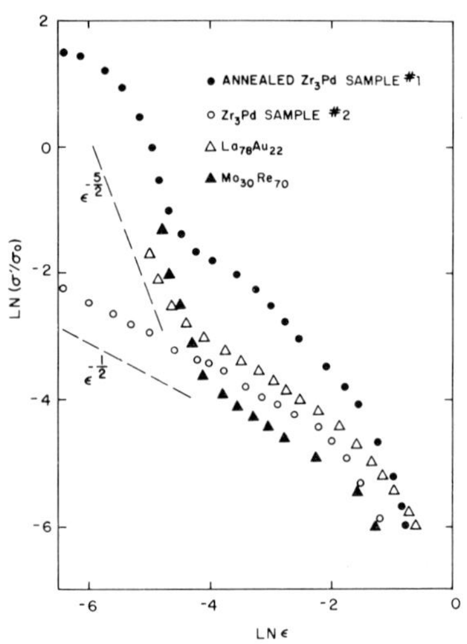

Figure 22.5 $\ln(\sigma'/\sigma_0)$ as a function of $\ln \varepsilon$ for four amorphous superconductors, where ε is $(T - T_{co})/T_{co}$. Data on $La_{78}Au_{22}$ and $Mo_{30}Re_{70}$ are taken from Johnson and Tsuei[49]. Data are to be compared with dashed lines for $\ln \varepsilon < -4$

while sample #1 follows the AL prediction near T_c but with a slightly enhanced σ' in the region $-6 < \ln \varepsilon < -1$. The transition width ΔT_c defined by the 0 per cent and 90 per cent points on the resistivity transition curve is about 4 mK for these samples. Both samples show no detectable crystallization from X-ray diffraction. However by annealing these samples at 280 °C for 10 hours, well below the estimated crystallization temperature of about 480 °C, onset of crystallization is observed in sample #1. This is accompanied by an increase of ΔT_c to around 30 mK. This suggests the presence of quenched-in inhomogeneities (probably crystallites) in this sample. Its paraconductivity resembles very much the behaviour of most bulk superconductors studied[45]. By introducing a distribution function which describes the T_c variation in the sample, the following proportionality can be derived[50] for T near T_{c0}

$$\frac{\sigma'_{AL}}{\sigma'_{AL}(T_{c0})} \propto \frac{\gamma^2 M_2}{(T - T_{c0})^{5/2}} \tag{22.10}$$

where T_{c0} is the temperature at which the sample resistivity vanishes, γ is the characteristic width of the T_c distribution function, and M_2 is the second moment of the distribution function. A straight line showing the relation $\ln(\sigma'_{AL}/\sigma_0) \sim (T - T_{c0})^{-5/2}$ is shown in *Figure 22.5*. On the other hand, statistical fluctuation in composition can also affect $\sigma'(T)$. However, the effect only becomes significant for $\log \varepsilon < -10$, far below our present temperature resolution.

In summary, for samples with a high degree of homogeneity in the amorphous state, the paraconductivity follows the AL prediction near T_c and it gradually decreases below the AL curve away from T_c. For samples with a lower degree of homogeneity, divergence in σ' is observed near T_c. Data on amorphous $La_{78}Au_{22}$ and $Mo_{30}Re_{70}$ taken from Johnson and Tsuei[49] are included in *Figure 22.5*. For homogeneous samples, the transition width ΔT_c due to paraconductivity alone can be written in the following form

$$\Delta T_c = 0.1811\, \rho_0^2 T_c{}^2 H'_{c2}(T_c) \tag{22.11}$$

where ρ_0 (Ωcm) is the normal state resistivity and $H'_{c2}(T_c)$ (Oe K^{-1}) is the upper critical field slope at T_c. It is estimated that ΔT_c for the otherwise homogeneous $La_{78}Au_{22}$ and $Mo_{30}Re_{70}$ samples should be about 1.3 and 12 mK respectively. The experimental values are an order of magnitude higher, indicating that the samples are rather inhomogeneous. While for the Zr_3Pd and Zr_3Rh alloys $\Delta T_c \simeq 3.5$ and 5 mK respectively, in reasonable agreement with observation. Thus simple measurement of the paraconductivity of amorphous superconductors might prove to be a very convenient way of identifying homogeneous samples.

22.3.2 Upper critical field

Let us be reminded that amorphous superconductors are type II superconductors. One notices that for inhomogeneous superconductors the characteristic distribution in T_c on a spatial scale of about $10\xi_0$ ($T \simeq 0.99\, T_{c0}$) is approximately 30 mK. The comparison with theoretical predictions of $H_{c2}(T)$ is important in the region $T \lesssim 0.4 T_c$ corresponding to $\xi \lesssim 1.35\xi_0$. On such a scale, the degree of inhomogeneity is expected to be even larger. The effect on H_{c2} might be significant. Indeed enhanced H_{c2} above the Ginzburg–Landau–Abrikosov–Gorkov (GLAG) prediction in amorphous superconductors has been reported[52]. Summarizing the existing data on $\sigma'(T)$ and $H_{c2}(T)$, one notices that enhanced σ' above the AL curve goes hand in hand with enhanced H_{c2} above the GLAG curve. *Figure 22.6* shows data on homogeneous amorphous Mo–Ge film[53] following the GLAG curve and 'inhomogeneous'[52] $Mo_{30}Re_{70}$ (σ' shown in *Figure 22.5*) film deviating from theoretical prediction at decreasing temperature. An extreme case of inhomogeneity is illustrated in phase-separated amorphous Pb–Sb–Au alloys with divergent H_{c2} values[54]. In addition, H_{c2} data on homogeneous and inhomogeneous Zr–Pd samples[50] are shown in *Figure 22.7*. Thus far, there exists cumulative evidence that anomalous H_{c2} behaviour does originate from inhomogeneities in the samples. Various models have been proposed to explain the critical field curves. They all incorporate distributions in H_{c2} and T_c[53,55,56].

One then can pose the question: What are the forms and origins of inhomogeneities in glassy superconductors? First, the presence of inhomogeneities on

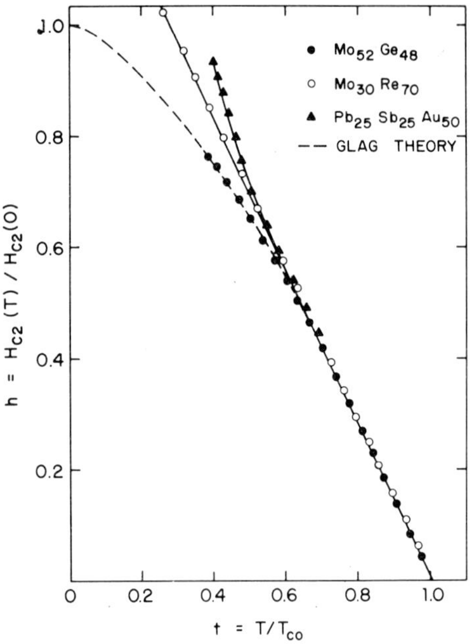

Figure 22.6 Reduced upper critical field $h = H_{c2}(T)/H_{c2}(0)$ as a function of reduced temperature. $H_{c2}(0)$ is the GLAG value at $T = 0$. Data on ○, $Mo_{30}Re_{70}$; ●, $Mo_{52}Ge_{48}$, and ▲, $Pb_{25}Sb_{25}Au_{50}$ are taken from Tenhover *et al.*[52], Carter *et al.*[53], and Kim and Johnson[54] respectively. The broken line represents GLAG theory

the scale of ξ_0 has been observed in low-angle X-ray scattering experiments[57]. For the alloy systems discussed here, they undergo eutectic crystallization into phases with different compositions from the amorphous hosts. For example, evidence of crystalline La_2Au and α–Zr embedded in amorphous La–Au and Zr–Pd alloys was routinely observed. If significant crystallization were not suppressed during the supercooling process, there would be a tendency for the composition of the regions in the

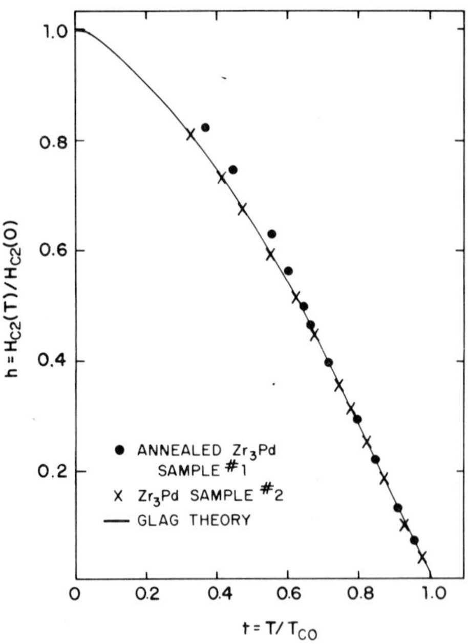

Figure 22.7 Reduced upper critical field $h = H_{c2}(T)/H_{c2}(0)$ as a function of reduced temperature for ●, annealed sample 1 and ×, as-quenched sample 2 or Zr_3Pd. The full line represents GLAG theory. Critical values are: Sample 1, $T_c = 2.68\,K$, $dH_{c2}(T_c)/dT = 26\,kG/K$; Sample 2, $T_c = 3.04\,K$, $dH_{c2}(T_c)/dT = 24.0\,kG/K$

neighbourhood of these crystallites to fluctuate. Another form of inhomogeneity might occur, for example, in alloys where several intermetallic compounds or a sizeable homogeneity range for an intermetallic compound exists in the composition of interest. Then, there would be a tendency towards microscopic ($\lesssim \xi_0$) incipient decomposition. This happens especially in ternary alloys where a large choice of short range order is available. Examples of macroscopic phase separation ($> \xi_0$) had also been reported in several amorphous ternary alloys[58]. Systematic studies are needed to correlate upper critical field behaviour with the different forms of inhomogeneities suggested here. More generally, the present problem is also related to the upper critical field behaviour in high-T_c metastable superconductors of technological importance[53].

22.3.3 Low-angle X-ray scattering

The scattering of X-rays and neutrons at small angles provides a useful method of determining the degree of inhomogeneity in samples of metallic glass[59,60]. The scattering intensity function appropriate to a line (point) collimation geometry is called $J(k)$ $[I(k)]$ and can be expressed in electron units per atom of sample[61]. The Fourier transforms of $J(k)$ or $I(k)$ give the autocorrelation function for electron density fluctuations. As an example of the usefulness of this method, *Figure 20.8* shows the scattering intensity $J(k)$ for an as-quenched and neutron irradiated sample of $(Mo_{0.6}Ru_{0.4})_{82}B_{18}$ metallic glass[62,63]. The irradiated sample was exposed to a flux of 10^{19} neutrons/cm^2 with average neutron energy of 1 MeV. Following irradiation, the superconducting transition was observed to sharpen[62] and the shape of the $H_{c2}(T)$ curve was found to deviate from linearity more strongly than that of the as-quenched sample[55]. *Figure 22.8* shows that the degree of homogeneity of the sample is altered by

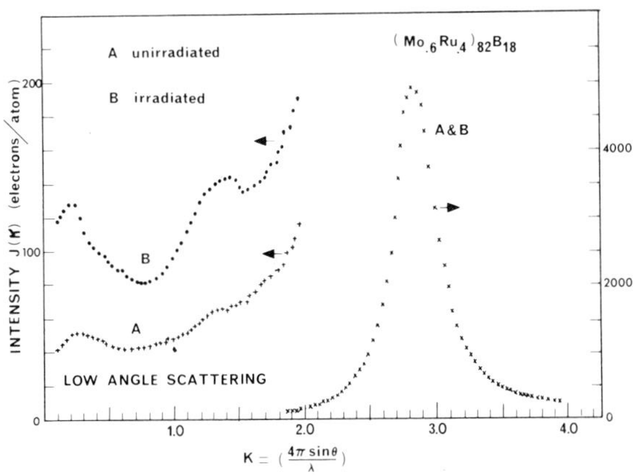

Figure 22.8 Low-angle X-ray scattering intensity as a function of wavenumber K [plotted as $(4\pi \sin \theta)/\lambda$] for A, as-quenched and B, irradiated amorphous $(Mo_{0.6}Ru_{0.4})_{82}B_{18}$

neutron irradiation. The overall increase in the level of small angle scattering suggests an increase in atom-scale defects and irregularities following irradiation. The development of several oscillatory features in the small angle scattering suggest that these atomic scale electron density fluctuations are correlated in real space. This in fact implies that a degree of short atomic range order develops following irradiation. The changes in superconducting properties are almost certainly connected with these changes in structure.

22.4 Relaxation effects

22.4.1 Dependence of T_c on annealing temperature and time

It was first observed in amorphous Zr_3Rh that T_c decreases according to $\log t$ upon annealing for a length of time t well below the glass temperature[64]. This is typical of a thermal relaxation process governed by a distribution in the activation energy. This is well known in the thermal activation of domains in ferromagnets[65] and the decay of remanent magnetization in spin glasses[66]. Given sufficiently long annealing time, a saturation value of T_c should be reached for a particular alloy if crystallization and phase separation do not occur. By comparing the trends towards saturation, one might be able to answer some of the questions concerning the metastability of amorphous alloys. Like spin glasses[67], structural glasses are highly degenerate systems with many local free energy minima. It is interesting to find out if the value of saturation T_c depends on the quenching condition or annealing temperature, or both. Dependence of magnetic and transport properties on short time annealing in metallic glasses has been studied by various authors[68,69]. It usually takes about 100 to 1000 hours of annealing to observe the saturation effect. Therefore, caution must be taken to avoid crystallization and possible gaseous contamination. When superconductivity is used as a probe, a superconducting transition width of several millikelvin is used as a criterion for a single amorphous phase. To detect a small quantity of a second phase, proper structural probes such as electron microscopes and low-angle X-ray scattering are available. However, it would be too tedious and time consuming to examine a sample at each annealing temperature and time. From our discussion on paraconductivity, samples with transition widths of several millikelvins obey the AL curve near T_c, characteristic of a homogeneous phase.

Since the as-quenched samples inevitably contain voids (defects), annealing would relax the quenched-in strains and cause the collapse and redistribution of voids. The latter are responsible for the irreversible and reversible phenomena respectively observed in metallic glasses upon annealing, as discussed by Egami[69]. The superconducting transition temperature is rather sensitive to the presence of strains[70]. The decrease in T_c upon annealing is due to the 'hardening' of phonon modes as the strains are relaxed. Annealing results of two sets of amorphous Zr_3Rh samples

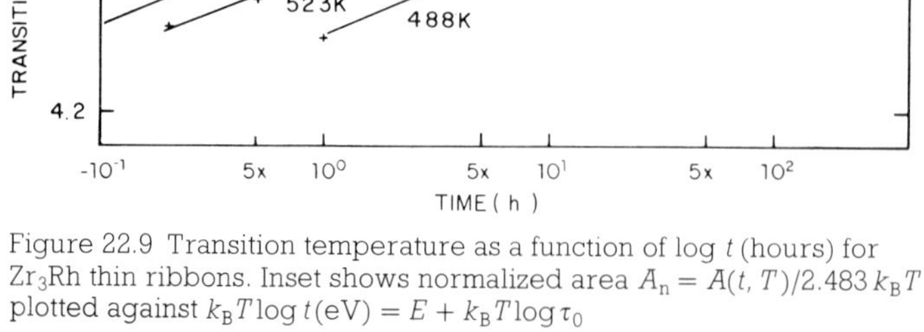

Figure 22.9 Transition temperature as a function of $\log t$ (hours) for Zr_3Rh thin ribbons. Inset shows normalized area $A_n = A(t, T)/2.483\, k_B T$ plotted against $k_B T \log t\,(\text{eV}) = E + k_B T \log \tau_0$

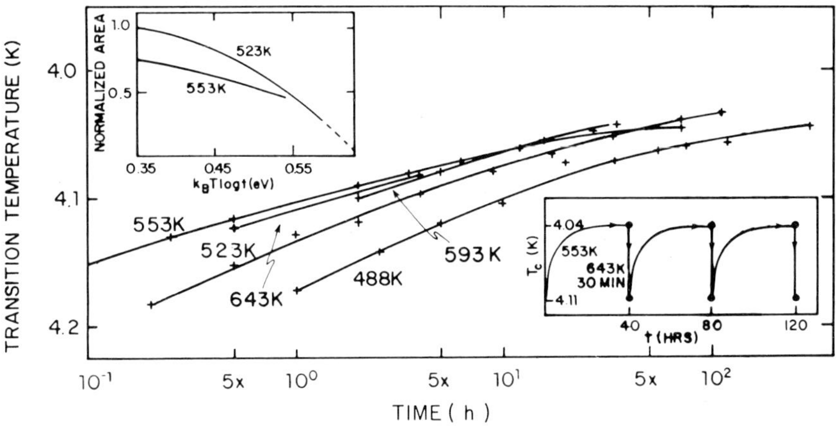

Figure 22.10 Transition temperature as a function of log t (hours) for Zr_3Rh wide ribbons. Lower right inset shows reversibility in T_c upon cyclic annealing at 280 °C (553 K) for 40 hours followed by 370 °C (643 K) for 30 minutes. The T_c value at 370 °C for 30 minutes, however, is not the saturation T_c at that temperature

quenched at different rates are shown in *Figures 22.9* (thin ribbon) and *22.10* (wide ribbons). Other technical details are given by Elmquist and Poon[71]. Saturation trends are more apparent in the wide ribbon samples. The extrapolated saturation T_c values for this ribbon at different annealing temperatures lie within a range of about 20—30 mK, while the overall decrease in T_c upon annealing is about 270 mK, an order of magnitude greater than the range of saturation T_c. In contrast, these two values can be of the same order of magnitude in amorphous ferromagnets[68,69]. The difference can be understood qualitatively. The Curie temperature, which depends essentially on the nearest neighbour interactions, is sensitive to the chemical short range order. The latter can be varied reversibly upon cyclic annealing at different temperatures[69,72]. On the other hand, superconductivity depends on the bulk phonon frequency as well as the short range order. Thus annealing changes T_c by both altering the vibrational modes and rearrangement of local structures in amorphous superconductors. It is suggested that the major decrease in T_c (irreversible) upon annealing is due to the strain effect, while chemical short range order of the equilibrium phases has only a minor effect on T_c. From T_c cycling (*see* inset of *Figure 22.10*), the saturation regimes correspond to the equilibrium degenerate states at different temperatures. Saturation is not achieved even at high annealing temperatures for the thin ribbon samples. The extrapolated saturation T_c values are expected to be at least 60 mK lower than those in the wide ribbon samples. Whether this is due to the degree of contamination or multiplicity of states introduced at different quenching conditions is at present not understood.

To facilitate the discussion of strain effect on T_c, the phenomenological model originated by Gaunt[73] is recalled. For small change in T_c, the approximation $\Delta T_c \propto \Delta S$ (S denotes strain) can be used. The relaxation process is represented by a distribution in activation energy. The change in T_c is found to be given by

$$\frac{\mathrm{d}\,\Delta T_c}{\mathrm{d}\log t} \propto -0.368 S_0 A(t, T) \tag{22.12}$$

where S_0 is the strain in the as-quenched state, and $A(t, T)$ is an integral of the distribution function. If the function is flat over a certain energy range, then equation 22.12 is reduced to

$$\frac{d\,\Delta T_c}{d\log t} \propto -0.914 S_0 k_B T$$

In the insets of *Figures 22.9* and *22.10* are plotted the normalized $A(t, T)$ for $T = 553$ K and 523 K as a function of $k_B T \log t$. The thin ribbon samples have broader activation spectra than the wide ribbon samples. The former samples have a faster quenching rate. It would be of interest to study these two sets of samples from a structural point of view. Finally, it is interesting to mention that, so far, the T_c of amorphous superconductors have been found to decrease upon annealing. These include

$Zr_{100-x}Pd_x(20 < x < 40)$[41,71]; $Zr_{100-x}Ni_x(20 < x < 35)$[41];
$Zr_{100-x}Fe_x(15 < x < 20)$[41]; Zr_3Rh[64,71]; $La_{70}Cu_{30}$[74]; V_3Si; Nb_3Ge[75]

On the contrary, Curie temperatures in metallic glasses have been observed to change both positively and negatively upon annealing[68].

22.4.2 Pressure dependence of T_c

So far, there are only several published reports on the pressure dependence of superconductivity in amorphous superconductors. Fasol *et al.* [76] and Willer *et al.*[77] investigated amorphous $La_{78}Au_{22}$[76], $Cu_{40}Zr_{60}$ and $Pd_{30}Zr_{70}$[77] alloys up to quasihydrostatic pressures of 13.7 and 10.8 GPa respectively. They found that the rates of T_c variation dT_c/dP were comparable with their crystalline 'counterparts' of lanthanum and zirconium. Recently, the pressure dependence of T_c in amorphous $La_{80}Ga_{20}$, $La_{80}Al_{20}$, and Mo–Ru–B alloys has been reported for low pressures[78]. Applying high pressure to as-quenched samples would also lead to the collapse of voids and the rearrangement of atoms. It was estimated[77] that the change in volume at 10 GPa is about 5 per cent. On the other hand, annealing of as-quenched Zr_3Rh samples only led to a much smaller change in volume. However, the change in T_c upon annealing these samples is comparable with that in the high pressure experiment. Therefore, the pressure dependence of the Hopfield–McMillan parameter η and Coulomb pseudopotential μ^* must be important. One can then compare the pressure dependence of these local electronic parameters in amorphous and crystalline structures. It would also be interesting to study $T_c(P)$ in the thermally 'relaxed' state. By combining thermal annealing and pressure, one can study the reversibility of amorphous phases at high pressure.

22.4.3 Amorphous ribbons quenched at elevated substrate temperature

In addition to the quenching speed, one can also vary the substrate temperature in liquid quenching. In *Figure 22.11*, the annealing curves for amorphous Zr_3Pd quenched at room temperature are taken from Poon *et al.*[79]. Both the 130°C (403 K) and 310°C (583 K) samples were subject to annealing at 180°C (453 K). The annealing behaviour of the 130 °C sample follows the room temperature sample. For the 310 °C sample, T_c decreases by a negligible amount for the initial 18 minutes of annealing. Further annealing gradually yields data points closer to the 180 °C (453 K) curve. It is assumed that the initial state of the preannealed 310 °C sample is identical to the state defined by $\log t$ at the same T_c value on the annealing curve. In this case, the initial time on the 180 °C isotherm would be given by $t_0 \simeq 42$ min. Hence, the relevant time scale to be used would be $t = t_0 + t_a$ with $t_0 = 42$ min. The adjusted data are shown in *Figure 22.11*. One then sees that both the 130 °C and 310 °C samples follow closely the annealing curve for the room temperature sample. The reduced initial T_c for the 310 °C

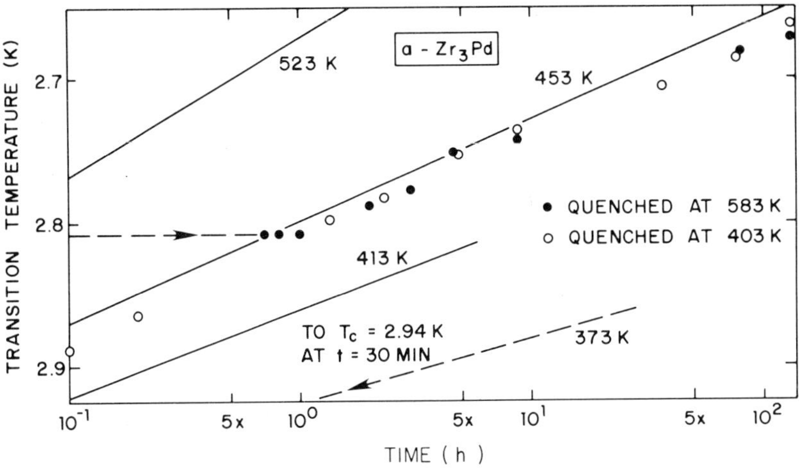

Figure 22.11 Data points show transition temperature as a function of log t (hours) for samples of α-Zr_3Pd quenched at elevated substrate temperatures: ●, 583 K; ○, 403 K, followed by annealing at 453 K. Solid lines are taken from Elmquist and Poon[71]. Horizontal dashed line connects initial state to annealing curve

sample results from postquenching annealing at a higher temperature. If it were assumed that the sample was annealed for about 1—2 minutes at the postquenching temperature T_{pq} before it was sufficiently cooled to a temperature ineffective in reducing T_c any further, the postquenching temperature could be determined. It has been known that in single roller melt spinning[80,81] the ribbon stays on the wheel for about 10^{-4} s at the end of which the postquenching period begins. On the time scale t, $t = 0$ would be the time at which glass transition for the bulk section has completed and cooling begins. Then it takes even less than 10^{-4} second for subsequent cooling to reach T_{pq}. For further cooling, one should start from t_0 on the postquenching isotherm. One then obtains $\Delta T_c \propto - T_{pq} \log(t_a/t_0)$. Thus the temperature T_{pq} can be determined by extrapolating the horizontal dashed line in *Figure 22.11* towards the 1—2 min region to obtain a temperature range for T_{pq}. It is found that $T_{pq} \propto$ 553—590 K. Thus it is estimated that the temperature at which the ribbon leaves the wheel is quite close to the substrate temperature (583 K) for Zr_3Pd. For the Zr_3Rh alloys, the data of which are not shown[79], a ribbon was quenched at 260°C (533 K) and subsequently annealed at 215 °C (473 K). Following the aforementioned scheme, T_{pq} was estimated to be about 553—573 K, somewhat higher than the substrate temperature (553 K).

Thus by making use of the superconducting properties of these alloys, postquenching temperatures are estimated and are found to be near the substrate wheel temperatures. The effective cooling time for the samples is about 1–2 minutes which is no worse than in the case of thin films. Therefore if substrate temperatures close to the glass temperatures were used so that glass formation would be hindered, then new metastable phases could be expected. It can then be compared with thin film metastable materials quenched at elevated substrate temperatures.

22.5 Conclusion and future prospects

The electromagnetic properties and T_c systematics of amorphous superconductors can be understood on a phenomenological level using the conventional alloy theories of type II superconductors. In lieu of rigorous theories, one always refers to Anderson's theorem[82] as the basis for understanding the occurrence of superconductivity in this class of material. As noted in section 22.2, a first-principles calculation of electron–phonon interaction in amorphous metals will be extremely difficult. So far, only special cases have been treated, such as in the low energy regime. Computer simulation of electron–phonon interaction is definitely needed to cover the entire phonon spectrum. More extensive experimental studies on amorphous transition metal alloys will also help further to understand metastable superconductors in general. It is suggestive to focus attention on simpler systems such as binary alloys so that experimental results can be readily compared with theoretical predictions.

More recently, superconductivity has been employed as a probe to study the physical properties of amorphous metals. The low-energy excitations of the two level systems in amorphous superconductors have been investigated by acoustic and specific-heat measurements[83,84]. The thermal relaxation studies of samples prepared under different conditions might unravel the different states of an amorphous alloy. The idea of almost degenerate free energy minima separated by low energy barriers in spin glasses might find its application in the study of metastability of amorphous states. With the aid of EXAFS[85] and EDXD[86] techniques, it may be hoped that this problem can be understood on a structural basis. High pressure studies of amorphous superconductors yield direct information on the local atomic properties such as the Hopfield–McMillan parameter and bulk modulus. Flux pinning studies in glassy superconductors have enhanced our present understanding of the mechanism of flux pinning and the dynamics of vortex lattice[87]. The physical properties (sometimes anomalous) mentioned here, however, depend on the degrees of homogeneities of the samples. Thus, it is conceivable that the study of 'inhomogeneous' amorphous superconductors will become an important subject in the near future.

Since most amorphous refractory alloys have rather high crystallization temperatures and are electrochemically inert, they might constitute a new class of useful material. It is interesting that amorphous refractory superconductors with transition temperatures near that of lead ($\sim 7.2\,K$) can readily be synthesized. The latter and its alloys are among the key components in today's superconducting thin film technology. Thus, at present it is premature to put limits on the potential applications of amorphous superconductors.

Acknowledgements

The author thanks W. L. Johnson for contributing his data to the section on low-angle X-ray scattering. This research is supported in part by the University of Virginia Pratt Fund and a grant from the Research Corporation.

References

1. Buckel, W. and Hilsch, R., *Z. Phys.*, **138,** 109 (1954); **138,** 118 (1954)
2. Bergmann, G., *Phys. Rep.*, **27C,** 161 (1976)
3. Collver, M. M. and Hammond, R. H., *Phys. Rev. Lett.*, **30,** 92 (1973)
4. Johnson, W. L., Poon, S. J. and Duwez, P., *Phys. Rev.*, **B11,** 150 (1975); Johnson, W. L. and Poon, S. J., *J. appl. Phys.*, **46,** 1787 (1975)
5. *See* for example, Johnson, W. L., in *'Glassy Metals I'* Eds. Güntherodt, H.-J. and Beck, H., Topics in Applied Physics, vol. 46, p. 191, Springer-Verlag, New York (1981); Tsuei, C. C., in *'Superconducting Material Science'*, Eds. Foner, S. and Schwartz, B., p. 735, Plenum Press, New York (1981)
6. Allen, P. B. and Dynes, R. C., *Phys. Rev.*, **B12,** 905 (1975)
7. Leavens, C. R., *J. Phys. F: Metal Phys.*, **7,** 1911 (1977)
8. McMillan, W. L., *Phys. Rev.*, **167,** 331 (1967)

9. Faber, T. E. and Ziman, J., *Phil. Mag.*, **11**, 153 (1965))
10. Busch, G. and Güntherodt, H.-J., *'Solid State Physics'*, Eds. Ehrenreich, H., Seitz, F. and Turnbull, D., p. 235, Academic Press, New York (1974)
11. Jäckle, J. and Froböse, K., *J. Phys. F: Metal Phys.*, **10**, 471 (1980)
12. Meisel, L. V. and Cote, P. J., *Phys. Rev.*, **B23**, 5834 (1981)
13. Bergmann, G., *Phys. Rev.*, **B3**, 3797 (1971)
14. Poon, S. J. and Geballe, T. H., *Phys. Rev.*, **B18**, 233 (1978)
15. Suck, J.-B., Rudin, H., Güntherodt, H.-J. and Beck, H., *J. Phys. C: Solid State Physics*, **13**, L1045 (1980)
16. Von Heimendahl, L., *J. Phys. F: Metal Phys.*, **9**, 161 (1979)
17. Harrison, W. A., *'Pseudopotentials in the Theory of Metals'*, Benjamin, New York (1966)
18. Barisic, S., Labbe, J. and Friedel, J., *Phys. Rev. Lett.*, **25**, 919 (1970); Barisic, S., *Phys. Rev.*, **B5**, 932 (1971)
19. Poon, S. J., *Solid St. Communs*, **34**, 659 (1980)
20. Kimhi, D. B. and Geballe, T. H., *Phys. Rev. Lett.*, **45**, 1039 (1980)
21. Ziman, J. M., *'Electrons and Phonons'*, chap. 5, Clarendon Press, Oxford (1960)
22. Tsuei, C. C., Johnson, W. L., Laibowitz, R. L. and Viggiano, J. M., *Solid St. Communs*, **24**, 615 (1977); Johnson, W. L., Tsuei, C. C., Raider, S. I. and Laibowitz, R. L., *J. appl. Phys.*, **50**, 424 (1979)
23. Pippard, A. B., *Phil. Mag.*, **46**, 1104 (1955): *J. Phys. Chem. Solids*, **3**, 175 (1957)
24. Poon, S. J. and Carter, W. L., *Solid St. Communs*, **35**, 249 (1980)
25. Koch, C. C., Kroeger, D. M., Scarbrough, J. O. and Giessen, B. C., *Phys. Rev.*, **B22**, 5213 (1980)
26. Babic, E., Ristic, R., Miljak, M., Scott, M. G. and Gregan, G., *Solid St. Communs*, **39**, 139 (1981)
27. Minnigerode, B. and Samwer, K., *Physica (B +* C), **108**, 1217 (1981); Samwer, K., doctoral dissertation, Georg-August-Universität Göttingen (1981)
28. Carter, W. L., Poon, S. J., Marshall, A. F. and Barbee, T. W., *Bull, Am. Phys. Soc.*, **25**, 385 (1980)
29. Johnson, W. L., Poon, S. J., Durand, J. and Duwez, P., *Phys. Rev.*, **B18**, 208 (1978)
30. Inoue, A. and Masumoto, T., Supplement to *Scientific Report* of Research Institute of Tohoku University (Sendai, Japan), A, March, 1980, p. 165
31. Strudthoff, C-H, Reichelt, J., and Freyhardt, H. C., *Physica* (B + C), **107**, 393 (1981)
32. Johnson, W. L. and Williams, A. R., *Phys. Rev.*, **B20**, 1640 (1979)
33. Varma, C. M. and Dynes, R. C., *'Superconductivity in d- and f-Band Metals'*, Ed. Douglass, D. H., Plenum Press, New York (1976)
34. Graebner, J. E., Golding, B., Schultz, R. J., Hsu, F. S. L. and Chen, H. S., *Phys. Rev. Lett.*, **39**, 1480 (1977)
35. Early, S. R., Carter, W. L., Poon, S. J. and Geballe, T. H., *Bull. Am. Phys. Soc.*, **25**, 432 (1980)
36. Zwicknagl, G., *Z. Phys.*, **B40**, 23 (1980–1981); **B40**, 31 (1980–1981)
37. Miedema, A. R., *J. Phys. F: Metal Phys.*, **4**, 120 (1974)
38. Lustfield, H., *Solid St. Communs*, **17**, 437 (1975)
39. Oelhafen, P., Hauser, E. and Güntherodt, H.-J. *Solid St. Communs*, **35**, 1017 (1980)
40. Amamou, A. and Johnson, W. L., *Solid St. Communs*, **35**, 765 (1981)
41. Poon, S. J., unpublished results
42. Tenhover, M. and Johnson, W. L., *Physica (B +* C), **108**, 1221 (1981)
43. Rapp, O., personal communication (1981)
44. Amamou, A. and Krill, G., *Solid St. Communs*, **28**, 957 (1978)
45. Aslamazov, L. G. and Larkin, A. I., *Phys. Lett.*, **A26**, 238 (1968); *Soviet Phys. Solid State*, **10**, 875 (1968)
46. Maki, K., *Theor. Phys.*, **39**, 897 (1968); *J. Low Temp. Phys.*, **1**, 513 (1969)
48. Skocpol, W. J. and Tinkham, M., *Rep. Prog. Phys.*, **38**, 1049 (1975)
49. Johnson, W. L. and Tsuei, C. C., *Phys. Rev.*, **B13**, 4827 (1976); Johnson, W. L., Tsuei, C. C. and Chaudhari, P., *Phys. Rev.*, **B17**, 2884 (1978)
50. Poon, S. J., *Phys. Rev.*, **B**, (1982)
51. Keck, B. and Schmid, A., *Solid St. Communs*, **17**, 799 (1975)
52. Tenhover, M., Johnson, W. L. and Tsuei, C. C., *Solid St. Communs*, **38**, 53 (1981)
53. Carter, W. L., Poon, S. J., Hull, J. W. and Geballe, T. H., *Solid St. Communs*, **39**, 41 (1981)
54. Kim, C. O. and Johnson, W. L., *Phys. Rev.*, **B23**, 143 (1981)
55. Clemens, B. M., Tenhover, M. and Johnson, W. L., *Physica* (B + C), **107**, 319 (1981)
56. Agyeman, K., Tsuei, C. C. and Chaudhari, P., personal communication (1981)
57. Johnson, W. L., personal communication (1981)
58. Köster, U. and Herold, U., in *Glassy Metals I'*, Eds Güntherodt, H. J. and Beck, H., Topics in Applied Physics, vol. 46, p. 225, Springer-Verlag, New York (1981)
59. Cargill, G. S. (III), *'Solid State Physics'*, Eds. Seitz, F. and Turnbull, D., vol. 30, p. 227, Academic Press, New York (1975)

60. Nold, E., Steeb, S., Lamparter, P. and Rainer Harback, G., *J. Phys. (Paris)*, **C8,** 186 (1980)
61. Guinier, A. and Fournet, G., in *'Small Angle Scattering of X-Rays'*, Wiley, New York (1955)
62. Kramer, E. A., Johnson, W. L. and Cline, C., *Appl. Phys. Lett.*, **35,** 815 (1979)
63. Johnson, W. L., in *'Amorphe Metallische Werkstoff'*, Proc. 14th Metals Conf. of DDR, Dresden, p. 183 (1981)
64. Drehman, A. J. and Johnson, W. L., *Phys. Status. Solidi, (a)* **52,** 499 (1979)
65. Street, R. and Woolley, J. C., *Proc. Phys. Soc.*, **A62,** 562 (1949); **69,** 1189 (1956)
66. Guy, C. N., *J. Phys. F: Metal Physics*, **8,** 1309 (1978); *J. appl. Phys.*, **50,** 7308 (1979)
67. Walker, L. R. and Walstedt, R. E., *Phys. Rev.*, **B22,** 3816 (1980)
68. *See,* for example, papers presented at the Twenty-Sixth Annual Conference on Magnetism and Magnetic Materials, *J. appl. Phys.*, **52,** 1859 (1981)
69. Egami, T., *Ann. New York Acad. Sci.*, **371,** 238 (1981)
70. Suhl, H., Matthias, B. T., Hecker, S. and Smith, J. L., *Phys. Rev. Lett.*, **45,** 1707 (1980)
71. Elmquist, R. E. and Poon, S. J., *Solid St. Communs*, **41,** 221 (1982)
72. Greer, A. L. and Leake, J. E., in *'Proc. 3rd Int. Conf. on Rapidly Quenched Metals'*, Ed. Cantor, B., vol. I, p. 299, The Metals Society, London (1978)
73. Gaunt, P., *Phil. Mag.*, **34,** 775 (1976)
74. Esquinazi, P., de la Cruz, M. E. and de la Cruz, F., *Physica*, (B + C), **108,** 1215 (1981)
75. Tsueli, C. C., *Appl. Phys. Lett.*, **33,** 262 (1978)
76. Fasol, G. Schilling, J. S., Tsuei, C. C. and Zimmer, F., *J. Phys. F; Metal Phys.*, **8,** L257 (1978)
77. Willer, J., Fritsch, G. and Lüscher, E., *Appl. Phys. Lett.*, **36,** 859 (1980)
78. Missell, F. P. and Frota-Pessôa, S., *Solid St. Communs*, **40,** 379 (1981); Bindilatti, V., Missell, F. P. and Tenhover, M., *Physica* (B + C), **108,** 391 (1981)
79. Poon, S. J. Dunn, P. L., Elmquist, R. E. and Smith, L. M., *Solid St. Communs*, **35,** 249 (1982)
80. Leontic, B., Lukatela, J., Babic, E. and Ocko, M., in ref. 72, vol. 1, p. 41
81. Walter, J. L., ref. 72, vol. 1, p. 30
82. Anderson, P. W., *J. Phys. Chem. Solids*, **11,** 26 (1959)
83. Black, J. L., in ref. 58, vol. 46, p. 167
84 *See,* for example, papers presented at the Sixteenth International Conference on Low Temperature Physics, *Physica* (B + C) **107,** (1981)
85. Wong, J., ref. 58, p. 45
86. Egami, T., ref. 58, p. 25
87. Kes, P. H. and Tsuei, C. C., *Phys. Rev. Lett.*, **47,** 1930 (1981)

Chapter 23

Thermal properties of amorphous metallic alloys

D. G. Onn

Department of Physics, University of Delaware, Newark, USA

23.1 Introduction

For many years the only amorphous metallic alloys available were evaporated or sputtered thin films, many compositions requiring cryoquenching for stability. The low mass and the presence of a substrate made measurement of thermal properties such as specific heat, C_P, and thermal conductivity, k, difficult. Even after the first synthesis of metallic glasses by the piston-and-anvil technique many foils, each individually prepared, were required for accurate measurement of C_P or k. With the advent of melt-spinning techniques larger samples of uniform character have become available and reports on thermal properties are rapidly increasing in number. In addition, recent advances in microcalorimetry now permit C_P of about 1 mg samples to be determined accurately.

In this chapter we discuss primarily the specific heat, C_P, and the thermal conductivity, k, of amorphous metallic alloys below about 100 K. C_P may be written as the sum of several contributions:

$$C_P = C_L + C_E + C_M + C_{EX} \tag{23.1}$$

We will first review in section 23.2 the phonon or 'lattice' term, C_L, which in amorphous alloys is distinctly different from crystalline alloys. We will then survey in section 23.3 recent results on C_E, the electronic contribution for 'simple' metals, which arises from electrons in states near the Fermi energy E_F and provides values of the electronic density of states at E_F, $N(E_F)$. In magnetic alloys (section 23.4) the specific magnetic state (spin-glass, ferromagnet, etc.) determines C_M, the magnetic contribution to C_P, but C_E for magnetic alloys is also important since it can define limits on $N(E_F)$ for the existence of each magnetic regime. Finally we observe in section 23.5 that for superconducting alloys C_P in the normal and superconducting states provides most of the parameters that describe the onset of the superconducting state.

These various contributions to C_P are not all present in every alloy. Where multiple contributions do occur separation, using a number of techniques, is almost always unambiguous. C_{EX} in equation 23.1 covers contributions to C_P not already classified, such as nuclear hyperfine effects.

The thermal conductivity, k, of a normal solid at low temperatures consists of phonon, k_p, and electron, k_E, contributions so that we can write:

$$k = k_p + k_E \tag{23.2}$$

k_p is of primary interest in amorphous materials and k_E may be subtracted by assuming the law of Wiedemann and Franz, or may be reduced by the occurrence of superconductivity. The Wiedemann–Franz law relates k_E and the electrical resistivity, ρ, provided that the electron scattering contributions to both have the same energy dependence:

$$\rho k_E = \frac{1}{3}\left(\frac{\pi^2 k_B^2}{e^2}\right)T \tag{23.3}$$

Measurements of ρ are easier to undertake than measurements of k.

Emphasis in this chapter will be on recent experimental results with references provided to relevant theoretical models. Experimental techniques are not covered since they are described by most authors or are developments of standard methods. The phrase 'metallic glasses' will be confined, as far as possible, to substrate-free alloys, while 'amorphous alloys' will cover all such materials. For brevity we will use [Fe] to indicate element concentrations.

A number of comprehensive reviews, conference proceedings and texts of the physical properties and applications of amorphous metallic alloys, primarily metallic glasses, have appeared recently. In addition to the present chapter, they include the comprehensive survey by Chen[1], Proceedings of the second, third and fourth International Conferences on Rapidly Quenched Metals[2,3,4] and texts edited by Gilman and Leamy[5], by Guntherodt and Beck[6] and by Hasegawa[7]. Several more specialized reviews on phonon, electronic, magnetic and superconducting properties will be referred to in the appropriate sections below.

23.2 Phonon thermal properties

23.2.1 Background

Although the thermal properties of insulating glasses had been investigated for many years, their study received a major stimulus with the discovery of a dominant linear contribution to C_L below about 2 K by Zeller and Pohl[8]. This term, and the accompanying thermal conductivity k_p (which varies with T^2) were attributed to low energy excitations which are describable by 'two-level systems' (TLS) with which the phonons interact. There have been several comprehensive reviews of both the experimental and theoretical situation in recent years[8–15]. We will not attempt to repeat here the detail provided by those reviews and their associated reference lists. We will provide a brief summary of the salient experimental features and relevant theoretical models as a background for the involvement of phonon behaviour in the electronic, magnetic and superconducting thermal properties of amorphous metal alloys discussed in sections 23.3 to 23.5.

It is now firmly established that C_L and k_p have a universal characteristic temperature dependence for all amorphous solids, insulating or conducting, that is quite distinct from that for crystalline solids. *Figure 23.1* shows, schematically, C_L/T^3 and k_p, both on bilogarithmic axes, for $0.1 \lesssim T \lesssim 100$ for typical amorphous and crystalline solids. Two distinct temperature regions are apparent, $T \lesssim 2$ K and $T \gtrsim 2$ K.

23.2.2 Temperatures less than about 2 K

In this region C_L is dominated by a phonon term that is very close to linear in T, which we refer to as (aT). This term is much smaller than the electronic linear contribution ($C_E = \gamma T$: section 23.3.1). It causes the upturn in C_L/T^3 below about 2 K seen in *Figure 23.1(a)*. Below about 1 K k_p increases approximately as T^2 in contrast to the T^3 dependence for crystalline materials [*Figure 23.1(b)*].

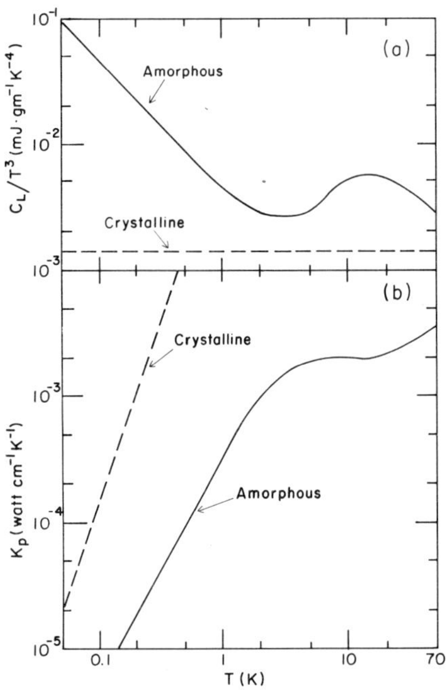

Figure 23.1 (*a*) C_L/T^3 as a function of T for typical amorphous and crystalline solids. (*b*) Phonon thermal conductivity k_p as a function of T for typical amorphous and crystalline solids. The common temperature axes of these graphs emphasizes the correlation of features discussed in the text

The indices for C_L and k_p are rarely integer, ranging from 0.8 up to 1.3 for C_L and from 1.8 to 2 for k_p. The integer powers are often assumed if the temperature range of measurement is limited.

The most accurate studies of these contributions for conductors have been made in superconducting amorphous alloys far below the transition temperature T_c where the electronic contributions C_E and k_E are much reduced and C_L and k_p are dominant.

Explanation of the (aT) contribution to C_L and of the $k_p \sim T^2$ behaviour requires a density of states which is almost energy independent. The actual densities of states estimated from (aT) for insulating and metallic glasses is within a factor of three. The slight deviations of indices from integer values can be accounted for by a slight energy dependence in the density of states.

The low energy excitations giving rise to these effects in C_L and k_p are considered to be localized 'two-level systems' (TLS) with tunnelling permitted between the levels. Coupling constants describing the interaction of the TLS with phonons can be determined from k_p and from acoustic studies and are found to be slightly higher in insulating glasses than in metallic.

The consistency of TLS related properties for all types of glass implies that TLS are fundamental entities associated with the amorphous state. Despite the great success of the TLS model, however, there is still no definite indentifiable physical identity for the TLS. Most recent suggestions focus on relatively large clusters of atoms which may be situated in nearly degenerate configurations, but experimental tests of this hypothesis have not yet been established.

23.2.3 Temperatures greater than about 2 K

Above about 2 K C_L/T^3 passes through a minimum [*Figure 23.1(a)*] above which a rounded peak occurs followed by a decrease to values of C_L/T^3 close to those of crystalline materials. These variations in C_L/T^3 are much greater than those observed in most crystalline materials. As C_L/T^3 passes through its maximum value, typically between about 5 and 20 K, k_p levels off to a near 'plateau' at about the same temperature. Above this plateau no consistent pattern for k_p has been identified. Recently similar correlations between the C_L/T^3 peak and the plateau in k_p have been observed in low dimensional crystalline materials which also show[16] TLS-like effects below 2 K.

Presentation and analysis of C_P results in this region tend to be governed by the upper temperature limit of measurement. Data acquisition often ceases about 6 K and, since C_P for metal alloys contains an electronic contribution it may be expressed as:

$$C_P = \gamma' T + \beta(0)T^3 + \delta T^5 \tag{23.4}$$

where $\gamma' (= \gamma + a)$ includes the small linear phonon term.

From $\beta(0)$ an 'effective Debye temperature' $\theta_D(0)$ can be obtained by use of the low temperature limit ($T = 0$) of the relation:

$$\beta(T) = \frac{12\pi^4 R}{5}\left(\frac{1}{\theta_D(T)}\right)^3 \tag{23.5}$$

This same relation can be used to obtain a temperature dependent Debye temperature $\theta_D(T)$ which is proportional to the inverse of $(C_L/T^3)^{1/3}$ and incorporates the T^5 or other temperature-dependent contributions.

For insulating glasses $\beta(0)$ consistently exceeds the value estimated from sound velocity measurements. In metallic glasses the situation is less clear since in binary alloys agreement with sound velocity based values is good[17] while in transition metal–metalloid glasses discrepancies occur[18,19]. In all cases however, the values of $\theta_D(0)$ are less than in crystalline materials of comparable compositions.

The T^5 term is often attributed to phonon dispersion effects. From trends in crystalline materials there is known to be an empirical relationship between $\theta_D(0)$ and δ which is also followed for amorphous metal alloys enabling the conclusion that the vibrational properties of the two phases can be considered to be the same with only a change in average force constants attributed to densification[20]. The explanation of this change must reside in differences in the material microstructures.

Determination of C_L/T^3 [or equivalently of $\theta_D(T)$ from equation 23.5] is unfortunately rare in amorphous alloys above about 10 K, so that correlation with the T^5 term, and with other physical properties, is limited. The T^5 term no longer provides an adequate description above about 10 K and Einstein contributions of one, or more, characteristic energies ($k\theta_E$) are used to fit the rounded hump[21]. The ratio $\theta_E/\theta_D(0)$ for various materials is remarkably consistent, while the number of modes involved is typically only 0.03 to 0.06 per cent which is quite close to the number involved in the TLS model, suggesting a possible common origin[22].

The complexity of trends in C_L above 10 K is illustrated in *Figure 23.2* where recent results on $\theta_D(T)$ are shown for $Fe_{40}Ni_{40}[PB]_{20}$ alloys as the P/B ratio is varied. The changes in both $\theta_D(0)$ and $\theta_D(T)$ are dramatic but correlation with associated structural changes and changes in other physical properties is still needed and no

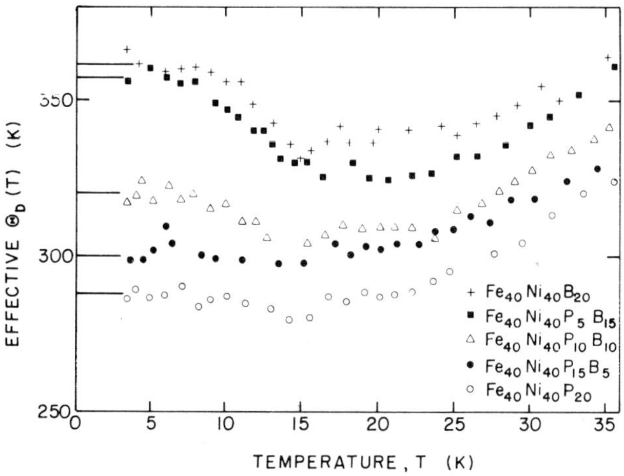

Figure 23.2 The effective Debye temperature $\theta_D(T)$ obtained from equations 22.4 and 22.5 for $Fe_{40}Ni_{40}M_{20}$ alloys with M varying from boron to phosphorus. The horizontal lines below about 5 K are the low temperature limits of the Debye temperature $\theta_D(0)$. Note that a similar figure given by Onn[18] has the key to the compositions reversed

effective explanation is yet available[18]. Unfortunately, the dominant C_E contribution in these non-superconducting alloys prevents correlation between the trends in C_L/T^3 and possible changes in the (aT) phonon term.

The occurrence of the plateau in k_p at the same temperature as the peak in C_L/T^3 for many amorphous materials strongly suggests a common origin, with the dispersion causing the peak in C_L/T^3 modifying the phonon scattering to create the strong frequency dependence giving rise to the plateau in k_p[16]. A theoretical model[17] based upon force constant variations results in enhanced T^3 and T^5 terms in C_L. However, as is the case for the TLS the physical entity responsible for the trends in C_L and k_p above 2 K remains unknown.

23.2.4 Annealing studies

Some insight into the nature of TLS and the states affecting C_L and k_p above 2 K may be obtained from extensive systematic annealing studies, associated with structural studies, but few such surveys have been undertaken to date.

Below 2 K structural relaxation annealing of an alloy is known[23] to enhance k_p between 0.1 K and 30 K. In an alloy deposited on a substrate the TLS contribution was halved and the power dependence changed from 0.55 to 0.85, implying a reduction in TLS states and a more uniform energy distribution after a structural relaxation anneal[24].

Above 2 K structural relaxation annealing decreases $\theta_D(0)$ in most, but not all cases. Crystallization annealing is more consistent, increasing $\theta_D(0)$ by about a factor of 1.3 ± 0.15.

The apparent inconsistency in changes in $\theta_D(0)$ with structural relaxation annealing response may arise from lack of measurements to a high enough temperature to determine $\theta_D(T)$ rather than just $\theta_D(0)$. For transition metal–metalloid alloys it has been shown that $\theta_D(T)$ is changed considerably in addition to the changes in $\theta_D(0)$, though no definitive model for the changes in terms of structure modifications is yet established[18]. In sharp contrast, it is now established that annealing, whether for structural relaxation or crystallization, has almost no effect on the electronic specific heat C_E in most materials (section 23.3.4).

23.2.5 Debye temperature and alloy stability

There is some indication from the limited data presently available, of a correlation between Debye temperature and alloy stability. This is shown most effectively by recent measurements[25] of the Fe–B and Fe–P alloy series which show a pronounced minimum in $\theta_D(0)$ at the most stable alloy composition points of $Fe_{80}M_{20}$. There is a similar minimum at the stable composition point in the Ag–Cu–Ge alloys[26] and at $Pd_{80}Si_{20}$ [27].

23.3 Electron thermal properties

23.3.1 Background

Determination of C_p combined with modern photoemission techniques provide valuable complementary insights into the electronic structure of amorphous alloys.

C_E, the contribution to C_P from electrons in states near E_F, is given by:

$$C_E = \tfrac{1}{3}\pi^2 k_B^2 (1 + \lambda p) N_0(E_F) T = \gamma T \tag{23.6}$$

where $N_0(E_F)$ is the bare density of electronic states at E_F. This is enhanced by the electron–phonon interaction λ_p to the experimentally determined value $N_\gamma(E_F)[=(1+\lambda_p)N_0(E_F)]$. Precise determination of γ gives $N_\gamma(E_F)$ to within a few percent throughout an alloy series. If λ_p is small, monotonic, or well known from another source, such as T_c for a superconductor (section 23.5.1), then $N_0(E_F)$ is well determined.

Complementary photoemission studies (ultraviolet: UPS; X-ray: XPS, etc.) provide a direct and semi-quantitative picture of the complete energy dependence of the bare density of state $N_0(E)$. Photoemitted electron intensities are not directly convertible to accurate $N_0(E)$ curves however, and the energy resolution is limited (about 0.3 to 1 eV). Also many orbitals contribute to $N_0(E)$ and $N_0(E_F)$ and a complex deconvolution process is needed to obtain partial density of states functions[28]. Recent reviews of electronic structure of amorphous alloys discuss photoemission results in detail[29,30].

$N(E_F)$ values can also be obtained from magnetic susceptibility $[N_\chi(E_F)]$ and, for superconductors, from critical field temperature dependence $[N_H(E_F)]$ (section 23.5.1). A definitive model of electronic structure will draw upon, and must explain, values from all sources in addition to related data such as the Hall coefficient R_H and the electrical resistivity ρ.

A great stimulus for C_E determination was the proposal of Nagel and Tauc[31] that metallic glass stability was enhanced by a reduction in electronic energy due to a minimum in $N(E)$ near E_F. The metallic glass Fermi surface could interact with a peak $S(q_p)$ in the structure factor $S(q)$ creating a minimum in $N(E)$ in analogy with the Fermi surface–Brillouin zone interaction in crystalline alloys which gives rise to band-gaps. So far no such minima in $N(E)$ have been identified through C_E studies. This may be because of the stringent conditions of a rigid-band model alloy and that $q_p = 2k_F$. An observed minimum in $N_\gamma(E_F)$ in $[Fe]_{80}[PB]_{20}$ alloys is probably not due to a Nagel–Tauc condition (section 23.4.4).

23.3.2 Simple amorphous alloys

In 'simple' amorphous metals, elemental or alloy, which involve no transition metals, the absence of long-range structural order will lead to a free-electron-like Fermi surface, in contrast to crystalline alloys where structure has a considerable influence on $N(E)$ due to Fermi surface–Brillouin zone effects. Bergmann[32] has reviewed the situation for cryoquenched simple metal amorphous films and more recent reviews support the same conclusions[31,33].

We will illustrate these conclusions with two very recent studies which followed the above reviews. C_P for the simple bivalent metallic glass $Mg_{70}Zn_{30}$ has a γ value within a few percent of the free-electron value and, since contributions from λ_p and (aT) are very small for this alloy, it appears that a free-electron model for $N_0(E_F)$ is applicable[34]. Magnetic susceptibility measurements[35] give $N_\chi(E_F)$ in agreement with $N_\gamma(E_F)$. Recent precision studies of R_H and ρ suggest small but consistent deviations from the free-electron model as [Zn] is varied from 23 to 35 atom %, though no definite cause for these deviations is yet identified[36].

Free-electron like behaviour is also inferred[26] for the amorphous Hume-Rothery (H-R) metallic glasses $(Ag_{0.5}Cu_{0.5})_{1-x}Ge_x(0.2 \lesssim x \lesssim 0.3)$. Experimentally γ values exceed free-electron values by only 2—3 per cent in distinct contrast to the case of the crystalline H-R alloys where each of seven different phases shows a unique dependence explained in terms of a specific Fermi-surface–Brillouin zone interaction[37]. While R_H of the amorphous H-R alloys is consistent with free-electron behaviour, $N_\chi(E_F)$ values are not since χ has the incorrect sign.

These two studies, combined with research covered in the reviews mentioned above confirm the general applicability of the free-electron model, for simple metals, though they also high-light the continuing inconsistencies when all electronic parameters are to be reconciled with that model.

23.3.3 Pd–Si–(Cu) alloys

Although palladium is a transition metal we discuss C_E of amorphous Pd–Si and Pd–Si–Cu alloys here because the experimental γ's are less than 15 per cent of the crystalline palladium value, suggesting a reduced d-band contribution. The interpretation of this marked change provides an excellent example of the contribution of various techniques to understanding electronic structure of amorphous alloys.

The drop in γ suggests that electrons from silicon have filled the d-band. However, γ is still higher than calculated for free-electron-like behaviour[17,21,38]. Although a recent theoretical model[29], combined with UPS data for $Pd_{84}Si_{16}$, suggests that the Pd d-electrons still contribute to $N_0(E_F)$, this conclusion does not fully explain all of the available experimental information.

An extensive composition survey of γ for Pd–Si(–Cu) alloys was originally designed to find a Nagel–Tauc minimum[38]. However, subsequent (e/a) values obtained from R_H were only 0.8 to 1.3 rather than 1.48 to 1.66 calculated from full valencies[39]. Thus the necessary $q_p = 2k_F$ condition of Nagel and Tauc was not met, which is just as well since a maximum in γ was found for the most stable composition [Si_{19}]! At the same composition a slow upwards trend in R_H is broken by a discontinuous increase of about 30 per cent. In contrast, measurements of χ showed a steady decrease in $N_\chi(E_F)$ for the same composition range. Thus the trends of γ, R_H and $N_\chi(E_F)$ seem inconsistent.

Mizutani and Massalski[39] concluded that there was an effective valence change for silicon at about [Si_{19}] with associated Pd 4d/Si 3p hybridization, a result that is consistent with photoemission studies of $Pd_{81}Si_{19}$ which show such a hybridization peak giving a local maximum in $N_0(E)$ near E_F. Johnson and Tenhover[30] point out that the seeming inconsistency in the electronic parameters can be reconciled with this model by recognizing that each experimental parameter contains a different weighting of electronic states. While $N_\gamma(E_F)$, which has a maximum at $[Si_{19}]$, contains the sum of all contributions to $N_0(E_F)$, R_H will tend to give more weight to *s*/*p* electrons and thus increase with [Si], while χ and UPS are influenced most by *d*-states which will decrease with increase in [Si].

23.3.4 Annealing effects

The effect of annealing on C_E should provide insight into the differences in the electronic states of amorphous and crystalline alloys of identical composition. However, annealing for structural relaxation or crystallization, even to a single crystalline phase, shows little, if any change in γ. The small changes observed are comparable with the reduction or disappearance of an (aT) TLS term (section 23.2.2) or may be attributed to small changes in λ_p due to the phonon spectrum changes which occur upon annealing (section 23.2.4). The near constancy of γ on crystallization is consistent with photoemission studies which show $N_0(E)$ to be remarkably similar in general shape for amorphous and crystalline phases of the same alloy for several alloy classes[28,40,41]. Crystalline and amorphous Nb_3Ge have γ values differing by a factor of 2.5. This difference can be related to superconducting properties as described below on p. 467.

23.4 Thermal properties of magnetic alloys

23.4.1 Background

The magnetic specific heat C_M is determined by the magnetic state (spin-glass, ferromagnetic, etc.) of the alloy and includes 'crystal-field effect' contributions in rare-earth alloys (section 23.4.6). The concentration and temperature dependence of C_M complement magnetic measurements such as $M(T)$ and $\chi(T)$ by providing thermodynamic information, including the entropy S_M associated with magnetic ordering, and thermal values of the spin-wave stiffness coefficient, D.

Trends in C_E throughout an alloy series also provide values of $N_\gamma(E_F)$ which may have critical limits for specific magnetic states (section 23.4.4). λ_p is small and monotonic through most magnetic alloy series so that $N_\gamma(E_F)$ closely reflects trends in $N_0(E_F)$. C_L for magnetic alloys was discussed in section 23.2. The close relationship between phonon behaviour and Invar properties will be discussed in section 23.2.2.

Measurement of thermal conductivity k in magnetic amorphous alloys has been confined to a few materials where the primary concern was with phonon properties as described in section 23.2.

In addition to the chapters on magnetic properties in this text and section 4 of Chen[1], a comprehensive review of magnetic properties has been given by Luborsky[42].

The most extensive studies of C_M for an alloy series to date have been in the readily made [FeNi][PB] series with the metalloids ranging from 12 to about 25 atom %. The thermal properties of these alloys were recently reviewed[18]. As the Fe/Ni ratio is increased a range of magnetic regimes from paramagnetic, through spin-glass and spin-cluster glass to ordered ferromagnetic occurs.

23.4.2 Dilute magnetic alloys

$Ni_{80}M_{20}$ and alloys of comparable composition are paramagnetic but close enough to the critical concentration for ferromagnetism that spin fluctuation effects are seen in C_M. C_M may be described by the model of Hahn and Wohlfarth[43]:

$$C_M = cA' \tag{23.7}$$

where A' is a constant and c a magnetic cluster concentration which can be estimated independently from magnetic measurements. For $Ni_{80}P_{14}B_6$ the values of c from both sources agree within about 30 per cent[44].

As [Fe] is increased up to about 8 atom %, the $[P_{14}B_6]$ alloys show spin-glass behaviour though this state is not yet confirmed for the $[B_{20}]$ alloys series. C_P data for two alloys[45] with low [Fe] are shown in *Figure 23.3*. An apparently linear C_M occurs up to about $0.8T_0$ where T_0 is the spin-glass transition temperature defined from the cusp in the $\chi(T)$ curve. For most spin-glasses, crystalline or amorphous, C_M can be described by:

$$C_M = AT \qquad (0.2T_0 \lesssim T \lesssim 0.8T_0) \tag{23.8}$$

There is increasing evidence, experimental and theoretical, that equation 23.8 does not necessarily reflect a fundamental property of all spin-glasses. If the predominant interaction is the RKKY interaction then scaling conditions ensure that $C_M/c \sim T_0/c$ so that A in equation 23.8 is independent of concentration c.

For the alloys in *Figure 23.3* the entropy S_M is only about 30 per cent of the expected value of $S_M = \mathrm{R}\ln(2J + 1)$ where J is the spin value obtained from magnetic measurements. Such reduced S_M values are typical for spin-glasses, implying that there is much residual disorder as T approaches 0 K. The entropy distribution obtained from

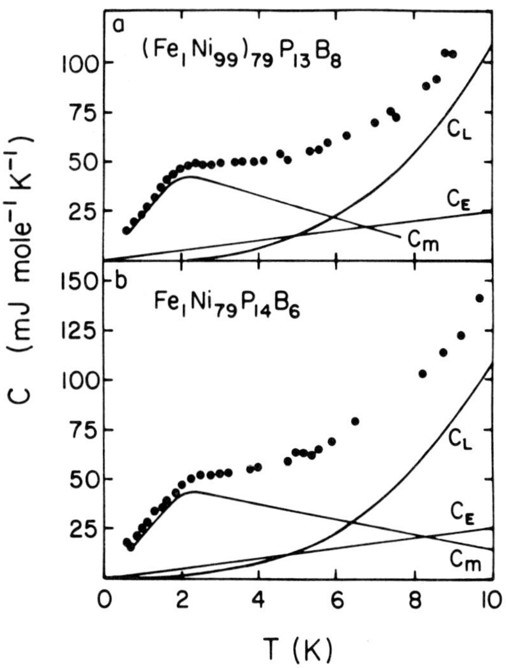

Figure 23.3 C_P for two spin-glass amorphous alloys containing about 1 atom % of iron. Separation into C_L, C_E and C_M is shown, C_E being estimated from the paramagnetic alloy $Ni_{80}P_{14}B_6$ with C_L and C_M then being separable through their different temperature dependences

C_M shows 60 per cent to be above T_0 implying that there is considerable short range order above T_0, again a typical spin-glass result.

C_M measurements for the metallic spin-glass $Zr_{30}Cu_{60}Fe_{10}$ down to 0.1 K show that $C_M = AT + BT^2$ provides a better description below 1 K than does equation 23.8. This conclusion agrees with recent results for crystalline spin-glasses with about 1 atom % of magnetic concentration[46,47]. Reinterpretation of the data on crystalline spin-glasses suggests $C_M \sim T^{3/2} (T \lesssim 0.2\, T_0)$ possibly because of the establishment of spin-wave like behaviour in finite regions of enhanced spin concentration[48]. Such a C_M could also describe the data on ZrCuFe.

Developments of the Edwards–Anderson spin-glass model have lead to a theoretical linear C_M considered as arising from magnetic two-level systems (MTLS)[49]. A computer calculation based on the RKKY interaction leads to a C_M that appears linear for a limited temperature range[50] but not down to $T = 0$ K. This agrees qualitatively with typical C_M data.

For a further extensive discussion of the spin-glass problem the reader is referred to recent reviews[51,52].

23.4.3 Concentrated magnetic alloys

For more concentrated spin-glasses or spin-cluster glasses such as for [Fe] $\gtrsim$ 8 atom % in the $[P_{14}B_6]$ alloys, the peak in C_M is concealed by C_L which is rising as T^3. However, for $T \lesssim 0.5\, T_0$, equation 23.8 still describes C_M. The vales of A up to $[Fe_{30}]$ are shown on bilogarithmic axes in *Figure 23.4*[18,53]. The smooth line is A estimated from the fraction of spins remaining isolated in a random distribution for a specific [Fe], and agrees well with the data up to about $[Fe_8]$. From the discussion above this suggests that RKKY interactions between the remaining isolated spins determine C_M at low [Fe] while at higher [Fe] the data exceed the smooth curve because of added contributions from pairs and larger spin-clusters. C_M thus provides a means of establishing limits of spin-glass and spin-cluster glass behaviour though a quantitative model of C_M for this region is still needed.

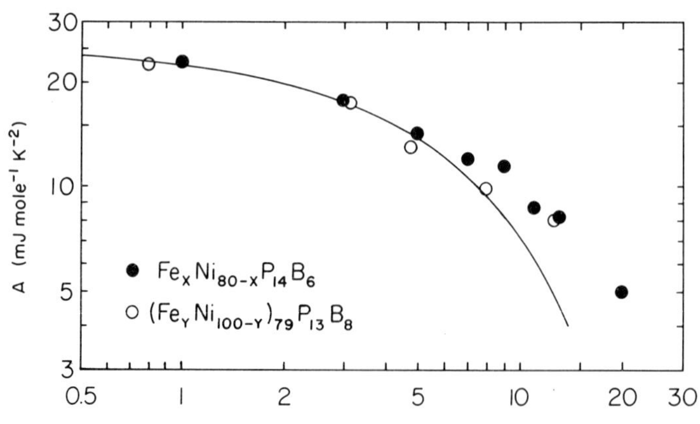

Figure 23.4 Comparison of the experimental linear magnetic contribution C_M (data points: ●, $Fe_xNi_{80-x}P_{14}B_6$; ○, $(Fe_yNi_{100-y})_{79}P_{13}B_8$) with the linear magnetic contribution estimated from the number of residual isolated spins in a random lattice (solid line). The excess C_M above about 10 K is attributed to contributions from spin-pairs and larger clusters

C_P of two sputtered amorphous MnSi samples have been reported[54]. The alloys were identified as spin-cluster glasses. Only 13 per cent of the expected S_M was found associated wth the rounded anomaly in C_M which peaked just above the ordering temperature of around 22 K. At low temperatures C_M was found to vary approximately with $T^{3/2}$, and was attributed to spin-waves in finite magnetic clusters.

Recently C_P for $Fe_{90}Zr_{10}$ was found to contain a linear term far in excess of C_E [55]. Spin-glass behaviour was observed[55,56], from magnetic studies, to develop below a ferromagnetic Curie temperature at about 220 K. Clear separation of the spin-glass contribution from possible Invar related properties is not yet possible for this alloy however.

Above [Fe_{30}] the [$P_{14}B_6$] alloys are characterized as ferromagnetic so that as T approaches 0 K the spin wave theory gives:

$$C_M = \alpha T^{3/2} = 0.113 k_B (k_B T/D)^{3/2} \tag{23.9}$$

D is the spin-wave stiffness coefficient also obtainable from neutron scattering and from $M(T)$. Values of D from C_M and $M(T)$ are in good general agreement[57–60,25] though in the past discrepancies with D values obtained from neutron scattering have been reported. In contrast to $Fe_{90}Zr_{10}$, $Co_{90}Zr_{10}$ is a well-behaved ferromagnet[55] down to $T = 1.5$ K with a small ferromagnetic C_M.

23.4.4 $N_\gamma(E_F)$ for magnetic alloys

Values of γ for the [FeNi][$P_{14}B_6$] and the [FeNi][B_{20}] alloys series are compared in *Figure 23.5*. Lack of a universal shape independent of metalloid suggests that a rigid-band model (RBM) is not applicable to $N(E)$ in these alloys. In fact at the nickel-rich end γ decreases with increase in [P] while at the centre [$Fe_{40}Ni_{40}$] it increases with increase in [P][18,53]. At the iron-rich end there is a minimum in γ near the $P_{14}B_6$ composition[58,60,24]. This is not necessarily a Nagel–Tauc minimum since an RBM does not apply to these alloys. The use of an RBM to analyse magnetic measurements[42] is not inconsistent with the conclusions that an RBM is unable to describe the trends in γ with composition, since magnetic measurements reflect the area below $N(E)$ rather than the $dN(E)/dE$ variations observed as γ varies with composition.

An attempt to describe $N(E_F)$ trends for [FeNi][PB] alloys has been made in terms

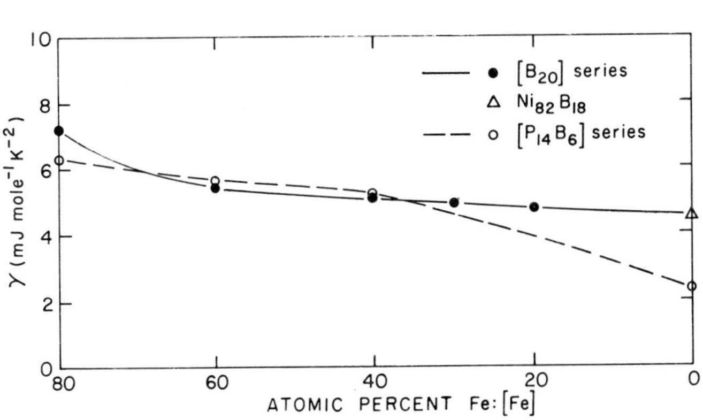

Figure 23.5 The electronic specific heat coefficient γ for $[FeNi]_{80}M_{20}$ amorphous alloys with [B_{20}] and [$P_{14}B_6$] metalloids. The different composition dependence of γ is inconsistent with a rigid-band model for $N(E)$

of the split-band model (SBM) proposed by Berger[61,62]. While trends similar to those of *Figure 23.5* are obtained the resulting band-shapes are not entirely consistent with the $N(E)$ shapes obtained for transition metal–metalloid alloys by photoemission[63].

It is apparent from *Figure 23.5* that the lowest $N_0(E_F)$ (or at least γ) values occur for the spin-glass (nickel-rich) region, and, as more magnetic and C_P data are obtained a precise correlation of magnetic state and $N_0(E_F)$ or $N_\gamma(E_F)$ should be possible for this alloy series and will be extended to others.

23.4.5 Invar alloys

Recognition of characteristic Invar behaviour[64] motivated C_P measurements in the binary amorphous Fe–B and Fe P alloys. Alloys with low [B] have dominant Invar behaviour. For those alloys γ increases rapidly while $\theta_D(0)$ rises close to the value for crystalline iron[65]. Values of D obtained from C_M are consistent with those from $M(T)$ and are very low (~ 50 meV Å) in the Invar region. The trends in γ and D are consistent with those in Invar crystalline Fe–Ni alloys but the rise in $\theta_D(0)$ is in direct contrast to the phonon-softening thought to be a signature of, if not a contributor to, Invar behaviour in crystalline alloys.

XPS results imply a decrease in $N_0(E_F)$ in amorphous $Fe_{80}B_{20}$ compared with crystalline iron[63], while a theoretical model of $N_0(E_F)$ for the amorphous Fe–B alloys gave only 50 per cent of the $N_\gamma(E_F)$ obtained from the experimental γ values[66]. The enhancement in γ occurs where D is decreasing rapidly and the magnetic moment is also decreasing, suggesting unstable ferromagnetism. Nevertheless the spin-wave C_M (varying as $T^{3/2}$) is still clearly defined in addition to the enhanced γ.

Matsuura and Mizutani[25] conclude that the high $\theta_D(0)$ value for the Invar amorphous Fe–B alloys provides a key to understanding Invar characteristics in both amorphous and crystalline alloys and that the enhanced γ term could be magnetic in origin. At present, it is only possible to conclude that magnetic effects, changes in $N_0(E_F)$ or in λ_p may be responsible for the enhanced γ of Invar alloys. A similar enhancement of γ has also been reported in the Invar alloy $Fe_{90}Zr_{10}$ in association with spin-cluster or spin-glass behaviour (p. 461)[55].

23.4.6 Rare earth alloys

The magnetic and related thermal properties of rare-earth based amorphous alloys are determined by the competition between exchange interactions and crystalline field effects (CEF). Harris *et al.*[67] proposed a Hamiltonian which combines exchange effects with a random uniaxial anisotropy from CEF:

$$\mathscr{H} = -\Sigma J_{ij}\vec{S}_i \cdot \vec{S}_j - D\Sigma(\vec{n}_i \cdot \vec{S}_i)^2 \qquad (D > 0) \qquad (23.10)$$

$\vec{S}$ is the total angular momentum, $\vec{n}$ the direction of the local anisotropy D at site i. A review of theoretical and experimental work based on this model has appeared[68]. We concentrate here primarily on C_M results. It has been shown that the above Hamiltonian cannot lead to long-range order but that it describes spin-glasses well[69]. For low temperatures the resulting C_M is given in a molecular field H_m approximation by:

$$C_M = Nk^2T/g\mu_B JH_m \qquad (23.11)$$

A large linear term (~ 155 T mJ mol^{-1} K^{-2}) attributable to this contribution was first observed in amorphous DyCu[70]. The estimated H_m was 82 kOe. A later study of extended compositions of DyCu confirmed that random crystalline anisotropy dominates C_M, but some form of long-range order could not be ruled out for all compositions[71].

Amorphous Ce–Au and Ce–Cu alloys show typical spin-glass peaks in $\chi(T)$ but 50 kOe is needed to suppress the peaks. C_M for these alloys was precisely determined by comparison with non-magnetic amorphous La–Au and H_m found to be 180 kOe from equation 23.11. Below 0.5 K a T^3 contribution to C_M was attributed to collective magnetic excitations[72].

$Pr_{21}Ag_{79}$ amorphous alloy is paramagnetic at the lowest temperatures allowing observation of pure CEF effects. C_M was determined to be large and almost constant and attributed to CEF excitations with a distributive rather than a uniaxial anisotropy[73]. This model gives a plateau in C_M in good agreement with the data. In $Sm_{21}Ag_{79}$ in contrast, C_M passes through a broad maximum near 4 K with associated S_M expected for a Kramers ion in the absence of CEF excitations. Exchange interaction effects dominate in this alloy since the Curie–Weiss constant is 8 K.

For the *s*-state ion Gd^{3+} random single-ion anisotropy should not be important. A spin-glass cusp in $\chi(T)$ occurred at 17.5 K in amorphous $GdAl_2$ while C_M varied as $T^{3/2}$ up to about 10 K with a broad maximum [74] just above T_0. As in similar cases described above, the $T^{3/2}$ term was attributed to collective excitations in spin-clusters with ferromagnetic behaviour.

Magnetic studies of rare-earth based magnetic amorphous alloys have been more extensive than thermal studies. The limitations on C_M measurements arise from sample preparation difficulties for the relatively large samples needed, but more thermodynamic information would complement and clarify many of the magnetic measurements already made.

23.5 Thermal properties of superconducting alloys

23.5.1 Background

According to BCS theory, developed for weak-coupling superconductors ($\lambda_p < 0.5$), the transition temperature T_c may be expressed as[75]:

$$T_c = 1.13\theta_D(0)e^{-1/N_0(E_F)\lambda_p} \tag{23.12}$$

which shows the relative dominance of λ_p and $N_0(E_F)$ over $\theta_D(0)$ which appears only in the pre-factor. Below T_c the electronic specific heat C_{ES} is ideally described by:

$$C_{ES} = 8.5\gamma T_c \exp(-2\Delta_0/k_B T) \tag{23.13}$$

where Δ_0 is the superconducting energy gap parameter which is related to T_c in the BCS limit by:

$$\frac{2\Delta_0}{k_B T_c} = 3.53 \tag{23.14}$$

At T_c BCS theory also predicts a discontinuity in C_E given by:

$$\Delta C_E = C_{ES} - C_{EN} = 1.43\gamma T_c \tag{23.15}$$

If the numerical factors (3.53, 1.43) of equations 23.14 and 23.15 are greatly exceeded then the superconductor is intermediate or strong-coupled depending upon the value of λ_p obtained.

A more quantitative insight into the role of λ_p is provided by the McMillan equation which, though it was developed originally for crystalline niobium, has found a wider spectrum of applications in the absence of effective alternative descriptions:[76]

$$T_c = \frac{\theta_D(0)}{1.45} \exp\left(-\frac{1.04(1+\lambda_p)}{\lambda_p - \mu^*(1+0.62\lambda_p)}\right) \qquad (23.16)$$

where μ^* is the effective Coulomb coupling constant usually taken to be 0.1 to 0.15. Modifications of equation 23.16 have been developed[77] which extend accurately to higher values of λ_p.

λ_p itself is defined by:

$$\lambda_p = \frac{N_0(E_F)\langle I^2\rangle}{M\langle\omega^2\rangle} \qquad (23.17)$$

where $\langle I^2\rangle$ is the average squared electron-ion matrix element, M the ionic mass and $\langle\omega^2\rangle$ a mean square phonon frequency defined by McMillan[76] and accessible from tunnelling data.

Information on most of the factors in the above descriptions of superconductivity can be obtained from C_P. In the normal state:

$$C_P = \gamma T + \beta(T)T^3 \qquad (23.18)$$

providing values of $N_\gamma(E_F)$, $\theta_D(0)$ and $\theta_D(T)$. The last may be useful in characterizing the phonon spectrum in the absence of other sources of phonon data. From T_c and $\theta_D(0)$ the McMillan equation 23.16 provides λ_p so that $N_0(E_F)$ can be obtained from $N_\gamma(E_F)$. All of the factors contributing to equation 23.17 are thus accessible if reasonable estimates of $\langle\omega^2\rangle$ are made in the absence of tunnelling data.

In addition, equations 23.14 and 23.15 can be used to characterize the nature of the coupling which should, of course, be compatible with the value of λ_p obtained, and also to give Δ_0.

For comparison, values of $N_x(E_F)$ can be obtained from interpretations of χ measurements, though agreement with $N_\gamma(E_F)$ is not universal. Better consistency is found for $N_\gamma(E_F)$ and $N_H(E_F)$, the density of states value obtained from the temperature dependence of the upper critical field $(\mathrm{d}H_{c2}/\mathrm{d}T)_{T_c}$, the electrical resistivity ρ and density d since[78]:

$$\gamma_H = \frac{2.23}{\rho d}\left(\frac{\mathrm{d}H_{c2}}{\mathrm{d}T}\right)_{T_c} \text{ mJ mol}^{-1}\text{ K}^{-2}$$

23.5.2 Elements and simple alloys

Superconductivity is found in only a few elemental amorphous metals, prepared generally by cryoquenching and often requiring small amounts of a second component to stabilize the amorphous phase. A review of their properties and of early studies on amorphous alloys was provided by Bergmann[32]. Recent reviews which encompass transition metal alloys also refer to advances in simple metal alloys[33,74] as does Chapter 22.

As we described on p. 457 the Fermi surface of simple metals is free-electron like, the absence of long-range order eliminating band effects which, in crystalline metals, cause deviations from the free-electron model. Johnson has shown, using a simple jellium model, that T_c/Ω_p where Ω_p is the ionic plasma frequency is a simple function of the valence Z, while λ_p is proportional to Z. Known simple metal superconductors fit these universal relationships very well[33,79].

23.5.3 Transition metal alloys

In transition metal alloys the *d*-band structure dominates the superconducting properties. The alloys for which thermal studies have been undertaken fall into two principal categories:

1. Transition metal–metalloid alloys such as [MoRu][PB].
2. Early transition metal–late transition metal alloys such as Cu–Zr, Ni–Zr and La–Au.

New microcalorimetry techniques can be applied to approximately 1 mg sputtered samples of, for instance, Mo–Ru binary alloys but few data are yet available[80]. Several reviews of superconductivity with emphasis on transition metal alloys have appeared recently[30,33,79,81]. In this section we will emphasize conclusions reached primarily through C_P measurements.

Even for the transition metal–metalloid alloys C_P studies have been relatively infrequent owing to the low mass of the samples generally available, and $N_\chi(E_F)$ and $N_H(E_F)$ have been used to estimate the density of electron states, the latter providing good agreement with $N_\gamma(E_F)$ when checked. Values of $\theta_D(0)$ are not yet known for many alloys, nor is $\theta_D(T)$ which, as we discussed on p. 455, is known to be very sensitive in the magnetic 3*d* transition metal–metalloid alloys to phosphorus/boron ratio and to annealing state.

A C_P survey of $(Mo_{0.60}Ru_{0.40})_{100-x}B_x$ alloys for $14 < x < 22$ has shown a monotonic variation of γ with x, with an increase of almost a factor of 2 between $x = 22$ and $x = 14$ as T_c increases[82] from 4.7 to 6.25 K. The ratios of equations 23.14 and 23.15 are within the BCS limits while $\lambda_p (= 0.69)$ suggests weak to intermediate coupling. The Debye temperature is discontinuous at $x = 16$, suggesting possible structural changes but shows a minimum at $x = 20$ consistent with the stability minimum mentioned on p. 456.

The values of γ and T_c extrapolated to $x = 0$ are quite close[80] to those of amorphous $Mo_{58}Ru_{42}$, for which T_c is 8.3 K. The binary alloy is, however, analysed to be strong-coupling with $\lambda_p = 0.84$. In addition the binary alloy has a $\theta_D(0)$ almost 40 K lower than the metalloid-containing alloys.

The analysis of the above alloys is consistent with the conclusion that in many cases [the possible exceptions of Cu–Zr and La–(Ga, Au) alloys are discussed below] superconductivity in amorphous transition metal alloys follows the model of Varma and Dynes developed for crystalline transition metal alloys[83]. They show that for the factors included in the definition of λ_p (equation 23.17) there is an approximate relationship:

$$\frac{\langle I^2 \rangle}{M\langle \omega^2 \rangle} \sim W(1 \mp S) \tag{23.20}$$

where W is an estimated *d*-band width and S an overlap integral for atomic *d*-orbitals on nearest neighbours. This ratio should be constant for a given alloy series so that λ_p should be proportional to $N_0(E_F)$ only.

For alloys containing 4*d* transition metals Johnson[30] shows that λ_p is proportional to $N_0(E_F)$ with proportionality constants depending on whether E_F lies in the upper or lower half of the 4*d* band, as shown in *Figure 23.6*. Thus it appears that amorphous transition metal superconductors with metal components from the same series behave in a very similar manner to crystalline alloys of similar composition.

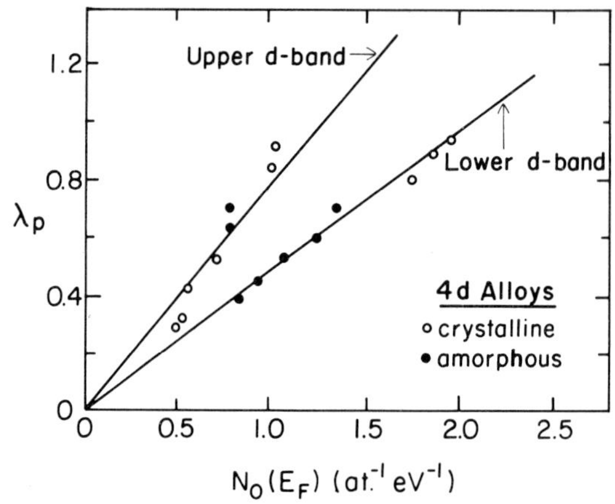

Figure 23.6 Proportionality of λ_p and $N_0(E_F)$ for the 4d alloys; ○, crystalline and ●, amorphous (after Johnson and Tenhover[30])

Inter-series alloys may not follow these same trends as is shown by the Zr_xCu_{100-x} alloy series (*Figure 23.7*) for which C_P is known for a wide range of x in both superconducting and normal states[84–89]. Consistency between results from various sources is very good. An added advantage of this series is that it has been used to determine the linear (aT) phonon contribution below 1 K so that very accurate corrections to C_E can be made[90].

The $\theta_D(0)$ values for this series are always below 230 K but are scattered with x. No T^5 contribution is found up to 6 K. The values of γ increase monotonically with [Zr] and exceed those of crystalline zirconium and copper. They are shown in *Figure 23.7* where they are compared with the $N_0(E_F)$ values obtained after reduction $[N_0(E_F) = N_\gamma(E_F)/(1+\lambda_p)]$ with λ_p from the McMillan equation[88]. Since $N_0(E_F)$ is almost independent of composition λ_p varies strongly with composition. This implies,

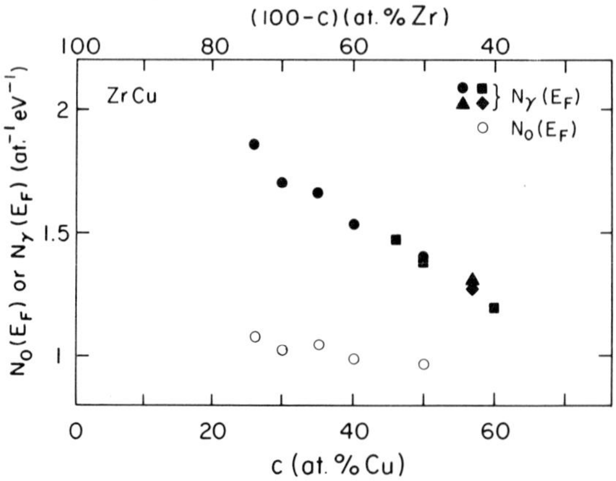

Figure 23.7 Concentration dependence of $N_\gamma(E_F)$ and $N_0E_F)$ for the amorphous Zr–Cu alloys showing the near constancy of $N_0(E_F)$ in the superconducting alloy range (after refs. 84—89)

from equation 23.17, that $\langle I^2 \rangle / M \langle \omega^2 \rangle$ is not constant for this alloy series in contrast to the Varma–Dynes relation (equation 23.20). Furthermore, it has been noted that for amorphous alloys $1/M\langle \omega^2 \rangle$ can be expected to be almost constant so that the changes in λ_p with [Zr] must be caused by strong variation of $\langle I^2 \rangle$ with alloying[88,89].

This variation in $\langle I^2 \rangle$ is then attributed to changes in hybridization between the *s*/*p* and *d* states of the zirconium atoms as a function of average atomic separation which will itself vary with [Zr][88,89]. This conclusion is consistent with a similar $\langle I^2 \rangle$ trend in amorphous noble metal alloys inferred from T_c changes alone[91] and with the increase of T_c with pressure in Zr–Cu metallic glasses[92].

$La_{100-x}M_x$ alloys where M is Au ($x = 0.20$, 0.24) or Ga ($0.16 < x < 0.28$) are superconducting. C_P surveys showed that they are weak to intermediate coupling, and that T_c is almost proportional[92,93] to λ_p. λ_p decreases monotonically as x increases, but $\theta_D(0)$ shows a compositional discontinuity similar to that seen in the [MoRu][PB] alloys discussed above. While $N_\gamma(E_F)$ decreases with increase of x this is attributed mostly to the λ_p change and after correction the bare density of states $N_0(E_F)$ is independent of γ within experimental error. As for the inter-series Cu–Zr alloys it is possible that $\langle I^2 \rangle$ variations are responsible for the changes in λ_p in the [La][Au, Ga] inter-series alloys.

Evidence that the role of $\langle I^2 \rangle$ may be different in amorphous and crystalline alloys is provided by an early study[94] of C_P for amorphous Nb_3Ge which was compared with crystalline Nb_3Ge. While the crystalline alloys have the highest known T_c (~ 23 K), the amorphous phase T_c is only about 1.4 K. $\theta_D(0)$ is reduced from 302 K in the crystalline to 222 K in the amorphous alloy, following the general pattern discussed in section 23.2.4, while γ is reduced from 30 to 12 mJ mol^{-1} K^{-2}, which is far greater than the usual small changes of γ between crystalline and amorphous phases (section 23.3.4). The amorphous alloy was concluded to be a weak-coupling superconductor while the crystalline alloy is strong-coupling.

By using the McMillan equation and well justified assumptions on the nature of the phonon spectrum it was shown from equation 23.17 that $\langle I^2 \rangle_c / \langle I^2 \rangle_a = 3$. This lowering in $\langle I^2 \rangle$ in the amorphous state was attributed to dehybridization between Nb 4*d* and Ge 3*p* orbitals, a conclusion consistent with XPS of both phases which shows such dehybridization near E_F for the amorphous alloy.

In most other cases for which C_P has been determined the compositions are unique or have narrow limits, so that trends in λ_p, $N_\gamma(E_F)$, $N_0(E_F)$ and $\theta_D(0)$ cannot be followed and conclusions are necessarily limited.

$Zr_{70}Pd_{30}$ was the first superconducting amorphous alloy prepared in bulk form studied primarily for information on the phonon (aT) contribution below T_c (section 23.2.3). It also proved to be a well-behaved weak-coupling superconductor with the numerical factors of equations 23.14 and 23.15 within about 15 per cent of the BCS values[95].

Thermal properties of sputtered alloys close to the $Zr_{70}M_{30}$ composition for *M* of silver, nickel, platinum and copper also show weak-coupling behaviour[96]. Annealing of the $Zr_{76}Ni_{24}$ alloys decreased $\theta_D(0)$, T_c and γ, the last by a factor too great for an (aT) contribution change. The $N_\gamma(E_F)$ change was attributed to a change in λ_p arising from the softening of the phonon spectrum[97].

Selected alloys of copper or nickel with zirconium, titanium or niobium showed only $Ni_{36}Zr_{64}$ to be superconducting above 1.5 K but showed remarkable consistency of γ values, all being[98] close to 4.0 mJ mol^{-1} K^{-2}. For the zirconium-based alloys this is not surprising since XPS has shown $N_0(E_F)$ to be dominated by zirconium states[99] and upon substitution of zirconium by titanium or niobium it is concluded that a similar pattern is maintained.

$La_{78}Zn_{22}$ has been shown to be strong-coupling in the amorphous phase with T_c

of 4.15 K but when crystallized to the single phase hexagonal $La_{0.78}Zn_{0.22}$ it becomes a BCS weak-coupling superconductor[100] with T_c of 2.23 K. Since $\theta_D(0)$ and $N_0(E_F)$ both increase upon crystallization the decrease in T_c is attributed entirely to changes in λ_p owing to differences in short-range order. The relation of weak coupling (crystalline) to strong coupling (amorphous) is the inverse of that discussed above for Nb_3Ge.

Superconducting amorphous alloys seem to be very well described by the relationships and parameters outlined in section 23.5 which were initially obtained for crystalline materials. With a few exceptions they are weak to intermediate coupled. For the series of 4*d* band alloys λ_p is determined primarily by $N_0(E_F)$ as in crystalline alloys. Interseries alloys appear to have a constant $N_0(E_F)$ with λ_p determined by changes in $\langle I^2 \rangle$ with composition. As more superconducting alloys are found and the range of present compositions extended C_P studies will continue to play a dominant role. When combined with photoemission studies and eventually with tunnelling data, further clarification of the factors contribution to λ_p, and hence to T_c should result.

23.6 Conclusions

We have seen that one technique, C_P measurement, can provide valuable and often unique information on many aspects of amorphous metallic alloys: phonon, electronic, magnetic and superconducting. Thermal conductivity provides valuable complementary information.

Measurements of C_P and k stimulated the great body of recent work on low-energy excitations in amorphous materials, leading to the concept of TLS. There remain many unanswered questions in the phonon areas, however, particularly at temperatures above 2 K. For the entire temperature range, the nature of the physical entity responsible for the low energy excitations remains unidentified. A Debye temperature minimum may be identified for several alloy series at the most stable compositions.

For simple metals a free-electron model provides an excellent description of the electronic properties and is directly verified by determination of C_E, while in more complicated situations, such as the Pd–Si–Cu alloys a study of C_E must be combined with ρ, R_H and photoemission results before even a tentative electronic model is proposed.

C_P provides, through C_M, information complementary to direct magnetic measurements, while C_E for magnetic alloys sets limits on $N_\gamma(E_F)$ for various magnetic regimes. In superconducting alloys the parameters obtained from C_P in both the normal and the superconducting states provide almost all of the parameters needed to fit existing models of superconductivity and enable conclusions to be reached on contributions to λ_p and hence the T_c value.

Acknowledgements

Preparation of this review was supported in part by the National Science Foundation. I thank the many colleagues who made information available, in many cases before publication, and to those who have contributed to my education in the area of amorphous metallic alloys.

References

1. Chen, H. S., *'Glassy Metals', Rep. Prog. Phys.*, **43,** 353 (1980)
2. *'Proceedings of the 2nd International Conference on Rapidly Quenched Metals'*, Eds. Grant, N. J. and Giessen, B. C., (Sect. I: Massachusetts Institute of Technology Press, Cambridge, Mass.); (Sect. II: *Mater. Sci. Engng*, **23**) (1976)
3. *'Proceedings of the 3rd International Conference on Rapidly Quenched Metals'*, Ed. Cantor, B., The Metals Society, London (1978)
4. *'Proceedings of the 4th International Conference on Rapidly Quenched Metals'*, Sendai, Japan, 1981 Eds. Masumoto, T. and Suzuki, K., Japan Institute of Metals, Sendai (1982)
5. *'Metallic Glasses'*, Eds. Gilman, J. J. and Leamy, H. J., American Society for Metals, Metals Park, Ohio (1978)
6. *'Glassy Metals I'*, Eds. H. J. Guntherodt and H. Beck, Springer-Verlag, Berlin, (1982)
7. *'The Magnetic, Chemical and Structural Properties of Glassy Metallic Alloys'*, Ed. Hasegawa, R., Chemical Rubber Co. Press, Boca, Florida (1982)
8. Zeller, R. C. and Pohl, R. O., *Phys. Rev.*, **B4,** 2029 (1971)
9. Pohl, R. O. and Salinger, G. L., *Ann. N.Y. Acad. Sci.*, **279,** 150 (1976)
10. Lohneysen, H. V., *Phys. Rep.*, **79,** 163 (1981)
11. Pohl, R. O., in *'Amorphous Solids: Low Temperature Properties'*, Ed. W. A. Phillips, Springer-Verlag, Berlin (1981)
12. Anderson, A. C., in ref. 11
13. Bhattacharya, A., *Contemp. Phys.*, **22,** 117 (1981)
14. Black, J. L., in ref. 6, p. 167
15. Bellessa, G., *J. Phys. (Paris)*, **41,** C8, 723 (1980)
16. Ackerman, D. A., Moy, D., Potter, R. C., Anderson, A. C. and Lawless, W. N., *Phys. Rev.*, **B23,** 3886 (1981)
17. Golding, B., Bagley, B. G. and Hsu, F. S. L., *Phys. Rev. Lett.*, **29,** 68 (1972)
18. Onn, D. G., *J. appl. Phys.*, **52,** 1788 (1981)
19. Malozemoff, A. P., Chang. P. H. and Grimsditch, M., *J. appl. Phys.*, **50,** 5896 (1979)
20. Mizutani, U. and Massalski, T. B., *J. Phys. F: Metal Phys.*, **10,** 1093 (1980)
21. Chen, H. S. and Haemmerle, W. H., *J. Non-cryst. Solids*, **11,** 161 (1972)
22. Montgomery, C. G., *J. Low Temp. Phys.*, **39,** 13 (1980)
23. Matey, J. R. and Anderson, A. C., *J. Non-cryst. Solids*, **23,** 129 (1977)
24. Ravex, A., Lasjaunias, J. and Bethoux, O., *Physica*, **107B,** 397 (1981)
25. Matsuura, M. and Mizutani, U., in ref. 4
26. Mizutani, U. in ref. 4
27. Mizutani, U. Hartwig, K. T., Massalski, T. B. and Hopper, R. W., *Phys. Rev. Letts.*, **41,** 661 (1978)
28. Riley, J. D., Ley, L., Azoulay, J. and Terakura, K., *Phys. Rev.*, **B20,** 776 (1979)
29. Guntherodt, J. H., *et al.*, *J. Phys. (Paris)*, **41,** C8, 381, 1980
30. Johnson,W. L. and Tenhover, M. in ref. 7
31. Nagel, S. R. and Tauc, J., *Phys. Rev. Lett.*, **35,** 380 (1975)
32. Bergmann, G., *Phys. Rep.*, **27,** 159 (1976)
33. Johnson, W. L., *J. Phys., (Paris)*, **41,** C8, 731 (1980)
34. Mizutani, U. and Mizoguchi, T., *J. Phys. F: Metal Phys.*, **11,** 1385 (1981)
35. Giessen, B. C., Calka, A., Raman, R. and Sellmeyer, D. J., *'Amorphous Magnetism II'* Eds. Levy, R. A. and Hasegawa, R., p. 197, Plenum Press, New York (1977)
36. Matsuda, T. and Mizutani, U., in ref. 4.
37. Massalski, T. B. and Mizutani, U., *Prog. mater. Sci.*, **22,** 151 (1978)
38. Massalski, T. B., Mizutani, U., Hartwig, K. T. and Hopper, R. W., ref. 3, vol. 2, p. 81; *Phys. Rev. Lett.*, **41,** 66 (1978)
39. Mizutani, U. and Massalski, T. B., *Phys. Rev.*, **B21,** 3180 (1980)
40. Nagel, S. R., Fisher, G. B., Tauc, J. and Bagley, B. G., *Phys. Rev.*, **B13,** 3284 (1976)
41. Amamou, A. and Krill, G., *Solid St. Communs*, **33,** 1087 (1980)
42. Luborsky, F. E., in *'Ferromagnetism'*, Ed. Wohlfarth, E. P., Ch. 6, North-Holland Publ. Co., Amsterdam (1981)
43. Hahn, A. and Wohlfarth, E. P., *Helv. Phys. Acta*, **41,** 857 (1968)
44. Onn, D. G., Antoniuk, T. H., Donnelly, T. A., Johnson, W. P., Egami, T., Prater, J. T. and Durand, J., *J. appl. Phys.*, **49,** 1730 (1978)
45. Long, T. C., Donnelly, T. A., Durand, J. and Onn, D. G., ref. 3, vol. 2, p. 69
46. Fogle, W. H., Ho, J. C. and Phillips, N. E., *J. Phys. (Paris)*, **39,** C6, 901 (1978)
47. Martin, D. L., *Phys. Rev.*, **B21,** 1906 (1980)
48. Thomson, J. O. and Thompson, J. R., *J. Phys. F: Metal Phys.*, **11,** 247 (1981)

49. Dasgupta, C., Ma, S. K. and Hu, C. K., *Phys. Rev.*, **B20,** 3837 (1979)
50. Walker, L. R. and Walstedt, R. E., *Phys. Rev.*, **B22,** 3816 (1980)
51. Mydosh, J. A., *J. Magnetism magn. Mater.*, **7,** 237 (1978)
52. Beck, P. A., *Prog. mater. Sci.*, **23,** 1 (1978)
53. Onn, D. G., Broomhall, J. R. and Krause, J. K., to be published
54. Hauser, J. J., Hsu, F. S. L., Kammlott, G. W. and Waszczak, J. V., *Phys. Rev.*, **B20,** 3391 (1979)
55. Obi, Y., Wang, L. Q., Motsay, R. and Onn, D. G., *J. appl. Phys.*, **53,** 2304 (1982)
56. Hiroyoshi, H. and Fukamichi, K., *J. appl. Phys.*, **53,** 2226 (1982)
57. Onn, D. G., Johnson, W. D., Gleeson, P. F., Donnelly, T. A., Egami, T. and Liebermann, H. H., *J. Phys. C:* **10,** L639 (1977)
58. Donnelly, T. A., Egami, T. and Onn, D. G., *Phys. Rev.*, **B20,** 1211 (1979)
59. Onn, D. G., Sundermier, A. and Krause, J. K., *J. appl. Phys.*, in press
60. Matsuura, M., Mizutani, U. and Yazawa, Y., *J. Phys. F: Metal Phys.*, **11,** 1393 (1981)
61. Berger, L., American Institute of Physics Conf. Proc. No. 34, p. 355 (1976)
62. Berger, L., *Physica,* **91B,** 31 (1977)
63. Cartier, E. , Baer, Y., Laird, M. and Guntherodt, H. J., *J. Phys. F: Metal Phys.*, **10,** L21 (1980)
64. Fukamichi, K., Kikuchi, K., Arakawa, S. and Masumoto, T., *Solid St. Communs,* **23,** 955 (1977)
65. Gupta, K. P., Cheng, C. H. and Beck, P. A., *J. Phys. Chem. Solids,* **25,** 73 (1964); Dixon, M., Hoare, F. E. and Holden, T. M., *Proc. R. Soc.*, **A303,** 339 (1968)
66. Fujiwara, T., *J. Phys. F: Metal Phys.*, (in press)
67. Harris, R., Plische, M. and Zuckermann, M. J., *Phys. Rev. Lett.*, **31,** 160 (1973)
68. Cochrane, R. W., Harris, R. and Zuckermann, M. J., *Phys. Rep.*, **48,** 1 (1979)
69. Pelcovitz, R. A., Pytte, E. and Rudnick, J., *Phys. Rev. Lett.*, **40,** 476 (1978)
70. Coey, J. M. D. and von Molnar, S., *J. Phys. Lett.*, **39,** L-327 (1978)
71. von Molnar, S., Guy, C. N., Gambino, R. J. and McGuire, T. B., *J. Magnetism magn. Mater.*, **15—18,** 1391 (1980)
72. Lohneysen, H. V., Schink, H. J., Felsch, W., Samwer, K. and Schröder, H., *Physica,* **107B,** 631 (1981)
73. Garoche, P., Fert, A., Veyssie, J. J. and Boucher, B., *J. Magnetism magn. Mater.*, **15—18,** 1397 (1980)
74. Coey, J. M. D., von Molnar, S. and Gambino, R. J., *Solid. St. Communs,* **24,** 167 (1977)
75. Bardeen, J., Cooper, L. N. and Scheiffer, J. R., *Phys. Rev.*, **108,** 1175 (1957)
76. McMillan, W. L., *Phys. Rev.*, **167,** 331 (1968)
77. Allen, P. B. and Dynes, R. C., *Phys. Rev.*, **B12,** 905 (1975)
78. Hake, R. R., *Phys. Rev.*, **158,** 356 (1966)
79. Johnson, W. L., in *'Glassy Metals I'*, Ed. Guntherodt, H. J. and Beck, H., p. 191, Springer-Verlag, Berlin (1981)
80. Early, S. R., Hellman, F., Marshall, J. and Geballe, T. H., *Physica,* **107B,** 327 (1981)
81. Johnson, W. L., *J. appl. Phys.*, **50,** 1557 (1979)
82. Hopkins, S. T. and Johnson, W. L., to be published
83. Varma, C. M. and Dynes, R. C., *'Superconductivity in d- and f-band Metals'*, Ed. Douglass, D. H., Plenum Press, New York (1976)
84. Mizoguchi, T., von Molnar, S. and Cargill, C. S. (III), in ref. 35, p. 513
85. Lohneysen, H. V., Platte, M., Sander, W., Schink, H. J., von Minnigerode, G. and Samwer, K., *J. Phys. (Paris)* **41,** C8, 745 (1980)
86. Garoche, P., Calvayrac, Y. and Veyssie, J. J., *J. Phys. (Paris)* **41,** C8, 766 (1980)
87. Ravex, A., Lasjaunias, J. C. and Bethoux, O., *Physica,* **107B,** 395 (1981)
88. von Minnigerode, G. and Samwer, K., *Physica,* **108B,** 1217 (1981)
89. Samwer, K. and Lohneysen, H. V., *Phys. Rev. B,* in the press
90. Schink, H. J., Lohneysen, H. V. and Sander, W., *Physica,* **107B,** 389 (1981)
91. Nast, P., Samwer, K. and von Minnigerode, G., *Z. Phys.*, **B38,** 89 (1980)
92. Willer, J. , Fritsch, G. and Lüscher, E. *Appl. Phys. Lett.*, **39,** 1580 (1977)
93. Shull, W. H., Naugle, D. G., Poon, S. J. and Johnson, W. L., *Phys. Rev.*, **B18,** 3263 (1978)
94. Tsuei, C. C., von Molnar, S. and Coey, J. M., *Phys. Rev. Lett.*, **41,** 664 (1981)
95. Graebner, J. E., Golding, B., Schutz, R. J., Hsu, F. S. L. and Chen, H. S., *Phys. Rev. Lett.*, **39,** 1480 (1977)
96. Ravex, A., Lasjaunias, J. C. and Bethoux, O., *Physica,* **107B,** 395 (1981)
97. Ravex, A., Lasjaunias, J. C. and Bethoux, O., *Physica,* **107B,** 397, (1981)
98. Moody, D. E. and Ng, H. K., Institute of Physics Conference Series, No. 55, p. 631 (1981)
99. Amamou, A. and Krill, G., *Solid St. Communs,* **28,** 957 (1978)
100. Bethoux, O., Laborde, O., Lasjaunias, J. C. and Ravex, A., *J. Phys. (Paris),* **41,** C8, 754 (1980)

Chapter 24

Chemical properties

K. Hashimoto

The Research Institute for Iron, Steel and Other Metals, Tohoku University, Sendai, Japan

24.1 Introduction

Amorphous alloys are new materials from the chemical point of view. The amorphous structure is thermodynamically metastable. However, they are free from defects associated with the crystalline state such as grain boundaries, dislocations and stacking faults. Furthermore, rapid quenching from the molten state to prepare amorphous alloys prevents solid state diffusion during quenching. Thus they are also free from defects such as second phases, precipitates and segregates, which are generally formed by solid state diffusion during slow cooling or heat treatment. The amorphous alloys are, therefore, regarded as ideally chemically homogeneous alloys with respect to the surface-related characteristics such as corrosion and catalysis. On the other hand, the amorphous alloys consist at least of two components and often contain large quantities of metalloids necessary for the formation of the amorphous structure. Such a complicated composition also greatly affects their chemical properties.

Before rapidly quenched amorphous alloys became available, Hoar[1] had stated that the formation of amorphous passive films would enhance the corrosion resistance. Okamoto[2] suggested that the most important parameter controlling the corrosion resistance of stainless steel is the amorphous nature of the passive film. Revesz and Kruger[3] have pointed out that the match of two structures (passive film and substrate alloy) with rigidly defined but different long-range order is poor and that a quasi-perfect solid–solid interface can be obtained only when either a quasi-perfect epitaxial relationship exists between single crystals or at least one side of the interface is non-crystalline. The amorphous alloys are, of course, non-crystalline, on which amorphous passive films can easily be formed. In this manner, the high corrosion resistance, particularly against localized corrosion attack, can be provided by their chemically homogeneous single phase nature along with the characteristic that they are always chemically homogeneous even when alloying elements beneficial in increasing the corrosion resistance are added in large quantities.

Their ability to form the homogeneous single phase of complicated composition with a sufficiently high mechanical strength will also be effective in preparing an alloy with the characteristic provided by alloying with large amounts of specific elements, such as a high catalytic activity for a specified chemical or electrochemical reaction.

24.2 Corrosion-resistant amorphous alloys

24.2.1 Corrosion resistance

In general, an amorphous metal–metalloid alloy containing only one metallic component possesses a lower corrosion resistance than the pure metal constituting the amorphous alloy. For instance, the corrosion rate of amorphous iron–metalloid alloys without a second metallic element is higher than that of crystalline pure iron[4–6]. However, the corrosion rate is decreased by addition of almost all second elements[7–12], such as titanium, zirconium, vanadium, tantalum, chromium, molybdenum, tungsten, cobalt, nickel, copper, ruthenium, rhodium, palladium and platinum. In particular, the addition of chromium is quite effective. Passivation occurs spontaneously for $Fe_{72}Cr_8P_{13}C_7$ alloy in 2M-hydrochloric acid at room temperature[13] and for $Fe_{75}Cr_3X_2P_{13}C_7$ alloys containing 2 atom % of second metallic element, X, by anodic polarization in 1M-hydrochloric acid at room temperature[14]. They do not suffer pitting corrosion even by anodic polarization in those solutions up to the transpassive region of chromium. The addition of chromium is also effective in improving the corrosion resistance of amorphous cobalt–[15,16] and nickel[17]–metalloid alloys.

The combined addition of chromium and molybdenum is further effective in increasing the corrosion resistance[5]. As shown in *Figure 24.1*, some amorphous Fe–Cr–Mo–metalloid alloys passivate spontaneously even in 12M-hydrochloric acid at 60°C[18]. *Figure 24.2* summarizes[19] the critical concentrations of chromium and molybdenum necessary for spontaneous passivation in hydrochloric acids of various

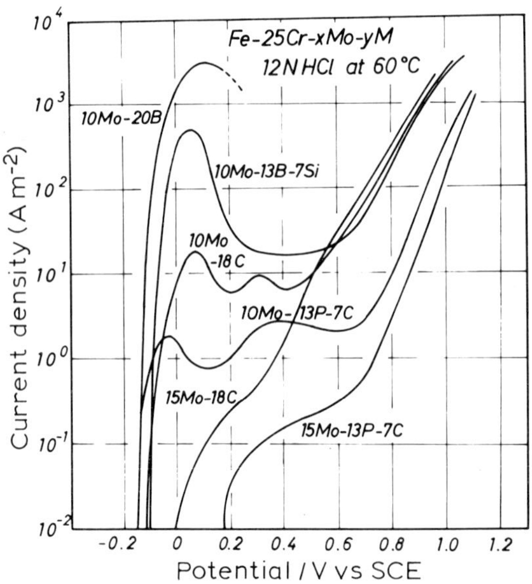

Figure 24.1 Potentiodynamic anodic polarization curves of amorphous $Fe_{75-x-y}Cr_{25}Mo_x(metalloid)_y$ alloys measured in 12M-hydrochloric acid at 60 °C. $Fe_{40}Cr_{25}Mo_{15}P_{13}C_7$ and $Fe_{42}Cr_{25}Mo_{15}C_{18}$ passivate spontaneously, but the other alloys require anodic polarization for passivation (after Hashimoto *et al.*[18])

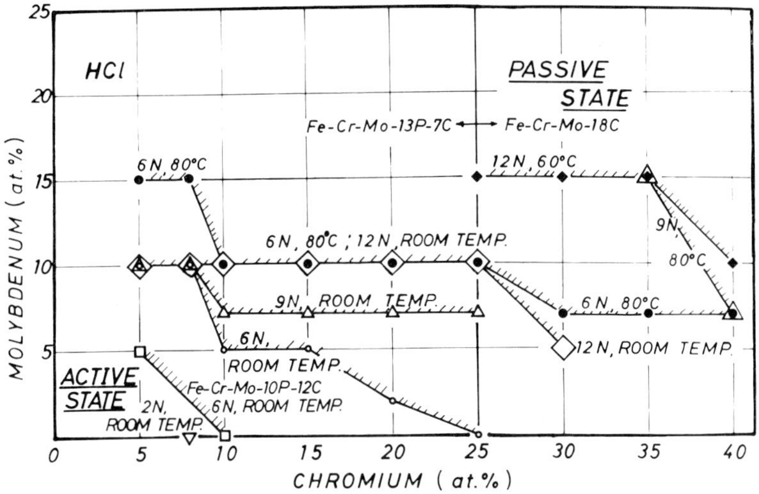

Figure 24.2 Critical concentrations of chromium and molybdenum necessary for spontaneous passivation of amorphous $Fe_{75-x-y}Cr_{25}Mo_x(metalloid)_y$ and $(Fe,Cr,Mo)_{82}C_{18}$ alloys in hydraulic acid of various normalities (N) and at various temperatures (after Kobayashi *et al.*[19])

concentrations and temperatures. Such a high corrosion resistance of the amorphous alloys is in marked contrast with the fact that any commercial, corrosion-resistant, crystalline metal except tantalum cannot passivate spontaneously even in dilute hydrochloric acids such as 1M-hydrochloric acid at room temperature.

In contrast to some amorphous metal–metalloid alloys, amorphous metal–metal alloys are not unusually corrosion-resistant. Their corrosion resistance never exceeds that of the crystalline pure metallic component of the alloys with the highest corrosion resistance or with the highest passivating ability[20]. Amorphous Fe–Si alloys prepared by sputtering resemble amorphous metal–metal alloys from the corrosion point of view; although the amorphous $Fe_{70}Si_{30}$ alloys shows about one order of magnitude lower corrosion rate in comparison with the crystalline $Fe_{70}Si_{30}$ alloy in dilute sulphuric acid, the increase in the silicon content of crystalline binary Fe–Si alloys from 20 to 30 atom % decreases the corrosion rate by a factor of 50 owing to passivation by silicon[21]. These results lead to the conclusion that the corrosion resistance of the metal–metal alloys is primarily determined not by the fact that they are in the amorphous state but by the presence of passivating element.

With respect to oxidation of amorphous alloys, there have so far been no findings peculiar to the amorphous state: Oxidation of amorphous $Pd_{81}Si_{19}$ alloys within the temperature range between room temperature and 160 °C at about 10^{-5} torr of air gives rise to the formation of a silicon-enriched protective oxide film about 10 nm thick. This is characteristic of the presence of silicon[22]. The addition of 14 atom % of chromium to an amorphous Fe–Ni–P–B alloy decreases the oxidation rate at 100–250 °C in dry and wet air[23]. The weight gain of $Cu_{60}Zr_{40}$ alloy as a result of reaction with oxygen and/or water contained in nominally pure argon at 25–250 °C is greater than the weight gain in oxygen or air at 1 atm because the formation of a ZrO_2 layer in higher partial pressures of oxygen acts as a barrier against further oxidation[24]. All of these results are not peculiar to the amorphous state.

24.2.2 Passivity: protective film

A decrease in the active dissolution rate by alloy additions is often responsible for an improvement of the corrosion resistance of various amorphous alloys, regardless of the activity of cathodic reactions, as will be described later. The extremely high corrosion resistance of amorphous metal–metalloid alloys containing chromium results from spontaneous passivation even in very corrosive solutions such as hot concentrated hydrochloric acids[13].

According to X-ray photoelectron spectroscopic studies, passive films on various amorphous and crystalline alloys containing chromium consist[17,25–30] mainly of hydrated chromium oxyhydroxide, $CrO_x(OH)_{3-2x},nH_2O$. The formation of a passive hydrated chromium oxyhydroxide film is responsible for spontaneous passivation also in hot concentrated hydrochloric acid as shown[18] in *Figure 24.3*. As can be seen from

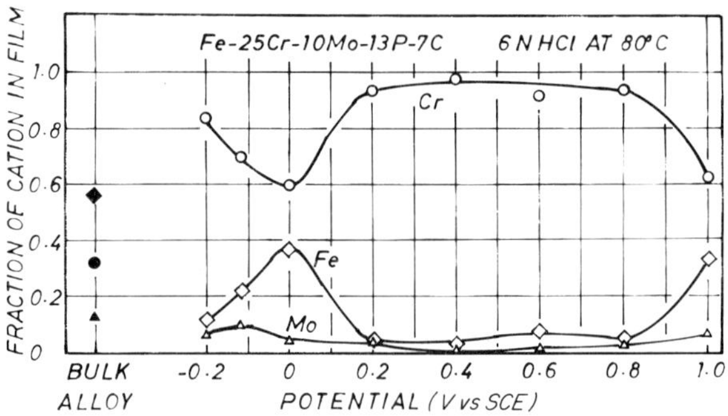

Figure 24.3 Analytical results by X-ray photoelectron spectroscopy for fractions of cations in the surface film formed on the amorphous $Fe_{45}Cr_{25}Mo_{10}P_{13}C_7$ alloys by potentiostatic polarization for 1 hour in 6M-hydrochloric acid at 80 °C (after Hashimoto *et al.*[18])

the example that the pitting potential of austenitic stainless steels is raised with increasing concentration of hydrated chromium oxyhydroxide in the surface film[31], the higher the concentration of hydrated chromium oxyhydroxide in the passive film, the higher is the protective quality of the film[32]. The amorphous metal–metalloid alloys have an ability extraordinarily to concentrate the beneficial chromic ions in the passive film, and accordingly have the high ability for spontaneous passivation with a consequent superior corrosion resistance.

24.2.3 Passivation rate

When passivation occurs, the amorphous metal–metalloid alloys are extremely stable. The high corrosion resistance and high passivating ability of amorphous metal–metalloid alloys are partly attributable to the high reactivity of their unfilmed surface. In general, active dissolution preceding passive film formation is required in aggressive solutions for accumulation of beneficial metallic ions and/or OH^- ions at the alloy–solution interface. *Figure 24.4* exhibits a good example of the correlation between active dissolution and passivation[33]. Polarization of amorphous $Pd_{55}Ti_{16}P_{19}$ alloy at a high anode current density in hot concentrated sodium chloride solutions results in evolution of chlorine gas along with severe corrosion. On galvanostatic polarization at 140 mA/cm^2 the alloy showed initially a low potential without chlorine evolution and

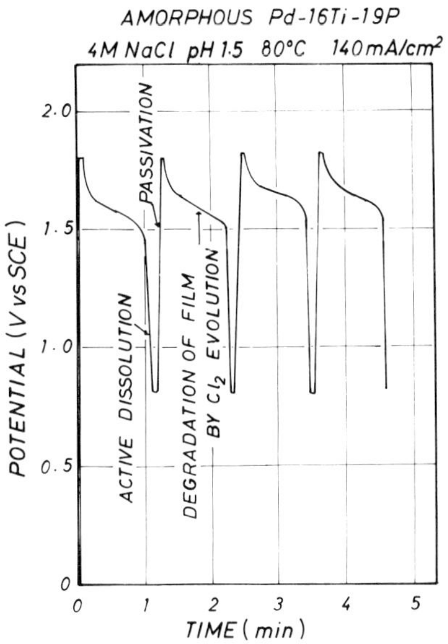

Figure 24.4 Changes in potential, showing repetition of active dissolution, passivation and depassivation during galvanostatic polarization of amorphous $Pd_{55}Ti_{16}P_{19}$ alloy at 140 mA/cm² in 4M-sodium chloride at pH 1.5 and 80 °C (after Hare *et al.*[33])

the solution in the vicinity of the alloy electrode turned brown, suggesting active dissolution of the alloy. Subsequently, an abrupt rise in the potential occurs and brown colouring of the solution simultaneously ceased, suggesting passivation. At the same time gas began to be evolved vigorously and this lasted for several tens of seconds, being accompanied by a gradual decrease in the potential. However, gas evolution led to the degradation of the passive film by nascent chlorine and to resumption of active dissolution with a consequent cessation of gas evolution which resulted in a sharp decrease in the potential. Nevertheless, active dissolution again led to passivation. In this manner, active dissolution, passivation and chlorine evolution were repeatedly observed. It can be said that active dissolution leads to passivation even in a very corrosive solution where breakaway of passivity occurs by nascent chlorine. Active dissolution preceding passivation provides concentration of passivating species at the alloy–solution interface. Consequently, when alloys which are able to passivate are exposed to an aggressive solution, the faster the active dissolution, the faster is the passivation and the higher the concentration of beneficial metallic ions in the passive film[34,35].

The active dissolution rate of amorphous metal–metalloid alloys without a stable surface film is essentially higher than that of crystalline metals[36]. The rapid active dissolution of the unfilmed surface of amorphous metal–metalloid alloys is responsible for the high concentration of beneficial chromic ions in the passive films on various amorphous alloys and for the high corrosion resistance and passivating ability.

24.2.4 Chemical homogeneity

In general, the corrosion resistance of almost all metallic materials depends on the composition, stability and uniformity of the surface film as well as the kinetics of film formation. When a given metal possesses a superior characteristic, that is, the rapid formation of the passive film containing a high concentration of beneficial species with a high stability, the corrosion resistance will depend upon the uniformity of the passive film. In this connection, the homogeneous nature of amorphous alloys ensures the formation of a uniform, highly protective passive film and hence the high corrosion resistance.

Amorphous Fe–Mo–metalloid alloys passivate in 1M-hydrochloric acid by anodic polarization owing to the formation of a passive hydrated iron oxyhydroxide film[10,37], which is the same as that formed on iron in 0.5M-sulphuric acid. This suggets that the passive hydrated iron oxyhydroxide film is stable in 1M-hydrochloric acid. Nevertheless iron cannot passivate in 1M-hydrochloric acid by anodic polarization, and suffers pitting corrosion. It has been found that a passive hydrated chromium oxyhydroxide film is formed on a ferritic 19Cr stainless steel in 1M-hydrochloric acid even when pitting occurs by anodic polarization[30]. Accordingly, hydrated iron oxyhydroxide may be formed on commercially pure, crystalline iron by anodic polarization in 1M-hydrochloric acid, but because crystalline iron is not chemically homogeneous, hydrated iron oxyhydroxide cannot cover the entire surface of the iron specimen in this acid and hence localized attack of chemically heterogeneous sites seems to prevent passivation of iron.

It has been known that the high corrosion resistance of amorphous metal–metalloid alloys disappears on heat treatment for crystallization[36,38–40]. The passivity of amorphous $Fe_{70}Cr_{10}P_{13}C_7$ and $Fe_{42}Ni_{36}Cr_{14}P_{12}B_6$ alloys in 1M-hydrochloric acid disappears as soon as the crystalline metastable phase I is formed in the amorphous matrix[40]. In contrast, low temperature annealing of amorphous $Fe_{32}Ni_{36}Cr_{14}B_{12}Si_6$ alloy at 100 or 200 °C for 1 hour improves the corrosion resistance in 10% iron(III) chloride owing to the homogenization of the alloy[41]. This has never been found for amorphous phosphorus-bearing alloys and seems characteristic of metal–metalloid alloys without phosphorus. On the other hand, heat treatment of amorphous $Fe_{80}B_{20}$ and $Fe_{40}Ni_{20}B_{20}$ alloys at 247 °C for ten days does not lead to crystallization but to embrittlement, and significantly increases the active current density in sulphate solutions possibly owing to short range decomposition into two amorphous phases[42].

It is interesting to clarify whether a dominant factor in decreasing the corrosion resistance is the chemical heterogeneity or the structural heterogeneity. According to Anthony and Cline[43], rapidly quenched crystalline 304 stainless steel after laser surface-melting possesses a higher corrosion resistance than solution-treated conventional 304 stainless steel since various compositional fluctuations are removed by rapid quenching from the molten state. Similarly, crystalline $Ti_{60}Ni_{40}$ and $Ti_{38.5}Ni_{61.5}$ prepared by rapid quenching of the alloy melts consist of a single TiNi type phase with fine grains and have high corrosive resistance and passivating ability in 1M-hydrochloric acid, whereas crystalline ordinary $Ti_{60}Ni_{40}$ and $Ti_{38.5}Ni_{61.5}$ alloys are composed of a two-phase mixture and are hardly passivated by anodic polarization in that acid[44]. On the other hand, vacuum annealing of amorphous $Fe_{70}Cr_{10}P_{13}C_7$ alloy at 350 °C for 3.5 months gives rise to the formation of a crystalline supersaturated solid solution phase, whose average composition has been thought to be the same as that of the surrounding amorphous matrix. Nevertheless, the passive current density in 1M-sulphuric acid increases by two orders of magnitude owing to the formation of the supersaturated solid solution phase in the amorphous matrix[36].

As described above, rapidly quenched, single phase alloys show significantly high corrosion resistance, even if they are crystalline. On the other hand, the formation of the crystalline phase with the same composition as that of the surrounding amorphous matrix by heat treatment considerably decreases the corrosion resistance. Since rapid quenching prevents the formation of various compositional fluctuations, crystal defects in the rapidly quenched crystalline alloys may be chemically clean and may not act as preferential active surface sites with respect to corrosion. The chemically homogeneous single phase nature without these compositional fluctuations is essential for the high corrosion resistance of amorphous alloys.

24.2.5 Effects of alloy additions

Metalloids

As can be seen from *Figure 24.1* the passivating ability and hence the corrosion resistance of amorphous metal–metalloid alloys are significantly different with different metalloids. The beneficial effect of the metalloid in improving the corrosion resistance decreases in the order phosphorus, carbon, silicon and boron[18,45]. Metalloids affect both the kinetics of passivation and the composition of passive film[26,41].

As mentioned above, the reactivity of unfilmed amorphous metal–metalloid alloys is generally higher than crystalline metals, and decreases in the above-mentioned order for alloys[46]. The amorphous phosphorus-bearing alloys dissolve most actively when they are not covered by a stable surface film, and hence they have the highest ability to accumulate beneficial species in the passive film. This is responsible for the high corrosion resistance of amorphous phosphorus-bearing alloys. Therefore, the corrosion resistance and passivating ability of amorphous metal–metal alloys is remarkably improved by the addition of a small amount of phosphorus[20,44].

Metalloid elements in the amorphous alloys also affect the composition of the surface film and hence the protective quality of the film. Phosphorus dissolves as P^{5+} into aqueous solutions[6] and is hardly contained in the stable passive film[25,26,32], although the content of phosphorus in the surface film as phosphate increases when the protective quality of the film is not sufficiently high, such as in the surface film formed in the active or transpassive region or formed on low chromium alloys[26,32,47,48]. Consequently, phosphorus contained in the amorphous metal–metalloid alloys is most effective for the rapid formation of a passive hydrated oxyhydroxide film containing a high concentration of beneficial species.

Carbon in amorphous alloys does not greatly accelerate the formation of passive films but is not contained in the surface film since metal carbonates are generally soluble in aqueous solutions. Accordingly, carbon is the second most effective metalloid in improving the corrosion resistance of amorphous alloys if the alloys contain a sufficient amount of passivating element.

In contrast, the ability of silicon and boron to accelerate the formation of passive films is not high and they tend to be contained in the surface film as silicate and borate[26]. Even if chromic ion comprises almost 100 per cent of cations in the surface film, the formation of chromium borate or chromium silicate inevitably lowers the concentration of hydrated chromium oxyhydroxide in the surface film. Thus their beneficial effect for increasing the corrosion resistance is lower than that of phosphorus and carbon.

Metallic elements

Addition of almost all metallic elements to amorphous metal–metalloid alloys improves the corrosion resistance[7–12]. The alloying additions can be classified into three groups depending upon their roles in increasing the corrosion resistance. Chromium forms its own passive film and hence is the most effective alloying element in improving the corrosion resistance of amorphous metal–metalloid alloys. Titanium also forms its own passive film. When the same amount of chromium or titanium is added to the amorphous $Ni_{40}P_{20}$ alloy[49], the open circuit corrosion resistance and passivating ability of the chromium-bearing alloy in 1M-hydrochloric acid is significantly higher than that of the titanium-bearing alloys, although the titanium-containing alloys are more stable than the chromium-bearing alloys at high potentials since the former do not suffer transpassive dissolution, being different from the latter alloys[49].

The second group of alloying additions includes metallic elements more active than the main metallic component of alloys. In amorphous iron–metalloid alloys, those are vanadium, niobium, molybdenum, tungsten, etc. Their passive potential region is lower than that of iron and the surface films consisting of their cations are less stable than the passive film of iron or chromium in the passive region of iron or chromium. Accordingly, the content of their cations in the passive hydrated iron or chromium oxyhydroxide film is considerably lower[18,19,27,29,30,37,50,51], except for the passive film formed on the alloys without sufficiently high passivating ability[10]. Nevertheless, they are beneficial for passivation of alloys. When the dissolution rate of alloys is high, as the alloys are polarized in the active region, the corrosion products of these alloying elements cover the alloy surface since they dissolve more slowly than other alloy constituents[29,37,50]. The formation of the corrosion product film, which acts as a diffusion barrier against further dissolution of alloys, lowers the active dissolution rate of alloys. Furthermore, when the alloy contains passivating elements such as chromium or iron, the formation of corrosion product films of the alloying elements assists the accumulation of the passivating species between the corrosion product film and the underlying alloy and assists passivation. Once passivation occurs with a consequent lowering of active dissolution rate, the condition for rapid supply of alloying elements to renew the corrosion product film fails and hence the corrosion product film dissolves into the solution because of the low stability of corrosion products of alloying elements in the passive region of the alloys. In this manner, these alloying elements do not constitute the stable passive film but assist passivation and prevent localized corrosion by mechanical and/or chemical breakdown of passive films on both amorphous and crystalline alloys containing passivating elements.

In this connection, the excess additions of these elements are, sometimes, detrimental. For example, as shown[51] in *Figures 24.5* the addition of 5 atom% of molybdenum to amorphous $Fe_{55}Cr_{25}B_{20}$ alloy results in spontaneous passivation in 6M-hydrochloric acid at room temperature. However, an increase in the molybdenum content up to 10 atom% significantly increases the passive current density. The excess addition of molybdenum leads to incorporation of molybdenum species in the passive film. Since molybdenum species are unstable in the passive region of alloys owing to transpassive dissolution, the protective quality of the passive film is lower than that of the film without the unstable species.

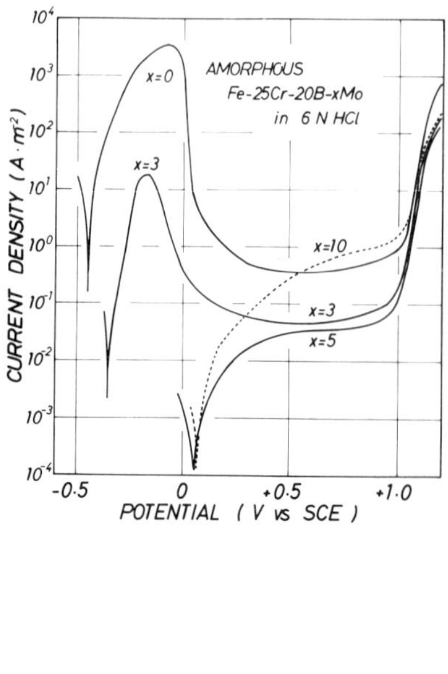

Figure 24.5 Potentiodynamic anodic polarization curves of amorphous $Fe_{55-x}Cr_{25}Mo_xB_{20}$ alloys in 6M-hydrochloric acid at room temperature (after Asami *et al.*[51])

The third group of alloy additions includes metallic elements more noble than the main metallic component of the amorphous iron–metalloid alloys. These are cobalt, nickel, copper, ruthenium, rhodium, palladium, platinum, etc. These elements are chemically stable and hence their cations are poorly contained in the surface film[50]. Therefore, corrosion of the alloys leads to the selective enrichment of those noble metals in the metallic state under the surface film, which decreases the anodic activity and dissolution rate of the alloys and increases the cathodic activity. Consequently, the enrichment of noble metals in the surface of underlying alloys increases the open circuit corrosion resistance and passivating ability.

24.2.6 Localized corrosion resistance

It has widely been found[5,9–11,13–20,32,34–41] that amorphous alloys do not suffer pitting corrosion even by anodic polarization in strongly acidic solutions containing chloride. In particular, it is typical of amorphous alloys[9,10,37] that the amorphous Fe-Mo-metalloid alloys passivate by forming the passive hydrated iron oxyhydroxide film by anodic polarization in 1M-hydrochloric acid, where crystalline iron and Fe–Mo alloys dissolve actively by anodic polarization without passivation, being accompanied by pitting. The chemically homogeneous single phase nature of amorphous alloys without compositional fluctuations is primarily responsible for the high localized corrosion resistance. However, since crevice corrosion does not require surface heterogeneity for initiation, the susceptibility of amorphous alloys to crevice corrosion is also dependent upon the passivating ability of the alloys.

The first finding of localized corrosion of amorphous alloys has been reported for cold-deformed specimens. Localized shear bands are generally observed by deformation of amorphous alloys. After the deformation marks are removed by electropolishing from the amorphous $Pd_{77.5}Cu_6Si_{16.5}$ alloy specimen, previously deformed about 11 per cent in compression, the shear bands reappear on electroetching. This has been interpreted in terms of the preferential etching of slip bands[52,53]. Such behaviour has also been found for amorphous $Ni_{55}Pd_{35}P_{10}$ alloy[54]. In contrast, the preferential etching of slip bands has not been observed for amorphous chromium-bearing alloys. According to Devine and his coworkers[55–57] plastic deformation of amorphous $Fe_{30}Ni_{35}Cr_{15}P_{14}B_6$ alloy to a total strain of 3.6 per cent does not lead to the preferential etching of slip bands during anodic polarization in 0.17 M-sodium chloride and has no effect on the passive corrosion rate nor on the pitting corrosion resistance. On the other hand, cold-rolling of this alloy to a 32 per cent reduction in thickness results in pitting corrosion because of crevice attack at the intersections of surface microcracks which are formed by cold-rolling[57].

Diegle has studied crevice corrosion of amorphous $Fe_xNi_{30}Cr_{50-x}P_{14}B_6$ alloys containing various concentrations of chromium: Cold rolling of the alloy with 16 per cent chromium to a 38 per cent reduction in thickness considerably increases the current densities in both the active and passive regions in 1M-sodium chloride at pH 7 with a consequent initiation of pitting corrosion at the intersections of surface microcracks. However, during potentiostatic polarization for 1 hour at 0.6 V (SCE), which is significantly anodic to promote pitting and crevice corrosion, the current density decreases rather than increases, the latter being expected if pitting and crevice corrosion grow deeply. At the end of polarization a backward potential sweep leads to a negative current in the active region of the alloy. These results indicate that localized corrosion occurs as a result of the formation of tight crevices due to severe cold-rolling, but pits thus formed readily repassivate by potentiostatic polarization at the high potential, suggesting a high passivating ability of the alloy. Furthermore, when the

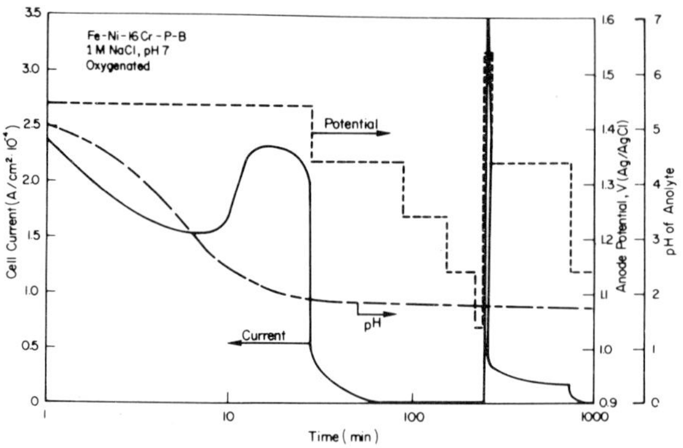

Figure 24.6 Crevice-cell current and potential of amorphous $Fe_{31}Ni_{30}Cr_{16}P_{15}B_6V_2$ alloy and anolyte pH in the instrumental crevice in oxygenated 1M-sodium chloride of pH 7 at 22 °C (after Diegle[59])

amorphous $Fe_xNi_{30}Cr_{50-x}P_{14}B_6$ alloys are placed inside an artificial crevice, crevice corrosion hardly occurs unless the specimens are polarized at high potentials[58]. As shown in *Figure 24.6*, potentiostatic polarization[59] of amorphous $Fe_{34}Ni_{30}Cr_{34}P_{14}B_6$ alloy inside the crevice at 1.4 V (SCE) results in initiation of crevice corrosion and pH decrease inside the crevice but a decrease in applied potential by 0.1 V leads to a decrease in crevice cell current essentially to zero. The cell current reappears by polarization at 1.5 V (SCE) but decreases by potential decrease to 1.3 V (SCE). Thus crevice corrosion starts at very high potentials and the difference between the crevice initiation potential and the crevice protection potential is less than 0.1 V. Consequently, very corrosive conditions are required for the onset of crevice corrosion for the amorphous alloys in comparison with crystalline steels of similar chromium contents. The amorphous alloys repassivate by a slight potential decrease, since they have considerably higher passivating ability primarily because of the beneficial effect of metalloids, as mentioned previously.

24.2.7 Stress corrosion cracking and hydrogen embrittlement

Environmentally induced embrittlement of amorphous alloys has also been studied. Stress corrosion cracking of crystalline alloys occurs owing to active path corrosion, often being associated with crystallographic slip planes or grain boundaries. Since amorphous alloys are free from those, they may not suffer stress corrosion cracking based on active path corrosion. Pampillo[60], in reporting the first stress corrosion experiment, stated that amorphous $Ni_{49}Fe_{29}P_{14}B_6Al_2$ alloy failed in 3.5 M-sodium chloride on exposure for 117 hours at 26 °C and a constant load of 75 per cent of the fracture stress in air. Because of the thin specimens, the mechanical properties measured for amorphous alloy ribbons tend to scatter owing to the presence of small notches. To obtain reproducible results both the surfaces and edges of the specimen must be carefully polished before the mechanical test. In particular, the above-mentioned alloy did not contain chromium and should be readily corroded in the sodium chloride solution. Accordingly if the corrosion notch was formed at the edge of the specimen, mechanical failure would easily occur under such a high applied stress. In this regard, there is no other report confirming stress corrosion cracking based on active path corrosion for amorphous alloys.

On thc other hand, some amorphous alloys are susceptible to hydrogen embrittlement when they are corroded under applied tensile stress or when charged with hydrogen[61–64]. Since hydrogen embrittlement occurs as a result of the absorption of atomic hydrogen produced by cathodic reactions of corrosion, the resistance to amorphous alloys against hydrogen embrittlement under free corrosion conditions in acidic solutions is improved with increasing corrosion resistance, and is significantly affected by metalloid elements in the alloys[64]. Increasing contents of chromium and/or molybdenum in amorphous iron–metalloid alloys decreases the susceptibility to hydrogen embrittlement. Phosphorus is the most effective in improving the resistance of corrosion-resistant metal–metalloid alloys against hydrogen embrittlement and the alloys containing only carbon as a metalloid are most susceptible to hydrogen embrittlement.

The amorphous metal–metalloid alloys are able to absorb a high concentration of hydrogen by cathodic charging[61,65], but the absorbed hydrogen is rapidly released from the alloys by storage in air at room temperature[61]. The diffusivity of hydrogen in various amorphous alloys, such as $Fe_{80}P_{13}C_7$[62], $Fe_{70}Cr_{10}P_{13}C_7$[62], $Fe_{32}Ni_{36}Cr_{14}P_{12}B_6$[65], $Fe_{42}Ni_{40}P_{12}B_6$[65] and $Ni_{81}P_{19}$[65] is within the range $6–8 \times 10^{-11}$ $cm^2\,s^{-1}$ and is smaller than that in iron or steels.

The fracture surface produced by hydrogen embrittled samples shows shell-like, parabolic lines near the crack origin. These seem to be Wallner lines which are characteristic of the fracture surface of completely brittle solids[66]. The crack propagation rate of amorphous Fe–Cr–Mo–C alloys calculated on the basis of Wallner lines is within the range 1100–1400 m/sec, suggesting instantaneous failure of hydrogen-embrittled alloys.

24.3 Catalysis

24.3.1 Electrocatalysis

Electrolysis of sodium chloride solution

The anode materials suitable for the electrolysis of sodium chloride solutions should have a high electocatalytic activity for chlorine evolution and a very low activity for oxygen evolution which competes with the former reaction. In this regard, palladium has an outstanding characteristic, superior to all other platinum group metals in dilute sodium chloride solution at room temperature, but is not practically used because of its low corrosion resistance in hot concentrated sodium chloride solution. Since amorphous alloys are characterized by the fact that alloying with suitable elements provides extremely high corrosion resistance, the use of amorphous palladium alloys was attempted to improve the corrosion resistance of palladium without harming its superior electrode characteristics for the electrolysis of sodium chloride solution[33,67].

In hot concentrated sodium chloride solution, such as 4M-NaCl with pH 4 and at 80 °C, amorphous $Pd_{81}P_{19}$ and $Pd_{80}Si_{20}$ alloys dissolved actively without showing passivation. However, the dissolution current of the amorphous $Pd_{81}P_{19}$ alloy was lower than that of crystalline pure palladium and the addition of various metallic elements to the amorphous $Pd_{81}P_{19}$ alloy significantly improved the corrosion resistance. *Figure 24.7* shows potentiostatic polarization curves of amorphous $Pd_{41}Ir_{40}P_{19}$ alloy and various crystalline metals[67]. The anodic dissolution current is observed in the potential region lower than 0.9 V (SCE). The current increase in the potential range higher than about 0.9 V corresponds to chlorine evolution. It was

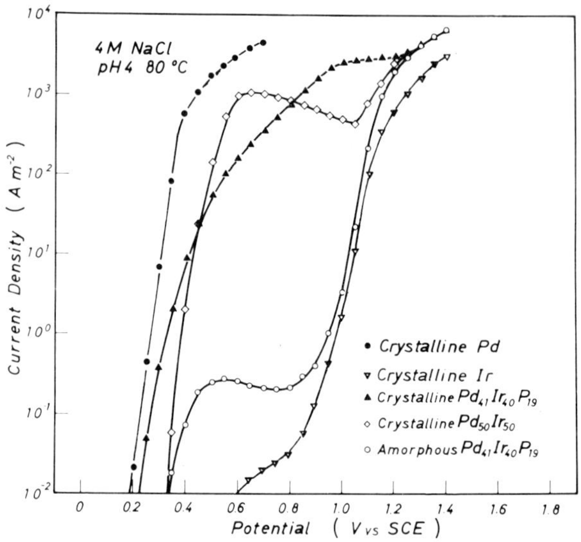

Figure 24.7 Potentiostatic anodic polarization curves of amorphous $Pd_{41}Ir_{40}P_{19}$ alloy and crystalline $Pd_{41}Ir_{40}P_{19}$ and $Pd_{50}Ir_{50}$ alloys measured in 4M-sodium chloride of pH 4 at 80 °C (after Hara *et al.*[67])

difficult to polarize crystalline palladium in the high potential region because of high active dissolution current. Two crystalline alloys show significantly high dissolution current and barely passivate by anodic polarization. On the contrary, the amorphous alloy and crystalline iridium exhibit a low passive current density and a subsequent steep current increase corresponding to gas evolution. The activity of the amorphous alloy for gas evolution is considerably higher than that of crystalline iridium.

When various metallic elements were added to the amorphous $Pd_{81}P_{19}$ alloy the corrosion rate in the gas evolution region was very low owing to the formation of a protective passive film in which cations of the alloy additions were highly concentrated[67]. *Table 24.1* shows the current densities[67] measured in 4M-sodium chloride and 1M-sodium sulphate solutions at pH 4 and at 80 °C for the comparison of

TABLE 24.1 Catalytic activities of various electrodes for chlorine and oxygen evolution in solutions of pH 4 and at 80 °C

Electrode	*Current Density*/A cm^{-2} *at* 1.15 V (SCE) 4M-NaCl	$1M-Na_2SO_4$
Amorphous alloy		
$Pd_{41}Ir_{40}P_{19}$	2000	5
$Pd_{46}Ir_{30}Ti_5P_{19}$	3000	2
$Pd_{31}Ir_{30}Ti_{10}P_{19}$	1500	5
$Pd_{46}Ir_{30}Rh_5P_{19}$	700	4
$Pd_{46}Ir_{30}Pt_5P_{19}$	860	17
$Pd_{41}Ir_{30}Ru_{10}P_{19}$	2200	28
Graphite	41	20
RuO_2/Ti	1700	340

catalytic activities of the alloys for chlorine and oxygen evolution. The current densities of graphite and RuO_2/Ti electrodes are also shown for comparison, the latter electrode being widely used as 'the dimensionally stable anode' in the chlorine–alkali industry. Several amorphous alloys have considerably higher activity for chlorine evolution and lower activity for oxygen evolution than the RuO_2/Ti electrode.

Consequently, the amorphous alloys have sufficiently high corrosion resistance and catalytic activity for chlorine evolution and very low activity for oxygen evolution in hot concentrated sodium chloride solution as a result of the transformation to the amorphous structure and alloying with passivating elements.

Electrode for fuel cell

An attempt was made to apply amorphous alloys to the fuel electrode in the methanol–air fuel cell[68], by using their unique characteristics; that is, the ability to make homogeneous single phase alloys consisting of high concentrations of multicomponents and having high corrosion resistance. Since various as-quenched amorphous alloys tested were not superior to platinized platinum, a surface-activation treatment was applied to increase the effective surface area. This treatment consisted of electrodeposition of zinc on the alloys, diffusion of zinc into the alloys by heat treatment, and leaching of zinc into hot 6M-potassium hydroxide.

As shown in *Figure 24.8*, some of the surface-activated amorphous Pd–P alloys

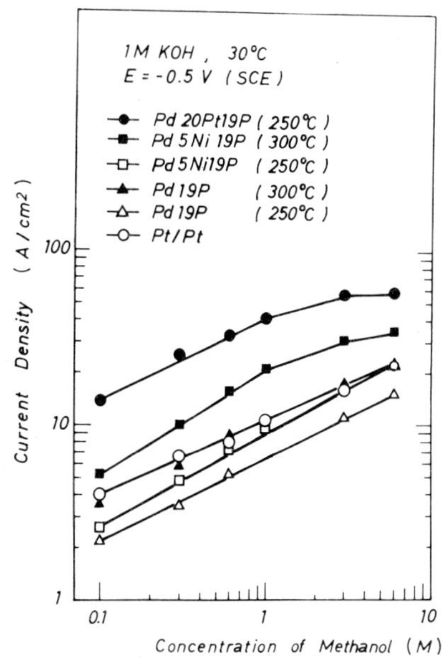

Figure 24.8 Change in current density of oxidation of methanol with concentration of methanol in 1M-potassium hydroxide during potentiostatic polarization of surface-activated amorphous palladium-based alloys and platinized platinum at −0.5 V (SCE) at 30 °C. The temperature of heat treatment for diffusion of zinc into alloys in the surface-activation treatment is shown in parentheses: ●, $Pd_{51}Pt_{20}P_{19}$ (250 °C); ■, $Pd_{76}Ni_5P_{19}$ (300 °C); □, $Pd_{76}Ni_5P_{19}$ (250 °C); ▲, $Pd_{81}P_{19}$ (300 °C); △, $Pd_{81}P_{19}$ (250 °C); ○, platinized platinum (after Kawashima and Hashimoto[68])

showed a considerably higher activity for oxidation of methanol and its derivatives and their catalytic activity was higher and more stable than that of the platinized platinum and surface-activated crystalline palladium. The amorphous structure was particularly suitable to apply the surface-activation treatment since zinc diffused uniformly into the alloys, being different from crystalline metals in which zinc was apt to diffuse along grain boundaries with a consequent grain-boundary embrittlement after leaching.

24.3.2 Catalytic synthesis of hydrocarbons

From the catalytic point of view the amorphous state seems to be more active than the crystalline state because of a high density of imperfections. Hence amorphous

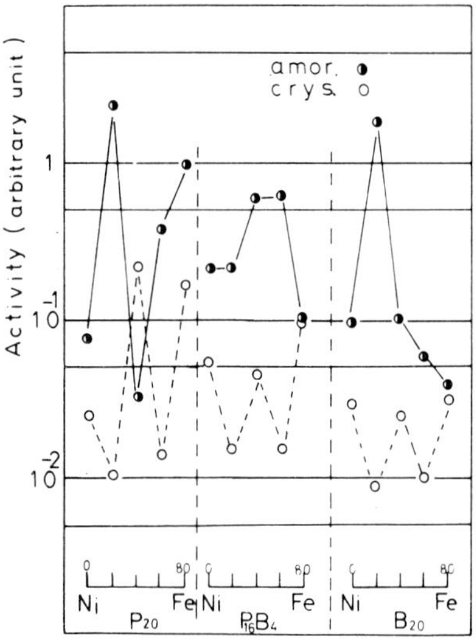

Figure 24.9 Change in catalytic activity for hydrogenation of carbon monoxide at 255 °C as a function of composition of amorphous alloy catalysts (after Yokoyama *et al.*[69,70])

Fe–Ni–metalloid alloys have been used as the catalyst to produce C_1 to C_5 hydrocarbons by hydrogenation of carbon monoxide[70,71].

After pretreatment in hydrogen or a mixture of carbon monoxide and helium at 320 °C, the amorphous alloys showed a stable activity for conversion of carbon monoxide into hydrocarbons. As shown in *Figure 24.9*, the catalytic activity[70] of amorphous alloys was always higher than that of the crystalline alloys with an exception of the $Fe_{40}Ni_{40}P_{20}$ alloy. The activation energy for the depletion of carbon monoxide was 100 ± 4 kJ/mole, regardless of alloy composition and of the amorphous or crystalline state. Accordingly, the amorphous catalysts have a high density of the active sites with the same nature as the active sites on the crystalline catalysts. The amorphous catalysts are also superior to the crystalline catalysts with respect to production of higher hydrocarbons ranging from C_2 to C_5.

24.4 Concluding remarks

As has been briefly described in this book the amorphous alloys possess a variety of very interesting chemical properties. They are a new type of metal with extraordinarily high resistance to various types of corrosion on the basis of their chemical homogeneity and high passivating ability. Their commercial utilization as corrosion-resistant metals is encouraging. For this purpose it may be necessary to solve some technical problems, such as their preparation to obtain a large surface area, the ability to weld them, etc.

The catalytic properties are indeed a novel field. There are a number of chemical and electrochemical reactions which require catalysts. These reactions are waiting to be examined in regard to the applicability of new amorphous metal catalysts.

References

1. Hoar, T. P., *J. Electrochem. Soc.*, **117**, 17C (1970)
2. Okamoto, G., *Corrosion, Sci.*, **13**, 471 (1973)

3. Revesz, A. G. and Kruger, J., *'Passivity of Metals', Corrosion Monograph Series,* Eds. Frankenthal, R. P. and Kruger, J., p. 137, The Electrochemical Society, Princeton, N.J. (1978)
4. Masumoto, T., Hashimoto, K. and Naka, M., *'Proc. 3rd Int. Conf. Rapidly Quenched Metals',* p. 435, The Metals Society, London, (1978)
5. Naka, M., Hashimoto, K., Inoue, A. and Masumoto, T., *J. Non-cryst. Solids,* **31,** 347 (1979)
6. Cadet, P., Keddam, M. and Takenouti, H., *Proc. 4th Int. Conf. Rapidly Quenched Metals,* Eds. Masumoto, T. and Suzuki, K., vol. II, p. 1447, The Japan Institute of Metals, Sendai (1982)
7. Hashimoto, K., Naka, M. and Masumoto, T., *Sci. Rep. Res. Insts Tôhoku Univ.,* **A26,** 48 (1976)
8. Naka, M., Hashimoto, K. and Masumoto, T., *Sci. Rep. Res. Insts Tôhoku Univ.* **A26,** 283 (1977)
9. Naka, M., Hashimoto, K. and Masumoto, T., *J. Non-cryst. Solids,* **29,** 61 (1978)
10. Hashimoto, K., Asami, K., Naka, M. and Masumoto, T., *Sci. Rep. Res. Insts Tôhoku Univ.,* **A27,** 237 (1979)
11. Naka, M., Hashimoto, K. and Masumoto, T., *J. Non-cryst. Solids,* **31,** 355 (1979)
12. Kovacs, P., Farkas, J., Kiss, L., Lovas, A. and Tompa, K., ref. 6, vol. II, p. 1471
13. Kobayashi, K., Hashimoto, K. and Masumoto, T., *Sci. Rep. Res. Insts Tôhoku Univ.,* **A29,** 284 (1981)
14. Hashimoto, K., Naka, M., Noguchi, J., Asami, K. and Masumoto, T., ref. 3, p. 156
15. Naka, M., Asami, K., Hashimoto, K. and Masumoto, T., ref. 4, p. 449
16. Naka, M., Hashimoto, K. and Masumoto, T., *Sci. Rep. Rep. Insts Tôhoku Univ.,* **A27,** 246 (1979)
17. Hashimoto, K., Kasaya, M., Asami, K. and Masumoto, T., *Corrosion Engng (Boshoku Gijutsy),* **26,** 445 (1977)
18. Hashimoto, K., Kobayashi, K., Asami, K. and Masumoto, T., *'Metallic Corrosion',* Proc. 8th Int. Cong. Metallic Corrosion, vol. I, p. 70, DECHEMA, Frankfurt/Main (1981)
19. Kobayashi, K., Asami, K. and Hashimoto, K., ref. 6, p. 1443
20. Naka, M., Hashimoto, K. and Masumoto, T., *J. Non-cryst. Solids,* **30,** 29, (1978)
21. Brusic, V., MacInnes, R. D. and Aboaf, J., ref. 3, p. 170
22. Ley, L. and Riley, J. D., *'Proc. 7th Int. Vac. Cong. and 3rd Int. Conf. Solid Surfaces,* Vienna, p. 2031 (1977)
23. Hunderi, O. and Bergersen, R., *Corrosion Sci.,* **22,** 135 (1982)
24. Bigot, J., Calvayrac, Y., Harmelin, M., Chevellier, J.-P. and Quivy, A., ref. 6, p. 1463
25. Asami, K., Hashimoto, K., Masumoto, T. and Shimodaira, S., *Corrosion Sci.,* **16,** 909 (1976)
26. Hashimoto, K., Asami, K., Naka, M. and Masumoto, T., *Corrosion Engng, (Boshoku Gijutsu),* **28,** 271 (1979)
27. Hashimoto, K. and Asami, K., ref. 3, p. 749
28. Asami, K., Hashimoto, K. and Shimodaira, S., *Corrosion Sci.,* **18,** 151 (1978)
29. Hashimoto, K., Asami, K. and Teramoto, K., *Corrosion Sci.,* **19,** 3 (1979)
30. Hashimoto, K. and Asami, K., *Corrosion Sci.,* **19,** 251 (1979)
31. Asami, K. and Hashimoto, K., *Corrosion Sci.,* **19,** 1007 (1979)
32. Hashimoto, K., Naka, M., Noguchi, J., Asami, K. and Masumoto, T., ref. 3, p. 156
33. Hara, M., Hashimoto, K. and Masumoto, T., *Electrochim. Acta,* **25,** 1215 (1980)
34. Hashimoto, K., Naka, M., Asami, K. and Masumoto, T., *Corrosion Engng (Boshoku Gijutsu),* **27,** 279 (1978)
35. Naka, M., Hashimoto, K. and Masumoto, T., *J. Non-cryst. Solids,* **34,** 257 (1979)
36. Hashimoto, K., Osada, K., Masumoto, T. and Shimodaira, S., *Corrosion Sci.,* **16,** 71 (1976)
37. Hashimoto, K., Naka, M., Asami, K. and Masumoto, T., *Corrosion Sci.,* **19,** 165 (1979)
38. Diegle, R. B. and Slater, J. E., *Corrosion,* **32,** 155 (1976)
39. Kulik, T., Baszkiewicz, J. Kaminski, M., Latuszkieeicz, J. and Matyja, H., *Corrosion Sci.,* **19,** 1001 (1979)
40. Naka, M., Hashimoto, K. and Masumoto, T., *Corrosion,* **36,** 679 (1980)
41. Diegle, R. B., ref. 6, p. 1457
42. Kapusta, S, and Heusler, K. E., *Z. Metallk.,* **72,** 785 (1981)
43. Anthony, T. R. and Cline, H. E., *J. appl. Phys.,* **49,** 1248 (1978)
44. Naka, M., Asami, K., Hashimoto, K. and Masumoto, T., *'Titanium '80',* Proc. 4th Int. Conf. on Titanium, Eds. Kimura, H. and Izumi, O., p. 2677, ASM — AIME (1981)
45. Naka, M., Hashimoto, K. and Masumoto, T., *J., Non-cryst. Solids,* **28,** 403 (1978)
46. Hashimoto, K., Naka, M., Asami, K. and Masumoto, T., *Corrosion Engng. (Boshoku Gijutsu),* **27,** 279 (1978)
47. Baer, R. D. and Thomas, M. T., *J. Vac. Sci. & Technol.,* **18,** 722 (1981)
48. Thomas, M. T. and Baer, R. D., ref. 6, p. 1453
49. Naka, M., Hashimoto, K. and Masumoto, T., *Sci. Rep. Res. Insts Tôhoku Univ.,* **A28,** 156 (1980)
50. Hashimoto, K., Asami, K., Naka, M. and Masumoto, T., *Corrosion Sci.,* **19,** 857 (1979)
51. Asami, K., Naka, M., Hashimoto, K. and Masumoto, T., *J. Electrochem. Soc.,* **127,** 2130 (1980)

52. Pampillo, C. A., *Scripta Met.*, **6,** 915 (1972)
53. Pampillo, C. A. and Chen, H. S., *Mater. Sci. Engng,* **13,** 181 (1974)
54. Takayama, S. and Maddin, R., *Acta metall.,* **23,** 9434 (1975)
55. Devine, T. M. and Wells, L., *Scripta Met.,* **10,** 309 (1976)
56. Lee, D. and Devine, T. M., *Proc. 2nd Int. Conf. Rapidly Quenched Metals, Section I,* Eds. Grant, N. J. and Giessen, B. C., p. 393, The Massachusetts Institute of Technology, Cambridge, Mass. (1976)
57. Devine, T. M., *J. Electrochem. Soc.,* **124,** 38 (1977)
58. Diegle, R. B., *Corrosion,* **35,** 250 (1979)
59. Diegle, R. B., *Corrosion,* **36,** 362 (1980)
60. Pampillo, C. A., *J. Mater. Sci.,* **10,** 1194 (1975)
61. Kawashima, A., Hashimoto, K. and Masumoto, T., *Corrosion Sci.,* **16,** 935 (1976)
62. Nagumo, M. and Takahashi, T., *Mater. Sci. Engng,* **23,** 257 (1976)
63. Viswanadhan, R. K., Green, J. A. S. and Montague, W. G., *Scripta Met.,* **10,** 229 (1976)
64. Kawashima, A., Hashimoto, K. and Masumoto, T., *Corrosion,* **36,** 577 (1980)
65. Latanision, R. M., Compeau, C. R. and Kurkela, M., Proc. Troiano Symp. Hydrogen Embrittlement and Stress Corrosion Cracking, in press.
66. Kawashima, A., Hashimoto, K. and Masumoto, T., *Scripta Met.,* **14,** 41 (1980)
67. Hara, M., Hashimoto, K. and Masumoto, T., ref. 6, p. 1423
68. Kawashima, A. and Hashimoto, K., ref. 6, p. 1427
69. Yokoyama, A., Komiyama, H., Inoue, H., Masumoto, T. and Kimura, H. M., *J. Catal.,* **68,** 355 (1981)
70. Yokoyama, A., Komiyama, H., Inoue, H., Masumoto, T. and Kimura, H. M., *Scripta Met.,* **15,** 365 (1981)
71. Yokoyama, A., Komiyama, H., Inoue, H. Masumoto, T. and Kimura, H. M., ref. 6, p. 1419

Chapter 25

Atomic diffusion in amorphous alloys

B. Cantor

Dept. of Metallurgy and Science of Materials, University of Oxford, Oxford, UK

R. W. Cahn

Laboratoire de Métallurgie Physique, Université de Paris-Sud, Centre D'Orsay, Paris, France

25.1 Introduction

Amorphous alloys possess a variety of desirable electrical, magnetic, mechanical and other properties. As described elsewhere in this book, the properties of an amorphous alloy are determined by its atomic structure, and both are very sensitive to heat-treatment. Processes such as structural relaxation, phase separation, and crystallization involve significant rearrangements in atomic structure, and therefore cause significant changes in amorphous alloy properties. In general, these processes are controlled by diffusion of atoms through the structure of the amorphous alloy. Understanding atomic diffusion in amorphous alloys is therefore essential if we are to make full use of the technological potential of these materials, first so as to develop heat treatments suitable for optimizing amorphous alloy properties during manufacture, and secondly so as to ensure safe working temperatures which prevent degradation of properties during service life.

The main objectives of this chapter are as follows:

1. To describe the techniques which have been used to investigate diffusion in amorphous alloys
2. To compare the diffusion data which have been obtained for different alloy systems
3. To discuss how diffusion is affected by thermal and mechanical treatment of the amorphous alloys
4. To consider possible atomic mechanisms of diffusions in amorphous alloys

25.2 Experimental techniques

For a given amorphous alloy matrix and a given diffusing atomic species, the fundamental property is the diffusion coefficient D, defined by Fick's law as the constant of proportionality between the mass flux of diffusing atoms dm/dt, the area through which they diffuse A, and the concentration gradient of diffusing atoms normal to this area dc/dx, with c in units of mass per unit volume[1]:

$$dm/dt = -DA\, dc/dx \tag{25.1}$$

When Fick's law and the law of conservation of mass are applied to every volume element of a general concentration field $c(x, y, z, t)$, the resulting differential equation describes the time evolution of the concentration field

$$dc/dt = \text{div}(D \,\text{grad}\, c) \tag{25.2}$$

When D is effectively constant over the concentration field, this reduces to the standard form of the diffusion equation:

$$dc/dt = D \nabla^2 c \tag{25.3}$$

There are two important solutions to the diffusion equation. In the thin film solution[1], a mass m of the diffusing species is concentrated initially in an infinitely thin layer, either at the surface of a semi-infinite slab of matrix, or in the interior of an infinite slab of matrix. At subsequent times, the concentration profile normal to the layer $c(x, t)$ is given by a Gaussian distribution:

$$c(x, t) - c_0 = [c(x_0, t) - c_0] \exp(-w^2) \tag{25.4}$$

where c_0 is the initial uniform concentration of the diffusing species in the matrix, x_0 is the initial position of the thin layer, $w = (x - x_0)/s\sqrt{2}$, s is the standard deviation of the Gaussian distribution given by $(2Dt)^{1/2}$, and $c(x_0, t)$ is the maximum concentration in the profile, which is at x_0 and can be evaluated by integrating equation (25.4) over the matrix volume and setting the result equal to m. In the thick film solution[1], two semi-infinite slabs with different initial concentrations of the diffusing species are brought into contact and allowed to interdiffuse across the common interface. At subsequent times, the concentration profile normal to the interface is given by an error function equation:

$$c(x, t) - c_1 = (c_2 - c_1)(1 - \text{erf}\, w) \tag{25.5}$$

where c_1 and c_2 are the initial uniform concentrations in the two slabs, and w is defined as before, with x_0 the initial position of the interface. *Figure 25.1* shows typical concentration profiles resulting from the thin and thick film solutions in equations 25.4 and 25.5.

In almost all the direct measurements of diffusion coefficients which have been made in amorphous alloys, the procedure has been to ion-implant or sputter-coat the diffusing atoms on to the amorphous alloy surface, measure the concentration profile as a function of time at a given temperature, and compare the results with the thin or thick film solution of the diffusion equation. Measurements of hydrogen diffusion are rather different, and are described separately later (p. 499). Ion-implanted profiles correspond to a thin film solution in an infinite slab as in *Figure 25.1(a)*, thin sputter-coated profiles correspond to a thin film solution in a semi-infinite slab as in *Figure 25.1(b)*, and thick sputter-coated profiles correspond to a thick film solution as in *Figure 25.1(c)*. Several important points need to be made about this general procedure:

1. Fick's law assumes that diffusion is only caused by a concentration gradient, and ignores the effect of other forces such as gravity, electric field etc. These effects can usually be avoided, but serious problems arise when there is a strong chemical interaction between the diffusing species and one or more components of the matrix. The mass flux is then determined by the gradient of chemical potential, and D as defined by equation 25.1 can even become negative. Unfortunately, very little can be done about this problem in amorphous alloys, because there are no available data for chemical potentials or other thermodynamic properties.
2. The diffusion equation and the thin and thick film solutions are strictly only valid

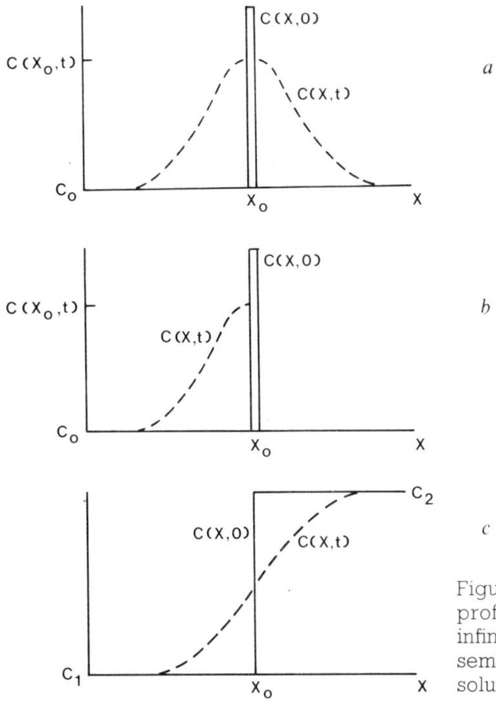

Figure 25.1 Schematic concentration profiles for (*a*) thin film solution in an infinite matrix; (*b*) thin film solution in a semi-infinite matrix; and (*c*) thick film solution

if the range of concentrations is small, so that D is effectively constant over the whole concentration profile. This requires small values of m in the thin film solution and $c_2 - c_1$ in the thick film solution. Ideally, therefore, accurate techniques for measuring concentrations are necessary to determine the details of a diffusion profile between narrow concentration limits.

3. Impurity diffusion is when the diffusing species is not a component of the amorphous alloy matrix, i.e. $c_0 = 0$ in the thin film solution, and $c_1 = 0$ in the thick film solution. Self-diffusion is when the diffusing species is also a component of the amorphous alloy matrix, and self-diffusion coefficients are important in controlling processes such as structural relaxation and crystallization. Unfortunately, self-diffusion coefficients are particularly difficult to measure because diffusion profiles must be determined against a significant background concentration in the amorphous alloy matrix.
4. A particularly suitable method for determining self-diffusion coefficients in amorphous alloys is from the interdiffusion of two isotopes of the diffusing species. The background matrix concentration can be removed with a measurement technique sensitive to different isotopes, and variation of isotope ratio at constant alloy composition ensures that D remains effectively constant over the diffusion profile.
5. The extent of the diffusion concentration profile can be estimated by setting $w = 1$ to define a characteristic diffusion distance x_c:

$$x_c^2 = 4Dt \qquad (25.6)$$

In the thin film solution, for instance, x_c is the distance over which the concentration of the diffusing species falls to e^{-1} of its maximum value. In amorphous alloys, heat-treatments are limited to short times at low temperatures to avoid crystallization, and diffusion coefficients and diffusion distances are correspondingly small. Exceptional spatial resolution is then required to measure diffusion profiles. *Figure 25.2* shows characteristic diffusion distances over the range of measured diffusion coefficients in amorphous alloys, 10^{-16}—$10^{-24}\,m^2\,s^{-1}$, and the range of accessible annealing times, 10^2—10^6 s. Also included in *Figure 25.2* is a band which represents a range of diffusion coefficients and annealing times which lead to the onset of crystallization in different amorphous alloys. Diffusion distances are evidently always limited to less than about 0.1 μm. Since the spatial resolution of concentration measurements must be significantly smaller than the diffusion distance, *Figure 25.2* indicates that 1—10 nm resolution is necessary for diffusion coefficient measurements in amorphous alloys.

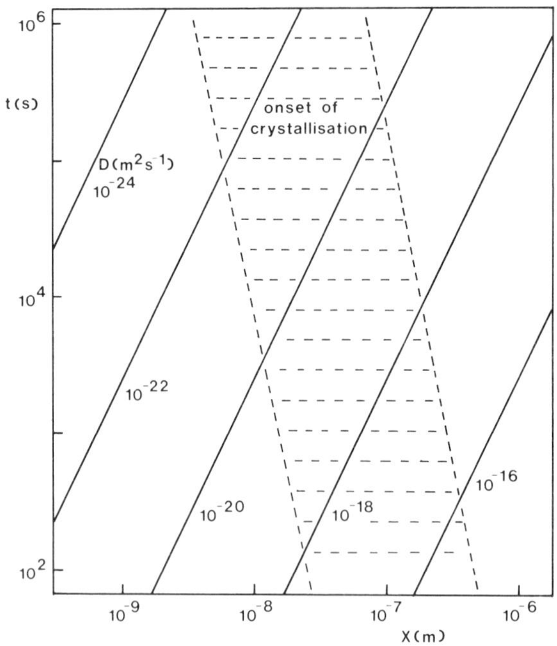

Figure 25.2 Characteristic diffusion distances x_c over the range of measured diffusion coefficients D in amorphous alloys, and the range of accessible annealing times t. The shaded band indicates the range of diffusion coefficients and annealing times which lead to the onset of crystallization in different amorphous alloys

Ion-erosion is one method for obtaining suitable depth resolution when measuring diffusion profiles in a thin layer below the surface of an amorphous alloy. In the first experiment on diffusion in amorphous alloys, Gupta *et al.*[2] measured the diffusion coefficient of silver in amorphous $Pd_{81}Si_{19}$, by ion-implanting ^{110}Ag radioactive isotope below the amorphous alloy surface, diffusion-annealing, then eroding the surface gradually by bombardment with argon ions, and measuring the silver concentration profile from the radioactivity of the material removed as a function of time. With an almost identical technique, Valenta *et al.*[3,4] used ^{59}Fe and ^{32}P

radioactive isotopes to determine iron and phosphorus self-diffusion coefficients in amorphous $Fe_{40}Ni_{40}P_{14}B_6$. Luborsky and Bacon[5] measured the silicon self-diffusion coefficient in amorphous $Fe_{82}B_{12}Si_6$ with a similar procedure, sputter-coating a thick silicon layer on the amorphous alloy surface, diffusion annealing, then gradually eroding the surface by bombardment with argon ions, and measuring the silicon surface concentration as a function of time by Auger electron spectrometry. In secondary ion mass spectrometry, the surface of a specimen is again ion-eroded, and a mass spectrometer is used to analyse the composition of the material removed. This method discriminates between different isotopes of the same element. Cahn *et al.*[6] prepared amorphous $Fe_{40}Ni_{40}B_{20}$ specimens containing the naturally-occurring $^{10}B/^{11}B$ isotope ratio of 0.20, by sputter-coating with a thick layer of the same alloy containing an enriched isotope ratio of 0.96. After diffusion-annealing, secondary ion mass spectrometry was used to determine the self-diffusion coefficient of boron. Edelin and Têtе[7] also used secondary ion mass spectrometry to measure concentration profiles of phosphorus diffused from the gas phase into amorphous $Fe_{80}B_{20}$.

To obtain good depth resolution without destroying the specimen by ion-erosion, the only alternative method is to use a high energy beam of particles which can penetrate below the surface of the amorphous alloy. In Rutherford back-scattering spectrometry, a specimen is irradiated with a monoenergetic beam of alpha-particles, the alpha-particles are preferentially scattered by heavy nuclei in the specimen, and the energy spectrum of scattered alpha-particles can be used to determine the concentration of scattering nuclei as a function of depth below the specimen surface. This technique is suitable for detecting heavy elements in a matrix of substantially lower atomic weight, and was used by Chen *et al.*[8] to measure ion-implanted gold diffusion profiles in amorphous $Pd_{77.5}Cu_6Si_{16.5}$, and by Kijek *et al.*[9,10] to measure thin and thick sputter-coated gold diffusion profiles in amorphous $Ni_{59.5}Nb_{40.5}$. With an almost identical method, Akhtar *et al.*[11,12] measured gold diffusion coefficients in amorphous $Pd_{77.5}Cu_6Si_{16.5}$, $Fe_{40}Ni_{40}B_{20}$, $Ni_{64}Zr_{36}$ and $Ni_{33.3}Zr_{66.7}$ and platinum and lead diffusion coefficients in amorphous $Ni_{33.3}Zr_{66.7}$. High energy particles can also be used to investigate diffusion of light elements, if the nuclei can be made to undergo a suitable nuclear reaction. Kijek *et al.*[10,13] measured the diffusion coefficient of ion-implanted boron in amorphous $Ni_{59.5}Nb_{40.5}$, by irradiating the amorphous alloy surface with high energy protons, detecting alpha-particles emitted from the nuclear reaction $^{11}B + p \rightarrow {}^{8}Be + \alpha$, and determining the concentration profile of ^{11}B below the surface from the number and energy of emitted alpha-particles as a function of incident proton energy. The very high profiling sensitivity of this technique allowed diffusion coefficients to be measured over a wider temperature range than that achieved by other methods. Birac and Lesueur[14] used a similar method, with high energy neutrons and the nuclear reaction $^{6}Li + n \rightarrow {}^{3}H + \alpha$, to investigate lithium diffusion in ion-implanted amorphous $Pd_{80}Si_{20}$.

Ion-implantation and sputter-coating both have disadvantages as methods of manufacturing diffusion couples. The main disadvantage of ion-implantation is the possibility of damaging the amorphous alloy structure, and locally enhancing the diffusion coefficient. On the other hand, it can be difficult to ensure that there is no oxide barrier between a sputter-coating and the amorphous alloy substrate, and sputter-coated diffusion profiles often extend over a large range of concentrations, which is undesirable when diffusion coefficients are obtained from the thin and thick film solutions to the diffusion equation. There are also disadvantages with both ion-erosion and high energy particle techniques for measuring diffusion profiles. Ion-erosion techniques are destructive of the specimen, topological profiling of the eroded surface can cause errors in depth measurements, and differential sputtering rates can cause errors in composition analysis. On the other hand, measured energy spectra from high energy particle methods are often difficult to deconvolute so as to obtain concentration profiles. A general, critical review of profiling methods has been provided by Coburn and Kay[15].

Several indirect methods have been used to investigate diffusion in amorphous alloys. In a series of iron-based amorphous alloys, Köster and Herold[16–18] estimated self-diffusion coefficients from measurements of crystal growth rates during crystallization, by assuming a specific model for the mechanism and kinetics of crystal growth. Further similar results have been obtained by Morris[19] and Tiwari *et al.*[20]. Using iron-based amorphous alloys, Baer *et al.*[21] estimated phosphorus self-diffusion coefficients from measured rates of phosphorus segregation to a free surface, again by assuming a specific kinetic law for the segregation process. An elegant technique, sensitive to very small diffusion coefficients, was used by Rosenblum *et al.*[22] and Greer *et al.*[23]. Specimens were manufactured from alternate thin sputtered layers of two amorphous alloys, and interdiffusion of the two alloys was monitored from the decay in intensity of the small-angle X-ray peak corresponding to the interlayer separation.

Indirect estimates of self-diffusion coefficients should be treated with some caution, because additional assumptions have to be made about the kinetics of the process being studied. In general, the mechanism and kinetics of a process are usually established from known diffusion data rather than vice versa. This is a particularly serious problem with the data derived from crystal growth measurements, since the kinetics of phase transformations can vary widely, and are sometimes independent of long range atomic diffusion[24]. Even when crystal growth rates are determined by long range diffusion, the relationship can be complex, and it is by no means always clear which is the species whose diffusion is rate-controlling[24]. For these reasons, the rest of this chapter concentrates on diffusion coefficients measured directly from composition profiles, although indirect data are also considered where appropriate.

25.3 Diffusion coefficients in amorphous alloys

With the exception of hydrogen diffusion, which is discussed separately (p. 499) *Figure 25.3* shows all the measurements which have been made of diffusion coefficients in amorphous alloys on an Arrhenius graph of log D against inverse temperature $1/T$. For comparison, *Figure 25.3* also includes low temperature extrapolations of typical radioactive tracer diffusion coefficients which have been measured in crystalline metals[25], with self-diffusion represented by fcc iron, and subsitutional and interstitial solute diffusion represented by gold in fcc nickel and boron in fcc iron, respectively. The amorphous alloy diffusion coefficients are spread in a wide band midway between the extrapolated lines for typical substitutional and interstitial diffusion processes in crystalline metals. As expected, amorphous alloy diffusion coefficients decrease with decreasing temperature in very roughly the same way as found in crystalline metals.

Kijek *et al.*[26] plotted diffusion data in amorphous alloys on an Arrhenius graph with the inverse temperature scale normalized to the glass transition temperature of the amorphous alloy T . The data fell into two groups separated by several orders of magnitude, which was interpreted by Kijek *et al.*[26], as indicating two different diffusion mechanisms in amorphous alloys. Unfortunately, this conclusion is much less clear when the more recent measurements[3–5,9–12,21–23] of diffusion in amorphous alloys are also included. *Figure 25.4* shows the data from *Figure 25.3* re-plotted as log D against inverse normalized temperature T_g/T, and the amorphous alloy diffusion coefficients are again spread in a wide band, with no clear separation into two groups. To compare

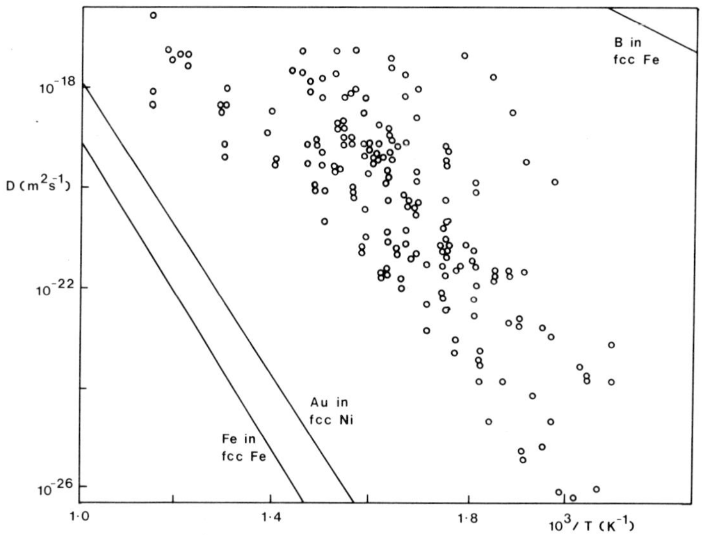

Figure 25.3 Arrhenius graph of all measured diffusion coefficients in amorphous alloys

different alloy systems consistently, the diffusion coefficient should also be normalized by replacing D with $D/a^2\nu$, where a is a characteristic atomic distance in the amorphous alloy structure, and ν is a characteristic atomic vibration frequency[27]. For all the amorphous alloys represented in *Figures 25.3* and *25.4*, $a^2\nu$ is in the range 3×10^6–$5 \times 10^6\ m^2\,s^{-1}$, with a taken as the mean Goldschmidt radius of the amorphous alloy components, and ν calculated from $\nu = kT_g/h$, where k is Boltzmann's constant and h is Planck's constant. Because of the narrow range of values for $a^2\nu$, normalizing D has almost no effect on the graph in *Figure 25.4*.

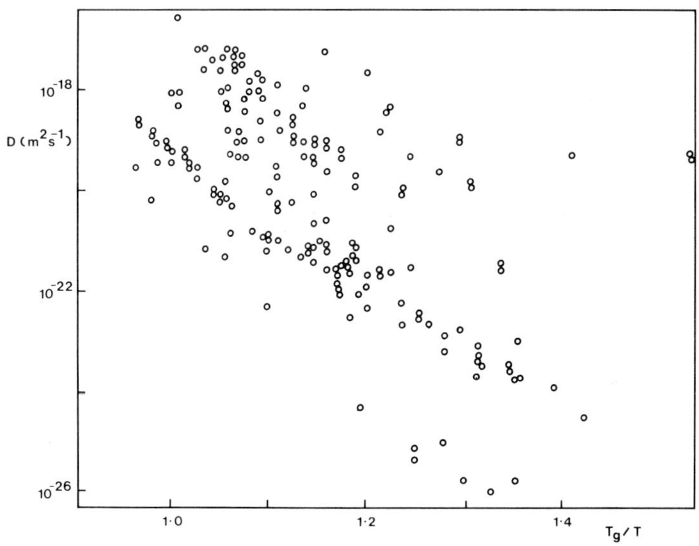

Figure 25.4 Normalized Arrhenius graph of all measured diffusion coefficients in amorphous alloys

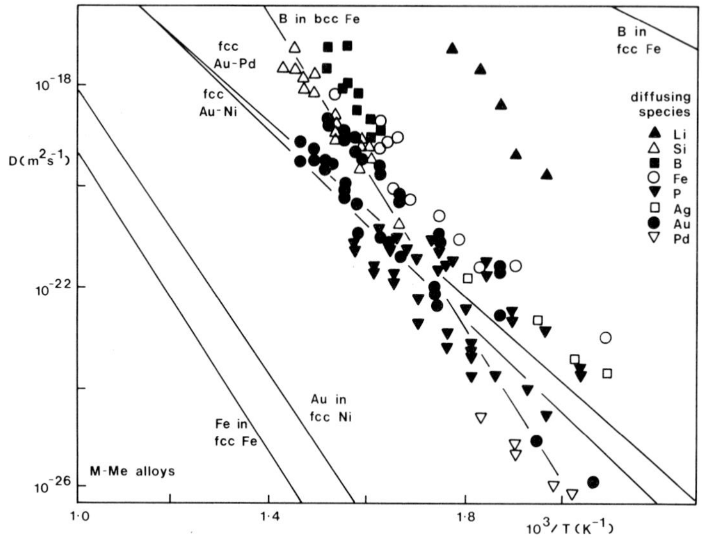

Figure 25.5 Arrhenius graph of measured diffusion coefficients in metal–metalloid amorphous alloys

Figure 25.4 does not show the two distinct groups of data found by Kijek *et al.*[26] mainly because of a series of recent measurements[9–12] of heavy metal diffusion coefficients in metal–metal amorphous alloys such as Ni–Nb and Ni–Zr. It seems reasonable therefore to consider separately the data for metal–metalloid (M–Me) and metal–metal (M–M) amorphous alloys. *Figures 25.5 and 25.6* show measured diffusion coefficients in M–Me amorphous alloys, plotted on conventional and normalized Arrhenius graphs respectively; similarly, *Figures 25.7* and *25.8* show measured diffusion

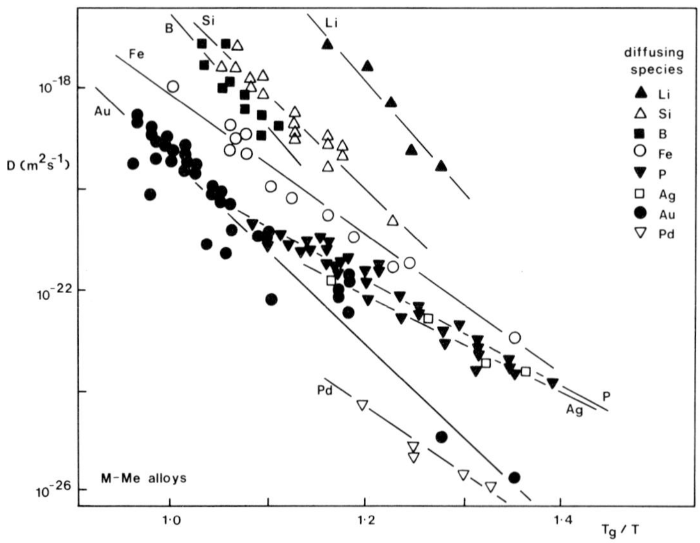

Figure 25.6 Normalized Arrhenius graph of measured diffusion coefficients in metal–metalloid amorphous alloys

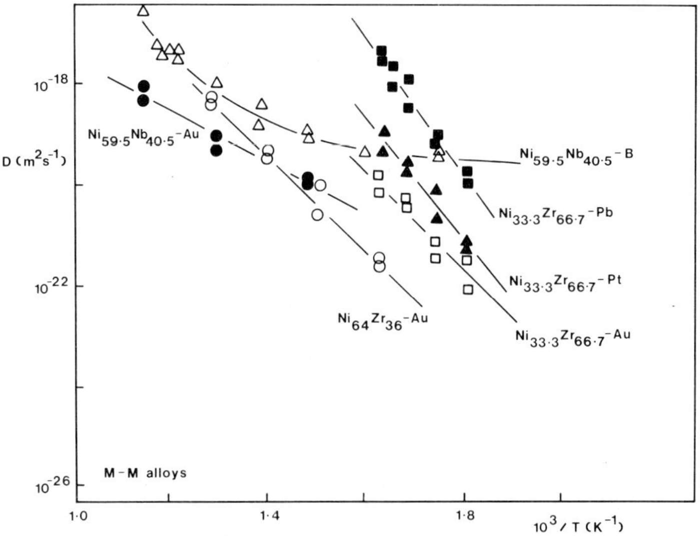

Figure 25.7 Arrhenius graph of measured diffusion coefficients in metal–metal amorphous alloys

coefficients in M–M amorphous alloys, plotted on conventional and normalized Arrhenius graphs, respectively. For reasons discussed below, *Figures 25.5—25.8* include data obtained at the higher annealing temperatures, when the amorphous alloy was either above[8,11] T_g or was partially crystallized during the diffusion-anneal[2,5,6,8], but exclude data for fully crystallized specimens[3,4,12]. Also, for reasons discussed below, they include data for relaxed rather than as-quenched amorphous alloys, when significant differences were found between the two measurements[8,22,23].

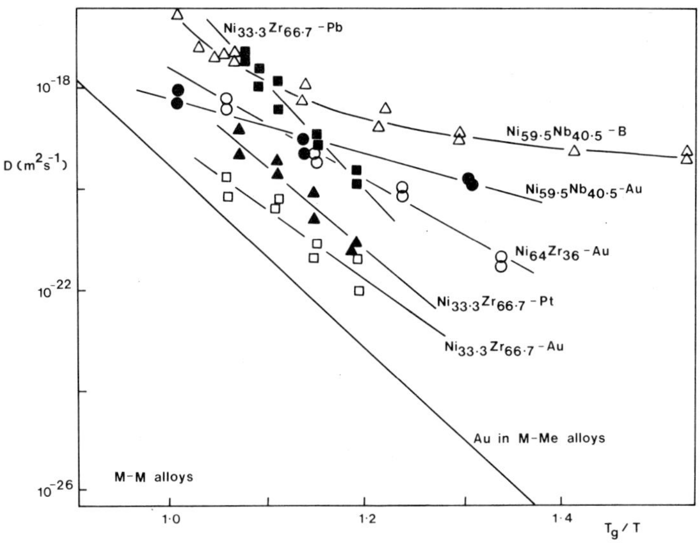

Figure 25.8 Normalized Arrhenius graph of measured diffusion coefficients in metal–metal amorphous alloys

25.3.1 Diffusion in metal–metalloid amorphous alloys

The normalized Arrhenius graph in *Figure 25.6* shows a consistent pattern for diffusion coefficients in M–Me amorphous alloys, with a different line for each of the different diffusing species. There is good agreement, well within experimental error, between the three sets of data for phosphorus diffusion[4,7,21], and also between the three sets of data for gold diffusion[8,11]. The indirectly measured[22] interdiffusion coefficients in $Pd_{80}Au_7Si_{13}$–Fe are on the same line as the three sets of direct measurements for gold diffusion[8,11]; and the diffusion coefficients estimated[17] indirectly from crystal growth rates in Fe–B and Fe–B–C are on the same line as the direct measurements[3] of iron self-diffusion in $Fe_{40}Ni_{40}P_{14}B_6$. Thus, *Figure 25.6* suggests that the indirect interdiffusion and crystal growth rate measurements are controlled by diffusion of the largest atom in the amorphous alloy structure, i.e. gold for interdiffusion[22] of $Pd_{80}Au_7Si_{13}$–Fe, palladium for interdiffusion[22] of both $Pd_{85}Si_{15}$–$Fe_{85}B_{15}$[22] and $Pd_{80}Si_{20}$–$Fe_{80}B_{20}$[23], and iron for crystal growth[17] in Fe–B and Fe–B–C. This interpretation is physically plausible and agrees well with other data in *Figure 25.6*. Köster and Herold[16–18] suggested that their crystal growth measurements represented diffusion of boron or carbon, but it seems more likely that the crystal growth process is controlled by the slower diffusion of the large iron atoms.

When the indirect interdiffusion and crystal growth measurements are interpreted in this way, all of the data in *Figure 25.6* are self-consistent. At a given fraction of T_g, the diffusion coefficients in metal–metalloid (M–Me) amorphous alloys depend only upon the nature of the diffusing species, and increase progressively from the larger, heavy atoms such as gold and palladium, through to the smaller, light atoms such as boron, silicon, and lithium. This indicates a progressive change in diffusion mechanism from the large atoms such as gold to the light atoms such as lithium, rather than the two distinct mechanisms suggested by Kijek *et al.*[26]. The more recent measurements of iron and silicon diffusion[3,5] prevent the data from separating into two distinct groups. For an element X, *Figure 25.6* shows that the diffusion coefficient in all M–Me amorphous alloys is given by:

$$D_x = D_{0x} \exp(-B_x T_g/T) \tag{25.7}$$

where D_x and B_x depend only on X. This is equivalent to saying that diffusion in M–Me amorphous alloys shows Arrhenius behaviour, with a frequency factor D_{0x} which depends only on the diffusing species, and an activation energy Q which is proportional to the glass transition temperature $Q = B_x k T_g$ for a given X.

With the exception of data for lithium, all the measured diffusion coefficients in M–Me amorphous alloys are grouped into a surprisingly narrow band on the conventional Arrhenius graph in *Figure 25.5*. As noted previously for some of the data[9,17], this band of diffusion coefficients is quite close to extrapolated Arrhenius lines for boron diffusion in bcc Fe, and interdiffusion in fcc Au–50%Ni and Au–50%Pd alloys[25]. The band of data in *Figure 25.5* seems to indicate that diffusion coefficents in M–Me amorphous alloys depend only on the annealing temperature. This is almost certainly incorrect. It is physically unlikely that diffusion is independent of both the diffusing species and the amorphous alloy matrix. In any case, there is no good reason for excluding lithium, and the scatter of data within the band is significantly greater than experimental error. The apparent dependence of diffusion coefficient only on temperature conflicts with equation 25.7 and the interpretation of *Figure 25.6* discussed above. Thus, it is worth noticing that the three sets of data for phosphorus diffusion[4,7,21], and the three sets of data for gold diffusion[8,11] are in each case more closely grouped on the normalized Arrhenius graph in *Figure 25.6* than on the conventional Arrhenius graph in *Figure 25.5*.

25.3.2 Diffusion in metal–metal amorphous alloys

Compared with M–Me amorphous alloys, there have been relatively few investigations of diffusion in metal–metal (M–M) amorphous alloys. The measured diffusion coefficients in M–M amorphous alloys are of roughly comparable magnitude with those found in M–Me amorphous alloys, and again show an Arrhenius or near-Arrhenius dependence on temperature. However, the conventional and normalized Arrhenius graphs in *Figures 25.7* and *25.8* both show considerable scatter in the data for M–M amorphous alloys, with no consistent pattern like the normalized M–Me amorphous alloy data in *Figure 25.6*. It is noteworthy that the data[10,13] for boron diffusion in $Ni_{59.5}Nb_{40.5}$ extend over a far wider temperature range than other measurements, and show unmistakable evidence of a non-linear Arrhenius plot.

Perhaps not surprisingly, small boron atoms diffuse[9,10,13] faster than gold atoms at a given temperature in amorphous $Ni_{59.5}Nb_{40.5}$. A more surprising result is that diffusion coefficients in amorphous $Ni_{33.3}Zr_{66.7}$ increase[12] in the order Au, Pt, Pb, although gold and platinum have similar atomic size and lead is slightly larger. Impurity diffusion coefficients in crystalline metals increase with increasing difference in melting point between the diffusing species and the matrix[28]. There seems to be no clear physical explanation of this behaviour, although an interpretation may perhaps be derived from the well-known relationship between self-diffusion coefficients and melting points in pure metals[29,30]. With melting points of 1064 °C for Au, 1772 °C for Pt, 327 °C for Pb and 1140 °C for crystalline NiZr, Akhtar *et al.*[12] suggest that a similar effect may explain the diffusion coefficients in amorphous $Ni_{33.3}Zr_{66.7}$.

The diffusion coefficient of gold has been measured in six amorphous alloys[8,9,11,12,22] including the interdiffusion measurements in $Pd_{80}Au_7Si_{13}$–Fe, more than for any other diffusing species. To compare the gold diffusion data, *Figure 25.8* includes the Arrhenius line for gold diffusion in M–Me amorphous alloys, from *Figure 25.6*. At a given fraction of T_g, gold diffuses faster in M–M amorphous alloys than in M–Me amorphous alloys. As discussed in the previous section, gold diffusion in M–Me amorphous alloys is independent of the alloy at a given fraction of T_g. However, at a given fraction of T_g in M–M amorphous alloys, the gold diffusion coefficient increases with increasing T_g, from $Ni_{33.3}Zr_{66.7}$ with $T_g = 660$ K, through to $Ni_{64}Zr_{36}$ with $T_g = 823$ K, and $Ni_{59.5}Nb_{40.5}$ with $T_g = 881$ K [9,11,12]. In *Figure 25.8*, the two lines for gold diffusion in $Ni_{33.3}Zr_{66.7}$ and $Ni_{64}Zr_{36}$ have the same slope, but the line for gold diffusion in $Ni_{59.5}Nb_{40.5}$ has a different slope. Thus, to extend equation 25.7 to include M–M amorphous alloys, both D_{0x} and B_x must become functions of T_g.

25.3.3 Thermal and mechanical effects

This section considers the effects of crystallization, structural relaxation, plastic deformation and irradiation on diffusion coefficients in amorphous alloys.

Partial crystallization

Gupta *et al.*[2] and Luborsky and Bacon[5] found no effect of slight crystallization during diffusion-annealing on the measured diffusion coefficients of silver in $Pd_{81}Si_{19}$ and silicon in $Fe_{82}B_{12}Si_6$, respectively. However, Cahn *et al.*[6] reported that partial crystallization affects diffusion of boron in $Fe_{40}Ni_{40}B_{20}$, and Chen *et al.*[8] reported that partial crystallization or heating above T_g without crystallization both affect diffusion of gold in $Pd_{77.5}Cu_6Si_{16.5}$. Careful examination of the results of Cahn *et al.* and Chen *et al.* does not support their interpretation, and in each case a single line can be drawn within experimental error through all the data, including partially crystallized specimens. As pointed out by Luborsky and Bacon[5], no effect is expected when the fraction of crystals is small.

Complete crystallization

Valenta *et al.*[3,4] found significantly slower diffusion of phosphorus and iron in $Fe_{40}Ni_{40}P_{14}B_6$, when the amorphous alloy was annealed at a high enough temperature to cause extensive crystallization. On the other hand, Akhtar *et al.*[12] found significantly faster diffusion of platinum in $Ni_{33.3}Zr_{66.7}$ after complete crystallization. These results are not incompatible. As described elsewhere in this book, there is a wide range of crystallization mechanisms in different amorphous alloys, which can lead to a wide variety of different crystal structures and grain sizes. Thus, no general rule is likely, and diffusion coefficients should be expected sometimes to increase and sometimes to decrease after crystallization. It is worth noticing that some amorphous alloys crystallize on a very fine scale, so that diffusion in the crystallized alloy can be enhanced by a large area of grain and interphase boundaries.

Relaxation

Reported data for the effect of relaxation of the amorphous alloy on diffusion coefficients show considerable apparent disagreement. Chen *et al.*[8] reported that a relaxation heat-treatment close to T_g caused the gold diffusion coefficient in $Pd_{77.5}Cu_6Si_{16.5}$ to fall by several orders of magnitude, and similar results were obtained in interdiffusion experiments on thin amorphous alloy films[22,23]. However, Luborsky and Bacon[5] and Baer *et al.*[21] detected no effect of relaxation on the diffusion of silicon in $Fe_{82}B_{12}Si_6$ and phosphorus in $Fe_{32}Ni_{36}Cr_{14}P_{12}B_6$ respectively. Kijek *et al.*[9,10] and Akhtar *et al.*[11,12] reported little or no effect of relaxation with a variety of different amorphous alloys and different diffusing species. The measured diffusion coefficients were consistently lower in the relaxed alloy than in the as-quenched alloy, but the difference was always small and usually within experimental error, so any effect of relaxation was slight. Because of the previous conflicting data, Akhtar *et al.*[11] repeated the experiments of Chen *et al.*[8] on gold diffusion in amorphous $Pd_{77.5}Cu_6Si_{16.5}$, using the same relaxation heat-treatment, same diffusion-annealing heat treatments, and same Rutherford back-scattering measurement technique. Unlike Chen *et al.*, however, Akhtar *et al.* found no significant difference between gold diffusion coefficients in relaxed and as-quenched amorphous $Pd_{77.5}Cu_6Si_{16.5}$.

Akhtar *et al.*[1] explained their failure to reproduce the results of Chen *et al.*[8] as a consequence of different methods for producing the amorphous alloy, and the same explanation can be used to resolve all the apparent disagreement about the effect of relaxation on diffusion in amorphous alloys. Luborsky and Bacon[5], Baer *et al.*[21], Kijek *et al.*[9,10], and Akhtar *et al.*[11,12] used melt-spinning to prepare the various amorphous alloys. Direct temperature measurements show a high initial cooling rate in melt-spun ribbons, but loss of contact with the substrate causes a sharp drop in cooling rate at a temperature about 500 K below the melt temperature[31,32]. For most amorphous alloys, this temperature is quite close to T_g, so melt-spun amorphous alloys can cool quite slowly through T_g and experience some autorelaxation during the quenching process. In contrast, Chen *et al.* prepared amorphous $Pd_{77.5}Cu_6Si_{16.5}$ by a variant of the gun splat-quenching technique, in which a thin film of liquid alloy is spread over the surface of a copper substrate. With this technique, the cooling rate remains high over most of the temperature range from the melt temperature to room temperature[33,34]. In addition, although not stated by Chen *et al.*, it is likely that the diffusion experiments were performed on the smooth, lower surface of the amorphous alloy specimens, where the cooling rate is particularly high. Thus, the amorphous $Pd_{77.5}Cu_6Si_{16.5}$ of Chen *et al.* was cooled very rapidly through T_g, so the amorphous alloy atomic structure was particularly far from internal equilibrium, and particularly susceptible to subsequent relaxation heat treatments. A similar unrelaxed and non-equilibrium amorphous alloy structure is likely in the sputtered films used for the interdiffusion experiments[22,23]. Substantially different diffusion coefficients have been found on upper and lower surfaces of as-quenched melt-spun amorphous alloys[7,35], which again indicates that diffusion is sensitive to the precise thermal history of the amorphous alloy during quenching.

The different reported effects of relaxation are therefore quite well explained by the different methods of producing the amorphous alloy. Diffusion is very sensitive to relaxation[8,22,23] in amorphous alloys which are very rapidly cooled through T_g, but autorelaxation in melt-spun amorphous alloys makes the diffusion coefficient independent[5,21] or nearly independent[9–12] of subsequent relaxation heat-treatments. Since there is little or no effect of relaxation in the more slowly cooled amorphous alloys, the time constant for the relaxation process must be quite short.

Plastic deformation

Akhtar *et al.*[12] measured the diffusion coefficient of gold in amorphous $Ni_{33.3}Zr_{66.7}$ at four different temperatures, using as-quenched, relaxed, and plastically deformed specimens. At each temperature, the diffusion coefficients were in the order deformed > as-quenched > relaxed. The differences were small and within the experimental error, but the consistency of the results does suggest that relaxation causes a slight decrease in diffusion coefficient, which is reversed by plastic deformation. This is consistent with amorphous alloy density measurements[36] which show that relaxation causes a slight decrease in mean atomic volume which is reversed by plastic deformation. Thus, the gold diffusion coefficient can be correlated with the mean atomic volume of the amorphous alloy matrix.

Irradiation

Cahn *et al.*[37] measured gold diffusion coefficients before and after fast neutron irradiation of amorphous $Ni_{64}Zr_{36}$. Unexpectedly, they found that irradiation reduced the diffusion coefficient even though the mean atomic volume is usually increased by irradiation. Cahn *et al.*[37] argued that irradiation-enhanced chemical short range order caused a reduction in atomic volume with a corresponding reduction in the gold diffusion coefficient.

25.4 Diffusion of hydrogen in amorphous alloys

Hydrogen, the smallest atom, is in a class by itself as a diffusing species. The first measurement of hydrogen diffusion in an amorphous alloy was made by an indirect method. Berry and Pritchet[38], while investigating internal friction in amorphous $Nb_{75}Ge_{25}$, traced one of the internal friction peaks to unintentionally absorbed hydrogen, and subsequently identified similar peaks in six other amorphous alloys intentionally charged with hydrogen[39,40]. The internal friction peak was a Gorsky peak: a hydrogenated strip was excited into flexural oscillation, with damping originating from diffusion of hydrogen across the thickness of the strip. From the variation of damping with temperature, a hydrogen diffusion coefficient could be deduced. A detailed study[39] of Gorsky damping in hydrogenated amorphous $Pd_{80}Si_{20}$ led to the Arrhenius plot shown in *Figure 25.9*, with the very low activation energy of

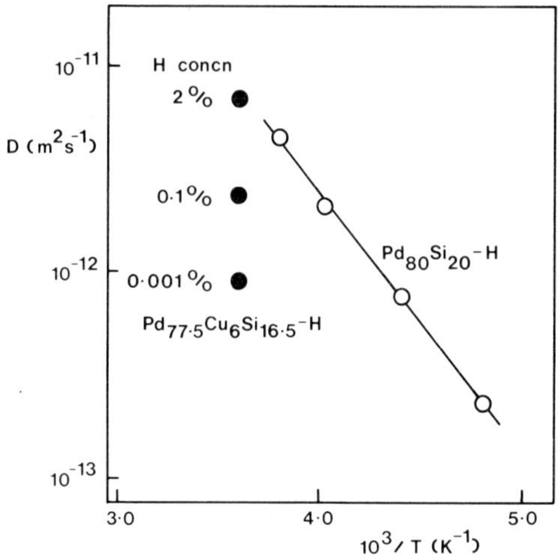

Figure 25.9 Arrhenius graph of hydrogen diffusion coefficients in amorphous Pd–Si and Pd–Cu–Si alloys, showing a strong variation with hydrogen concentration (from Berry and Pritchett[39] and Kirchheim *et al.*[41])

0.25 eV. Hydrogen diffusion in other amorphous alloys[40] led to activation energies in the range 0.33—0.60 eV, but in no case was the hydrogen concentration known.

Independently, Kirchheim *et al.*[41] used an electrolytic method to measure directly the solubility and diffusion coefficient of hydrogen in amorphous $Pd_{77.5}Cu_6Si_{16.5}$. A pulse of hydrogen was released electrolytically into one surface of an amorphous alloy sheet, and arrival at the other surface was monitored electrochemically. Some of the results of Kirchheim *et al.* are included in *Figure 25.9.* The diffusion coefficient is a function of hydrogen concentration, and at 2 atom % agrees well with Berry and Pritchet's data for the closely related $Pd_{80}Si_{20}$ amorphous alloy. The experiments of Kirchheim *et al.* provide the first observation of a diffusion coefficient in an amorphous alloy depending on solute concentration, with hydrogen diffusing more slowly in dilute solutions.

A further indirect approach to measuring hydrogen diffusion in amorphous alloys is to monitor the motional narrowing of nuclear magnetic resonance lines. Bowman *et al.*[42,43] have used this method to study hydrogen diffusion in amorphous $Pd_{33.3}Zr_{66.7}$, hydrogenated up to a concentration of 0.5 atom % of hydrogen. The activation energy for diffusion was 0.25 eV between 270 and 365 K, rising at high temperature to reach about 0.7 eV in the range 430—490 K. Earlier results[40] showed similar characteristics in hydrogenated amorphous Cu–Ti alloys. Hydrogen diffusion in hydrogenated Cu–Ti and Pd–Zr amorphous alloys is between one and two orders of magnitude faster than in the corresponding crystalline hydrides. All measured hydrogen diffusion coefficients in amorphous alloys are very much higher than for other diffusing species, with much lower activation energies.

25.5 Diffusion mechanisms in amorphous alloys

There is almost no direct evidence about mechanisms of atomic diffusion in amorphous alloys, so the discussion in this section is necessarily rather speculative. As shown elsewhere in this book, amorphous and crystalline alloys have similar densities and similar local arrangements of atoms. It seems sensible therefore to consider whether atomic models of diffusion in crystalline alloys can also be applied to amorphous alloys. For random walk diffusion in a crystalline lattice, the diffusion coefficient can be expressed by an Arrhenius equation[24]:

$$D = D_0 \exp(-Q/kT)$$
$$= \alpha f d^2 \exp(\Delta S/K) \exp(-Q/kT) \quad (25.8)$$

where α is a geometric factor including correlation effects, f is the vibration frequency of diffusing atoms, d is the diffusion jump distance, and ΔS and Q are the activation entropy and energy of an individual diffusion jump. In an amorphous structure, unlike a crystalline lattice, the values of α, d, ΔS and Q are likely to vary from one diffusion jump to another, so there is no guarantee that equation 25.8 is valid. However, *Figures 25.3—25.9* show that diffusion coefficients in amorphous alloys agree quite well with the Arrhenius law in equation 25.8. This indicates, rather surprisingly, that for a given diffusing species and a given amorphous alloy, the diffusion process is often controlled by a single mechanism over a considerable range of temperatures.

Vacancy migration is the most common mechanism of both self diffusion and impurity diffusion in crystalline metals and alloys[24]. In a crystal lattice, a vacancy is defined as an unoccupied lattice site, but this definition is clearly meaningless in an amorphous structure. Without reference to a crystal lattice, a vacancy can only be defined as a space in the amorphous alloy structure of atomic or near-atomic dimensions. There is then no formal distinction between a vacancy and an interstitial site, and a vacancy is nothing more than a particularly large interstitial site.

Bennett *et al.*[44] have reported that when a vacancy is created artificially by removing one atom from a dense random packed computer model of an amorphous alloy structure, atomic vibrations quickly redistribute the excess of space over a large volume. This seems to show that vacancies are unstable in amorphous alloys, and it has frequently been assumed that diffusion cannot take place by vacancy migration[3,4,8,14,26,45]. Unlike a crystal lattice however, different atomic positions in an amorphous structure are not equivalent, and Vitek[46] has pointed out that the stability of the artificial vacancy created by Bennett *et al.* should depend upon which atom is removed. In any case, as described above, a vacancy in an amorphous structure is not characterized by the absence of an atom, and stable vacancies should be found as large interstitial sites in the amorphous structure.

A simple Boltzmann expression gives the number of vacancies in internal equilibrium N_v as:

$$N_v = N \exp(-U/kT) \quad (25.9)$$

where N is the number of possible sites for a vacancy, and U is the excess internal energy of a vacancy. If anything, an amorphous alloy might then be expected to contain more vacancies than an equivalent crystalline alloy. An amorphous structure is 'softer' so that U should be slightly smaller, there are more interstitial sites than atoms so that N should be slightly larger, and in any case, an as-quenched amorphous alloy is not in internal equilibrium and contains excess free volume which can partly be distributed as excess vacancies. If a vacancy in an amorphous alloy can decompose into groups of smaller interstitial sites, then the statistical mechanics of the distribution of interstitial sites become quite complex[47], but the principle of detailed chemical balancing still indicates that the number of vacancies is given by equation 25.9.

There have been several investigations of the structure and size of interstitial sites in soft sphere dense random packed models of amorphous alloy structures[48–51]. *Figure 25.10* shows a typical size distribution of interstitial sites in a model constructed by Finney and relaxed under a Morse potential[48,49]. Most interstitial sites are surrounded by distorted tetrahedral and octahedral groups of atoms with more tetrahedra and

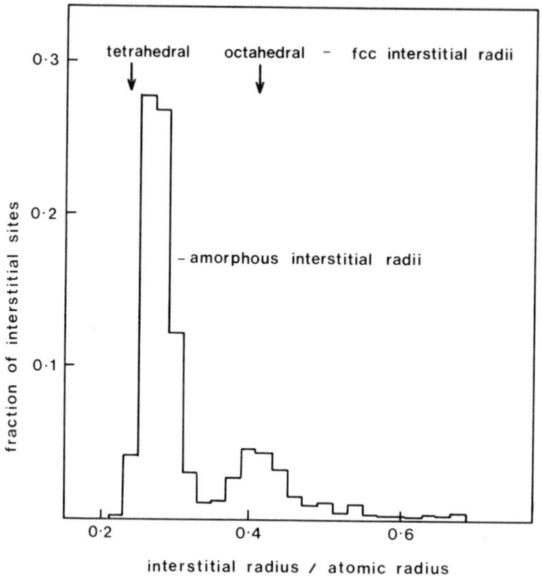

Figure 25.10 Size distribution of interstitial sites in Finney model after relaxation under a Morse potential (after Finner and Wallace[48])

fewer octahedra than in an equivalent close packed crystal. Mean tetrahedral and octahedral interstitial sizes are similar to the regular tetrahedral and octahedral interstitial sizes in a close packed crystal. However, the tail of the octahedral distribution contains a few large interstitial sites, up to a maximum diameter about 0.7 times the atomic diameter. The size of these larger interstitial sites cannot be very different from the relaxed size of a vacancy in a crystal lattice, so it might be argued that these are the vacancies predicted by equation 25.9. It is perhaps a semantic question whether the larger interstitial sites can reasonably be described as vacancies. However, migration of the larger interstitial sites is clearly a possible diffusion mechanism, analogous to vacancy migration in a crystalline lattice.

Several previous investigators have suggested that diffusion in amorphous alloys must take place by some form of cooperative motion of groups of adjacent atoms[3,4,8,14,26,52]. No detailed process has yet been described, but cooperative motion of a group of atoms is clearly more likely near one of the larger interstitial sites shown in *Figure 25.10*. In fact, a cooperative diffusion mechanism is not really different from the migration of interstitial sites described above. Cooperative atomic motion can only cause atomic diffusion if there is some redistribution of interstitial sites and, conversely, the migration of an interstitial site requires the cooperative motion of the surrounding atoms. As with vacancy migration in crystals, it may be more convenient to regard the diffusion process in amorphous alloys as a redistribution of the space between the atoms, rather than the atomic motion itself. Interstitial migration may take place by a relatively simple process, either by direct interchange of position between adjacent large and small interstitial sites, or by decomposition of a large interstitial site into several smaller ones, followed by reconstitution at a different position in the amorphous structure. In either case, the collective motion of the surrounding atoms is difficult to describe, which explains why no detailed cooperative diffusion mechanism has been suggested previously. As discussed elsewhere in this book, metalloid atoms in M–Me amorphous alloys are thought to occupy some of the larger interstitial sites in *Figure 25.10*, so the mechanism of interstitial migration should be sensitive to amorphous alloy composition. For a given diffusing species, *Figure 25.6* and equation 25.7 indicate that this effect is equivalent to using a normalized scale of temperature.

Table 25.1 shows best-fit activation energies Q and frequency factors D_0 for the amorphous alloy diffusion data in *Figures 25.3—25.9*. In general, activation energies for diffusion in amorphous alloys are close to the activation energies found for vacancy migration in crystalline alloys, supporting the suggestion that diffusion in amorphous alloys is often controlled by the migration of large, vacancy-like interstitial sites. The D_0 values in *Table 25.1* vary by more than 20 orders of magnitude, and Fernengel *et al.*[53] have suggested that there is a rough correlation between high D_0 and Q values in different systems.

In crystalline alloys, small atoms are often dissolved interstitially, by fitting into the interstitial sites within the crystal lattice. Solute diffusion can then take place by an interstitial mechanism, with the small diffusing atoms moving between adjacent interstitial sites[1,24]. A similar interstitial diffusion mechanism is also possible in an amorphous alloy, with small solute atoms migrating through the interstitial sites in the amorphous alloy matrix. Hydrogen almost certainly diffuses by an interstitial mechanism in amorphous alloys. At a given fraction of T_g, *Figure 25.6* shows that other diffusion coefficients in M–Me amorphous alloys increase progressively with decreasing size of the diffusing atoms, and as suggested previously by Kijek *et al.*[26], this seems to indicate a progressive change in diffusion mechanism, from cooperative atomic motion and interstitial site migration for large atoms such as gold, to an interstitial diffusion mechanism for small atoms such as lithium.

With an interstitial diffusion mechanism, the larger interstitial sites in an amorphous alloy are again likely to be important, acting as trapping centres for the diffusing atoms[26,49,50]. Interstitial diffusion is then controlled by the binding energy

TABLE 25.1 Activation energies Q and frequency factors D_0 for diffusion in amorphous alloys (experiments based upon crystal growth kinetics are not included)

Amorphous alloy	*Diffusing species*	D_0/m² s⁻¹	Q/eV	*Ref.*
$Fe_{40}Ni_{40}B_{20}$	B	10^{12}	3.6	6
	Au	2×10^{-4}	2.1	11
$Pd_{77.5}Cu_6Si_{16.5}$	Au	10^{-8}	1.7	8
	Au	2×10^{-5}	1.8	11
	H	$1—3 \times 10^{3}$*	0.25—0.5*	41
$Fe_{40}Ni_{40}P_{16}B_4$	P	10^{10}	3.1	4
	P	2×10^{10}	3.4	53
	Fe	10^{-3}	2.0	3
$Ni_{33.3}Zr_{66.7}$	Au	2×10^{-4}	2.0	11, 12
	Pt	9×10^{0}	2.4	12
	Pb	8×10^{4}	2.7	12
$Pd_{81}Si_{19}$	Ag	2×10^{-10}	1.3	2
$Pd_{80}Si_{20}$	Li	10^{10}	2.8	14
	H	3×10^{-7}	0.25	39
$Fe_{80}B_{20}$	P	10^{-4}	1.9	7
$Fe_{32}Ni_{36}Cr_{14}P_{12}B_6$	P	10^{-7}	1.8	21
$Fe_{82}B_{12}Si_6$	Si	5×10^{2}	2.8	5
$Ni_{64}Zr_{36}$	Au	10^{-6}	1.9	11
$Pd_{33.3}Zr_{66.7}$	H	—	0.25	42

*Depending on H concentration.

between the large interstitial sites and the diffusing atoms, rather like solute–vacancy or solute–dislocation binding in crystalline alloys. Interstitial diffusion coefficients should therefore be strongly dependent on composition, as the larger interstitial sites become filled with increasing solute content. The importance of solute trapping at large interstitial sites was first suggested by Kijek *et al.*[26], and then studied in detail by Ahmadzadeh and Cantor[49,50]. In any amorphous structure, the interstitial site distribution such as shown in *Figure 25.10* is equivalent to a distribution in energy wells or traps for interstitial solute atoms, and there is a corresponding distribution in saddle point energies between adjacent interstitial sites[49]. To allow for the resulting variety of individual diffusive jumps, Ahmadzadeh and Cantor[50] modelled interstitial diffusion by Monte Carlo methods in a dense random packed amorphous structure and found significant correlation effects between consecutive diffusive jumps. Kirchheim[54] used some of these ideas to explain the concentration dependence of interstitial hydrogen diffusion in amorphous alloys[41]. At low concentrations, hydrogen solute atoms are trapped at large interstitial sites, but at higher concentrations, the additional hydrogen atoms must occupy less favourable interstitial sites and can diffuse with a lower activation energy. Kirchheim[54] analysed the problem quantitatively in terms of Fermi–Dirac statistics, obtaining good agreement with the form of the observed concentration dependence of the hydrogen diffusion coefficient[41].

Acknowledgements

We thank the UK Science and Engineering Research Council for financial support.

References

1. Shewmon, P. G., *'Diffusion in Solids'*, McGraw-Hill, New York (1963)
2. Gupta, D., Tu, K. N. and Asai, K. W., *Phys. Rev. Lett.*, **35,** 796 (1975)
3. Valenta, P., Maier, K., Kronmüller, H. and Freitag K., *Phys. Status Solidi,* **106,** 129 (1981)
4. Valenta, P., Maier, K., Kronmüller, H. and Freitag, K., *Phys. Status Solidi,* **105,** 537 (1981)
5. Luborsky, F. E. and Bacon, F., in *'Proc. 4th Int. Conf. on Rapidly Quenched Metals'*, Eds. Masumoto, T. and Suzuki, K., Japan Institute of Metals, Sendai, p. 561 (1982)
6. Cahn, R. W., Evetts, J. E., Patterson, J., Somekh, R. E. and Kenway-Jackson, C., *J. mater. Sci.*, **15,** 702 (1980)
7. Edelin, G. and Tête, C., *Scripta Met.*, **15,** 739 (1981)
8. Chen, H. S., Kimerling, L. C., Poate, J. M. and Brown, W. L., *Appl. Phys. Lett.*, **32,** 461 (1978)
9. Kijek, M., Akhtar, D., Cantor, B. and Cahn, R. W., vol. 1, p. 573, in ref. 5
10. Kijek, M., Palmer, D. W. and Cantor, B., to be published
11. Akhtar, D., Cantor, B. and Cahn, R. W., *Acta metall.*, **30,** 1571 (1982)
12. Akhtar, D., Cantor, B. and Cahn, R. W., *Scripta Met.*, **16,** 417 (1982)
13. Kijek, M., Ahmadzadeh, M. and Cantor, B., in *'Proc. Int. Conf. on Amorphous Metals'*, Central Research Institute for Physics, Budapest (1980)
14. Birac, C. and Lesueur, D., *Phys. Status Solidi,* **A36,** 247 (1976)
15. Coburn, J. W. and Kay, E., *CRC Crit. Rev. Solid St. Sci.*, **4,** 561 (1974)
16. Köster, U. and Herold, U., *J. Phys. C,* **8,** 352 (1980)
17. Köster, U., Herold, U., Hillenbrand, H. G. and Denis J., *J. mater. Sci.*, **15,** 2125 (1980)
18. Köster, U. and Herold, U., p. 587 in ref. 5
19. Morris, D. G., *Acta metall.*, **29,** 1213 (1981)
20. Tiwari, R. S., Ranganathan, S. and von Heimendahl, M., *Z. Metallk.*, **72,** 563 (1981)
21. Baer, D., Pederson, L. R. and Thomas, M. T., *Mater. Sci. Engng,* **48,** 283 (1981)
22. Rosenblum, M. P., Spaepen, F. and Turnbull, D., *Appl. Phys. Lett.*, **37,** 184 (1980)
23. Greer, A. L., Lin, C. J. and Spaepen, F., p. 567 in ref. 5
24. Christian, J. W., *'The Theory of Transformations in Metals and Alloys'*, Pergamon, Oxford (1975)
25. Smithells, C. J. and Brandes, E. A., *'Metals Reference Book'*, 5th edition, Butterworths, London (1976)

26. Kijek, M., Ahmadzadeh, M., Cantor, B. and Cahn, R. W., *Scripta Met.*, **14,** 1337 (1980)
27. Ashby, M. F., personal communication
28. LeClaire, A. D., *Prog. Metal Phys.*, **1,** 306 (1949)
29. Tiwari, G. P., *Z. Metallk.*, **72,** 211 (1981)
30. Brown, A. M. and Ashby, M. F., *Acta metall.*, **28,** 1085 (1980)
31. Cantor, B., in *Rapidly Solidified Amorphous and Crystalline Alloys,* Eds Kear, B. H., Giessen, B. C. and Cohen, M., Materials Research Society, Boston, p. 317 (1982)
32. Hayzelden, C., Rayment, J. J. and Cantor, B., *Acta metall.*, in the press
33. Jones, H., *Rep. Prog. Phys.*, **36,** 1425 (1973)
34. Predecki, P., Mullendore, A. W. and Grant, N. J., *Trans. AIME,* **233,** 1581 (1965)
35. Akhtar, D., personal communication
36. Pratten, N. A., Scott, M. G. and Cahn, R. W., to be published
37. Cahn, R. W., Toloui, B., Akhtar, D. and Thomas, M., p. 749 in ref. 5
38. Berry, B. S. and Pritchet, W. C., *'Proc. 3rd Int. Conf. on Rapidly Quenched Metals',* Ed. Cantor, B., vol. 2, p. 21, The Metals Society, London (1978)
39. Berry, B. S. and Pritchet, W. C., *Phys. Rev.*, **B24,** 2299 (1981)
40. Berry, B. S. and Pritchet, W. C., *Scripta Met.*, **15,** 637 (1981)
41. Kirchheim, R., Sommer, F. and Schluckebier, G., *Acta metall.*, **30,** 1059 (1982)
42. Bowman, R. C., Rosker, M. J. and Johnson, W. L., to be published
43. Bowman, R. C. and Maeland, A. J., *Phys. Rev.*, **B24,** 2328 (1981)
44. Bennett, C. H., Chaudhari, P., Moruzzi, V. and Steinhardt, P., *Phil. Mag.*, **A40,** 485 (1979)
45. Spaepen, F., in ref. 38, vol. 2, p. 253
46. Vitek, V., personal communication
47. Cantor, B. and Ramachandrarao, P., vol. 1, p. 291 in ref. 5
48. Finney, J. L. and Wallace, J., *J. Non-cryst. Solids,* **43,** 165 (1981)
49. Ahmadzadeh, M. and Cantor, B., *J. Non-cryst. Solids,* **43,** 189 (1981)
50. Ahmadzadeh, M. and Cantor, B., vol. 1, p. 591 in ref. 5
51. Yamamoto, R., Shibuta, H., Mihara, T., Haga, K. and Doyama, M., in ref. 5
52. Kiritani, M. and Hamada, T., vol. 1, p. 583 in ref. 5
53. Fernengel, W., Kronmüller, H., Rapp, M. and He, Y., *J. appl. Phys.*, in the press.
54. Kirchheim, R., *Acta metall.*, **30,** 1069 (1982)

Chapter 26

Amorphous metal powder: production and consolidation

S. A. Miller

General Electric Company, Schenectady, New York, USA

26.1 Introduction

The preparation of amorphous metallic alloys by direct quenching from the melt has been an active research area for over 20 years[1–3]. The resultant amorphous alloys have demonstrated many unique and promising properties, such as high strength[4], high hardness[5], low magnetic hysteresis losses[6] and excellent corrosion resistance[7]. Despite the potential economic benefits, however, glassy metals largely remain subjects for laboratory investigation, and only in very few instances have been employed in commercial application[8,9].

The lack of development originates in part from the geometric limitations imposed on the as-cast alloy by the very high quench rates (approximately 10^6 K/s) required in order to suppress crystallization during solidification. The requisite high cooling rate must be provided by heat conduction through the sample (usually an alloy of relatively low thermal conductivity) and by conduction or convection away from the sample by the cooling substrate. These considerations severely restrict the dimensions of a glassy specimen as quenched from the melt. That is, the sample must be extremely thin in at least one dimension and have a high surface area to volume ratio. Thus, glassy alloys of interest have only been available in the form of splats, thin foils, ribbons and, recently, powder.

The above geometries are not suitable for many engineering applications which require the product to be more complex in configuration and relatively large in all three dimensions, as for example, a low hysteresis transformer core fabricated from a ferromagnetic glassy alloy. Thus, much research has been directed towards developing techniques for preparing three-dimensionally large glassy alloys. These include plasma spraying, cold compaction studies, explosive compaction, warm die compaction, warm extrusion and dynamic compaction.

Of the previously mentioned as-cast configurations of amorphous metals, powder, because of its more nearly equiaxed geometry, is the preferred form for consolidation. The deformation processes which occur during the consolidation of powder are characterized by large amounts of relative motion between particles and the flow of individual particles. The high degree of interfacial shear that results helps to disrupt oxide films providing clean surfaces across which good interparticle bonding can occur. Flat geometries, such as splats and ribbon characterized by large width and length to thickness ratios, tend to lay up flat against one another, producing little intersurface shear. To attain the same degree of bonding as powder, these geometries usually require processes characterized by a higher degree of macroscopic deformation, such as extrusion employing a large reduction ratio. As a result, attempts to consolidate amorphous ribbons and flake have been only partially successful. This helped stimulate the current interest in technology to produce a glassy powder.

This section will review processes currently developed for the production of amorphous powder (process technology, alloys investigated, etc.), the applicable theory governing the high cooling rate technology of atomized powder and the current status of consolidation studies to produce three-dimensionally large amorphous structures from powder and ribbon.

26.2 Powder production processes

Present amorphous powder production techniques may be broadly categorized as being one of two types. In the first process, the powder particles result from the quenching and solidification of discrete droplets of liquid metal (atomization techniques). In the second process, the powder results from the chopping or comminuting of ribbons or filaments of amorphous alloy (comminution techniques). Apparatus outlines are shown in *Figures 26.1—26.5*.

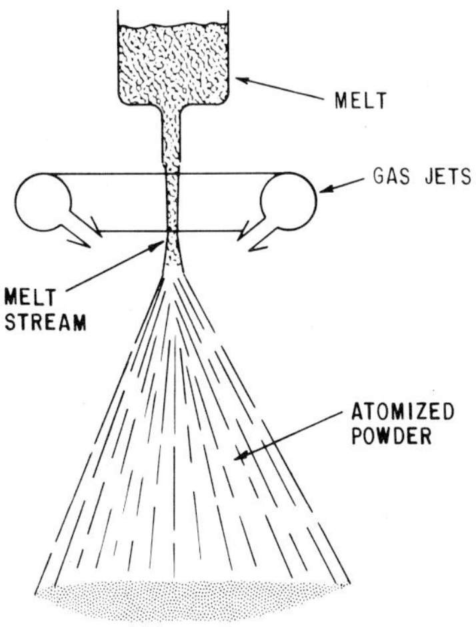

Figure 26.1 Gas or water atomization

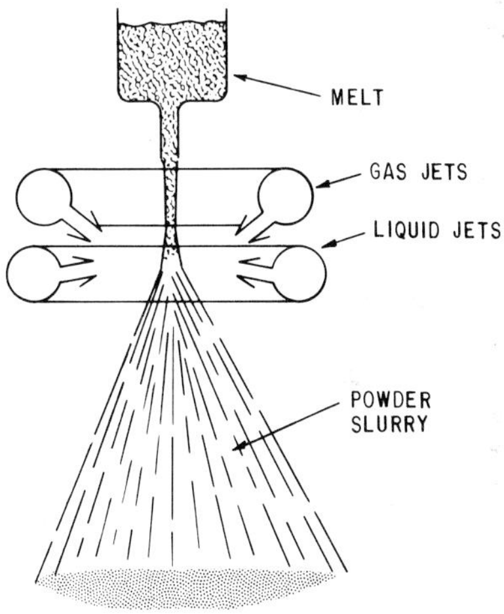

Figure 26.2 Gas–liquid atomization

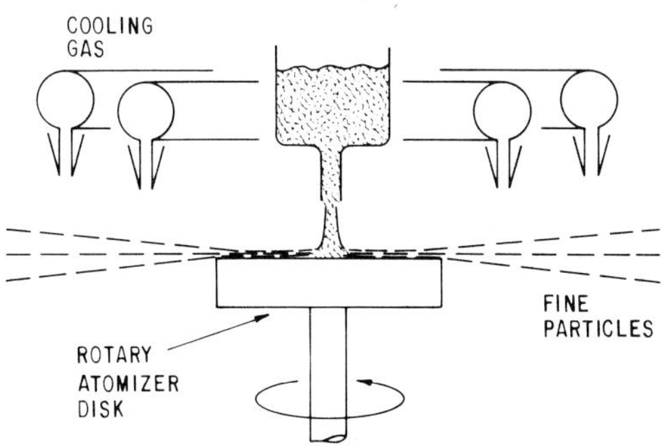

Figure 26.3 Rotating disk atomization

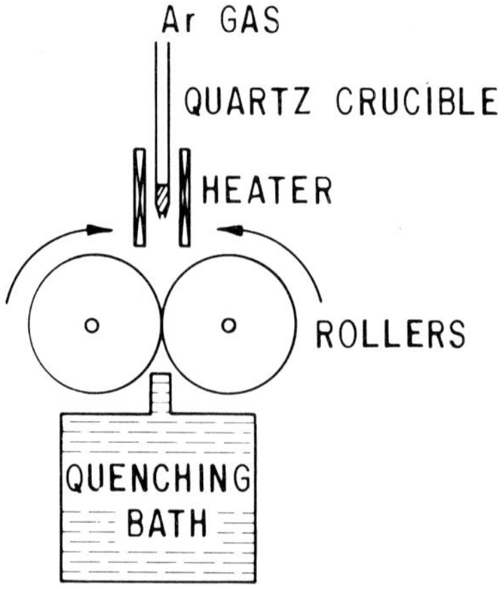

Figure 26.4 Roller atomization

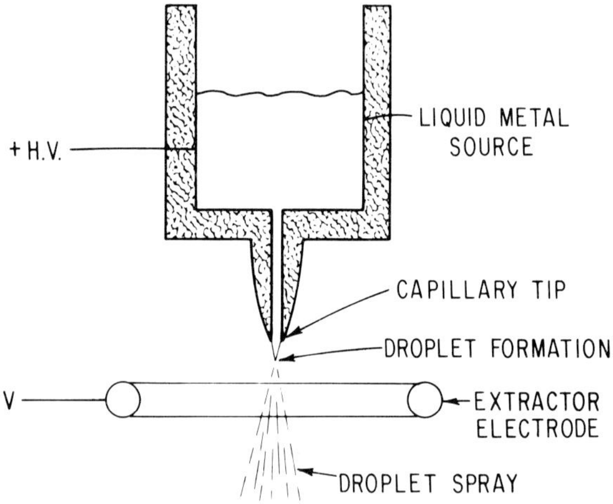

Figure 26.5 Electrohydrodynamic atomization

26.2.1 Ultrasonic gas atomization

This process is similar in concept to conventional gas atomization where a thin stream of the metal is impinged upon by high velocity gas[10–14]. The gas both disrupts the stream into droplets and acts as the quenching medium for the droplets. Geometric arrangements in gas atomization vary from a finite number of discrete gas jets[15–16] to an annular jet[17,18] which completely surrounds the melt stream (*Figure 26.1*). The principal difference between ultrasonic and conventional gas atomization is that in the former the gas jets are pulsating[19,20] at a frequency of approximately 80 kHz while in the latter the gas jets flow in a continuous manner. The pulsations are provided by the incorporation of Hartmann shock tubes in the gas passages of the atomizing die. It is claimed that droplet formation is a one-step process, with individual droplets being sheared off the melt stream by the impinging ultrasonic pulses.[21]

Amorphous $Cu_{60}Zr_{40}$ powder has been prepared by using helium gas at 8.1 MPa. XRD analysis indicates particle sizes below 50 μm to be fully amorphous and to have increasing crystalline content as the size approaches 125 μm. Above 125 μm, the powder is essentially crystalline. The estimated cooling rate for 20 μm powder is 10^5 K/s. The particle morphology is spherical (*Figure 26.6*), typical of inert gas atomized powder[22]

26.2.2 Water atomization

The geometric arrangement of water atomization devices (*Figure 26.1*) is conceptually identical with that of gas atomizers[23,24]. However, as the name implies, water replaces gas as the atomizing-quenching medium. The quench is rapid, and in fact, solidification interrupts the atomization process as the presence of many dumbell (ligaments solidified in the process of separating into droplets) and irregularly shaped particles indicate (*Figure 26.6*). Using jet velocities of 70–140 m/s, Yamaguchi and Narita were able to produce smaller than 20 μm powder of $Fe_{69}Si_{17}B_{14}$ and $Fe_{74}Si_{15}B_{11}$ containing an amorphous phase[25,26]. While no size range was determined to be completely amorphous, they estimated by magnetic saturization–temperature measurements that approximately 80 per cent volume fraction of the smaller than 20 μm particles and 20 per cent volume fraction of the 50 μm particles consisted of an amorphous phase.

26.2.3 Gas–liquid atomization quenching

High pressure gas and liquid jets are focused on the liquid metal stream (*Figure 26.2*)[27,28]. The process is established so that primary atomization of the metal is performed by the gas, and enhanced by the liquid. There is evidence that the small particle sizes (10—15 μm) are gas quenched, but for larger sizes the liquid is the principal agent producing the high quench rate. To date argon at 4.2 MPa and water at 1.6 MPa have been employed as the atomizing quenching agents to produce amorphous $Cu_{60}Zr_{40}$, $Fe_{75}Si_{15}B_{10}$ and $Fe_{81.5}Si_{14.5}B_4$ powder. The $Cu_{60}Zr_{40}$ powder is completely amorphous for sizes less than 105 μm, and retains a large amorphous-volume fraction for powders in excess of 600 μm. The larger than 600 μm powder contains a combination of fully amorphous particles, particles consisting of an amorphous matrix with crystalline nuclei and particles exhibiting an equilibrium solidification microstructure[29]. Because the Fe–Si–B alloys have a higher critical quench rate, they could be made amorphous in particle sizes up to only 20 μm. Above 20 μm in size, the particles were essentially completely crystalline[28]. The powder consists of both spherical and highly irregular particles (*Figure 26.8*). The process offers some degree of control over particle geometry by the relative placement of the gas and liquid jets. The process cooling rate for 20 μm particles has been estimated at about 10^6 K/s by use of the existing cooling rate correlations for Al–Cu alloys[29].

26.2.4 Centrifugal atomization

Pratt and Whitney has developed a centrifugal atomization process for producing rapidly quenched powder (*Figure 26.3*). In this process the melt stream is directed on to a rapidly rotating (~2500 rev/s) concave disk (diameter ~8.4 cm). The melt is accelerated by the disk to near rim speed and, under the action of the centrifugal forces present, flows to the disk periphery. It is believed that atomization occurs at the disk edge by direct droplet formation (by-passing the usual intermediate step of ligament formation) because of the melt flow rates and high melt surface tensions involved. After the metal droplets are flung off the disk they pass through a transverse array of helium gas jets which provide the quench[30,31,32].

The process has been employed to produce amorphous powder from $Ni_{75}B_{17}Si_8$

and several iron-rich Fe–Si–B alloys. While it is not yet known what size powder can be solidified in a fully amorphous state, particle sizes as large as 150 μm have some amorphous phase present[33,34]. The powder morphology is generally spherical, similar to gas atomized powder. It is calculated[35] that 20 μm particles are quenched at a rate of 10^5—10^6 K/s.

26.2.5 Roller atomization

Ishii *et al.* have applied the roller atomization process developed by Singer[36,37] to the production of amorphous powder and flake (*Figure 26.4*). Atomization is accomplished by feeding the liquid metal into the nip of a rapidly (1000—5000 r.p.m.) rotating roller. As the liquid metal emerges from the far side of the rollers (the rolls are coated with a carbon paste to reduce heat transfer and prevent solidification) it cavitates and is thrown off as droplets. The droplets are quenched in a water bath, located within 25 mm of the roller nip. Amorphous $Pd_{78}Cu_6Si_{16}$ and partially amorphous $Ni_{75}Si_8B_{17}$, and $Co_{72.5}Si_{12.5}B_{15}$ have been produced. The amorphous phase content of the powder increases with decreasing particle size and reached a maximum of 85 per cent for powder particles smaller than 37 μm $Ni_{75}Si_8B_{17}$. At present the process is limited by its inability to produce small liquid metal droplets, as only 2—3 per cent of the powder produced is below 37 μm in diameter[38]. The quench rate of the system is estimated at approximately 10^5 K/s, based on the critical cooling rates required to retain an amorphous phase in the alloys investigated[39].

In a variation of the process, the water quench is replaced by a solid substrate. Using this procedure amorphous flakes of $Co_{73}Si_{12}B_{15}$ have been prepared[38].

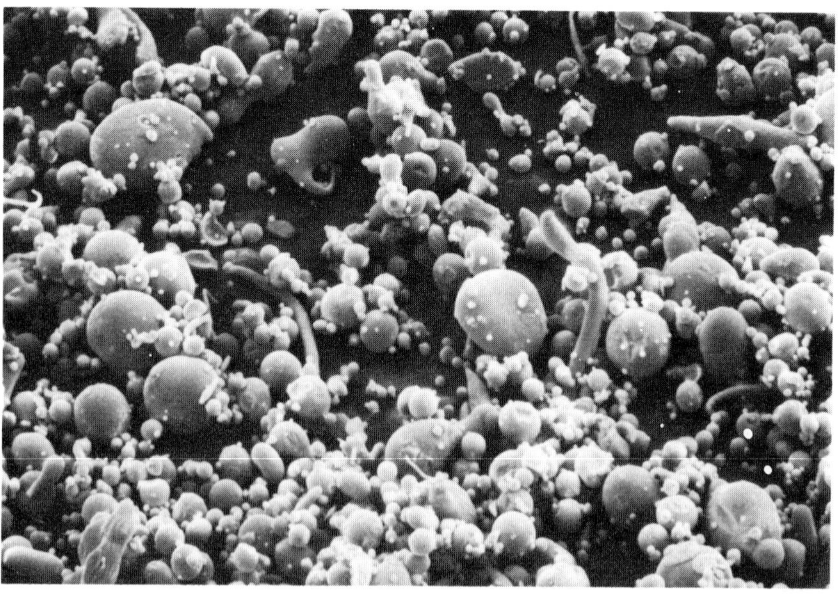

Figure 26.6 Micrograph showing typical morphology of powder produced by ultrasonic gas atomization ($Cu_{60}Zr_{40}$) (×600)

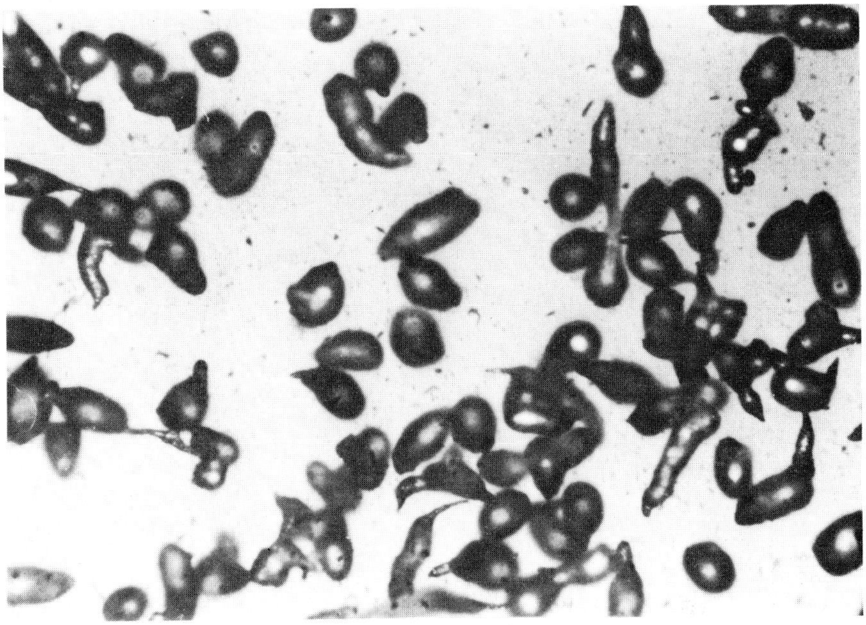

Figure 26.7 Micrograph of $Fe_{74}Si_{15}B_{11}$ powder produced by water (× 600)

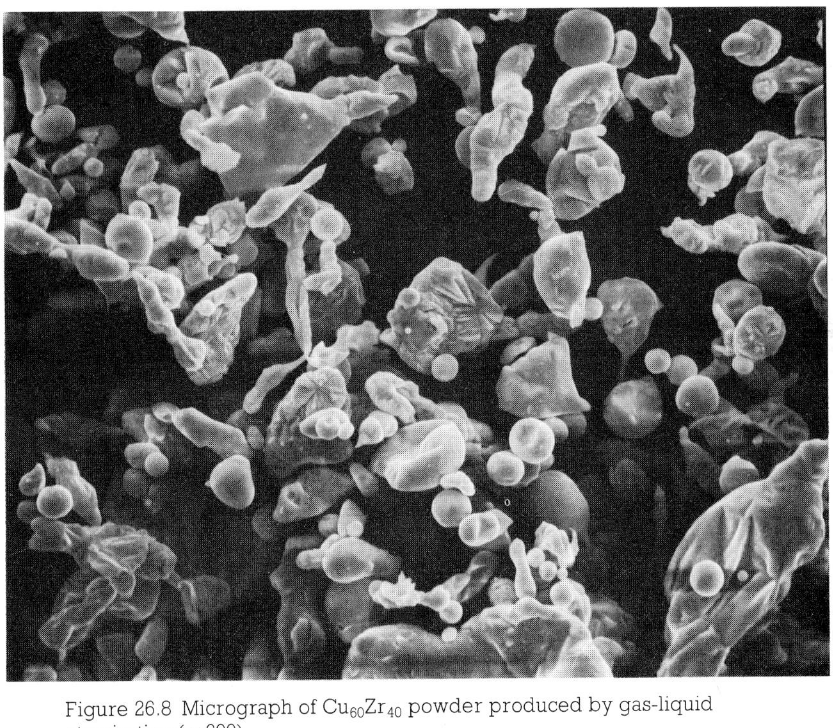

Figure 26.8 Micrograph of $Cu_{60}Zr_{40}$ powder produced by gas-liquid atomization (× 600)

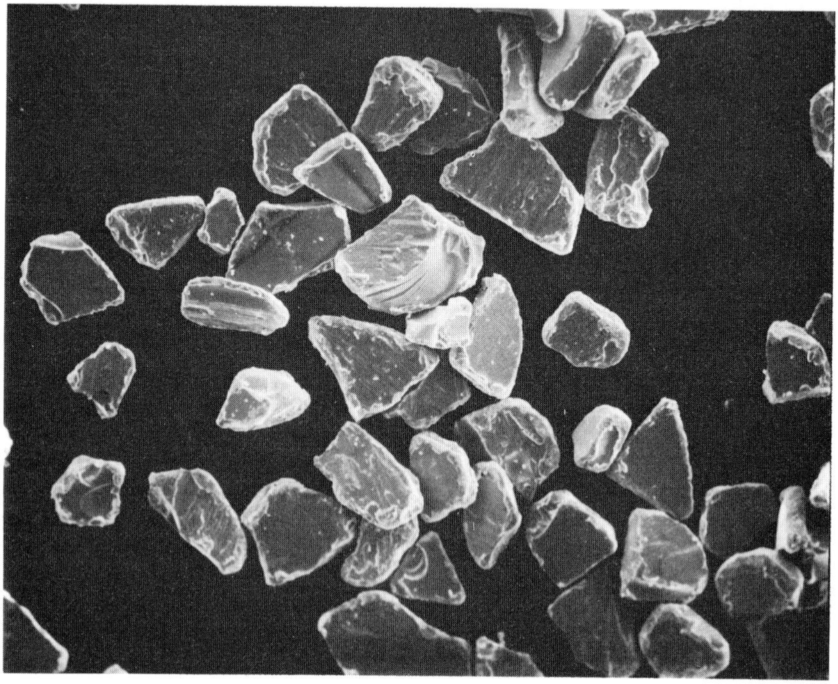

Figure 26.9 Micrograph of powder produced by comminution of ribbon ($\times 600$)

26.2.6 Spark erosion

This technology for producing rapidly solidified powder is an extension of electrodischarge machining (EDM)[40–44]. The alloy which is to be made into a powder is employed as the electrodes in an EDM unit. The powder is formed by each discharge event, when a small portion of the electrodes is either melted or vaporized[45–46] and propelled into the surrounding dielectric where it is quenched.

A broad range of ferromagnetic alloys have been converted into amorphous powder by this technique. Dodecane, benzene, water, liquid nitrogen and liquid argon have been employed as the dielectric fluid[40–48]. Dodecane, by virtue of its higher vaporization point, has provided the best quench rate. Quench rates, estimated by secondary dendrite arm spacings of Rene 95, appear to be in excess of 10^6 for $20\,\mu m$ powder and approach 10^9 for $5\,\mu m$ powder[49]. Powders produced are typically 0.5 to $30\,\mu m$ in size (although particles as large as $50\,\mu m$ have been observed to be fully amorphous and have a spherical geometry). Production rates are low[46], approximately 3.5 g/h.

26.2.7 Electrohydrodynamic atomization

Intense electric fields ($>10^5$ V/cm) can be employed to disrupt the surface of a liquid metal into fine droplets. A schematic diagram of the process is shown in *Figure 26.5*. A small diameter (50—125 μm) melt source is provided by a heated capillary tube. The field is generated by applying a high voltage (3—20 kV) across the melt and the extractor electrode located in front of the capillary tip. Atomization occurs when the applied electrostatic stresses exceed the surface tension of the melt meniscus, causing a droplet to be emitted from the capillary tip. Electrohydrodynamic atomization can provide greatly decreased mean droplet sizes and tighter size distributions about the mean than conventional atomization techniques. Amorphous $Fe_{40}Ni_{40}P_{14}B_6$ powder has been prepared approximately 0.1 μm in size. The calculated quench rate for 0.1 μm powder is 10^6 K/s. In its present stage of development as a laboratory unit, powder rates are about 20 grams per day[50–53].

Comminuted powder

The powder is fabricated by first casting an amorphous ribbon via conventional melt spinning technology[54–58]. If not inherently brittle, the ribbon is then embrittled; i.e., by annealing either in an inert gas or in hydrogen to charge hydrogen into the alloy. The ribbon can then be converted into powder by hammer or ball milling[59]. For conventionally sized powder ($>$10—20 μm diameter) this technique provides a higher quench rate than atomized powder[60]. Additionally, a relatively uniform quench rate that is independent of particle size can be attained when process design eliminates the entrapment of gas pockets between the ribbon and rotating chill block[61]. The quench rate of atomized powder is, of course, directly affected by particle size[62]. The powder morphology is that of irregularly shaped polygons (*Figure 26.9*). Because of its embrittled state, the powder is not suitable for consolidation processing in which the glassy state is retained but is finding use as precursor powder for forging billets with a microcrystalline structure[63,64] or as a coating material applied by plasma spraying. The plasma sprayed coating benefits from the original homogeneity of the powder and in certain cases the coating itself will solidify with an amorphous structure[64].

26.2.9 Dispersion techniques

The above techniques have all been characterized by their cooling rates. This approach was taken partially because it is the accepted convention for rapidly solidified powder processes and partially because it is the process parameter used in the above techniques to obtain the large degree of undercooling necessary to prevent crystallization during solidification[65,66]. However, other techniques which do not require high cooling rates have been developed which produce powder with amorphous structures. These low cooling rate techniques produce large undercooling of the melt by eliminating the presence of local heterogeneous nucleation sites. Droplets free from sites are generated by subdividing the melt into more droplets than there are nucleation sites. Under the proper conditions, the droplets without nucleation sites will then form an amorphous phase during solidification. A typical laboratory droplet dispersal process might emulsify 1 gram of molten metal in 5 cm^3 of carrier fluid. For droplets in the 5—20 μm size range, this results in approximately 10^8 droplets from one gram of metal[67]. Amorphous powder of $Cu_{29}Te_{71}$ has been prepared by this technique[68]. In the shot tower approach, droplets solidify while falling in an inert gas filled column. This process has produced amorphous $Pd_{82}Si_{18}$ powder larger than 800 μm in diameter[69,70].

26.2.10 Splat process

These processes produce high quench rates by droplet solidification against a solid substrate. One technique forms the droplets by ejecting a melt stream at an acute angle on to the inside surface of a rotating drum[71]. Impact with the drum disintegrates the stream and secondary impact with the drum by the droplets formed causes them to solidify. In a second approach, droplets are generated by a rotary atomization technique and quenched against a rotating substrate[72].

26.3 Process theory

26.3.1 Atomization techniques

Atomization processes to produce powders may be reduced to two sequential steps; liquid metal droplet formation and droplet quenching.

Gas atomization, because of the complexity of the process, is only understood qualitatively. The impinging gas jets disrupt the metal stream into an inverted hollow cone which continuously forms, disintegrates and re-forms during the atomizing process[73]. The wall of the cone, a thin sheet, then breaks up into tubular ligaments and the ligaments into droplets under the dynamic forces present. The disruption process has been verified by several investigators using high speed photography[74,75].

Attempts to develop quantitative relations for droplet size as a function of process parameters have not provided consistent results[76]. A number of correlations exist, each dependent on the atomizing geometry and parameters employed. Some of the existing correlations are presented in references 76—80. Agreement does exist that the important factors in obtaining an increased yield of fines are[73–80]:

1. Decreasing the resistance of the melt to atomization. Fine particle sizes are favoured by low melt surface tension (σ) and viscosity (η). These can be altered by alloy composition, superheat, and the use of surfactants (surface tension only).
2. Increasing the kinetic energy of the gas stream at the atomization point. This is accomplished by reduced gas jet impingement distances, increased gas velocity (gas pressure and nozzle design) and increased gas density.
3. Increasing the ratio of the mass flow rate of the gas to the mass flow rate of the metal (M_m).

Studies have lead to the conclusion that water atomization occurs by both shear and splash drop formation mechanisms[81–84]. However, as with gas atomization, several different expressions relating the mean droplet size (Dp) to atomizing parameters have been derived[81–85]. It is generally agreed that water velocity is the most important process parameter with the mean particle size being inversely proportional to water velocity or inversely proportional to the water velocity component normal to the melt stream[83]. Friedmann *et al.* have determined that the mean droplet size produced during rotary atomization process is given by

$$Dp = K\frac{M_m^{0.2}}{(\omega^2 r)^{0.3}}\cdot\frac{\sigma^{0.1}\mu^{0.2}}{\rho^{0.5}} \tag{26.1}$$

where ω, r, ρ and K are disk angular velocity, disk radius, metal density and a process constant, respectively[86]. This expression has been shown to be valid over a wide range of atomization parameters.

Heat removal from atomized droplets, assuming that they are not directed against a solid substrate, occurs by both convection and radiation. Earlier work has shown that heat loss by radiation from small droplets in a convective environment is at least an order of magnitude less than convective losses[51]. Considering convection heat transfer only, and assuming Newtonian cooling (there is some uncertainty as to whether the upper limit on the Biot number is 0.01 or 0.001 for this assumption to be valid[87,51]), the cooling rate of a small diameter sphere in a convective medium has been shown to be reasonably described by

$$\frac{\mathrm{d}T}{\mathrm{d}t} = \frac{6k_f(T_p - T_f)}{D_p^2\rho_p\bar{C}_p}(2.0 + 6Re^{0.5}Pr^{0.3}) \quad (26.2)$$

Re = Reynold's Number = $vD\rho_f/\mu_f$ and Pr = Prandtl Number = $C_{pf}\mu/k_f$, where k_f, T_f, ρ_f, μ_f and C_{pf} are the cooling fluid thermal conductivity, temperature, density, viscosity and specific heat; D_p, T_p, ρ_p and $\bar{C}_p$ are the metal droplet diameter, temperature, density and average specific heat over the temperature range of interest[87–89].

The above expression shows that the ability to generate fine droplet sizes in the atomization stage is the most important factor in attaining high quench rates as the maximum attainable quench rate is inversely proportional to the square of the particle diameter. Additionally, in any given lot of convectively cooled powder, there will be a large variation in attained cooling rate simply by virtue of particle size. It is believed the application of this expression to a sphere in a liquid medium is valid if the relative velocity of the sphere with respect to the medium is sufficient to prevent local boiling[53].

26.3.2 Comminuted powder

Process parameters and their importance to solidification in melt spinning ribbon (a precursor for comminuted powder) will not be discussed here. They are treated in detail in Chapter 3.

26.4 Consolidation

A major driving force for the development of techniques to produce amorphous powder has been the goal to produce a three-dimensionally large amorphous product. Particle geometries that are approximately equiaxed, as compared with splats or ribbons, have an inherent advantage in producing good interparticle bonding during consolidation by deformation processing because of the high degree of interparticle shear that occurs during densification. The shear acts to abraid surface oxides and expose uncontaminated surfaces for bonding.

Three dimensional specimens have been prepared by plasma spraying, explosive and dynamic compaction, swaging, and warm deformation processes. Plasma sprayed deposits of amorphous $Cu_{60}Zr_{40}$ did not have attractive mechanical properties. The deposits were porous, of low strength, and brittle[90].

High energy rate forming techniques, explosive and dynamic compaction, have been used successfully to consolidate powder while retaining the amorphous state. Both techniques require an intense pressure pulse (0.1—10 GPa) of short duration ($\sim 10^{-6}$ s) to accomplish interparticle welding. Cline *et al.*[91,92] have successfully consolidated $Pd_{78}Cu_6Si_{16}$ and $Fe_{40}Ni_{40}P_{14}B_6$ into rods by explosive techniques. No compacted specimen properties were reported. Researchers at the Minsk Powder Metallurgy Institute have explosively compacted amorphous fibres (prepared by a modified melt extraction process)[93] of $Fe_{48}Ni_{48}B_4$ and $Fe_{40}Ni_{40}P_{16}B_4$ into plates measuring 200 × 50 × 3 mm in size. Crystallization studies showed the plates to retain the amorphous nature of the starting powder. While plate density was only 92—98 per cent of theoretical density (*Figure 26.10*), it was concluded the density could be improved by process parameter optimization and the use of finer powder[94]. However, others have reported local delamination and crystallization (due to melting and resolidification) in explosively fabricated specimens of melt spun ribbon[95].

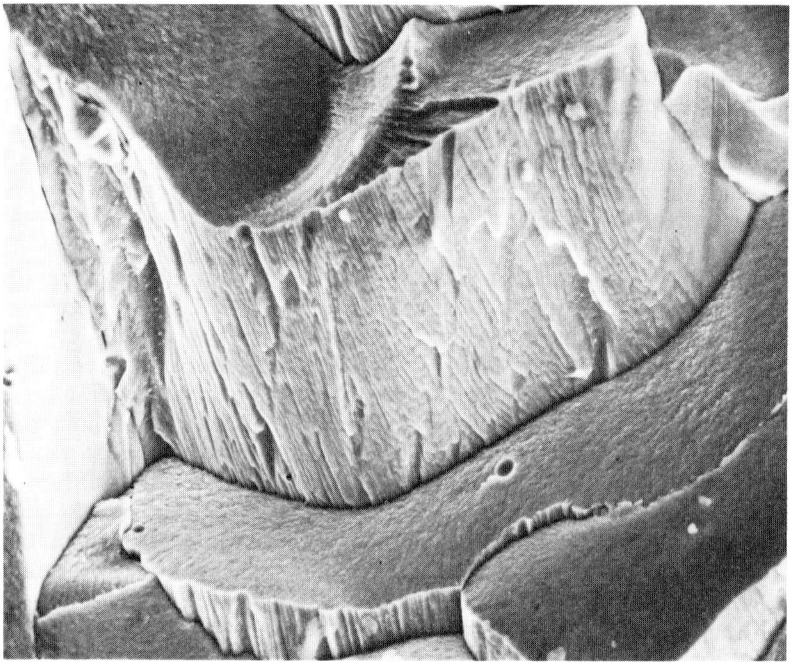

Figure 26.12 Micrograph of a tensile fracture surface of a specimen prepared by warm extrusion of a Fe–B–Si alloy. Note that the fracture surface follows ribbon boundaries (× 800)

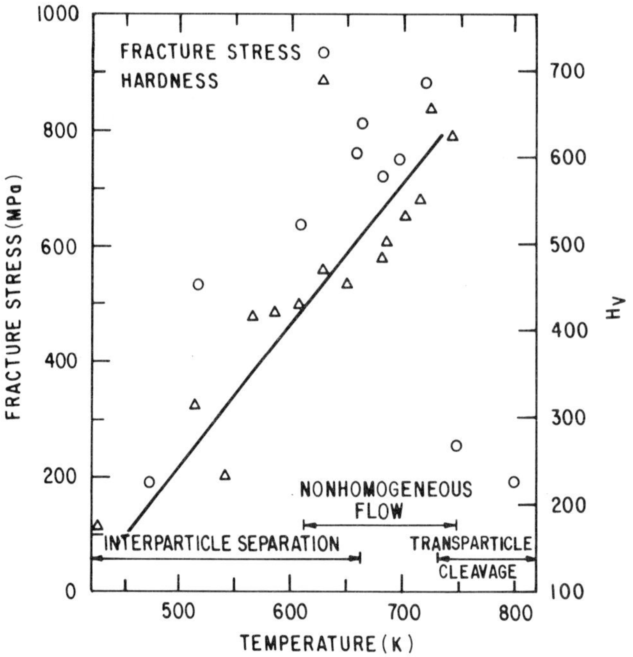

Figure 26.13 Billet hardness, fracture strength and fracture mechanism as a function of billet compaction temperature

structures without degrading the mechanical properties of the original alloy. This degradation has been attributed to tempering embrittlement (dynamic compaction), insufficient material flow (warm die compaction of ribbon), and incomplete interparticle bonding due to the presence of surface oxides on the starting powder (warm die compaction of powder). Studies of the magnetic properties of consolidated parts show that in all cases the magnetic properties are severely degraded owing to the introduction of high stresses and shear bonds. These could not be annealed out. The degree of degradation disappeared as the magnetostriction of the alloy approached zero[107].

26.5 Conclusion

In the past three years, there has been a rapid growth in the number of ways to fabricate amorphous alloy powder. All depend on attaining a large degree of undercooling, before solidification, in order to prevent crystallization. To accomplish this, most processes use high cooling rate technology. Some of the processes show the potential to be scaled up (or are already scaled up) to the pilot plant level. Others provide an excellent laboratory source of amorphous powder.

Research in the area of consolidation processing with the goal of producing three-dimensionally large structures is still in its infancy. At present, the data are limited and what data exist are incomplete in scope. In order to accurately assess a compaction technique, a more complete determination of final compact properties than is presently being reported is necessary. Properties investigated should include those already reported for the as-cast state. A brief list might comprise hardness, fracture strength, ductility, toughness, density, amorphous phase retention, as determined by electron diffraction studies, etc.

Additionally as-cast specimens (i.e., ribbon) should be given processing histories similar to that of the consolidated material to determine the effect of the process on intrinsic material properties, separately from the problem of consolidated properties. Finally, comparison and analysis of as-cast properties, properties as affected by the processing history, and compact properties can bring about an understanding of process phenomena that will hasten process development and the attainment of optimal properties in the final consolidated structure.

References

1. Klement, W. (jun.), Willens, R. H. and Duwez, P., *Nature,* **187,** 809 (1960)
2. Salli, I. V., *J. Inorg. Chem. Acad. Sci. USSR,* **3,** 136 (1958)
3. Duwez, P. and Willens, R. H., *TMS — AIME,* **227,** 362 (1963)
4. Davis, L. A., in *'Metallic Glasses',* p. 190, American Society for Metals, Metals Park, Ohio (1963)
5. Li, J. C. M., in *'Ultrarapid Quenching of Liquid Alloys',* Ed. Herman, H., p. 326, Academic Press, New York, (1981)
6. Gyorgy, E. M., in ref. 4, p. 275
7. Hashimoto, K. and Masumoto, T., in ref. 5, p. 291
8. Chen, H. S., Leamy, H. J. and Miller, C. E., in *A. Rev. mater. Sci.,* Eds. Huggins, R. A., Bube, R. H. and Vermilyea, D. A., p. 363, Annual Reviews, Inc., Palo Alto, CA (1980)
9. Polk, D. E. and Giessen, B. C., in ref. 3, p. 1
10. Beddow, J. K., *'The Production of Metal Powders by Atomization',* Heyden and Sons Ltd., London (1980)
11. Lawley, A., *Int. J. Powder Met. Powder Tech.,* **13,** 169 (1977)
12. Lubanska, H., *J. Metals,* **22,** 45 (1970)
13. Klar, E. and Shafer, W. M., in *'Powder Metallurgy for High Performance Applications',* Eds. Burke, J. J. and Wiess, V., p. 57, Syracuse University Press, Syracuse, NY, 57 (1972)

14. Mehrotra, S. P. and Khedkar, P. Y., *Trans. Indian Inst. Metals,* **33,** 361 (1980)
15. Feldman, K., *U.S. Pat.,* 3 340 334 (1967)
16. Clark, I. S. and Pargeter, J. K., *U.S. Pat.,* 4 272 463 (1981)
17. Hegmann, W. G., *U.S. Pat.* 3 588 951 (1971)
18. Lev, M. B., Pavlovskaya, E. I., Shibryaev, B. F. and Barkan, B. L., *Soviet Powder Met. Metal Ceram.,* **2,** 153 (1964); translated from *Poroshkavaya Metall.,* **2,** 89 (1964)
19. Grant, N. J., in *'Rapid Solidification Processing, Principles and Technologies',* Eds. Mehrabian, R., Kear, B. H. and Cohen, M., p. 230, Claitor's Publishing Division, Baton Rouge, LA (1977)
20. Anand, V., Kaufman, A. J. and Grant, N. J., in *'Rapid Solidification Processing, Principles and Technologies, II',* Eds. Mehrabian, R., Kear, B. H. and Cohen, M., p. 273, Claitor's Publishing Division, Baton Rouge, LA (1980)
21. Lawley, A., *J. Metals,* **33,** 13 (1981)
22. Grant, N. J. and Domalavaghe, P. K., (Massachusetts Institute of Technology), personal communication (1982)
23. Dain, R. J., *U.S. Pat.,* 4 191 516, (1980)
24. Grandzol, R. J. and Tallmadge, J. A., *Int. J. Powder Met.,* **11,** 103 (1975)
25. Yamaguchi, T. and Narita, K., *Appl. Phys. Lett.,* **33,** 498 (1978)
26. Yamaguchi, T., in *'Amorphous Electronic Materials and Their Application',* p. 78, Science Forum Press, Tokyo, Japan (1981)
27. Miller, S. A. and Murphy, R. J., *Scripta Met.,* **13,** 673 (1979)
28. Miller, S. A. and Murphy, R. J., in ref. 20, p. 385
29. Miller, S. A., PhD Thesis, Northeastern University, Boston, MA (1980)
30. Glickstein, M. R., Patterson, R. J. (jun.) and Schockey, N. E., in ref. 19, p. 46
31. Thompson, C. C., *U.S. Pat.,* 4 284 394
32. Holiday, P. R., Cox, A. R. and Patterson, R. J., (jun.), in ref. 19
33. Patterson, R. J. (jun.), in Proc. Fall Annual Meeting AIME (1981) to be published
34. Patterson, R. J., (jun.), United Technologies Corp., West Palm Beach, FL, personal communication (1982)
35. Glickstein, M. R., Patterson, R. J. (jun.) and Schockey, N. E., in ref. 19, p. 46
36. Singer, A. R. E., Roche, A. D. and Day, L., *Powder Metall.,* **2,** 81 (1980)
37. Singer, A. R. E. and Roche, A. D., in *'Modern Developments in Powder Metallurgy',* p. 127, Metal Powder Industries Federation, Princeton, NJ (1977)
38. Ishii, H., in *'Proc. 4th Int. Conf. on Rapidly Quenched Metals,* Eds. Masumoto, T. and Suzuki, K., Japan Institute of Metals, Sendai (1982)
39. Ishii, H., Naka, M. and Masumoto, T., *Sci. Rep. Res. Insts. Tôkohu Univ.,* **29,** 343 (1981)
40. Svedberg, T., *Nova Acta R. Soc. Scient. upsal.,* **2,** 14 (1904)
41. Bucklow, T. A. and Drain, L. E., *J. scient. Instrum,* **41,** 614 (1964)
42. Ananthararaman, T. A., *Z. Metallk.,* **61,** 760 (1970)
43. Ruppersberg, H. and Bold, H. J., *Metallurgia,* **26,** 34 (1972)
44. Filimonenko, V. N., Gurland, J. and Richman, M. H., *Metallography,* **2,** 125 (1969)
45. Yamaguchi, T. and Narita, K., *IEEE Trans. Magn.,* **MAG-13,** 1621 (1977)
46. Berkowitz, A. E. and Walter, J. L., in ref. 20, p. 294
47. Berkowitz, A. E., Walter, J. L. and Wall, K. F., *Phys. Rev. Lett.,* **46,** 1484 (1981)
48. Liebermann, H. H. and Walter, J. L., in *Proc. Mater. Res. Soc.,* to be published (1981)
49. Walter, J. L. (General Electric Co., Schenectady) personal communication (1982)
50. Mahoney, J. F., Perel, J. and Vickers, K. E., *U.S. Pat.,* 4 264 641 (1981)
51. Perel, J., Mahoney, J. F.,Kalensher, B. E., Vickers, K. E. and Mehrabian, R., in ref. 19, p. 258
52. Perel, J., Mahoney, J. F., Duwez, P. and Kalensher, B. E., in ref. 20, p. 287
53. Perel, J., Mahoney, J. F., Kalensher, B. E. and Mehrabian, R., in *Advances in Metal Processing,* Eds. Burke, J. J., Mehrabian, R. and Weiss, V., p. 79, Plenum Publishing Corp., New York (1981)
54. Kavesh, S., in ref. 3, p. 36
55. Pond, R. and Maddin, R., *Metals Trans.,* **A245,** 2475 (1969)
56. Pond, R. B., *U.S. Pat.* 2 825 108 (1958)
57. Narasimhan, M. C., *U.S. Pat.* 4 142 571 (1980)
58. Liebermann, H. H., *IEEE Trans. Magn.,* **MAG-15,** 1393 (1979)
59. Ray, R., *U.S. Pat.* 4 290 808 (1981)
60. Panchanathan, V., Ray, R., Tsai, C. L. and Whang, S., in ref. 48
61. Huang, S. C., (General Electric Co., Schenectady) personal communication (1982)
62. Ray, R., *Metals Prog.,* (1982), to be published
63. Ray, R., *J. mater. Sci.,* **16,** 2924 (1981)
64. Ray, R., *J. mater. Sci.,* **16,** 2927 (1981)
65. Chen, H. S. and Jackson, K. A., in ref. 3, p. 74

66. Cohen, M., Kear, B. H. and Mehrabian, R., in ref. 20, p. 1
67. Perepezko, J. H., in ref. 20, p. 56
68. Rasmussen, D. H. and Loper, C. R., *U.S. Pat.*, 4 042 374 (1977)
69. Drehman, A. J. and Turnbull, D., in ref. 48
70. Drehman, A. J. and Turnbull, D., *Scripta. Met.*, **15**, 543 (1981)
71. Ray, R., *U.S. Pat.*, 4 221 587 (1980)
72. Schlienger, H. P., *U.S. Pat.*, 3 963 812 (1976)
73. Lawley, A., in ref. 3, p. 91
74. See, J. B. and Johnson, G. H., *Powder Tech.*, **21**, 119 (1978)
75. See, J. B., Runkle, J. and King, T. B., *Metals Trans.*, **4**, 2669 (1973)
76. Rao, K. P. and Mehrotra, S. P., in *'Modern Developments in Powder Metallurgy'*, p. 113, Metal Powder Industries Federation and American Powder Metallurgy Institute, Princeton, NJ (1980)
77. Bradley, D., *J. Phys.*, **14**, 1724 (1973)
78. Lubanska, H., *J. Metals*, **2**, 45 (1970)
79. Mehrotra, S. P., *Powder Metall. Int.*, **13**, 80 (1981)
80. Schmitt, H., *Powder Metall. Int.*, **11**, 18 (1979)
81. Grandzol, R. J., PhD Thesis, Drexel University, Philadelphia (1973)
82. Dunkley, J. J., *Powder Metall. Int.*, **10**, 38 (1978)
83. Tallmadge, J. A., in *'Powder Metallurgy Processing, New Techniques and Analysis'*, Eds. Kuhn, H. A. and Lawley, A., p. 1, Academic Press, New York (1978)
84. Small, S. and Bruce, T. J., *Int. J. Powder Metall.*, **4**, 7 (1968)
85. Gummeson, P. U., in ref. 13, p. 27
86. Friedman, S. J., *Chem. Engng. Prog.*, **48**, 181 (1952)
87. Mehrabian, R., Hsu, S. C., Levi, C. G. and Kou, S., in ref. 53, p. 13
88. Levi, C. G. and Mehrabian, R., *Metals Trans.*, **11B**, 21 (1980)
89. Jones, H., *J. Phys. D: Appl. Phys.*, **4**, 1657 (1971)
90. Giessen, B. C., Madhava, N. M., Murphy, R. J., Ray, R. and Surette, J., *Metals Trans.*, **8A**, 346 (1977)
91. Cline, C. F. and Hopper, R. W., *Scripta Met.*, **11**, 1137 (1977)
92. Cline, C. F., Mahler, J., Milton, F., Kuhl, W. and Hopper, R., in ref. 19, p. 380
93. Maringer, R. E. and Mobley, C. E., *J. Vac. Sci. Technol.*, **11**, 1067 (1974)
94. Roman, O. V., Gorobtsov, V. G., Mitin, B. S. and Vasilyev, V. A., in ref. 38
95. Clauer, A. H., Raman, R. V., Carbonara, R. S. and Maringer, R. E., in ref. 20, p. 339
96. Morris, D. G., *Metal Sci.*, **14**, 215 (1980)
97. Raybould, D., Morris, D. G. and Cooper, G. A., *J. mater. Sci.*, **14**, 2523 (1979)
98. Morris, D. G., in ref. 20, p. 372
99. Morris, D. G., in ref. 38
100. Smith, J. S., Perepezko, J. H., Rasmussan, D. H. and Loper, C. R. (jun.), *U.S. Pat.*, 4 282 034 (1981)
101. Spaepen, F. and Turnbull, D., in ref. 3, p. 114
102. Stempin, J. L. and Wexell, D. R., *U.S. Pat.*, 4 298 382 (1981)
103. Liebermann, H. H., in ref. 20, p. 393
104. Liebermann, H. H. (Allied Chemical Co.), personal communication (1982)
105. Miller, S. A. and Murphy, R. J., in ref. 38
106. Pourahimi, S., M. S. Thesis, Northeastern University, Boston (1980)
107. Luborsky, F. E. (General Electric Co., Schenectady), personal communication (1982)

Index of amorphous alloys

Elements of alloy arranged alphabetically except that the metalloids, B, Be, C, P and Si, are placed last and the transition metals Fe, Co and Ni, are placed before other elements.

Subject index